Clinical 3D Dosimetry in Modern Radiation Therapy

IMAGING IN MEDICAL DIAGNOSIS AND THERAPY

Series Editors: Andrew Karellas and Bruce R. Thomadsen

Published titles

Quality and Safety in Radiotherapy
Todd Pawlicki, Peter B. Dunscombe, Arno J. Mundt, and Pierre Scalliet, Editors
ISBN: 978-1-4398-0436-0

Adaptive Radiation Therapy
X. Allen Li, Editor
ISBN: 978-1-4398-1634-9

Quantitative MRI in Cancer
Thomas E. Yankeelov, David R. Pickens, and Ronald R. Price, Editors
ISBN: 978-1-4398-2057-5

Informatics in Medical Imaging
George C. Kagadis and Steve G. Langer, Editors
ISBN: 978-1-4398-3124-3

Adaptive Motion Compensation in Radiotherapy
Martin J. Murphy, Editor
ISBN: 978-1-4398-2193-0

Image-Guided Radiation Therapy
Daniel J. Bourland, Editor
ISBN: 978-1-4398-0273-1

Targeted Molecular Imaging
Michael J. Welch and William C. Eckelman, Editors
ISBN: 978-1-4398-4195-0

Proton and Carbon Ion Therapy
C.-M. Charlie Ma and Tony Lomax, Editors
ISBN: 978-1-4398-1607-3

Physics of Mammographic Imaging
Mia K. Markey, Editor
ISBN: 978-1-4398-7544-5

Physics of Thermal Therapy: Fundamentals and Clinical Applications
Eduardo Moros, Editor
ISBN: 978-1-4398-4890-6

Emerging Imaging Technologies in Medicine
Mark A. Anastasio and Patrick La Riviere, Editors
ISBN: 978-1-4398-8041-8

Cancer Nanotechnology: Principles and Applications in Radiation Oncology
Sang Hyun Cho and Sunil Krishnan, Editors
ISBN: 978-1-4398-7875-0

Image Processing in Radiation Therapy
Kristy Kay Brock, Editor
ISBN: 978-1-4398-3017-8

Informatics in Radiation Oncology
George Starkschall and R. Alfredo C. Siochi, Editors
ISBN: 978-1-4398-2582-2

Cone Beam Computed Tomography
Chris C. Shaw, Editor
ISBN: 978-1-4398-4626-1

Computer-Aided Detection and Diagnosis in Medical Imaging
Qiang Li and Robert M. Nishikawa, Editors
ISBN: 978-1-4398-7176-8

Cardiovascular and Neurovascular Imaging: Physics and Technology
Carlo Cavedon and Stephen Rudin, Editors
ISBN: 978-1-4398-9056-1

Scintillation Dosimetry
Sam Beddar and Luc Beaulieu, Editors
ISBN: 978-1-4822-0899-3

Handbook of Small Animal Imaging: Preclinical Imaging, Therapy, and Applications
George Kagadis, Nancy L. Ford, Dimitrios N. Karnabatidis, and George K. Loudos Editors
ISBN: 978-1-4665-5568-6

IMAGING IN MEDICAL DIAGNOSIS AND THERAPY

Series Editors: Andrew Karellas and Bruce R. Thomadsen

Published titles

Comprehensive Brachytherapy: Physical and Clinical Aspects
Jack Venselaar, Dimos Baltas, Peter J. Hoskin, and Ali Soleimani-Meigooni, Editors
ISBN: 978-1-4398-4498-4

Handbook of Radioembolization: Physics, Biology, Nuclear Medicine, and Imaging
Alexander S. Pasciak, PhD., Yong Bradley, MD., J. Mark McKinney, MD., Editors
ISBN: 978-1-4987-4201-6

Monte Carlo Techniques in Radiation Therapy
Joao Seco and Frank Verhaegen, Editors
ISBN: 978-1-4665-0792-0

Stereotactic Radiosurgery and Stereotactic Body Radiation Therapy
Stanley H. Benedict, David J. Schlesinger, Steven J. Goetsch, and Brian D. Kavanagh, Editors
ISBN: 978-1-4398-4197-6

Physics of PET and SPECT Imaging
Magnus Dahlbom, Editor
ISBN: 978-1-4665-6013-0

Tomosynthesis Imaging
Ingrid Reiser and Stephen Glick, Editors
ISBN: 978-1-138-19965-1

Ultrasound Imaging and Therapy
Aaron Fenster and James C. Lacefield, Editors
ISBN: 978-1-4398-6628-3

Beam's Eye View Imaging in Radiation Oncology
Ross I. Berbeco, Ph.D., Editor
ISBN: 978-1-4987-3634-3

Principles and Practice of Image-Guided Radiation Therapy of Lung Cancer
Jing Cai, Joe Y. Chang, and Fang-Fang Yin, Editors
ISBN: 978-1-4987-3673-2

Radiochromic Film: Role and Applications in Radiation Dosimetry
Indra J. Das, Editor
ISBN: 978-1-4987-7647-9

Clinical 3D Dosimetry in Modern Radiation Therapy
Ben Mijnheer, Editor
ISBN: 978-1-4822-5221-7

Clinical 3D Dosimetry in Modern Radiation Therapy

Edited by

Ben Mijnheer

CRC Press is an imprint of the
Taylor & Francis Group, an **informa** business

CRC Press
Taylor & Francis Group
6000 Broken Sound Parkway NW, Suite 300
Boca Raton, FL 33487-2742

Printed on acid-free paper

International Standard Book Number-13: 978-1-4822-5221-7 (Hardback)

Library of Congress Cataloging-in-Publication Data

Names: Mijnheer, Ben, editor.
Title: Clinical 3D dosimetry in modern radiation therapy / [edited by] Ben Mijnheer.
Other titles: Imaging in medical diagnosis and therapy ; 28.
Description: Boca Raton : Taylor & Francis, 2017. | Series: Imaging in medical diagnosis and therapy ; 28 | Includes bibliographical references.
Identifiers: LCCN 2017023203 | ISBN 9781482252217 (hardback : alk. paper)
Subjects: | MESH: Neoplasms–radiotherapy | Radiometry–methods | Radiotherapy, Conformal–methods | Radiation Oncology
Classification: LCC RC271.R3 | NLM QZ 269 | DDC 616.99/40642–dc23
LC record available at https://lccn.loc.gov/2017023203

Visit the Taylor & Francis Web site at
http://www.taylorandfrancis.com

and the CRC Press Web site at
http://www.crcpress.com

Contents

Series Preface

Since their inception over a century ago, advances in the science and technology of medical imaging and radiation therapy are more profound and rapid than ever before. Further, these disciplines are increasingly cross-linked as imaging methods become more widely used to plan, guide, monitor, and assess treatments in radiation therapy. Today, the technologies of medical imaging and radiation therapy are so complex and computer driven that it is difficult for the people (physicians and technologists) responsible for their clinical use to know exactly what is happening at the point of care, when a patient is being examined or treated. The people best equipped to understand the technologies and their applications are medical physicists, and these individuals are assuming greater responsibilities in the clinical arena to ensure that what is intended for the patient is actually delivered in a safe and effective manner.

The growing responsibilities of medical physicists in the clinical arenas of medical imaging and radiation therapy are not without their challenges, however. Most medical physicists are knowledgeable in either radiation therapy or medical imaging, and expert in one or a small number of areas within their disciplines. They sustain their expertise in these areas by reading scientific articles and attending scientific talks at meetings. In contrast, their responsibilities increasingly extend beyond their specific areas of expertise. To meet these responsibilities, medical physicists periodically must refresh their knowledge of advances in medical imaging or radiation therapy, and they must be prepared to function at the intersection of these two fields. The challenge is to accomplish these objectives.

At the 2007 annual meeting of the American Association of Physicists in Medicine in Minneapolis, this challenge was the topic of conversation during a lunch hosted by Taylor & Francis Group and involving a group of senior medical physicists (Arthur L. Boyer, Joseph O. Deasy, C.-M. Charlie Ma, Todd A. Pawlicki, Ervin B. Podgorsak, Elke Reitzel, Anthony B. Wolbarst, and Ellen D. Yorke). The conclusion of this discussion was that a book series should be launched under the Taylor & Francis banner, with each volume in the series addressing a rapidly advancing area of medical imaging or radiation therapy of importance to medical physicists. The aim of this series would be for each volume to provide medical physicists with the information needed to understand technologies driving a rapid advance and their applications to safe and effective delivery of patient care.

Each volume in the series is edited by one or more individuals with recognized expertise in the technological area encompassed by the book. The editors are responsible for selecting the authors of individual chapters and ensuring that the chapters are comprehensive and intelligible to someone without such expertise. The enthusiasm of volume editors and chapter authors has been gratifying and reinforces the conclusion of the Minneapolis luncheon that this series of books addresses a major need of medical physicists.

The series *Imaging in Medical Diagnosis and Therapy* would not have been possible without the encouragement and support of the series manager, Lu Han, of Taylor & Francis Group. The editors and authors, and most of all I, are indebted to his steady guidance of the entire project.

— William R. Hendee
Founding Series Editor
Rochester, Minnesota

Preface

This book's title, *Clinical 3D Dosimetry in Modern Radiation Therapy*, needs some clarification with respect to its scope. Radiation dosimetry can have a range of meanings, depending on context. In a narrow sense, it might be taken to mean dose *measurement*, which could be all of absolute or relative dose measurements, in phantom or *in vivo*, in standard or technique developmental situations. In a clinical radiation oncology context, it is often also used to mean dose *determination* via computation, evaluation of dose distributions for treatment planning, and verification of radiation delivery. Dose determination also critically depends on dose measurements providing the baseline dataset of beam characteristics and machine performance required to execute each treatment. Equally it depends on measurements, typically from image guidance, for determining the relevant patient characteristics during therapy.

In some countries, dosimetrists may be employed under that specific job title to carry out some of these activities, including quality assurance (QA) of several stages in the preparation and delivery of a specific patient treatment. In some other countries, these activities may be carried out by other professionals, for example, specialized radiation therapists or medical physicists. In practice, in radiation oncology departments and processes, there is a seamless continuum between the many steps involving dose measurement and dose determination using a variety of professional staff.

For the purposes of this book, clinical dosimetry has been taken in the widest sense that covers a range of aspects of the radiation oncology process that can affect the dose absorbed in tissues in the patient: (1) dose measurement and requisite instrumentation; (2) dose delivery performance of the equipment (including calibration and QA) and of other treatment-related procedures; (3) dose prediction (e.g., treatment planning dosimetry); and (4) verification of dose delivery using dosimeters or data from image guidance.

As to *Clinical 3D Dosimetry in Modern Radiation Therapy*, the rationale for a book on this topic is based on the rapid recent expansion of new radiotherapy techniques, delivering more sophisticated three-dimensional (3D) (and increasingly for some treatment sites, four-dimensional [4D]) radiation dose distributions. Therefore, more sophisticated 3D (and 4D) dosimetry methods are demanded to support that push for more precision and accuracy. In principle, dosimetry has always been 3D in a real-world sense, and certain aspects of 3D dosimetry have been used in clinical practice for several decades, such as in the commissioning of accelerators and treatment planning systems. This has until relatively recently been largely based on point (0D), one-dimensional (1D), and 2D practical approaches such as the use of ionization chambers and planar films or diode arrays. The major recent advances in the field have demanded new and innovative direct 3D or semi-3D dosimetric tools and methods to accurately characterize and validate hard- and software, and to determine and verify patient- and organ-specific doses. This book aims to bring together these developments and information to cover the state-of-the-art of the accuracy, instrumentation, methods, and clinical as well as preclinical applications of 3D dosimetry in modern radiation therapy.

Thus, in summary, the topic is wide, but for the purposes of this book the main focus is on dose measurement, although dose calculations are also discussed in a restricted manner. Dose calculation algorithms will not be discussed in detail; however, the verification of the algorithms and their clinical application is an important part of the book. The theoretical physical aspects of the characteristics of dose computation models and measurement devices can be found in many textbooks and will only be discussed briefly. In this book, we focus on new developments of measurement techniques and 3D dosimetry of modern radiotherapy techniques including those of new image-guided treatment modalities.

This book is divided into five sections dealing with various aspects related to clinical 3D dosimetry. In the section entitled "Introduction," after summarizing the main topics discussed in the different chapters in this book, the clinical need for accurate 3D dosimetry is elucidated from different points of view. In the second section on "Instrumentation," experts in the use of the many different types of radiation detectors currently applied for dosimetry describe the specific application of these detectors to 3D dosimetry. The emphasis is on the clinical application of these detectors with a brief overview of the unique characteristics of the detectors of importance to 3D dosimetry. In the third section on "Measurement and Computation," various 3D and 4D dosimetry methods required for special treatment techniques, both already routinely applied or at the developmental stage, are described. In the fourth section of the book on "Clinical Applications," a range of 3D dosimetry techniques are discussed extensively for a large variety of treatment techniques and disease sites including intensity-modulated radiotherapy (IMRT), volumetric-modulated arc therapy (VMAT), brachytherapy, and proton/carbon ion therapy. In the final section, entitled "Emerging Technological Developments," 3D dosimetry techniques used for emerging technological developments in the field of radiotherapy are introduced.

The target audiences are primarily physicists involved in clinical radiotherapy. These include medical physicists working in radiotherapy departments, medical physics residents, and graduate students training in the field of radiotherapy physics. In addition, the book is intended to provide state-of-the-art information on relevant dosimetry methods for basic scientists and other researchers working on the development of new radiation detectors or sources. Finally, some chapters of the book are suitable as background information for radiation oncologists and radiation therapists, and may serve as teaching material for radiation oncology residents and other students and trainees.

Acknowledgments

The idea for this book originated about four years ago. Since that time both original editors (Ben Mijnheer and David Thwaites) had many exciting discussions about the concept of the book, its outline and content, finding authors and discussing with them about their chapters and developing the chapter outlines. Also, both spent a lot of time in responding to the questions and comments up to the point that draft chapters started to come in and the editing process began. Unfortunately, around that time, it became clear that it was not possible for David Thwaites to continue to put in the time to this project that he had expected and intended. He therefore had to withdraw as a coeditor, which was a setback for both of us and for the book timing. The editor (BM) very much appreciates that David has helped enormously to initiate and support such a book, for which we felt a clear need in our clinical practice and research community.

Editor

Ben Mijnheer got his PhD at the University of Amsterdam, Netherlands, on a study concerning standardization of neutron dosimetry. After working on a neutron therapy project in the Netherlands Cancer Institute (NKI) in Amsterdam, he joined this institution as a clinical physicist. Later, he became head of the Physics Department, professor at Inholland University of Applied Sciences, Hoofddorp, Netherlands, and is until now involved in research projects and teaching activities at the NKI. He served for many years on the editorial board of *Radiotherapy and Oncology* and *Medical Physics Journal*, is (co-)author of about 250 articles and chapters in books, and was supervisor of about 25 PhD theses. He was involved in the organization of the physics program of many international scientific meetings, and faculty member of numerous courses dealing with various aspects of radiotherapy for medical physicists, radiation oncologists, and radiation therapists, both at the national and international level. He received the ESTRO Breur Gold Medal and Emmanuel van der Schueren Award.

Contributors

Parham Alaei
Department of Radiation Oncology
University of Minnesota
Minneapolis, Minnesota

Michael B. Altman
Department of Radiation Oncology
Washington University
St. Louis, Missouri

Jerry J. Battista
Departments of Oncology and Medical Biophysics
Western University
London, Ontario, Canada

Glenn S. Bauman
Division of Radiation Oncology
Department of Oncology
Western University
London, Ontario, Canada

James L. Bedford
Joint Department of Physics
The Institute of Cancer Research and The Royal Marsden NHS Foundation Trust
London, England, UK

Bryan P. Bednarz
Department of Medical Physics
University of Wisconsin
Madison, Wisconsin

Jeremy T. Booth
Northern Sydney Cancer Centre
Royal North Shore Hospital
St. Leonards, New South Wales, Australia

and

School of Physics
University of Sydney
Sydney, New South Wales, Australia

Elke Bräuer-Krisch
Experiments Division
ESRF-The European Synchrotron
Grenoble, France

Indrin J. Chetty
Department of Radiation Oncology
Henry Ford Hospital
Detroit, Michigan

Gye Won (Diane) Choi
Department of Radiation Physics
The University of Texas MD Anderson Cancer Center
Houston, Texas

Catharine H. Clark
Department of Medical Physics
St. Luke's Cancer Centre
Royal Surrey County Hospital
Guildford, Surrey, UK

and

Metrology for Medical Physics
National Physical Laboratory
Teddington, UK

Emma Colvill
Sydney Medical School
University of Sydney
Sydney, New South Wales, Australia

J. Adam M. Cunha
Department of Radiation Oncology
University of California
San Francisco, California

Joanna E. Cygler
Medical Physics Department
The Ottawa Hospital Cancer Centre
and
Department of Radiology
University of Ottawa
and
Department of Physics
Carleton University
Ottawa, Ontario, Canada

Indra J. Das
Department of Radiation Oncology
New York University Langone
Medical Center
New York, New York

Yves De Deene
Department of Engineering
Faculty of Science and Engineering
Macquarie University
Sydney, New South Wales, Australia

Christopher L. Deufel
Department of Radiation Oncology
Mayo Clinic
Rochester, Minnesota

Nesrin Dogan
Radiation Oncology Department
Sylvester Comprehensive Cancer
Center
University of Miami
Miami, Florida

Simon Duane
Division of Acoustics and Ionizing
Radiation
National Physical Laboratory
Teddington, United Kingdom

David S. Followill
Department of Radiation Physics
The University of Texas MD Anderson
Cancer Center
Houston, Texas

Paolo Francescon
Department of Medical Physics
San Bortolo Hospital
Vicenza, Italy

Dietmar Georg
Division Medical Radiation Physics
Department of Radiation Oncology
Medical University of Vienna
Vienna, Austria

Peter B. Greer
Department of Radiation Oncology
Calvary Mater Newcastle
Waratah, New South Wales,
Australia

and

Mathematical and Physical
Sciences
University of Newcastle
Newcastle, New South Wales,
Australia

Rebecca M. Howell
Department of Radiation Physics
The University of Texas MD Anderson
Cancer Center
Houston, Texas

Mohammad Hussein
Department of Medical Physics
St. Luke's Cancer Centre
Royal Surrey County Hospital NHS Foundation Trust
Guildford, Surrey, UK

Geoffrey S. Ibbott
Department of Radiation Physics
The University of Texas MD Anderson Cancer Center
Houston, Texas

Titania Juang
Department of Radiation Oncology
Stanford Cancer Center
Stanford, California

Paul Keall
Sydney Medical School
University of Sydney
Sydney, New South Wales, Australia

Eric E. Klein
Department of Radiation Oncology
Washington University
St. Louis, Missouri

Tomas Kron
Department of Physical Sciences
Peter MacCallum Cancer Centre
Melbourne, Australia

Stephen F. Kry
Department of Radiation Physics
The University of Texas MD Anderson Cancer Center
Houston, Texas

Hannah Jungeun Lee
The University of Texas MD Anderson Cancer Center
UT Health Graduate School of Biomedical Sciences
Houston, Texas

Boyd McCurdy
Division of Medical Physics
Department of Physics & Astronomy
CancerCare Manitoba and University of Manitoba
Winnipeg, Manitoba, Canada

Moyed Miften
Department of Radiation Oncology
University of Colorado School of Medicine
Aurora, Colorado

Ivaylo B. Mihaylov
Radiation Oncology Department
Sylvester Comprehensive Cancer Center
University of Miami
Miami, Florida

Jean M. Moran
Department of Radiation Oncology
University of Michigan
Ann Arbor, Michigan

Mark Oldham
Department of Radiation Oncology
Duke University Medical Center
Durham, North Carolina

Hugo Palmans
Department of Medical Physics
EBG MedAustron GmbH
Wiener Neustadt, Austria
and
Division of Acoustics and Ionizing Radiation
National Physical Laboratory
Teddington, United Kingdom

Katia Parodi
Department of Experimental Physics—Medical Physics
Ludwig-Maximilians-Universität München
Munich, Germany

Chester Reft
Department of Radiation and Cellular Oncology
University of Chicago
Chicago, Illinois

Mark J. Rivard
Department of Radiation Oncology
Tufts University School of Medicine
Boston, Massachusetts

Donald A. Roberts
Department of Radiation Oncology
University of Michigan
Ann Arbor, Michigan

Yvonne Roed
Department of Physics
University of Houston
Department of Radiation Physics
The University of Texas MD Anderson Cancer Center
Houston, Texas

Emiliano Spezi
School of Engineering
Cardiff University
Cardiff, Wales, UK

Matthew T. Studenski
Radiation Oncology Department
Sylvester Comprehensive Cancer Center
University of Miami
Miami, Florida

Jonathan Sykes
Radiation Oncology Medical Physics
Sydney West Cancer Network—Blacktown Hospital
Blacktown, New South Wales, Australia

Jacob Van Dyk
Departments of Oncology and Medical Biophysics
Western University
London, Ontario, Canada

Frank Verhaegen
Maastro Clinic
Maastricht, the Netherlands

Zhifei Wen
Department of Radiation Physics
The University of Texas MD Anderson Cancer Center
Houston, Texas

David Westerly
Department of Radiation Oncology
University of Colorado School of Medicine
Aurora, Colorado

Suk Whan (Paul) Yoon
Department of Radiation Oncology
Duke University Medical Center
Durham, North Carolina

Kelly C. Younge
Department of Radiation Oncology
University of Michigan
Ann Arbor, Michigan

Section I
Introduction

Introduction and Overview of the Contents of the Book

Ben Mijnheer

1.1 Dosimetric Issues in Modern Radiation Therapy

In the last two decades of the twentieth century, radiotherapy showed a fundamental change in approach by the transition from conventional two-dimensional radiotherapy (2DRT) to three-dimensional conformal radiotherapy (3DCRT). In 2DRT, tumor volume and critical structures were drawn on orthogonal simulator films or on a few computed tomography (CT) images, simple setups with a few fields were used, treatment planning was performed with isodose plans in only one or a few planes, and generally broad margins were used. When implementing 3DCRT, tumor volume and critical structures were drawn on slice-by-slice CT or magnetic resonance (MR) images in combination with beam's eye views created from digitally reconstructed radiographs. Furthermore, complex setups of four or more fields with rigid immobilization were used, 3D treatment planning was introduced with 3D visualization and 3D plan analysis, while tight margins became general practice. The additional efforts for a clinic when starting 3DCRT have been elucidated in many reports published by national and international organizations such as the International Commission on Radiation Units and Measurements (ICRU, 1993, 1999) and the International Atomic Energy Agency (IAEA, 2008). The consequences with respect to dosimetric issues when introducing 3DCRT were enormous. The possibilities of the newly developed 3D treatment planning systems (TPSs) had to be well understood and extensive sets of commissioning and validation measurements were required to ensure the safe introduction of 3DCRT techniques in the clinic. Capabilities of a 3D TPS that had to be tested included the use of irregularly shaped treatment fields, 3D dose calculation algorithms, display of 3D anatomy and 3D dose distributions, and treatment plan evaluation tools. For all these activities, new dosimetric approaches were introduced, often based on the use of the same type of dosimeters as applied in conventional 2DRT, but measurements were now required in many more points in multiple planes.

Gradually, 3DCRT became more complicated with the use of techniques having small fields shaped by multileaf collimators (MLCs) or special cones, for instance, for stereotactic radiotherapy. With the introduction of intensity-modulated radiotherapy (IMRT) and volumetric-modulated arc therapy (VMAT), as well as with the clinical use of proton and carbon ion beams, numerous other new possibilities for creating complex treatment plans became available. All these new treatment modalities needed, however, additional efforts in assessing the accuracy of the various steps in the radiotherapy process as discussed in many reports (e.g., Ezzell et al., 2003; ESTRO, 2008; ICRU, 2010). This book aims to discuss the many aspects related to 3D, or even 4D, that is, including the time variation, dosimetry techniques needed to determine very accurately the 3D dose distributions involved in modern radiotherapy using these novel planning and delivery tools. It is intended to explain the many issues involved in the design and clinical implementation of these newly developed 3D dosimetry tools required for the introduction of new treatment technology. It should enable readers to select the most suitable dosimetry approaches, as well as equipment and methods for their application, based on numerical data, examples, and case studies provided.

In the first section of this book, diverse aspects related to the accuracy that is required and can be achieved in modern radiotherapy are elucidated. The need for accurate 3D dosimetry is clarified in Chapter 2 from different points of view, starting with the physical aspects followed by the biological and clinical context for accuracy and precision in contemporary radiation therapy. After discussing the basic metrology terminology and the relationship between dosimetry and geometry, a historical perspective on accuracy in radiotherapy is given, followed by a summary of recent data on uncertainties associated with 3D dosimetry and 3D dose delivery. The drive for dosimetric accuracy needs, however, to be considered in context with the inherent uncertainties associated with complex biological environments. It is shown in Chapter 2 that in many cases, biological and clinical uncertainties may predominate over uncertainties associated with the physical delivery of therapeutic radiation. These biological and clinical sources of uncertainties are associated with underlying tissue biology and with target delineations as determined by human observers. The impact of systematic and random uncertainties on clinical response has been used to formulate accuracy requirements for planning target volumes and organs at risk. Furthermore, it is discussed how these uncertainties can influence the design and interpretation of clinical trial outcomes incorporating radiation treatments. Strategies to deal with uncertainties in the clinic, such as quality management, clinical audit, and risk analysis, are also briefly discussed in this chapter. On the basis of this review of uncertainties in modern radiotherapy, recommendations regarding accuracy requirements are given, especially in the context of 3D dosimetry, which were partially taken from a recent report published by the IAEA (2016).

1.2 Instrumentation

In the second section of the book, experts in the use of the many different types of radiation detectors currently applied for dosimetry describe the specific application of these detectors to 3D dosimetry. The emphasis is on the clinical use of these detectors with an overview of the unique characteristics of the detectors of importance to 3D dosimetry for specific applications. The latest developments of dosimeters that can be used for applications in modern radiotherapy have recently also been discussed by Kron et al. (2016), with special attention to spatial resolution and dimensions for measurement as sorting criteria.

Detectors for reference dosimetry are reviewed in Chapter 3. After discussing a number of properties of reference-class ion chambers, the traceability of a calibration of these chambers, that is, the link to primary standards of absorbed dose, is discussed in detail. Special attention is paid to the use of various types of calorimeters for reference dose measurements. Details of the absorbed dose calibration and measurement process, including the calibration in flattening filter-free (FFF) beams, as well as a number of practical issues related to reference dose measurements, are also given in this chapter. New developments in reference dose measurements of treatment modalities for which it is difficult to meet the reference conditions described in dosimetry protocols, such as in photon beams in the presence of strong magnetic fields, are also briefly discussed in this chapter.

Point detectors (0D devices) are also very often used in the IMRT/VMAT/4D era, for instance, for measuring input data for a TPS and data sheets for quality assurance (QA) purposes as well as for validation of software/hardware/plans at multiple points. Some of these point detectors can also be configured into arrays to provide simultaneous multidimensional dose information. In Chapter 4, dosimetric characteristics and properties of commercially available point detectors, divided into active and passive detectors, are described. The specific application of a dosimeter depends upon the type of measurement: *in vivo* or *in vitro*, in-field or out-of-field, single point or multipoint, and whether an immediate measurement result is required or if off-line analysis may be performed. Some types of detectors, such as thermoluminescent dosimeters, supply also information on the linear energy transfer (LET) of particle beams. The information provided in Chapter 4 may give guidance to medical physicists in selecting a specific detector for a specific purpose.

The complex dose distributions produced by today's treatment equipment and delivery techniques require advanced dosimetry systems to provide confidence that the delivered distribution is consistent with the planned distribution. 3D dosimetry techniques are valuable to enable acquisition of volumetric information with a single irradiation. 3D dosimetry can be performed using a large variety of dosimeters, which can be divided into full-3D, pseudo-3D, or semi-3D systems. The only types of dosimeter currently available that are able to measure a full-3D dose distribution are polymer gel dosimeters and radiochromic 3D detectors. In Chapters 5 and 6, the rationale and methods of using these types of 3D dosimeters are discussed in detail, with particular attention to the preparation and dose readout of the various systems. The advantage of these dosimeters over pseudo-3D and semi-3D dosimetry systems is that they are tissue equivalent and can be molded into an anthropomorphic shape. They are therefore used as benchmarking tool for the commissioning of treatment plans of specific treatment techniques, and several examples of such applications of full-3D dosimetry are presented. After discussing the uncertainties in the dose determination due to variation in the preparation of these dosimeters and in the readout procedure, guidelines for the implementation of polymer gel dosimetry and radiochromic 3D detectors are provided.

In addition to their original use for patient setup verification, electronic portal imaging devices (EPIDs) are also used for many other purposes including dosimetric applications. Some EPID-based approaches are able to reconstruct the 3D dose distribution in the patient anatomy from EPID images and can therefore be considered as pseudo-3D dosimetry systems. Amorphous silicon (a-Si)–type EPIDs possess useful dosimetric characteristics such as the (almost) linearity of the response with dose and dose rate, good long-term stability, high spatial resolution, and real-time readout. These characteristics, and their applications in radiation therapy, have been described in detail in Chapter 7. Technical challenges to the routine use of a-Si EPIDs, such as improvements to signal acquisition and clinical software, are also explained. The remaining discussion in this chapter focuses on updating the literature review on EPID dosimetry provided by van Elmpt et al. (2008) to include publications between 2008 and mid-2016, numbering more than 100 additional references.

In Chapter 8, the use of multidimensional (2D and semi-3D) dosimetry systems is discussed. These systems, whether in a 2D or other array of detectors, provide real-time feedback at the time of measurements, and an overview of the basic characteristics of planar and semi-3D systems is provided. It is essential that these multidimensional detectors are adequately characterized for clinical use, and guidelines for performing these types of measurements are given. However, when superior spatial resolution is required, passive detector systems such as radiochromic films are often the preferred method, and their dosimetric characteristics are also provided in this chapter. Furthermore, novel 2D or semi-3D dosimetry systems are discussed in this chapter, including detectors to measure Cherenkov radiation during delivery, plastic scintillators, which can be set up in arrays similar to other multidimensional systems, and devices that provide real-time feedback of a patient's plan delivery based on the exit fluence from the collimator of the linear accelerator (linac).

1.3 Measurement and Computation

In the third section of the book, various dose measurement and dose computation techniques required for special treatment methods, both already routinely applied or at the developmental stage, are described. These techniques differ because of the use of particular field sizes, varying from very small to total body irradiation conditions, are changing with time (4D approaches), or apply special treatment machines, including proton and carbon ion facilities. Quantifying differences of measurements and computations such as Monte Carlo (MC) calculations are also discussed in this section of the book.

Many modern irradiation techniques have changed the paradigm on the limit of radiation fields using very small fields for patient treatment; up to subcentimeter dimension. In large fields, dosimetric parameters are well defined and can be accurately measured. However, with shrinking field size, lateral electron equilibrium cannot be established and traditional dosimetry techniques cannot be utilized. Manufacturers provide many types of detectors whose characteristics in small fields are, however, often not well known. Chapter 9 provides detailed information to better understand the complexity of using detectors in small fields and its implications for dosimetry purposes. Data are given for a number of detectors for their use in small fields generated by linacs and other treatment units as a function of field size and beam energy. Classes of detectors having favorable characteristics such as water equivalence, small volume, and minimum field perturbation have been identified and should therefore be encouraged for use in small fields.

Most external beam radiotherapy treatments are delivered with accelerators having C-type gantries. Chapter 10 discusses the development of these C-shaped RT treatment units, their possibilities, as well as their limitations. Characteristics of other types of modern radiotherapy equipment, also able to deliver special treatment techniques, are compared with these types of accelerators. This chapter then discusses a number of improvements in the design of radiotherapy treatment units that are still possible, such as improvement of dose delivery of very small or very large fields, and improvement of target visualization and motion management. The final section of the chapter gives comments on health economics and a general outlook on issues such as access to RT worldwide, the cost of RT equipment including optimization of features required in a department, as well as computerization and automation.

Chapter 11 describes how to perform 4D dosimetry by measuring the dose in 3D while the detector position is changed with time during dose delivery. The rationale for performing time-dependent dose measurements is that human anatomy during a radiotherapy treatment may change with time. Currently available 4D dosimetry systems to mimic target volume and/or normal tissue motion during radiotherapy treatments are discussed. They are still approximations of real treatment situations but allow for assessment, commissioning, and quality assurance of treatment techniques involving motion management. This chapter includes 4D dosimetry of photon beam delivery methods in which anatomic motion is implicit and covered in margins (motion inclusive), as well as delivery methods in which anatomic motion is explicitly accounted for, that is, gating and real-time adaptive radiotherapy.

Radiation therapy with proton and carbon ion beams is a rapidly emerging treatment modality due to its superior beam characteristics compared to photon beam radiotherapy, allowing in principle excellent sparing of healthy tissue for the same, or better, target coverage. In Chapter 12, aspects of the interaction of light ions with matter that make ions distinct from photon and electron beams are summarized, followed by an overview of different detector technologies to determine the absorbed dose to water at a point or in 3D in light-ion beams. The main issues with respect to 3D dosimetry that are discussed in this chapter are the water equivalence and energy dependence of detectors. Due to the LET variation in the depth dose distributions, special attention is given to the energy dependence of the various detectors currently used. Ion chambers can in general be considered the gold standard for relative dosimetry in light-ion beams, but their response nevertheless exhibits a small energy dependence that is worth considering when aiming for more detailed and better quantitative data, for instance, for the determination of experimental relative biological effectiveness (RBE) values. Other detectors can exhibit substantial response quenching as a function of energy and should therefore always be benchmarked against ion chambers prior to clinical use. In order to better understand radiation-induced biological effects resulting from light-ion beam irradiation, microdosimetric and nanodosimetric detectors are also briefly discussed in this chapter.

The impact of accurate dose distributions on patient clinical outcomes is of ultimate importance. The MC method performs calculation of photon and electron tracks within the treatment unit and patient tissues very accurately, and as such imitates how radiation is physically delivered to the patient, as discussed in a comprehensive way in many textbooks (e.g., Verhaegen and Seco, 2013). In Chapter 13, it is shown that MC techniques are increasingly used in assessing the accuracy, verification, and calculation of 3D dose distributions. After giving an overview of MC-based photon and electron transport codes, application of MC in 3D photon and electron dosimetry is elucidated. MC techniques also aid in perturbation calculations of detectors used in small and highly conformal radiation beams. Furthermore, it is shown that research evaluating dose–volume effect relationships may help in elucidating the benefit of the more accurate dose distributions afforded by the MC method on observed clinical outcomes.

Given the complexity and large numbers of variables involved in modern radiation therapy, it is often necessary to compare calculated and/or measured 3D dose distributions. In Chapter 14, a number of quantitative methods are discussed how to perform these comparisons, including the use of dose difference and distance-to-agreement methods, gamma evaluation, and region of interest analysis metrics. The advantages and disadvantages of each of these methods, applied in different situations, are presented. Issues related to extracting the clinical meaning of observed differences when verifying patient treatment plans in a phantom, that is, defining tolerances tied to clinical outcomes, are elucidated. Another subject discussed in this chapter stems from the need to assign physical dose difference and distance tolerances, as well as thresholds for the number of points in a comparison that pass a given test. Further discussion on this issue is presented in Chapter 17 when presenting patient-specific pretreatment 3D dose verification methods.

1.4 Clinical Applications

In the fourth section of the book, clinical applications of the various 3D dosimetry approaches are extensively discussed for a large variety of treatment techniques and disease sites delivered using IMRT, VMAT, brachytherapy, and light-ion therapy. With the implementation of IMRT and VMAT, low-dose regions outside the treatment volume received more attention than in the past, while the impact of imaging dose for modern RT techniques also became more important with the introduction of image-guided radiotherapy (IGRT). These topics are therefore also discussed in this chapter in addition to a large variety of issues related to 3D dosimetry of the target volume and nearby organs at risk.

Before clinical application of a specific treatment technique can start, extensive testing of the delivery and planning equipment used for that treatment is a prerequisite. Chapters 15 and 16 discuss in detail the acceptance testing, commissioning, and QA of linear accelerators, and commissioning and QA of treatment planning systems, respectively. In Chapter 15, a number of different approaches and techniques for acceptance testing, commissioning, and regular linac QA are presented. A large variety of equipment needed for this purpose is discussed in this chapter, which includes water tank systems, phantoms, detectors, and devices for scanning data, point dose, and array measurements, for use in photon as well as electron beams. Guidance for acceptance testing and the types and quantities of measurements needed for commissioning of linacs is also provided. The last part of the chapter discusses the final step in an acceptance testing and commissioning process of a linac, which is the compilation and documentation of collected data into reports and data sheets to be used for regular QA measurements.

The increased complexity of a modern TPS requires a comprehensive commissioning and QA program of the system, which should incorporate all constraints imposed by the rapidly evolving new technologies. Chapter 16 summarizes the measurements, testing, and validation of the dosimetric aspects of a modern TPS. It includes a description of experimental techniques to assure a complete and accurate beam model, the commissioning process, the validation of state-of-the-art 3D dose calculation algorithms, the dosimetric effects of immobilization devices and couch, and guidelines for performing routine QA tests and end-to-end tests for many types of external beam irradiations including IMRT, VMAT, and stereotactic treatments. A number of examples of these tests are then provided, starting from the simplest possible setup for 3D conformal radiation therapy to complex IMRT setups.

It was the general belief that performing a thorough commissioning program of both the treatment machine and the TPS, as elucidated in Chapters 15 and 16, respectively, would allow the safe introduction of advanced 3DCRT techniques in the clinic. Patient-specific pretreatment verification of 3DCRT was generally limited to an independent monitor unit (MU)/dose calculation, whereas measurements with 2D or semi-3D devices for individual patient treatments were not a common procedure. With the introduction of IMRT this situation changed. The higher complexity of the dose calculation and dose delivery of IMRT treatments compared to 3DCRT was the main reason that with the introduction of IMRT patient-specific pretreatment verification started. Chapters 17 and 18 give an overview of the various approaches in which patient-specific QA is performed.

Chapter 17 starts with discussing the rationale behind pretreatment verification and what errors can be detected in this way and which not. The evolution of pretreatment 2D and 3D dose verification, the various measurement-based techniques, and tolerance and acceptance levels are then presented in a comprehensive way. After discussing measurement-based techniques, the increasing use of calculation-based techniques and their limitations are elucidated. The chapter ends with a discussion of the impact of recent developments in RT, for instance, the use of FFF photon beams and adaptive radiotherapy, on future QA concepts.

A number of dosimetric errors cannot be detected by pretreatment QA, such as those which occur randomly and those which are associated with patient setup or anatomy changes. These errors can only be discovered using *in vivo* dosimetry approaches and are discussed in Chapter 18. After elucidating the relationship between *in vivo* dosimetry and pretreatment verification, details about the different measurement techniques for *in vivo* dosimetry are provided with emphasis on 3D verification using EPIDs. For the clinical implementation, tolerance and action levels have also to be defined, as in the case of pretreatment verification, which needs additional input of other clinical staff. A number of examples are then given of what has been observed in clinical practice, followed by recent developments and future approaches of *in vivo* dosimetry such as real-time verification.

Since the planning and delivery process of an advanced RT treatment comprise a series of interdependent steps, it is extremely difficult to perform QA for each individual step. For that reason, the entire treatment process must be evaluated using what is known as an end-to-end QA test. For RT treatments, an end-to-end QA test is an audit methodology that tests whether all components in the treatment process function in a manner such that the desired radiation dose is delivered accurately only to the intended spatial location. It is therefore different compared to the patient-specific pretreatment procedures outlined in Chapter 17, which verify only part of the treatment process. In Chapter 19, end-to-end approaches applied in North America (Imaging and Radiation Oncology Core [IROC] Houston), Western Europe (Institute of Physics and Engineering in Medicine [IPEM]), and Australia (Australian Clinical Dosimetry Service [ACDS]) by external groups are presented. The differences in approaches taken to end-to-end testing across the world, due to historical, resource, and demand reasons, are elucidated. Details are provided about many issues involved in end-to-end testing by these three approaches such as clinical trial credentialing, phantom-detector design, audit acceptance criteria, and practical logistics. Results, historical and trending, as well as accomplishments in reducing errors are also provided.

The content of this book is mainly devoted to 3D dosimetry in external beam radiation therapy (EBRT) because there exist already many textbooks on brachytherapy physics that include information dedicated to 3D dosimetry (e.g., Venselaar et al., 2012). It was thought, however, that it would be good to include in this book also a chapter elucidating specifically the differences between 3D dosimetry in EBRT and in brachytherapy. Chapter 20 starts with discussing 3D data considerations in brachytherapy, which are related to using lower photon energies and the placement of the radioactive sources inside or near the tumor. Building upon these concepts, it describes the basis for calculating and measuring brachytherapy dose distributions in 3D, covers the practical aspects of performing image-guided 3D treatment planning, and covers the importance of techniques for commissioning applicators and validation of treatment delivery with *in vivo* dosimetry. It is shown that the measurement of brachytherapy dose distributions is more challenging than for EBRT, mainly due to the higher dose gradients and larger detector response corrections at lower photon energy. For these and other reasons, the methods for validating brachytherapy treatment delivery are currently inferior to and less commonly used than those used in EBRT.

Dose outside the treatment volume is detrimental to the patient and may be particularly concerning under special circumstances such as the treatment of pregnant patients or of patients having an implantable electronic device. Knowledge of the low radiation doses outside the treatment volume is also of growing concern because the risk of late effects from secondary radiation may be more evident today than in the past. Consequently, properly assessing nontarget doses, as a means to document and minimize them, is an increasingly important issue. In Chapter 21, it is clarified that measuring this dose poses many unique challenges because the radiation field outside the treatment volume is much different from that inside it. The average energy is lower, the dose rate is lower, and the dose distribution is dissimilar; all these issues must be considered when conducting out-of-field measurements. Details are provided in this chapter about the use of various types of detectors for measuring the dose outside the treatment volume in EBRT. Particular attention is given to dose, or dose equivalent, measurements of neutrons in or on a phantom or patient, because of the complexity of energy spectrum variations of the neutrons in relation to the energy response of the dosimeter.

Most previous chapters in this book are focused on the instrumentation and measurement techniques for dosimetry of therapeutic external beams or around radioactive sources. In Chapter 22, a variety of measurement techniques are reviewed for characterizing the radiation dose from x-ray imaging systems used in radiation therapy for localizing the target volume and nearby organs at risk. Similar systems are used for imaging the patient prior to beam delivery, to ensure accurate patient alignment, or during beam delivery to monitor intrafraction motion. Many x-ray imaging systems are used to perform image guidance and they all lead to additional dose to the patient. It is therefore important to quantify this dose in order to justify the risks of using x-ray imaging against the benefits for a particular IGRT protocol. Because imaging procedures produce a highly inhomogeneous 3D dose distribution in a patient, special measurement techniques are required to determine the imaging dose. In this chapter, the measurement of imaging dose is reviewed for CT and cone-beam computed tomography (CBCT) using various detector systems and measurement quantities. Dose measurement techniques for QA purposes, for planar kilovoltage (kV) imaging and for megavoltage (MV) portal imaging, as well as results of these measurements for various treatment techniques, are discussed in a comprehensive way. At the end of the chapter, calculation techniques of the imaging dose and application of these dose calculations, for instance, for combining dose from RT and imaging, are provided.

Radiation therapy with proton and carbon ion beams is a rapidly emerging modality. However, the scanned beam delivery of these beams poses challenges for accurate characterization of the pencil beams and in-field dosimetric verification. In Chapter 12, specific requirements and characteristics of detectors to be used for reference and 3D dosimetry in light-ion beams are discussed. Chapter 23 addresses the complex requirements to the clinical 3D dosimetry of proton and carbon ion beam therapy related to the wide range of pencil-beam parameters and the dynamic way of dose delivery. After discussing the dosimetric measurements for pencil-beam characterization and TPS basic data generation, tests are described for TPS commissioning using treatment plans of different complexity. Whereas these dosimetric measurements enable 3D dose calculations, additional depth-dependent fluence energy spectra measurements are needed to characterize the complex RBE dependence with radiation quality, particularly in carbon ion beams. Methods are then described to verify dosimetrically light-ion beam therapy plans of patient-specific treatments prior to the first treatment session, similar to the current practice of IMRT and VMAT. Accurate verification of the actual dose delivered to the patient or, at least, *in vivo* confirmation of the beam range, is still an unmet challenge of light-ion beam therapy. Several methods are discussed in this chapter including positron-emission tomography (PET) and the detection of high-energy photons promptly emitted during dose delivery. The chapter ends with discussing neutron dosimetry and spectrometry measurements, which is particularly important to account for out-of-field secondary effects during light-ion beam therapy.

1.5 Emerging Technological Developments

In the final section of this book, 3D dosimetry aspects related to emerging technological developments in the field of radiotherapy are revealed. A modern trend in radiation biology preclinical research is to mimic the radiation conditions of human radiotherapy as closely as possible. This has led to various devices to allow image-guided irradiations using very small kV x-ray beams, which is the topic discussed in Chapter 24. These novel radiation platforms, however, pose new challenges for mechanical targeting accuracy, dosimetric accuracy, and imaging characteristics. After presenting compensation mechanisms for systematic mechanical errors, absolute dosimetry calibration of these kV x-ray beams using existing dosimetry protocols is discussed. 3D dosimetry methods of small field precision irradiators is challenging due to the sharp beam penumbras and steep drop in beam output for these small fields. Characteristics of several dosimeters used for accurate measurements in points, planes, and volumes, in realistic irradiation geometries, exposed to small beams of kV photons, are then presented. Several methods to perform dose calculations in small animal specimens are also discussed, ranging from simple analytical models, superposition–convolution approaches, to MC simulation. Furthermore, issues related to focal spot, portal dosimetry, and QA of these devices are discussed in this chapter. Future developments are briefly described as the radiation platforms will increase their degrees of freedom, for instance by adding variable collimators and motion-gated irradiation.

Stereotactic synchrotron radiotherapy (SSRT) and microbeam radiation therapy (MRT) are novel approaches in radiation therapy to treat brain tumors and potentially other tumors using synchrotron radiation. In Chapter 25, the medical physics aspects of these treatments are discussed, which are challenging due to the very small field sizes used, having sometimes a width of 25 μm by several centimeters in height, down to 50 μm × 50 μm spot sizes for pencil beams. An important other feature of these synchrotron sources is the extremely high dose rate. A number of examples are given of micrometer-sized field dosimetry for the specific application in MRT elucidating the MRT-specific medical physics problems related to the combination of dose range, dose rate, low- to medium-energy photons, and extremely high resolution. Characteristics of various types of dosimeters used in synchrotron beams, including radiochromic film, Si-based single and multiple strip detector systems, and radiochromic plastic dosimeters are then discussed in detail. The chapter ends with discussing various issues related to preclinical research in MRT, which has demonstrated great potential of spatially fractionated radiation therapy using microscopically small beam sizes.

In Chapter 26, it is shown that in the emerging field of MR IGRT, the presence of strong magnetic fields can affect the performance of most conventional dosimetry systems. For instance, ion chamber design and construction play important roles in the magnitude of magnetic field effects on their response. Even more important is the orientation of the ion chamber axis to the magnetic field; the influence of the magnetic field decreases considerably when the chamber axis is parallel to the magnetic field. However, in this chapter, it is also shown that several novel 3D dosimeters perform well in the presence of magnetic fields and can provide quantitative dose distributions in a volumetric manner. While the available data are preliminary, these results indicate the potential for 3D dosimeters, including both gels and radiochromic polyurethane, to provide reliable measurements in clinically relevant circumstances.

References

ESTRO (2008) Guidelines for the verification of IMRT. *ESTRO*, Booklet No. 9. (Eds. Mijnheer B and Georg D, Brussels, Belgium: European Society for Radiotherapy and Oncology).

Ezzell G et al. (2003) Guidance document on delivery, treatment planning, and clinical implementation of IMRT: Report of the IMRT Subcommittee of the AAPM Radiation Therapy Committee. *Med. Phys.* 30: 2089–2115.

IAEA (2008) Transition from 2-D radiotherapy to 3-D conformal and intensity modulated radiotherapy. TECDOC-1588. (Vienna, Austria: International Atomic Energy Agency).

IAEA (2016) Accuracy requirements and uncertainties in radiation therapy. Human Health Series No. 31. (Vienna, Austria: International Atomic Energy Agency).

ICRU (1993) Prescribing, recording, and reporting photon beam therapy. Report 50. (Bethesda, MD: International Commission on Radiation Units and Measurements).

ICRU (1999) Prescribing, recording, and reporting photon beam therapy (Supplement to ICRU Report 50). Report 62. (Bethesda, MD: International Commission on Radiation Units and Measurements).

ICRU (2010) Prescribing, recording, and reporting photon-beam intensity-modulated radiation therapy (IMRT). Report 83. (Bethesda, MD: International Commission on Radiation Units and Measurements).

Kron T, Lehmann J and Greer PB (2016) Dosimetry of ionising radiation in modern radiation oncology. *Phys. Med. Biol.* 61: R167–R205.

van Elmpt W, McDermott LN, Nijsten S, Wendling M, Lambin P and Mijnheer B (2008) Literature review of electronic portal imaging for radiotherapy dosimetry. *Radiother. Oncol.* 88: 289–309.

Venselaar J, Meigooni AS, Baltas D and Hoskin PJ (2012) *Comprehensive Brachytherapy: Physical and Clinical Aspects.* (Boca Raton, FL: CRC Press).

Verhaegen F and Seco J (2013) *Monte Carlo Techniques in Radiation Therapy.* (Boca Raton, FL: CRC Press).

2

Accuracy Requirements for 3D Dosimetry in Contemporary Radiation Therapy

Jacob Van Dyk, Jerry J. Battista, and Glenn S. Bauman

2.1 Introduction

The level of accuracy required in radiation therapy is primarily driven by the clinical response of diseased and normal tissue to radiation and the realistic acceptance of what dosimetric accuracy can be practically achievable using today's instrumentation. The clinical aspects relate to dose–response curves that tend to be very steep for both tumors and normal tissues such that small changes in dose have the potential of producing large changes in radiobiological response. Moreover, the curves may be displaced such that a strong dose *gradient* is essential to achieving a good tumor response without inducing a major negative side effect in adjacent normal tissues and organs. This is indeed the *raison d'être* and goal of contemporary intensity-modulated radiation therapy (IMRT). The rapid evolution of more conformal dose delivery combined with the transition of image guidance from two-dimensional (2D) radiography and ultrasound to three-dimensional (3D) computed tomography (CT) and magnetic resonance imaging (MRI) is also adding potential for dose escalation and improving the therapeutic ratio between tumor and normal tissue response. A wider exploitation of dose–response curves, including novel dose fractionation schemes, will lead to new dimensions in spatiotemporal optimization of radiotherapy. Hence, a review of the required accuracy in radiation therapy seems timely if these goals are to be achieved realistically and safely with new technology and optimization strategies. This is the primary focus of this chapter.

The International Atomic Energy Agency (IAEA) has recently indicated four major considerations for setting realistic goals for accuracy requirements in radiation therapy (IAEA, 2016; van der Merwe et al., 2017):

1. Radiobiological: steepness of dose–response curves and the impact of dose accuracy on clinically based outcomes such as tumor control and normal tissue complications.
2. Clinical variability in defining the target volume and organs at risk (OAR) with today's imaging capabilities and limitations.
3. Statistical considerations for establishing and interpreting the measurement of clinical outcomes in clinical trials.
4. Dose accuracy that is practically achievable with the current generation of dosimetric instruments and dose modeling software.

In this chapter, we address the many aspects of accuracy requirements, especially as related to 3D dosimetry. The meaning of 3D dosimetry was described in the preface with the main focus being on dose measurement, although dose calculations used in treatment planning are also considered here in a more restricted manner. In addition, some of the nontechnical limitations influencing dosimetry accuracy requirements are briefly discussed. A large component of these considerations is tempered by what accuracy is practically available and achievable routinely based on practical experience in clinical physics. Clearly, even if the clinical requirements call for a required accuracy of 1% but only 5% is practically measurable, then setting a *requirement* of 1% becomes academic and totally impractical. Thus, a large component of this chapter reviews the literature describing what has been practically achieved in 3D dosimetry with today's hardware and software tools. Similarly, a technical ability to measure dose with an accuracy of 1% becomes irrelevant if other inaccuracies (such as target volume delineation or knowledge of fundamental radiobiology or dose–response effects) are limiting.

2.1.1 Interplay of Dose and Geometric Accuracy

In the most general sense, dose is a scalar function of a four-dimensional (4D) space, i.e., $D(x, y, z, t)$ when we consider the dynamic nature of radiation delivery and a patient with a changing anatomy over the course of therapy. Therefore, there are two main components in assessing the physical aspects of accuracy requirements in radiation therapy dosimetry. The first relates to accuracy in the magnitude of delivered dose (D) "at a point" and the second relates to the spatial location of such a point in the *patient coordinate space* (x, y, z) and irradiation time (t). The accuracies in dose and in geometry are not independent, since spatial distortion in the patient's tissue space can often impact dose when a dose gradient is present. In other words, we can consider a tissue voxel as a mobile collection of cells that can move in the applied radiation fields. However, for the sake of this discussion, these two features will be treated individually since the underlying characteristics and quality controls required to maintain accuracy can be quite different. For example, a calculation of a dose distribution in heterogeneous tissue requires a trustworthy physical model of radiation absorption and scattering transport, while tissue movement may be mitigated by radiation gating techniques.

In the 2D era, until the 1980s, accuracy considerations related primarily to the question of how well we could calibrate a radiation beam or source in a reference phantom, with major uncertainty in ascertaining if this accuracy could then be preserved in 3D patient tissues. Spatial accuracy considerations have now become much more important as we have moved into the 3D era of treatment planning with patient-specific, CT scan-based, anatomical data such that we can derive information about the current location and *in vivo* densities of small elements (voxels) of tumors and normal tissues, and new methods of beam collimation delivery incorporating steep dose gradients. Spatial accuracy of these voxel locations has become even more relevant as online image guidance has become available on the therapy machine providing many instances of evolving anatomy, i.e., the "anatomy of the day." Thus, image guidance in radiation therapy provides the basis for improved 3D retargeting and 3D dose delivery, and a potential to enhance accuracy of dose at the moving tissue voxel level. If the virtual dose-image grid used in treatment planning is mapped reliably onto the patient's coordinate space, the dose accumulated by each tissue voxel over multiple dose fractions is within grasp. There is clearly a strong role here for accurate deformable image-dose registration algorithms that have been developed in computer sciences and are being adopted and evaluated for oncology applications. Several questions remain

1. How accurately do we need to know the *in vivo* dose to tissue voxels?
2. Which voxels are more important?
3. What accuracy can we realistically attain, taking into account the propagation of uncertainty across multiple clinical processes with quality assurance (QA) check points?

2.1.2 Dimensionality of Dose Measurements

While modern technologies provide 3D and some 4D capabilities, both in dose delivery and in dose measurement, some of the fundamental dosimetry and commissioning procedures still rely on measurements taken in subdimensions of the full dosimetry space. Basic beam calibrations are still performed with a "0D" ionization chamber at a reference point in a water (or water-like) phantom (Figure 2.1a). The challenge is to port this absolute dose accuracy into knowledge of the full 3D/4D dose distribution in the patient (Figure 2.1b).

Point measurements (0D) still remain as the fundamental procedure for determining the reference absolute dose rate, traceable to a national standards laboratory, and are used for the dose determination in the treatment of patients to achieve the clinically prescribed dose and dose distribution. For commissioning of treatment planning systems (TPSs), we often also measure dose profiles across the radiation beam at multiple depths. These profiles individually provide a one-dimensional set of data that can be used for setting dose calculation parameters. If film is used, or any other measurement capability that provides an array of doses in a single plane, then we have a 2D data set. Volumetric chemical dosimetry in gels or plastics, read out by CT or magnetic resonance, provides a new capability of generating a dose data set in 3D using a single irradiation as discussed in

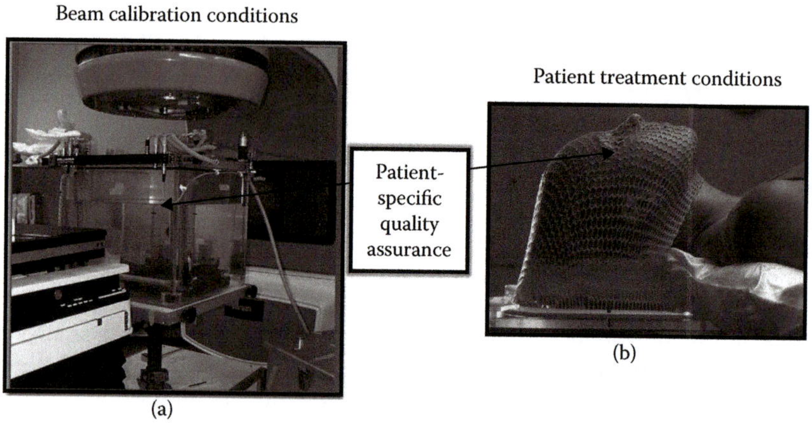

Beam calibration conditions

Patient treatment conditions

Patient-specific quality assurance

(a)

(b)

Figure 2.1

(a) Beam calibration and machine commission geometry (Courtesy of T. Kron). (b) Patient treatment space (Courtesy of L. Verhey). The radiation dosimetry challenge is to determine the 3D or 4D dose distribution in the heterogeneous patient, based on dose samples determined in the calibration conditions and quality assurance procedures. One of the goals of this chapter is to review the uncertainties associated with phantom and patient irradiation conditions.

Chapters 5 and 6. If motion is incorporated during 3D dose measurements to simulate breathing or other motion effects, then we achieve 4D dosimetry, which is discussed in detail in Chapter 11. For example, Poole et al. (2011) describe a hybrid dosimetry system to allow for the simultaneous determination of spatio-temporal dependence of dose during a dynamic radiation delivery.

Each of these methods of dose determination offers different dose and spatial accuracy and their limitations in accuracy and precision need to be understood. Note that patient-specific QA procedures (Ezzell et al., 2003; Galvin et al., 2004; Hartford et al., 2012) using ancillary phantom measurements fall within the central portion of Figure 2.1. On the other hand, *in vivo* dose measurements using surface, intracavitary, or transit dosimetry with electronic portal imaging devices (EPIDs) and CT data (Mijnheer et al., 2013b; Van Uytven et al., 2015), fall closer to the right side of the figure. In summary, all forms of calibration and QA procedures are aimed at narrowing the uncertainty gap region and minimizing the potential degree of discrepancy between the two types of irradiation illustrated in Figure 2.1.

2.1.3 Dosimetry within an Evolving Technology

An interesting historical note is that the accuracy in absolute dose delivery has not always improved with new radiation treatment technology. By way of example, in the 2D era, the International Commission on Radiation Units and measurements (ICRU) Report 24 (ICRU, 1976) indicated that striving for a 5% accuracy in dose delivery was reasonably achievable. As we moved into the IMRT era, study groups were developing criteria of acceptability for some treatment circumstances that were as large as 7% in dose and 4 mm in spatial accuracy. For head and neck cases, only 69% of the institutions passed such wide criteria in the early days of implementation, although that has improved to 91% in recent years (Molineu et al., 2013). Thus, major technological transitions to more complex procedures carry a risk of increasing treatment uncertainty, if not significant treatment errors during the early adoption and learning phases (ICRP, 2009).

A counter example is in order. A review by Van Dyk and Battista (2014) of the impact of the specific use of computers on the accuracy of radiation dose delivery concluded that, as a result of computer applications, we are now better able to track changes in internal anatomy of the patient before, during, and after treatment. This has yielded the most significant advance to the knowledge of *in vivo* dose distributions propagated in the patient over the weeks of therapy. Furthermore, a much richer set of 3D/4D coregistered dose-image data is becoming available for retrospective analyses of radiobiological tissue response and clinical outcomes. The impact of computer simulations on absolute beam calibration improvements has yielded changes in dose-delivery accuracy of the order of 1%–4%, primarily due to fundamental, calibration-related parameters that can now be determined with Monte Carlo techniques in reasonable computation times.

2.1.4 Chapter Objective

In view of the major hardware and software changes in dose-delivery technologies and techniques (e.g., image-guided adaptive radiation therapy [IGART], IMRT, volumetric-modulated arc therapy [VMAT]) in the last decade and in view of a wider range of dosimetry and QA procedures, this chapter reviews recent data on uncertainties associated with 3D dosimetry and 3D/4D dose delivery. Based on this review, a concise summary is presented on the state-of-the-art accuracy in contemporary radiation oncology. The current state is also placed in the context of other sources of inaccuracy in radiation treatment to provide perspective on other limiting factors in the effective delivery of radiation.

2.2 Terminology of Metrology

Before we describe the uncertainties encountered in the current era of radiotherapy, we define some key concepts. Recent reports (Thwaites, 2013; Van Dyk et al., 2013; IAEA, 2016) have included clearer definitions of metrology terminology largely based on recommendations of international organizations such as the *Bureau International des Poids et Mesures* (BIPM) and the International Organization for Standardization (ISO). A few definitions are reiterated here in view of the importance of these terms in the context of this chapter and book.

Conceptually, *accuracy* is the closeness of agreement between a measured quantity and its "true value." This determination generally falls within the domain of national standards procedures achieving a consensus on ground truth. It is especially important to have a universal dose calibration to avoid *systematic* dose offsets across multicenter clinical trials to maintain validity of intercomparing clinical outcomes internationally. Conceptually, *precision* is the closeness of agreement of the results when the same measurement is made repeatedly in real or virtual experiments. In real experiments, this usually refers to reproducibility of the experimental setup, irradiation conditions, and random noise in the detection system (e.g., Poisson fluctuations in a radiation counter). In virtual simulations of radiation particles, it normally refers to the intrinsic statistical fluctuations of a scored quantity such as energy fluence or dose, resulting from using a finite number of particle histories in Monte Carlo simulations, for example.

Numerically, *uncertainty* characterizes the dispersion of values that can be obtained for a particular procedure when it is performed repeatedly. It is intended to reflect the confidence in accuracy or precision in numerical terms. For repeated measurements or simulations, the results are represented by a statistical distribution, which can be summarized by standard statistical quantities such as mean, mode, standard deviation (SD), and variance. Uncertainty statements are most often given by the SD of a measured quantity. This is also referred to as the *standard uncertainty*, which is often represented by the symbol u.

Combined standard uncertainty is the standard uncertainty of a quantity that is composed of multiple components, each of which has individual uncertainty (u_a, u_b, etc.). It is usually obtained by taking the square root (SQRT) of the individual uncertainties added in quadrature, if we assume that the contributing uncertainties are normally distributed and independent:

$$u_{\text{total}} = \text{SQRT}\ (u_a^2 + u_b^2 + u_c^2 + \ldots) \tag{2.1}$$

where u_{total} is the total uncertainty and u_a, u_b, u_c, and so on, are individual uncertainties contributing to the overall uncertainty.

The *expanded uncertainty* is the standard uncertainty multiplied by a *coverage factor, k*, such that an increasing k-value attributes greater confidence that the correct value lies within the resultant extended range. The expanded uncertainty is given by U:

$$U = k \times u \tag{2.2}$$

Often k is greater than 1 for higher levels of confidence so that the values obtained by an independent identical experiment or simulation will lie within the specified range with a defined probability. For k equal to 1, 2, or 3, there is approximately a 68%, 95%, or 99% probability, respectively, that the mean of the measured quantity lies between $\pm U$. Often we will see uncertainties quoted "at the $k = 2$ level" or "with $k = 2$." Alternatively, the results are sometimes reported as corresponding percent confidence intervals (CIs) or error bars in tabular or graphical data, respectively.

Note that quantitative statements of uncertainty can apply to accuracy and/or precision. In absolute dose calibrations, the uncertainty statement is issued by the standards calibration laboratory. In local measurements of doses in a clinical setting, the precision is determined by ambient conditions and uncertainty is determined by the experimentalist based on the procedures and instruments used. If an absolute dose is inferred by using a set of local instrument readings made with a calibrated device, the overall uncertainty statement must include the intrinsic calibration uncertainty and the precision of the experiment. Furthermore, error propagation algorithms to determine overall uncertainty are required if the dose quantity of interest (e.g., output factors) relies on combinations of independently measured dose values (e.g., ratio of readings).

Uncertainties also fall into two categories, based on the methods used to determine their values: *Type A uncertainties* are those that are evaluated by repeated procedures and statistical assessment methods described above. *Type B uncertainties* are those that are determined by means other than statistical methods, often using intelligent estimates based on expert scientific judgments. In the past, Type A and Type B uncertainties have frequently been loosely referred to as random and systematic errors; however, it is now well recognized that there is no correspondence between these terms and this practice should be discontinued.

2.3 Review of Recent Reports on Achievable Accuracy

2.3.1 Historical Reports and Perspective

One of the most frequently cited reports on accuracy requirements in radiation therapy is the 1976 report by the ICRU (1976), which stated that "*the available evidence for certain types of tumor points to the need for an accuracy of ± 5% in the delivery of an absorbed dose to a target volume if the eradication of the primary tumor is sought. Some clinicians have requested even closer limits such as ± 2%, but at the present time it is virtually impossible to achieve such a standard.*" Since 1976, there have been multiple other updated reports analyzing accuracy requirements (Brahme, 1984; Dutreix, 1984; Svensson, 1984; Mijnheer et al., 1987; Brahme, 1988; Wambersie, 2001) with the general conclusion being that we should aim for an overall accuracy of ±5% in the determination and delivery of dose to tumors and normal tissues although in some cases "*an absorbed dose delivery of 3.5% is proposed even though it is known that in many cases larger values are acceptable and in a few special cases an even smaller value should be aimed at*" (Mijnheer et al., 1987). Most of these analyses were performed in the 2D era during the 1980s. All of these uncertainty statements implicitly refer to uncertainty specified by one SD (i.e., $k = 1$).

2.3.2 Recent Reports on Accuracy and Uncertainty in Radiation Therapy

The most recent comprehensive report on accuracy and uncertainties in radiation therapy has been produced by a consultants' group at the IAEA (IAEA, 2016; van der Merwe et al., 2017). This report not only reviews the uncertainties associated with the radiation beam calibration process but incorporates those associated with the dose calculation algorithms used in radiation treatment planning software. Furthermore, it analyses every step in the radiation treatment process and develops propagated uncertainty estimates based on collective recent published data. The report also gives a summary of typical uncertainty estimates for different components of the radiation treatment chain for both external beam radiation therapy (EBRT) and brachytherapy.

The first conclusion of this report indicates that "All forms of radiation therapy should be applied as accurately as reasonably achievable (AAARA), technical and biological factors being taken into account." To illustrate this with extreme examples, accuracy requirements for curative, small-field, stereotactic radiotherapy are quite different from those for total body or total nodal irradiation for bone marrow transplantation, or those for single fraction palliative radiation treatment.

Another recent publication (Van Dyk et al., 2013) provided an overview of accuracy and uncertainty considerations both from a historical perspective and from a perspective of the use of modern technologies. The authors considered uncertainties in two major categories: (1) human-related (patient or personnel) uncertainties and (2) technology- or dose-related uncertainties. They provide comprehensive summaries of uncertainty, both for external beam and brachytherapy, associated with each step of the therapy process ranging from diagnosis and clinical evaluation to treatment dose delivery. *In vivo* dosimetry for external beam therapy and posttreatment imaging with recalculation of dose distributions (e.g., post insertion for brachytherapy) is also discussed. A publication in the proceedings of the 7th International Conference on 3D Radiation Dosimetry (Thwaites, 2013) examines historical data based on 3D conformal radiation therapy (3DCRT) techniques and provides an update for more recent IMRT techniques. A more recent paper (Seravalli et al., 2015) provides a comprehensive evaluation of treatment accuracy but only for a specific treatment technique, intracranial stereotactic radiotherapy; however, the emphasis of this paper is on geometric uncertainties, specifically to aid in the determination of margins that should be used between the gross tumor volume (GTV) and the planning target volume (PTV). Indeed, there are many similar publications in the literature on the generation of treatment margins and treatment margin "recipes," as is discussed in more detail in Section 2.6.1.

2.3.3 Review of Recommendations from Equipment Commissioning and Quality Assurance Reports

Multiple reports have been produced on the commissioning and QA of various technologies associated with radiation treatment, including CT simulators (Aird and Conway, 2002; Mutic et al., 2003), TPSs (Van Dyk et al., 1993; Fraass et al., 1998; IAEA, 2004; Van Dyk, 2008), and linear accelerators (Ezzell et al., 2009; Klein et al., 2009; Bissonnette et al., 2012). TPSs are at the core of the radiation treatment process and provide a 3D display of the dose distribution within the context of the 3D anatomy of the patient. It is the accuracy of the dose computation algorithm used in the TPS that is often the main consideration for dose uncertainty estimates in the radiation treatment process. However, the entire chain of procedures merits attention as opposed to a solitary component. Figure 2.2 provides a summary of the steps in the modern radiation treatment process along with a brief statement of some of the major uncertainty considerations. It should be noted that the anatomy is usually sampled before the commencement of treatment and it progressively undergoes systematic change during therapy over many weeks. In some cases, changes occur over much shorter time frames, as is the case for moving lung tumors driven by respiration. Such variability introduces uncertainty that has largely been ignored until the introduction of image guidance and beam gating. Recent advances in adaptive radiotherapy (ART) include considerations for recomputing and accumulating daily dose distributions, accounting for progressive changes in anatomy detected by onboard CT or portal dosimetry systems (Van Dyk et al., 2010; Battista et al., 2013; Mijnheer et al., 2013a). This includes the possibility of replanning treatment of the dose distributions when poor convergence to the original clinical goals becomes apparent in the anatomy or associated dose distortions.

2.3.3.1 Treatment Machine Commissioning

Prior to commencement of the patient treatment process, the radiation machine must be commissioned for clinical usage. One of the major components of this procedure is absolute beam calibration of dose rate (i.e., determining dose per monitor unit or time). Various recent reports have reviewed beam calibration uncertainties (Andreo, 2011; Thwaites, 2013; IAEA, 2016) with the general conclusion that the dose to a reference point in a water phantom can be determined to an accuracy ranging between 1% and 2% (with $k = 1$) depending on beam energy

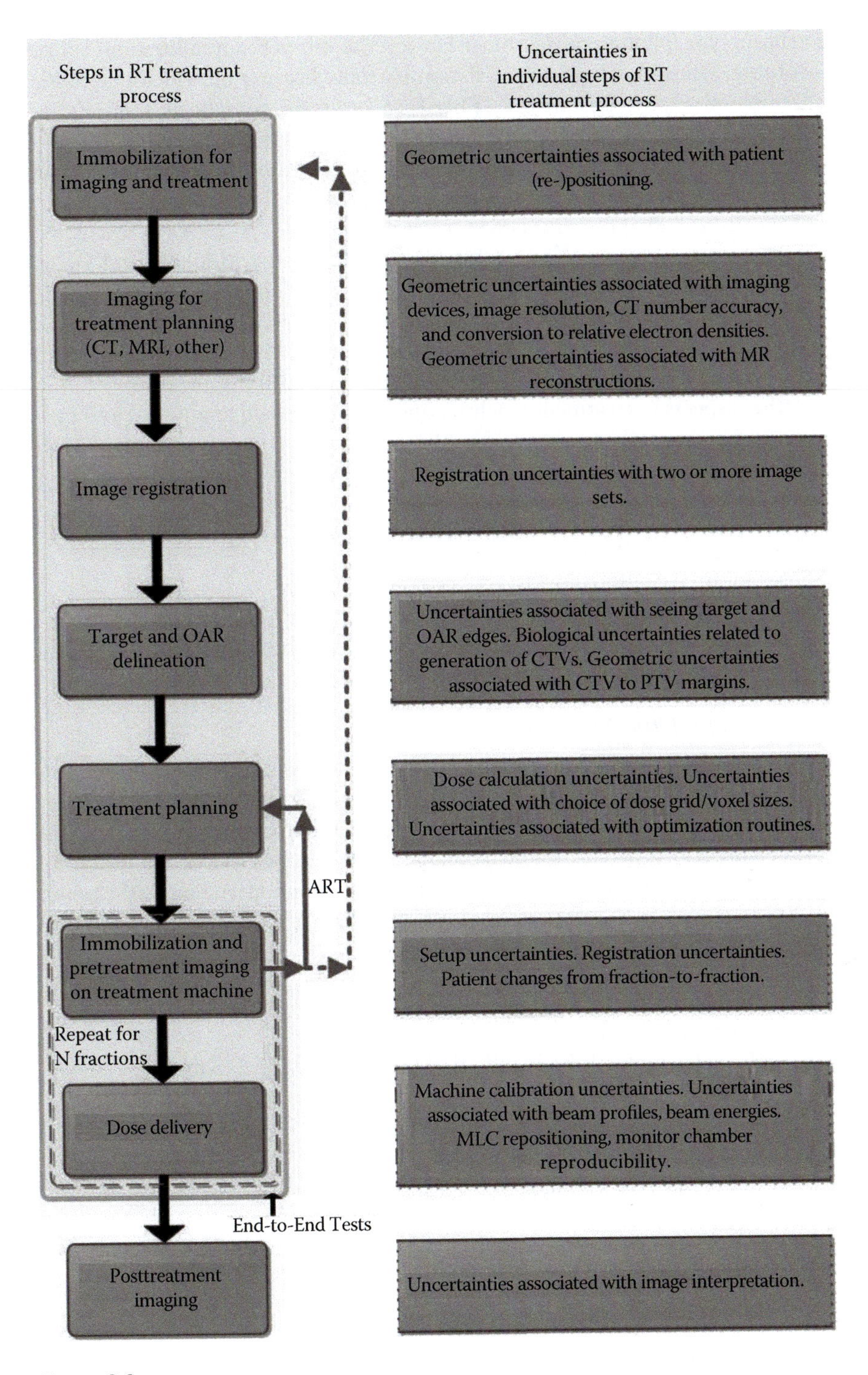

Figure 2.2

Steps in the overall radiation treatment process and major uncertainties associated with each step. The slightly shaded region shows the components that are tested with full end-to-end tests. The two external arrows (with solid line and dashed line) from "immobilization and pretreatment imaging" show feedback options for replanning during adaptive radiation therapy (ART).

and beam type (photons or electrons). However, audits of beam calibrations across institutions show much greater variation partially because of detector selection (thermoluminescent dosimeters [TLDs] or optically stimulated luminescent dosimeters [OSLDs]) and precision, and partially because of institutional variations in calibration procedures. In some cases, these uncertainties relate to actual errors in the interpretation or implementation of calibration protocols. The IAEA TLD mail-in audit program shows uncertainties of 2.2%–3.5% (IAEA, 2016). Similar results were summarized for multiple studies by Mijnheer and Georg (2008). The Imaging and Radiation Oncology Core Houston (IROC-Houston), formerly known as the Radiological Physics Center (RPC), experienced a spread in machine "output" data of ~2.5% (1 SD) over the last 15 years for both photons and electrons with an SD of ~1.7% between 2009 and 2010 (IAEA, 2016).

Other aspects of treatment machine uncertainties relate to physical and geometrical features of the treatment machine components (Moran and Ritter, 2011) (at the $k = 1$ level):

- Jaw positioning accuracy: <1 mm
- Wedges: 2%/2 mm
- Multileaf collimator (MLC) positioning: ≤1 mm
- MLC transmission: up to several percent of central beam dose
- Table tops/couch tops: depends on angle, energy, and position. Up to 20% in dose due to attenuation through couch support elements

2.3.3.2 Patient Immobilization

With a commissioned machine, the next major component of assessing uncertainties is the entire treatment planning process from patient immobilization, 3D imaging, treatment planning, and finally data transfer to the radiation treatment machine (as in Figure 2.2). A detailed comparison of immobilization devices and the corresponding uncertainties demonstrates significant variations depending on the tumor site, the treatment technique, and immobilization device used. The following summarizes mean setup uncertainties for different body sites (Meeks, 2011):

- Head and neck: 3 ± 1 mm
- Spine: 2–3 ± 2 mm
- Lung (stereotactic body radiation therapy [SBRT]): 2–5 mm
- Breast: 2 ± 3 mm
- Abdomen: 2–6 mm
- Prostate: 3–5 mm
- Intracranial stereotactic: 1–2 ± 1 mm

Some of these uncertainties can be reduced at the treatment machine by the use of onboard image-guidance procedures and appropriate retargeting.

2.3.3.3 Imaging for Treatment Planning

Image quality for CT scanners is characterized by image noise, signal-to-noise ratio, and spatial resolution, for example. The in-plane resolution is related to detector size and sampling, and the pixel size as well as the image reconstruction filter used. In-plane resolution is typically 1 mm for a diagnostic CT scanner. The longitudinal resolution (i.e., slice thickness) could be 1–2 mm but is often 5 mm. Such resolutions tend to be similar for MRI with improved soft-tissue contrast but geometric distortions due to tissue composition and magnetic field inhomogeneity can range up to 15 mm depending on the location and patient size (Chen et al., 2006; Glide-Hurst et al., 2015). For positron-emission tomography (PET) and single-photon emission tomography (SPECT), the spatial resolutions degrade significantly to about 3–7 mm, but image contrast reflects metabolic activity through highly sensitive molecular and functional imaging.

2.3.3.4 Delineation of Target and OAR

It is well recognized that target volume delineation on various imaging modalities results in one of the largest uncertainties in the entire radiation treatment process (Hamilton and Ebert, 2005). An excellent meta-study (Jameson et al., 2010) has been provided which reviewed 69 articles and included 10 different body sites. Most of these articles included the quantification of both inter- and intraobserver variations in target volume delineation. As summarized in the IAEA report (IAEA, 2016), these uncertainties can range between 5 and 50 mm for target definition and 5 and 20 mm for OARs. These are substantial variations and reflect the limitations in imaging and their interpretation.

2.3.3.5 Image Registration

Since quality of tumor imaging is both dependent on clinical site and tumor pathology, a variety of 3D imaging techniques is used to segment the target and surrounding normal tissues and critical organs. However, dose computations during treatment planning are almost universally based on CT images, because the pixel values (in Hounsfield units [HUs]) reflect kilovoltage x-ray attenuation coefficients that mostly depend on electron density (electrons per cm^3) with some perturbation due to atomic composition especially in bone (see appendix of Battista and Bronskill, 1981). For other imaging modalities, there is a need for registration procedures to connect with the CT-derived electron density map; the assignment of regional bulk densities is sometimes sufficient (Kim et al., 2015). In their simplest forms, registration procedures can be simple translation and rotation of images to obtain the best match in anatomy of the two image sets. In a more complex approach, deformable registration can be used to align and register the voxels across two sets of patient anatomy with tissue losses or displacement. The uncertainties associated with image registration are dependent on the algorithm used and its software implementation with a lower limit of ~0.5 mm for rigid registration (Bissonnette et al., 2012) and about 2 mm for deformable registration; however, larger uncertainties can exist for some deformable registration algorithms with suboptimal performance or input parameters (Brock et al., 2011). The registration errors could misplace tissue voxel pairs leading to regional point registration uncertainties. These are indeed "geometric errors" that propagate into "dose errors" especially in regions of steep dose gradients. When registration algorithms are further used to track dose accumulation across treatment fractions, the user must be cognizant of these possible registration pitfalls and downstream propagation of uncertainty.

2.3.3.6 Computer-Aided Treatment Planning

Computerized radiation TPSs yield the 3D dose distributions upon which clinical treatment decisions are made; they are therefore recognized as being at the heart and along the critical path of the radiation treatment process. They are used by the various professionals involved in radiation treatment planning, including medical physicists, radiation oncologists, dosimetrists, and radiation therapists, through a local area network. The systems provide software for target and normal tissue delineation, beam's eye views with digitally reconstructed radiographs (DRRs) for treatment geometry verification, design of treatment technique, dose optimization, and machine control information (e.g., MLC). Over the years, various national and international reports have been developed that describe the purchase, acceptance, and commissioning and QA components of a TPS (Van Dyk et al., 1993, 1999; IAEA, 2004, 2007, 2008; Van Dyk, 2008). Recently, an American Association of Physicists in Medicine (AAPM) Medical Physics Practice Guideline has been published on commissioning and QA of dose calculation algorithms for both photons and electrons (Smilowitz et al., 2015). Chapter 16 reviews the commissioning and QA procedures for a modern TPS, providing tests that should be performed prior to initial clinical use of a TPS, as well as before the periodic testing of an existing TPS.

Earlier reports on criteria of acceptability as part of the commissioning process gave numerical values that related primarily to the accuracy of dose calculations (e.g., Van Dyk et al., 1993) with geometric accuracy being considered in high-dose gradient regions. The more recent AAPM and IAEA reports investigated many other aspects of TPSs as well (Fraass et al., 1998; IAEA, 2004).

A significant issue that relates to the commissioning of TPSs is the determination of measured data and the accuracy of detectors that are used to generate these input data sets. If the measured data have significant inaccuracies or uncertainties "at source," then these could have a major bearing on the quality of the beam-fitting parameters used subsequently by dose calculation algorithms. Many of these parameters also are derived from combinations of several dose measurements, such as percent depth dose (PDD) and tissue-phantom ratios (Mitchell et al., 2009) that require pairs of readings, each with uncertainty. Examples have been published, especially as related to small-field dosimetry. Depending on the detector used, doses measured in small fields could be in error by more than 50% (Alfonso et al., 2008; Kron et al., 2013b) as discussed in Chapter 9. Another study demonstrated that depending on the choice of detectors, there is a potential for large errors when effects such as partial volume averaging, perturbation and differences in material properties of detectors are not taken into account (Azangwe et al., 2014). These factors not only affect the doses determined in small fields but also have a significant impact on the subsequent determination of doses near the beam edge or penumbra region. If these foundational measurements are inaccurate, then it is likely that the beam modeling for different conditions will also be inaccurate as a result of inappropriate fitting, extrapolations, and subsequent error propagation.

An audit of 60 TPSs was performed under the guidance of the IAEA in eight European countries yielding 190 datasets (Gershkevitsh et al., 2014). Eight different case scenarios were evaluated with agreement criteria for each ranging from 2% to 5% in dose depending on the complexity of the treatment technique. The results were grouped by calculation algorithm type and the mean relative difference between the calculated and measured doses (Δ%) was determined along with uncertainty estimates corresponding to two SDs. As expected, the mean and SD were smaller for physics-rich transport algorithms, which primarily use point kernel convolution/superposition models, and account for tissue density variation and lateral scattering in 3D including scattered photon and recoil electron transport (mean 0.5%, 2 SD 3.3%). Simpler algorithms use a pencil beam convolution and equivalent radiological path length restricted to primary beam rays to account for tissue heterogeneities; they ignore lateral electron transport and this results in greater uncertainty (mean 0.8%, 2 SD 5.6%). The differences were particularly pronounced for points within the lung material with differences up to 17% for some of the pencil beam models when used with higher energy small x-ray beams. These results are summarized in Figure 2.3.

Similar types of results were observed by the RPC (Davidson et al., 2008). More recently, a retrospective analysis was performed of 304 irradiations of the RPC thorax phantom at 221 different institutions as part of credentialing for Radiation Therapy Oncology Group (RTOG) clinical trials; the irradiations were all done using 6 MV beams (Kry et al., 2013). Pencil beam algorithms overestimated the dose delivered to the center of the target by 4.9% on average. Surprisingly, convolution/superposition algorithms and the anisotropic analytic algorithm (AAA) also showed a systematic overestimation of the dose to the center of the target, by 3.7% on average. It is possible that these discrepancies originate from a combination of inadequate commissioning parameters and algorithm implementation "short cuts" used to achieve faster calculations. In contrast, the Monte Carlo algorithm produced better results that agreed with measurement within 0.6% on average. It is encouraging that there was no difference observed between the accuracy for IMRT and 3DCRT techniques. With recent advances in computational speed using graphical processing units (GPUs) (Pratx and Xing, 2011; Su et al., 2014), it is predicted that well-commissioned Monte Carlo models will be the method of choice in the near future.

Modern TPSs use automated dose optimization software to determine the MLC configurations to provide the optimum external beam fluence distribution. A similar process is used to determine optimum source position and dwell times in brachytherapy. The user-specified input data include dose and volume constraints for both the target(s) and OARs that are subject to contouring variations discussed previously. This has implications when a dose–volume histogram

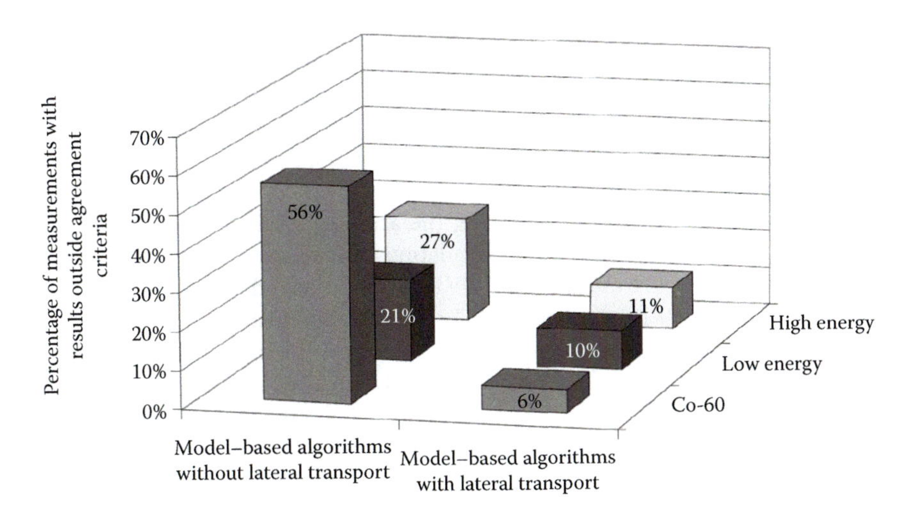

Figure 2.3

Percentage of measurement points with results lying outside of dose agreement criteria for different algorithm types (3D versus 2D convolution) and beam energy groups. (Adapted from Gershkevitsh E et al., *Acta Oncol*, 53, 628–636, 2014.)

(DVH, see Section 2.4.1) is judged during approval or review of treatment plans; incorrect contouring can lead to distorted DVH curves and displace optimization constraint points. In addition, many of the modern TPS optimization routines now also consider radiobiological estimations of tumor control probability (TCP) and normal tissue complication probability (NTCP). Uncertainties associated with these quantities are significantly larger compared to physical dose and volume uncertainties since these quantities are not only dependent on the dose–volume input data but also on the limited understanding of the multifactorial radiobiological response of cancerous and normal tissue cells (at minimum, the four "R"s of radiobiology). The most comprehensive collated summary of normal tissue responses can be found in the Quantitative Analyses of Normal Tissue Effects in the Clinic (QUANTEC) papers (Bentzen et al., 2010; QUANTEC, 2010).

2.4 Relationship Between Dose and Geometry

2.4.1 Dose–Volume Histogram

As CT scanning became available for radiation treatment planning in the late 1970s, volumes of both targets and normal tissues were readily quantifiable. This allowed the correlation of dose levels within specific contoured volumes of interest and resulted in the distilling of a 3D dose plan into a single curve using the concept of DVH (Drzymala et al., 1991), a concept used previously in image processing for display contrast optimization. A DVH is obtained by plotting the (relative or absolute) volume of the structure of interest against (relative or absolute) local dose values (for a differential DVH) or against a (relative or absolute) dose value that is equal to or greater than that particular dose level (for an integral or cumulative DVH). Most commonly, when the literature refers to DVHs, the cumulative DVHs are understood unless otherwise specified. The advantage (and disadvantage) of DVHs is their simplicity in collapsing 3D dose-space information versus scrolling through the entire 3D dose distribution and their utility for guiding the dose optimization process based on clinical goals, i.e., DVH constraint points. The disadvantage is that DVHs lose information about the 3D spatial location of the specific dose values; all eligible doses from anywhere within a segmented volume contribute equally to a DVH point in the histogram. The accuracy of DVHs is dependent upon the computed dose accuracy and the accuracy of the contoured volume in a region, as interpreted from the imaging modality by the radiation oncologist.

In one of the first reports evaluating the accuracy of DVHs (Panitsa et al., 1998), the consistency of the DVH calculations was determined by comparing data taken from the TPS and a control DVH recomputed independently from the 3D dose distributions for the same case. For six different institutions, they examined key DVH parameters such as D_{min}, D_{max}, dose in the structure, and volume of the structure. They did this in both low- and high-dose gradient regions with the results indicating variations to a maximum of about 3% in maximum dose determinations in low-dose gradient regions and up to 27% in volume in high-dose gradient regions. These results were dependent on the institution's TPS, the beam energy, and the choice of dose grid calculation parameters.

Craig et al. (1999) developed a special phantom which allowed for a direct determination of the DVH from a well-defined volume and well-defined dose distribution. The results showed mean differences in relative volumes for specific doses of 1.0%–1.5% with maximum differences ranging between 3% and 5%.

However, in a separate phantom test containing well-defined absolute volumes, uncertainties as large as 17 cm^3 were observed in a volume of 125 cm^3.

A more recent analysis (Nelms et al., 2015) used DICOM RT structure sets imported into two different TPSs: (1) to test for the accuracy of various DVH parameters such as total volume, D_{max}, D_{min}, and various doses to percent volumes, and (2) to determine volume errors along the DVH curves. Significant deviations were found for the different treatment planning algorithm implementations. The errors were related to inadequate 3D dose grid sampling and inconsistencies in the implementation of "endcapping" of 3D structures. As noted previously, there is a tacit assumption that the regions are accurately contoured and this is normally only done "slice by slice" rather than in a 3D view.

These reports indicate that the accuracy of DVH curves relies on the accuracy of the dose calculation algorithm, choice of dose grid spacing, volume delineation limited by observer bias, and imaging resolution (especially CT slice spacing). In addition to these controllable components to DVH accuracy, the DVH histogram binning used by the system can introduce variation; this aspect is not normally controllable by the user. Several authors have developed the concept of DVH "bands" to describe the range of dose–volume uncertainties that can occur. For example, one group (Trofimov et al., 2012) presented a method to visualize the variability when accounting only for dose delivery-related uncertainties. Another group (Fattori et al., 2014) used DVH bands to investigate variations due to setup errors and range uncertainties in image-guided carbon ion radiotherapy of head chordomas. Another group (Cutanda Henriquez and Vargas Castrillon, 2010) has developed CIs in DVHs to account for uncertainties associated with the dose computation *per se*. These were found particularly useful when optimizing and approving a treatment plan while being cognizant of inherent uncertainties.

The application of DVHs has been further extended by mapping NTCPs onto regions of dose–volume space with statistical considerations of risk (Kupchak et al., 2008). The authors plotted a series of DVHs derived from a set of clinically approved plans for prostate cancer and illustrated the wide "bands" of acceptability. They proceeded to generate technique-specific maps that highlight high-risk zones in DVH space, a feature which is advantageous over fixation on single-point constraints. The maps also provide a visualization tool to help select robust treatment plans and open the possibility for improving the efficiency of biologically based plan optimization by focusing on more critical segments of DVH curves. While such maps can help minimize risk of normal tissue complications, they did not include the intrinsic uncertainties associated with developing the DVH plots. A more recent application of such risk maps has been published with examples for stereotactic ablative and radiosurgery applications (Asbell et al., 2016).

2.4.2 Gamma Index

During the 1990s, there was a realization that scoring of local dose discrepancies between calculated and measured dose distributions was too stringent and the distance-to-agreement (DTA) of neighboring dose points was needed to account for small geometric offsets between data set coordinates. Thus, when commissioning a TPS, new criteria of acceptability were developed which recommended not only an allowable percentage local dose deviations in low-dose gradient regions but greater variations in high-dose gradient regions if neighboring dose points could be found within an acceptable nearby distance. This relaxed criterion was implemented to account for inevitable positional uncertainties in comparing experimental data and computed data (e.g., Van Dyk et al., 1993). The question of how both of these uncertainties (i.e., dose and distance) could be addressed in a composite efficient assay was addressed by the concept of the *gamma index*, γ (Low et al., 1998; Cutanda Henriquez and Vargas Castrillon, 2011; Low et al., 2011). If the dose difference tolerance is ΔD and the spatial tolerance is ΔR in Euclidean distance, then the gamma (γ) index map is defined by

$$\gamma(x,y,z)=\min\sqrt{\frac{\delta D^2}{\Delta D^2}+\frac{\delta R^2}{\Delta R^2}} \tag{2.3}$$

where δD and δR are the actual differences in local dose and distance (ideally found in 3D space) compared to a closely matching dose neighbor point (ΔD, ΔR) in the reference dataset. In practice, the dose differences (δD and ΔD) are often set as percentages of a reference dose (e.g., at isocenter). The selected local gamma value is the one that minimizes the square root value after searching the neighborhood voxels. In other words, the gamma value is the minimum value of the vector length in (δD, δR) space, for a neighborhood around the test point (x, y, z). Typical tolerance criteria used in IMRT are ΔD = 3% (of isocentric dose) and ΔR = 3 mm. A point passes the check if this index is less than or equal to 1. ΔD and ΔR individually are no longer strict tolerances since dose difference could be greater than ΔD for a point passing the gamma test and DTA could also be greater than ΔR for a point with a gamma less than 1. However, if the dose difference is greater than ΔD and DTA is greater than ΔR at the same point, the gamma test fails (Low et al., 1998). At the same time, the value of γ at a point where the test has not passed is a measure of the severity of the failure. The gamma index has been generalized to 3D with DTA being computed within a 3D image search (Wendling et al., 2007; Pulliam et al., 2014). While originally intended for efficiently comparing pairs of dose distributions, the gamma index concept has also been used for planar dose assessments used in patient-specific QA using diode arrays in Cartesian or radial configuration or film placed in a body-like phantom. When the calculations are complete, a map of gamma values can be displayed to show regions of unacceptable discrepancy and possible concern. Note that published reports of gamma calculations do not always specify if the neighborhood search is done only within a plane (2D) or in 3D with consideration of adjacent dose slices. Subsampling of the neighborhood could lead to inherent uncertainty and possible errors in gamma pass rates (Schreiner, 2011).

Various authors have correlated gamma pass rates with other dosimetric verification metrics, as discussed more extensively in Chapters 14 and 17, including DVH deviations because these are used in treatment plan approvals (Zhen et al., 2011). A recent study (Jin et al., 2015) found poor correlation between the gamma pass rate performed in 2D and 3D using ArcCheck®, Sun Nuclear (www.sunnuclear.com), for pretreatment VMAT and measured dosimetric errors. DVH-based metrics (3DVH® software, Sun Nuclear) were found to be more clinically oriented and informative. Another group (Carrasco et al., 2012) found a lack of correlation between the gamma index and DVH violations, regardless of regional or global tests and 2D or 3D assessments. Some of the tests yielded false positives or false negatives in per-beam gamma analyses. Similar low correlations between gamma index and DVH analyses have been shown by other groups (Nelms et al., 2011, 2013; Stasi et al., 2012; Coleman and Skourou, 2013).

In summary, using 3%/3 mm thresholds for the entire dose matrix showed low sensitivity, specificity, and questionable predictive power for patient-specific pretreatment QA. One approach to addressing this limitation is to incorporate inherent uncertainty in the gamma analysis (Cutanda Henriquez and Vargas Castrillon, 2011). Thus, false positive rejections due to known dose uncertainty do not occur and there is no need to inflate tolerances. The method is based on rules of uncertainty propagation and helps define more rigorous pass/fail criteria, based on experimental information. Another approach (Stojadinovic et al., 2015) has been described by dividing dose distributions into four distinct regions: the high-dose or umbra region, the high-dose gradient or penumbra region, the medium-dose region, and the low-dose region. Different gamma passing criteria are defined for each type of region. This has been described as the "divide and conquer" gamma analysis method. This approach revealed a much poorer agreement between calculated and measured dose distributions with large local point dose differences within different dose regions compared to the use of the global gamma analysis approach of AAPM Task Group 119 (Ezzell et al., 2009), i.e., the poorer agreement demonstrating more sensitive analysis compared to the global analysis.

It is clear from the above discussion that while the gamma index analysis provided an efficient comparator tool for 3D dose distributions (it's original purpose), extended application to patient-specific QA of modern radiation techniques could lead to a false assurance that a plan is clinically acceptable.

2.5 Review of Dose Auditing Results

One source of information about accuracy and uncertainties in the radiation treatment process can be obtained by performing independent, peer-review audits of various steps in the radiation treatment process. The aims of such external audits are to support the development and improvement of quality and safety in radiation therapy, to avoid treatment errors, to provide confidence in the accuracy of dosimetric calibration, and to provide a communal process for centers to ensure that new machines and processes can be validated before widespread clinical use (Ibbott and Thwaites, 2015). In addition, such audits are prerequisites for many institutions in order to participate in multi-institution clinical trials (Ibbott et al., 2013; Kron et al., 2013a). Dosimetry audits can be performed internally by different individuals and/or different dose measuring tools in the same department, or they can be performed independently by external individuals from other departments or organizations involved in quality checks of the dosimetry procedures. A simple audit process can involve single-point beam calibration measurements using mailed dosimeters such as TLDs or OSLDs. More complex auditing procedures include measurements under nonreference conditions, on-site visits with phantoms and measurements for specialized techniques (Letourneau et al., 2013), and end-to-end tests (Ibbott et al., 2013; Seravalli et al., 2015), which consider the entire treatment process as discussed in detail in Chapter 19. A classification of different auditing procedures has been described by the Australian Clinical Dosimetry Service (ACDS) (Kron et al., 2013a) and is summarized in Table 2.1 (see also Chapter 19).

Two major groups involved in dosimetry audit programs include the IAEA (Izewska et al., 2003) and IROC-Houston (Aguirre et al., 2002). In addition, there are various national (Hoornaert et al., 1993; Nisbet et al., 1998; Kroutilikova et al., 2003; Rassiah et al., 2004; Hourdakis and Boziani, 2008;

Table 2.1 Classification of Different Auditing Protocols as Described by the Australian Clinical Dosimetry Service

Dosimetry Level	Types of Tests	Detector Type	Mode	System Checked	Comments
Level I	Dose output under reference conditions	TLD, OSLD	Remote	Every radiation beam	Identical to IROC-H (RPC) audit
Level IB	Dose output under reference conditions	Ionization chamber	On-site	Every radiation beam	Offered for new centers prior to opening
Level II	Dose distribution in physical phantoms	Detector array	Remote	Planning system	Can include tissue inhomogeneities and allows for clarification of Level III findings
Level III	Anthropomorphic phantom end-to-end	Ionization chamber, radiochromic film	On-site	Entire treatment chain	Treatment specific—most relevant for clinical trials

Source: Adapted from Kron T et al., J. Phys. Conf. Ser., 444, 012014, 2013a.

Note: IROC-H, Imaging and Radiation Oncology Core Houston; OSLD, optically stimulated luminescent dosimeter; RPC, Radiological Physics Center; TLD, thermoluminescent dosimeter.

Williams et al., 2012; Clark et al., 2015) and regional organizations (Roue et al., 2004; Palmer et al., 2011), as well as clinical trials study groups (Hansson et al., 1993; Gomola et al., 2001) performing such audits.

In addition to single-point-dose calibration audits, recent years have seen multi-institutional audits performed for a variety of conditions, technologies, or new treatment procedures including measurements under nonreference conditions (Izewska et al., 2007), TPS verification (Davidson et al., 2008; Gershkevitsh et al., 2014; Dunn et al., 2015), IMRT (Budgell et al., 2011; Molineu et al., 2005, 2013; Clark et al., 2009), rotational intensity-modulated radiation therapy (Clark et al., 2014), intraoperative radiation therapy (Eaton et al., 2013), and high-dose rate (HDR) brachytherapy (Roue et al., 2007; Palmer et al., 2013). Not only is dose checked through these auditing processes but special procedures have also been developed for geometric assessments (Roue et al., 2006).

The uncertainty of the mail-in OSLD system for measuring the dose output of accelerators has been evaluated to be 1.5% at the $k = 1$ level (Ibbott and Thwaites, 2015). Consequently, the IROC-Houston's measurement of an institution's output can be stated at an uncertainty of less than 5% using a 99% CI and this has been established as the threshold for calibration acceptability (Ibbott and Thwaites, 2015).

2.6 Accuracy Requirement Considerations

2.6.1 Clinical Considerations

Ultimately, therapeutic applications involve the interaction of radiation with both tumor and normal tissue cells in dynamic, complex biologic environments. As such, the drive for dosimetric accuracy needs to be considered in context with the inherent uncertainties associated with these biologic environments and in many cases, biological uncertainties may predominate over uncertainties associated with the physical delivery of therapeutic radiation. In this section, we discuss sources of uncertainties associated with underlying tissue biology, the uncertainties associated with target delineations as determined by human observers, and how these uncertainties can influence the design and interpretation of clinical trial outcomes incorporating radiation treatments. For an in-depth discussion of these issues, the reader is referred to comprehensive publications (Van Dyk et al., 2013; IAEA, 2016).

2.6.1.1 Radiobiological Considerations

At the most fundamental level, radiation response is determined largely by the intrinsic radiosensitivity and recovery of normal and tumor cells. This intrinsic radiosensitivity can be characterized by cell survival curves that plot proportion of cells surviving as a function of radiation dose and parameters like SF2 which measures the proportion of cells surviving after a standard radiation dose fraction (in this case 2 Gy) (Fowler, 2010). Even in these simple *in vitro* systems, however, a wide range of intrinsic radiosensitivities may be seen between tumor and normal tissues and, importantly, even within a single tumor type. For example, Williams et al. (2008) demonstrated a nearly 10-fold range in SF2 across tumor cell types and a fourfold range within a given tumor cell type. Other authors have described differential radiosensitivity between stem-cell-like and nonstem-cell-like tumor cells, and postulate that radiocurability is contingent upon the radiosensitivity of stem-cell-like tumor cell populations (Gerweck and Wakimoto, 2016). Thus, even in simple model systems considerable variation in radiosensitivity exists depending on the tumor cell type, tumor cell population examined, and experimental conditions. Nevertheless, radiation response of such simple systems can be approximated by mathematical models such as the linear-quadratic (LQ) model (Moiseenko, 2004), and such models can be useful for designing bioequivalent radiation dose prescriptions (Fowler, 2010). Within the commonly used LQ model, cellular response is commonly characterized by the parameter α/β, the dose at which the survival versus dose curve for a cell line transitions from a predominantly linear response to a predominantly quadratic response. Normal, so called "late reacting" tissues (e.g., rectum) and some tumors (e.g., prostate cancer) have been postulated to have an α/β in the low range of experimental measurement (α/β = 1–5 Gy) whereas, so called "acute reacting" tissues (e.g., mucosa) and most tumors (Wilson et al., 2003) are associated with an α/β in the higher range of experimental measurement (α/β = 5–10 Gy). Figures 2.4a and b are display hypothetical plots for prostate cancer (2.4a) and lung cancer (2.4b) comparing biologically effective dose (BED) as a function of uncertainty in α/β and uncertainty in delivered dose (± 10%). Such plots demonstrate that, for tumors with a low α/β and greater fractional size (as commonly used for hypofractionated and stereotactic treatments), effects of uncertainties in α/β assumptions can overwhelm effects of dosimetric uncertainty. For tumors with high α/β values, effects are more balanced. Such *Gedanken* experiments suggest that for certain clinical scenarios (Song et al., 2006), increased effort in understanding the underlying tumor radiobiology may be more pressing than concerns around improving delivered dose accuracy. In other clinical scenarios (hypofractionated treatments of lung cancer), there is less influence of radiobiological uncertainties and more influence of increased uncertainties accruing in dose delivery due to physiologic motion. In this case, the accurate computation

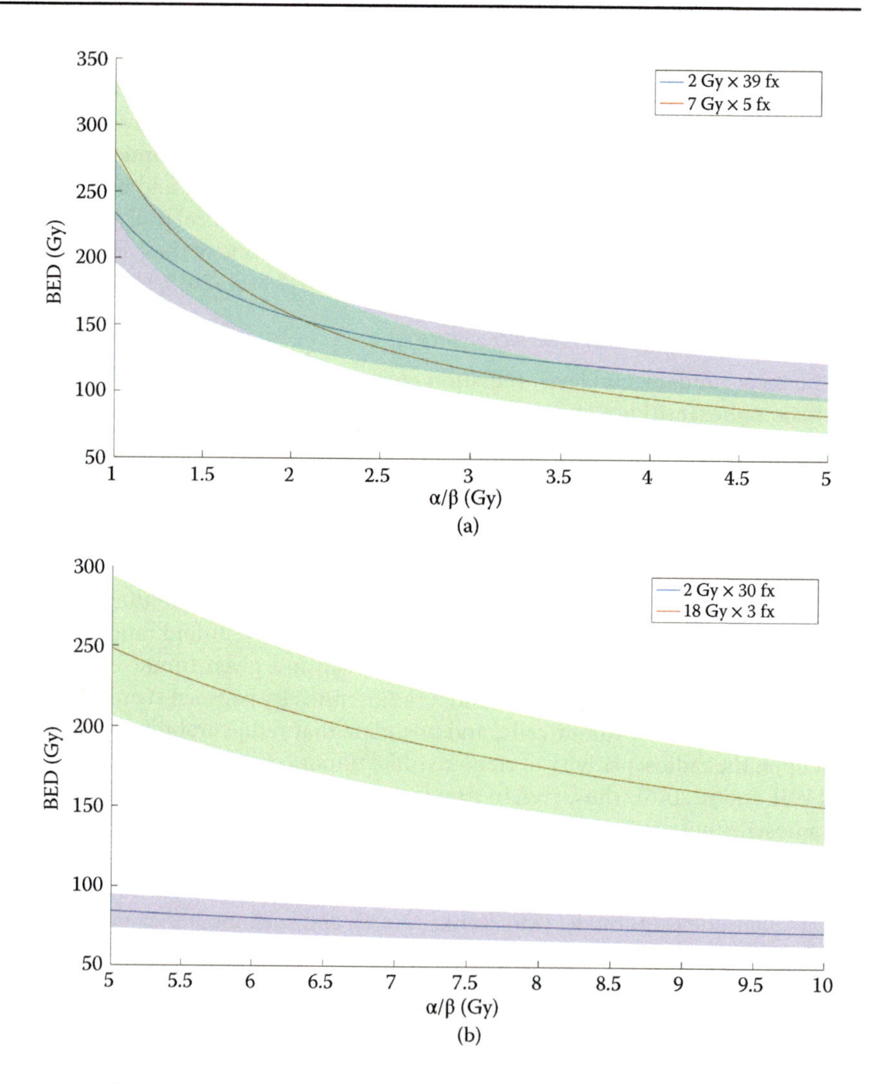

Figure 2.4

Examples of the relative contribution of radiobiologic uncertainty versus dosimetric uncertainty (±10%) for treatment scenarios for (a) prostate and (b) lung cancer by plotting biologically effective dose (BED) versus α/β. (a) For prostate cancer, where α/β is proposed to be low (<5 Gy), radiobiologic uncertainty (regarding the α/β ratio) has a greater effect (represented by the solid lines) on estimated BED than dosimetric uncertainty (±10% dose/fraction uncertainty represented by the shaded area above and below the lines) for both conventional (2 Gy/day × 39 fractions, blue line) and stereotactic (7 Gy/day × 5 fractions, red line) treatment. (b) For lung cancer, at conventional fractionation (2 Gy/day × 30 fractions, blue line) and an assumed high α/β (5–10 Gy), dosimetric accuracy has a greater influence, whereas for stereotactic treatments (18 Gy/day × 3 fractions, red line) both radiobiologic and dosimetric uncertainty contribute significantly to variations in the estimated BED.

or measurement of delivered dose is needed in the context of 4D dose delivery. Compounding this uncertainty is the lack of validation of models like the LQ formulation at larger doses per fraction (>6 Gy) where novel biologic phenomenon (such as vascular ablation) may play a role in addition to direct tumor cell kill (Fowler, 2010).

In a clinical context, such uncertainties are compounded further by the fact that both normal and tumor cells exist within more complex organized entities *in vivo*, influenced by the surrounding microenvironment and physiological conditions. Under these circumstances, modeling of clinical outcomes as a function of dose across individuals and populations has been attempted for various tumor sites and scenarios. Such modeling is complicated by the heterogeneity of patient populations because of underlying variation in tumor and normal tissue radiosensitivity, variations in tumor burden, clinical scenarios (radical versus adjuvant treatment, treatment of GTVs versus elective tumor volumes), and modifiers of radiation response like variations in blood flow/oxygenation to tissues and chemotherapeutic drugs. In general, radiation dose–response can be characterized by sigmoidal dose–response curves for tumor and normal tissue. The dose–response is steepest around the linear component of the dose–response curve located at D_{50} (dose required to achieve 50% response). Gamma50 (γ_{50}) is commonly used as the associated slope to describe the steepness of the dose–response curve in this region. Again, a range of γ_{50} is described with late responding normal tissues exhibiting a larger (steeper) γ_{50} (2–6) compared to tumors (1.5–2.5). As described elsewhere (IAEA, 2016), a population γ_{50} tends to be reduced by factors that increase heterogeneity across subjects such as variation in tumor stage and burden and adjuvant versus primary treatment scenarios. Conversely, γ_{50} tends to be increased (steeper dose response) by better characterized patient populations, clinical trial inclusion criteria, and more accurate dosimetry, factors that decrease heterogeneity. Modeling a required dosimetric accuracy in the setting of such clinical dose–response uncertainty suggests that systematic uncertainties and random uncertainties in dose delivery should be maintained at less than 1%–2% and <3%, respectively, and that the PTV volume receiving <90% of the dose should be kept at <6%–12% depending on the assumptions regarding γ_{50}. Such modeling also suggests that dosimetric accuracy should be more rigorous in clinical trials of radiotherapy, reflecting the steeper γ_{50} and thus greater effect of changes in delivered dose on clinical response associated with the typically more homogeneous patient populations selected by stricter inclusion and exclusion criteria associated with clinical trials.

Such dose–response considerations become more complex when defining normal tissue dose-volume effects. In addition to heterogeneity between patients and among populations studied, defining normal tissue toxicity endpoints is complicated by the limitations of current toxicity scales like the Common Toxicity Criteria Adverse Event scale, which "lumps" individual signs and symptoms under broad categories. This introduces heterogeneity in toxicity endpoints that can reduce statistical power and/or affect estimates of parameters like γ_{50} for normal tissues and organs and inconsistencies in reporting toxicity endpoints, especially among populations analyzed retrospectively. In addition, relationships of late effects may be based on historical treatment patterns that do not reflect current technology and practice such as use of combination chemoradiotherapy, IMRT including VMAT, and altered fractionation schemes. The consensus QUANTEC document summarizes the best available clinical data, discusses limitations, and provides direction for future research (Bentzen et al., 2010). In a related article, Jaffray et al. (2010) discussed the need for robust systems that relate the true accumulated dose distribution in normal tissues to clinical outcomes. They highlight issues specific to OAR dose–volume analyses including development of automated segmentation methods, tissue mapping algorithms to account for normal tissue deformation, defining optimal schedules for imaging monitoring, and development of informatics tools for analyses and uncertainty estimates of dose–volume relationships. Similarly, the benefits of "big data" analysis allowing consideration of dosimetric, image-based, clinical, socioeconomic, and biological factors in improving our understanding of normal tissue response has been highlighted (Deasy et al., 2010).

2.6.1.2 Target Volume and OAR Definition

The overall goal of dosimetric measurements is to estimate the accuracy of delivery of radiation dose as measured against a desired radiation delivery plan computed with software. For most radical (curative) radiation treatment courses, such plans are developed on the basis of volumetric imaging (typically CT) with the subsequent delineation and segmentation of tissue volumes representing various tumor targets (GTV, clinical target volume [CTV], internal target volume [ITV], PTV) and OAR volumes (including OAR and planning organ at risk volumes [PRVs]) (ICRU, 2010). These volumes are defined by human observers and are subject to inter- and intraobserver variability in delineating boundaries of these volumes within the limitations of available image contrast. Thus, the standard against which a delivered plan is adjudicated is, itself, bounded by levels of uncertainty associated with decisions made by imperfect observers (e.g., target boundaries, assigned dose-volume constraints, target prioritization). For example, Weiss and Hess (2003) in a systematic review of uncertainties in the delineation of GTV and CTV noted that such delineations are subject not only to the interpretation of digital imaging information but also available background clinical information, as well as the experience of the individual and collective oncology experience regarding patterns of tumor growth and spread. They noted in the literature review that among a variety of disease sites, contoured volumes could vary between 1.3- and 2-fold among observers, and that the ratio of the common volume between the total inclusive volumes among observers ranged from 0.4 to 0.75 depending on the clinical scenario. Factors such as tumor site, imaging modality and protocol, and use of multimodality imaging influenced volume segmentation as did the specialty of the observers, with radiation oncologists typically contouring larger volumes than diagnostic radiologists. As well, CTV variation in general was noted to be larger than GTV variation. They concluded that for some clinical scenarios interobserver variability (IOV) is a major and sometimes dominant factor contributing to geometric inaccuracy. Similarly, Segedin and Petric (2016) in a literature review highlighted the variability even in the metrics used to characterize IOV. They grouped such measurements into broad categories such as mean volumes, ratios of smallest and largest delineated volumes, coefficients of variation, concordance measurements that characterize the geometric relationship of common and encompassing volumes (i.e., conformity index, Dice coefficient), and local variation measurements (distance between observer and reference contours or surfaces such as Hausdorff distance [HD]). Like Weiss and Hess, Segedin and Petric noted a wide variation in volume (V) among observers that varied with tumor site with the largest variations noted for esophagus (V_{max}/V_{min} up to 6), head and neck (V_{max}/V_{min}, 18), lung (V_{max}/V_{min}, 7), and lymphoma (V_{max}/V_{min}, 15). Smaller variation was noted for other clinical scenarios such as prostate delineation (V_{max}/V_{min}, 1.2–1.6), brain tumor (V_{max}/V_{min}, 1.3–2), or breast lumpectomy cavity (V_{max}/V_{min}, 2). Larger uncertainties were noted for those sites where there was more clinical uncertainty regarding decisions on inclusion of elective nodal volumes (Diez et al., 2007) or where tumor boundaries may be more difficult to distinguish (e.g., lung). They estimated the effects of volume delineation uncertainty resulted in dosimetric uncertainties of 10%–20% due to undercontouring of the CTV/GTV or overcontouring of the OARs.

These uncertainties have also been modeled prospectively, typically in the context of credentialing and QA protocols for clinical trials. For example, in a prospective trial of hippocampal sparing whole brain radiotherapy, Gondi et al. (2015) undertook both "dry runs" and real-time review for the QA of delineation of the hippocampi by radiation oncologists at participating sites. They used the HD as a quality metric and compared variation in the HD between expert contours (generated by reference radiation oncologists leading the study) and local contours. An HD <7 mm was judged to be acceptable and in the dry run, 8/113 observers failed on the first attempt, with subsequent submissions being deemed acceptable. During the study, real-time review of at least the first three cases treated occurred and after three successful consecutive submissions, observers were allowed to proceed without real-time review. Among the real-time review, however, a higher failure rate (24%) was noted and interestingly this rate did not decrease with progressive experience (23% failure rate among cases undergoing posttreatment review), suggesting that complex contouring tasks such as hippocampal contouring may require significantly greater experience than four to six cases before a low-error state was reached. In an analysis of contouring variation associated with PET-based planning for non–small lung cancer in the RTOG 1106 trial, Cui et al. (2015) noted volume differences between observer volumes and reference volumes (simultaneous truth and performance level estimation [STAPLE] consensus volumes) of -28% to 7.7% (mean –5.9%) with a mean intersurface distance of 2.55–4.56 mm and Dice coefficients of 86%–92%. A questionnaire among participants attempted to elicit the factors contributing to contouring variation and factors identified for tumor and OAR included CT and PET image quality, drawing precision with contouring tools, imperfect contrast due to suboptimal window and level settings, variation in standard uptake value (SUV) threshold, observer effort, delineation of tumor from atelectasis, mediastinal vessels and other nontumor anatomy, inclusion of lymph nodes, variation in protocol interpretation, and observer's experience, knowledge, and judgment. They noted large drops in estimated TCP with undercontouring of PTV volumes when the extent of undercontouring exceeded 10%–20% compared to consensus contours.

Such uncertainties in delineation and their downstream dosimetric effects can be mitigated through training interventions to reduce IOV (Vinod et al., 2016). In a systematic review of interventions to reduce IOV, Vinod and coworkers identified 56 studies and found benefits within four broad categories of interventions: written guidelines and protocols (reduced IOV in 7/9 studies), training (8/9 studies), use of automated segmentation tools (6/78 studies), and use of alternative imaging modalities. They found that the benefit of the use of alternative imaging (i.e., PET or MRI) varied by clinical scenario with CT assisting in the definition of seroma cavities (versus fluoroscopy) for breast cancer, PET assisting in reducing IOV in lung cancer, rectal cancer, and lymphoma, and MRI reducing IOV for OAR definition in head and neck cancer. They also highlighted the new uncertainties and possible ambiguity that can be introduced with the introduction of *multimodality* imaging including registration error, geometric distortion, variable viewing parameters and thresholding for alternative imaging, and emphasized the benefits of interdisciplinary image interpretation and peer review. Similarly, Jeraj et al. (2015) reviewed the potential benefits and pitfalls of molecular imaging for radiation therapy planning and identified a spectrum of complexity for molecular imaging in terms of registration, segmentation, target definition, and motion management issues. In particular, the use of "dose painting" to intensify treatment to tumor subvolumes was discussed. While theoretically attractive, such strategies are subject to the same IOV associated with other GTV delineations along with added uncertainty of interpreting the biologic significance and histopathologic correlations of differential imaging signals on PET and MRI (van Baardwijk et al., 2006). Other systematic reviews of cancer-specific issues for multi-imaging and target delineation interventions are available (see, e.g., van Baardwijk et al., 2006; Ippolito et al., 2008; Gwynne et al., 2012; Yang et al., 2013). In addition, system-level guidelines such as the Royal Australian and New Zealand College of Radiologists (RANZCER) Quality Guidelines for Volume Delineation in Radiation Oncology are available and seek to address these issues with multidimensional recommendations for improving contouring quality (Faculty of Radiation Oncology, 2015). Table 2.2 summarizes the RANZCER guidelines.

Another approach to addressing IOV is to incorporate these uncertainties as part of a broader definition of the PTV. For example, Rasch et al. (2005) characterized target delineation as a form of systematic error that should be treated similarly to other geometric errors. Factors contributing to target delineation are discussed for prostate and head and neck cancers and include imaging factors such as acquisition parameters, spatial resolution, type of modality, and inherent tissue contrast. For nasopharynx cancers, they estimated a 1 SD error in GTV definition of 4.4 mm with overall interobserver agreement of 36% with CT alone (uncertainty is greatest in the craniocaudal direction) and found the combination of MRI plus CT improved observer performance (1 SD decreases to 3.3 mm and increases to 64% agreement). For prostate cancer, they also observed benefits to GTV delineation with combined MRI plus CT with GTV volumes 30%–40% smaller than with CT alone, and better agreement at the apex where soft-tissue contrast is limited on CT alone. Overall, they estimated a systematic margin error of 2 mm based on contouring contributing to an overall systematic error of 2–4 mm and a calculated PTV margin of 6–9 mm. For head and neck cancers, a 3-mm 1 SD GTV delineation error was estimated for an overall total systematic error of 3.5 mm and an overall calculated PTV margin of 8–10 mm.

Finally, the important function that peer review can serve in reducing variability in treatment plans has been noted and is becoming a routine component of the radiation treatment process in many institutions. For example, in a Canada-wide survey, Caissie et al. (2016), noted that all radiation oncology programs reported incorporating peer review for at least a subset of treated cases; a majority (53%) reported peer review for 80% or more of curative treatment plans. All radiation oncology programs surveyed attached a strong importance to peer review (7 or higher than that on a 10-point scale where 10 is extremely important) and felt that peer review was important to extremely important in reducing practice variation (8 or higher than that on a 10-point scale). These impressions are borne out in a systematic review of the literature (Brunskill et al., 2017), including reports on the impact of QA on treatment plans encompassing 11 studies and a total of 11,491 patient cases, and peer review led to a change in treatment plans in 10.8% of cases reviewed. They noted major changes in about 2% of plans. The most

Table 2.2 Quality Guidelines for Volume Delineation in Radiation Oncology

Education and Training

1. Radiation oncologists are encouraged to attend volume delineation workshops to ensure maintenance of professional standards.

Standard Contouring Protocols

1. It is recommended that radiation oncologists use standard contouring protocols and atlases for target and organ at risk volume delineation.

Delineation of Volumes

1. The radiation oncologist should have access to adequate hardware and software to enable high-precision contouring.
2. The planning system(s) used should have preset window levels to allow optimal viewing of imaging of different body sites and of different imaging modalities.
3. It is recommended that departments have standardized nomenclature for naming of volumes for radiation therapy planning. This includes both tumor volumes and normal tissues. Standard color coding of the major volumes delineated (GTV, CTV, ITV, PTV) is encouraged.

Gross Tumor Volume

1. Multimodality imaging used to aid in volume delineation should ideally be performed in the treatment position contemporaneously with the simulation CT.
2. The accuracy of image fusion performed by the radiation therapists should be checked by the radiation oncologist prior to contouring. Volumes may need to be adjusted to account for any misregistration.
3. Radiation oncology departments should develop strong links with radiology and nuclear medicine departments in order to help interpretation of multimodality imaging used for volume delineation in radiation therapy.

Clinical Target Volume

1. It is recommended that published or web-based atlases be used where available for CTV delineation.

Internal Target Volume

1. It is recommended that departments have a defined protocol for deriving ITV for the relevant tumor site and method of assessment.

Planning Target Volume

1. It is recommended that departments have defined collaborative protocols for deriving PTV for individual tumor sites with reference to specific immobilization and verification techniques.
2. Measurement and analysis of local data such as setup errors and online versus offline shifts are recommended to help determine appropriate PTV margins.

Review of Volumes

1. Radiation therapy target volumes should be reviewed by other radiation oncologist(s) using the Peer-Review Audit Tool, prior to a patient commencing treatment. As many cases as practical should be audited with a particular focus on patients undergoing definitive radiation therapy.
2. Clinical radiologist involvement in review of target volumes is encouraged.
3. In larger departments where there is site-subspecialization, peer review should be performed within these site-specific groups. Smaller departments with ≤3 full-time radiation oncologists are encouraged to link up with other departments for the purpose of peer review including checking of target volumes.

Radiation Therapy Techniques

1. Contouring protocols are strongly recommended for highly conformal radiation therapy techniques. For specialized techniques such as stereotactic radiation therapy or brachytherapy, review of contouring by a second radiation oncologist (or clinical radiologist) or consensus contouring is recommended.

Source: Adapted from Faculty of Radiation Oncology, Royal Australian and New Zealand College of Radiologists, Quality guidelines for volume delineation in radiation oncology. https://www.ranzcr.com/college/document-library/quality-guidelines-for-volume-delineation-in-radiation-oncology

Note: CTV, clinical target volume; GTV, gross tumor volume; ITV, internal target volume; PTV, planning target volume.

common changes recommended were target volume delineation (45% of plans), dose prescription or written directives (24%), or normal tissue delineation (7.5%).

2.6.1.3 Impact of Radiation Variation on Clinical Outcomes

As outlined above, clinical trials have a more homogeneous patient population being evaluated and, therefore, demand a greater dosimetric accuracy due to "tighter" γ_{50} with less statistical blurring. Clinical trials also offer the opportunity for formal integration and evaluation of the introduction of QA and improvement processes. For example, as part of a prospective study of EBRT and brachytherapy for cervical cancer, centers were required to submit a "dummy run" as part of credentialing for participation in the study (Kirisits et al., 2015). A subsequent dummy run by participating centers revealed that over half of participating centers failed the initial dummy run; 13/16 center deviations were in brachytherapy parameters (most commonly GTV delineation variation) and 11/16 deviations (most commonly GTV or CTV delineation variation) were noted in external beam deviations. More experienced centers reporting a historical volume of cervical brachytherapy cases greater than 30 cases per annum were associated with better compliance. In another prospective study of radiotherapy for cervical cancer (Eminowicz et al., 2016), the study organizers conducted a systematic review of contouring guidelines and generated a consensus atlas for use in the study. They examined compliance with protocol defined before and after introduction of the standardized atlas and noted an improvement from 25%–59% to 50%–83% compliance with contouring of four key structures (primary CTV, nodal CTV, bladder, and rectum). They also noted an improvement in an overall contouring quality score from an average of 1.8/4 to 2.7/4 after introduction of the atlas into the standardized radiotherapy QA package used for the study. Similarly, Mason et al. (2016) reported on delivery and target volume delineation accuracy in the context of the ProtecT trial comparing surgery, EBRT, and active surveillance for early prostate cancer. They noted delivered target dose variations of −4.4% to 0.2% as assessed by standardized QA measurements among ProtecT sites. They examined contouring quality among a random sample of three patients from each participating site (approximately 10% of the 554 patients enrolled on the radiation arm) and scored contouring quality for prostate, seminal vesicles, bladder, and rectum. Overall quality was judged as "satisfactory" in 78%, "acceptable variation" in 10%, "unacceptable variation, unlikely clinically significant" in 11%, and "unacceptable variation, possible clinical significance" in 1%. Of the significant variation (n = 3), all were in variations in the prostate contours; most of the acceptable variations were in seminal vesicle contours. Compliance with planning constraints was rated favorably with 80% demonstrating two or fewer (among 13) planning constraint variations and satisfactory PTV coverage was achieved in 49/54 cases examined.

At the other end of the spectrum, studies have reported on the clinical impact of deviation from protocol-specified treatment. For example, Ohri et al. (2013) identified eight studies (four pediatric, four adult; two lung, one sarcoma, three brain, one head and neck, one pancreatic cancer) eligible for inclusion in a meta-analysis of effects of protocol deviation on clinical outcomes. In a random effects model, deviations were associated with a statistically significant decrease in overall survival (Ohri et al., 2013) (1.74, 95% CI 1.28–2.35, $p < .001$), with a similar magnitude of effect seen for secondary outcomes for treatment failure (1.79, 95% CI 1.15–2.78, $p = .009$). Bekelman et al. (2012) referenced guideline documents for improving clinical trial QA in radiotherapy trials. They noted that the effect sizes seen in their meta-analysis surpassed those typically seen in successful trials of novel therapies. This identifies a "clear and present danger" in missing clinically significant treatment effects in the presence of inadequate QA in trials with a radiotherapy treatment component. This effect is most evident in one of the included studies, a Trans-Tasman Radiation Oncology Group (TROG) Phase III study comparing radiation plus cisplatin to radiation plus tirapazamine (Peters et al., 2010). In this study, posttreatment review revealed a noncompliance rate of 25% with almost half of the noncompliant plans (12% overall) exhibiting deviations to the extent that they were judged to carry a high risk of negative impact on tumor control. Compliance correlated with the number of patients enrolled (<5 patients enrolled were associated with a 30% noncompliance rated versus 5% among centers enrolling >20 patients). Patients with major deficiencies had inferior overall survival (50% versus 70%) and inferior locoregional failure free survival (54% versus 78%). This negative impact on survival outcomes of 20%–25% was more than twice the difference in survival that the trial was designed to detect (10% difference).

2.6.1.4 Concluding Remarks on Clinical Considerations

While a full discussion of the clinical sources of uncertainty is beyond the scope of this chapter (indeed of this book), it is important to appreciate major sources of clinical uncertainty and their relative magnitude in the context of overall treatment uncertainty, especially as related to 3D dosimetry. A "consumer's guide" to common cancer situations and the major clinical and dosimetric issues is supplied in Table 2.3 as a qualitative overview of some of these issues.

2.6.2 Physical Considerations

From a physical perspective, many reports have been written on uncertainty estimation and their categorization aligns favorably with the procedures of Figure 2.2. One report (Andreo and Nahum, 2007) classified uncertainties in EBRT into four major groups according to the following:

1. Absolute dose determination at the reference point in a water phantom (i.e., beam calibration) performed with a standard laboratory calibrated ionization chamber according to a code of practice or dosimetry protocol.
2. Procedures involving relative beam dosimetry (field size output dependence, influence of beam modifiers like wedges, secondary collimation, and blocks, etc.), performed with any type of detector suitable for measurements in water or in a plastic phantom.
3. The calculation of the dose delivered to the patient (monitor units or irradiation time to deliver the prescribed dose) and its distribution, usually performed with a computerizedTPS.
4. The process of treatment delivery throughout a complete treatment course, which accounts for daily variations both in patient and machine setup, patient movements, and machine instability during several weeks of treatment.

For the last group, the current trend in applying image guidance is to implement ART as needed, if unacceptable deviations are detected during the course of treatment. The situation is often corrected by repositioning the patient but in

Table 2.3 "Consumer's Guide" on Uncertainties for Common Cancer Situations in Radiation Therapy and the Major Clinical and Dosimetric Issues

	Clinical Sources of Uncertainty • Imaging defined boundaries • Radiobiology (BTV) • Interobserver variation (BTV/CTV) • Practice variation (CTV/PTV)				Dosimetric Focus				
Site	BTV	GTV	CTV	PTV	Dose Gradients	Dose Painting	Deformable Dosimetry	4D Dosimetry	Notes
SRS/SBRT	+	+	++	+	+++	+	–	+/–	Large fraction RBE; managing motion in lung SBRT
CNS	++	+	++	+	+++	++	–	–	BTV-based on mpMRI/PET
Head and neck	+++	+	+	+	+++	+++	+++	–	BTV-based on PET ART
Lung	+++	+++	+	+++	++	+	++	+++	Respiratory motion BTV-based on PET
Breast	+	++	+	+	+	+	+	+	GTV definition in partial breast RT
Lymphoma	++	+	+	+	+	+	+++	+	PET-based prescriptions
Rectal	+	++	+++	+	+	+	+	+	Movement from 3D to IMRT techniques
Cervix	++	+++	+++	+	+++	++	+++	+	MRI/PET GTV definition, mobile GTV, deform OAR
Bladder	–	+	+	+	+	+	+++	–	Highly variable CTV volumes
Prostate	+++	++	++	+	++	++	++	+	Defining intraprostatic boost, nodes

Note: ART, adaptive radiation therapy; CNS, central nervous system; 4D, four-dimensional; MRI, magnetic resonance imaging; OAR, organ at risk; PET, positron-emission tomography; SBRT, stereotactic body radiation therapy.

BTV: Biologic target volume defined on functional or metabolic imaging (multiparametric MRI or PET).

GTV/CTV/PTV: Gross tumor volume/clinical target volume/planning target volume.

Dose gradients: Measuring steep dose gradients required for OAR sparing.

Dose painting: Measuring inhomogeneous dose distributions within CTV/GTV to differentially boost biologic subvolumes.

Deformable dosimetry: Estimating dose accumulation in deformable OAR (i.e., rectum) or changing GTV (i.e., Head and neck, lymphoma).

4D dosimetry: Measuring dose when motion management strategies are used (i.e., gating, tumor tracking).

some cases, a refresh of the treatment replan is required and the above step is reiterated.

A recent report (Thwaites, 2013) discusses "optimum" uncertainty values as those representative of "achievable" values. In this chapter, we review uncertainty levels that have been reported and to use those as a guide to what is possible. The major recent references in this context include Palta and Mackie (2011), Thwaites (2013), Van Dyk et al. (2013), and IAEA (2016). An abbreviated summary of sample data for the most typical routine (noncomplex) external beam treatment techniques are shown in Table 2.4. Similarly, data for typical brachytherapy scenarios are shown in Table 2.5.

Various reviews have summarized *in vivo* dosimetry detectors, their advantages, and disadvantages as well as the estimated uncertainties in dose determination (IAEA, 2013; Mijnheer, 2013; Mijnheer et al., 2013a). A summary of these uncertainties is shown in Table 2.6.

In the context of direct 3D dosimetry, whereby a full dose distribution is obtained in a single irradiation, several systems have been described using polymer or radiochromic absorbers. Water radiolysis causes polymerization or chemical changes that can be read out with either MRI or optical/x-ray CT scanning techniques. An overview of the underlying principles of gel response has been presented elsewhere (Schreiner and Olding, 2009). The current generation of gel materials and their performance is updated in this book (Chapters 5 and 6). Typically, these methods yield a dose accuracy better than 5% at submillimeter spatial resolution within a one liter cylindrical volume, with variable dose sensitivity and linearity. Three-dimensional dosimetry avoids the spatial "patching" of results obtained by several dosimeters placed at different locations and obtained at different times. As an example of 3D IMRT dose verification, Figure 2.5 shows a radiochromic cylinder placed within the IROC-Houston (formerly RPC) head and neck phantom. Gel results obtained by optical CT scanning (VISTA, Modus Medical Devices, London, Ontario Canada) were compared against TPS-computed dose distributions, radiochromic film, and TLD in critical planes. Gamma analysis (3D) yielded a pass rate of 97% for thresholds of 5%/3mm. Absolute dose agreement with film and TLD was within 5%. To date, gel and plastic dosimeters have been applied mainly to research labs and "niche" dose verification situations during TPS commissioning or introduction of a new radiotherapy technique where traditional dosimeter results were in question because of poor tissue-equivalence or spatial resolution. An example is the application of gel dosimetry to small beams of radiosurgery (Babic et al., 2009). End-to-end testing may evolve with continuing developments in dosimeter materials and scanners more suitable for a clinical physics environment where process efficiency is paramount.

There is a growing commercial availability of radiochromic gels (Modus Medical Devices Inc., London, Ontario, Canada), radiochromic plastics (Heuris Pharma LLC, Skillman, New Jersey), and of turnkey-dedicated optical scanners (Modus Medical Devices Inc., London, Ontario, Canada). Application to patient-specific QA has been more limited; diode arrays in planar or cylindrical geometry provide more convenience and patient throughput, but it is recognized that these devices only sample a subset of the available 3D dose space. A full description of various dosimetry tools and techniques that can be used for evaluating the 3D quality of IMRT plans is given in a report from AAPM Task Group 120 (Low et al., 2011) and in Chapter 9. Gel dosimetry is rapidly evolving with promising

Table 2.4 Uncertainties (1 SD) Associated with Various Physical Components of the EBRT Process

Process and Quantity				Uncertainty
Dose at calibration point in water				1.5%–3.0%
Dose as determined by auditing program (TLD, OSLD)				1.5%–3.0%
Treatment machine-related uncertainties				
• Lasers				1–2 mm
• Relative dose ratios				2%
• Beam monitor stability				2%
• Machine jaw positioning				<1 mm
• Wedges				2 mm
• MLC static or dynamic position				≤1 mm
• MLC transmission				Several %
• Table/couch top attenuation				Up to 20%
Patient positioning				<1–15 mm
In vivo dosimetry				3%–5%
Treatment planning imaging-related uncertainties				
Imaging modality	MR	PET	US	CT
• Image geometry	<1–15 mm	<2 mm	<1 mm	<2 mm
• Image resolution	<1 mm	4–7 mm	0.3–3 mm	<1 mm
• CT number accuracy				±20 HU
• Imaging dose	0	8 mSv	0	1–4 cGy
Electronic portal imaging				
• Imaging geometry				1–2 mm
• Imaging resolution				<1 mm
• Imaging dose				~2 cGy
Image guidance-related uncertainties				
• Imaging modality	HT MV CT		MV CBCT	kV CBCT
• Image geometry	1–2 mm		1 mm	1 mm
• Imaging resolution	1.6 mm		2 mm	<1 mm
• CT number accuracy	30 HU		80 HU	±20–100 HU
• Imaging dose	1–3 cGy		5–10 cGy	5–25 cGy
TPS dose calculation				
• Central axis data				2%
• Off-axis, high-dose, low-dose gradient				2%
• High-dose gradient				2–4 mm
• Low-dose, low-dose gradient				5%
• Buildup region				50%
• Nonunit density tissues				2%–20%
EBRT end-to-end phantom				3%–10%/2 mm
EBRT end-to-end patient				5%–10%/5 mm

Source: Adapted from IAEA, Accuracy requirements and uncertainties in radiation. Human Health Series No. 31, Vienna, Austria: International Atomic Energy Agency, 2016.

Note: CT, computed tomography; EBRT, external beam radiation treatment; MLC, multileaf collimator; MR, magnetic resonance; OSLD, optically stimulated luminescent dosimeter; PET, positron-emission tomography; SD, standard deviation; TLD, thermoluminescent dosimeter; TPS, treatment planning system, HT, helical tomotherapy; MV, megavoltage; kV, kilovoltage; CBCT, cone-beam CT; US, ultrasound; HU, Hounsfield unit

Table 2.5 Uncertainties (1 SD) Associated with Various Physical Components of the Brachytherapy Radiation Treatment Process

Process and Quantity	Uncertainty
Dose at reference point in water	
• Air kerma strength in clinic	1.3%
TPS dose calculation	
• In water, compared with published data	2%
• Tissue inhomogeneities	10%[a]
Dose delivery	
• HDR	
• Source calibration	1.5%
• Source position	1 mm
• Temporal accuracy	<0.5%
• Dose delivery	3.4%
• LDR/MDR	
• Source calibration	1.3%
• Linear uniformity	<5%
• Source position	2 mm
• Temporal accuracy	1 s
• Dose delivery	4.4%

Source: Adapted from IAEA, Accuracy requirements and uncertainties in radiation. Human Health Series No. 31, Vienna, Austria: International Atomic Energy Agency, 2016.

Note: HDR, high-dose rate; LDR, low-dose rate; MDR, medium-dose rate; SD, standard deviation; TPS, treatment planning system.

[a] For high-energy photon sources, these values are likely to be much smaller.

Table 2.6 Summary of Estimated Dose Uncertainties (1 SD) for Different *in vivo* Dosimetry Detectors

Detector	Estimated Dose Uncertainty (%)
Diode	1.5–3[a]
MOSFET	2–5[a]
TLD	2–3[a]
OSLD	2–3[a]
Radiographic film	3[b]
Radiochromic film	3[c]
EPID	1.5–3[a]

Source: Adapted from Mijnheer, *Modern Technology of Radiation Oncology: A Compendium for Medical Physicists and Radiation Oncologist,* Medical Physics Publishing, Madison, WI, 2013.

Note: EPID, electronic portal imaging device; MOSFET, metal–oxide–semiconductor field-effect transistor; OSLD, optically stimulated luminescent dosimeter; SD, standard deviation; TLD, thermoluminescent dosimeter.

[a] Lower values are applicable for dosimeters that are regularly calibrated and have well-known correction factors.

[b] Assumes a well-maintained processor.

[c] Assumes following a strict readout protocol.

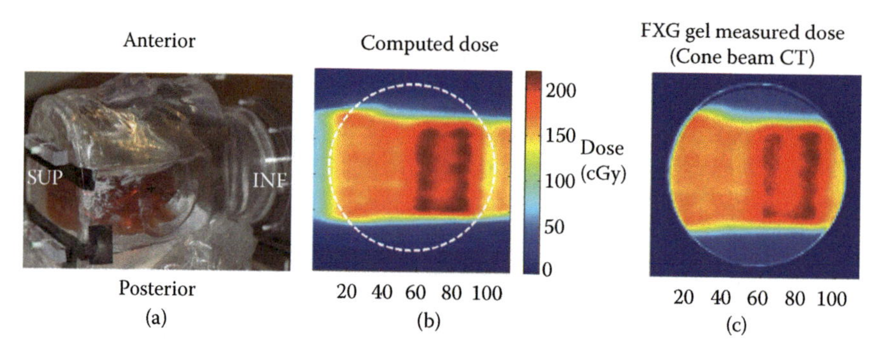

Figure 2.5

(a) Three-dimensional dose verification using a ferrous-xylenol orange-gelatin gel (FXG) placed in the Radiological Physics Center (RPC) head phantom. (b) The computed dose shown in the middle panel (Pinnacle software, Philips Medical) for an extracted coronal plane is compared with (c) optical computed tomography (CT) readout. (Adapted from Babic S et al., *Int. J. Radiat. Oncol. Biol. Phys.*, 2008.)

impact on clinical 3D dosimetry and QA for new radiation devices and software (see also Chapters 5 and 6).

2.6.3 Technology-Related Versus Patient-Related Uncertainties

Uncertainties associated with decision-making are clearly dependent upon the state of radiation oncology knowledge, interpretation, and practical limitations. Thus, decisions regarding disease staging, target volume definition, and dose prescription, for example, have the potential for human-related inaccuracies with potential clinical outcome consequences. These have been analysed by comparing inter- and intraperson variations in decision-making given the same patient information. These types of variations in clinical judgment and practice were discussed earlier. However, another component of uncertainty relates to the physical features of the patient and the changes that might occur in the patient during a course of therapy both intrafraction, such as patient breathing effects, and interfraction, such as tumor shrinkage and patient weight loss. These types of changes are time dependent and incorporated in target volume and OAR margin "recipes." Target volume recipes determine the margin required between the CTV and the PTV (Tanderup et al., 2010). It is interesting to note that these geometric prescriptions account for both machine- and patient-related uncertainties. They also incorporate both systematic and random uncertainties where it is recognized that the random uncertainties occur as a variation in performing the same procedure whereas systematic uncertainties occur when it is recognized that a set of results deviate by a consistent offset from the true value of the measurement. An example of a frequently used target volume margin recipe is given as follows (van Herk, 2011):

$$\mathrm{Margin}_{\mathrm{CTV-PTV}} = 2.5\sum + 0.7\sigma \tag{2.4}$$

where Σ is the systematic uncertainty and σ is the random uncertainty in target delineation. It is clear from this formulation that the systematic uncertainty has a relatively greater effect on the total margin size in comparison to random uncertainties. A detailed description of how such uncertainties can be determined from clinical practice protocols in individual cancer centers has been given by the joint report with the UK Royal College of Radiologists (The

Royal College of Radiologists et al., 2008). The report also describes how various contributing uncertainties can be combined. Assuming that each of the individual uncertainty components is normally distributed, the combined *systematic* uncertainty is given by

$$\Sigma_{Total}^{2} = \Sigma_{Delineation}^{2} + \Sigma_{Motion}^{2} + \Sigma_{Transfer}^{2} + \Sigma_{Patient\_setup}^{2} \tag{2.5}$$

where $\Sigma_{Delineation}$ is the systematic uncertainty due to target volume delineation representing the difference between the defined volume using an imaging technology and the "ideal" CTV, Σ_{Motion} is the systematic uncertainty due to change in target position and shape between delineation during treatment planning and treatment execution, possibly due to tumor regression or growth, or organ filling or distention, $\Sigma_{Transfer}$ is the uncertainty that accumulates when transferring image coordinate data from initial localization (e.g., CT simulation) through the TPS and onto the linear accelerator. This uncertainty is usually measured with a phantom and may include geometric imaging, TPS, and linac geometric uncertainties.

$\Sigma_{Patient\_setup}$ describes the setup uncertainties not accounted for by the transfer uncertainties including patient shape changes and relative changes of target position vis-à-vis surface reference marks over the course of treatment.

The components contributing to the *random* uncertainty include $\sigma_{Patient\_setup}$, the random uncertainty due to patient setup uncertainty and σ_{Motion}, the random variation in organ position and shape (except breathing since it is not random and can be addressed separately). These random variations can also be combined in quadrature similar to Equation 2.5. It is the onboard image-guidance procedures that can provide valuable quantitative information on both systematic and random uncertainties.

In principle, the same concepts can be applied to brachytherapy. In practice, this is more complex because the brachytherapy sources are generally located within the patient and the relationship between source locations and target or OAR volumes is complex because the sources can move along with the anatomy displacements and distortions.

2.6.4 End-to-End Tests

The above discussion regarding uncertainty propagation to generate treatment margins only addresses issues related to geometric uncertainties. As indicated earlier, there are two components to the radiation therapy uncertainty analysis, one being dose related and the other being geometry related. One approach to determining the combination of both uncertainties is to perform end-to-end tests, which includes all the stages of the radiation treatment process from machine calibration, imaging for therapy planning, transfer of data to the treatment planning computer, performing the actual treatment planning, transferring the resultant data to the therapy machine, and performing the actual treatment over a course of weeks (Figure 2.2). End-to-end tests have been performed by executing all these steps with anatomical phantom-containing targets, OARs, and dosimeters to measure the actual dose delivered. Such tests have been performed and reported by various clinical trials audit organizations for external beam therapy (Ibbott et al., 2006; Molineu et al., 2013) and for brachytherapy

(Palmer et al., 2015) as discussed in detail in Chapter 19. While all the major steps of the physical radiation process are considered, rigid phantoms are clearly limited and do not incorporate the uncertainties of nonrigid anatomy in real patients. More realistic simulations using Monte Carlo techniques of real patient "instances" on the treatment machine have been proposed (Van Dyk et al., 2010).

Such tests might also be performed with patient image data taken during therapy and concomitant with *in vivo* dosimetry on (or in) the patient. The *in vivo* dosimetry capability is possible for x-ray beams through the use of an EPID and dose reconstruction using a CT data set for back projection of fluence (Mijnheer et al., 2013b; Van Uytven et al., 2015). These authors concluded that EPID-based *in vivo* dosimetry, in combination with in-room imaging, is a fast and accurate tool for 3D *in vivo* verification for more complex technologies such as IMRT and VMAT delivery. This technique could verify a 3D conformal plan, using 8-point-dose measurements, with a resulting difference in dose of 1.3% ± 3.3% (1 SD) compared with the reconstructed dose (van Elmpt et al., 2009).

2.6.5 Uncertainty Propagation across Dose Fractions

While the end-to-end tests discussed above, including the *in vivo* measurements, describe uncertainties associated with the various steps of the radiation treatment process, they do not describe the overall uncertainty in dose delivery associated with all the fractions over the total treatment process for real patients. One way of modeling the total uncertainty is to determine the patient anatomy for each treatment fraction using onboard CT imaging and then determining the cumulative dose distribution to the patient using deformable dose determination, registering each voxel "of the day" in a patient to a reference (planning) image set. An example of this process has been performed by our group for a multifraction course for the treatment of cancer of the prostate (Van Dyk et al., 2010; Battista et al., 2013). This methodology was used to examine a range of scenarios for IGART of prostate cancer, including different schedules for CT imaging, patient repositioning, and dose replanning (Song et al., 2005). Our conclusions were that the frequency of adaptive interventions depends on the target margins used during IMRT optimization. The application of adaptive CT target margins determined one week into therapy minimizes the need for subsequent dose replanning. This is one example of how dose accumulation for a full course of therapy can be simulated to guide decisions about adapting or altering the treatment parameters depending on patient changes. Such studies could be expanded to include machine-related or other uncertainties using Monte Carlo propagation methods.

One of the questions that arises with daily image guidance is when to adjust the treatment to account for patient changes since daily replanning and adaptation is resource intensive and still remains impractical for routine online application at the present time. In the prostate study described above, it was noted that the D_{95} values dropped away from the intended dose prescription with less-frequent image guidance, especially for the small 5 mm margin. Daily repositioning and replanning restores agreement with the D_{95} of the original treatment plan. For NTCPs, the average NTCP values were in the range of 3%–6% for all IGART scenarios evaluated, compared with 8% obtained during initial treatment planning, although some outlier patients had NTCP values of >10%. This demonstrates that uncertainties in the total radiation treatment process are dependent on individual patient changes throughout a course of treatment and requires personalized due diligence in terms of IGART.

In another study from our group for lung cancer treatments (Woodford et al., 2007), it was found that 40% of the 17 patients evaluated had ≥30% reduction in GTV during the course of treatment with the average GTV change observed over 30 fractions being 38%, ranging from 12% to 87%. The conclusion was that if the GTV decreases by >30% at any point in the first 20 fractions of the 30 fraction treatment course, adaptive planning is appropriate to further improve the therapeutic ratio. It was also noted that adaptive planning can yield significant reductions in cumulative doses to OARs.

Another group (Kwint et al., 2014) developed a "traffic light protocol," an alarm system for adaptive treatment decision-making for lung cancer patients, which has four levels of action: red (immediate action before treatment), orange (action before next fraction), yellow (no action required), and green (no change). These authors observed that 128 out of 177 patients (72%) had intrathoracic anatomical changes with a maximum level of red, orange and yellow in 12%, 36% and 24%, respectively. The action levels were based on the location of the GTV with respect to the CTV, for example, the red level occurs when the GTV falls outside the PTV due to intrathoracic anatomical changes. Of course, such changes, if not corrected, would result in a clear geographic miss and inadequate dose to the GTV.

There has been some discussion in the literature about not using the PTV concept at all but rather incorporating the treatment-related uncertainties directly into the dose calculation by means of convolution or other methods (Craig et al., 2001; Gordon and Siebers, 2009). This has been called coverage-based or probabilistic treatment planning. This has been further extended with the use of dose coverage histogram criteria (Gordon et al., 2010; Xu et al., 2015).

Other developments in treatment planning include the consideration of uncertainties in such a way that the resultant treatment plan is minimally affected by the uncertainties in the overall treatment process. This has become known as robust treatment planning and optimization (Heath et al., 2009; Fredriksson, 2012; Li et al., 2015) and has been especially addressed for proton-based treatment planning.

One of the aspects of estimating dose uncertainties in an IGART course relates to the additional untargeted dose that is delivered as a result of the frequent CT image-guidance procedures (Ding et al., 2008). Dependent on the dose prescription and the number of images taken for image guidance, tissue doses could be increased by about 3% compared to the prescription at the target site. This could translate to about 10% of the prescription for dose to bone; skin doses could be about 5% of the prescription dose (Alaei et al., 2010). Clearly, the additional incidental imaging doses cannot be ignored if we are to make an accurate statement of the total dose actually delivered to the patient (Nelson and Ding, 2014) during a full course of radiotherapy. Ideally, such doses would be incorporated as part of the treatment plan and cumulative dose distribution, assuming the CT dose distributions can be accurately superimposed over the course of treatment (Alaei et al., 2010). In Chapter 22, issues related to the radiation dose from x-ray imaging systems used in radiation therapy are discussed in a comprehensive way.

2.7 Summary of Accuracy Requirement Considerations

As shown by the content of this chapter, the determination of the accuracy and uncertainties associated with the estimation of the actual dose delivered to all tissues and organs within the patient is a nontrivial task. The following is a concise summary of many of the clinical and physical uncertainty considerations associated with patient dose determination.

2.7.1 Clinical Considerations

- The radiation treatment process consists of a complex series of steps each with their own technology content and corresponding subprocess uncertainties that propagate collectively.
- Patient (re)positioning can be done to an accuracy of 1–6 mm depending on the target site, the immobilization device, and the image-guidance procedure that is used.
- Under some conditions, uncertainties associated with radiobiological parameters such as the α/β value may have a greater impact on treatment-related uncertainties compared to uncertainties in dose delivery. For example, for tumors with a low α/β and a higher fraction size (as commonly used for hypofractionated and stereotactic treatments), the effects of uncertainties in α/β assumptions overwhelm the effects of dosimetric uncertainty.
- Clinical trials demonstrate that patients with major protocol deficiencies have inferior clinical outcome (e.g., overall survival and inferior locoregional failure free survival), in many cases of similar (but negative) magnitude to the benefits reported for successful new treatments.
- Target and OAR delineation still remains as one of the larger uncertainties in the overall radiation treatment process with uncertainties ranging from 5 to 50 mm for target definition and 5–20 mm for OAR delineation. Improved imaging resolution will have limited impact compared with enhanced training on the interpretation of visible targets seen on CT, MRI, and PET images. Interventions to reduce IOVs are the subject of ongoing research. Documents like the RANCZER guidelines provide practical suggestions for improving target volume delineation.
- Target volume delineation recipes to generate the CTV to PTV margin generally incorporate systematic and random uncertainties as related to setup uncertainties as well as patient-related deformation changes. The systematic uncertainties have a greater impact on the total uncertainty compared to random uncertainties but need to be assessed at the local institution level since they can be related to imaging interpretation or setup protocols.

2.7.2 Physical Considerations

- The determination of dose to water in a beam can potentially be performed with an accuracy of 1%–2% ($k = 1$), generally using a standards laboratory calibrated ionization chamber.
- Audits of radiation beams using mailed dosimeters such as TLDs or OSLDs can be done with an accuracy of 1%–2% and demonstrate agreement of multiple institutions to within 2%–3.5%.
- Uncertainty in both dose and geometry need to be considered in radiation therapy uncertainty analysis. Multiple radiation dose detectors, ranging from 0D (point detector) to 3D sampling, are very useful for evaluating treatment dose distributions, with each having their own level of uncertainty in dose determination.

- The mechanical components (jaw settings, MLC settings, wedges, and gantry rotation) can generally be done with an accuracy of 1 mm or 1°.
- The in-plane imaging resolution for CT scanning is approximately 1 mm with the longitudinal resolution depending on the slice thickness. Similar resolutions exist for MRI, but PET and SPECT have resolutions of approximately 3–7 mm.
- CT data are still optimal to provide electron density information to an uncertainty level that does not deteriorate accuracy of dose computation algorithms. Density substitutions (e.g., in segmented MRI zones) can be used in certain situations after validation of dose predictions versus CT-based results.
- For image registration, rigid registration "errors" have a lower uncertainty limit of about 0.5 mm while deformable registrations can have an uncertainty of several millimeters to centimeters. The impact on dose depends largely on the local dose gradients in the region of voxel misplacements.
- Dose computations using TPSs can generally comply with a gamma-specified accuracy of 3%/3 mm although lung dose calculations still remain a challenge and results are strongly dependent on the type of algorithm used, and the energy and size of the beam; uncertainties are still varying well beyond the 4% range under some conditions. Monte Carlo calculations offer better accuracy when commissioned with due diligence and will become available for routine treatment planning in the near future.
- Gamma analysis has become a very useful tool for efficiently surveying the quality of 3D dose distributions; however, it is recognized that there is the potential for a lack of sensitivity and specificity to clinically meaningful parameters under some conditions. Various proposals have been made to provide more sensitive and clinically relevant analyses.
- Robust treatment planning can be performed to select plans that minimize the impact of uncertainties from the total treatment process.
- The ICRU, in its Report 83 on IMRT (ICRU, 2010), made the following recommendation for the accuracy of absorbed-dose delivery: "With these considerations, it is recommended that for a low-gradient (<20%/cm) region, the difference between the measured (or independently computed) absorbed dose and the treatment planning absorbed dose, normalized to the absorbed-dose prescription (e.g., D_{50}%) should be no more than 3.5%. For high-gradient (>20%/cm) regions, the accuracy of DTA should be 3.5 mm."
- DVH accuracy is dependent on the "hidden" algorithm and binning used by the TPS supplier and the user's choice of imaging parameters such as CT slice thickness and dose grid parameters. DVH accuracy is also influenced by the inaccuracies in volume segmentation.
- Independent audits by peers are extremely useful in support of quality and safety in radiation therapy, to avoid treatment errors and to provide confidence in the accuracy of dosimetry.
- Summaries of typical achievable accuracies are shown in Tables 2.4 and 2.5 for external beam and brachytherapy, respectively.
- Uncertainties are generally propagated by summation in quadrature under the assumption that the individual uncertainties are independent and normally distributed.

- End-to-end experimental tests and computer simulations provide a means of evaluating the accuracy and uncertainties of the overall treatment process including beam calibration, CT scanning, planning, dose optimization, and treatment. End-to-end tests are recommended for evaluating any new treatment technique or technology, initially internally but eventually reaffirmed by an external independent review.
- *In vivo* dosimetry provides a means of evaluating the treatment accuracy in a personalized manner, especially if it can be done by CT-based dose reconstruction within the specific patient. Such results from one group demonstrate an uncertainty of about 3% (1 SD) to the target in addition to providing a means of detecting treatment errors early, before the completion of a multifraction course of therapy.

2.7.3 Broad Considerations

- The total uncertainty of the dose delivered to a patient for a multifraction course of therapy requires the use of daily imaging and deformable registration to accumulate daily doses on a reference image. Computer simulations of the procedures per tumor site can be used to design the optimal image-guidance schedules or to test "alarms" for effective replanning decisions.
- A single statement of required treatment accuracy is a gross oversimplification. The best single statement that can be made comes from the first recommendation of the IAEA report (IAEA, 2016), "*All forms of radiation therapy should be applied as accurately as reasonably achievable (AAARA), technical and biological factors being taken into account.*"
- Ideally, each department should determine local uncertainties associated with the dosimetric and treatment procedures for all of its major techniques. These can then be folded into practice guidelines and refined over time as the techniques evolve. Computer simulations of each subprocess and of the overall chain of processes can provide valuable guidance for assuring effective and efficient optimal cancer therapy in light of inevitable uncertainties.
- As uncertainties are collectively reduced, it is anticipated that the full promise of new radiation technology advances will be achieved for the benefit of cancer patients. The contents of this book on clinical 3D dosimetry go a long way to achieving this objective.

References

Aguirre JF et al. (2002) Thermoluminescence dosimetry as a tool for the remote verification of output for radiotherapy beams: 25 years of experience. In: Proceedings of the International Symposium on Standards and Codes of Practice in Medical Radiation Dosimetry, IAEA-CN-96/82, pp. 191–199. (Vienna, Austria: International Atomic Energy Agency).

Aird EG and Conway J (2002) CT simulation for radiotherapy treatment planning. *Br. J. Radiol.* 75: 937–949.

Alaei P, Ding G and Guan H (2010) Inclusion of the dose from kilovoltage cone beam CT in the radiation therapy treatment plans. *Med. Phys.* 37: 244–248.

Alfonso R et al. (2008) A new formalism for reference dosimetry of small and nonstandard fields. *Med. Phys.* 35: 5179–5186.

Andreo P (2011) Accuracy requirements in medical radiation dosimetry. In: *International Symposium on Standards, Applications and Quality Assurance in Medical Radiation Dosimetry*, pp. 1–23. (Vienna, Austria: International Atomic Energy Agency).

Andreo P and Nahum A (2007) Supplementary details on codes of practice for absolute dose determination. In: *Handbook of Radiotherapy Physics: Theory and Practice*, pp. 385–427. (Eds. Mayles P, Nahum A and Rosenwald J-C, Boca Raton, FL: Taylor and Francis).

Asbell SO, Grimm J, Xue J, Chew MS and LaCouture TA (2016) Introduction and clinical overview of the DVH risk map. *Semin. Radiat. Oncol.* 26: 89–96.

Azangwe G et al. (2014) Detector to detector corrections: A comprehensive experimental study of detector specific correction factors for beam output measurements for small radiotherapy beams. *Med. Phys.* 41: 072103.

Babic S, Battista J and Jordan K (2008) Three-dimensional dose verification for intensity-modulated radiation therapy in the Radiological Physics Center head-and-neck phantom using optical computed tomography scans of ferrous xylenol-orange gel dosimeters. *Int. J. Radiat. Oncol. Biol. Phys.* 70: 1281–1291.

Babic S et al. (2009) Three-dimensional dosimetry of small megavoltage radiation fields using radiochromic gels and optical CT scanning. *Phys. Med. Biol.* 54: 2463–2481.

Battista JJ and Bronskill MJ (1981) Compton scatter imaging of transverse sections: An overall appraisal and evaluation for radiotherapy planning. *Phys. Med. Biol.* 26: 81–99.

Battista JJ et al. (2013) Dosimetric and radiobiological consequences of computed tomography-guided adaptive strategies for intensity modulated radiation therapy of the prostate. *Int. J. Radiat. Oncol. Biol. Phys.* 87: 874–880.

Bekelman JE et al. (2012) Redesigning radiotherapy quality assurance: Opportunities to develop an efficient, evidence-based system to support clinical trials—Report of the National Cancer Institute Work Group on Radiotherapy Quality Assurance. *Int. J. Radiat. Oncol. Biol. Phys.* 83: 782–790.

Bentzen SM et al. (2010) Quantitative Analyses of Normal Tissue Effects in the Clinic (QUANTEC): An introduction to the scientific issues. *Int. J. Radiat. Oncol. Biol. Phys.* 76(3 Suppl): S3–S9.

Bissonnette JP et al. (2012) Quality assurance for image-guided radiation therapy utilizing CT-based technologies: A report of the AAPM TG-179. *Med. Phys.* 39: 1946–1963.

Brahme A (1984) Dosimetric precision requirements in radiation therapy. *Acta Radiol. Oncol.* 23: 379–391.

Brahme A (1988) Accuracy requirements and quality assurance of external beam therapy with photons and electrons. *Acta Oncol.* Suppl. 1: 5–76.

Brock KK, Al-Mayah A and Velec M (2011) Uncertainties in deformable registration. In: *Uncertainties in External Beam Radiation Therapy*, pp. 403–442. (Eds. Palta JR and Mackie TR, Madison, WI: Medical Physics Publishing).

Brunskill K et al. (2017) Does peer review of radiation plans affect clinical care? A systematic review of the literature. *Int. J. Radiat. Oncol. Biol. Phys.* 97: 27–34.

Budgell G et al. (2011) A national dosimetric audit of IMRT. *Radiother. Oncol.* 99: 246–252.

Caissie A et al. (2016) A pan-Canadian survey of peer review practices in radiation oncology. *Pract. Radiat. Oncol.* 6: 342–351.
Carrasco P et al. (2012) 3D DVH-based metric analysis versus per-beam planar analysis in IMRT pretreatment verification. *Med. Phys.* 39: 5040–5049.
Chen Z et al. (2006) Investigation of MR image distortion for radiotherapy treatment planning of prostate cancer. *Phys. Med. Biol.* 51: 1393–1403.
Clark CH et al. (2009) Dosimetry audit for a multi-centre IMRT head and neck trial. *Radiother. Oncol.* 93: 102–108.
Clark CH et al. (2014) A multi-institutional dosimetry audit of rotational intensity-modulated radiotherapy. *Radiother. Oncol.* 113: 272–278.
Clark CH et al. (2015) Radiotherapy dosimetry audit: Three decades of improving standards and accuracy in UK clinical practice and trials. *Br. J. Radiol.* 88: 20150251.
Coleman L and Skourou C (2013) Sensitivity of volumetric modulated arc therapy patient specific QA results to multileaf collimator errors and correlation to dose volume histogram based metrics. *Med. Phys.* 40: 111715.
Craig T, Battista J et al. (2001) Considerations for the implementation of target volume protocols in radiation therapy. *Int. J. Radiat. Oncol. Biol. Phys.* 49: 241–250.
Craig T, Brochu D and Van Dyk J (1999) A quality assurance phantom for three-dimensional radiation treatment planning. *Int. J. Radiat. Oncol. Biol. Phys.* 44: 955–966.
Cui Y et al. (2015) Contouring variations and the role of atlas in non-small cell lung cancer radiation therapy: Analysis of a multi-institutional preclinical trial planning study. *Pract. Radiat. Oncol.* 5: e67–e75.
Cutanda Henriquez F and Vargas Castrillon S (2010) Confidence intervals in dose volume histogram computation. *Med. Phys.* 37: 1545–1553.
Cutanda Henriquez F and Vargas Castrillon S (2011) A probability approach to the study on uncertainty effects on gamma index evaluations in radiation therapy. *Comput. Math. Methods Med.* 2011: 861869.
Davidson SE et al. (2008) Technical note: Heterogeneity dose calculation accuracy in IMRT: Study of five commercial treatment planning systems using an anthropomorphic thorax phantom. *Med. Phys.* 35: 5434–5439.
Deasy JO et al. (2010) Improving normal tissue complication probability models: The need to adopt a "data-pooling" culture. *Int. J. Radiat. Oncol. Biol. Phys.* 76(3 Suppl): S151–S154.
Diez P, Hoskin PJ and Aird EG (2007) Treatment planning and delivery of involved field radiotherapy in advanced Hodgkin's disease: Results from a questionnaire-based audit for the UK Stanford V regimen vs ABVD clinical trial quality assurance programme (ISRCTN 64141244). *Br. J. Radiol.* 80: 816–821.
Ding GX, Duggan DM and Coffey CW (2008) Accurate patient dosimetry of kilovoltage cone-beam CT in radiation therapy. *Med. Phys.* 35: 1135–1144.
Drzymala RE et al. (1991) Dose-volume histograms. *Int. J. Radiat. Oncol. Biol. Phys.* 21: 71–78.
Dunn L et al. (2015) National dosimetric audit network finds discrepancies in AAA lung inhomogeneity corrections. *Phys. Med.* 31: 435–441.
Dutreix A (1984) When and how can we improve precision in radiotherapy? *Radiother. Oncol.* 2: 275–292.

Eaton DJ et al. (2013) A national dosimetry audit of intraoperative radiotherapy. *Br. J. Radiol.* 86: 20130447.
Eminowicz G et al. (2016) Improving target volume delineation in intact cervical carcinoma: Literature review and step-by-step pictorial atlas to aid contouring. *Pract. Radiat. Oncol.* 6: e203–e213.
Ezzell GA et al. (2003) Guidance document on delivery, treatment planning, and clinical implementation of IMRT: Report of the IMRT Subcommittee of the AAPM Radiation Therapy Committee. *Med. Phys.* 30: 2089–2115.
Ezzell GA et al. (2009) IMRT commissioning: Multiple institution planning and dosimetry comparisons, a report from AAPM Task Group 119. *Med. Phys.* 36: 5359–5373.
Faculty of Radiation Oncology, Royal Australian and New Zealand College of Radiologists. 2015. Quality guidelines for volume delineation in radiation oncology. https://www.ranzcr.com/college/document-library/quality-guidelines-for-volume-delineation-in-radiation-oncology
Fattori G et al. (2014) Dosimetric effects of residual uncertainties in carbon ion treatment of head chordoma. *Radiother. Oncol.* 113: 66–71.
Fowler JF (2010) 21 years of biologically effective dose. *Br. J. Radiol.* 83: 554-568.
Fraass B et al. (1998) American Association of Physicists in Medicine Radiation Therapy Committee Task Group 53: Quality assurance for clinical radiotherapy treatment planning. *Med. Phys.* 25: 1773–1829.
Fredriksson A (2012) A characterization of robust radiation therapy treatment planning methods from expected value to worst case optimization. *Med. Phys.* 39: 5169–5181.
Galvin JM et al. (2004) Implementing IMRT in clinical practice: A joint document of the American Society for Therapeutic Radiology and Oncology and the American Association of Physicists in Medicine. *Int. J. Radiat. Oncol. Biol. Phys.* 58: 1616–1634.
Gershkevitsh E et al. (2014) Dosimetric inter-institutional comparison in European radiotherapy centres: Results of IAEA supported treatment planning system audit. *Acta Oncol.* 53: 628–636.
Gerweck LE and Wakimoto H (2016) At the crossroads of cancer stem cells, radiation biology, and radiation oncology. *Cancer Res.* 76: 994–998.
Glide-Hurst CK et al. (2015) Initial clinical experience with a radiation oncology dedicated open 1.0T MR-simulation. *J. Appl. Clin. Med. Phys.* 16(2): 218–240.
Gomola I et al. (2001) External audits of electron beams using mailed TLD dosimetry: Preliminary results. *Radiother. Oncol.* 58: 163–168.
Gondi V et al. (2015) Real-time pretreatment review limits unacceptable deviations on a cooperative group radiation therapy technique trial: Quality assurance results of RTOG 0933. *Int. J. Radiat. Oncol. Biol. Phys.* 91: 564–570.
Gordon JJ et al. (2010) Coverage optimized planning: Probabilistic treatment planning based on dose coverage histogram criteria. *Med. Phys.* 37: 550–563.
Gordon JJ and Siebers JV (2009) Coverage-based treatment planning: Optimizing the IMRT PTV to meet a CTV coverage criterion. *Med. Phys.* 36: 961–973.
Gwynne S et al. (2012) Imaging for target volume delineation in rectal cancer radiotherapy-a systematic review. *Clin. Oncol. (R. Coll. Radiol.)* 24: 52–63.
Hamilton CS and Ebert MA (2005) Volumetric uncertainty in radiotherapy. *Clin. Oncol. (R. Coll. Radiol.)* 17: 456–464.

Hansson U et al. (1993) Mailed TL dosimetry programme for machine output check and clinical application in the EORTC radiotherapy group. *Radiother. Oncol.* 29: 85–90.

Hartford AC et al. (2012) American College of Radiology (ACR) and American Society for Radiation Oncology (ASTRO) Practice guideline for intensity-modulated radiation therapy (IMRT). *Am. J. Clin. Oncol.* 35: 612–617.

Heath E, Unkelbach J and Oelfke U (2009) Incorporating uncertainties in respiratory motion into 4D treatment plan optimization. *Med. Phys.* 36: 3059–3071.

Hoornaert MT et al. (1993) A dosimetric quality audit of photon beams by the Belgian Hospital Physicist Association. *Radiother. Oncol.* 28: 37–43.

Hourdakis CJ and Boziari A (2008) Dosimetry quality audit of high energy photon beams in Greek radiotherapy centers. *Radiother. Oncol.* 87: 132–141.

Ibbott GS, Haworth A and Followill DS (2013) Quality assurance for clinical trials. *Front. Oncol.* 3: 311.

Ibbott GS, Molineu A and Followill DS (2006) Independent evaluations of IMRT through the use of an anthropomorphic phantom. *Technol. Cancer Res. Treat.* 5: 481–487.

Ibbott GS and Thwaites DI (2015) Audits for advanced treatment dosimetry. *J. Phys. Conf. Series* 573: 012002.

IAEA (2004) Commissioning and quality assurance of computerized planning systems for radiation treatment of cancer. Report TRS-430. (Vienna, Austria: International Atomic Energy Agency).

IAEA (2007) Specification and acceptance testing of radiotherapy treatment planning systems. TECDOC-1540. (Vienna, Austria: International Atomic Energy Agency).

IAEA (2008) Commissioning of radiotherapy treatment planning systems: Testing for typical external beam treatment techniques. TECDOC-1583. (Vienna, Austria: International Atomic Energy Agency).

IAEA (2013) Development of procedures for in vivo dosimetry in radiotherapy. Human Health Reports No. 8. (Vienna, Austria: International Atomic Energy Agency).

IAEA (2016) Accuracy requirements and uncertainties in radiation therapy. Human Health Series No. 31. (Vienna, Austria: International Atomic Energy Agency).

ICRP (2009) Preventing accidental exposures from new external beam radiation therapy technologies, Publication 112. (Oxford, UK: International Commission on Radiation Protection, Pergamon Press).

ICRU (1976) Determination of absorbed dose in a patient irradiated by beams of X or gamma rays in radiotherapy procedures. Report 24. (Washington, DC: International Commission on Radiation Units and Measurements).

ICRU (2010) Prescribing, recording, and reporting photon-beam intensity-modulated radiation therapy (IMRT). Report 83. (Bethesda, MD: International Commission on Radiation Units and Measurements).

Ippolito E et al. (2008) IGRT in rectal cancer. *Acta Oncol.* 47: 1317–1324.

Izewska J et al. (2003) The IAEA/WHO TLD postal dose quality audits for radiotherapy: A perspective of dosimetry practices at hospitals in developing countries. *Radiother. Oncol.* 69: 91–97.

Izewska J et al. (2007) A methodology for TLD postal dosimetry audit of high-energy radiotherapy photon beams in non-reference conditions. *Radiother. Oncol.* 84: 67–74.

Jaffray DA et al. (2010) Accurate accumulation of dose for improved understanding of radiation effects in normal tissue. *Int. J. Radiat. Oncol. Biol. Phys.* 76(3 Suppl): S135-S139.

Jameson MG et al. (2010) A review of methods of analysis in contouring studies for radiation oncology. *J. Med. Imaging Radiat. Oncol.* 54: 401–410.

Jeraj R, Bradshaw T and Simoncic U (2015) Molecular imaging to plan radiotherapy and evaluate its efficacy. *J. Nucl. Med.* 56(11): 1752–1765.

Jin X, Yan H, Han C, Zhou Y, Yi J and Xie C (2015) Correlation between gamma index passing rate and clinical dosimetric difference for pre-treatment 2D and 3D volumetric modulated arc therapy dosimetric verification. *Br. J. Radiol.* 88: 20140577.

Kim J et al. (2015) Dosimetric evaluation of synthetic CT relative to bulk density assignment-based magnetic resonance-only approaches for prostate radiotherapy. *Radiat. Oncol.* 10: 239.

Kirisits C et al. (2015) Quality assurance in MR image guided adaptive brachytherapy for cervical cancer: Final results of the EMBRACE study dummy run. *Radiother. Oncol.* 117: 548–554.

Klein EE et al. (2009) Task Group 142 report: Quality assurance of medical accelerators. *Med. Phys.* 36: 4197–4212.

Kron T, Haworth A and Williams I (2013a) Dosimetry for audit and clinical trials: Challenges and requirements. *J. Phys. Conf. Ser.* 444: 012014.

Kron T, Taylor M and Thwaites D (2013b) Small field dosimetry. In: *The Modern Technology of Radiation Oncology: A Compendium for Medical Physicists and Radiation Oncologists*, Vol. 3, pp. 245–300. (Ed. Van Dyk J, Madison, WI: Medical Physics Publishing).

Kroutilikova D, Novotny J and Judas L (2003) Thermoluminescent dosimeters (TLD) quality assurance network in the Czech Republic. *Radiother. Oncol.* 66: 235–244.

Kry SF et al. (2013) Algorithms used in heterogeneous dose calculations show systematic differences as measured with the Radiological Physics Center's anthropomorphic thorax phantom used for RTOG credentialing. *Int. J. Radiat. Oncol. Biol. Phys.* 85: e95–e100.

Kupchak C, Battista J and Van Dyk J (2008) Experience-driven dose-volume histogram maps of NTCP risk as an aid for radiation treatment plan selection and optimization. *Med. Phys.* 35: 333–343.

Kwint M et al. (2014) Intra thoracic anatomical changes in lung cancer patients during the course of radiotherapy. *Radiother. Oncol.* 113: 392–397.

Letourneau D, McNiven A and Jaffray DA (2013) Multicenter collaborative quality assurance program for the province of Ontario, Canada: First-year results. *Int. J. Radiat. Oncol. Biol. Phys.* 86: 164–169.

Li Y et al. (2015) Selective robust optimization: A new intensity-modulated proton therapy optimization strategy. *Med. Phys.* 42: 4840–4847.

Low DA et al. (1998) A technique for the quantitative evaluation of dose distributions. *Med. Phys.* 25: 656–661.

Low DA et al. (2011) Dosimetry tools and techniques for IMRT. *Med. Phys.* 38: 1313–1338.

Mason MD et al. (2016) Radiotherapy for prostate cancer: Is it 'what you do' or 'the way that you do it'? A UK perspective on technique and quality assurance. *Clin. Oncol. (R. Coll. Radiol.)* 28: e92–e100.

Meeks SL (2011) Immobilization from rigid to non-rigid. In: *Uncertainties in External Beam Radiation Therapy*, pp. 45–67. (Eds. Palta JR and Mackie TR, Madison, WI: Medical Physics Publishing).

Mijnheer B (2013) In vivo dosimetry. In: *Modern Technology of Radiation Oncology: A Compendium for Medical Physicists and Radiation Oncologists*, Vol. 3, pp. 301–336. (Ed. Van Dyk J, Madison, WI: Medical Physics Publishing).

Mijnheer BJ, Battermann JJ and Wambersie A (1987) What degree of accuracy is required and can be achieved in photon and neutron therapy? *Radiother. Oncol.* 8: 237–252.

Mijnheer B et al. (2013a) In vivo dosimetry in external beam radiotherapy. *Med. Phys.* 40: 070903.

Mijnheer B et al. (2013b) 3D EPID-based *in vivo* dosimetry for IMRT and VMAT. *J. Phys. Conf. Series* 444: 012011.

Mijnheer B and Georg D (Eds.) (2008) *Guidelines for the verification of IMRT. ESTRO Booklet No. 9*. (Brussels, Belgium: European Society for Radiotherapy and Oncology).

Mitchell DM et al. (2009). Assessing the effect of a contouring protocol on post-prostatectomy radiotherapy clinical target volumes and interphysician variation. *Int. J. Radiat. Oncol. Biol. Phys.* 75: 990–993.

Moiseenko V (2004) Effect of heterogeneity in radiosensitivity on LQ based isoeffect formalism for low alpha/beta cancers. *Acta Oncol.* 43: 499–502.

Molineu A et al. (2005) Design and implementation of an anthropomorphic quality assurance phantom for intensity-modulated radiation therapy for the Radiation Therapy Oncology Group. *Int. J. Radiat. Oncol. Biol. Phys.* 63: 577–583.

Molineu A et al. (2013) Credentialing results from IMRT irradiations of an anthropomorphic head and neck phantom. *Med. Phys.* 40: 022101.

Moran JM and Ritter T (2011) Limits of precision and accuracy of radiation delivery systems. In: *Uncertainties in External Beam Radiation Therapy*, pp. 215–232. (Eds. Palta JR and Mackie TR,Madison, WI: Medical Physics Publishing).

Mutic S et al. (2003) Quality assurance for computed-tomography simulators and the computed-tomography-simulation process: Report of the AAPM Radiation Therapy Committee Task Group No. 66. *Med. Phys.* 30: 2762–2792.

Nelms BE et al. (2013) Evaluating IMRT and VMAT dose accuracy: Practical examples of failure to detect systematic errors when applying a commonly used metric and action levels. *Med. Phys.* 40: 111722.

Nelms B et al. (2015) Methods, software and datasets to verify DVH calculations against analytical values: Twenty years late(r). *Med. Phys.* 42: 4435–4448.

Nelms BE, Zhen H and Tome WA (2011) Per-beam, planar IMRT QA passing rates do not predict clinically relevant patient dose errors. *Med. Phys.* 38: 1037–1044.

Nelson AP and Ding GX (2014) An alternative approach to account for patient organ doses from imaging guidance procedures. *Radiother. Oncol.* 112: 112–118.

Nisbet A, Thwaites DI and Sheridan ME (1998) A dosimetric intercomparison of kilovoltage x-rays, megavoltage photons and electrons in the Republic of Ireland. *Radiother. Oncol.* 48: 95–101.

Ohri N, Shen X, Dicker AP, Doyle LA, Harrison AS and Showalter TN (2013) Radiotherapy protocol deviations and clinical outcomes: A meta-analysis of cooperative group clinical trials. *J. Natl. Cancer Inst.* 105: 387–393.

Palmer A et al. (2011) Analysis of regional radiotherapy dosimetry audit data and recommendations for future audits. *Br. J. Radiol.* 84: 733–742.
Palmer AL et al. (2015) A multicentre 'end to end' dosimetry audit for cervix HDR brachytherapy treatment. *Radiother. Oncol.* 114: 264–271.
Palmer AL et al. (2013) Design and implementation of a film dosimetry audit tool for comparison of planned and delivered dose distributions in high dose rate (HDR) brachytherapy. *Phys. Med. Biol.* 58: 6623–6640.
Palta JR and Mackie TR (Eds.) (2011) AAPM Medical Physics Monograph No. 35. In: *Uncertainties in External Beam Radiation Therapy*. (Madison, WI: Medical Physics Publishing).
Panitsa E, Rosenwald JC and Kappas C (1998) Quality control of dose volume histogram computation characteristics of 3D treatment planning systems. *Phys. Med. Biol.* 43: 2807–2816.
Peters LJ et al. (2010) Critical impact of radiotherapy protocol compliance and quality in the treatment of advanced head and neck cancer: Results from TROG 02.02. *J. Clin. Oncol.* 28: 2996–3001.
Poole C et al. (2011) A hybrid radiation detector for simultaneous spatial and temporal dosimetry. *Australas. Phys. Eng. Sci. Med.* 34: 327–332.
Pratx G and Xing L (2011) GPU computing in medical physics: A review. *Med. Phys.* 38: 2685–2697.
Pulliam KB et al. (2014) Comparison of 2D and 3D gamma analyses. *Med. Phys.* 41: 021710.
QUANTEC (2010) Quantitative Analyses of Normal Tissue Effects in the Clinic (QUANTEC). *Int. J. Radiat. Oncol. Biol. Phys.* 76(3 Suppl): S1–S160.
Rasch C, Steenbakkers R and van Herk M (2005) Target definition in prostate, head, and neck. *Semin. Radiat. Oncol.* 15: 136–145.
Rassiah P et al. (2004) A thermoluminescent dosimetry postal dose inter-comparison of radiation therapy centres in Malaysia. *Australas. Phys. Eng. Sci. Med.* 27: 25–29.
Roue A et al. (2006) The EQUAL-ESTRO audit on geometric reconstruction techniques in brachytherapy. *Radiother. Oncol.* 78: 78–83.
Roue A et al. (2004) The EQUAL-ESTRO external quality control laboratory in France. *Cancer Radiother.* 8(1 Suppl): S44–S49.
Roue A et al. (2007) Development of a TLD mailed system for remote dosimetry audit for (192)Ir HDR and PDR sources. *Radiother. Oncol.* 83: 86–93.
The Royal College of Radiologists, Society and College of Radiographers, Institute of Physics and Engineering in Medicine, National Patient Safety Agency, and British Institute of Radiology (2008) On target: Ensuring geometric accuracy in radiotherapy. (London, UK: The Royal College of Radiologists). https://www.rcr.ac.uk/docs/oncology/pdf/BFCO(08)5_On_target.pdf.
Schreiner LJ and Olding T (2009) Gel dosimetry. In: *Clinical Dosimetry Measurements in Radiotherapy—AAPM 2009 Summer School*, pp. 979–1025. (Eds. Rogers DWO and Cygler JE, Madison, WI: Medical Physics Publishing).
Schreiner LJ (2011) On the quality assurance and verification of modern radiation therapy treatment. *J. Med. Phys.* 36: 189–191.
Segedin B and Petric P (2016) Uncertainties in target volume delineation in radiotherapy—Are they relevant and what can we do about them? *Radiol. Oncol.* 50: 254–262.

Seravalli E et al. (2015) A comprehensive evaluation of treatment accuracy, including end-to-end tests and clinical data, applied to intracranial stereotactic radiotherapy. *Radiother. Oncol.* 116: 131–138.

Smilowitz J et al. (2015) AAPM Medical Physics Practice Guideline 5.a: Commissioning and QA of treatment planning dose calculations—Megavoltage photon and electron beams. *J. Appl. Clin. Med. Phys.* 16(5): 14–34.

Song W et al. (2005) Image-guided adaptive radiation therapy (IGART): Radiobiological and dose escalation considerations for localized carcinoma of the prostate. *Med. Phys.* 32: 2193–2203.

Song WY et al. (2006) Evaluation of image-guided radiation therapy (IGRT) technologies and their impact on the outcomes of hypofractionated prostate cancer treatments: A radiobiologic analysis. *Int. J. Radiat. Oncol. Biol. Phys.* 64: 289–300.

Stasi M et al. (2012) Pretreatment patient-specific IMRT quality assurance: A correlation study between gamma index and patient clinical dose volume histogram. *Med. Phys.* 39: 7626–7634.

Stojadinovic S et al. (2015) Breaking bad IMRT QA practice. *J. Appl. Clin. Med. Phys.* 16(3): 154–165.

Su L et al. (2014) ARCHERRT—A GPU-based and photon-electron coupled Monte Carlo dose computing engine for radiation therapy: Software development and application to helical tomotherapy. *Med. Phys.* 41: 071709.

Svensson H (1984) Quality assurance in radiation therapy: Physical aspects. *Int. J. Radiat. Oncol. Biol. Phys.* 10(1 Suppl): 59–65.

Tanderup K et al. (2010) PTV margins should not be used to compensate for uncertainties in 3D image guided intracavitary brachytherapy. *Radiother. Oncol* 97: 495–500.

Thwaites D (2013) Accuracy required and achievable in radiotherapy dosimetry: Have modern technology and techniques changed our views? *J. Phys. Conf. Series* 444: 012006.

Trofimov A et al. (2012) Visualization of a variety of possible dosimetric outcomes in radiation therapy using dose-volume histogram bands. *Pract. Radiat. Oncol.* 2: 164–171.

van Baardwijk A et al. (2006) The current status of FDG-PET in tumor volume definition in radiotherapy treatment planning. *Cancer Treat. Rev.* 32: 245–260.

van der Merwe D et al. (2017) Accuracy requirements and uncertainties in radiotherapy: A report of the International Atomic Energy Agency. *Acta Oncol.* 56: 1–6.

Van Dyk J (2008) Quality assurance of radiation therapy planning systems: Current status and remaining challenges. *Int. J. Radiat. Oncol. Biol. Phys.* 71: S23–S27.

Van Dyk J, Barnett RB and Battista JJ (1999) Computerized radiation treatment planning systems. In: *The Modern Technology of Radiation Oncology: A Compendium for Medical Physicists and Radiation Oncologists*, Vol. 1, pp 231–286. (Ed. Van Dyk J, Madison, WI: Medical Physics Publishing).

Van Dyk J et al. (1993) Commissioning and quality assurance of treatment planning computers. *Int. J. Radiat. Oncol. Biol. Phys.* 26: 261–273.

Van Dyk J et al. (2010) Impact analysis of image-guided radiation therapy: Description of a multi-fraction dose propagation model. In: Proc XVIth International conference on the use of computers in radiation therapy (ICCR). Paper 12383. (Ed. Sonke JJ, Amsterdam, the Netherlands: The Netherlands Cancer Institut-Antoni van Leeuwenhoek Ziekenhuis).

Van Dyk J and Battista J (2014) Has the use of computers in radiation therapy improved accuracy in dose delivery? *J. Phys. Conf. Series* 489: 012098.

Van Dyk J, Battista JJ and Bauman GS (2013) Accuracy and uncertainty considerations in modern radiation oncology. In: *The Modern Technology of Radiation Oncology: A Compendium for Medical Physicists and Radiation Oncologists*, Vol. 3, pp. 361–415. (Ed. Van Dyk J, Madison, WI: Medical Physics Publishing, Madison).

van Elmpt W et al. (2009) 3D in vivo dosimetry using megavoltage cone-beam CT and EPID dosimetry. *Int. J. Radiat. Oncol. Biol. Phys.* 73: 1580–1587.

van Herk M (2011) Margins and margin recipes. In: *Uncertainties in External Beam Radiation Therapy*, pp. 169–190. (Eds. Palta JR and Mackie TR, Madison, WI: Medical Physics Publishing).

Van Uytven E et al. (2015) Validation of a method for in vivo 3D dose reconstruction for IMRT and VMAT treatments using on-treatment EPID images and a model-based forward-calculation algorithm. *Med. Phys.* 42: 6945–6954.

Vinod SK et al. (2016) A review of interventions to reduce inter-observer variability in volume delineation in radiation oncology. *J. Med. Imaging Radiat. Oncol.* 60: 393–406.

Wambersie A (2001) What accuracy is required and can be achieved in radiation therapy? (Review of radiobiological and clinical data). *Radiochim. Acta* 89: 255–264.

Weiss E and Hess CF (2003) The impact of gross tumor volume (GTV) and clinical target volume (CTV) definition on the total accuracy in radiotherapy theoretical aspects and practical experiences. *Strahlenther. Onkol.* 179: 21–30.

Wendling M et al. (2007) A fast algorithm for gamma evaluation in 3D. *Med. Phys.* 34: 1647–1654.

Williams JR et al. (2008) A quantitative overview of radiosensitivity of human tumor cells across histological type and TP53 status. *Int. J. Radiat. Biol.* 84: 253–264.

Williams I et al. (2012) The Australian Clinical Dosimetry Service: A commentary on the first 18 months. *Australas. Phys. Eng. Sci. Med.* 35: 407–411.

Wilson EM et al. (2003) Validation of active breathing control in patients with non-small-cell lung cancer to be treated with CHARTWEL. *Int. J. Radiat. Oncol. Biol. Phys.* 57: 864–874.

Woodford C et al. (2007) Adaptive radiotherapy planning on decreasing gross tumor volumes as seen on megavoltage computed tomography images. *Int. J. Radiat. Oncol. Biol. Phys.* 69: 1316–1322.

Xu H, Gordon JJ and Siebers JV (2015) Coverage-based treatment planning to accommodate delineation uncertainties in prostate cancer treatment. *Med. Phys.* 42: 5435–5443.

Yang TJ et al. (2013) Tumor bed delineation for external beam accelerated partial breast irradiation: A systematic review. *Radiother. Oncol.* 108: 181–189.

Zhen H, Nelms BE, and Tome WA (2011) Moving from gamma passing rates to patient DVH-based QA metrics in pretreatment dose QA. *Med. Phys.* 38: 5477–5489.

Section II
Instrumentation

3

Detectors for Reference Dosimetry

Simon Duane and Ben Mijnheer

3.1 Introduction

The term "absolute" is widely used in dosimetry, but it has two distinct meanings. Clinical physicists commonly use the term absolute to emphasize that a measurement is not relative; for instance, the calibration of machine output requires such an absolute measurement, while the acquisition of percentage depth dose and beam profile data involves relative measurements. In some dosimetry calibration laboratories, the term absolute is reserved for measurements made with a primary standard such as a calorimeter; the clinical physicist's "absolute" measurement may instead be referred to as a "reference" measurement.

There is room for further confusion, since the measurement conditions specified in a reference dosimetry protocol would normally be referred to as "reference" conditions, and any measurement following the protocol would normally be referred to as a "reference" dose measurement. The confusion arises if an absolute measurement, whether or not it is made using a primary standard, is made under nonreference conditions. Such a measurement may be "absolute," and is possible provided that an appropriate correction is made for the change in sensitivity of a dosimeter when it is used under nonreference conditions rather than reference conditions; however, it is not a "reference" measurement.

This chapter is concerned with measurements that are absolute as opposed to relative, it includes some discussion of absorbed dose primary standards, but the main concern is the accurate measurement of absorbed dose using a calibrated ion chamber under reference conditions.

3.2 Detectors Used for Reference Dose Measurements

3.2.1 Requirements for a Reference Dosimetry System

Reference dosimetry requires a detector that has a calibration coefficient traceable to a standards dosimetry laboratory, which provides the conversion of the detector signal to absorbed dose to water. These reference detectors are generally reserved for the output calibration of radiation therapy machines. Most radiotherapy departments have only one reference dosimetry system with a calibration traceable to a standards dosimetry laboratory. However, calibration factors for other detectors can be obtained through a transfer calibration procedure (e.g., IAEA, 2000; Abdel-Rahman et al., 2009).

In a reference dose measurement, the detector reading is multiplied by its calibration coefficient to give absorbed dose to water. The sensitivity of a reference detector must remain stable over the time that elapses between calibration and use, and it must be possible to correct readings for the effect on detector sensitivity of any change in ambient circumstances such as temperature, pressure, and so on. In a relative measurement, the detector is used to determine the ratio of absorbed dose, for example, at two different points in the same field, or at the same point in two different fields. There is no need for the detector to show good long-term stability, and ambient circumstances can generally be assumed not to change significantly between the two readings. The sensitivity of a silicon diode, for example, varies with temperature in a nontrivial way. Consequently, the uncertainty in correcting for temperature dependence would be unacceptable in reference dosimetry, but becomes negligible in a relative measurement because the correction would cancel in the dose ratio. Air-filled ion chambers can meet the requirements of reference dosimetry, provided they show good dimensional stability and have a large enough sensitive volume to generate an adequate signal. Further requirements are summarized below. Other detectors such as alanine, synthetic single crystal diamond, and plastic scintillator detectors, which are discussed in detail in Chapter 4, offer some promise, but are unlikely to replace the use of air-filled ion chambers for reference measurements of absorbed dose at a point. Currently most, if not all, absorbed dose protocols use ion chambers for reference dosimetry (e.g., Almond et al., 1999; IAEA, 2000; McEwen et al., 2014).

3.2.2 Reference-Class Detectors

In Chapter 4, characteristics of point detectors and their use for dosimetry in modern radiation therapy are discussed. Most of these detectors are not suitable for reference dosimetry and only ion chambers are recommended for this purpose. Even not all types of ion chambers are appropriate for reference dosimetry, and a specification of the characteristics of reference-class ion chambers to be used for the measurement of absorbed dose in megavoltage (MV) photon beams are provided in the addendum to the AAPM TG-51 protocol (McEwen et al., 2014).

The aspects of chamber performance identified in that addendum as being crucial to determining reference-class behavior are chamber stabilization, leakage current, polarity correction, recombination correction, and long-term stability. Table 3.1 gives specifications for these properties for reference-class ion chambers.

Only cylindrical chambers are recommended for reference dosimetry in high-energy photon beams, as elucidated in the addendum to the AAPM TG-51 protocol (McEwen et al., 2014), and in the International Atomic Energy Agency (IAEA) Code of Practice (IAEA, 2000). For a large number of chamber types, data required for reference dosimetry are provided in these documents. Plane-parallel chambers cannot be used for reference dose measurements in MV photon beams because their long-term stability is worse and the chamber-to-chamber variation is larger than for cylindrical chambers. However, parallel-plate chambers are useful for relative dosimetry in high-energy photon beams, for instance, for measurements in the buildup region.

Plane-parallel chambers are recommended for reference dose measurements for all electron beam qualities. In the IAEA Code of Practice, it is furthermore stated that they *must* be used for electron beams with incident energies lower than 10 MeV, while according to the AAPM TG-51 protocol, this threshold is 6 MeV. For electron beams with energies higher than these thresholds, cylindrical chambers *may* be used if appropriate gradient (effective point of measurement) corrections are taken into account.

Ion chambers for dose measurements in low-energy x-ray beams are also of the plane-parallel type. These chambers must have an entrance window consisting of a thin membrane as clarified in the low-energy x-ray dosimetry protocols (IAEA, 2000; Ma et al., 2001). Cylindrical chambers are recommended for reference dose measurements in medium-energy x-ray beams.

Table 3.1 Specification of a Reference-Class Ionization Chamber for Megavoltage Photon-Beam Dosimetry

Measurand[a]	Specification
Chamber settling	Should be less than a 0.5% change in chamber reading per monitor unit from beam-on for a warmed up machine to stabilization of the ionization chamber.
P_{leak}	<0.1% of chamber reading ($0.999 < P_{leak} < 1.001$).
P_{pol}	<0.4% correction ($0.996 < P_{pol} < 1.004$). <0.5% maximum variation in P_{pol} with energy (total range)
$P_{ion} = 1 + C_{init} + C_{gen} D_{pp}^{b}$	
General	P_{ion} should be linear with dose per pulse.
Initial	Initial recombination should be <0.2%, that is, $C_{init} < 0.002$, for the TG-51 reference conditions.[c]
Polarity dependence	Difference in initial recombination correction between opposite polarities should be <0.1%.
Chamber stability	Should exhibit less than a 0.3%[d] change in calibration coefficient over the typical recalibration period of 2 years.

Source: Reproduced from McEwen M. et al., *Med. Phys.*, 41, 041501, 2014.

[a] Refer to McEwen (2010) for details on how each parameter was evaluated.

[b] Both initial and general recombination need to be considered.

[c] Value derived from data presented by McEwen (2010).

[d] This value is derived from calibration data from dosimetry calibration laboratories.

3.3 Traceability to Primary Standards of Absorbed Dose to Water

3.3.1 The International Measurement System

Reference dose measurements, where a protocol is used to convert the measured signal to D_w, the absorbed dose to water, should be accurate, reproducible, and traceable to assure tumor control and mitigate normal tissue complications. High accuracy is important because it directly influences all patient treatments with that calibrated beam. General issues related to the accuracy of dose measurements in radiotherapy are discussed in Chapter 2, while the accuracy of reference dose measurements is described in this chapter, with the emphasis on traceability to primary standards of absorbed dose.

In radiation dosimetry, primary standards dosimetry laboratories (PSDLs) developed primary standards for radiation measurement. Primary standards are instruments of the highest metrological quality, which permit determination of the unit of a quantity according to its definition, the accuracy of which has been verified by comparison with standards of other institutions of the same level, that is, with those of the BIPM, the International Bureau of Weights and Measures (Bureau International des Poids et Mesures) in Paris, France, and other PSDLs. It should always be possible to link the calibration of a detector used for reference dosimetry in a hospital back to the national primary standard of absorbed dose. This traceability consists of a chain of cross calibrations, linking one instrument to another, back to the primary standard, and helps to ensure that reference dose measurements made with different instruments across a country are compatible. International compatibility relies on a measurement infrastructure in which the various national primary standards are regularly compared in a process that is coordinated by the BIPM. In a recent comprehensive report on accuracy and uncertainties in radiation therapy, the international measurement system and the relationship between the BIPM, PSDLs, secondary standards dosimetry laboratories (SSDLs), and users have been elucidated (IAEA, 2016; van der Merwe et al., 2017).

3.3.2 Calorimeters

Because the quantity of interest in reference dosimetry is D_w, water calorimeters have been developed as primary standards to measure D_w in x-ray and electron beams (e.g., Ross and Klassen, 1996; Seuntjens and Duane, 2009). Water calorimeters determine D_w by measuring the temperature rise in water as a result of energy deposition during irradiation of a specific water volume. Graphite calorimeters have also been developed as primary absorbed dose standard because they do not need a heat defect correction, necessary because of chemical reactions in irradiated water (e.g., DuSautoy, 1996). Also the sensitive volume of a water calorimeter may increase with the duration of the measurement, which can be avoided with a graphite calorimeter (see Chapter 12). However, a conversion procedure is required for graphite calorimeters to determine absorbed dose to water, resulting in a somewhat larger total uncertainty in D_w determinations. Water calorimeters are the primary standards for absorbed dose in photon beams in PSDLs in the United States, Canada, Germany, the Netherlands, and Switzerland, while graphite calorimeters are used for that purpose for photon and electron beams in the United Kingdom, France, Australia, and Italy. Ion chamber calibrations are carried out in these PSDLs with a standard uncertainty of about 0.43% and 0.56% for water and graphite calorimeters, respectively.

With the installation of new treatment modalities, there is an increasing need that D_w measurements with primary standards can be performed on-site to calibrate detectors. Portable calorimeters have been developed for this purpose for photon and electron beams (e.g., McEwen and Duane, 2000), as well as for light-ion beams (e.g., Palmans et al., 2004); issues related to the latter type of calorimeter are discussed in detail in Chapter 12. Recently VSL, the PSDL in the Netherlands, has developed a new transportable water calorimeter serving as a primary D_w standard for ^{60}Co and MV photons including magnetic resonance imaging (MRI) incorporated treatment equipment (de Prez et al., 2016). Special attention was paid to its operation in different beam geometries and beam modalities including the application in magnetic fields (Figure 3.1).

Another interesting development is the construction of probe-type calorimeters (Duane et al., 2012) such as the one shown in Figure 3.2 (Renaud et al., 2017). In the latter publication, it was demonstrated that photon-beam output measurements using the Aerrow, the ionization chamber-sized graphite calorimeter, were in agreement with chamber-based clinical reference dosimetry data within combined standard uncertainties. These devices may be used by clinical physicists as a local absorbed dose standard for high-energy photon beams, even in dosimetrically challenging situations such as in intensity-modulated radiation therapy (IMRT) and magnetic fields.

Table 3.2 shows the estimated combined standard uncertainty in D_w at the reference depth in water in MV photon beams (IAEA, 2016). For most hospitals, this value will vary between 1.2% and 1.5%, while a somewhat better accuracy can be obtained if the ion chamber is calibrated in a PSDL using an accelerator having the same beam quality as the one used in the hospital.

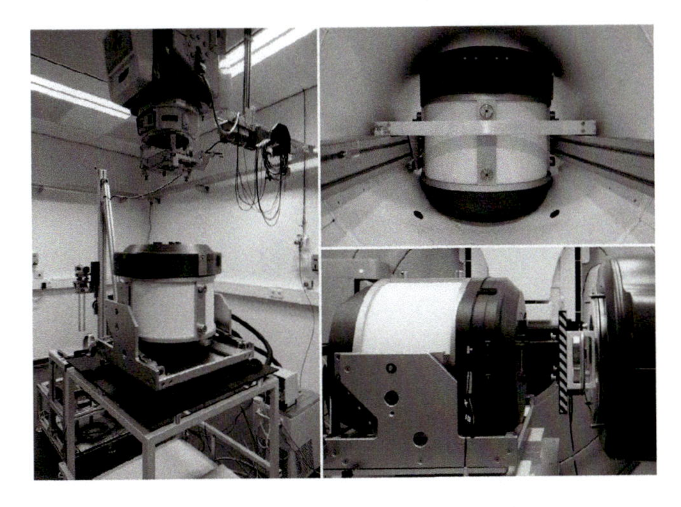

Figure 3.1

The calorimeter in the vertical VSL ^{60}Co beam (left), inside the bore of an Elekta Atlantic MRI-linac combination at UMC Utrecht, the Netherlands (top right) and in a horizontal beam orientation in front of an Elekta Versa HD accelerator at the Netherlands Cancer Institute in Amsterdam, the Netherlands (bottom-right). (Reproduced from de Prez L. et al., *Phys. Med. Biol.*, 61, 5051–5076, 2016.)

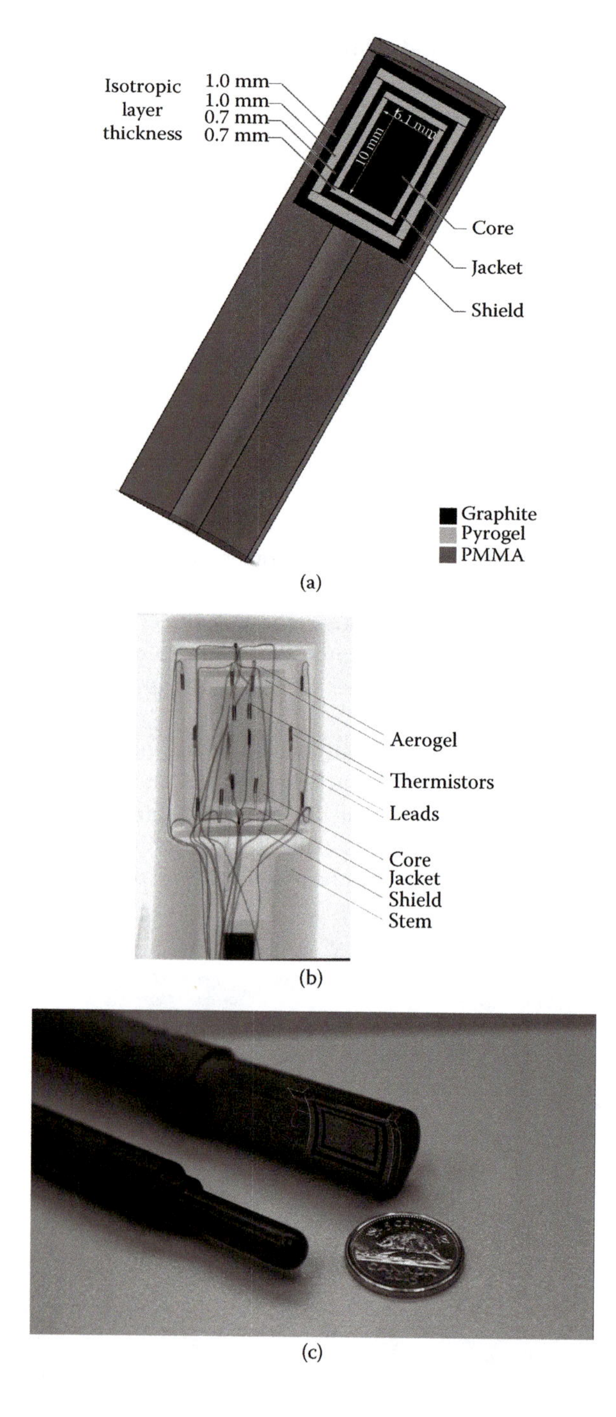

Figure 3.2

(a) A cross-sectional schematic diagram of the Aerrow design and (b) a digitally reconstructed radiograph of a microcomputed tomography scan of the prototype calorimeter showing multiple embedded thermistors and leads. (c) The comparable size of the Aerrow to that of a Farmer-type ionization chamber is illustrated by the Exradin A12 positioned alongside the probe calorimeter (internal Aerrow structure is shown as a blended rendering) and a 5 cent coin (21 mm wide) for scale. PMMA, poly-methyl-methacrylate. (Reproduced from Renaud J. et al., *Med. Phys.*, 2017.)

Table 3.2 Estimated Combined Standard Uncertainty in D_w at the Reference Depth in Water in Megavoltage Photon Beams

Physical Quantity or Procedure	Relative Standard Uncertainty (%)			
	SSDL Co-60	PSDL Co-60	PSDL Co-60 and Accelerator	PSDL Accelerator
Step 1: Standards laboratory				
$N_{D,w}$ calibration of the secondary standard	0.5	–	–	–
Long-term stability of the secondary standard	0.1	–	–	–
$N_{D,w}$ calibration of the used dosimeter at the standards laboratory	0.4	0.5	0.5	0.5
Combined uncertainty of Step 1	0.6	0.5	0.5	0.5
Step 2: Hospital				
Long-term stability of user dosimeter	0.3	0.3	0.3	0.3
Establishment of reference conditions	0.4	0.4	0.4	0.4
Dosimeter reading relative to timer or beam monitor	0.6	0.6	0.6	0.6
Correction for influence quantities	0.4	0.4	0.4	0.4
Beam quality correction	1.0[a]	1.0[a]	0.7[b]	–
Combined uncertainty of Step 2	1.3	1.3	1.1	0.9
Combined standard uncertainty in D_w (Steps 1 and 2)	1.5	1.4	1.2	1.0

Source: Reproduced from IAEA, Accuracy requirements and uncertainties in radiation therapy. Human Health Series No. 31. Vienna, Austria: International Atomic Energy Agency, 2016.

Note: $N_{D,w}$ = absorbed dose to water calibration coefficient, D_w—absorbed dose to water. PSDLs, primary standards dosimetry laboratories; SSDLs, secondary standards dosimetry laboratories.

[a] Calculated values.

[b] Measured values normalized to ^{60}Co.

3.4 Absorbed Dose Calibration and Measurement

3.4.1 Calibration of MV Photon Beams

The sensitivity of a given detector, in terms of absorbed dose to water, depends on the properties of the radiation field at the position of the detector in the phantom. For an air-filled ion chamber used in MV photon beams, the variation in sensitivity can approach 5%. Inevitably the radiation field is perturbed by the presence of the detector and, for definiteness, the calibration coefficient is defined so that the measurement is of absorbed dose at a point in the undisturbed phantom, that is, when the detector is removed. It is not practical to specify the radiation field completely in terms of the distributions in energy and in angle of electron and photon fluences. Instead, dosimetry protocols (e.g., Almond et al., 1999; IAEA, 2000; McEwen et al., 2014) specify reference conditions under which measurements must be made. The conventional conditions for MV photon beams are the following:

1. A unidirectional beam is normally incident on a full scatter rectangular water phantom.
2. The beam is collimated to produce a 10 cm × 10 cm field in the plane of measurement.

3. The measurement point is on the central axis at a specified depth, usually 5 or 10 g/cm^2.
4. The beam has a nominally flat profile.
5. Beam quality is parameterized in terms of its penetrating power using a quality index such as $\%d_{10,X}$ or $TPR_{\frac{20}{10}}$ as defined in dosimetry protocols.
6. Ion chamber readings are corrected to standard air density, for example, 20°C and 1013.25 mbar, to 50% relative humidity, and for saturation.

Of these conditions, 1–4 are intended to control the proportion of scattered to primary photons, and to ensure that there is at least transient charged particle equilibrium (CPE) at the point of measurement. The use of a beam quality index (5) is intended to deal with most of the variation in sensitivity of a given ion chamber from one MV beam to another. The correction for saturation in (6) is required in case the dose rate differs between calibration and use. After applying the corrections listed in (6), readings taken with a reference-class ion chamber can show long-term consistency of the order 0.1%. This level of consistency is significantly better than the combined standard uncertainty when the readings are expressed in terms of absorbed dose to water, which is usually not better than 0.5%, and dominated by the uncertainty of calibrations in the traceability chain and the uncertainty of the primary standard itself, as discussed above.

Conditions (2) and/or (4) cannot be met in beams produced by certain machines, and in this case it is necessary to specify a machine-specific reference field (Alfonso et al., 2008) as discussed in Section 3.4.3. Even when the absorbed dose calibration coefficient is expressed in terms of a beam quality index as specified in (5), there remains a small residual variation in ion chamber sensitivity from one MV beam to another. This variation may, for instance, be up to 0.5% between flattening filter free (FFF) and flattened beams. Although small, this variation illustrates that beam quality index and beam quality are distinct concepts. It turns out that the field size specification in (2) is crucial in the measurement of the beam quality index, because both $\%d_{10.X}$ and $TPR_{\frac{20}{10}}$ are strongly dependent on field size. However the beam quality, that is, the energy spectrum of the electron fluence, is much less sensitive to field size, and so the chamber calibration coefficient is also relatively insensitive to field size, provided that the field does not qualify as a small field.

3.4.2 Calibration of FFF Beams

Fano's theorem (Bouchard et al., 2012) is a great help in understanding the response of ion chambers under conditions of CPE. An analysis based on cavity theory shows why the chamber cavity does not significantly perturb the equilibrium electron fluence, so that the water-to-air mass-stopping power ratio accounts for almost all of the observed quality dependence of ion chamber sensitivity. Full CPE is never achieved in practice because of the attenuation of primary photons, but lateral CPE exists in the interior of a large flat field and is enough to prevent disequilibrium effects provided the chamber is entirely contained within the flat part of the field.

In the case of FFF beams, there is a lateral dose gradient that increases with distance from the central axis, and this creates a small amount of lateral disequilibrium in the electron fluence. As a result, the chamber cavity slightly perturbs the electron fluence, even on the central axis of the reference field. The effect is qualitatively identical to what happens when a small field is reduced to the point where the penumbrae begin to overlap an air-filled ion chamber used on the central axis. This fluence perturbation combines with volume averaging to further reduce the response of the ion chamber, compared to its response in a flat field of the same beam quality. This should be considered whenever a reference ion chamber is calibrated in a flat beam and then used in an FFF beam, particularly if the cavity length is comparable to that of a Farmer-type chamber.

The associated higher dose rates and doses per pulse of FFF beams improve treatment delivery efficiency but may need larger corrections for recombination and polarity behavior of ion chambers. In a recent study, several models of small-volume ion chambers have been shown to meet reference-class requirements with respect to ion recombination and polarity, even in a high dose rate environment (Hyun et al., 2017). However, the results of his study also emphasize the need for careful reference detector selection, and indicate that ion chambers ought to be extensively tested in each beam of interest prior to their use for reference dose measurements.

3.4.3 Calibration of Beams of Other Types of Treatment Modalities

There is an increasing number of treatment modalities for which it is impossible to meet the conditions under which reference dose measurements must be made in the way described in dosimetry protocols. For reference dosimetry of photon fields delivered by a Gamma Knife, CyberKnife, or Tomotherapy unit, machine-specific reference fields are used (Alfonso et al., 2008). The absorbed dose to water can then be derived from a detector reading by the introduction of a correction factor accounting for the detector's difference in dose response between the conditions of field size, geometry, phantom material, and beam quality of the conventional reference field and the machine-specific reference field. Details of this approach and values for that correction factor for a number of detectors are provided in Chapter 9.

As shown in Chapter 26, the emerging field of MR-guided radiotherapy requires the presence of strong magnetic fields that can affect the performance of most conventional dosimetry systems. In that chapter, the magnitude of the effect of a magnetic field on the response of a number of detectors can be found, while in an earlier section of this chapter, the use of a water calorimeter for reference dosimetry in an MR-Linac is described (Figure 3.1). Reference dosimetry using ion chambers can be performed in the presence of strong magnetic fields with the use of magnetic field correction factors. Recently, data have been reported by O'Brien et al., (2016) for a number of cylindrical ion chamber models in a 1.5 T magnetic field. These authors showed that chamber magnetic field correction factors are smaller than 1% for reference-class Farmer-type chambers when the chambers are aligned parallel with the magnetic field lines, but can reach 4%–5% depending on the orientation. It was also observed that the $TPR_{\frac{20}{10}}$ beam quality specifier is robust in the presence of magnetic fields. However, the $\%d_{10,X}$ beam quality index cannot be measured directly because of SSD restrictions and changes in the dose distribution in the build-up region, but can be derived from the $TPR_{\frac{20}{10}}$ value using a conversion factor. O'Brien and colleagues also observed that even small air gaps around an ionization chamber altered the reading of the instrument, and therefore recommended that nonwater phantoms should not be used for reference dose measurements.

Absolute calibration of beams of low-energy x-rays used in small animal irradiation platforms is discussed in Chapter 24. As shown in that chapter, the reference dose measurements described in low-energy x-ray dosimetry protocols (IAEA, 2000; Ma et al., 2001) are not directly applicable to the potentially very small fields from these novel irradiation platforms. These dosimetry protocols recommend a calibration field size of 10 cm × 10 cm, which is impractical for precision animal irradiators. Instead, often a field of 4 cm × 4 cm is used, requiring various correction and conversion factors, some of which need to be inter-/extrapolated from the tables provided in these dosimetry protocols. A number of issues related to dose measurements in small-field animal irradiators are also valid for measurements in synchrotron microbeams, as explained in Chapter 25.

As discussed in Chapter 12, calorimeters are in principle also in light-ion beams the primary instruments for reference dose measurements. Although for photon and electron beams there is sufficient knowledge of the chemical heat defect to ensure a high accuracy, limited information is available of the chemical heat defect in light-ion beams due to its complicated linear energy transfer (LET) dependence. The more common instruments to perform reference dosimetry in light-ion beams are therefore air-filled ion chambers, having only small variations of their response as a function of energy in the clinical energy range. However, the energy dependence of the mean energy required to produce an ion pair in air, the correction for fluence perturbation, and the variation of the stopping power ratio water-to-air with LET cannot be ignored if a high accuracy is required, as will be elucidated in Chapter 12.

Several groups have developed calorimeters for absorbed dose rate measurements close to high dose rate ^{192}Ir brachytherapy sources (Sarfehnia and Seuntjens, 2010; Guerra et al., 2012; Sander et al., 2012). Although such an approach gives an improvement in the combined standard uncertainty in the absorbed dose rate to water at a distance of 1 cm, the brachytherapy community is reluctant to change their practice of using an air kerma-based calibration method. The reason for this is that for specifying source strength using calorimetry, several changes would be needed to clinical treatment planning systems. Also, the decreased uncertainties are not substantial when the big picture of clinical procedures is considered. Finally, it is not that easy to deliver such an absorbed dose standard for sources other than ^{192}Ir.

3.5 Practical Issues Related to Absolute Dose Measurements

3.5.1 Beam Quality Correction

Calibration of a machine's output will ideally be made with an ion chamber that has been cross-calibrated in the same user beam, in order to avoid the need for a quality-dependent correction. However, that cross-calibration will require the use of a reference ion chamber, which will usually have been calibrated in another beam, and its calibration may require a quality-dependent correction. That correction factor, k_{Q,Q_0}, may be available as a function of a beam quality index such as $\%d_{10,X}$ or $TPR_{\frac{20}{10}}$ as discussed in the various dosimetry protocols. The required quality index must have been measured for the user beam and also be known for the calibration beam. However, the quality index is not sufficient to completely specify beam quality and additional care may be required in transferring the calibration of a detector from one beam to another, as discussed in a previous section of this chapter.

3.5.2 Ion Chamber Corrections and Influence Quantities

The effect of ion recombination is reduced if the polarizing voltage is increased; however, at some point charge multiplication sets in and leads to nonlinear behavior in a Jaffe plot of the inverse of the chamber reading against the inverse of the polarizing voltage (e.g., McEwen, 2010). This is not a disadvantage in relative dosimetry, because the charge multiplication is essentially independent of dose rate and so cancels when taking the ratio. However in reference dosimetry, it is essential to correct for ion recombination, and the validity of the correction should be demonstrated by using a Jaffe plot to show that the operating voltage is within the linear region. For this reason, one might choose a smaller polarizing voltage when a chamber is used for reference dosimetry compared to relative dosimetry. The polarity effect is usually small (or negligible) in photon-beam measurements, but should be checked if the chamber used has a very small volume.

The chamber must be vented in order for the air pressure in the cavity to reach equilibrium with the measured ambient pressure. A chamber will usually reach thermal equilibrium in a reasonably short time when used in water. This process may be slower in a solid phantom, if used, and a solid phantom must be left to reach a stable temperature before starting measurements.

Thimble chambers have better dimensional stability than parallel-plate (electron) chambers. Graphite-walled chambers can have particularly good long-term stability, but if a waterproof chamber is used directly in water, there may be a risk of the thimble distorting through absorption of water if the chamber is left in water for an extended period.

3.5.3 Phantoms Used for Absolute Dose Measurements

Absolute dose calibration is generally performed with a traceable calibrated ionization chamber and electrometer in a scanning water system or a small calibration water phantom. The latter types of water phantom are generally preferred due to their ease of setup and their ability to accommodate larger ion chambers such as Farmer-type chambers, as discussed in Chapter 15.

The TG-51 absorbed dose protocol (AAPM, 1999) as well as the IAEA code of Practice (IAEA, 2000) have made the use of liquid water as a phantom material for reference dosimetry mandatory for reasons of phantom material reproducibility. Also, the more recent addendum to the TG-51 protocol (McEwen et al., 2014) upholds the recommendation that a water phantom *must* be used for the output calibration of high-energy photon beams. As mentioned in that addendum, the main advantage of solid phantoms is in the ease of setup, but this is countered by the uncertainty in the correction factor to convert from dose in plastic to dose in water. Despite the developments in the formulation of water equivalent phantoms that have been documented in the literature, for example, Seuntjens et al. (2005), the additional uncertainty in using such materials still negates any ease-of-use issue.

3.6 Summary and Future Developments

In this chapter, detectors used for reference dosimetry are reviewed. After discussing a number of properties of reference-class ion chambers, the traceability of a calibration of these chambers, that is, the link to primary standards of absorbed dose is discussed in detail. Special attention is paid to the role of calorimeters in reference dosimetry. Details of the absorbed dose calibration and measurement process, including the calibration in FFF beams, as well as a number of practical issues related to reference dose measurements, are also given in this chapter.

New developments can be anticipated in reference dose measurements of treatment modalities for which it is difficult to meet the conditions described in dosimetry protocols. For instance, for reference dosimetry in light-ion beams, as well as in photon beams in the presence of strong magnetic fields, improvement of existing procedures might be expected when new data become available. An interesting development is also the use of probe-type calorimeters for reference dosimetry purposes.

References

Abdel-Rahman W et al. (2009) Clinic based transfer of the N(D,w)(Co-60) calibration coefficient using a linear accelerator. *Med. Phys.* 36: 929–938.

Alfonso R et al. (2008) A new formalism for reference dosimetry of small and nonstandard fields. *Med. Phys.* 35: 5179–5186.

Almond P et al. (1999) AAPM's TG-51 protocol for clinical reference dosimetry of high-energy photon and electron beams. *Med. Phys.* 26: 1847–1870.

Bouchard H, Seuntjens J and Palmans H (2012) On charged particle equilibrium violation in external photon fields. *Med. Phys.* 39: 1473–1480.

de Prez L et al. (2016) A water calorimeter for on-site absorbed dose to water calibrations in ^{60}Co and MV-photon beams including MRI incorporated treatment equipment. *Phys. Med. Biol.* 61: 5051–5076.

Duane S et al. (2012) An absorbed dose calorimeter for IMRT dosimetry. *Metrologia* 49(5): S168–S173.

DuSautoy AR (1996) The UK primary standard calorimeter for photon-beam absorbed dose measurement. *Phys. Med. Biol.* 41: 137–151.

Guerra AS et al. (2012) A standard graphite calorimeter for dosimetry in brachytherapy with high dose rate ^{192}Ir sources. *Metrologia* 49(5): S179–S183.

Hyun MA et al. (2017) Ion recombination and polarity corrections for small-volume ionization chambers in high-dose-rate, flattening-filter-free pulsed photon beams. *Med. Phys.* 44: 618–627.

IAEA (2000) Absorbed dose determination in external beam radiotherapy: An international code of practice for dosimetry based on standards of absorbed dose to water. Technical Reports Series No. 398. (Vienna, Austria: International Atomic Energy Agency, IAEA).

IAEA (2016) Accuracy requirements and uncertainties in radiation therapy. Human Health Series No.31. (Vienna, Austria: International Atomic Energy Agency).

Ma C-M et al. (2001) AAPM protocol for 40–300 kV x-ray beam dosimetry in radiotherapy and radiobiology. *Med. Phys.* 28: 868–893.

McEwen M (2010) Measurement of ionization chamber absorbed dose k_Q factors in megavoltage photon beams. *Med. Phys.* 37: 2179–2193.

McEwen M et al. (2014) Addendum to the AAPM's TG-51 protocol for clinical reference dosimetry of high-energy photon beams. *Med. Phys.* 41: 041501.
McEwen MR and Duane S (2000) A portable calorimeter for measuring absorbed dose in the radiotherapy clinic. *Phys. Med. Biol.* 45: 3675–3691.
O'Brien DJ et al. (2016) Reference dosimetry in magnetic fields: Formalism and ionization chamber correction factors. *Med. Phys.* 43: 4915–4927.
Palmans H et al. (2004) A small-body portable graphite calorimeter for dosimetry in low-energy clinical proton beams. *Phys. Med. Biol.* 49: 3737–3749.
Renaud J et al. (2017) Aerrow: A probe-format graphite calorimeter for absolute dosimetry of high-energy photon beams in the clinical environment. *Med. Phys.* (in press).
Ross CK and Klassen NV (1996) Water calorimetry for radiation dosimetry. *Phys. Med. Biol.* 41: 1–29.
Sander T et al. (2012) NPL's new absorbed dose standard for the calibration of HDR ^{192}Ir brachytherapy sources. *Metrologia* 49(5): S184–S188.
Sarfehnia A and Seuntjens J (2010) Development of a water calorimetry-based standard for absorbed dose to water in HDR ^{192}Ir brachytherapy. *Med. Phys.* 37: 1914–1923.
Seuntjens J and Duane S (2009) Photon absorbed dose standards. *Metrologia* 46: S39–S58.
Seuntjens J et al. (2005) Absorbed dose to water reference dosimetry using solid phantoms in the context of absorbed-dose protocols. *Med. Phys.* 32: 2945–2953.
van der Merwe D et al. (2017) Accuracy requirements and uncertainties in radiotherapy: A report of the International Atomic Energy Agency. *Acta Oncol.* 56: 1–6.

4

Point Detectors for Determining and Verifying 3D Dose Distributions

Chester Reft

4.1 Introduction

Radiation detectors that provide point measurements have a long and extensive use in radiotherapy. These measurements include, but are not limited to, in and outside of the radiation field, in-phantom and *in vivo*, as well as absolute and relative dose measurements. With the increasing complexity in radiotherapy to deliver radiation to patients safely and efficiently, measuring dose is becoming both more difficult and important. Although many radiation detectors provide point dose measurements, they can also be utilized to obtain three-dimensional (3D) dose distributions, as well as provide dose information at regions of interest at various anatomical sites such as eyes, skin, and scrotum. Due to the many different types of dose measurements, there is no detector that can meet all the dosimetric requirements in a radiotherapy clinic. Besides absolute reference dose measurements, there are many treatment procedures such as stereotactic radiosurgery (SRS), total body irradiation (TBI), total skin electron therapy (TSET), and in general skin doses that require dose verification. Many of these treatment procedures require detectors with particular properties for accurate dose measurements such as sensitivity, energy dependence, angular dependence, accuracy, reproducibility, and physical dimensions. Therefore, clinics treating a variety of anatomical sites using various treatment procedures will require an assortment of radiation detectors for *in vivo* and in-phantom measurements.

Detectors can be broadly classified into two categories: active (real-time) or passive. Active detectors provide an instantaneous reading while passive detectors require some time postirradiation for a reading. The type of detector used will depend upon the particular dosimetric application. In this chapter, the various types of commercially available detectors used in radiotherapy are described along with their particular dosimetric characteristics and clinical applications.

4.2 Use of Detectors in Radiation Therapy

4.2.1 Types of Detectors Used in Radiation Therapy

Active detectors such as ionization chambers and diodes provide an immediate "real-time" reading. These types of detectors are frequently used because of their immediate readout and favorable dosimetric properties such as robustness, reproducibility, and stability.

Passive detectors such as thermoluminescent detectors (TLDs) and film require some time following irradiation to provide a stable signal. This time delay can vary from a few minutes to a day depending upon the detector type. The time delay is due to allowing the detector to stabilize such as for TLDs and optically stimulated luminescence detectors (OSLDs), or for processing the detector as for film. The time delay can either be an advantage or disadvantage depending on the measurement requirement. An example of an advantage is in audits of beam output while a disadvantage would be in verifying the dose prior to the next treatment of a patient undergoing TBI.

4.2.2 Characteristics and Properties

4.2.2.1 Accuracy

The accuracy of a detector measurement is the degree of closeness of the measured value to the actual value. In radiation dosimetry, it can refer to either measuring absolute (reference dose) or relative dose. The accuracy of a detector measurement is important in reference dose measurements where a protocol is used to convert the measured signal to absorbed dose. General issues related to the accuracy of dose measurements in radiotherapy are discussed in Chapter 2, while the use of detectors for reference dosimetry is described in Chapter 3. It can also be important in some specialized relative dosimetric applications such as small-field measurements where a number of detector perturbation factors can affect the accuracy of the measurement as discussed in Chapter 9. However, since most dosimetric measurements are relative measurements where the ratio of detector responses is required, accuracy of the detector response is less important. Other factors such as reproducibility and sensitivity are generally more important.

4.2.2.2 Reproducibility (Precision)

The reproducibility or precision of a detector refers to the closeness of the agreement between the results of measurements of the same quantity performed under identical radiation conditions. It can be described quantitatively in terms of the spread or dispersion of the measurements. This differs somewhat from repeatability which refers to the closeness of agreement between consecutive measurements obtained under the same radiation conditions using the same equipment. The reproducibility of a detector can vary from a few tenths of a percent (one standard deviation [1SD]) for ionization chambers and diodes, to 3% for TLDs and OSLDs. For relative dosimetry, the reproducibility and sensitivity of the detector can be the most important factors affecting the measurements.

4.2.2.3 Stability

The stability of a detector is related to its robustness, i.e., how it maintains its dosimetric characteristics over time with use and handling. Ionization chambers and diodes are generally very stable while TLDs and film are less stable requiring careful handling. For ionization chamber measurements, the time evolution of its response to radiation to achieve a stable reading can vary. As described in the literature (McEwen et al., 2014), Farmer-type chambers have a short time to stabilize while scanning-type chambers take slightly longer. Microchambers can require long irradiation times to stabilize their response (McEwen, 2010), whereas solid-state detectors such as diodes exhibit short equilibration times (Pierret, 2002).

4.2.2.4 Energy Response

Since radiotherapy utilizes polyenergetic radiation, the ideal radiation detector would have a response independent of energy. For megavoltage radiation, many detectors such as diodes, film, TLDs, and OSLDs have a response relatively independent of energy. This is because the ratio of mass-energy absorption coefficients and mass-energy stopping powers of their sensitive material relative to water varies slowly with energy (e.g., see www.physics.nist.gov/PhysRefData). However, for diodes, this energy dependence in megavoltage photon beams is affected by the material surrounding the silicon chip which usually contains some high-Z material (Saini and Zhu, 2007). The response of ionization chambers varies less than 10% for energies from ^{60}Co to 25 MV (AAPM, 1983). However, in the kilovoltage region where the atomic number of the detector becomes important due to increased photoelectron interactions, all detectors exhibit energy dependence. The magnitude of this energy dependence is related to the atomic number of the sensitive material. However, even for megavoltage radiation, the energy dependence of the detector becomes important for measurements where lower energy scatter radiation is increased such as outside of the radiation field and at large phantom depths.

4.2.2.5 Recombination Effects

The response of solid-state type detectors, such as TLDs, OSLDs, radiophotoluminescent detectors (RPDs), and plastic scintillators, is relatively independent of dose rate or dose per pulse (Tochilin and Goldstein, 1966; Beddar et al., 1992) and therefore exhibit only minor recombination effects. Although diodes are also solid-state devices, one study reported a variation in the dose rate correction factor of 6% over the clinical range of interest for the investigated diodes (Shi et al., 2003). Ionization chambers exhibit recombination effects from less than 1% to a few percent (AAPM, 1983) under most irradiation conditions, whereas for liquid ionization chambers it can vary up to 8% depending on the polarizing voltage and intensity of the radiation (Anderson and Tolli, 2010). The recombination effects can be determined by altering the polarizing voltage across the chamber and plotting the inverse of the chamber reading as a function of the inverse of the polarizing voltage; the so-called Jaffe plot. A linear relation indicates that the chamber is behaving correctly and the effects of recombination can be evaluated by varying the polarizing voltage (McEwen, 2010).

4.2.2.6 Water Equivalence

The water equivalence of a detector refers to how well the dosimetric properties of the detector's sensitive material match those of the measurement medium. Water-equivalent materials are most commonly used for the measuring medium. The dosimetric quantities determining the material's water equivalence are its density, atomic number, mass-energy absorption coefficient, and mass-energy stopping power relative to water. Any variation of these quantities from water will introduce a perturbation into the measurement.

Table 4.1 provides a representative summary of commercially available detectors and some of their properties used in radiotherapy.

Table 4.1 Representative Summary of Commercially Available Point Detectors and Some of Their Properties

Type	Sensitive Width (cm)	Thickness (cm) or Volume (cm^3)	Water Proof	*In Vivo*
Active				
SFD[a] (photon diode)	0.25	0.003	Y	Y
SRS diode	0.06	0.025	Y	Y
Electron diode	0.25	0.003	Y	Y[d]
Edge diode	0.11	0.003	Y	N
Synthetic diamond (CVD)[b]	0.22	0.0001	Y	Y[d]
Plastic scintillator	0.10	0.3	Y	Y
Farmer-type ionization chamber	0.60	0.6	N	N
Miniature ionization chamber	0.55	0.125	Y	N
Microionization chamber	0.60	0.009	Y	N
PinPoint ionization chamber	0.29	0.019	Y	N
Liquid ionization chamber	0.25	0.002	Y	N
Parallelplate—advanced Markus	0.50	0.02	Y	N
Parallel plate—Roos	1.5	0.035	Y	N
Parallel-plate microchamber	0.60	0.002	Y	N
Parallel-plate chamber	2.0	0.62	Y	N
MOSFET	0.2	0.13	Y	Y
Passive				
TLD—microcubes	0.10	0.10	N	Y[d]
TLD—chips	0.31	0.015 → 0.089	N	Y[d]
OSLD	0.70	0.03[c]	Y	Y[d]
RPD	0.55	0.10	N	Y[d]

Note: MOSFET, metal-oxide-silicon field-effect transistor; OSLD, optically stimulated detector; RPD, radiophotoluminescent detector; SRS, stereotactic radiosurgery; TLD, thermoluminescent detector.

[a] Stereotactic field diode.

[b] Chemical vapor deposition.

[c] For bare Al_2O_3:C film.

[d] Applicable for out-of-field measurements.

4.2.3 Dosimetric Applications

4.2.3.1 Absolute or Relative Dosimetry

Absolute or reference dosimetry requires a detector that has a calibration factor traceable to a national standards laboratory which provides a quantity to convert the detector signal to an absorbed dose. These detectors are generally reserved for the output calibration of radiation therapy machines. Currently, all absorbed dose protocols use ionization chambers as discussed in Chapter 3. Most dosimetry measurements in radiotherapy are relative measurements where the quantity of interest is determined by taking the ratio of detector readings. The important detector properties for these types of measurements are its reproducibility, sensitivity, energy dependence, and water equivalence. Therefore, any of the detectors listed in the table with these properties or with appropriate correction factors for the radiation of interest can provide accurate relative measurements.

4.2.3.2 Small-Field Dosimetry

A small field is generally defined as that field size which does not meet the conditions for lateral charged particle equilibrium, and therefore depends upon the energy of the radiation. For 6 MV, field sizes less than 3×3 cm^2 are considered small fields (International Atomic Energy Agency [IAEA], 2017). Small-field dosimetry places stringent requirements on detectors. Not only the dimensions of the detector are important but also its density, atomic number, and water equivalence all affect the measurements (Das et al., 2006). A thorough discussion on small-field dosimetry is presented in Chapter 9.

4.2.3.3 In Vitro/In Vivo

Many point dose measurements are performed *in vitro*, i.e., in water equivalent-type phantoms to measure various dosimetric quantities. These include, but are not limited to: dose measurements in-phantom and in-air such as determination of wedge factors, tissue–phantom ratios, collimator scatter factors, and out-of-field measurements to estimate the dose to critical organs for a patient treatment. These types of measurements are typically performed with ionization chambers or diodes. *In vivo* measurements are performed to assess the dose to the patient either in-field or out-of-field. *In vitro* and *in vivo* measurements, as well as the particular detector requirements for each type of measurement, are discussed in Sections 4.4 and 4.5.

4.2.3.4 Measuring Linear Energy Transfer in Particle Beams

With the increasing number of proton machines entering radiotherapy clinics, and the potential introduction of carbon ion radiotherapy, more patients will potentially be treated with these modalities. A current topic of interest is the spatial variation of linear energy transfer (LET) along the path of the protons. Most proton facilities use a fixed value of 1.1 for the relative biological effectiveness (RBE) for prescribing dose. However, there is increasing evidence that the LET increases in the distal region of the spread-out Bragg peak (SOBP) (Chaudhary et al., 2014). Since LET is one of the factors contributing to the RBE, its spatial variation can impact patient treatment. A number of studies show that some types of TLDs can be used to estimate the LET for high LET radiation (Loncol et al., 1996; Schoner et al., 1999). Usually, the total integrated charge under the TLD glow curve is used to determine the absorbed dose. However, in LiF and CaF_2, the high temperature peak shows an increased sensitivity to high LET radiation relative to the main lower temperature peak (Hoffmann and Prediger, 1983). This is illustrated in Figure 4.1 showing glow curves for TLD-700 irradiated with beams having different LET. Also, by analyzing the decay of the signal from irradiated OSLDs, information on the LET of proton therapy beams can be obtained (Granville et al., 2016).

The high-temperature ratio (HTR) of the high-temperature peak height (P2) to the low-temperature peak height (P1) can now be used to estimate the LET via calibration in radiation beams of known LET as described in the literature (Loncol et al., 1996). As an example of these types of measurements, Figure 4.2 shows the HTR measured in the SOBP and in the distal fall-off region of the Bragg peak in a clinical proton beam. The R_{90} is the distance to the 90% percentage depth dose (PDD) for the various proton energies, and the negative values are distances beyond the 90% PDD.

Also, by determining the signal in each peak and from the calibration of the TLDs for the low and high LET radiation, the high and low-LET absorbed dose components can be obtained (Busuoli et al., 1970). Thus, there is the potential for using these types of TLDs to measure the absorbed dose and the LET in-phantom and *in vivo*.

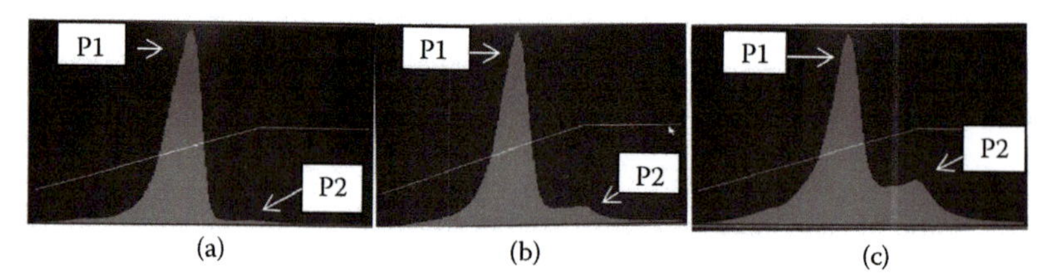

Figure 4.1

(a) Glow curve for TLD-700 irradiated to 15 cGy with low linear energy transfer 6 MV photons, (b) 250 MeV protons, and (c) therapy neutrons produced by 50.5 MeV protons on a Be target. TLD, thermoluminescent detector.

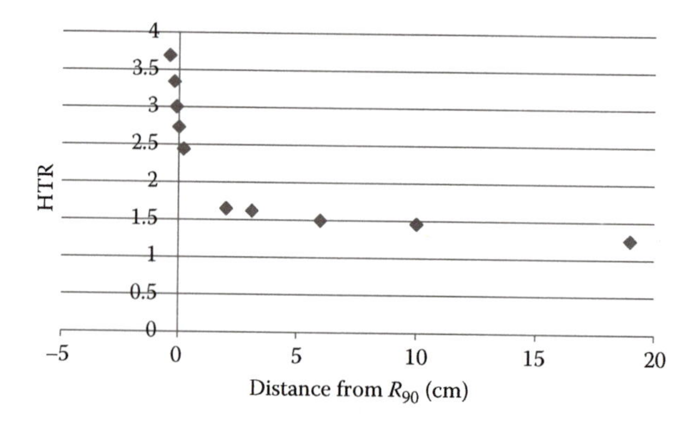

Figure 4.2

Spatial variation of the ratio HTR of the height of the high temperature peak to the low temperature peak as a function of the distal R_{90} value determined with TLD-700 measurements along proton beams with varying energy. TLD, thermoluminescent detector.

4.2.4 3D Dosimetric Applications

Although the detectors discussed in this chapter provide point dose measurements, they can be configured to provide multiple point dose measurements. Ionization chambers, diodes, diamond detectors, and plastic scintillators are used in water tanks to measure various dosimetric quantities such as percent depth doses, tissue–phantom ratios, and beam profiles. Metal-oxide-silicon field-effect transistors (MOSFETs) detectors can provide measurements at a single point but also simultaneously at five different locations. Multiple passive detectors such as TLDs and OSLDs can be positioned at several locations in a phantom or *in vivo* to measure absorbed dose. In Section 4.5, there is a discussion on the use of these detectors for 3D dosimetry.

4.3 Calibration and Reference Dosimetry

As mentioned in Section 4.2.3, all absorbed dose protocols use ionization chambers for reference dosimetry. Compliant radiotherapy treatment machines are those that meet the requirements for distance, field size, and energy prescribed in various protocols such as presented in the AAPM TG-51 report (Almond et al., 1999) and in IAEA Report TRS-398 (IAEA, 2000). Most radiotherapy machines satisfy these requirements. Noncompliant therapy machines do not adhere to one or more of the requirements prescribed in these absorbed dose protocols. These include CyberKnife, TomoTherapy, and Gamma Knife units. The absolute and relative dosimetry involved for these machines is the subject of the IAEA and American Association of Physicists in Medicine produced Code of Practice (IAEA, 2017).

Ionization chambers used for reference dosimetry require a calibration factor with traceability to a national calibration facility. Most facilities only have one ionization chamber meeting this requirement. However, calibration factors for other ionization chambers can be obtained through a transfer calibration procedure. This procedure involves the comparison of the calibrated with the uncalibrated chamber in a water-equivalent phantom irradiated in a megavoltage photon beam as described in the literature (Abdel-Rahman et al., 2009). The specific detector requirements and procedures to convert the measured signal to absorbed dose are discussed in Chapter 3.

4.4 Relative Dosimetry

For relative dosimetry, the important properties of a detector are its sensitivity and reproducibility although its energy dependence can be an issue in regions of low-energy scattered radiation such as in the beam penumbra, in large fields, and at large depths. The following section provides a discussion on commonly used detectors used in radiotherapy.

4.4.1 Real-Time Detectors

For many measurements of dosimetric quantities such as percent depth doses, tissue–phantom ratios and beam profiles, active or real-time detectors are most efficient. These detectors are used in a water phantom in a scanning mode to efficiently measure various dosimetric quantities as discussed in Chapter 15. In photon and proton beams, ionization chambers are most commonly used because of their stability, small energy dependence, and various dimensions that can be optimized for specific measurements. The near-water-equivalent plastic scintillators and synthetic diamond detectors can also be used for these measurements. However, their relatively high cost may limit their use in many radiotherapy clinics. Although MOSFETs provide real-time measurements, they are not practical for scanning applications. For point dose measurements, such as output factors and wedge factors, any of the detectors listed in Table 4.1 are acceptable. One caveat is the performance of small-field measurements, which imposes additional requirements that limit the use of some of these detectors. Small-field dosimetry, and detectors recommended for these measurements, is the subject of Chapter 9.

For measurements in electron fields, electron diodes are generally the detector of choice for both scanning-type and point dose measurements because of their small size, known effective point of measurement, minimum perturbation factor, and relative energy independence for clinical electron beams. Plastic scintillators and the new synthetic diamond detectors have similar properties in electron beams, but their higher cost may limit their use. Ionization chambers can be used but corrections are required for the change in the mass-energy stopping power ratio of water-to-air with depth, change in fluence with depth, and the effective point of measurement. Well-guarded parallel-plate ionization chambers have a fluence correction factor of unity and a well-defined point of measurement (Almond et al., 1994) and only need correction for the change in mass-stopping power ratio with depth. For specific point dose measurements, such as cone and block field factors (Gerbi et al., 2009), all of the detectors listed in the table are acceptable detectors.

4.4.2 Passive Detectors

Passive detectors and MOSFETs are not practical for scanning measurements and are therefore limited to point dose measurements. Their small dimensions and relative energy independence to megavoltage radiation make them, however, useful for measurements in phantoms containing various heterogeneities such as bone, lung, and metallic implants (Ding et al., 2007). TLDs are often used in measuring the absorbed dose in phantoms for radionuclides used in both low-dose rate (LDR) and high-dose rate (HDR) brachytherapy (Moura et al., 2015). Since TLDs, OSLDs, and RPDs require no direct electrical connections, they are ideal for *in vivo* measurements both in and outside the radiation field. Although MOSFET detectors require some cables attached to the patient, there is, however, similar to diodes, no active electrical connection and multiple individual measurements may be obtained simultaneously with a specific type of commercial device (Best Medical, Kanata, Ontario, Canada). Unlike other passive detectors, measurements can be repeated without removing the detectors to read out the signal. The requirements and use of these detectors for patient measurements within the radiation field are discussed in the following section. In Chapter 21, the dose outside the treatment volume in external beam therapy is discussed.

4.5 Specific Requirements for *In Vivo* Dosimetry for 3D Dosimetry

The requirements for measuring the absorbed dose to patients limit the number of detectors available for these measurements. Detectors that require an active electrical connection such as ionization chambers are not recommended because of the possibility of producing an electrical shock to the patient. Also, other restrictions such as sensitivity and energy dependence of the detector become a factor if an out-of-field measurement is required. Table 4.1 identifies those detectors generally recommended for out-of-field *in vivo* dose measurements. Detailed information about the use of detectors for *in vivo* dosimetry in external beam radiotherapy can be found elsewhere (IAEA, 2013; Mijnheer et al., 2013).

4.5.1 Detectors and Their Characteristics Recommended for *In Vivo* Point Dose Measurements within the Treatment Field

4.5.1.1 Thermoluminescent Detectors

TLDs have a long history for use as *in vivo* dose measurements in radiation protection and radiotherapy applications due to their small size and favorable dosimetry properties. They come in various compositions, sizes, and shapes such as chips, rods, microcubes, and powder. Their effective point of measurement is at the center of the TLD which depends upon its thickness, which varies from 0.015 to 0.089 cm. The most commonly used TLD material is LiF which has an atomic number of 8.2 close to that of water, 7.4, and exhibits a linear dose–response relationship over a wide dose range from 10 μGy to about 1.5 Gy. Their response is relatively energy independent for therapy photon, electron, and proton beams and shows only a small angular dependence (McKinlay, 1981). By following a strict annealing procedure, they can be reused any number of times without a loss in sensitivity. By using a number of these detectors, the dose can be measured at various locations in the patient treatment field. Due to their small dimensions, they can be placed in body cavities such as the mouth or rectum. Their disadvantages are that they are relatively labor intensive to use, require about a day for their response to stabilize, lose signal in their readout, and require determination and maintenance of individual sensitivities to reduce their measurement uncertainty below ±2% (1SD).

4.5.1.2 Optically Stimulated Luminescent Detectors

OSLDs have very similar dosimetric properties as TLDs. The major differences are the density (3.7 g/cm^3), the atomic number (11.1), and the signal response (stabilization after about 10 minutes postirradiation). They can also be reread multiple times because the readout decreases the signal by about 0.05%, while optical bleaching is used to remove the effects of radiation (Jursinic, 2007). The commercially available nanoDot OSLDs (Landauer, Inc., Glenwood, Illinois) have an effective point of measurement of 0.08 cm (Zhuang and Olch, 2014), which includes the thickness of the plastic cover. They are also easier to handle and identify than TLDs because they are encapsulated in a labeled plastic holder. As with TLDs, individual sensitivities should be determined to reduce the measurement uncertainty below ±2% (1SD). However, unlike TLDs where annealing eliminates all the radiation effects, optical bleaching cannot clear the deep traps. For accumulated doses above about 15 Gy, there is an increased background signal from the deep traps that changes the detector sensitivity (Jursinic, 2010).

4.5.1.3 Radiophotoluminescent Detectors

RPDs are comprised of silver-activated metaphosphate glass with dimensions 0.85 cm in length and 0.15 cm in diameter. They have properties similar to OSLDs: relative energy independence to megavoltage radiation, linear dose response, and good reproducibility. They are, however, read out using a pulsed N_2 laser system (Araki et al., 2004). One major difference compared to OSLDs is that they can be annealed to clear all the trapping centers, so there is no buildup of residual signal and therefore no change in sensitivity. They are commercially available (Asahi Techno Glass Corp, Shiuoka, Japan). Their small dimensions allows for simultaneously obtaining a number of *in vivo* dose measurements at various anatomical sites. As with TLDs, individual sensitivity factors are necessary to reduce their measurement uncertainty below 2% (1SD).

4.5.1.4 Diodes

Silicon diodes have many favorable dosimetric properties for in-field patient measurements. They provide a real-time response, high sensitivity due to their high density (2.3 g/cm^3) combined with a low energy to produce an electron–hole pair (3.6 eV), high spatial resolution, and small recombination effect. However, as a point detector, they only provide a measurement at one location. For *in vivo* measurements they are located on the entrance or exit surface of the patient and encapsulated in a tissue-equivalent material to measure the dose at maximum depth. Detailed discussion on their characteristics and clinical use may be found in the AAPM Report "Diode in vivo dosimetry for patients receiving external beam radiotherapy" (Yorke et al., 2005) and the ESTRO publication "Practical guidelines for the implementation of *in vivo* dosimetry with diodes in external radiotherapy with photon beams (entrance dose)" (Huyskens et al., 2001). The specifications of a number of commercially available diodes have been published (Saini and Zhu, 2002). The diode response depends on the dose per pulse, energy and angular orientation in the radiation beam, and temperature, which changes by about 0.3% per degree centigrade (Saini and Zhu, 2002). Since diodes are generally calibrated at around 22°C, their sensitivity could increase 3%–4% when performing dose measurements on patients with a body temperature of about 37°C. Because diode response is affected by the radiation dose per pulse, their sensitivity could vary by 4% over the clinical dose range (Saini and Zhu, 2004). Although silicon has an atomic number of 14, this is only of minor concern for megavoltage photons and electrons. The angular dependence of diodes can vary by 10% depending upon their internal composition (Heydarian et al., 1996; Jursinic, 2009), which would require corrections for measurements off-axis.

4.5.1.5 Metal-Oxide-Silicon Field-Effect Transistors

MOSFETs are active detectors with dosimetric properties similar to diodes. Their dosimetric characteristics have been described at several places in the literature (e.g., Ramaseshan et al., 2004). Their major advantages are their small size, linear dose response, relative energy independence for megavoltage radiation, reusability without any annealing process, read out repeatability, and relatively inexpensive equipment. Since they do not require connection to the readout system during irradiation, there are no cables required in the treatment room. Their advantages for *in vivo* dose measurements are their small size, and therefore little radiation attenuation, and providing simultaneously dose measurements at multiple separate anatomical sites. They have shown to be useful for TBI procedures (e.g., Scalchi et al., 2005; Briere et al., 2008) by providing separate dose measurements along the length of the patient. Recently, a special type of MOSFET has also become available for skin dose measurements (Jong et al., 2014). Their disadvantages are similar to diodes: temperature, field size, dose per pulse, and angular dependence. Modern MOSFETs have been developed to be independent of temperature. However, unlike diodes, they have a limited lifetime of about 100 Gy attributed to the complete filling of the hole traps resulting in a saturation of the radiation-induced response (Cygler, 2009).

4.5.1.6 Synthetic Single Crystal Diamond Detectors

The newly developed synthetic single crystal diamond detector produced by chemical vapor deposition is now commercially available (microDiamond 60019, PTW, Freiburg, Germany). It is an improvement over the original natural diamond detector in that it is operated in the photovoltaic mode which does not require a bias voltage, and is relatively independent of the dose per pulse and dose rate of the radiation beam (Di Venanzio et al., 2013). Its major advantages are its small dimensions, near water equivalence due to its atomic number ($Z = 6$) close to water ($Z = 7.4$), and a reported temperature dependence of only 0.08%/°C (De Angelis et al., 2002) compared to diodes of about 0.3%/°C. However, unlike a diode it requires a preirradiation of about 5 Gy to stabilize its response to radiation. Although it can be used for *in vivo* measurements, it only provides a single-point dose measurement, while the preirradiation requirement and its high cost will limit its use to in-phantom measurements, particularly for small-field dosimetry.

4.5.1.7 Plastic Scintillators

Plastic scintillators have many favorable dosimetric properties such as their relative water equivalence due to their atomic number ($Z \approx 6$) and density ($\approx$1.05 g/cm^3) close to water, angular independence, linear dose response, small dimensions, and high sensitivity (Beddar, 1994). A polystyrene-based scintillator became recently commercially available (Standard Imaging, Middleton, Wisconsin). Although it can be used for *in vivo* measurements, it only provides point dose measurements, the optical fiber has a minimum bend radius of 6 cm, and the high cost associated with requiring a two channel electrometer will probably limit its use to in-phantom dosimetry.

4.5.2 In-phantom Measurements

All of the following detectors can be used for in-phantom measurements. Whether dosimetric functions or point dose measurements are required will dictate the type of detector to use.

4.5.2.1 Dosimetric Quantities

For dosimetric quantities such as percent depth doses and beam profiles, all active detectors, except MOSFETs, operating in the scanning mode in a water phantom are acceptable. Due to their over-response to low-energy scattered radiation, diodes are not recommended for percent depth dose measurements in photon beams. However, energy-compensated diodes have been developed that decrease their over-response to lower energy photons (McKerracher and Thwaites, 2006). For measurements in areas of noncharged particle equilibrium such as in the buildup region of photon and electron beams, well-guarded parallel-plate ionization chambers with minor corrections for their dimensions are recommended (Gerbi and Khan, 1990). Passive detectors are generally used in water-equivalent phantoms containing various types of heterogeneities such as bone, lung (Carrasco et al., 2004), or metal implants for point dose measurements at various locations in an inhomogeneous phantom. Due to their favorable dosimetric characteristics, small and variable dimensions, TLDs are most often used for these measurements.

4.5.2.2 Verification of Patient Treatment Plans

Patient-specific verification of treatment plans for intensity-modulated radiation therapy (IMRT) and volumetric-modulated arc therapy (VMAT) are commonly required. These measurements can be performed with the two-dimensional and 3D dosimetry systems described in Chapters 8 and 17. Although in principle all of the active and passive detectors mentioned in Table 4.1 can be used to provide a point dose measurement in a phantom to compare with the treatment planning calculation, ionization chambers are most commonly used due to their high sensitivity and accuracy.

4.5.3 In-field Patient Dose

In vivo dose measurements place additional limitations to the use of detectors such as their nonobtrusiveness and nonactive electrical connection. Therefore, all detectors mentioned in the table, except ionization chambers, can be used for patient dose measurements. The type of detector to use depends upon whether a real-time point dose measurement is required, the available detectors, and the personnel experience in the specific center. To verify a 3D treatment plan, multiple dose point measurements can be obtained with MOSFET detectors or with a number of passive detectors positioned at various anatomical sites. The following sections provide clinical examples of in-field patient dose measurements.

4.5.3.1 Skin Dose Measurements

The International Commission on Radiological Protection (ICRP) recommends that skin dose measurements should be performed at a depth of 7 mg/cm^2 (ICRP, 1991). This limits the use of point detectors to TLDs and OSLDs. The effective depth of the detector, d_{eff}, is typically taken at its center which is calculated by the following relation:

$$d_{\text{eff}} = 0.5 \times \rho \times \text{detector thickness}$$

where ρ is the physical density.

By using different thickness of TLDs on the patient surface, the dose measurements can be extrapolated to estimate the skin dose (Kron et al., 1996). The commercially available nanoDot OSLDs have an effective measurement depth of 0.08 cm (Zhuang and Olch, 2014), which includes the encapsulation material, and therefore, can provide an estimate of the skin dose. Two clinical examples of the importance of knowing the skin dose are in the treatment of the chest wall and in TSET.

4.5.3.2 Entrance Dose (d_{max}) Measurements

Measurements of the entrance dose, i.e., the dose at the depth of dose maximum, d_{max}, are required for a number of purposes such as evaluating the output of the therapy unit, to compare the measured dose with the calculated dose, and to verify the delivery of the correct number of monitor units. These measurements are obtained with the detector located on the patient surface under sufficient buildup material with the effective detector depth at d_{max}. MOSFETs and diodes provide a real-time measurement, while OSLDs can provide measurements 10 minutes postirradiation, while TLDs are generally read out the following day. For 3D conformal and IMRT treatments, the dose at d_{max} can be determined at multiple anatomical sites with MOSFETs and diodes, or by using a number of the passive detectors: TLDs, OSLDs, or RPDs. The entrance dose for patients undergoing IMRT for prostate cancer was measured with MOSFETs fitted with brass buildup caps (Varadhan et al., 2006). Diodes are also used for IMRT treatment verification. To develop a practical approach for the routine verification of IMRT delivery doses, diodes were used to measure the entrance dose (Vinall et al., 2010; Kadesjo et al., 2011).

4.5.3.3 Intracavitary Dose Measurements

There are a number of anatomical locations where *in vivo* dose measurements can be obtained for external beam radiotherapy and brachytherapy procedures to compare with treatment planning calculations. These sites include mouth and esophagus for head-and-neck treatments and the vagina, rectum, and anal canal for pelvic treatments. For treating gynecological patients with LDR and HDR brachytherapy procedures, measurements can be made to determine the rectal dose to avoid an over- or underdose of the prescription dose. These measurements require relatively nonobtrusive, small detectors that can be sealed from body fluids. MOSFETs, TLDs, OSLDs, or RPDs are generally used because of their small size and providing multiple dose measurements, while diodes have been used to provide a single-point dose measurement.

TLDS are often used to measure the midline dose along the length of a patient undergoing TBI (e.g., Pacyna et al., 1997; Duch et al., 1998). OSLDs also provide accurate measurements at various treatment sites for patients undergoing TBI with ^{60}Co (Mrcela et al., 2011). The use of diodes to measure entrance and exit dose values at various positions on a patient during TBI has also been reported (Patel et al., 2014). Since TBI conditions are quite different compared to the typical calibration conditions for the detectors, their calibration and correction factors must be carefully evaluated before performing these dose measurements.

Due to their high-dose gradients, IMRT measurements are more difficult to perform. However, with careful planning and detector positioning, some studies report good agreement between measurements and calculations. TLDs inserted into a flexible nasopharyngeal tube into the nasopharynx were used in studies measuring the dose during IMRT of head-and-neck patients (Engstrom et al., 2005; Gagliardi et al., 2009). MOSFETs attached to a custom-made mouth plate were used to measure the dose to the oral cavity for patients treated for head-and-neck cancer (Marcie et al., 2005).

Brachytherapy treatments produce isodoses with high-dose gradients that require detectors with high spatial resolution. TLDs and diodes are generally used for these measurements. For instance, TLDs were positioned within the rectum to measure the rectal dose along the length of the implant for patients treated for uterine and cervical carcinoma with ^{192}Ir HDR (Kapp et al., 1992). In another study, four energy-compensated diodes were used to measure the dose at various locations in the rectum (Alecu and Alecu, 1999). Detailed information about the use of detectors for *in vivo* dosimetry in brachytherapy can be found elsewhere (Tanderup et al., 2013).

4.5.3.4 Intraoperative Radiation Therapy

Intraoperative radiation therapy (IORT) treatment procedures typically use electrons with energies from 6 to 12 MeV or ^{192}Ir to deliver a single high dose of radiation to the treatment area following surgery in the operating room. Since a single high-dose fraction is delivered, many institutions require verification of the dose using an appropriate detector. Since these measurements are performed in a sterile environment, the detector must be sterilized and the response of the detector to these high doses must be known. The most commonly used detectors for these measurements are TLDs, diodes, and MOSFETs (Biggs et al., 2011). Since OSLDs and RPDs have properties similar to TLDs, they could also be used for these measurements. All of these detectors can provide multiple dose readings within the treatment field.

For the treatment of prostate cancer with IORT using 9 MeV electrons, MOSFETs were inserted into sterile catheters and placed in either the rectum lumen or into the bladder through the urethra to verify the dose and patient setup (Soriani et al., 2007). To verify the dose to the pancreas, MOSFETs were inserted in sterile plastic bags and placed over the pancreatic stump to measure the dose delivered by either 7 or 9 MeV electrons (Consorti et al., 2005). In the treatment of early-stage breast cancer with 4–10 MeV electrons, MOSFETs were inserted into sterile catheters and positioned in the radiation field to measure the entrance dose at various locations in the treatment field (Ciocca et al., 2006). For patients receiving IORT with ^{192}Ir HDR for breast cancer, skin dose measurements were obtained with TLD chips sterilized in thin polyethylene bags (Perera et al., 2005). In another study of patients receiving IORT for breast cancer using a 50 kV x-ray source (Carl Zeiss, Oberkochen, Germany), TLD rods prepared in heat-sealed plastic envelopes and further sealed in sterile Tegaderm film (3M, St Paul, Minnesota) were used to measure the skin dose (Eaton et al., 2011).

4.5.4 Out-of-Field Point Dose Measurements

Current treatment planning systems do not provide accurate dose calculations outside of the treatment fields (Howell et al., 2010). Therefore, *in vivo* measurements are often required to estimate the dose to organs at risk during radiotherapy such as the contralateral breast, scrotum, and eye. Also, there are an increasing number of patients receiving radiotherapy with various implanted electronic devices such as cardiovascular implantable electronic devices, which are sensitive to radiation (Last, 1998; Hurkmans et al., 2005) and may require a measurement of the dose it receives during treatment. The particular problems and detector requirements associated with these measurements such as their sensitivity, energy dependence, and nonobtrusiveness are the subject of Chapter 21.

4.6 Summary and Future Developments

As discussed in the previous sections, all of the detectors listed in Table 4.1 have particular uses in radiation dosimetry as point detectors. However, some of the detectors such as TLDs, OSLDs, RPDs, and MOSFETs can be configured into arrays to provide multidimensional dose information. The specific application of all the dosimeters depends upon a number of factors such as whether an immediate measurement is required, *in vivo* or *in vitro* measurement, in-field or out-of-field measurement, or multipoint measurements. For absolute or reference dosimetry the important characteristic of the detector is its accuracy, whereas most clinical dosimetry involves a relative measurement where the sensitivity and reproducibility of the detector are important. With the extension of radiotherapy to protons and carbon ion beams where their LET becomes important because of their relation to RBE, *in vivo* and *in vitro* measurements of LET are important in affecting treatment planning. Analysis of TLD and OSLD curves offer the potential to provide not only dose measurements but also information on the particle beam's LET, particularly in and beyond the SOBP. Since most clinical radiation therapy departments have a limited budget for dosimetry equipment, it is important that their medical physicists carefully consider the dosimetry requirements of their department in selecting radiation detectors. The purpose of this chapter is to provide the dosimetric characteristics of the commercially available detectors to give some guidance to medical physicists in selecting detectors. The choice of detectors will not only depend on the types of treatment equipment but also on the spectrum of patient treatment procedures.

References

AAPM (1983) A protocol for the determination of absorbed dose from high-energy photon and electron beams. *Med. Phys.* 10: 741–771.

Abdel-Rahman W et al. (2009) Clinic based transfer of the N(D,W)(Co-60) calibration coefficient using a linear accelerator. *Med. Phys.* 36: 929–938.

Alecu R and Alecu M (1999) In-vivo rectal dose measurements with diodes to avoid misadministrations during intracavitary high dose rate brachytherapy for carcinoma of the cervix. *Med. Phys.* 26: 768–770.

Almond P et al. (1994) The calibration and use of plane-parallel ionization chambers for dosimetry of electron beams: An extension of the 1983 AAPM protocol report of AAPM radiation therapy Committee Task Group No. 39. *Med. Phys.* 21: 1251–1260.

Almond P et al. (1999) AAPM's TG-51 protocol for clinical reference dosimetry of high-energy photon and electron beams. *Med. Phys.* 26: 1847–1870.

Anderson J and Tolli H (2010) Application of the two-dose-rate method for general recombination for liquid ionization chambers in continuous beams. *Phys. Med. Biol.* 56: 299–314.

Araki F et al. (2004) Dosimetric properties of radiophotoluminescent glass rod detector in high-energy photon beams from a linear accelerator and cyberknife. *Med. Phys.* 31: 1980–1986.

Beddar AS (1994) A new scintillator detector system for the quality assurance of ^{60}Co and high-energy therapy machines. *Phys. Med. Biol.* 39: 253–263.

Beddar AS, Mackie TR and Attix FH (1992) Water-equivalent plastic scintillation detectors for high-energy beam dosimetry: II. Properties and measurements. *Phys. Med. Biol.* 37: 1901–1913.

Biggs P et al. (2011) Intraoperative electron beam irradiation: Physics and techniques. In: *Intraoperative Irradiation*, (Eds. Gunderson L, Willett C, Calvo F and Harrison L, New York, NY: Springer).

Briere TM et al. (2008) Patient dosimetry for total body irradiation using single-use MOSFET detectors. *J. Appl. Clin. Med. Phys.* 9(4): 200–205.

Busuoli G et al. (1970) Mixed radiation dosimetry with LiF (TLD-100). *Phys. Med. Biol.* 15: 673–681.

Carrasco P et al. (2004) Comparison of dose calculation algorithms in phantoms with lung equivalent heterogeneities under conditions of lateral electronic disequilibrium. *Med. Phys.* 31: 2899–2911.

Chaudhary P et al. (2014) Relative biological effectiveness variation along monoenergetic and modulated Bragg peaks of a 62-MeV therapeutic proton beam: A preclinical assessment. *Int. J. Radiat. Oncol. Biol. Phys.* 90: 27–35.

Ciocca M et al. (2006) Real-time in vivo dosimetry using micro-MOSFET detectors during intraoperative electron beam radiation therapy in early stage breast cancer. *Radiother. Oncol.* 78: 213–216.

Consorti R et al. (2005) In vivo dosimetry with MOSFETTs: Dosimetric characterization and first clinical results in intraoperative radiotherapy. *Int. J. Radiat. Oncol. Biol. Phys.* 63: 952–960.

Cygler J (2009) MOSFET Detectors in radiotherapy. In: *Clinical Dosimetry Measurements in Radiotherapy*, pp. 943–977. (Eds. Cygler J and Rogers D, Madison, WI: Medical Physics Publishing).

Das IJ, Ding GX and Ahnesjo A (2008) Small fields: Non-equilibrium radiation dosimetry. *Med. Phys.* 35: 206–215.

De Angelis C et al. (2002) An investigation of the operating characteristics of two PTW diamond detectors in photon and electron beams. *Med. Phys.* 29: 248–254.

Ding GX et al. (2007) Impact of inhomogeneity corrections on dose coverage in the treatment of lung cancer using stereotactic body radiation therapy. *Med. Phys.* 34: 2985–2994.

Di Venanzio C et al. (2013) Characterization of a synthetic single crystal diamond Schottky diode for radiotherapy electron beam dosimetry. *Med. Phys.* 40: 02172.

Duch MA et al. (1998) Thermoluminescence dosimetry applied to in vivo dose measurements for total body irradiation techniques. *Radiother. Oncol.* 47: 319–324.

Eaton DJ et al. (2011) In vivo dosimetry for single-fraction targeted intraoperative radiotherapy (TARGIT) for breast cancer. *Int. J. Radiat. Oncol. Biol. Phys.* 82: 819–824.

Engstrom PE, Haraldsson P, Landberg T, Hansen HS, Engelholm SA and Nystrom H (2005) In vivo dose verification of IMRT treated head and neck cancer patients. *Acta Oncol.* 44: 572–578.

Gagliardi FM, Roxby KJ, Engstrom PE and Crosbie JC (2009) Intra-cavitary dosimetry for IMRT head and neck treatment using thermoluminescent dosimeters in a naso-oesophageal tube. *Phys. Med. Biol.* 54: 3649–3657.

Gerbi BJ and Khan FM (1990) Measurement in the build-up region using fixed separation plane-parallel ionization chambers. *Med. Phys.* 17: 17–26.

Gerbi BJ et al. (2009) Recommendations for clinical electron beam dosimetry: Supplement to the recommendations of Task Group 25. *Med. Phys.* 36: 3242–3279.

Granville DA, Sahoo N and Sawakuchi GO (2016) Simultaneous measurements of absorbed dose and linear energy transfer in therapeutic proton beams. *Phys. Med. Biol.* 61: 1765–1779.

Heydarian M, Hoban PW and Beddoe AH (1996) A comparison of dosimetry techniques in stereotactic radiosurgery. *Phys. Med. Biol.* 41: 93–110.

Hoffmann W and Prediger B (1983) Heavy particle dosimetry with high temperature peaks of CaF_2:Tm and 7LiF phosphors. *Radiat. Prot. Dosim.* 6: 149–152.

Howell RM et al. (2010) Accuracy of out-of-field dose calculations by a commercial treatment planning system. *Phys. Med. Biol.* 55: 6999–7008.

Hurkmans CW et al. (2005) Influence of radiotherapy on the latest generation of pacemakers. *Radiother. Oncol.* 76: 93–98.

Huyskens D et al. (2001) Practical Guidelines for the Implementation of in vivo Dosimetry with Diodes in External Radiotherapy with Photon Beams (Entrance Dose). ESTRO Booklet No. 5. (Brussels, Belgium: European Society for Radiotherapy and Oncology, ESTRO).

IAEA (2000). Absorbed Dose Determination in External Beam Radiotherapy: An International Code of Practice for Dosimetry Based on Standards of Absorbed Dose to Water. Technical Reports Series No. 398. (Vienna, Austria: International Atomic Energy Agency, IAEA).

IAEA (2013) Development of Procedures for in vivo Dosimetry in Radiotherapy. Human Health Report No. 8. (Vienna, Austria: International Atomic Energy Agency, IAEA).

IAEA (2017) Dosimetry of Small Static Fields Used in External Beam Radiotherapy: An IAEA-AAPM International Code of Practice for Reference and Relative Dose Determination. Technical Reports Series No. 483. (Vienna, Austria: International Atomic Energy Agency, IAEA).

ICRP (1991) *1990 Recommendations of the International Commission on Radiological Protection.* ICRP Publication 60. (Oxford. UK: Pergamon Press).

Jong WL et al. (2014) Characterization of MOSkin detector for *in vivo* skin dose measurement during megavoltage radiotherapy. *J. Appl. Clin. Med. Phys.* 15(5): 120–132.

Jursinic PA (2007) Characterization of optically stimulated luminescent dosimeters for clinical dosimetric measurements. *Med. Phys.* 34: 4594–4604.

Jursinic PA (2009) Angular dependence of dose sensitivity of surface diodes. *Med. Phys.* 36: 2165–2171.

Jursinic PA (2010) Changes in optical stimulated luminescent dosimeter (OSLD) dosimetric characteristics with accumulated dose. *Med. Phys.* 37: 132–140.

Kadesjo N, Nyholm T and Olofsson J (2011) A practical approach to diode based in vivo dosimetry for intensity modulated radiotherapy. *Radiother. Oncol.* 98: 378–381.

Kapp KS et al. (1992) Dosimetry of intracavitary placements for uterine and cervical carcinoma: Results of orthogonal film, TLD, and CT-assisted techniques. *Radiother. Oncol.* 24: 137–146.

Kron T et al. (1996) TLD extrapolation for skin dose determination in-vivo. *Radiother. Oncol.* 41: 119–123.

Last A (1998) Radiotherapy in patients with cardiac pacemakers. *Brit. J. Radiol.* 71: 4–10.

Loncol T et al. (1996) Response analysis of TLD-300 dosimeters in heavy particle beams. *Phys. Med. Biol.* 41: 1665–1678.

Marcie S et al. (2005) In vivo measurements with MOSFET detectors in oropharynx and nasopharynx intensity-modulated radiation therapy. *Int. J. Radiat. Oncol. Biol. Phys.* 61: 1603–1606.

McEwen M (2010) Measurement of ionization chamber absorbed dose k_q factors in megavoltage photon beams. *Med. Phys.* 37: 2179–2193.

McEwen M et al. (2014) Addendum to the AAPM's TG-51 protocol for clinical reference dosimetry of high-energy photon beams. *Med. Phys.* 41: 041501.

McKerracher C and Thwaites D (2006) Notes on the construction of solid-state detectors. *Radiother. Oncol.* 79: 348–351.

McKinlay A (1981) *Thermoluminescence Dosimetry.* (Bristol, UK: Adam Hilger).

Mijnheer B et al. (2013) *In vivo* dosimetry in external beam radiotherapy. *Med. Phys.* 40: 070903.

Moura ES et al. (2015) Development of a phantom to validate high-dose-rate brachytherapy treatment planning systems with heterogeneous algorithms. *Med. Phys.* 42: 1566–1574.

Mrcela T et al. (2011) Optically stimulated luminescence in-vivo dosimetry for radiotherapy: Physical characterization and clinical measurements in ^{60}Co beams. *Phys. Med. Biol.* 56: 6065–6082.

Pacyna LG, Darby M and Prado K (1997) Use of thermoluminescent dosimetry to verify dose compensation in total body irradiation. *Med. Dosim.* 22: 319–324.

Patel RP et al. (2014) In vivo dosimetry for total body irradiation: Five-year results and technique comparison. *J. Appl. Clin. Med. Phys.* 15(4): 306–315.

Perera F et al. (2005) TLD skin dose measurements and acute and late effects after lumpectomy and high-dose-rate brachytherapy only for early breast cancer. *Int. J. Radiat. Oncol. Biol. Phys.* 62: 1283–1290.

Pierret RF (2002) Advanced semiconductor fundamentals, 2nd Ed. (New York, NY: Prentice Hall).

Ramaseshan KS et al. (2004) Performance characteristics of microMOSFET as an in vivo dosimeter in radiation therapy. *Phys. Med. Biol.* 49: 4031–4048.

Saini AS and Zhu TC (2002) Temperature dependence of commercially available diode detectors. *Med. Phys.* 29: 622–630.

Saini AS and Zhu TC (2004) Dose rate and SSD dependence of commercially available diode detectors. *Med. Phys.* 31: 914–924.

Saini AS and Zhu TC (2007) Energy dependence of commercially available diode detectors for in-vivo dosimetry. *Med. Phys.* 34: 1704–1711.

Scalchi P, Francescon P and Rajaguru P (2005) Characterization of a new MOSFET detector configuration for in vivo skin dosimeter. *Med. Phys.* 32: 1571–1578.

Schoner W, Vana N and Fugger M (1999) The LET dependence of LiF:Mg,Ti dosimeters and its application for LET measurements in mixed radiation field. *Radiat. Prot. Dosim.* 85: 263–266.

Shi J, Simon WE and Zhu TC (2003) Modeling the instantaneous dose rate of radiation detectors. *Med. Phys.* 30: 2509–2519.

Soriani A et al. (2007) Setup verification and in vivo dosimetry during intraoperative radiation therapy (IORT) for prostate cancer. *Med. Phys.* 34: 3205–3210.

Tanderup K et al. (2013). *In vivo* dosimetry in brachytherapy. *Med. Phys.* 40: 070902.

Tochilin E and Goldstein N (1966) Dose-rate and spectral measurements from pulsed x-ray generators. *Health Phys.* 12: 1705–1713.

Varadhan R et al. (2006) In vivo prostate IMRT dosimetry with MOSFET detectors using brass buildup caps. *J. Appl. Clin. Med. Phys.* 7(4): 22–32.

Vinall AJ et al. (2010) Practical guidelines for routine intensity-modulated radiotherapy verification: Pre-treatment verification with portal dosimetry and treatment verification with in vivo dosimetry. *Brit. J. Radiol.* 83: 949–957.
Yorke E et al. (2005) Diode in vivo Dosimetry for Patients Receiving External Beam Radiation Therapy. AAPM Report 87. (Madison, WI: Medical Physics Publishing).
Zhuang AH and Olch AJ (2014) Validation of OSLD and a treatment planning system for surface dose determination in IMRT treatments. *Med. Phys.* 41: 081720.

5 Polymer Gel Dosimetry

Yves De Deene

5.1 Introduction

In clinical polymer gel dosimetry (Baldock et al., 2010), a hydrogel is poured in a humanoid-shaped cast. The hydrogel contains monomers that upon exposure to ionizing radiation undergo a radiation-induced radical chain reaction. The radiation-induced polymerization reaction results in the formation of highly cross-linked microscopically small polymer aggregates that are entangled with the gelatin hydrogel matrix that keeps them in place. The polymer density is related to the amount of absorbed radiation.

Three-dimensional (3D) gel dosimetry has a unique role to play in safeguarding conformal radiotherapy treatments as this quality assurance (QA) technique can cover the full treatment chain, from imaging to treatment, and provides the radiation oncologist with the integrated dose distribution in 3D (Figure 5.1). 3D gel dosimetry can also be applied to benchmark new treatment strategies such as image-guided and tracking radiotherapy techniques. A major obstacle that has hindered the wider dissemination of polymer gel dosimetry in radiotherapy centers is a lack of confidence in the reliability of the measured dose distribution. Moreover, the workload associated with the implementation of accurate 3D polymer gel dosimetry in the clinical routine practice is substantial.

The advantage of polymer gel dosimeters is that they can be easily poured in anthropomorphic-shaped casts such as a head cast, a head-and-neck cast, and a pelvic cast. Also, large phantoms can be scanned. In addition, lung-equivalent gel dosimeters have been fabricated that are created by beating the hydrogel into a stable hydrogel foam (De Deene et al., 2013).

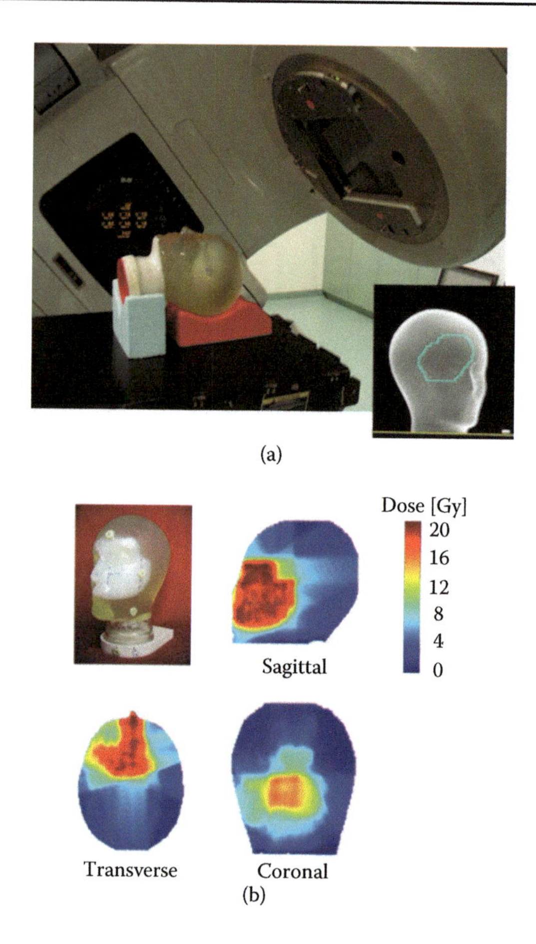

Figure 5.1

(a) Polymer gel dosimeter irradiated according to a conformal radiotherapy treatment. The white region is due to irradiation-induced polymerization. (b) Maps of absorbed radiation dose are obtained by use of high-accuracy quantitative R_2 imaging on a clinical magnetic resonance imaging (MRI) scanner.

5.2 Polymer Gel Dosimetry Principles

5.2.1 Basic Chemistry of Polymer Gel Dosimeters

Polymer gel dosimeters are hydrogels based on an organic gelling agent such as agarose or gelatin and contain vinyl monomers that on irradiation undergo a radiation-induced polymerization. The created polymer has an influence on several physical properties such as the nuclear magnetic resonance (NMR) relaxation times, the optical turbidity, the density, and the speed of sound, enabling dose readout through, respectively, magnetic resonance imaging (MRI), optical computed tomography (CT), x-ray CT, and ultrasound imaging.

Polymer systems for the use of radiation dosimetry were first suggested as early as 1954, where Alexander et al. (1954) discussed the effects of ionizing radiation on polymethyl methacrylate (PMMA). Subsequently, Hoecker and Watkins (1958) investigated the dosimetry of radiation-induced polymerization in liquids, and Boni (1961) used polyacrylamide as a gamma dosimeter. Kennan et al. (1992)

reported that the NMR longitudinal spin–lattice relaxation rate ($R_1 = 1/T_1$) of an aqueous solution of *N,N′*-methylenebisacrylamide (Bis) and agarose increases after irradiation. Gel systems based on the copolymerization of acrylamide and Bis monomers were first proposed by Maryanski et al. (1993, 1994). It was also found that a higher dynamic range was obtained with gelatin as gelling agent and for the NMR spin–spin relaxation rate ($R_2 = 1/T_2$). Later, various other comonomers have been suggested as potential candidates for polymer gel dosimeters (Pappas et al., 1999; Lepage et al., 2001a). Table 5.1 lists the most common monomers that have been applied in polymer gel dosimeters. In most polymer gel systems, the comonomers are combined with equal amounts of the cross-linker Bis monomer. The relatively high amount of cross-linker monomer results in highly cross-linked spherical polymer aggregates on irradiation. A schematic microstructure representation of a polymer gel is shown in Figure 5.2a before and

Table 5.1 R_2-dose Sensitivity, Half-Dose ($D_{1/2}$), and Saturation Values of Different Polymer Gel Dosimeters

Monomer	$D_{1/2}$ (Gy)	R_2-dose Sensitivity ($s^{-1} \cdot Gy^{-1}$)	R_{2sat}-R_{20} ($s^{-1} \cdot Gy^{-1}$)	Molecular Structure	Reference
Acrylamide (AAm)	5.5 (±0.1)	0.331 (±0.012)	4.2 (±0.4)	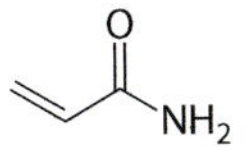 O, NH_2	Maryanski et al. (1993), Lepage et al. (2001a)
Acrylic Acid (AAc)	31.2 (±0.1)	0.358 (±0.006)	10.6 (±0.4)	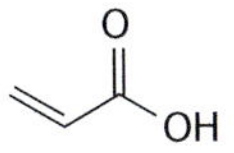 O, OH	Maryanski et al. (1996a), Lepage et al. (2001a)
Methacrylic Acid (MAc)	12.5 (±0.1)	1.193 (±0.048)	18.4 (±0.4)	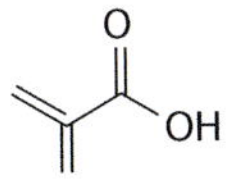 O, OH	Lepage et al. (2001a), De Deene et al. (2006a)
1-Vinyl-2-Pyrrolidone (VP)	23.6 (±0.1)	0.082 (±0.004)	13.7 (±0.4)	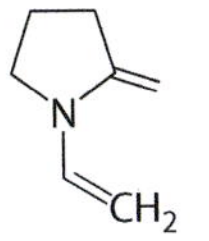 N, CH_2	Pappas et al. (1999), Lepage et al. (2001a)
2-Hydroxyethyl Acrylate (HEA)	5.5 (±0.1)	0.498 (±0.003)	4.2 (±0.4)	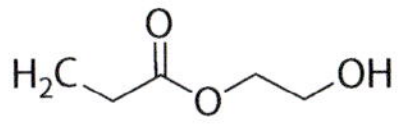 H_2C, O, O, OH	Baldock et al. (2000), Lepage et al. (2001a), Gustavsson et al. (2004a)
2-Hydroxyethyl Methacrylate (HEMA)	41.6 (±0.1)	0.046 (±0.002)	4.9 (±0.4)	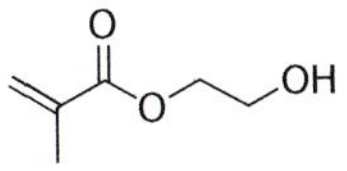 O, O, OH	Baldock et al. (2000), Lepage et al. (2001a)
N-isopropylacrylamide	10	0.13 (±0.012)	4.2 (±0.4)	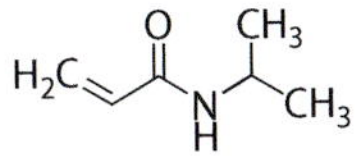 H_2C, O, N, H, CH_3, CH_3	Senden et al. (2006)
N,N′-methylenebisacrylamide (Bis)	N/A	N/A	N/A	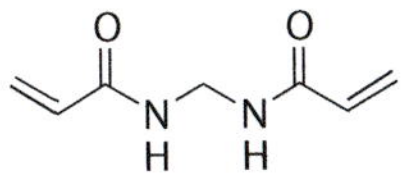 O, O, N, N, H, H	N/A

Note: The cross-linker monomer N,N′-methylenebisacrylamide is also shown. The values are for polymer gels where 50% of the comonomer is N,N′-methylenebisacrylamide.

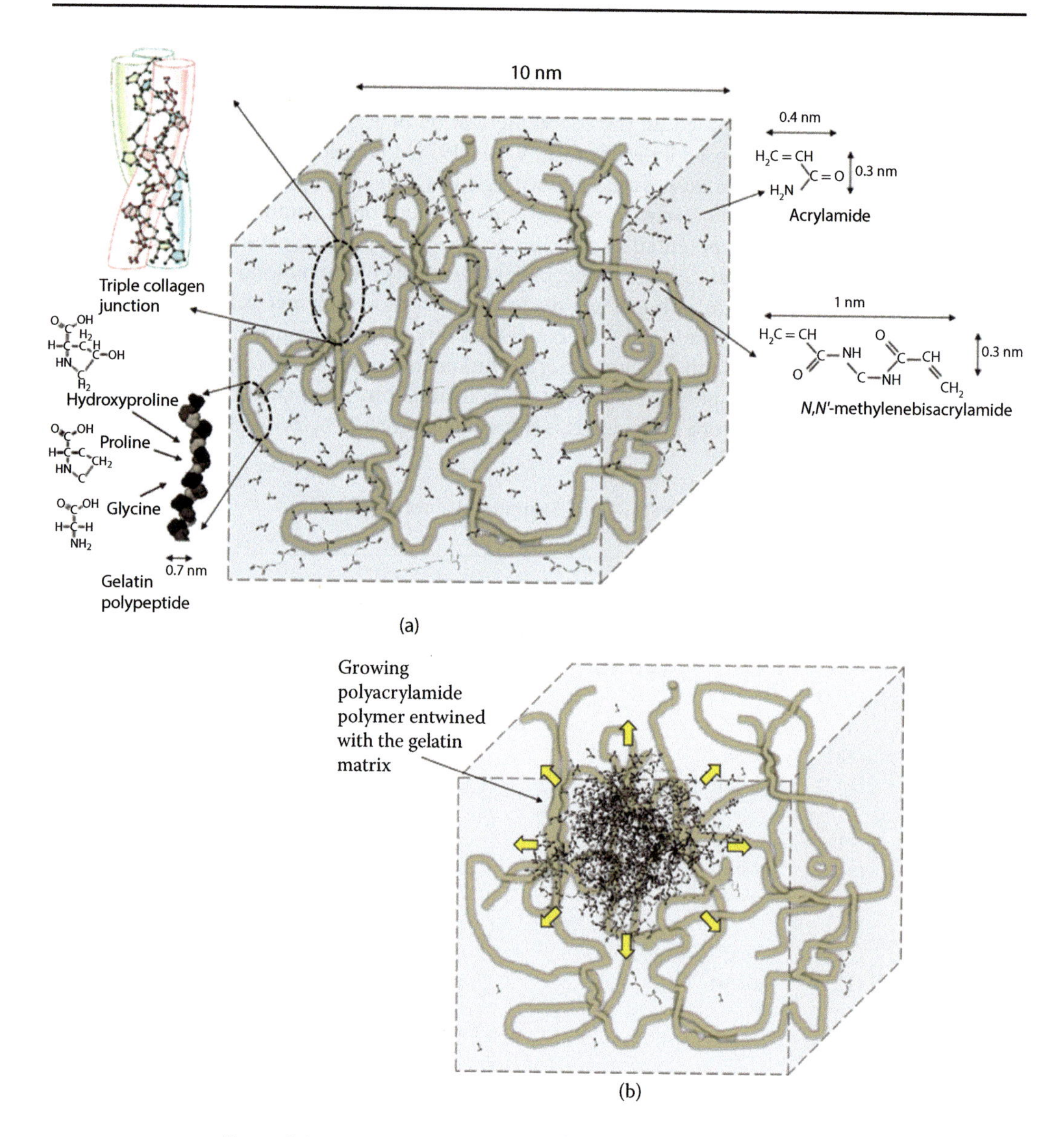

Figure 5.2

Schematic illustration of a polyacrylamide polymer gel (a) before and (b) after irradiation. The polyacrylamide aggregate that is created during radiation grows to a size in the order of 200–600 nm and is intertwined with the gelatin matrix. The gelatin matrix is made up of the gelatin polypeptide with tropocollagen junctions where three polypeptides meet.

Figure 5.2b after exposure to ionizing radiation. Based on the Mie–Debye theory of light scatter, it is estimated that the polymer aggregates in polyacrylamide gel (PAG) dosimeters reach a size in the order of 200–600 nm depending on the radiation dose (Maryanski et al., 1996b).

In some methacrylic acid (MAc)–based polymer gels, no cross-linker is added (De Deene et al., 2002a). It is believed that in these systems, the MAc grafts on the gelatin backbone. This assumption is based on the physical properties of polymer gel dosimeters irradiated to different doses such as the completely different characteristics of ultrasonic speed and elasticity modulus (Mather et al., 2002), the different characteristics of restricted molecular self-diffusion of the water molecules, the melting temperature of both gels, the chemical stability of the gels (De Deene et al., 2002a), and the dose–R_2 response curves obtained for different irradiation temperatures.

The fabrication of the first polymer gel dosimeters was however rather complicated by the fact that oxygen had to be expelled from the polymer gel as oxygen inhibits the radiation-induced polymerization through the creation of peroxides. This was achieved by bubbling the polymer gel solution with nitrogen for several hours. To avoid infiltration of oxygen in the polymer gel dosimeter while pouring the gel mixture in the final recipient, the fabrication procedure was performed in a glove box filled with nitrogen. The dissolved oxygen concentration in the gel needed to be less than 0.01 mg/L to avoid inhibition at low doses (De Deene et al., 2000a). The high sensitivity to oxygen also posed stringent requirements in terms of oxygen permeability on the containers in which the final polymer gel was poured (De Deene et al., 1998a). Containers made of glass and Barex® were found to be adequate to keep the polymer gel oxygen free.

In order to circumvent the devious fabrication procedure in an anoxic atmosphere, the use of antioxidants was proposed (Fong et al., 2001). These antioxidants are believed to compete with the monomers in capturing the radiation-induced oxygen radicals. Although preliminary results with these normoxic polymer gel dosimeters demonstrated a significant improvement in user-friendliness above their anoxic counterparts, it was later discovered that fresh oxygen infiltration still had an effect on the dose–R_2 response in the normoxic gel dosimeters and care is required in keeping the time between fabrication and irradiation to a minimum (De Deene et al., 2002a; Sedaghat et al., 2011b). To obtain an insight in the radiation chemistry of polymer gel dosimeters, basic chemical analysis (De Deene et al., 2000d, 2002a; Jirasek et al., 2006), pulse radiolysis (Kozicki, 2011), Raman spectroscopy (Jirasek and Duzenli, 2001; Jirasek et al., 2001; Rintoul et al., 2003), NMR spectroscopy (Rintoul et al., 2003), and reaction kinetic modeling (Zhang et al., 2002; Fuxman et al., 2003, 2005) have been performed. Some modifications to the standard gel composition have been made in an attempt to increase the sensitivity of the polymer gel dosimeters (Fernandes et al., 2008; Jirasek et al., 2009; Koeva et al., 2009; Jirasek et al., 2010; Zhu et al., 2010; Chain et al., 2011; Khadem-Abolfazli et al., 2013) and to reduce the toxicity of the polymer gel (Senden et al., 2006; Hsieh et al., 2013).

5.2.2 Radiation Chemistry in Polymer Gel Dosimeters

The water content of polymer gel dosimeters is generally in the order of 90%. To understand the physical mechanisms that take place in a gel dosimeter on irradiation, we can rely to a large extent on the physical processes that occur in water. Basic experimental observation is that for numerous solutions of different compounds in water the solute is not being affected directly by the radiation but indirectly by some entity or entities produced from water (Swallow, 1973). On irradiation, water molecules are dissociated in several highly reactive radicals and ions (Spinks and Woods, 1976; Freeman, 1987), a process named "radiolysis."

The cluster size of dissociated water products and the types of species that are created depend on the type of irradiation (linear energy transfer [LET]) and the energy. In the case of x-rays, gamma rays, and electrons, the products occur in "spurs." These prethermal events occur in femtoseconds (10^{-15} to 10^{-14} s).

For 6MV photons, the location of the dissociated products is within 1 nm from the path of the incident ionizing particle. The observation of these events is limited by intrinsic quantum uncertainties. From that moment onward, the probability that these reactive particles reach each other by Brownian motion and react with one another in the form of chain reactions increases with time. As a result, the action radius starts to grow. After 10^{-11} s, a local thermal equilibrium in the recombination of reactive particles is reached. With an average diffusion coefficient of the reactive particles of 4×10^{-9} m^2/s in water (Freeman, 1987), it can be estimated that after 10^{-11} s the quadratic average displacement of the particles from the point of creation is 0.28 nm, which is only one tenth of the intermolecular distance of the monomers in a typical (PAG) gel dosimeter. As the molecular diffusion coefficient of water in the hydrogel is only 15% lower than in pure water (De Deene et al., 2000d), it can be expected that the diffusion coefficient for the radiolytic products of water is in the same order of magnitude. After 10^{-8} s, the average quadratic displacement $\langle r^{\uparrow}2 \rangle = \sqrt{6Dt}$ amounts to 9 nm. The most present intermediates after 10^{-8} s are the hydrated electron (e_{aq}^-), the hydroxyl radical ($OH^\bullet$), and the hydroxonium ion (H_3O^+). These particles may react with the monomers. The hydrated electron reacts with the monomers by the formation of a radical anion that can be further neutralized by a proton (Panajkar et al., 1995). In summary, the decomposition of reactive intermediates can be written as a simplified reaction of which the reaction rate is proportional to the absorbed dose:

$$H_2O \xrightarrow{k_D} 2R^\bullet \tag{5.1}$$

The radicals initiate the polymerization of monomers or polymers containing a double bond by binding with an electron of the double bond:

$$R^\bullet + M_n \xrightarrow{k_I(n)} RM_n^\bullet \tag{5.2}$$

Initially, there will be no polymers in the gel and n will be equal to one. However, as the cross-linking monomers have two double bonds on the same molecule, there can be reactive double bonds in the cross-linking polymer. Hence, during the complete period of polymerization, there may be polymers M_n consisting of n monomer units that react with the radicals. Note the index n in Equation 5.2. No quantitative data were found in the scientific literature on the reaction of polymers with the radicals. The reaction constant $k_I(n)$ depends on the size of the polymers (i.e., the number of repetitive monomer units). It can be expected that the reaction rate will be smaller for larger polymers as the reactions are diffusion controlled (Bosch et al., 1998), and the larger the molecule, the higher the chance is that the reactive site on the molecule will be shielded (Tobita and Hamielec, 1990, 1992). This implies that the reaction rate k_I can be seen as a function of the number of monomer units n (Chernyshev et al., 1997). Note that on the molecular level, it is not only the size of the polymer that is determining the reaction rate but also the shape of the molecule and the location of the reactive groups (double bonds) on the polymer. However, on a macroscopic scale one may think of a statistical average of the different configurations of copolymers. The growth of polymer chains is a result of propagation chain reactions by which the created monomer and polymer radicals react further with other monomers or polymer chains.

The growth of polymer chains is a result of propagation chain reactions by which the created monomer and polymer radicals react further with other monomers or polymer chains:

$$RM_n^{\bullet} + M_{nm} \overset{k_p(n,m)}{\rightarrow} RM_{n+m}^{\bullet} \tag{5.3}$$

The general case in which a polymer radical with n monomer units reacts with a polymer of length m is illustrated. Termination of the polymerization reaction takes place by the combination of two radicals or by disproportionation. The growing polymer–radical may also terminate by transfer of the radical group to other molecules. Typical chain transfer constants $C_M = k_{trans}/k_p$ of radicals are of the order of 10^{-3} to 10^{-4} (Brandrup et al., 1999). The radical site on $M_n{}^*$ may undergo further reaction such as initiation of a new growing polymer chain. The chain transfer agent may be the growing polymer but also the gelatin biopolymer. The decrease of polymerization rate with increasing gelatin concentration provides some evidence of gelatin moderating the polymerization, possibly through chain transfer reactions or through scavenging of initiating fragments by the gelatin molecules (Lepage et al., 2001b).

If oxygen is present in the polymer gel during irradiation, peroxide radicals are created. These peroxide radicals will quickly react with other radicals leading to a termination. This explains the inhibition that may occur in the low-dose region of polymer gel dosimeters. Oxygen can be removed from the gel system by purging the gel solution with inert gasses such as nitrogen or argon gas (Maryanski et al., 1993). Another way to remove oxygen is by use of an antioxidant (Fong et al., 2001; De Deene et al., 2002a).

At high conversions of monomers, the viscosity of a polymerizing system becomes very high. This hinders termination by mutual interaction of growing chains but has less effect on the propagation reaction given by Equation 5.3, because diffusion of the small monomer molecules is not that much affected by the increased viscosity. As a result, the rate of polymerization shows an increase with high conversions (Swallow, 1973). This effect of autoacceleration (Chernyshev et al., 1997) is also called the gel effect or Trommsdorff effect. It has been reported that in systems in which the polymer precipitates from the solution by the creation of a heterogeneous gel system, the increase of viscosity takes place very rapidly even at low conversions (Chapiro, 1962). This effect has also been illustrated through mathematical models of dispersion radical polymerization kinetics (Chernyshev et al., 1997). The autoacceleration caused by a decrease in the termination rate is also responsible for the increasing size of the polymer aggregates with increasing dose as has been observed by optical turbidity spectra (Maryanski et al., 1996b). It is not completely clear yet if the nonlinear response in the low-dose region (seen from 0 to 1 Gy) (De Deene et al., 2002a) of most gel systems is a reflection of this sudden change, or if it is due to chemical kinetics with other molecular species in the gel. In several publications, this nonlinear behavior in the low-dose region is often ignored and a monoexponential saturation curve or linear fit is applied to the dose-R_2 plots of polymer gel dosimeters. With polymer gel dosimeters in which cross-linking copolymerization occurs (such as the acrylamide/*N*,*N*′-methylenebisacrylamide [AAm/Bis] system), the reaction kinetic models are more complex because of the differences in reactivity of the two comonomers (Lepage et al., 2001d; Fuxman et al., 2003) and the change in the reaction rate coefficients during the growth of the copolymer network. The different reaction rate of the comonomers leads to a shift in the instantaneous relative comonomer concentration (Baselga et al., 1988, 1989). The reaction rate of the copolymer structures is dependent not only on the number of monomer units but also on the cross-linking density and the shape of the polymer structures (Tobita and Hamielec, 1992).

According to Baselga et al. (1989), three different reaction steps can be observed in the cross-linking copolymerization of an AAm/Bis aqueous solution: a pregel step, gelation, and postgel reactions. In the pregel step, the cross-linked polymer particles are richer in Bis for both statistical and chemical reasons. At the gel point, the rate of reaction increases for both comonomers but the increment is larger for AAm. During gelation, the pregel particles are joined by chains, which are slightly richer in AAm than only according to the reactivity of both monomers. On the network formation during the post-gel phase characterized by slow cross-linking, only some hypotheses have been formulated such as the retardation by shielding of the radical group by the copolymer chains (Tobita and Hamielec, 1992; De Deene et al., 2000d) and reorganization of the polymer networks (Lepage et al., 2001d). Some studies have been performed on PAGs and aqueous solutions with different ratios of AAm and Bis. From Fourier-transform (FT) Raman spectroscopy studies, it is seen that the relative content of AAm and Bis has a significant influence on the consumption rate of both monomers (Jirasek et al., 2001). This is translated in a difference in dose sensitivity of gels with different compositions. Previously, it was reported that the dose sensitivity of PAG dosimeters is maximum for equal amounts (in weight) of monomer (AAm) and cross-linker (Bis) (Maryanski et al., 1997). This finding appeared to be independent of scanning temperature. It was also found that the saturation R_2 (the R_2 for very high doses) increases with increasing cross-linker fraction. In the study performed by Maryanski et al., it has been assumed that the dependence of dose sensitivity on cross-linker fraction reflects two opposing trends: an increase in sensitivity with cross-linker content up to 50 %C (%C is the relative content of cross-linker with respect to the total amount of comonomer in percentages of weight) due to greater NMR relaxivity of more cross-linked (rigid) polymer, whereas a decrease in sensitivity with increase in cross-linker content beyond 50 %C may be caused by lower reactivity of the cross-linker (Bis). The latter explanation has been contradicted by several studies using FT Raman spectroscopy in which it was found that the consumption rate of the Bis cross-linker monomer is twice as large as the AAm monomer (Baldock et al., 1998; Jirasek et al., 2001; Lepage et al., 2001d). The difference in consumption rate of comonomers makes that the relative fraction of monomer on cross-linker changes with dose. Thus, the polymer structures created at low doses differ from the structures created at higher doses. It has been proposed that the change in viscosity by structures rich in AAm on the one hand and the higher incidence of comonomer reacting with itself at high cross-linker (Bis) concentrations on the other hand, explains the dose sensitivity versus cross-linker concentration (Jirasek and Duzenli, 2001). These reactions have also been described in previous works on cross-linked PAGs (Gelfi and Righetti, 1981; Baselga et al., 1988, 1989; Tobita and Hamielec, 1992).

It is a necessary condition for the final accuracy of the polymer gel dosimeter that for any polymer gel composition, the dosimeter satisfies a few essential criteria. Essential characteristics of 3D radiation dosimeters are the following (De Deene, 2004):

- *Tissue equivalence*: The dosimeter should absorb the radiation in a similar manner as human soft tissue. The tissue equivalence is related to the electron density of the gel.
- *Spatial integrity:* The dose distribution captured by the dosimeter should be preserved for an extended time.
- *Temporal stability:* The R_2-dose response should not change over time.
- *Independence of the temperature during irradiation*: The R_2-dose response should not vary too much with temperature during irradiation as temperature differences in the order of 2°C are likely to occur between operator room and the linear accelerator treatment bunker.
- *Dose rate independence*: The R_2-dose response should not vary much with the rate at which the radiation dose is delivered in order to preserve a unique relation between radiation dose and the measured R_2 values.
- *Independence of the temperature during scanning:* During readout, temperature differences of 1°C–2°C are likely to occur as a result of temperature differences between operator room and scanner room and temperature fluctuations in the scanner bore.
- *Energy independence:* Ideally, the irradiation R_2-dose response of the 3D dosimeter should be the same for different types of irradiation and within the energy spectrum of the irradiation beams.

All polymer gel dosimeters that have been investigated exhibit excellent tissue equivalence and energy independence as a result of the high water content in the hydrogel. However, not all polymer gel dosimeters that have been proposed in the literature demonstrate satisfactory dose rate dependence and temperature dependence during irradiation and scanning. Despite their lower sensitivity, AAm-based gel (so-called PAGAT) dosimeters have superior characteristics that favor their reliability for dose measurements (De Deene et al., 2006a), but data on all the aforementioned characteristics of several other dosimeters are lacking.

5.2.3 MRI Contrast

The polymer aggregates that are created in a polymer gel dosimeter on radiation are highly restricted in their mobility because of their size and because they are entangled with, or grafted onto, the gelatin matrix. As a result of these restrictions, the NMR spin–spin relaxation time (T_2) is significantly decreased because of the static dipole–dipole coupling component (Bloembergen, 1948) and chemical proton exchange between water and the polymer (Zimmerman and Brittin, 1957; Gochberg et al., 2001; McConville et al., 2002). The R_2 ($1/T_2$)-dose response of the spin-spin relaxation rate in polymer gel dosimeters is more pronounced than that of the spin–lattice relaxation rate R_1 ($1/T_1$). It is shown in later studies that also other MRI contrasts, such as magnetization transfer (Lepage et al., 2002; De Deene et al., 2013) and chemical shift (Murphy et al., 2000), can be used to image polymer gel dosimeters.

To describe the effect of radiation-induced polymerization on the magnetic resonance relaxation rates R_1 and R_2, it is practical to consider different proton pools (i.e., ensembles of protons that can be considered as belonging to molecules that experience the same chemical environment). Three major groups of proton pools can be considered in a polymer gel dosimeter (Lepage et al., 2001c): (1) the proton pool of free and quasi-free protons (denoted as "mob"). These are the protons from free water molecules and monomers; (2) the proton pool of the growing polyacrylamide network (poly); and (3) the proton pool of the gelatin matrix (gel) and of the water molecules bound to the gelatin. It can be noted that in order to study other phenomena in more detail, a subdivision of these proton pools can be considered as well. In a study of the chemical stability of polymer gel dosimeters, the third pool is subdivided into two pools (De Deene et al., 2000d). According to the Bloembergen–Purcell–Pound (BPP) theory, the spin–spin relaxation of the different proton pools is determined by the rate of molecular "tumbling" and Brownian motion of the molecules that contain these protons (Bloembergen et al., 1948). This results in a change of the efficiency of dipolar coupling between neighboring protons and, as a result, in a change in the diphase rate of the spin-magnetic dipole moments. As this is directly correlated with the spin–spin relaxation, it follows that the relaxation rate of the proton pools is correlated with the mobility of the protons within these pools. The different proton pools are thus characterized by different relaxation rates. If the lifetimes of protons in the various environments are long compared to the characteristic correlation times of the environments, each environment has intrinsic relaxation rates that are independent of the specific lifetime value ($R_{2,mob}$, $R_{2,poly}$, $R_{2,gel}$). If, furthermore, the lifetimes are long compared to these relaxation times, the NMR signal is the same as the sum of the signals from isolated, nonexchanging environments. When this happens, the relaxation curves are multiexponential of which the population fractions of the different pools are determined by the coefficients of the different exponential components. This is the slow exchange case. On the other hand, when these lifetimes are short compared to the relaxation times but still long compared to the correlation times (the rapid exchange limit), the observed relaxation curve will be monoexponential with a relaxation rate that is the weighted average of the relaxation rates of the different proton pools in the sample (Zimmerman and Brittin, 1957):

$$R_2 = f_{mob} \cdot R_{2,mob} + f_{poly} \cdot R_{2,poly} + f_{gel} \cdot R_{2,gel} \tag{5.4}$$

For the R_2 measurements that have been performed on polymer gel dosimeters, the condition of fast exchange is satisfied. Before irradiation, the polymer proton pool is empty ($f_{poly} = 0$) whereas the mobile proton pool (f_{mob}) is at its maximum. On irradiation, the polymer proton pool (f_{mob}) starts to grow at the cost of the mobile proton pool. As a result, the relaxation rate will change proportional with the amount of converted monomers. The mobility of monomers is relatively high and thus also the mobility of water molecules that bind to the monomers by hydrogen bridges. However, on irradiation of the gel dosimeters, the molecular mobility is significantly reduced. As the mobility of the bound water molecules is reduced, the spin–spin relaxation is more effective, which is observed by an increase in the observed spin–spin relaxation rate (R_2). A comparison of the change in R_2 of gel dosimeters consisting of different monomers suggests that the change in relaxation rate cannot entirely be explained by the BPP theory. The dose sensitivity of different gel dosimeters is listed in Table 5.1.

From studies in which different water pools are selectively inverted (Edzes and Samulski, 1978), it is seen that cross-relaxation can occur between the different proton pools, for example, between protons of the polymer with protons of mobile water (Ceckler et al., 1992; Gochberg et al., 1998). The exchange of magnetization may occur by proton chemical exchange between bound water and free water and by magnetization transfer between nonexchangeable macromolecular protons and bound water. It has been shown that magnetization transfer can also be mediated by chemical exchange interactions (Kennan et al., 1996). It is shown that both chemical exchange and magnetization transfer are influenced by the pH of the system. As a result of the different interactions between the different proton pools, the relaxation rates of the different pools ($R_{2,mob}$, $R_{2,poly}$, $R_{2,gel}$) as they occur in Equation 5.4, are determined not only by the mobility of the molecules but also by the exchange rates of protons. As some monomers have acidic or alkaline functional groups, the overall R_2 relaxation rate also depends on the pH of the gel (Gochberg et al., 1998). From Table 5.1, it can be seen that the dose sensitivity of the different monomers is influenced by the functional group. The functional group determines both the polymerization rate of the monomers (inversely proportional to the half-dose value $D_{1/2}$) and the efficiency of cross-relaxation. The hydroxyl and amino groups serve as hydrogen-bonding sites (Ceckler et al., 1992). The hydroxyl group seems to be more efficient than the amino group in the exchange of magnetization. However, it is seen that the reaction rate of AAm in the PAG is much higher than that of acrylic acid. As a result, the dose sensitivity of both monomers is nearly the same, but the dose range of the acrylic acid gel is larger than that for the AAm-based gel. Although the alkyl group (in MAc and 2-hydroxyethyl methacrylate [HEMA]) does not have a large influence on the cross-relaxation efficiency, it has a significant effect on the polymerization rate of the monomers.

5.3 Imaging of Polymer Gel Dosimeters

All imaging techniques rely on the effect of the polymer created upon radiation on measurable physical properties. The precision and accuracy of the dosimeter is hence determined by both the polymer-induced physical effect and the imaging technique itself.

5.3.1 Magnetic Resonance Imaging

To obtain quantitative R_2 maps of the exposed polymer gel dosimeter, a series of T_2-weighted images is recorded from which an R_2 map is calculated. T_2-weighted images can be obtained by different methods (De Deene, 2013), but the most commonly used methods are based on spin echo acquisitions, either using a single-spin echo sequence twice (Figure 5.3a) with different echo times (TE) or using a multiple spin echo sequence with various echo times (Figure 5.3b).

In the case of two images acquired with a single-spin echo sequence with different echo times, the R_2 value can be calculated algebraically by the following equation:

$$R_{2,(i,j)} = \frac{1}{\mathrm{TE}_1 - \mathrm{TE}_2} ln\left(\frac{S_{(i,j)}(\mathrm{TE}_2)}{S_{(i,j)}(\mathrm{TE}_1)}\right) \tag{5.5}$$

where $R_{2,(i,j)}$ is the R_2 value for the pixel with coordinates (i,j), TE_1 and TE_2 are the echo times for both base images (Figure 5.3a), and $S_{(i,j)}(\mathrm{TE}_1)$ and $S_{(i,j)}(\mathrm{TE}_2)$ are the corresponding pixel intensities in pixel (i,j).

In the case of the multi-echo sequence, the R_2 map can be obtained by fitting an exponential T_2-decay function for each pixel coordinate in the base images on a pixel-by-pixel basis. It has been shown that in order to obtain an optimal signal-to-noise ratio (SNR) in the calculated R_2-image, the echo times need to be

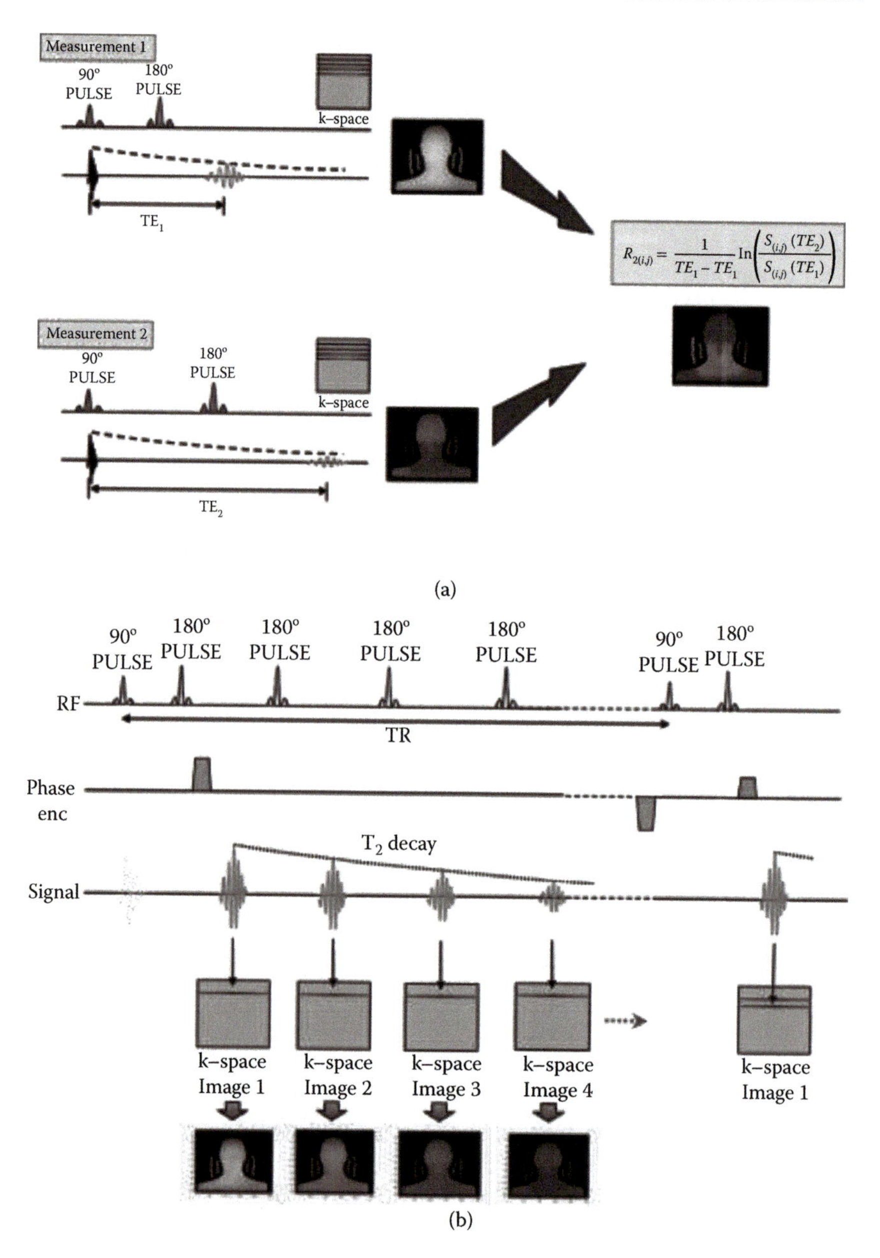

Figure 5.3

(a) Single-spin echo and (b) dual-spin echo pulse sequence for acquiring R_2 maps of the polymer gel dosimeter.

optimized in both the single-spin echo approach and multiple spin echo approach (De Deene et al., 1998b; De Deene and Baldock, 2002). An overview of MRI pulse sequences with their main characteristics is provided in Table 6.2 of Chapter 6.

5.3.2 Optical CT Scanning

Although an unirradiated polymer gel is transparent for visible light, the irradiated gel becomes increasingly turbid with increasing radiation dose as a result of the formation of polymer aggregates that have typical sizes in the order of the wavelength of visible light (Maryanski et al., 1996b). As a result, the dosimeter will Mie scatter visible light in many directions and the light in the primary direction will be attenuated proportional to the polymer density. The optical density spectra also depend strongly on the cross-linker concentration, as the cross-linker concentration determines the size of the polymer aggregates. It has also been found that the particle size distribution depends on the dose range with larger polymer aggregates being produced for dose increments in the high-dose regions (Maryanski et al., 1996b).

Similar to x-ray CT, a 3D image dataset of the radiation dosimeters can be obtained through optical CT scanning where the transmitted visible light is captured by optical detectors. Different kinds of optical CT scanners have been constructed to read out the 3D radiation dose distribution that is captured by the dosimeters. A comprehensive explanation of the theory behind optical CT scanning is given by Doran and Krstajic (2006).

The first generation of optical laser CT scanners (Figure 5.4a and b) consisted of a red laser and a set of moving mirrors that create a traveling laser beam through the dosimeter (Gore et al., 1996). To avoid deflection of the laser beam at the surface of the dosimeters, the dosimeter is immersed in an optically transparent square tank containing a fluid that has the same refractive index as the dosimeter. The refractive index matching fluid for polymer gel dosimeters consists of a mixture of water and glycerol. A single one-dimensional (1D) transmitted light projection through the phantom is acquired by moving the mirrors on either side of the fluid tank synchronically. The phantom is then rotated with a small angular increment and another 1D profile is acquired. After a full rotation, a cross-sectional optical density image can be reconstructed. Different slices can be scanned by moving the tank up or down with respect to the laser beam. Variations on this design have been published (Kelly et al., 1998; Oldham et al., 2001; Islam et al., 2003; Xu et al., 2004; Lopatiuk-Tirpak et al., 2008). The scanning beam optical laser CT scanners are able to scan a 1-L gel phantom with an isotropic resolution of 1 mm^3 in a few hours. The optical laser CT scanner with translating laser beam has been marketed by the company MGS Research (Madison, CT) under the name "OCTOPUS scanner."

A faster optical CT scanner makes use of a charge-coupled device (CCD) camera that records entire images for each angular increment of the phantom instead of transmission line profiles in the optical laser scanner (Figure 5.4c and d). In this cone-beam optical CT scanner (Wolodzko et al., 1999), the dosimeter phantom is placed between a diffuse light source and a pinhole camera. The pinhole camera receives the transmitted light from a cone. CT images are reconstructed using a cone-beam reconstruction algorithm. With the cone-beam optical CT scanner, an entire 3D volume can be scanned in less than 10 min. The cone-beam image reconstruction is in the order of a few minutes on a modern PC but has currently been sped up by use of parallel computing (graphics processor unit [GPU]—CUDA). A cone-beam optical scanner has been marketed by the company Modus Medical Devices (London, Ontario, Canada) under the name "Vista scanner."

To minimize image artifacts originating from secondary light scatter in the cone-beam scanner, a parallel beam scanner was introduced using a big lens to create a parallel beam of light (Doran et al., 2001) that projects a transmission image of the phantom on a diffuser screen which is captured by a CCD camera. This system has been further improved by replacing the lens and diffuser screen by two telecentric lenses (Krstajic and Doran, 2006) (Figure 5.4e and f). In this configuration, the second

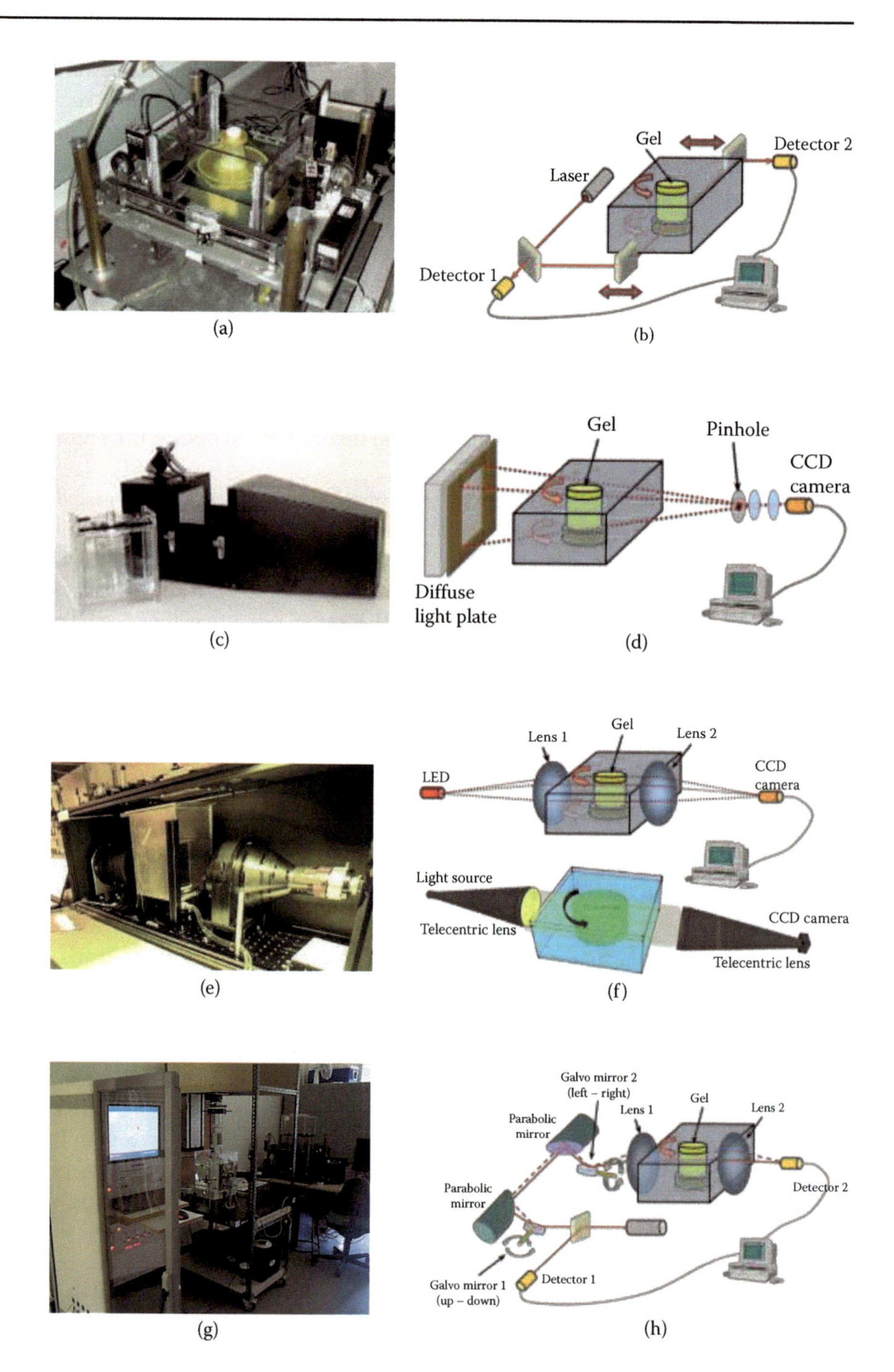

Figure 5.4

Different types of optical computed tomography (CT) scanners. (a and b) Moving mirror optical CT laser scanner. (c and d) Cone-beam optical CT scanner with diffuse light source and charge-coupled device (CCD) camera. (e and f) Parallel-beam telecentric CCD-based optical CT scanner. (g and h) Scanning optical laser CT scanner.

(Continued)

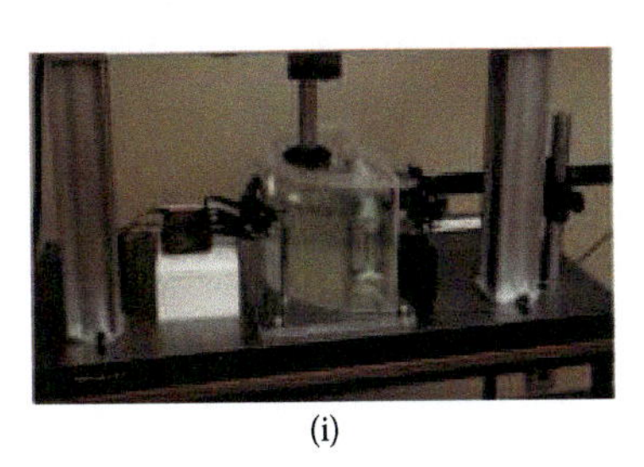

(i)

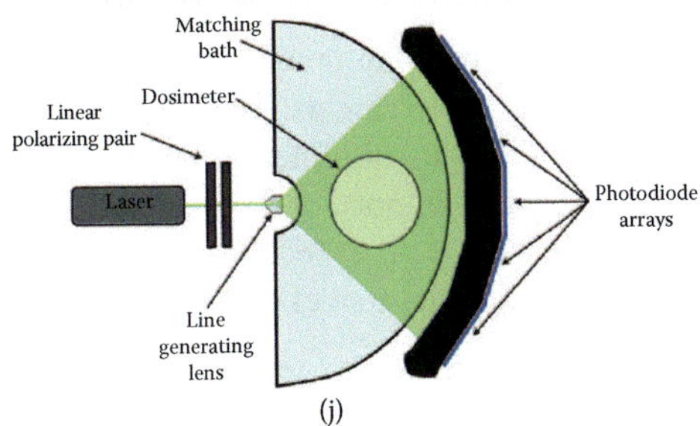

(j)

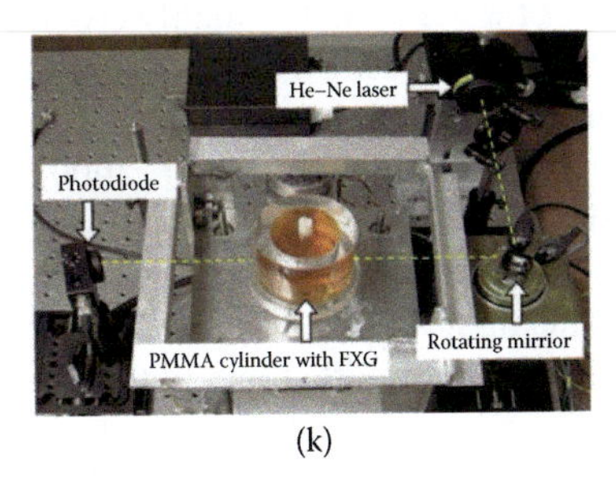

(k)

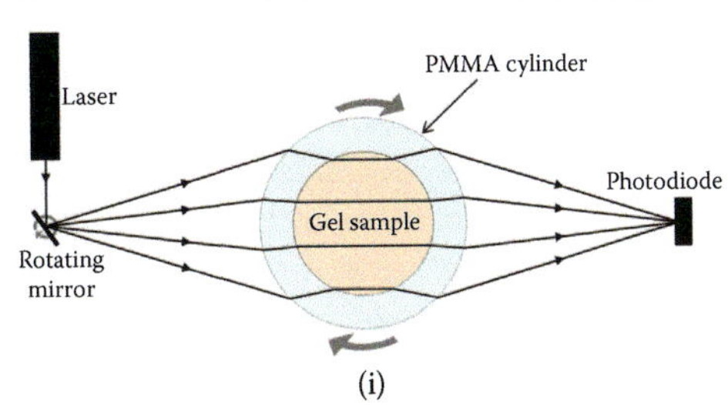

(i)

Figure 5.4 (Continued)

Different types of optical computed tomography (CT) scanners. (i and j) Fan-beam optical CT scanner with linear array photodetector. (k and l) Dry optical CT laser scanner. (FXG, ferrous xylenolorange gelatin gel; PMMA, polymethyl methacrylate.)

telecentric lens at the receiver side captures only rays of light that are orthogonal to the lens, hence filtering out any scattered light from other directions.

The first generation laser scanning systems have been improved by replacing the translating mirrors and vertical stage with the Galvano mirror that creates a sweeping beam of light that is incident on a large plano-convex lens (Figure 5.4g and h). The lens converges the laser beams in straight parallel rays that travel through the fluid tank and dosimeter phantom. The rays are captured by a second plano-convex lens that converges the beams in a photodetector. In the system introduced by Krstajic and Doran (2007), two Galvano mirrors are used for scanning in two directions. To compensate for nonuniform optical aberration in the vertical direction and to increase the possible phantom size, a modification was made by Vandecasteele and De Deene (2009, 2013a) where the vertically scanning Galvano mirror is replaced by a linear stage that moves the phantom vertically with respect to the laser beam.

In a fan-beam optical CT scanner (Campbell et al., 2013), a fan beam of light is produced which is collected by a circular array of detectors after passing through the fluid tank and dosimeter phantom (Figure 5.4i and j). To remove the need for a refractive index matching fluid, "dry" scanners (Maryanski and Ranade, 2001; Ramm et al., 2012) make use of the light-focusing effect of the cylindrical gel dosimeter phantom or cylindrical PMMA container that encloses the dosimeter phantom (Figure 5.4k and l).

The problem with optical scanning of polymer gel dosimeters is the diffuse light scattering by the irradiated dosimeter (Oldham et al., 2003). Light scattered in a direction different from the primary beam direction can end up in the detector, which may give rise to image artifacts. Although laser scanning systems are less susceptible to scattering artifacts than CCD-based cone-beam scanning systems (Olding et al., 2010), nonlinear perturbation of the detected light by scattered light is considered an important factor that contributes to the uncertainty in the final dose maps (Bosi et al., 2009). For that reason, 3D radiochromic dosimeter systems have been fabricated that exhibit far less light scattering (see Chapter 6).

5.3.3 X-Ray CT Scanning

The small radiation-induced change in electron density in polymer gel dosimeters results in a change in the linear attenuation coefficient of x-rays. The change in electron density is attributed to the expulsion of water from the highly cross-linked polymer aggregates (Trapp et al., 2002) leading to a redistribution of mass within the polymer gel system. The corresponding dose-dependent change in Hounsfield units (HUs) enables the use of x-ray CT scanning (Hilts et al., 2000). The dose sensitivity depends on the gel composition and ranges from 0.23 HU/Gy for a PAG without cross-linker to 1.43 HU/Gy for a PAG with 12% (w/w) total monomers and 50% cross-linker (Hilts, 2006). The highest sensitivity is found for large concentrations of monomers that can be obtained with the addition of cosolvents such as glycerol (Jirasek et al., 2009) or isopropanol (Jirasek et al., 2010). The dose-CT number sensitivity plot is generally not linear but is best approximated by a hyperbolic tangent function (Jirasek and Hilts, 2014).

5.3.4 Other Scanning Techniques

Acoustic properties such as propagation speed, ultrasonic absorption, and attenuation (Mather et al., 2003) have also been found to be affected by the formation of the polymer aggregates in the polymer gel dosimeters. The change in acoustic speed has been attributed to the polymer aggregates that affect both the elasticity modulus and the mass density (Mather et al., 2002) of the polymer gel dosimeter. An ultrasound scanner prototype has been constructed using an ultrasound transducer and a needle hydrophone (Mather and Baldock, 2003) but the image quality of these preliminary dose maps was not sufficient to be useful for clinical dosimetry.

Physical properties such as the young elasticity modulus and electrical conductivity are affected by the radiation-induced formation of polymer aggregates and create potential for other scanning techniques such as elastography (Oudry et al., 2009) and electrical impedance tomography (Kao et al., 2008).

5.4 Applications of Polymer Gel Dosimetry

5.4.1 Dose Verification of Conformal External Beam Therapy

Polymer gel dosimetry was developed with the aim of providing adequate 3D dosimetry in conformal radiotherapy (Ibbott et al., 1997; De Deene et al., 1998a; De Neve et al., 1999; De Deene, 2002) where steep dose gradients are encountered in three dimensions. With the first applications, gel dosimetry was considered as a time-consuming and labor-intensive dosimetry technique as the fabrication of polymer gel dosimeters could easily take up to several hours and the scanning required significant MRI physics expertise. Also these early polymer gel dosimeters were susceptible to several sources of uncertainty that were related to oxygen contamination and MRI scanning artifacts. These elements have made the use of polymer gel dosimetry as a dosimetric QA tool to be restricted to specialized academic centers, as it is still the case today, although much progress has been made in eliminating inaccuracies and the introduction of normoxic gel dosimeters that are easier to fabricate.

Polymer gel dosimetry still requires that a relatively strict procedure is followed in order to deliver reliable dose measurements. Even in specialized centers, polymer gel dosimetry is only used in the process of benchmarking and commissioning a new delivery technique, or in verifying class solutions. The unique feature of polymer gel dosimetry is that a humanoid-shaped dosimeter can be regarded as a "dummy patient" and can be taken through the entire treatment chain, from scanning to treatment (Figure 5.5). After irradiation, the measured dose distribution can be compared with the planned dose distribution. Several metrics can aid in this process such as gamma map evaluations (Low et al., 1998), difference maps, and dose–volume histograms (DVHs). If an intolerable deviation between measured and calculated dose distribution is detected, the source of the deviation should be determined and remedied before the treatment delivery technique is implemented in clinical routine.

Polymer gel dosimetry here serves as a top-level QA tool in determining the presence of any errors in the overall treatment delivery procedure. To determine the actual source of treatment errors, other dosimeters and QA checks can be performed. It can be noted that gel dosimeters can also indicate human errors in the treatment delivery such as setup errors and are therefore ideal training tools.

Until now, the biggest size in humanoid-shaped 3D dosimeters has been obtained with polymer gels, such as in the case of a whole abdominopelvic intensity-modulated arc therapy (IMAT) dose verification of an ovarium carcinoma (Duthoy et al., 2003). With these large phantom shapes, special consideration needs to be given to imaging artifacts that originate from radiofrequency (RF) nonuniformity (Vergote et al., 2004a) and temperature heterogeneity during scanning (De Deene and De Wagter, 2001).

Apart from the verification of entire intensity-modulated radiation therapy (IMRT) or IMAT treatments, gel dosimetry can also be used to investigate the dose distribution in extreme beam configurations. One such case where gel dosimetry showed to be very useful was in studying the effect of lead markers placed in a tray that was connected to the linac collimator, on the delivered dose. The tray with lead markers served as an isocenter position control device for radiation therapy setup verification using electronic portal images (De Deene, 2002). With the intention of studying the possibility of treatment without removing the lead markers, the disturbance of the dose distribution was studied by use of gel dosimetry. As an underdosage of 10% was found underneath the lead markers, it was concluded that the tray with lead markers needed to be removed before the treatment was given to the patient. Another case where gel dosimetry was found to be very flexible was in the study of leaf leakage of multileaf collimators, where the effect can be easily studied at different depths (De Deene, 2002).

In IMRT, typically several small beams are often used. When small beams cross a low-density structure such as an air cavity, electronic equilibrium may be disturbed, resulting in dose rebuild up in structures behind the low-density medium. The effect of tissue heterogeneities can be studied by inserting air cavities in a polymer gel dosimeter (De Deene, 2002). To study the effect of lung tissue on IMRT dose distributions, a thoracic phantom with lung cavities has been fabricated.

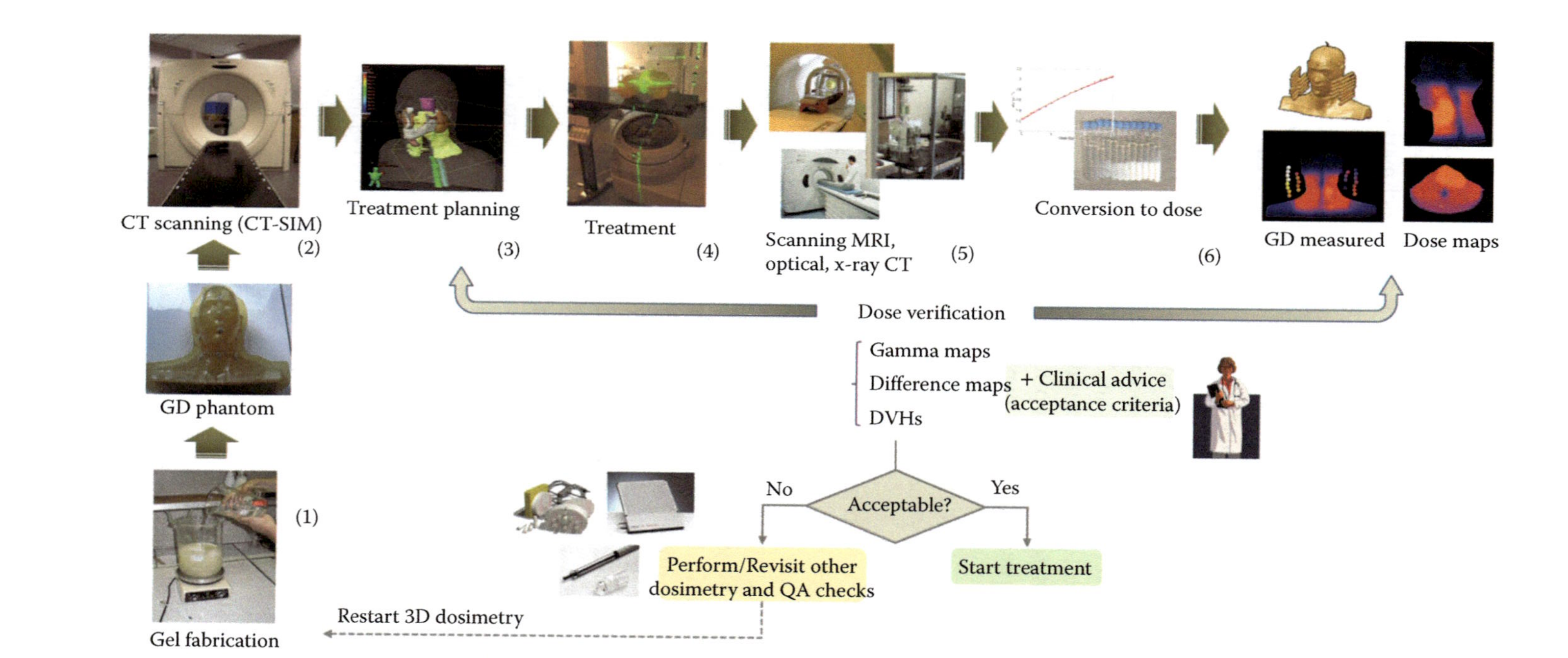

Figure 5.5 In a clinical dose verification experiment with three-dimensional (3D) dosimetry, the dosimeter phantom is taken through the entire treatment chain. Possible dosimetric or setup errors can occur at the different stages. 3D dosimetry fulfills an important role in benchmarking the complete treatment chain. (CT, computed tomography; GD, gel dosimeter; DVHs, dose–volume histograms; MRI, magnetic resonance imaging; QA, quality assurance.)

The effect of the lower density of the lungs on the dose distribution of a mediastinal tumor has been studied by performing 3D dosimetry in the scenario where the lung cavity was filled with air and in the scenario where the lung cavity was filled with water (Vergote et al., 2003). This study demonstrated the different performance of dose calculation algorithms in terms of accounting for radiation transport in low-density structures. It was concluded that by minimizing the beam path length through lung tissue, the effect of the heterogeneity of the thorax phantom on the dose to the planning target volume (PTV) could be minimized.

To measure the dose distribution in the lungs, a lung-tissue-equivalent polymer gel dosimeter has been proposed. These lung-tissue-equivalent dosimeters were fabricated by beating the gel into a hydrogel foam by use of a household mixer (De Deene et al., 2006b). It was found that these polymer hydrogel foams resemble lung tissue on the microscopic scale very well. An alternative imaging technique was needed to acquire the dose distribution in these low-density polymer gel foams. Interestingly, it was later shown that the average bubble size could be determined by use of T_2 relaxation rate dispersion measurements (Baete et al., 2008), and proton density maps were correlated with electron densities (De Deene et al., 2006b). This proof of principle has not been explored any further but opens a new window of possibilities for heterogeneous anthropomorphic 3D dosimetry.

5.4.2 Brachytherapy

Several studies have demonstrated the potential of 3D dosimetry of brachytherapy with both low-dose-rate and high-dose-rate sources (Baldock et al., 2010). However, additional caution is required when applying polymer gel dosimetry for brachytherapy. Additional artifacts may compromise the accuracy of the dose measurements. These artifacts are related to the insertion of a catheter in the gel dosimeter, which may disturb the oxygen concentration in the polymer gel dosimeter (De Deene et al., 2001), may cause susceptibility artifacts (De Deene et al., 2001), and may result in uncertainties caused by diffusion gradients in high-resolution MRI (Hurley et al., 2003). When point sources have been applied, the associated steep dose gradients may cause additional uncertainties in the dose distribution as a result of nonequilibrium diffusion reaction kinetics (De Deene et al., 2001).

5.4.3 High-LET Particle Irradiations

3D dosimetry is attractive for dose verification of proton (Heufelder et al., 2003; Gustavsson et al., 2004b) or other hadron (Ramm et al., 2000) treatments because of the steep dose gradient near the Bragg peak. However, all these studies demonstrate a significant LET dependence of the polymer gel dosimeters. Theoretical track structure calculations and FT Raman spectroscopy of proton beams demonstrate the decreased sensitivity of the dose response of polymer gel dosimeters for increasing LET (Jirasek and Duzenli, 2002). This is attributed to a larger density of polymer radical chains close to the proton track favoring a faster termination of the polymerization reactions. At high LET, radical recombination decreases the amount of polymerization initiation reactions.

5.4.4 Boron Neutron Capture Therapy

In boron neutron capture therapy (BNCT), it is assumed that boron-10 accumulates in the tumor after administration of a tumor-specific boron carrier. After redistribution of the carrier, the patient is irradiated with epithermal neutrons. A nuclear reaction of the boron-10 with the neutrons leads to the formation of ^{7}Li, an alpha particle and gamma radiation. Although the irradiation with epithermal neutrons results in background radiation, the localized radiation effect of the short-range alpha and ^{7}Li-particles leads to a significant therapeutic gain. An increase in dose has been reported in a PAG dosimeter doped with boron as compared to an undoped PAG dosimeter after irradiation with epithermal neutrons (Farajollahi et al., 2000), illustrating the potential of polymer gel dosimetry. However, more studies are needed to convert the NMR R_2-response to dose.

5.4.5 Dosimetry in Diagnostics

The diagnostic radiation dose distribution administered during x-ray CT scanning has been acquired with a MAGIC-type polymer gel dosimeter for a range of imaging protocols (Hill et al., 2005). In this study, dose profiles were acquired for different slice profiles. To achieve a measurable response, 50 accumulated single transaxial slices were acquired. Results were compared to dose measurements obtained with an ionization chamber, and parameters that were extracted from the MRI-measured polymer gel dose distributions where the computer-tomography dose index (CTDI) and the slice width dose profile (SWDP). The advantage of polymer gel dosimetry for x-ray CT is that with a single measurement, the regional variation of dose in the transaxial plane can be obtained in an anthropomorphic phantom.

5.5 Reliability of Polymer Gel Dosimetry

Uncertainties in measurements with 3D dosimeters are attributed to both dosimeter properties and scanning performance. In polymer gel dosimetry with MRI readout, discrepancies in dose response of large polymer gel dosimeters versus small calibration phantoms have been reported, which can lead to significant inaccuracies in the dose maps that are obtained by calibrating the R_2 values in the large humanoid-shaped 3D dosimeter with the R_2-dose relation extracted from the small calibration vials. The sources of error propagation in polymer gel dosimetry with MRI readout are well understood and it has been demonstrated that with a carefully designed scanning protocol, the overall uncertainty in absolute dose (maximum difference with respect to ionization chamber–measured dose) that can currently be obtained falls within 5% on an individual voxel basis, for a minimum voxel size of 5 mm^3 (De Deene and Vandecasteele, 2013).

5.5.1 Absolute Versus Relative Dosimetry

Several research groups have chosen to use polymer gel dosimetry in a relative manner by normalizing the dose distribution toward an internal reference dose within the gel dosimeter phantom. 3D dosimetry with optical scanning has also been mostly applied in a relative way, although in principle absolute calibration is possible.

5.5.2 Uncertainties

Uncertainties can be classified into type A and type B, where type A standard uncertainties are obtained from a probability density function derived from an observed frequency distribution, whereas type B standard uncertainties are obtained from an assumed probability density function that is based on the degree of belief that an event will occur. Type A standard uncertainties can be perceived as the inverse of precision (or the ability of a measurement to be consistently reproduced) and type B uncertainties as the inverse of accuracy (or the degree of conformity of a measured quantity or "measurand" to its actual, true, value). Type A standard uncertainties can be derived as the standard deviation on repetitive measurements whereas type B uncertainties can only be derived by comparison of the measured data with a "golden truth." The terminology of metrology has been more extensively discussed in Chapter 2.

A complicating factor in gel dosimetry is that the uncertainty applies to both dose and space. In a gel dosimetry dose verification experiment (and any radiation treatment), the spatial and dosimetric dimensions are interwoven. It is theoretically impossible to extract both dosimetric and spatial errors from a measured spatial dose distribution, for example, the result of a 3D gel dosimetry experiment. To comprise both spatial and dosimetric performance in one parameter, concepts such as the gamma index (Low et al., 1998) and maximum allowed dose difference (Steve et al., 2006) have been introduced.

To obtain the overall uncertainty of polymer gel dosimetry, also the fabrication and irradiation have to be included in the analysis. This can be achieved through a reproducibility study of the complete gel dosimetry experiment from gel fabrication to dose distribution analysis (Vandecasteele and De Deene, 2013b). It is imperative that in a clinical verification with gel dosimetry, to be considered as a single experiment, the uncertainty of the dose measured in each voxel comprises both the imprecision (type A uncertainties) and systematic inaccuracy (type B uncertainties).

Gel dosimetry involves different steps and errors can occur at each stage (Figure 5.5). (1) The polymer gel dosimeter is fabricated in a chemical laboratory and is stored until radiation. Any deviations in the chemical composition through inhomogeneous mixing, impurities in the recipients, or oxygen leaks in the recipient can give rise to discrepancies between calibration tubes and volumetric phantoms (Sedaghat et al., 2011a). Moreover, a difference in temperature course during storage may also affect the dose-R_2 response (De Deene et al., 2007). Most of these deviations are compensated by using calibration phantoms that are fabricated from the same batch of gel. (2) The gel dosimeter phantom is then scanned with CT and the treatment planning is optimized (3) on the scanned set for a virtual PTV and a set of critical organs. At this stage, just as with a patient, a reference coordinate system is allocated to the phantom by use of marker lines that are drawn on the phantom. In addition, stereotactic fixtures or fiducial markers (Meeks et al., 1999) can be placed on the phantom that will later be visible on the MRI, or optical and CT images in the case of optical or x-ray CT readout, respectively, for coregistration of the dose maps with the treatment-planning system (TPS). Positional setup errors are likely to result in deviations between the planned and measured dose distribution. These deviations are not intrinsic to the gel dosimeter but are indicative for errors that may also occur during the actual patient treatment. (4) On irradiation, a complex set of radiation-induced chemical reactions take place. On a molecular level, these reactions are probabilistic in nature but these uncertainties are negligible on the scale of typical imaging voxels. Other sources of uncertainty in dose reading that may occur during irradiation of the gel dosimeter are related to the dependence of the dose response on the temperature of the 3D dosimeter during irradiation, on the dose rate, and on the energy spectrum of the irradiation beam (De Deene et al., 2006a; Karlsson et al., 2007). It is important to realize that differences in dose rate (and energy) occur in the irradiated 3D volume even if the photon fluence rate (monitor unit [MU]/min) of the individual beams is kept constant. Also temporal and spatial chemical instability can give rise to dose uncertainties (De Deene et al., 2002b). During scanning (5), thermal measurement noise will add to the acquired images. Propagation of image noise into the fitted dose maps can be minimized by an optimal selection of scanning parameters and fit algorithm (De Deene et al., 1998b; De Deene and Baldock, 2002). Imaging artifacts can result in systematic dose and spatial uncertainties.

5.5.3 Type A Uncertainties

Stochastic variations (random uncertainties) occur at different stages of the dosimetry process. However, physicochemical variations are canceled out by the calibration procedure where gel samples of the same batch are applied. Therefore, irreproducibility during the fabrication procedure such as weighting the chemicals and temperature treatment of the gel during fabrication is not considered to contribute to type A uncertainties. During irradiation, sources of stochastic variations that may potentially affect the overall precision are variations in the dose delivery, variations in the temperature during irradiation, and stochastic variations in the positioning of the calibration phantoms. However, from experiments where 20 (calibration) samples were exposed to the same radiation dose, it is derived that these uncertainties are smaller than 1% (one standard deviation) of the nominal dose (De Deene and Vandecasteele, 2013). This level of uncertainty is further reduced in the fitted parameters as a result of the averaging effect on the measured data points (De Deene and Baldock, 2002). The resulting type A uncertainty is therefore considered as predominantly originating from thermal image noise during scanning.

The concept of dose resolution was introduced to evaluate *the intrinsic dosimetric precision* in terms of dose sensitivity and scanning SNR (Baldock et al., 2001). The dose resolution, D_Δ^p, is defined as the minimal detectable dose difference within a given level of confidence, p. The dose resolution is related to the standard deviation on dose, σ_D, by the following equation:

$$D_\Delta^p = \sqrt{2}\,.\,k_p\,.\,\sigma_D \tag{5.6}$$

For a 95% confidence level, the dose resolution becomes $D_\Delta^p = 2.77\sigma_D$.

The *relative dose resolution*, $D_{\Delta\%}^p$, is defined as the dose resolution relative to the operating dose range:

$$D_{\Delta\%}^p = \frac{D_\Delta^p}{(D_{\max} - D_{\min})} = \sqrt{2}\,.\,k_p \frac{\sigma_D}{(D_{\max} - D_{\min})} \tag{5.7}$$

If the dose maps are derived from quantitative NMR-R_2 maps, it can be shown that the relative dose resolution ($D_{\Delta\%}^p$) is equal to the relative R_2 resolution ($R_{2\Delta\%}^p$), which is defined in a similar way:

$$D_{\Delta\%}^p = \sqrt{2}\,.\,k_p \frac{\sigma_D}{(D_{\max} - D_{\min})} = \sqrt{2}\,.\,k_p \frac{\sigma_{R2}}{(R_{2_{\max}} - R_{2_{\min}})} = R_{2\Delta\%}^p \tag{5.8}$$

It should be emphasized that the dose resolution is related not only to the type of gel dosimeter but also to the scanning protocol (De Deene and Baldock, 2002). In some publications, the concept of dose resolution has been used as a criterion to compare different types of gel dosimeters. This can be misleading as most of these studies report on dose resolutions obtained with suboptimal scanning parameters. The concept of dose resolution is a practical metric to optimize the NMR sequence. In optimizing the NMR sequence, it is also important to take into account the number of slices that are required for the 3D dosimetry application.

5.5.4 Type B Uncertainties

An interbatch reproducibility study of polymer gel dosimetry with polyacrylamide gel with the antioxidant tetrakishydroxyphosphonium salt (PAGAT) gel dosimeters demonstrated that although the variation in the dose distribution between different experiments was less than 3%, a systematic deviation of more than 10% was found between the gel measured dose and the dose recorded with an ionization chamber at the isocenter (Vandecasteele and De Deene, 2013b). Also preceding studies had shown poor correspondence of absolute dose measurements with polymer gel dosimetry as compared to ionization chamber measurements. This has resulted in a relative calibration approach by several groups, where R_2 maps were converted to dose maps by a renormalization to the dose registered at the isocenter. The poor absolute accuracy has been attributed to a discrepancy in dose response between the small calibration samples and the large volumetric gel dosimeter phantom, which has been related to both physicochemical (Vandecasteele and De Deene, 2013c) and scanning factors (Vandecasteele and De Deene, 2013d).

5.5.4.1 Physicochemical Sources of Uncertainty

The influence of the temperature history during the storage period between fabrication and irradiation of the gel dosimeter on the dose response curve has been assessed for both PAGAT and methacrylic acid gel with antioxidant tetrakishydroxyphosphonium salt (MAGAT) gel dosimeters (De Deene et al., 2007). It was found that the dose-R_2 response was significantly more dependent on the storage temperature in MAGAT gel dosimeters as compared to PAGAT gel dosimeters. When both the volumetric gel dosimeter phantom and the calibration vials are placed in the fridge after fabrication, they will cool down at different rates as a result of the difference in thermal inertia. This has been estimated to result in an absolute dose uncertainty in the order of 5% with respect to the nominal dose range (systematic maximum dose deviation with respect to the maximum measured dose). However, if both calibration vials and volumetric gel dosimeter phantom are cooled down slowly after fabrication, for example, by placing them in a large water container, the absolute dose uncertainty is reduced to below 1%.

It was found that in some polymer gel dosimeters also the temperature during irradiation has an influence on the dose response (Jirasek et al., 2001). This is likely due to a temperature-dependent change in diffusivity of the monomers in the gel matrix and a change in the chemical reaction kinetics. A difference in temperature during irradiation of 3°C results in a maximum dose uncertainty of 1% (systematic maximum dose deviation with respect to the maximum measured dose).

After irradiation, the dose response changes over time as a result of postirradiation chemical instability (De Deene et al., 2000d). Theoretically, if calibration vials and the volumetric gel dosimeter would be irradiated at the same time, the chemical instability would affect both sets of gel phantoms in the same way so that no uncertainty in the calibrated dose values would be induced. However, as there is realistically a lead time of approximately 1 h between irradiation of calibration vials and volumetric gel dosimeter phantom, a dose uncertainty of 0.5% (systematic maximum dose deviation with respect to the maximum measured dose) can be expected when all phantoms are scanned simultaneously 1 day after irradiation. Immediately after irradiation, a redistribution of monomers may take place as a result of diffusion. These monomers can react with long-living polymer radicals resulting in spatial uncertainties in the registered dose distribution (De Deene et al., 2001, 2002b; Vergote et al., 2004b). As long as the maximum dose is lower than 20 Gy in the PAGAT gel dosimeter, the estimated uncertainty is less than 1%.

As radiation-induced polymerization reactions are regulated by complex reaction schemes involving initiation, propagation, cyclization, transfer, and termination (Fuxman et al., 2005), it is not surprising that the dose-R_2 response of polymer gel dosimeters is dependent on dose rate. Indeed, a significant dependence of the dose-R_2 response on the dose rate has been found in some polymer gel dosimeters (De Deene et al., 2006a). In PAGAT gel dosimeters, the dose rate dependence is minimal and would account for less than 1% uncertainty (systematic maximum dose deviation with respect to the maximum measured dose) in the dose distribution. However, the influence of the dose rate–dependent dose-R_2 response on the accuracy of dose verification with polymer gel dosimetry (PGD) should not be underestimated as it may lead to a depth-dependent dose response in other types of polymer gel dosimeters.

With the introduction of "normoxic" polymer gel dosimeters (Fong et al., 2001), using an antioxidant to scavenge oxygen from the gel, it was expected that problems related to oxygen infiltration in the gel would be solved. Although the procedure of fabrication has been significantly simplified, the issue of oxygen infiltration after closing the dosimeter is still present (De Deene et al., 2002a, 2006a; Sedaghat et al., 2011a,b). This also restricts the use of different cast materials and restricts the storage time before irradiation. In PAGAT gel dosimeters, reduced dose-R_2 sensitivity is also found for increasing amounts of antioxidant.

To avoid permeation of oxygen through the wall of the container, Barex or glass is often used as phantom material. It should be noted that some types of glass may contain heavy metals. These specific glass materials may result in a stronger attenuation of the incident beam and may also result in beam hardening. Some caution is therefore advised in selecting glass as cast materials. It was shown by Monte Carlo simulations that the effect of a borosilicate glass wall and backscatter of a layer of air did not have a significant effect on the delivered dose in a test vial (Michael et al., 2000). From an experimental study where a small glass vial was inserted into a larger volumetric gel dosimeter phantom, it was concluded that the effect of the glass container wall accounts for less than 1% uncertainty in dose (systematic maximum dose deviation with respect to the maximum measured dose) (Vandecasteele and De Deene, 2013c).

5.5.4.2 Readout-related Sources of Uncertainty

Imaging artifacts can cause uncertainties in both dose and space. Imaging artifacts can be machine related or object related. Machine-related artifacts originate from imperfections in the scanning device whereas object-related artifacts originate from the dosimeter itself. The most important machine-related MRI artifacts are attributed to eddy currents, stimulated echoes, B_1 field heterogeneity, imperfect slice profiles, and standing waves. These machine-related artifacts may depend on the gel dosimeter shape and make it difficult to make general statements on the accuracy of the dosimeter. A larger phantom or a phantom with sharp edges may perform differently than a smaller cylindrical- or spherical-shaped phantom. Standing waves can severely deteriorate the dose distribution in dosimeters with specific shapes and spatial dimensions but may be almost completely absent if the dosimeter phantom has a slightly different shape. Object-related MRI artifacts are mainly attributed to a temperature drift during scanning or molecular self-diffusion. MRI artifacts that may contribute to type B uncertainties have been summarized in Table 5.2.

In optical imaging, dosimetric artifacts are related to reflection and absorption by the recipient walls, off-axis positioning of the recipient, variation of the laser output, and photodetector and light-scattering by both impurities in the matching fluid, container, and by the polymer (Oldham and Kim, 2004; Xu et al., 2004; Doran and Krstajić, 2006; Simon, 2010).

Geometrical distortions in MRI may originate from a nonuniform static magnetic field of the MR scanner, gradient nonlinearity, and eddy currents that result in time-varying magnetic field deviations. The magnitude of the geometrical distortion is dependent on magnitude of the magnetic field deviation and the sampling (pixel) bandwidth, which is inversely correlated with the readout time of each echo (frequency encoding window) in the pulse sequence. A shorter spin echo readout time will result in a larger pixel bandwidth and corresponding smaller distortion for similar magnetic field nonuniformity. However this will go at the cost of SNR, hence dose resolution. The magnetic field of the scanner is generally expressed by the manufacturer in parts per million (ppm) of the main magnetic field within a certain volume of interest. The displacement of a pixel in the frequency-encoding direction can be easily calculated as follows:

$$\Delta x = \frac{\gamma \,.\, \Delta B_0}{2\pi \cdot \mathrm{BW_{pix}}} \cdot \left(\frac{\mathrm{FOV}_x}{N_x} \right) \tag{5.9}$$

Table 5.2 Overview of Important MRI Artifacts Classified by Two Criteria

Geometrical Distortions		Dose Inaccuracies	
Machine-related	**Object-related**	**Machine-related**	**Object-related**
B_0-field nonuniformity	Susceptibility differences	Eddy currents	Temperature drift
Gradient nonlinearity	Chemical shifts	Stimulated echoes	Molecular self-diffusion
Eddy currents	–	B_1-field nonuniformity	Standing waves and dielectric effects
–	–	Slice profile	Susceptibility differences

Note: MRI, magnetic resonance imaging.

where Δx is the shift in the frequency-encoding direction in metric units, γ is the gyromagnetic constant ($\gamma = 2.675310^8 s^{-1}T^{-1}$), ΔB_0 is the magnetic field inhomogeneity, BW_{pix} is the bandwidth per pixel ($BW_{pix} = 1/T_{ro}$), FOV_x is the field-of-view in the frequency-encoding direction, and N_x is the number of pixels in the frequency-encoding direction. As an example, a 1-ppm field deviation on a 3-T MRI scanner corresponds to a field inhomogeneity $\Delta B_0 = 3\,\mu T$ which results in a shift of approximately one pixel for a bandwidth per pixel of $BW_{pix} = 130\,Hz$ ($T_{ro} \cong 8\,ms$). Note that a higher spatial resolution (smaller pixel size) will result in a smaller spatial shift in metric units.

Similarly, a shift in the phase-encoding direction in pixel units is given as follows:

$$\Delta y = \frac{\gamma}{2\pi} \cdot \Delta B_0 \cdot t_{ph} \cdot \left(\frac{FOV_y}{N_y} \right) \tag{5.10}$$

where t_{ph} is the duration of the phase-encoding gradient, FOV_y is the field-of-view in the phase-encoding direction, and N_y is the number of pixels in the phase-encoding direction. The time of the phase-encoding gradient is usually significantly smaller than the readout time (in the order of 1 ms) and the distortion can therefore be ignored in standard MRI sequences (in the case of R_2 mapping: a multiple spin echo sequence).

The shift in the slice selective direction is given as follows:

$$\Delta z = \frac{\Delta B_o}{G_z} \tag{5.11}$$

where G_z is the gradient strength of the slice selective imaging gradient. For the previous example with a magnetic field inhomogeneity of 1 ppm and a slice selective magnetic field gradient strength of $G_z = 3\,mT/m$ the slice shift will also be 1 mm. Several QA phantoms have been proposed to measure the magnitude of spatial deformations (Price et al., 1990; Schad et al., 1992; Michiels et al., 1994). For observing in-plane distortions a pin-cushion phantom is often used. To account for errors in the construction of the phantom, the phantom is first scanned using CT. By overlaying the MRIs of the pin-cushion phantom with the CT images, a distortion map can be derived.

To study the effect of heterogeneous tissue structures on the dose distribution, air cavities or other materials with different electron densities are inserted in the gel phantom. These materials most often also have a different magnetic susceptibility (χ). This will result in susceptibility-related distortions in the base images and in the final parametric images. The magnitude of the geometrical distortion can be described with the same set of Equations 5.9 through 5.11. The magnetic field nonuniformity can be computed by numerically solving the Maxwell equations (Li et al., 1996) or can be measured with MRI using a dedicated sequence (Park et al., 1988). Several compensation strategies have been developed to correct the image distortions (Holland et al., 2010). Susceptibility-induced deformations have been also observed when a brachytherapy source guiding catheter is inserted inside a gel phantom (De Deene et al., 2001). In humanoid-shaped phantoms, sharp boundaries may also result in susceptibility-induced magnetic field distortions. These can also be compensated by placing the humanoid-shaped phantom in a larger cylindrical recipient filled with paramagnetic water doped with contrast agent. The contrast agent serves to lower the T_2 value beyond the first echo time in the multiple spin echo sequence in order to compensate artifacts from turbulent flow in the container.

Eddy currents may also invoke dose inaccuracies. The magnetic fields induced by eddy currents experience a certain decay time. The succession of many imaging gradients in a multiple spin echo sequence may lead to an increase in the magnetic field during the start of the imaging pulse sequence. This may lead to slice profile imperfections, which have an effect on the excitation history and spin magnetization pathways of stimulated echo components (De Deene et al., 2000b). The result is a change in the measured T_2-decay curve. As the eddy currents (and especially the induced magnetic field offset) are dependent on the imaging direction, the disturbance of the excitation history of stimulated echoes is different and therefore also the measured R_2 values. Eddy current effects can be minimized by playing out a gradient train before the start of the actual multi-spin echo acquisition (De Deene et al., 2000b). Before setting up an actual gel dosimetry experiment, it is important to investigate a possible orientation dependence of the acquired R_2 values. It is advisable to use a calibration plot that is derived from scans acquired with the same imaging parameters. Without eddy current compensation and using a noncorresponding calibration set, the dose uncertainty may be up to 8%. It should be noted that this value is strongly dependent on the MRI scanner. The uncertainty can be reduced to 1% by applying a calibration that is derived from the same set of images with calibration samples placed around the volumetric polymer gel phantom.

The imperfect slice profiles and associated stimulated echoes also have an effect on the signal decay of echoes in a multi-spin echo sequence. It is typically found that the signal intensity in the first two base images deviate significantly from the expected monoexponential T_2 decay. The first two base images are therefore often ignored in the fitting, or corrected (Fransson et al., 1993).

Different RF coils can be used to scan the dosimeter. However, RF coils should be chosen very carefully as the radiofrequency field of the coils is only uniform within a limited region. An imperfect excitation may occur outside the homogeneous region. As a result, the spin magnetization history in the multiple spin echo sequence may deviate and stimulated echoes will be created (De Deene et al., 2000c). When scanned with a circularly polarized transmit/receive head coil, the R_2 map was found to be uniform only over an area of 120 cm in the center of the coil, whereas the R_2 values (apparent dose values) decreased considerably near the edges of the coil. The R_2 map of the homogeneous phantom is much more uniform when scanned with the body coil. However, measuring R_2 using the body coil as both transmitter and receiver goes at the cost of SNR. Modern MRI scanners are equipped with several receive-only coils. When still transmitting with the body coil, a good compromise between homogeneity and SNR can be obtained. It is advisable to always scan a homogeneous (blank) phantom to assess the homogeneity in the reconstructed R_2 maps before scanning an irradiated gel phantom. An excitation B_1 field inhomogeneity of about 10% gives rise to a dose error in the order of 1.5%. R_2 maps can be corrected for B_1 field nonuniformities by use of acquired B_1-field maps (Vergote et al., 2004a; Vandecasteele and

De Deene, 2013d). These corrections are particularly useful for large gel dosimeter phantoms. In MR scanners with a field strength of more than 1.5 T, standing waves may occur. These standing waves can result in large B_1-field nonuniformities. The creation of standing waves can be avoided by changing the dielectric properties (electrical conductivity and relative dielectric permittivity) of the polymer gel dosimeter, for example, by adding salt to the gel.

Some of the most significant contributions to type B dose uncertainty are temperature nonuniformity and temperature variations in the gel dosimeter during scanning (De Deene and De Wagter, 2001; Vandecasteele and De Deene, 2013a,d). The dose-R_2 response of polymer gel dosimeters is very temperature sensitive. A temperature difference of only 1°C results in a dose error of 0.8 Gy at a dose of 10 Gy in a (6% T; 50% C) PAGAT gel. Temperature recordings in the scanner room have shown temperature fluctuations in the order of 1°C over a time span of a few hours, the typical measurement time of a polymer gel dosimeter. It has also been found that the temperature fluctuation is different in the small calibration vials than in the larger volumetric dosimeter, which results in significant dose uncertainties after calibration. Temperature increases in the gel dosimeter may also occur as a result of the absorption of RF energy from the excitation and refocusing pulses (De Deene and De Wagter, 2001). To better control the temperature during scanning, an active temperature-controlled experimental setup can be applied where the gel dosimeters are wrapped in thermal pads that are perfused with doped water in order to stabilize and homogenize the temperature in the dosimeter phantoms (Vandecasteele and De Deene, 2013d). The imaging sequence can be rendered less sensitive for nonuniform temperature drift by implementing a centric k-space recording scheme (De Deene and De Wagter, 2001). Until now, efforts to decrease the temperature sensitivity of the dose response by increasing the dose sensitivity of the gel dosimeters have proven to be unsuccessful (Berndt et al., 2015).

By implementing adequate compensation strategies and a robust imaging protocol, the overall type B uncertainty on absolute dose measurements can be reduced to below 5% (systematic maximum dose deviation with respect to the maximum measured dose).

5.6 Guidelines for the Implementation of Polymer Gel Dosimetry

Although most medical physicists possess a basic understanding of the principles behind MRI, the optimization of quantitative imaging sequences and protocols is often perceived as the work of MRI experts. Also the dose-related physicochemical properties of the gel dosimeters have a significant influence on the overall accuracy of 3D polymer gel dosimetry verification. Therefore, we provide a set of guidelines that can help in setting up a 3D polymer gel dosimetry experiment.

1. *Choose a polymer gel dosimeter*: The choice of a polymer gel dosimeter may depend on the dynamic range, but it is important to realize that several polymer gel dosimeters may exhibit temporal and/or spatial instability, or the response can be dose rate dependent or temperature dependent. It is advisable not to compromise on accuracy and therefore

choose a polymer gel dosimeter with high stability and minimal dose rate and temperature dependence.

2. *Choose an MRI sequence*: Until now, a multi-spin echo sequence has been the preferred MRI pulse sequence of choice for water-equivalent gel dosimeters. For lung-equivalent low-density polymer gel foams, the method of choice is a magnetization transfer weighted sequence. It is crucial to the accuracy of readout that the performance of the MRI pulse sequence has been thoroughly tested and optimized.
3. *Choose optimal sequence parameters for the polymer gel dosimeter*: The SNR for a particular polymer gel dosimeter can be maximized by an optimal choice of sequence parameters. In the case of the multi-echo pulse sequence, this comes down to optimizing the total echo time interval that is equivalent to an optimal number of spin echoes. When the number of spin echoes is restricted by the pulse sequence the echo time intervals should be optimized.
4. *Construct a cast for the humanoid-shaped dosimeter phantom*: A cast should be made of a material with minimal oxygen permeability, such as Barex or glass. Barex is thermoformable and can be casted using a vacuum-forming machine.
5. *Fill the cast with a blank (nonirradiated) gel*: A crucial step in assuring the accuracy of the gel dosimetry experiment is to evaluate the uniformity of the quantitative MRIs. This can easily be done by filling the cast with a uniform blank gel. The gel can be doped with some MRI contrast agent to obtain a similar relaxation rate as the polymer gel dosimeter.
6. *Scan the blank phantom with CT and perform treatment planning on the scanned data set*: The blank phantom can then be used as a template for the TPS.
7. *Scan the blank phantom and determine uniformity, distortion, and SNR*: The uniformity, SNR, and possible distortion can be extracted from the quantitative MRIs of the blank phantom using methods described in the literature.
8. *Fabricate the polymer gel*: While fabricating the polymer gel dosimeter, it is important that calibration samples are subject to the same conditions as the volumetric dosimeter phantom.
9. *Perform the polymer gel dosimetry experiment*: Much care is required to keep the temperature in the gel system stable at all stages of the experiment and not to expose the polymer gel dosimeter to too much ambient light.
10. *Compare the gel dosimetry–derived dose maps with those of a TPS*: Evaluation tools such as gamma maps, DVHs, and difference maps are helpful tools in demonstrating any discrepancies, but need to be interpreted within their own rights.

5.7 Future Developments

Although the fact that polymer gel dosimeters with MRI readout have been thoroughly benchmarked and useful clinical applications have been demonstrated, 3D dosimetry with polymer gel dosimeters have seen only a moderate dissemination in the radiotherapy community. As described previously, a significant reason for this is the painstaking procedure that is needed in order to acquire reliable dose maps. A major contributing factor for the meticulous measurement procedure is the temperature sensitivity of the dose-R_2 response of polymer gel dosimeters. Future research may lead to the development of polymer gel dosimeters that are less sensitive to temperature variations. To solve this challenge in a targeted manner, it is important that the fundamental NMR-related mechanism behind the temperature–R_2 relation in hydrogel systems becomes well understood. Another promising direction is the development of lung-tissue-equivalent 3D gel dosimeters using either a hydrogel foam or Styrofoam microbeads to decrease the average density of the polymer gel dosimeter. The introduction of small cavities results in magnetic susceptibility differences, which lead to diffusive dispersion of the MRI signal. Other MRI contrast mechanisms than spin–spin (T_2) relaxation may need to be explored in order to obtain reliable dose maps in these low-density gel dosimeters. In many hospitals, easy access to an MRI scanner for scanning polymer gel dosimeters is restricted. The development of dedicated low-cost permanent magnet and high-temperature superconducting magnet MRI systems may have a positive impact on the application of polymer gel dosimeters as these MRI systems can be readily installed in radiotherapy centers.

References

Alexander P, Charlesby A and Ross M (1954) The degradation of solid polymethylmethacrylate by ionizing radiation. *Proc. R. Soc. London A.* 223: 392–404.

Baete SH et al. (2008) Microstructural analysis of foam by use of NMR R2 dispersion. *J. Magn. Reson.* 193: 286–296.

Baldock C et al. (2010) Polymer gel dosimetry. *Phys. Med. Biol.* 55: R1–R63.

Baldock C et al. (2001) Dose resolution in radiotherapy polymer gel dosimetry: Effect of echo spacing in MRI pulse sequence. *Phys. Med. Biol.* 46: 449–460.

Baldock C et al. (2000) Different monomers for improved characteristics of polymer gel dosimeters. *Australas. Phys. Eng. Sci. Med.* 23: 158.

Baldock C et al. (1998) Fourier transform Raman spectroscopy of polyacrylamide gels (PAGs) for radiation dosimetry. *Phys. Med. Biol.* 43: 3617–3627.

Baselga J et al. (1989) Polyacrylamide gels—process of network formation. *Eur. Polym. J.* 25: 477–480.

Baselga J et al. (1988) Polyacrylamide networks—sequence distribution of crosslinker. *Eur. Polym. J.* 24: 161–165.

Berndt B et al. (2015) Do saccharide doped PAGAT dosimeters increase accuracy? *J. Phys. Conf. Ser.* 573: 012029.

Bloembergen N (1948) *Nuclear Magnetic Relaxation.* The Hague: Nijhoff M; the Netherlands: Springer.

Bloembergen N, Purcell EM and Pound RV (1948) Relaxation effects in nuclear magnetic resonance absorption. *Phys. Rev.* 73: 679–712.

Boni AL (1961) A polyacrylamide gamma dosimeter. *Radiat. Res.* 14: 374–380.

Bosch P et al. (1998) Kinetic investigations on the photopolymerization of di- and tetrafunctional (meth)acrylic monomers in polymeric matrices. ESR and calorimetric studies. 1. Reactions under irradiation. *J. Polym. Sci. Pol. Chem.* 36: 2775–2783.

Bosi SG et al. (2009) Modelling optical scattering artefacts for varying pathlength in a gel dosimeter phantom. *Phys. Med. Biol.* 54: 275–283.

Brandrup J et al. (1999) *Polymer Handbook.* (New York, NY: Wiley).

Campbell WG et al. (2013) A prototype fan-beam optical CT scanner for 3D dosimetry. *Med. Phys.* 40: 061712.
Ceckler TL et al. (1992) Dynamic and chemical factors affecting water proton relaxation by macromolecules. *J. Magn. Res.* 98: 637–645.
Chain JN et al. (2011) Cosolvent-free polymer gel dosimeters with improved dose sensitivity and resolution for x-ray CT dose response. *Phys. Med. Biol.* 56: 2091–2102.
Chapiro A (1962) *Radiation Chemistry of Polymeric Systems.* (New York, NY: Interscience Publishers, Wiley).
Chernyshev AV et al. (1997) A mathematical model of dispersion radical polymerization kinetics. *J. Polym. Sci. Pol. Chem.* 35: 1799–1807.
De Deene Y (2002) Gel dosimetry for the dose verification of intensity modulated radiotherapy treatments. *Z. Med. Phys.* 12: 77–88.
De Deene Y (2004) Essential characteristics of polymer gel dosimeters. *J. Phys Conf. Ser.* 3: 34.
De Deene Y (2013) How to scan polymer gels with MRI? *J. Phys. Conf. Ser.* 444: 012003.
De Deene Y et al. (2002a) A basic study of some normoxic polymer gel dosimeters. *Phys. Med. Biol.* 47: 3441–3463.
De Deene Y and Baldock C (2002) Optimization of multiple spin-echo sequences for 3D polymer gel dosimetry. *Phys. Med. Biol.* 47: 3117–3141.
De Deene Y and De Wagter C (2001) Artefacts in multi-echo T2 imaging for high-precision gel dosimetry. III. Effects of temperature drift during scanning. *Phys. Med. Biol.* 46: 2697–2711.
De Deene Y et al. (2000a) Validation of MR-based polymer gel dosimetry as a preclinical three-dimensional verification tool in conformal radiotherapy. *Magn. Reson. Med.* 43: 116–125.
De Deene Y et al. (2000b) Artefacts in multi-echo T2 imaging for high-precision gel dosimetry. I. Analysis and compensation of eddy currents. *Phys. Med. Biol.* 45: 1807–1823.
De Deene Y et al. (2000c) Artefacts in multi-echo T-2 imaging for high-precision gel dosimetry. II. Analysis of B-1-field inhomogeneity. *Phys. Med. Biol.* 45: 1825–1839.
De Deene Y et al. (1998a) Three-dimensional dosimetry using polymer gel and magnetic resonance imaging applied to the verification of conformal radiation therapy in head-and-neck cancer. *Radiother. Oncol.* 48: 283–291.
De Deene Y et al. (2000d) An investigation of the chemical stability of a monomer/polymer gel dosimeter. *Phys. Med. Biol.* 45: 859–878.
De Deene Y, Pittomvils G and Visalatchi S (2007) The influence of cooling rate on the accuracy of normoxic polymer gel dosimeters. *Phys. Med. Biol.* 52: 2719–2728.
De Deene Y, Reynaert N and De Wagter C (2001) On the accuracy of monomer/polymer gel dosimetry in the proximity of a high-dose-rate 192Ir source. *Phys. Med. Biol.* 46: 2801–2825.
De Deene Y et al. (1998b) Mathematical analysis and experimental investigation of noise in quantitative magnetic resonance imaging applied in polymer gel dosimetry. *Signal Process.* 70: 85–101.
De Deene Y and Vandecasteele J (2013) On the reliability of 3D gel dosimetry. *J. Phys. Conf. Ser.* 444: 012015.

De Deene Y, Vandecasteele J and Vercauteren T (2013) Low-density polymer gel dosimeters for 3D radiation dosimetry in the thoracic region: A preliminary study. *J. Phys. Conf. Ser.* 444: 012026.

De Deene Y et al. (2002b) Dose-response stability and integrity of the dose distribution of various polymer gel dosimeters. *Phys. Med. Biol.* 47: 2459–2470.

De Deene Y et al. (2006a) The fundamental radiation properties of normoxic polymer gel dosimeters: A comparison between a methacrylic acid based gel and acrylamide based gels. *Phys. Med. Biol.* 51: 653–673.

De Deene Y et al. (2006b) Three dimensional radiation dosimetry in lung-equivalent regions by use of a radiation sensitive gel foam: Proof of principle. *Med. Phys.* 33: 2586–2597.

De Neve W et al. (1999) Clinical delivery of intensity modulated conformal radiotherapy for relapsed or second-primary head and neck cancer using a multileaf collimator with dynamic control. *Radiother. Oncol.* 50: 301–314.

Doran SJ and Krstajić N (2006) The history and principles of optical computed tomography for scanning 3-D radiation dosimeters. *J. Phys. Conf. Ser.* 56: 45–57.

Doran SJ et al. (2001) A CCD-based optical CT scanner for high-resolution 3D imaging of radiation dose distributions: Equipment specifications, optical simulations and preliminary results. *Phys. Med. Biol.* 46: 3191–3213.

Duthoy W et al. (2003) Whole abdominopelvic radiotherapy (WAPRT) using intensity-modulated arc therapy (IMAT): First clinical experience. *Int. J. Radiat. Oncol. Biol. Phys.* 57: 1019–1032.

Edzes HT and Samulski ET (1978) The measurement of cross-relaxation effects in the proton NMR spin-lattice relaxation of water in biological systems: Hydrated collagen and muscle. *J. Magn. Res.* 31: 207–229.

Farajollahi AR et al. (2000) The potential use of polymer gel dosimetry in boron neutron capture therapy. *Phys. Med. Biol.* 45: N9–N14.

Fernandes JP et al. (2008) Formaldehyde increases MAGIC gel dosimeter melting point and sensitivity. *Phys. Med. Biol.* 53: N53–N58.

Fong PM et al. (2001) Polymer gels for magnetic resonance imaging of radiation dose distributions at normal room atmosphere. *Phys. Med. Biol.* 46: 3105–3113.

Fransson A et al. (1993) Properties of the phase-alternating phase-shift (PHAPS) multiple spin-echo protocol in MRI: A study of the effects of imperfect RF pulses. *Magn. Reson. Imag.* 11: 771–784.

Freeman GR (1987) *Kinetics of Nonhomogeneous Processes.* (New York, NY: Wiley).

Fuxman AM, McAuley KB and Schreiner LJ (2003) Modeling of free-radical cross-linking copolymerization of acrylamide and N,N'-methylenebis(acrylamide) for radiation dosimetry. *Macromol. Theor. Simul.* 12: 647–662.

Fuxman AM, McAuley KB and Schreiner LJ (2005) Modelling of polyacrylamide gel dosimeters with spatially non-uniform radiation dose distributions. *Chem. Eng. Sci.* 60: 1277–1293.

Gelfi C and Righetti PG (1981) Polymerization kinetics of polyacrylamide gels. I. Effect of different cross-linkers. *Electrophoresis* 2: 213–219.

Gochberg DF, Fong PM and Gore JC (2001) Studies of magnetization transfer and relaxation in irradiated polymer gels—interpretation of MRI-based dosimetry. *Phys. Med. Biol.* 46: 799–811.

Gochberg DF et al. (1998) The role of specific side groups and pH in magnetization transfer in polymers. *J. Magn. Res.* 131: 191–198.

Gore JC et al.(1996) Radiation dose distributions in three dimensions from tomographic optical density scanning of polymer gels. I. Development of an optical scanner. *Phys. Med. Biol.* 41: 2695–2704.

Gustavsson H et al. (2004a) Development and optimization of a 2-hydroxyethylacrylate MRI polymer gel dosimeter. *Phys. Med. Biol.* 49: 227–241.

Gustavsson H et al. (2004b) Linear energy transfer dependence of a normoxic polymer gel dosimeter investigated using proton beam absorbed dose measurements. *Phys. Med. Biol.* 49: 3847–3855.

Heufelder J et al. (2003) Use of BANG polymer gel for dose measurements in a 68 MeV proton beam. *Med. Phys.* 30: 1235–1240.

Hill B, Venning AJ and Baldock C (2005) A preliminary study of the novel application of normoxic polymer gel dosimeters for the measurement of CTDI on diagnostic x-ray CT scanners. *Med. Phys.* 32: 1589–1597.

Hilts M (2006) X-ray computed tomography imaging of polymer gel dosimeters. *J. Phys. Conf. Ser.* 56: 95–107.

Hilts M et al. (2000) Polymer gel dosimetry using x-ray computed tomography: A feasibility study. *Phys. Med. Biol.* 45: 2559–2571.

Hoecker FE and Watkins IW (1958) Radiation polymerization dosimetry. *Int. J. Appl. Radiat. Isot.* 3: 31–35.

Holland D, Kuperman JM and Dale AM (2010) Efficient correction of inhomogeneous static magnetic field-induced distortion in echo planar imaging. *Neuroimage* 50: 175–183.

Hsieh BT, Wu J and Chang YJ (2013) Verification on the dose profile variation of a 3-D—NIPAM Polymer Gel Dosimeter. *IEEE Trans. Nucl. Sci.* 60: 560–565.

Hurley C et al. (2003) The effect of water molecular self-diffusion on quantitative high-resolution MRI polymer gel dosimetry. *Phys. Med. Biol.* 48: 3043–3058.

Ibbott GS et al. (1997) Three-dimensional visualization and measurement of conformal dose distributions using magnetic resonance imaging of BANG polymer gel dosimeters. *Int. J. Radiat. Oncol. Biol. Phys.* 38: 1097–1103.

Islam KT et al. (2003) Initial evaluation of commercial optical CT-based 3D gel dosimeter. *Med. Phys.* 30: 2159–2168.

Jirasek AI et al. (2001) Characterization of monomer/crosslinker consumption and polymer formation observed in FT-Raman spectra of irradiated polyacrylamide gels. *Phys. Med. Biol.* 46: 151–165.

Jirasek A and Duzenli C (2002) Relative effectiveness of polyacrylamide gel dosimeters applied to proton beams: Fourier transform Raman observations and track structure calculations. *Med. Phys.* 29: 569–577.

Jirasek A and Hilts M (2014) Dose calibration optimization and error propagation in polymer gel dosimetry. *Phys. Med. Biol.* 59: 597–614.

Jirasek A, Hilts M, Berman A and McAuley KB (2009) Effects of glycerol co-solvent on the rate and form of polymer gel dose response. *Phys. Med. Biol.* 54: 907–918.

Jirasek A, Hilts M and McAuley KB (2010) Polymer gel dosimeters with enhanced sensitivity for use in x-ray CT polymer gel dosimetry. *Phys. Med. Biol.* 55: 5269–5281.

Jirasek A et al. (2006) Investigation of tetrakis hydroxymethyl phosphonium chloride as an antioxidant for use in x-ray computed tomography polyacrylamide gel dosimetry. *Phys. Med. Biol.* 51: 1891–1906.

Jirasek AI and Duzenli C (2001) Effects of crosslinker fraction in polymer gel dosimeters using FT Raman spectroscopy. *Phys. Med. Biol.* 46: 1949–1961.

Kao T-J et al. (2008) A versatile high-permittivity phantom for EIT. *IEEE Trans. Biomed. Eng.* 55: 2601–2607.
Karlsson A et al. (2007) Dose integration characteristics in normoxic polymer gel dosimetry investigated using sequential beam irradiation. *Phys. Med. Biol.* 52: 4697–4706.
Kelly RG, Jordan KJ and Battista JJ (1998) Optical CT reconstruction of 3D dose distributions using the ferrous-benzoic-xylenol (FBX) gel dosimeter. *Med. Phys.* 25: 1741–1750.
Kennan RP et al. (1992) Hydrodynamic effects and cross relaxation in cross linked polymer gels. In *Proceedings of the International Society for Magnetic Resonance in Medicine*, Issue S1, p. 1316; In *Eleventh Annual Scientific Meeting*, August 8–14, 1992, Berlin, Germany. Hoboken, NJ: Wiley.
Kennan RP et al. (1996) The effects of cross-link density and chemical exchange on magnetization transfer in polyacrylamide gels. *J. Magn. Reson. Ser.* B 110: 267–277.
Khadem-Abolfazli M et al. (2013) Dose enhancement effect of gold nanoparticles on MAGICA polymer gel in megavoltage radiation therapy. *Int. J. Radiat. Res.* 11: 55–61.
Koeva VI et al. (2009) Preliminary investigation of the NMR, optical and x-ray CT dose-response of polymer gel dosimeters incorporating cosolvents to improve dose sensitivity. *Phys. Med. Biol.* 54: 2779–2790.
Kozicki M (2011) How do monomeric components of a polymer gel dosimeter respond to ionising radiation: A steady-state radiolysis towards preparation of a 3D polymer gel dosimeter. *Radiat. Phys. Chem.* 80: 1419–1436.
Krstajic N and Doran SJ (2006) Focusing optics of a parallel beam CCD optical tomography apparatus for 3D radiation gel dosimetry. *Phys. Med. Biol.* 51: 2055–2075.
Krstajic N and Doran SJ (2007) Fast laser scanning optical-CT apparatus for 3D radiation dosimetry. *Phys. Med. Biol.* 52: N257–N263.
Lepage M et al. (2001a) Dose resolution optimization of polymer gel dosimeters using different monomers. *Phys. Med. Biol.* 46: 2665–2680.
Lepage M et al. (2001b) Modelling of post-irradiation events in polymer gel dosimeters. *Phys. Med. Biol.* 46: 2827–2839.
Lepage M et al. (2001c) The relationship between radiation-induced chemical processes and transverse relaxation times in polymer gel dosimeters. *Phys. Med. Biol.* 46: 1061–1074.
Lepage M et al. (2001d) C-13-NMR, H-1-NMR, and FT-Raman study of radiation-induced modifications in radiation dosimetry polymer gels. *J. Appl. Polym. Sci.* 79: 1572–1581.
Lepage M et al. (2002) Magnetization transfer imaging for polymer gel dosimetry. *Phys. Med. Biol.* 47: 1881–1890.
Li S et al. (1996) Three-dimensional mapping of the static magnetic field inside the human head. *Magn. Reson. Med.* 36: 705–714.
Lopatiuk-Tirpak O et al. (2008) Performance evaluation of an improved optical computed tomography polymer gel dosimeter system for 3D dose verification of static and dynamic phantom deliveries. *Med. Phys.* 35: 3847–3859.
Low DA et al. (1998) A technique for the quantitative evaluation of dose distributions. *Med. Phys.* 25: 656–661.
Maryanski MJ et al. (1994) Magnetic resonance imaging of radiation dose distributions using a polymer-gel dosimeter. *Phys. Med. Biol.* 39: 1437–1455.

Maryanski MJ, Audet C and Gore JC (1997) Effects of crosslinking and temperature on the dose response of a BANG polymer gel dosimeter. *Phys. Med. Biol.* 42: 303–311.

Maryanski MJ et al.(1993) NMR relaxation enhancement in gels polymerized and cross-linked by ionizing radiation: A new approach to 3D dosimetry by MRI. *Magn. Reson. Imag.* 11: 253–258.

Maryanski MJ et al. (1996a) Radiation therapy dosimetry using magnetic resonance imaging of polymer gels. *Med. Phys.* 23: 699–705.

Maryanski MJ and Ranade MK (2001) Laser microbeam CT scanning of dosimetry gels. In *Proceeding of SPIE , Volume 4320, Medical Imaging 2001: Physics of Medical Imaging, 17–22 February 2001, San Diego, CA, United States: SPIE. DigitalLibrary*, pp. 764–774.

Maryanski MJ, Zastavker YZ and Gore JC (1996b) Radiation dose distributions in three dimensions from tomographic optical density scanning of polymer gels. II. Optical properties of the BANG polymer gel. *Phys. Med. Biol.* 41: 2705–2717.

Mather ML and Baldock C (2003) Ultrasound tomography imaging of radiation dose distributions in polymer gel dosimeters: Preliminary study. *Med. Phys.* 30: 2140–2148.

Mather ML, Charles PH and Baldock C (2003) Measurement of ultrasonic attenuation coefficient in polymer gel dosimeters. *Phys. Med. Biol.* 48: N269–N275.

Mather ML et al. (2002) Investigation of ultrasonic properties of PAG and MAGIC polymer gel dosimeters. *Phys. Med. Biol.* 47: 4397–4409.

McConville P, Whittaker MK and Pope JM (2002) Water and polymer mobility in hydrogel biomaterials quantified by H-1 NMR: A simple model describing both T-1 and T-2 relaxation. *Macromolecules* 35: 6961–6969.

Meeks SL et al. (1999) Image registration of BANG gel dose maps for quantitative dosimetry verification. *Int. J. Radiat. Oncol. Biol. Phys.* 43: 1135–1141.

Michael GJ et al. (2000) Effects of glass and backscatter on measurement of absorbed dose in polyacrylamide gel (PAG) dosimeters. *Phys. Med. Biol.* 45: N133–N138.

Michiels J et al. (1994) On the problem of geometric distortion in magnetic resonance images for stereotactic neurosurgery. *Magn. Reson. Imag.* 12: 749–765.

Murphy PS et al. (2000) Proton spectroscopic imaging of polyacrylamide gel dosimeters for absolute radiation dosimetry. *Phys. Med. Biol.* 45: 835–845.

Oldham M and Kim L (2004) Optical-CT gel-dosimetry II: Optical artifacts and geometrical distortion. *Med. Phys.* 31: 1093–1104.

Oldham M et al. (2001) High resolution gel-dosimetry by optical-CT and MR scanning. *Med. Phys.* 28: 1436–1445.

Oldham M et al. (2003) Optical-CT gel-dosimetry I: Basic investigations. *Med. Phys.* 30: 623–634.

Olding T, Holmes O and Schreiner LJ (2010) Cone beam optical computed tomography for gel dosimetry I: Scanner characterization. *Phys. Med. Biol.* 55: 2819–2840.

Oudry J et al. (2009) Cross-validation of magnetic resonance elastography and ultrasound-based transient elastography: A preliminary phantom study. *Magn. Reson. Imag.* 30: 1145–1150.

Panajkar MS, Guha SN and Gopinathan C (1995) Reactions of hydrated electrons with N,N'-methylenebisacrylamide in aqueous-solution—A pulse-radiolysis study. *J. Macromol. Sci. Pure.* A32: 143–156.

Pappas E et al. (1999) A new polymer gel for magnetic resonance imaging (MRI) radiation dosimetry. *Phys. Med. Biol.* 44: 2677–2684.

Park HW, Ro YM and Cho ZH (1988) Measurement of the magnetic susceptibility effect in high-field NMR imaging. *Phys. Med. Biol.* 33: 339–349.

Price RR et al. (1990) Quality assurance methods and phantoms for magnetic resonance imaging: Report of AAPM nuclear magnetic resonance Task Group No. 1. *Med. Phys.* 17: 287–295.

Ramm D et al. (2012) Optical CT scanner for in-air readout of gels for external radiation beam 3D dosimetry. *Phys. Med. Biol.* 57: 3853–3868.

Ramm U et al. (2000) Three-dimensional BANG gel dosimetry in conformal carbon ion radiotherapy. *Phys. Med. Biol.* 45: N95–N102.

Rintoul L, Lepage M and Baldock C (2003) Radiation dose distribution in polymer gels by Raman spectroscopy. *Appl. Spectrosc.* 57: 51–57.

Schad LR (1992) Correction of spatial distortion in magnetic resonance angiography for radiosurgical treatment planning of cerebral arteriovenous malformations. *Magn. Reson. Imag.* 10: 609–621.

Sedaghat M, Bujold R and Lepage M (2011a) Investigating potential physicochemical errors in polymer gel dosimeters. *Phys. Med. Biol.* 56: 6083–6107.

Sedaghat M, Bujold R and Lepage M (2011b) Severe dose inaccuracies caused by an oxygen-antioxidant imbalance in normoxic polymer gel dosimeters. *Phys. Med. Biol.* 56: 601–625.

Senden RJ et al. (2006) Polymer gel dosimeters with reduced toxicity: A preliminary investigation of the NMR and optical dose-response using different monomers. *Phys. Med. Biol.* 51: 3301–3314.

Simon JD (2010) Imaging and 3-D dosimetry: Top tips for MRI and optical CT. *J. Phys. Conf. Ser.* 250: 012086.

Spinks JWT and Woods RJ (1976) *An Introduction to Radiation Chemistry.* (New York, NY: Wiley).

Steve BJ et al. (2006) On dose distribution comparison. *Phys. Med. Biol.* 51: 759–776.

Swallow AJ (1973) *Radiation Chemistry: An Introduction.* (London, UK: Longman).

Tobita H and Hamielec AE (1990) Cross-linking kinetics in polyacrylamide networks. *Polymer* 31: 1546–1552.

Tobita H and Hamielec AE (1992) Control of network structure in free-radical cross-linking copolymerization. *Polymer* 33: 3647–3657.

Trapp JV et al. (2002) Attenuation of diagnostic energy photons by polymer gel dosimeters. *Phys. Med. Biol.* 47: 4247–4258.

Vandecasteele J and De Deene Y (2009) Optimization of a fast optical CT scanner for nPAG gel dosimetry. *J. Phys. Conf. Ser.* 164: 012024.

Vandecasteele J and De Deene Y (2013a) Evaluation of radiochromic gel dosimetry and polymer gel dosimetry in a clinical dose verification. *Phys. Med. Biol.* 58: 6241–6262.

Vandecasteele J and De Deene Y (2013b) On the validity of 3D polymer gel dosimetry. I. Reproducibility study. *Phys. Med. Biol.* 58: 19–42.

Vandecasteele J and De Deene Y (2013c) On the validity of 3D polymer gel dosimetry. II. Physico-chemical effects. *Phys. Med. Biol.* 58: 43–61.

Vandecasteele J and De Deene Y (2013d) On the validity of 3D polymer gel dosimetry. III. MRI-related error sources. *Phys. Med. Biol.* 58: 63–85.
Vergote K et al. (2003) Application of monomer/polymer gel dosimetry to study the effects of tissue inhomogeneities on intensity-modulated radiation therapy (IMRT) dose distributions. *Radiother. Oncol.* 67: 119–128.
Vergote K et al. (2004a) Validation and application of polymer gel dosimetry for the dose verification of an intensity-modulated arc therapy (IMAT) treatment. *Phys. Med. Biol.* 49: 287–305.
Vergote K et al.(2004b) On the relation between the spatial dose integrity and the temporal instability of polymer gel dosimeters. *Phys. Med. Biol.* 49: 4507–4522.
Wolodzko JG, Marsden C and Appleby A (1999) CCD imaging for optical tomography of gel radiation dosimeters. *Med. Phys.* 26: 2508–2513.
Xu Y, Wuu CS and Maryanski MJ (2004) Performance of a commercial optical CT scanner and polymer gel dosimeters for 3-D dose verification. *Med. Phys.* 31: 3024–3033.
Zhang Q et al. (2002) Modeling of dose and dose rate dependence of changes in irradiated polyacrylamide gel dosimeters. *Med. Phys.* 29: 1208–1208.
Zhu XP et al. (2010) Improved MAGIC gel for higher sensitivity and elemental tissue equivalent 3D dosimetry. *Med. Phys.* 37: 183–188.
Zimmerman JR and Brittin WE (1957) Nuclear magnetic resonance studies in multiple phase systems: Lifetime of a water molecule in an adsorbing phase on silica gel. *J. Phys. Chem.* 61: 1328–1333.

6 Radiochromic 3D Detectors

Mark Oldham, Titania Juang, and Suk Whan (Paul) Yoon

6.1 Introduction

Radiation therapy (RT) is a principle treatment modality for many cancers and other diseases. Over half of all patients diagnosed with cancer will receive some form of RT as part of their treatment. In recent years, the sophistication and complexity of these treatments have increased dramatically. Two engines of innovation in particular can be discerned: (1) the discovery and development of mathematical tools to optimize the radiation fluence for a desired dose prescription and normal tissue tolerance (Webb, 1989; Bortfeld et al., 1994) and (2) associated computer control and electromechanical hardware developments that can implement the delivery of such precise intensity-modulated radiation therapy (IMRT) treatments (Convery and Rosenbloom, 1992; Stein et al., 1994). An example of a modern IMRT treatment for head-and-neck cancer is shown in Figure 6.1. This figure shows the extraordinary potential of IMRT to deliver complex distributions of dose that conform to a highly irregularly shaped lesion in the patient.

There are many assumptions in the mathematical modeling leading to the virtual computer representation of the patient's dose distribution, such as shown in Figure 6.1. The limitations of these models, and the potential for errors in RT treatment, have recently received significant attention (Ibbott, 2010; Moran et al., 2011a). In 2010, partly in response to several catastrophic events (e.g., Bogdanich, 2010a,b), the US Congress held hearings on the topic of how to improve safety in RT (Herman, 2011). Independently, the Imaging and Radiation Oncology Core Houston (IROC Houston, formerly the Radiological Physics Center [RPC])

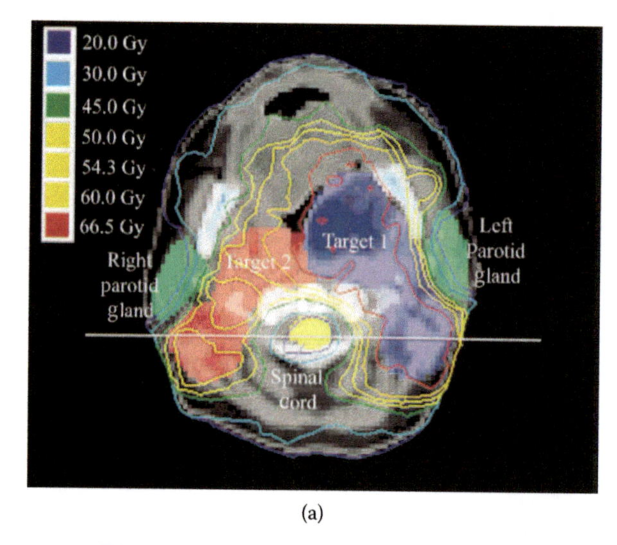

(a)

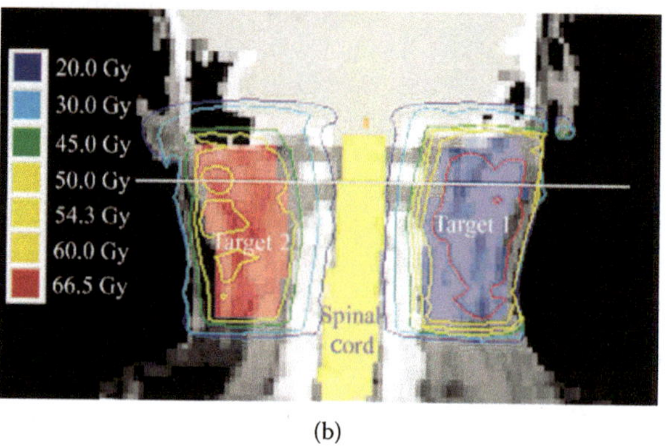

(b)

Figure 6.1

This figure shows a modern precision radiation therapy treatment plan. This precise scalloping of isodose lines around multiple targets (prescribed doses of 66.5 and 54.3 Gy, respectively), with conformal avoidance of the spinal cord and parotid glands, only became achievable with the advent of intensity-modulated therapy techniques. (a) Transverse section and (b) coronal section. White lines indicate slice locations. (Reprinted from Intensity-Modulated Radiotherapy Collaborative Working Group, *Int. J. Radiat. Oncol. Biol. Phys.*, 51, 880–914, 2001.)

reported unacceptably high failure rates for dosimetry credentialing in national clinical trials for several IMRT treatments: 18% for head-and-neck (7%/4 mm criteria) (Molineu et al., 2013) and 32% for spine (5%/3 mm criteria) (Ibbott, 2010). These failures were detected on a pair of orthogonal two-dimensional (2D) film planes normalized with thermoluminescent dosimeter (TLD) point measurements.

The concerns outlined earlier have led many to recognize an urgent need to radically strengthen the foundations of quality assurance (QA) in RT. There are, of course, many aspects to a comprehensive QA program (Moran et al., 2011b). A foundational aspect, however, is comprehensive end-to-end verification of the delivered dose in phantoms, normally performed during extensive commissioning measurements when the accuracy of a new technique is established prior to implementation in the clinic. Historically, in the absence of viable three-dimensional (3D) dosimetry systems, such measurements have typically been made at selected planes and points using film, TLDs, diodes, and ion chambers (Baldock et al., 2010). These low-sampling methods can catch planning and delivery errors, but there remains the possibility of undetected failure in regions that were not measured. This chapter focuses on a review of the development and current state-of-the-art of a suite of full-3D dosimetry techniques that can provide a more comprehensive solution to the problem of verification of complex RT treatments and 3D dose measurement in general. Also of interest (although not discussed further here) are *in vivo* pseudo-3D methods of transit dosimetry (e.g., Mans et al., 2010; Mijnheer et al., 2013). Although great strides have been made, the challenge of obtaining high accuracy has led to the clinical use of transit dosimetry more as an online check for gross errors as discussed in detail in Chapters 7 and 18.

6.2 Methods of 3D Dosimetry

The term 3D dosimetry is in common use, but is rather nonspecific and needs definition as pointed out in the introduction of this book. It is possible to obtain measurements in all three dimensions using many traditional dosimetry methods, such as film, TLDs, ion chambers, and diodes. For instance, in early IMRT verification studies, semi-3D dosimetry was achieved using stacks of film in anthropomorphic phantoms (Bortfeld et al., 1994; Bortfeld, 2006). Isotropic and high-resolution 3D full-3D dosimetry with these methods, however, is not readily achievable without a prohibitive amount of effort. Thus, the term "3D dosimetry" will be restricted in this chapter to refer solely to full-3D dosimetry systems that can directly perform high-resolution (2mm or less) isotropic dosimetry.

Some discussion of this definition was given in the form of the Resolution-Time-Accuracy-Precision (RTAP) criteria (Oldham et al., 2001). The RTAP represents a performance or capability goal for a 3D dosimetry verification system, to be viable for clinical use. Such a system should be able to deliver a 3D dosimetric analysis of a treatment plan with 1-mm isotropic spatial resolution, within 1 hour, with accuracy within 3% of the true value, and with 1% (1 standard deviation) precision. This definition rules out semi-3D dosimetry systems as the Delta^4 and the ArcCHECK, which have planes or surfaces of point detectors and interpolate measured doses to 3D (Letourneau et al., 2004, 2009; Feygelman et al., 2011). These approaches represent commercial and innovative momentum toward the ideal of a truly comprehensive 3D dosimetry system and have advantages in terms of efficiency and convenience, but are not discussed further here. More information about these systems can be found in Chapters 8 and 17.

Various full-3D dosimetry systems are in current use as shown in Figure 6.2. These systems are all chemical dosimetry systems, where a uniform mass of material exhibits a physical response to radiation that can be quantified by an imaging readout system. The vast majority of work to date has been performed with three classes of materials: polymer gels, radiochromic gels, and radiochromic plastics. Typical gel agents are water-based components like gelatin or agarose, while plastics have included polyurethane and silicone. These three material classes and their associated imaging methods for readout are described in the following sections. The term radiochromic indicates a material or substance that changes

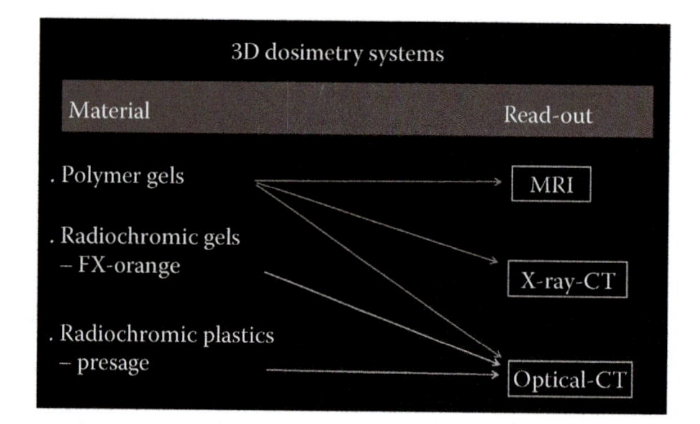

Figure 6.2

Three-dimensional (3D) dosimetry systems in current use are capable of directly measuring 3D dose isotropically with high resolution (2 mm or better) and approach the Resolution-Time-Accuracy-Precision capability criteria. CT; computed tomography; MRI; magnetic resonance imaging.

color (optical density [OD]) when exposed to ionizing radiation. Although there is a long history of radiochromic fluids and films (Niroomand-Rad et al., 1998), this review focuses on 3D radiochromic dosimetry methods. Recently, scintillation (Beddar et al., 2009) and Cherenkov radiation (Glaser et al., 2013) systems have been proposed as alternative methods for 3D dosimetry. These systems are in the preliminary phase of development and are not discussed further here, but some information can be found in Chapter 8.

6.2.1 Polymer Gel Dosimetry

The discovery of radiation-sensitive polymer gels initiated the field of 3D dosimetry, and is discussed at length in the accompanying Chapter 5. Here, we briefly highlight these origins in order to contrast with radiochromic techniques, which are the focus of this chapter. The first 3D dosimetry work employed agarose- or gelatin-based gels doped with acrylamide (AAm) monomers and *N,N*-methylenebisacrylamide (Bis) cross-linkers (Maryanski et al., 1993, 1994). These polyacrylamide gels (PAGs, commercially available as BANG [bis-acrylamide-nitrogen-gelatin] polymer gel dosimeters*) were highly tissue equivalent and exhibited a linear dose response (Baldock et al., 2010). The mechanism of radiosensitivity is radiation-induced free radical polymerization (Figure 6.3). Free radicals generated from water radiolysis in the gel during irradiation attack either AAm or Bis monomers to create reactive ends with an unpaired valence electron, which in turn attacks carbon–carbon double bonds (C=C) of a neighboring monomer in a polymerizing or cross-linking reaction. This transfers the reactive end to the newly incorporated monomer, which can then attack C=C of another monomer, forming a polymer radical. While AAm monomer only has one C=C, Bis has two C=C; hence, incorporating Bis in the growing polymer radical chain can result in cross-linking, where multiple adjacent polymer chains can link up with each other and form a network (Figure 6.3c). The series of radical reactions creates polymer microparticles until a termination reaction occurs. In the termination reaction, two radical species combine to form a stable bond and end the radical reaction.

* MGS Research Inc., Madison, CT.

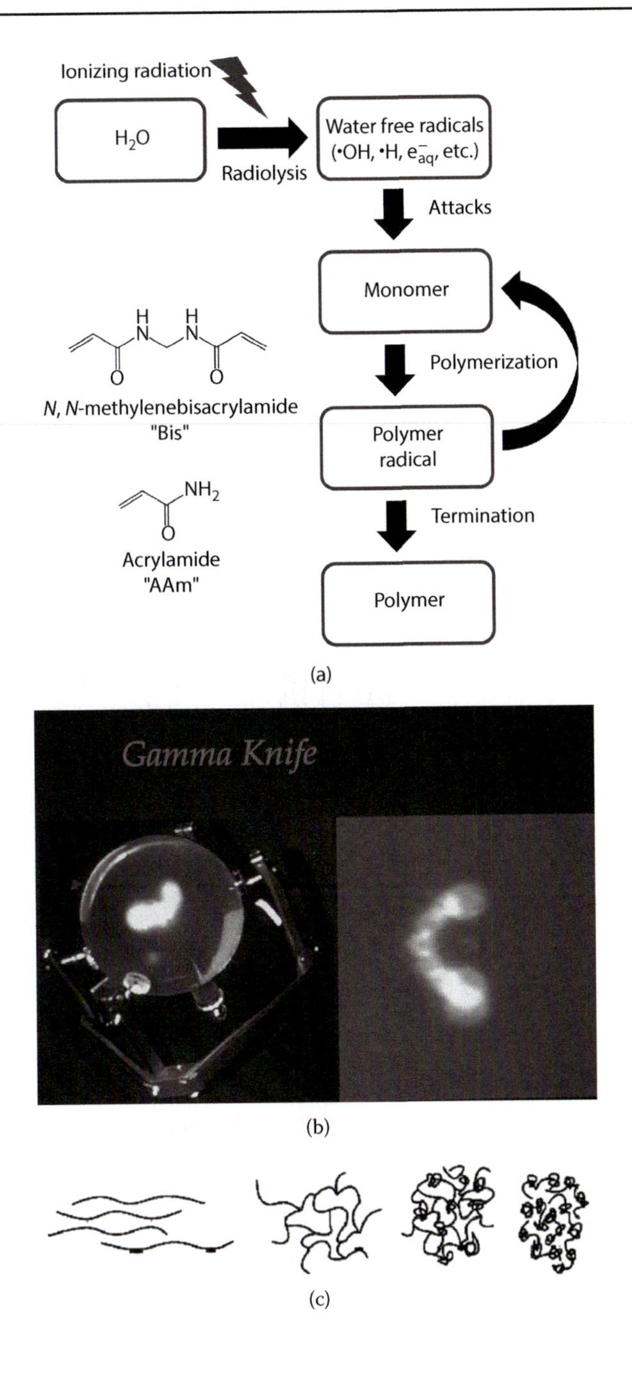

Figure 6.3

Polymer gel 3D dosimetry. (a) Schematic indication of the mechanisms of radiation-induced polymerization. (Adapted from Lepage M and Jordan K, *J. Phys. Conf. Ser.*, 250, 012055, 2010.) (b) Photograph (left) and magnetic resonance image (right) of a BANG polymer gel in a glass sphere after irradiation with a Gamma Knife treatment. (c) Schematic illustration of the process of how resultant polymer microparticles depend on gel composition (Baldock et al., 2010). The four images represent postirradiation gel structure corresponding to: monomer only (AAm); low initial Bis cross-linker fraction; high Bis fraction; and only Bis cross-linker (no monomer).

The polymerization reaction continues well after irradiation has finished, and dosimeters are often imaged ~24 hours postirradiation when the gel is relatively stable. Polymerization creates a cloudy or milky visual appearance within the polyacrylamide gel (PAG) as the result of light scattering from the radiation-induced polymer microparticles. The density of particles is proportional to the locally absorbed dose. The presence of these particles affects at least three physical aspects of the gel, which can be quantitatively imaged as relaxivity of water protons (magnetic resonance imaging [MRI]), attenuation of visible light (optical computed tomography [CT]), and physical density (x-ray CT). These methods are described in Section 6.2.4.

There have been several refinements of the polymer gel material system since first inception. These refinements have generally targeted improvements in several areas:

1. *Decreased sensitivity to oxygen:* The sensitivity of polymer gels decreases drastically when oxygen is present in the gel because of premature termination of radicalized polymer ends. The practical implications are that gels must be made in a deoxygenated environment within a glove box, and then set in an oxygen-impermeable container to prevent oxygen leaching into the gel with time. Investigators have typically used glass or Barex plastics. An alternative approach is to incorporate an oxygen-scavenging compound in the gel, such as ascorbic acid with copper. Such a "MAGIC" (Methacrylic and Ascorbic acid in Gelatin Initiated by Copper) gel was reported by Fong et al. (2001), and was shown to be compatible with normoxic conditions in the lab, without specialized equipment.
2. *Increased sensitivity:* The relative concentration of the cross-linker, Bis, in PAG dosimeters determines the dose sensitivity of the gel dosimeter; specifically, the PAG dosimeter dose sensitivity increases up until ~ 50% (by weight) of the monomers are cross-linkers (Bis), but decreases if a greater amount of Bis is used (Baldock et al., 2010). In MRI dose sensitivity studies (Lepage et al., 2001a), substituting the functional group $CONH_3$ on AAm with other functional groups (such as the hydroxyl group, which makes acrylic acid) drastically altered the dose sensitivity of the polymer gel, suggesting that the chemical structure can be used to alter dose sensitivity.
3. *Lower density lung-equivalent gels:* PAG gels are the only 3D dosimetry material to date that has proven amenable to manufacture in low density for lung-equivalent dosimeters (De Deene et al., 2006).

6.2.2 Fricke-Based Gel Dosimetry

The Fricke dosimeter, or ferrous sulfate dosimeter, has a long history of use as both absolute and relative dosimetry systems (Fricke and Morse, 1927; Schreiner, 2004). The standard Fricke dosimeter consists of small vials of ferrous solution containing Fe^{2+}. The high water content and low atomic number (Z) constituents render Fricke dosimeters tissue equivalent and water equivalent even for lower energy radiation fields (Kron et al., 1993). When irradiated, Fe^{2+} is oxidized to Fe^{3+}, and the quantification of the proportion of Fe^{3+}, which is proportional to absorbed dose, is determined by spectrophotometry (Scharf and Lee, 1962). Challenging but manageable aspects of the method are a temperature-sensitive reaction and limited dose-response sensitivity caused by the modest OD change of the Fe^{3+} ions at the probing light wavelength (224 nm). Imaging at such a short wavelength is feasible for optical cuvettes (path length 1 cm), but is not feasible for scaling up for large gel volumes due to overattenuation of the light.

3D Fricke-based dosimetry was first proposed by Gore and colleagues who showed that ferric (Fe^{3+}) ions have a greater influence on proton relaxation times than ferrous (Fe^{+2}) ions, enabling Fricke dosimetry by MRI (Gore et al., 1984; Appleby and Leghrouz, 1991). These works focused on imaging the dose distribution in the gel using MRI. The spin–lattice relaxation rate R_1 was found to vary linearly with dose (Podgorsak and Schreiner, 1992). R_1 relaxivity measurement was preferred to spin–spin relaxivity (R_2) for Fricke gels because of higher dynamic range (the unirradiated Fricke dosimeter exhibits low R_1) (De Deene, 2010). Diffusion of Fe^{3+} ions in the gel was identified as a key practical limitation, and requires the 3D distribution to be imaged within 1 hour (Chu et al., 2000; Tseng et al., 2002; de Pasquale et al., 2006). Early attempts to stabilize the Fricke solution in gels were performed by a number of investigators (Gore et al., 1984; Olsson et al., 1990, 1992; Hazle et al., 1991; Kron et al., 1993; Schreiner et al., 1994). Early work reported low spatial resolution (thick slices) owing to the challenge of obtaining low noise with high spatial resolution using MRI. Recent work looking at fast 3D MRI sequences, however, has shown promising results: 1 mm isotropic scans in a 20-minute imaging session (Cho et al., 2013).

The need for access to MRI machines is a restrictive aspect for many medical centers. Much effort was therefore devoted to developing Fricke gels that could be imaged by cheaper benchtop optical systems. The addition of a metal ion indicator, such as xylenol orange, leads to a visible color change in the presence of ferric (Fe^{3+}) ions (Gupta et al., 1982). This development allows the radiochromic response to be optically imaged over larger gel volumes using light of a longer wavelength (550 nm) and greater penetrating power. Studies by Appleby and Kelly demonstrated the feasibility of 3D optical CT dosimetry of Fricke–benzoic–xylenol (FBX) gels (see Figure 6.4) (Appleby and Leghrouz, 1991; Kelly et al., 1998). Their system used a scanning laser beam and was able to achieve submillimeter spatial resolution and an accuracy of reconstructed attenuation values within 2%, corresponding to dose measurement within 5% for a dose range of 1–10 Gy. An important advantage of the FBX dosimeter for optical CT dosimetry is the nature of the radiation-induced optical contrast. Use of optical contrast minimizes the amount of stray light in the dosimeter because the response to dose is light absorbing rather than light scattering. The practical implications are that fast broad-beam scanning systems become feasible (Olding et al., 2010; Olding and Schreiner, 2011). In practical terms, Fricke and FBX gels are relatively easy and convenient to make. Guidelines have been provided by Jordan (2010). There is no oxygen sensitivity and no toxicity of components. The main challenges relate to the need for an external casing to support the gel (largely a cost and convenience factor) and the time limitation requiring the gel to be read or imaged prior to the corrupting onset of diffusion.

6.2.3 Radiochromic Plastics

Prior to 2005, 3D dosimetry studies were conducted with water-based gel dosimeters. Alternative novel radiochromic plastic dosimeters were introduced in 2006 with the invention of "PRESAGE," a radiochromic polyurethane-based material by Adamovics and Maryanski (2006) that could be imaged with optical CT for 3D dosimetry. PRESAGE consists of a firm polyurethane matrix

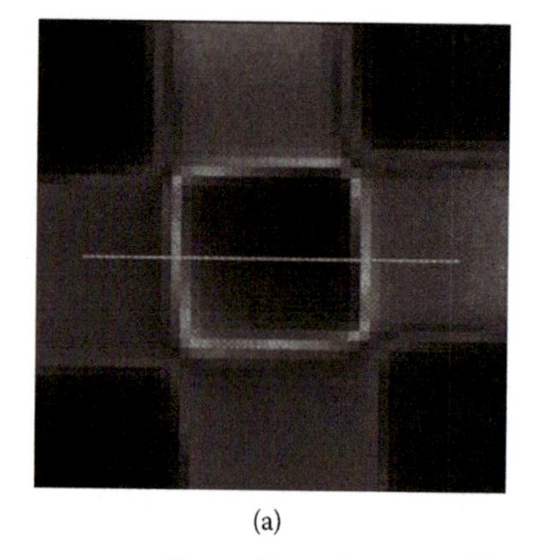
(a)

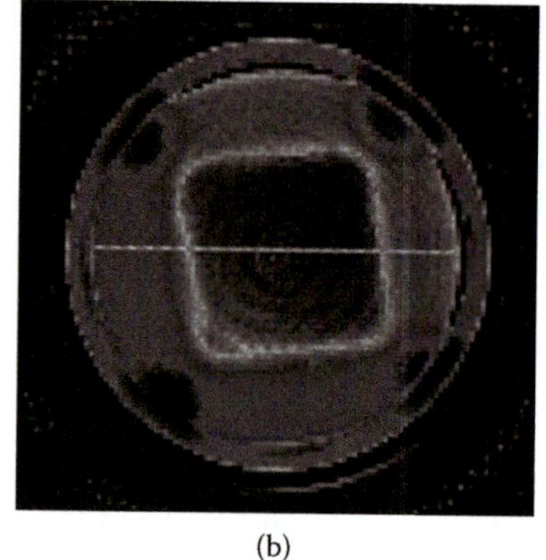
(b)

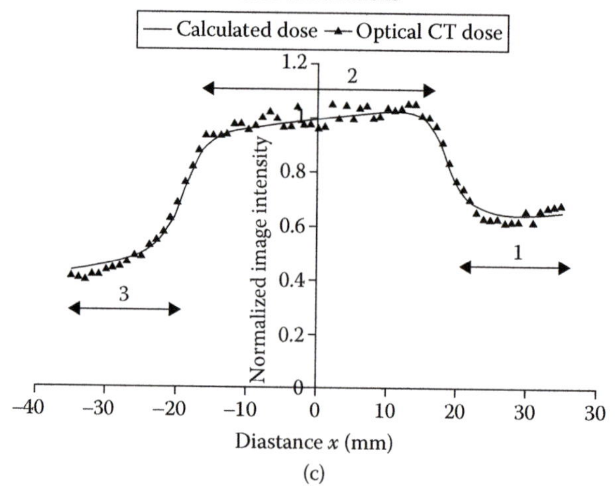

(c)

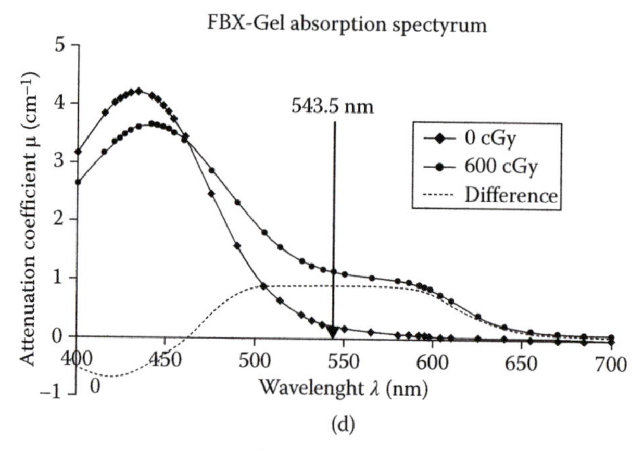

(d)

Figure 6.4

(a) Calculated dose distribution of a four-field box irradiation (6 MV; 4 cm × 4 cm field). (b) Corresponding optical CT scan through an irradiated Fricke-benzoic-xylenol (FBX) gel. (c) Line profile comparison along the line indicated in A and B. (d) The change in absorbance spectrum of an irradiated FBX dosimeter, indicating optimal readout wavelength of ~535 nm. (Adapted from Oldham M, In DJ Godfrey et al. [ed.], *Advances in Medical Physics*, Madison, WI, Medical Physics Publishing, 2014.)

doped with a trihalomethane or tetrahalomethane free radical initiator (such as trichloromethane, carbon tetrachloride, or carbon tetrabromide) and a triarylmethane leuco dye (leucomalachite green [LMG] or derivatives of LMG, shown in Figure 6.5). The polyurethane matrix is highly transparent, significantly more so than gelatin or agarose matrices due to an optimized curing process and degassing techniques that eliminate micrometer-scale inhomogeneities that can diffract and scatter light. Radiation-induced optical contrast is generated through the oxidation of colorless LMG to light-absorbing malachite green (MG) (Figure 6.6). PRESAGE thus also maintains the advantage of radiochromic gels in the minimization of stray light through an absorptive radiation-induced contrast mechanism. Unlike gels, however, the polyurethane substrate is a solid plastic and does not require an external casing to maintain shape. PRESAGE has proved exceptionally versatile in terms of casting in custom molds, including embedding

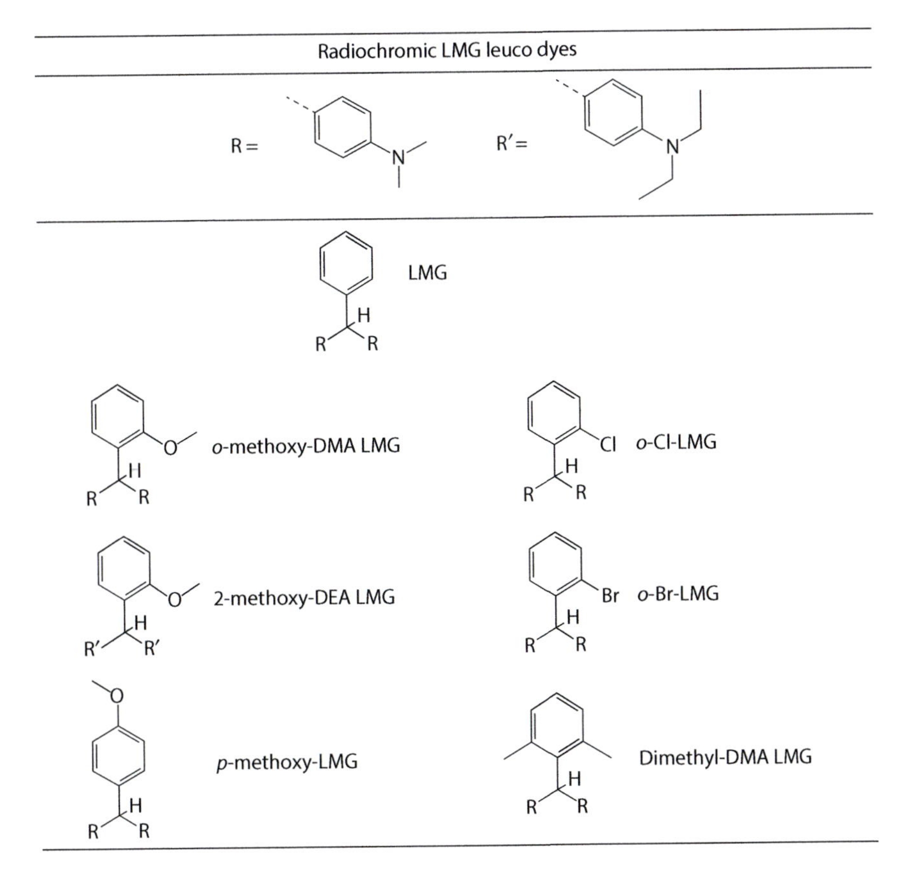

Figure 6.5

Chemical structure of leucomalachite (LMG) and LMG-derivative leuco dyes that have been used in radiochromic plastic dosimeters. The ortho- (*o*-) and para- (*p*-) prefixes indicate the position of the functional groups (e.g., methoxy (OCH_3) group) with respect to the central tetrahedral carbon. Ortho groups are attached to the second carbon away from the central carbon atom on the benzene ring, whereas para groups are attached to the opposite carbon on the benzene ring. Alternately, *o*-methoxy-LMG is called 2-methoxy-LMG to indicate position of the methoxy group on the benzene ring. DMA and DEA indicate dimethylamine and diethylamine, respectively, on the R or R′ group.

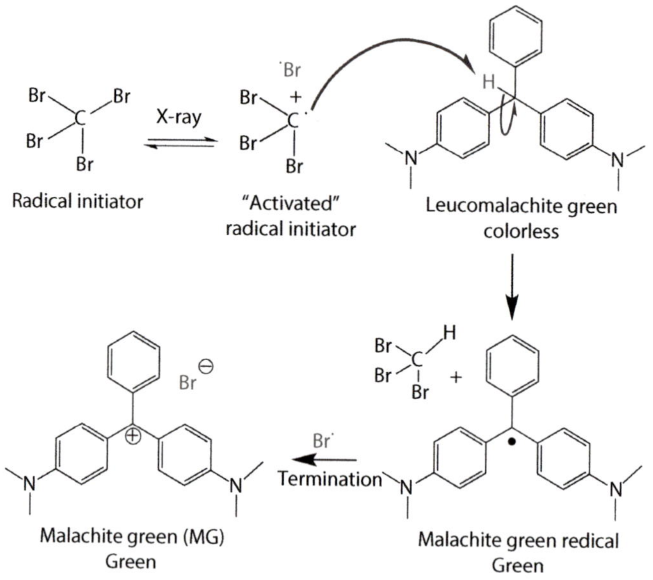

Figure 6.6

Oxidation reaction converting leucomalachite green (LMG), a colorless triarylmethane leuco dye used in radiochromic plastics, into its light-absorbing oxidation product, malachite green. The representative initiator pictured here is tetrabromochloride (CBr_4, a tetrahalomethane). All other trihalomethane and tetrahalomethane radical initiators react with LMG and LMG-derived leuco dyes in an analogous manner. (From Scheme 1 in Alqathami M et al., *Radiat. Phys. Chem.*, 85, 204–209, 2013.)

cavities and channels, including for brachytherapy measurements (Juang et al. 2013b; Adamson et al., 2014). Bache et al. (2015) have recently demonstrated the feasibility of both casting PRESAGE dosimeters as various 3D-printed anatomical shapes, and incorporating regions of different electron densities.

The chemical dose–response mechanism (Figure 6.6) is a step-by-step process that involves three main reactions: (1) activation of free radical initiators (trihalomethane or tetrahalomethane) via irradiation, (2) activation of LMG by the activated free radical initiators, and (3) termination leading to conversion of LMG to light-absorbing MG (Alqathami et al., 2013). Free radical initiators are heat sensitive, ultraviolet light sensitive, and radiation-sensitive molecules with a weak hydrogen–halogen bond. This weak bond is broken upon irradiation, generating radicals, and releasing a radical halogen. The generated radicals then activate LMG by attacking its center carbon and removing a hydrogen atom. This reaction is thought to be favored due to the stability of the resulting LMG radical; triarylmethane radicals are well known for stability, and last for weeks in solution and in crystalline forms (Gomberg, 1900; Griller and Ingold, 1976). The LMG radical is believed to then terminate with a radical termination reaction with radical halogen released during the first step, forming MG.

The radiochromic response of the PRESAGE dosimeter has been demonstrated by many authors to be linear for doses as high as 80 Gy (Adamovics and Maryanski, 2006; Guo et al., 2006b; Sakhalkar et al. 2009a; Wang et al., 2010). Several studies have reported negligible dependence on energy or dose rate in the range of 145 kVp to 18 MV (Adamovics and Maryanski, 2006; Guo et al., 2006a,b; Sakhalkar et al. 2009b). Although the majority of PRESAGE studies have been in the context of photon or electron irradiations, PRESAGE has also been studied for use in proton dosimetry. When used with protons, as with many materials, including PAGs and external beam therapy (EBT) film, under-response has been reported in the Bragg peak (Al-Nowais et al., 2009; Zhao et al., 2012).

The sensitivity and stability of the radiochromic response in PRESAGE can vary depending on the specific constituents used and their relative concentrations. For example, recent studies have shown that the composition of catalytic metal compounds in the polyurethane and the trihalomethane initiator influence dosimeter sensitivity (Alqathami et al., 2012a,b), and wide variability in postirradiation stability has been associated with relatively minor changes in dosimeter components as shown in Figure 6.7 (Juang et al., 2013b) and Table 6.1. The radiological and mechanical properties of PRESAGE can also be adjusted with variations in composition, including improving tissue equivalence and modifying the dosimeter's polyurethane matrix, for example, by replacing polyol with polyether or incorporating a plasticizer, to create dosimeters with markedly different hardness (Juang et al., 2013a,b). Although this allows a wide range of flexibility in customizing different PRESAGE formulations to address specific clinical and research applications, differences between formulations and variations between batches can also lead to markedly different dose–response characteristics. Some studies have reported conflicting results as to the stability of the radiochromic response and the relative 3D distribution (Yates et al., 2011), and substantial variations in sensitivities have been seen between batches of the same formulation, and even different volumes of the same batch (Jackson et al., 2013). These results suggest that caution should be exercised in assuming similar characteristics between different PRESAGE formulations, and that each individual batch of dosimeters should be evaluated independently prior to use.

Several chemical derivatives of LMG have been tested to enhance the sensitivity and stability of PRESAGE. For example, substituting two of the hydrogen atoms with methyl ($-CH_3$) groups in ortho-positions with respect to the center carbon, dimethyl-DMA LMG in Figure 6.5, tended to increase the temporal stability of PRESAGE (Juang et al., 2013b). The mechanism of radiochromic signal sensitization or stabilization is unclear. However, because the three aryl groups in the MG radical (shown bottom-right on Figure 6.6) are capable of delocalizing the unstable central unpaired electron throughout the structure (Griller and Ingold, 1976), it could be that different LMG formulations exhibit different degrees of radical stabilization and thus dose sensitivity. In addition, it was previously shown that the reactivity of the central carbon with water increased for cationic MG when hydrogen in the ortho- or para-position with respect to the central carbon (Figure 6.5) is substituted with an electron-attracting group such as –Br, –Cl, or –OH. This may explain why *p*-methoxy-LMG exhibits poor stability of the dose–response signal (MG) upon irradiation, and thus loses color quickly (Cigén et al., 1961).

Recent developments in the area of radiochromic plastic dosimeters have expanded the dose reporting chemical system used in PRESAGE to 3D silicone dosimeters (FlexyDos3D) (De Deene et al., 2015). These silicone dosimeters are formulated similarly to PRESAGE (with trichloromethane and LMG employed as the chemical radiochromic response system) except that an elastic silicone

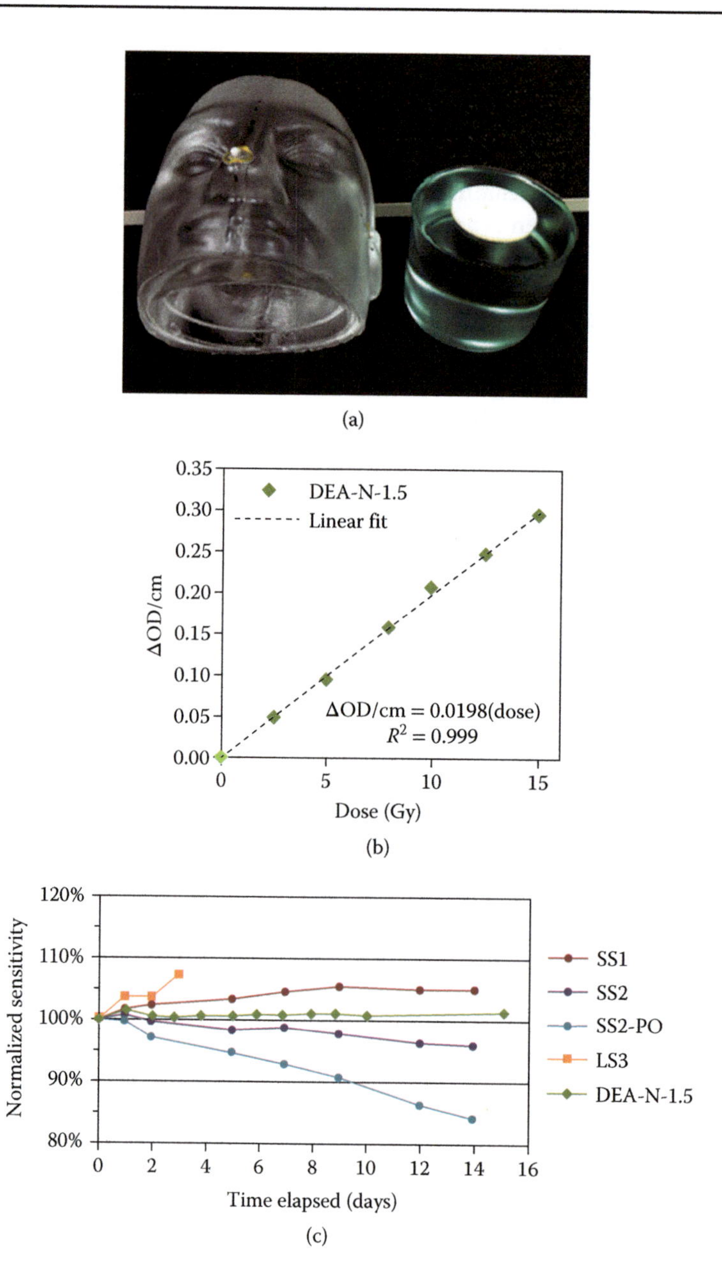

Figure 6.7

(a) Photograph of a cylindrical PRESAGE dosimeter (11 cm diameter) which inserts into a polyurethane head phantom for radiosurgery verification. (From Thomas A et al., *Med. Phys.*, 40, 121725, 2013a.) (b) Linearity of a PRESAGE formulation. (From Mein S et al., *Med. Phys.*, 42, 3492, 2015.) (c) Postirradiation temporal stability of radiation sensitivity of several PRESAGE variants illustrating how minor changes in formulation can yield substantially different effects. (Adapted from Juang T, Clinical and research applications in 3D dosimetry. PhD Thesis, Duke University, Durham, NC, 2015.) Details of each formulation are listed in Table 6.1, and all samples were kept in cold storage (3°C–10°C) between measurements.

Table 6.1 PRESAGE Formulations Shown in Figure 6.7

Formulation	Polyurethane	Leuco Dye	Initiator Compound(s)	Other Components
SS1 Juang et al. (2013b, 2014) *Also reported as:* D21 Niebanck (2012), Vidovic et al. (2014) *and* DX Jackson et al. (2015)	90.25% Smooth-On Crystal Clear® 206	2.0% LMG	0.50% CBr_4 0.25% $CBrCl_3$	7.0% $(CH_2)_5CO$
SS2 Juang et al. (2013b, 2014)	90.50% Smooth-On Crystal Clear 206	2.0% LMG	0.50% CBr_4	7.0% $(CH_2)_5CO$
SS2-PO	90.50% Polytek Poly-Optic®	2.0% LMG	0.50% CBr_4	7.0% $(CH_2)_5CO$
LS3 Juang et al. (2013b), Niebanck et al. (2013)	91.25% Smooth-On Crystal Clear 206	1.0% 2-methoxy-DMA LMG	0.75% CBr_4	7.0% $(CH_2)_5CO$
DEA-N-1.5 Mein et al. (2015)	91.00% Smooth-On Crystal Clear 206 + initiator additive	1.5% 2-methoxy-DEA LMG	0.50% CBr_4	2.0% DMSO 5.0% $C_4H_8O_2$

Source: Juang T., Clinical and research applications in 3D dosimetry. PhD Thesis, Duke University, Durham, NC, 2015.
Note: Percentages listed are by weight.

matrix is used instead of polyurethane. One prime advantage of this silicone radiochromic dosimeter is the relative ease of dosimeter fabrication, which can be performed in a laboratory environment without specialized equipment. Results so far have shown that this silicone dosimeter is tissue equivalent and energy independent, but exhibits dose-rate dependence and darkening of the unirradiated background over time, both of which potentially limit its application.

6.3 Methods of Dose Readout

6.3.1 Magnetic Resonance Imaging

It was noted earlier that MRI was the first imaging modality utilized for 3D gel dosimetry in the context of quantitative R_1 (spin–lattice) relaxivity of water molecules in Fricke gels. For Fricke gels, R_1 imaging is faster and has higher dynamic range than R_2 (spin–spin) relaxivity, owing to the low R_1 values in unirradiated gels. The mechanisms underlying the effects of paramagnetic molecules on relaxivity have been described elsewhere (Gore et al., 1984; Podgorsak and Schreiner, 1992). The main limitations to MRI 3D dosimetry of Fricke gels have been difficulty in generating low noise images for high spatial resolution, and difficulty in stabilizing the irradiated dose distribution. The latter effect originates in the fact that the small Fe^{3+} ions can easily diffuse in gelatin or agarose gels (Schreiner, 2004). The diffusion problem was solved with the introduction of MRI polymer gel 3D dosimetry, where the radiation-induced polymer cross-linked microparticles are unable to move in the gel substrate

(Maryanski et al., 1993). For polymer gels, R_2 was found to be the preferred readout quantity owing to a fast spin–spin exchange model (Maryanski et al., 1993; Kennan et al., 1996; Baustert et al., 2000; Lepage et al., 2001b,c; Ceberg et al., 2012). In the absence of oxygen contamination in the polymer gels, a linear dose response of R_2 is observed below onset of saturation effects (Oldham et al., 1998a; Baldock et al., 2010). Imaging magnetization transfer (MT) has also been found very useful for low density lung-equivalent gels (Lepage et al., 2002; De Deene et al., 2006; Baldock et al., 2010).

From the above discussion, we can distinguish three fundamental MRI parameters that are relevant for 3D dosimetry by MRI: spin–lattice relaxivity (R_1), spin–spin relaxivity (R_2), or MT. Whichever parameter is chosen, it is essential that the appropriate sequence is implemented to ensure accurate measurement. A list of relevant MRI sequences is shown in Table 6.2 (from De Deene, 2010). A number of limitations and challenges have been identified for MRI 3D dosimetry, leading to concern that expertise in MRI is a prerequisite for accurate 3D dosimetry (Vandecasteele and De Deene 2013b,c,d). Although some expertise is necessary, the prospect has been greatly simplified by published recommendations (De Deene, 2010) and guidelines presented in Chapter 5.

6.3.2 Optical CT Imaging

Optical CT appears to have arisen independently and nearly simultaneously in three different specialty fields (Gore et al., 1996; Maryanski et al., 1996; Winfree et al., 1996; Sharpe et al., 2002; Sharpe, 2004). The technique of optical CT is analogous to the more familiar x-ray CT, except with visible light as the photon source. In both optical and x-ray CT, line integrals of attenuation are acquired at various views through the object to be imaged. The main differences are the methods of producing and detecting either the x-rays or visible light and the scanning configurations. The relatively small size of 3D dosimeters makes it practical to rotate the dosimeter rather than the source and the detector. The same mathematics, for example, filtered back projection or inverse radon transform, can be used to reconstruct either dataset. 3D maps of the local x-ray or optical attenuation coefficients, are produced, and both methods are potentially susceptible to numerous artifacts including stray light, rings, beam hardening, attenuation, and motion (Oldham et al., 2003; Oldham and Kim, 2004). Several configurations of optical CT systems for 3D dosimetry are shown in Figure 6.8. It should be noted that there is complimentary overlap in the optical CT systems and discussion presented here and that in Chapter 5. Here we contrast the development of first-generation optical CT systems, designed for high-scatter polymer gel dosimeters, with later broad-beam systems designed for low-scatter radiochromic dosimeters. When compared to 3D dosimetry with MRI, the primary advantages of optical CT are substantially reduced cost, increased accessibility, and potentially higher accuracy and precision in shorter imaging times (Oldham et al., 2001).

The first optical CT system was developed for polymer gels, where the radiation-induced contrast is from light scattering polymer microparticles. The presence of scattered light poses challenges to obtain accurate optical tomographic attenuation maps in a manner analogous to the confounding effects of x-ray scatter in cone beam CT (Letourneau et al., 2005; Oldham et al., 2005). Accurate optical CT is only feasible if the stray light can be prevented from corrupting the line integral measurements in projection images. To counteract

Table 6.2 MR Imaging Sequences for 3D Dosimetry

Sequence Type	Conditions	Variable	Postprocessing	Availability	Spatial Accuracy
1. Quantitative R_1 imaging sequences ($R_1 = 1/T_1$)					
SE	TE short	TR (×2/×N)	Fit	c	Very good
Saturation recovery (SRGE/SRSE)	TR long, TE short	TM (×2/×N)	Fit	c	Good/very good
Inversion recovery (IRGE/IRSE)	TR long, TE short	TI (×2/×N)	Fit	c	Good/Very good
DESPOT	–	TI (×2/×N)	Fit	a	Good
LL, TOMROP	Fa small	TI (×2/×N)	Fit	a	Good
SSFP	TR >> T_2	FA (×2/×N)	Anal./Fit	b	Good
IR—Very fast acquisition (EPI, GRASE, HASTE)	TR long	TI (×2/×N)	Fit	b	Poor
2. Quantitative R_2 imaging sequences ($R_2 = 1/T_2$)					
Single SE	TR long	TE (×2/×N)	Anal./Fit	c	Very good
(FSE, TSE, RARE)	TR long	TE (×2/×N)	Anal./Fit	c	Good
(MSE, MC-SE)	TR long	[ΔTE (N)]	Fit	c	Very good
SSFP	TR << T_1 FA = 90°	[2 echoes]	Anal.	c	Good
3. Quantitative magnetization transfer (MT) imaging sequences					
MT pulse prepared spin echo imaging sequence	TR long	MT pulse amplitude	Anal.	b	Very good
Pulsed MT steady state	TR short	MT pulse amplitude	Anal.	b	Good
Stimulated echo preparation	–	TM	Anal.	a	Very good

DESPOT, Driven-equilibrium single-pulse observation of T_1; FSE, fast spin echo; LL, look-locker; MSE, multiple spin echo; SE, spin echo; SSFP, steady-state free precession.

Source: From De Deene Y, *J. Phys. Conf. Ser.*, 250, 012015, 2010.

Note: Overview of important quantitative MR imaging sequences for R_1 (= $1/T_1$), R_2 (= $1/T_2$) and magnetization transfer ratio (MTR) imaging. The variable that is changed to acquire different contrast weighted images is shown in the third column. If the variable is varied automatically within the sequence, the parameter is shown between squared brackets.

The meaning of the availability is as follows: ([a]) The imaging sequence is not provided by the manufacturer. The sequence should be developed in-house. ([b]) The sequence is available but significant changes to the imaging parameters are required. ([c]) The imaging sequence is readily available on all clinical MRI scanners.

this effect, the first optical CT system for 3D dosimetry, the "OCTOPUS" (Gore et al., 1996; Maryanski et al., 1996), used a first-generation scanning laser configuration operating in a translate-rotate manner: a laser beam is translated across a polymer gel dosimeter, and a line profile of attenuation is acquired with a photodiode detector. The dosimeter is then rotated incrementally to acquire

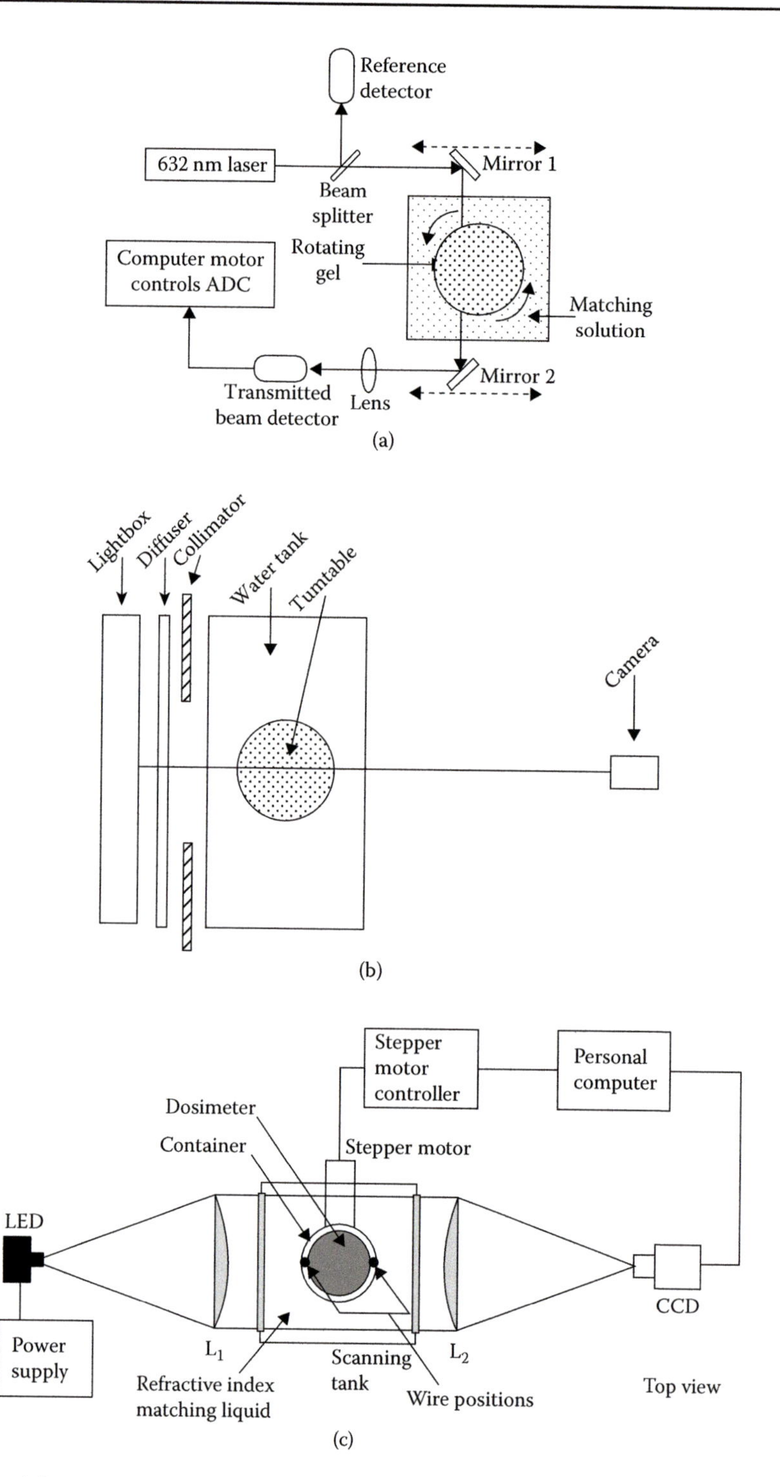

Figure 6.8

Optical CT scanning configurations: (a) first-generation scanning laser, (b) cone beam CT arrangement, (c) parallel beam arrangement, CCD, charge-coupled devices; LED, light-emitting diode. (Adapted from Oldham M, In DJ Godfrey et al. [ed.], *Advances in Medical Physics*, Madison, WI, Medical Physics Publishing.) *(Continued)*

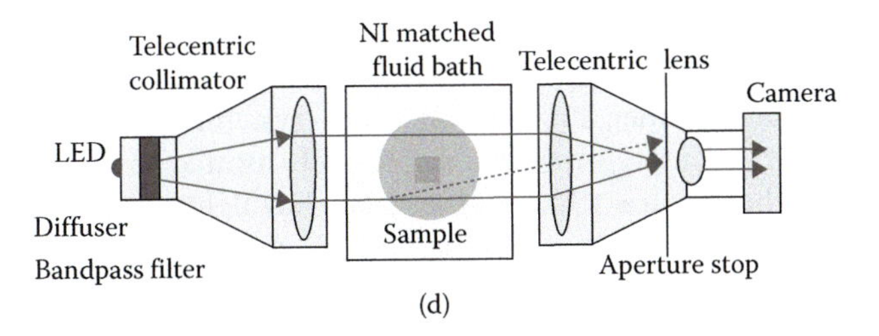

Figure 6.8 (Continued)

Optical CT scanning configurations: (d) fully bitelecentric system. (Adapted from Oldham M, In DJ Godfrey et al. [ed.], *Advances in Medical Physics*, Madison, WI, Medical Physics Publishing.)

the next profile. This first-generation scanning configuration can be effective at eliminating stray light from projection images through implementation of a collimating aperture that restricts light acceptance on the detecting photodiode. Also, by acquiring attenuation data of a single line profile at a time, it is impossible for stray light to contaminate neighboring line attenuation measurements in a projection. These systems can thus be used to image dose in dosimeters with high scattered light component like PAGs (Xu et al., 2004; Baldock et al., 2010). Several groups have investigated the performance of the OCTOPUS (Islam et al., 2003; Xu et al., 2004; Guo et al., 2006a; Lopatiuk-Tirpak et al., 2008). Variations on the scanning laser system were also explored (Kelly et al., 1998; Oldham et al., 2001).

The limitation of the first-generation laser scanning configurations is that long imaging times are needed to translate and rotate the dosimeter to acquire all necessary line profiles. Early works reported scanning times of several hours for large volumes (Gore et al., 1996; Oldham et al., 2001; Islam et al., 2003; Campbell et al., 2013). More recent scanning laser systems achieve faster speeds by up to a factor of 4 (Campbell et al., 2013; Qian et al., 2013). Still faster scanning speeds are possible by using rotating mirrors to translate the laser beams (Krstajic and Doran, 2007b). These latter systems require sophisticated motion control and data acquisition calibration and coordination due to the nonlinear laser motion. For smaller samples, the feasibility of the rotating mirror scanning system in reconstructing the dose in 3D had been demonstrated (Doran et al., 2010, 2013; Abdul Rahman et al., 2011). Despite the advances in scanning laser systems, it was not until the advent of broad-beam scanning systems that 3D dosimetry systems, which could yield results in clinically acceptable time frames as per RTAP criteria became feasible. The first prototype broad-beam optical CT scanners were proposed at the turn of the millennium (Wolodzko et al., 1999; Doran et al., 2001). Like many pioneering papers, the image quality from these first systems was not suitable for practical use, a result compounded by the fact that radiochromic low-scatter dosimeters were not yet readily available for optical CT. Other groups also developed in-house broad-beam systems with more sophisticated components and configuration (Doran et al., 2001; Babic et al., 2008; Sakhalkar and Oldham, 2008; Olding et al., 2010).

At the present time, two categories of fast broad-beam optical CT scanning configurations can be identified: parallel beam and cone beam (Figure 6.8). Cone beam optical CT systems have the advantage of lower cost, but the potential disadvantage of lower accuracy due to susceptibility to stray light artifacts (Oldham, 2006; Babic et al., 2009; Olding et al., 2010; Wuu and Xu, 2010; Olding and Schreiner, 2011). Parallel beam systems, and in particular telecentric systems, are more expensive due to sophisticated lenses, but have the capacity for higher accuracy by removing the stray light component (Krstajic and Doran, 2007a; Sakhalkar and Oldham, 2008). A debate has ensued as to the relative merits of these two approaches. Stray light artifacts have been reported when using cone beam optical CT scanners and polymer gels, as might be expected due to the high light scatter component (Olding et al., 2010; Olding and Schreiner, 2011). A continued increase in sophistication and image quality of broad-beam optical CT scanners is evident in the literature (Wolodzko et al., 1999; Doran et al., 2001; Krstajic and Doran, 2006, 2007a; Babic et al., 2008; Sakhalkar and Oldham 2008; Olding et al., 2010; Thomas and Oldham, 2010; Olding and Schreiner, 2011).

A current state-of-the-art system was introduced by Thomas et al. (2011b) consisting of a powerful, large field of view, matched bi-telecentric scanning system. This is the first broad-beam system that was commissioned and benchmarked for clinical use, and the high-quality data presented set a new standard for 3D dosimetry. Accompanying papers addressed minor corrections associated with stray light reflections, mostly within the lenses, and spectral artifacts (Thomas et al., 2011b,c). These corrections are negligible except under certain extreme conditions, such as small field output factor and percent depth dose commissioning.

6.3.3 Other 3D Dosimetry Readout Methods

While MRI and optical CT readout account for the majority of 3D dosimetry work conducted to date, other readout methods have been explored. The most promising is x-ray CT of polymer gels (Hilts et al., 2000, 2005; Trapp et al., 2001; Koeva et al., 2009). The underlying radiation-induced contrast mechanism is an increase of mass density of irradiated gel in the range of 1 mg/cm^3/Gy (or a few HU per Gy). With such a relatively low sensitivity, maintaining sufficient signal to noise with adequate spatial resolution becomes challenging. Improved sensitivity gel formulations and increasingly sophisticated filtering and imaging protocols have been developed (Jirasek et al., 2010, 2012; Chain et al., 2011; Kakakhel et al., 2011; Johnston et al., 2012). In a recent work (Johnston et al., 2012), it was shown that IMRT 3D verification was feasible by evaluating x-ray CT 3D dosimetry with gamma criteria (3%/3 mm), which yielded a 93.4% passing rate over the entire treated volume. Other readout techniques which have been investigated for PAG dosimeter include ultrasound (Mather and Baldock, 2003) and Raman imaging (Rintoul et al., 2003).

6.4 Applications of 3D Dosimetry

The immediate concern driving the early development of 3D dosimetry came from the field of RT. The key need was to find a way to verify complex therapeutic megavoltage (MV) dose distributions in a more comprehensive manner than possible with conventional planar (e.g., film) or point (e.g., ion chamber, TLD) detectors. As described earlier, tremendous progress has been achieved, and there are now several 3D dosimetry systems that can achieve and surpass the RTAP performance goal described in Section 6.2. It is not surprising, then,

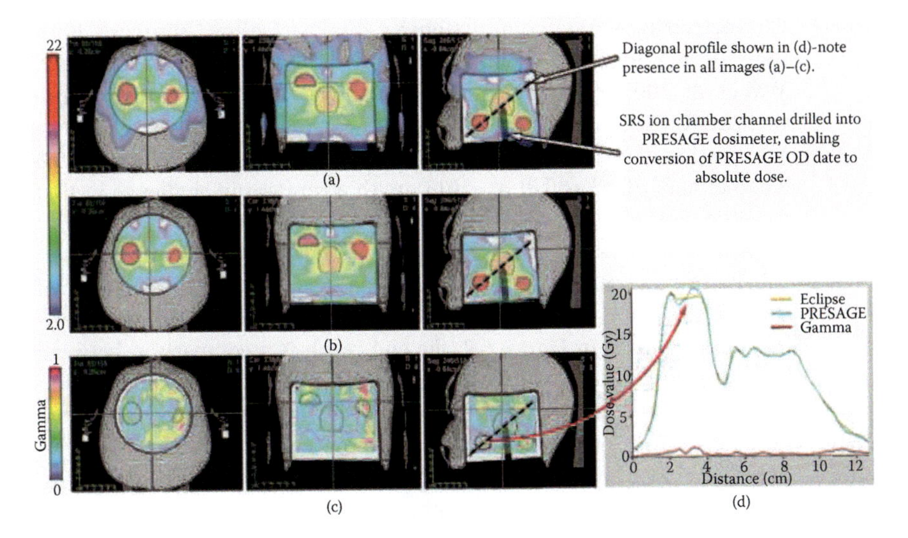

Figure 6.9

Clinical utility of three-dimensional (3D) dosimetry for investigating the accuracy of a new treatment technique: a single isocenter, multiple lesion, volumetric-modulated arc therapy radiosurgery treatment of intracranial brain lesions. (a) Eclipse planned dose distribution in axial, coronal, and sagittal planes. (b) PRESAGE measured 3D dose distribution. 1 mm isometric resolution, 15 minutes scan time. (c) 3D gamma comparison: 3%, 2 mm criteria. 3D pass rate inside dosimeter was 98.6%. (d) Diagonal dose profiles. SRS, stereotactic radiosurgery. (From Thomas A et al., *J. Phys. Conf. Ser.*, 444, 012049, 2013c.)

that the majority of 3D dosimetry applications to date relate to the verification of advanced therapy treatment techniques. Examples are listed below, and the utility of 3D dosimetry is further shown in Figures 6.9 through 6.11.

- Intracranial radiosurgery (Olsson et al., 1992; Oldham et al., 2001; Novotny et al., 2002; Sandilos et al., 2006; Babic et al., 2009; Clift et al., 2010; Wang et al., 2010; Gopishankar et al., 2011; Oldham et al., 2013; Thomas et al., 2013a,b).
- Stereotactic body radiotherapy (Cosgrove et al., 2000; De Deene et al., 2006; Brady et al., 2010; Olding et al., 2013).
- 3D conformal therapy (Ibbott et al., 1997; De Deene et al., 1998, 2000).
- IMRT (Oldham et al., 1998b; Gustavsson et al., 2003; Vergote et al., 2003; Wuu and Xu, 2006; Babic et al., 2008; Sakhalkar et al., 2009; Pavoni et al., 2012; Vandecasteele and De Deene, 2013a).
- Volumetric-modulated arc therapy (VMAT) (Vergote et al., 2004; Babic et al., 2008; Ceberg et al., 2010; Thomas et al., 2013a).
- Respiratory gating techniques (Ceberg et al., 2008; Lopatiuk-Tirpak et al., 2008; Brady et al., 2010; Thomas et al., 2013c).
- Image-guided radiation therapy (IGRT) (Thomas et al., 2013a).
- Deformable cumulative dosimetry (Niu et al., 2012; Yeo et al., 2012; Juang et al., 2013a).
- Patient-specific dose–volume estimations; a distinguishing advantage of 3D dosimetry is the capability of relating QA data to likely clinical significance for a patient (Figure 6.10) (Oldham et al., 2012).

- Characterizing and commissioning brachytherapy sources, both high dose rate (HDR) (Schreiner et al., 1994; McJury et al., 1999; Wai et al., 2009; Pierquet et al., 2010; Palmer et al., 2013) and low dose rate (LDR) (Adamson et al., 2012, 2013).
- Verification of proton deliveries (Back et al., 1999; Al-Nowais et al., 2009; Zhao et al., 2012; Kroll et al., 2013), with a remaining challenge in that most systems exhibit an under-response in the Bragg peak.

It is informative to consider how 3D dosimetry is utilized in the majority of cases listed above, and its corresponding strengths and limitations. In many verification studies, as shown in Figure 6.9, the end result is a comparison between a predicted (or planned) dose distribution against a measured 3D dose distribution. A straightforward qualitative comparison in the traditional three orthogonal planes can be effective at determining gross differences, and can be made visually, using either color wash dose maps (Figure 6.9), or comparing isodose lines (Figure 6.11). Often, however, a more quantitative evaluation is required, such as that provided by a gamma map analysis (Low and Dempsey, 2003) as shown in Figure 6.9c. The gamma metric contains two criteria: a dose-difference (DD) criterion, and a distance-to-agreement (DTA) criterion. The gamma map is constructed by evaluating the gamma metric at each point in the experimental distribution. The advantage is that the agreement between the two distributions can be characterized by a single number; typically the passing rate, or the percentage of points in the target distribution for which the gamma metric is less than unity. Gamma

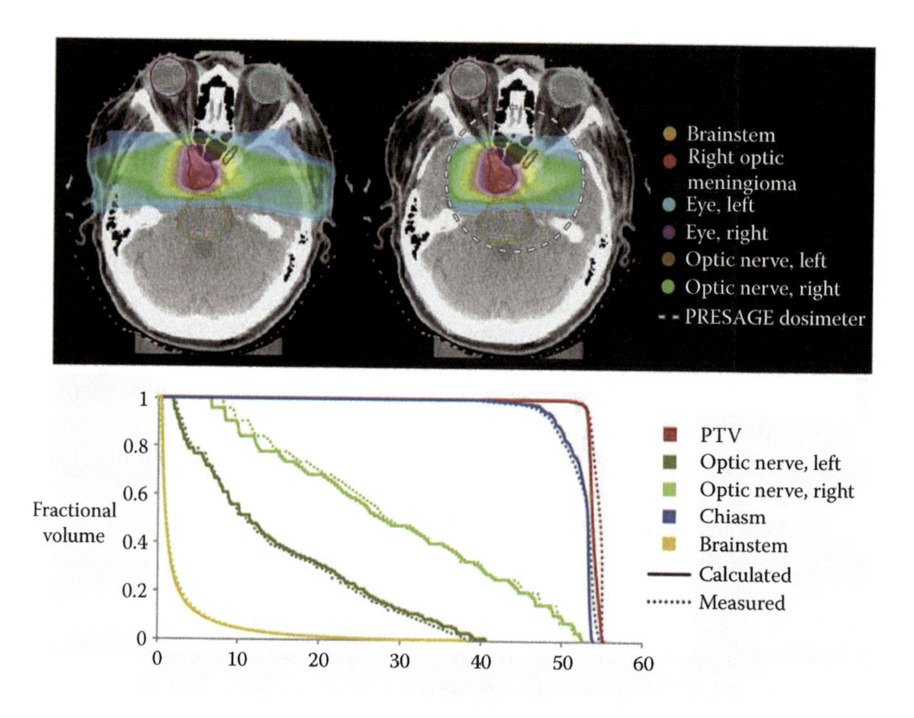

Figure 6.10

Illustration of an important advantage of three-dimensional (3D) dosimetry: the facility to relate quality assurance data to likely clinical significance for patients. (From Oldham M et al., *Int. J. Radiat. Oncol. Biol. Phys.*, 84, 540–546, 2012.)

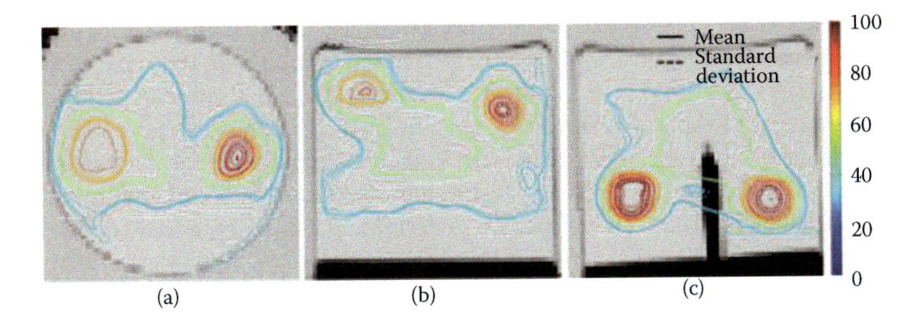

Figure 6.11

(a)–(c) Reproducibility and dosimetric agreement of four independent, end-to-end deliveries of the image-guided radiation therapy treatment shown in Figure 6.9. Solid lines indicate 95%, 90%, 80%, 70%, 50%, and 30% isodose lines, while dashed lines indicate one standard deviation. (From Thomas A et al., *J. Phys. Conf. Ser.*, 444, 012050, 2013b.)

pass rates can be calculated on individual planes (Ezzell et al., 2009) or in a full 3D calculation. The latter represents a comprehensive metric taking full account of the agreement between the two distributions in 3D. Whenever a gamma map analysis is presented, it is important to note the magnitudes of the criteria used, whether a minimum threshold dose level was included (above which gamma is calculated), and whether the dose difference criterion is a local or global percentage. The presence of significant noise in a distribution can also impact the usefulness of gamma evaluation (Low and Dempsey, 2003; Low et al., 2011). Common gamma criteria values in current use are 3%/3 mm, as proposed by the American Association of Physicists in Medicine (AAPM) Task Group (TG) 119 (Ezzell et al., 2009). These values, however, have been recently criticized as being too generous, and may allow clinically significant errors to pass undetected (Nelms et al., 2011, 2013). In practice, many authors now include a range of passing rates calculated with different criteria as discussed in Chapters 14 and 17.

The strengths of a comprehensive 3D verification capability are readily apparent. Oldham and colleagues recently described two key advantages for 3D dosimetry QA: treatment verification is comprehensive throughout the whole treated volume, and the clinical significance of deviations can be assessed through the generation of dose–volume histogram curves and dose overlays on the patient's anatomy (Oldham et al., 2012). Both steps represent developments that advance the clinical relevance of complex treatment QA. A limitation of cumulative 3D dosimetry verification, however, is that when deviations are identified, it is difficult to identify the cause of the deviation from the 3D measurement alone. Other dosimetry tools are required to isolate the origin of the deviation to, for example, a specific treatment field as discussed in Chapter 14.

In addition to the clinical applications described above, the techniques of 3D dosimetry have also found useful application in the preclinical world and in nonmedical specialties. A rapidly expanding field of RT and biological research involves the precise delivery of miniature RT dose distributions to small animal models. Precision and accuracy of delivery are key issues to achieving successful trial data. Examples include commissioning a microirradiator (Newton et al., 2011), and the evaluation of end-to-end accuracy of IGRT micro-RT treatment

(Rankine et al., 2013). Extremely high-resolution 3D dosimetry measurements have been demonstrated for use in characterizing small and novel x-ray systems (Doran et al., 2010, 2013). The dosimetry of small animal precision irradiators is the topic of Chapter 24, while in Chapter 25 the dosimetry of synchrotron x-ray microbeams is discussed. In the field of radiation protection and the nuclear industry, 3D dosimetry techniques have been proposed and investigated as efficient diagnostic methods to determine potential radiation contamination sources (Farfan et al., 2010, 2012; Oldham et al., 2010; Stanley et al., 2012).

6.5 Summary and Recent Developments

3D dosimetry has proved a highly innovative field of research and clinical development for over 20 years. The originating motivating factor of developing more comprehensive methods for verification of complex RT treatments is even more relevant today due to the seemingly endless increase in sophistication and complexity of modern RT treatments. Current areas of continued innovation and further development can be identified in three major areas: materials development, readout development, and novel applications. In materials development, the main challenges are to develop more stable dosimeters for remote credentialing activities and reusable dosimeters for improved economics within an institution (Juang et al., 2013b). Additionally, new classes of potential dosimeters have emerged from micelle technology (Jordan and Avvakumov, 2009; Nasr et al., 2013; Vandecasteele and De Deene, 2013a), and polymer gels are being developed with enhanced sensitivity for x-ray CT (Jirasek et al., 2010). The challenge of a 3D dosimetry material without linear energy transfer (LET) dependence is also an active area of development (Lopatiuk-Tirpak et al., 2012; Zhao et al., 2012). Another interesting recent development is the use of novel 3D printing technology to create anatomically accurate 3D dosimeters (Bache et al., 2015).

For 3D dosimetry readout, improved optical CT scanner designs and protocols that are faster, more accurate (with reduced scatter), cheaper, and higher resolution are being pursued (Olding and Schreiner, 2011; Thomas et al., 2011a), along with scanning configurations which require substantially less or no fluid matching (Doran and Yatigammana, 2012; Ramm et al., 2012; Rankine and Oldham, 2013; Miles, 2016). MRI sequences are under development to minimize problems due to sample heating while preserving low noise and higher spatial resolution (Vandecasteele and De Deene, 2013a,d). For x-ray CT readout, more sensitive formulations are being developed (Jirasek et al., 2012; Johnston et al., 2012).

A review of the field at this time reveals strong progress toward the long-sought goal of a reliable, practical, and cost-effective 3D dosimetry system that can match the RTAP criteria (Section 6.2). Several systems have demonstrated they can meet this goal, but are not yet realized in the commercial world. This is likely to occur in the next few years, and will represent an important component to strengthening core QA practices in RT, contributing to safer and more accurate treatment of our patients.

References

Abdul Rahman AT et al. (2011) Sophisticated test objects for the quality assurance of optical computed tomography scanners. *Phys. Med. Biol.* 56: 4177–4199.

Adamovics J and Maryanski MJ (2006) Characterisation of PRESAGE: A new 3-D radiochromic solid polymer dosimeter for ionising radiation. *Radiat. Prot. Dos.* 120: 107–112.

Adamson J et al. (2012) Commissioning a CT-compatible LDR tandem and ovoid applicator using Monte Carlo calculation and 3D dosimetry. *Med. Phys.* 39: 4515–4523.

Adamson J et al. (2013) Towards comprehensive characterization of Cs-137 seeds using PRESAGE® dosimetry with optical tomography. *J. Phys. Conf. Ser.* 444: 012100.

Adamson J et al. (2014) On the feasibility of polyurethane based 3D dosimeters with optical CT for dosimetric verification of low energy photon brachytherapy seeds. *Med. Phys.* 41: 071705.

Al-Nowais S et al. (2009) A preliminary analysis of LET effects in the dosimetry of proton beams using PRESAGE and optical CT. *Appl. Radiat. Isot.* 67: 415–418.

Alqathami M et al. (2013). Evaluation of ultra-sensitive leucomalachite dye derivatives for use in the PRESAGE® dosimeter. *Radiat. Phys. Chem.* 85: 204–209.

Alqathami M et al. (2012a) Optimizing the sensitivity and radiological properties of the PRESAGE® dosimeter using metal compounds. *Radiat. Phys. Chem.* 81: 1688–1695.

Alqathami M et al. (2012b) Optimization of the sensitivity and stability of the PRESAGE™ dosimeter using trihalomethane radical initiators. *Radiat. Phys. Chem.* 81: 867–873.

Appleby A and Leghrouz A (1991) Imaging of radiation dose by visible color development in ferrous-agarose-xylenol orange gels. *Med. Phys.* 18:309–312.

Babic S, Battista J and Jordan K (2008) Three-dimensional dose verification for intensity-modulated radiation therapy in the radiological physics centre head-and-neck phantom using optical computed tomography scans of ferrous xylenol-orange gel dosimeters. *Int. J. Radiat. Oncol. Biol. Phys.* 70: 1281–1291.

Babic S et al. (2009) Three-dimensional dosimetry of small megavoltage radiation fields using radiochromic gels and optical CT scanning. *Phys. Med. Biol.* 54: 2463–2481.

Bache S et al. (2015) Investigating the accuracy of microstereotactic-body-radiotherapy utilizing anatomically accurate 3D printed rodent-morphic dosimeters. *Med. Phys.* 42: 846–855.

Back SA et al. (1999) Ferrous sulphate gel dosimetry and MRI for proton beam dose measurements. *Phys. Med. Biol.* 44: 1983–1996.

Baldock C et al. (2010) Polymer gel dosimetry. *Phys. Med. Biol.* 55: R1–R63.

Baustert IC et al. (2000) Optimized MR imaging for polyacrylamide gel dosimetry. *Phys. Med. Biol.* 45: 847–858.

Beddar S et al. (2009) Exploration of the potential of liquid scintillators for real-time 3D dosimetry of intensity modulated proton beams. *Med. Phys.* 36: 1736–1743.

Bogdanich W (2010a) Radiation offers new cures, and ways to do harm. *New York Times*, January 23.

Bogdanich W (2010b) As technology surges, radiation safeguards lag. *New York Times*, January 26.

Bortfeld T (2006) IMRT: A review and preview. *Phys. Med. Biol.* 51: R363–R379.

Bortfeld T et al. (1994) Realization and verification of three-dimensional conformal radiotherapy with modulated fields. *Int. J. Radiat. Oncol. Biol. Phys.* 30: 899–908.

Brady SL et al. (2010). Investigation into the feasibility of using PRESAGE/optical-CT dosimetry for the verification of gating treatments. *Phys. Med. Biol.* 55: 2187–2201.

Campbell WG et al. (2013) A prototype fan-beam optical CT scanner for 3D dosimetry. *Med. Phys.* 40: 061712.

Ceberg S et al. (2010) RapidArc treatment verification in 3D using polymer gel dosimetry and Monte Carlo simulation. *Phys. Med. Biol.* 55: 4885–4898.

Ceberg S et al. (2008) Verification of dynamic radiotherapy: The potential for 3D dosimetry under respiratory-like motion using polymer gel. *Phys. Med. Biol.* 53: N387–N396.

Ceberg S et al. (2012) Modelling the dynamic dose response of an nMAG polymer gel dosimeter. *Phys. Med. Biol.* 57: 4845–4853.

Chain JN et al. (2011) Cosolvent-free polymer gel dosimeters with improved dose sensitivity and resolution for x-ray CT dose response. *Phys. Med. Biol.* 56: 2091–2102.

Cho NY et al. (2013) Isotropic three-dimensional MRI-Fricke-infused gel dosimetry. *Med. Phys.* 40: 052101.

Chu KC et al. (2000) Polyvinyl alcohol-Fricke hydrogel and cryogel: Two new gel dosimetry systems with low Fe3+ diffusion. *Phys. Med. Biol.* 45: 955–969.

Cigén R et al. (1961) Studies on derivatives of malachite green. Protolytic equilibria and reaction rate constants of o-methoxy malachite green. *Acta Chem. Scand.* 15: 1905–1912.

Clift C et al. (2010) Toward acquiring comprehensive radiosurgery field commissioning data using the PRESAGE/optical-CT 3D dosimetry system. *Phys. Med. Biol.* 55: 1279–1293.

Convery D and Rosenbloom M (1992) The generation of intensity-modulated fields for conformal radiotherapy by dynamic collimation. *Phys. Med. Biol.* 37: 1359–1374.

Cosgrove VP et al. (2000) The reproducibility of polyacrylamide gel dosimetry applied to stereotactic conformal radiotherapy. *Phys. Med. Biol.* 45: 1195–1210.

De Deene Y (2010) How to scan polymer gels with MRI? *J. Phys. Conf. Ser.* 250: 012015.

De Deene Y et al. (2000) Validation of MR-based polymer gel dosimetry as a preclinical three-dimensional verification tool in conformal radiotherapy. *Magn. Reson. Med.* 43: 116–125.

De Deene Y et al. (1998) Three-dimensional dosimetry using polymer gel and magnetic resonance imaging applied to the verification of conformal radiation therapy in head-and-neck cancer. *Radiother. Oncol.* 48: 283–291.

De Deene Y et al. (2015) FlexyDos3D: A deformable anthropomorphic 3D radiation dosimeter: Radiation properties. *Phys. Med. Biol.* 60: 1543–1563.

De Deene Y et al. (2006) Three dimensional radiation dosimetry in lung-equivalent regions by use of a radiation sensitive gel foam: Proof of principle. *Med. Phys.* 33: 2586–2597.

de Pasquale F et al. (2006) Ion diffusion modelling of Fricke-agarose dosemeter gels. *Radiat. Prot. Dos.* 120: 151–154.

Doran SJ and Yatigammana DN (2012) Eliminating the need for refractive index matching in optical CT scanners for radiotherapy dosimetry: I. Concept and simulations. *Phys. Med. Biol.* 57: 665–683.

Doran SJ et al. (2013) Establishing the suitability of quantitative optical CT microscopy of PRESAGE(R) radiochromic dosimeters for the verification of synchrotron microbeam therapy. *Phys. Med. Biol.* 58: 6279–6297.

Doran SJ et al. (2010) An investigation of the potential of optical computed tomography for imaging of synchrotron-generated x-rays at high spatial resolution. *Phys. Med. Biol.* 55: 1531–1547.

Doran SJ et al. (2001) A CCD-based optical CT scanner for high-resolution 3D imaging of radiation dose distributions: Equipment specifications, optical simulations and preliminary results. *Phys. Med. Biol.* 46: 3191–3213.

Ezzell GA et al. (2009) IMRT commissioning: Multiple institution planning and dosimetry comparisons, a report from AAPM Task Group 119. *Med. Phys.* 36: 5359–5373.

Farfan EB et al. (2010) RadBall technology testing in the Savannah River site's health physics instrument calibration laboratory. *J. Phys. Conf. Ser.* 250: 012080.

Farfan EB et al. (2012). Submerged RadBall(R) deployments in Hanford Site hot cells containing 137CsCl capsules. *Health Phys.* 103: 100–106.

Feygelman V et al. (2011) Evaluation of a new VMAT QA device, or the "X" and "O" array geometries. *J. Appl. Clin. Med. Phys.* 12(2): 146–168.

Fong PM et al. (2001) Polymer gels for magnetic resonance imaging of radiation dose distributions at normal room atmosphere. *Phys. Med. Biol.* 46: 3105–3113.

Fricke H and Morse S (1927) The chemical action of Roentgen rays on dilute ferrosulphate solutions as a measure of dose. *Am. J. Roentgenol. Radium Ther. Nucl. Med.* 18: 430–432.

Glaser AK et al. (2013) Three-dimensional Cerenkov tomography of energy deposition from ionizing radiation beams. *Opt. Lett.* 38: 634–636.

Gomberg M (1900) An instance of trivalent carbon: Triphenylmethyl. *J. Am. Chem. Soc.* 22: 757–771.

Gopishankar N, Watanabe Y and Subbiah V (2011) MRI-based polymer gel dosimetry for validating plans with multiple matrices in Gamma Knife stereotactic radiosurgery. *J. Appl. Clin. Med. Phys.* 12(2): 133–145.

Gore JC, Kang YS and Schulz RJ (1984) Measurement of radiation dose distributions by nuclear magnetic resonance (NMR) imaging. *Phys. Med. Biol.* 29: 1189–1197.

Gore JC et al. (1996) Radiation dose distributions in three dimensions from tomographic optical density scanning of polymer gels: I. Development of an optical scanner. *Phys. Med. Biol.* 41: 2695–2704.

Griller D and Ingold KU (1976) Persistent carbon-centered radicals. *Acc. Chem. Res.* 9: 13–19.

Guo P, Adamovics J and Oldham M (2006a) A practical three-dimensional dosimetry system for radiation therapy. *Med. Phys.* 33: 3962–3972.

Guo PY, Adamovics JA and Oldham M (2006b) Characterization of a new radiochromic three-dimensional dosimeter. *Med. Phys.* 33: 1338–1345.

Gupta BL et al. (1982) Use of the FBX dosemeter for the calibration of cobalt-60 and high-energy teletherapy machines. *Phys. Med. Biol.* 27: 235–245.

Gustavsson H et al. (2003) MAGIC-type polymer gel for three-dimensional dosimetry: Intensity-modulated radiation therapy verification. *Med. Phys.* 30: 1264–1271.

Hazle JD et al. (1991) Dose-response characteristics of a ferrous-sulphate-doped gelatin system for determining radiation absorbed dose distributions by magnetic resonance imaging (Fe MRI). *Phys. Med. Biol.* 36: 1117–1125.

Herman M (2011) Medical radiation: An overview of the issues. Hearing before the Subcommittee on Health of the Committee on Energy and Commerce, House of Representatives, 111th congress, 2nd session, Serial No. 111-100, US Government Printing Office, On behalf of the American Association of Physicists in Medicine (AAPM). Available at: https://www.gpo.gov/fdsys/pkg/CHRG-111hhrg76012/pdf/CHRG-111hhrg76012.pdf

Hilts M et al. (2000) Polymer gel dosimetry using x-ray computed tomography: A feasibility study. *Phys. Med. Biol.* 45: 2559–2571.

Hilts M, Jirasek A and C Duzenli (2005) Technical considerations for implementation of x-ray CT polymer gel dosimetry. *Phys. Med. Biol.* 50: 1727–1745.

Ibbott G (2010) QA in radiation therapy: The RPC perspective. *J. Phys. Conf. Ser.* 250: 0120001.

Ibbott GS et al. (1997) Three-dimensional visualization and measurement of conformal dose distributions using magnetic resonance imaging of BANG polymer gel dosimeters. *Int. J. Radiat. Oncol. Biol. Phys.* 38: 1097–1103.

Intensity-Modulated Radiotherapy Collaborative Working Group (2001) Intensity-modulated radiotherapy: Current status and issues of interest. *Int. J. Radiat. Oncol. Biol. Phys.* 51: 880–914.

Islam KT et al. (2003) Initial evaluation of commercial optical CT-based 3D gel dosimeter. *Med. Phys.* 30: 2159–2168.

Jackson J et al. (2013) An investigation of the dosimetric characteristics of a novel radiochromic 3D dosimeter. *Med. Phys.* 40: 215.

Jackson J et al. (2015) An investigation of PRESAGE® 3D dosimetry for IMRT and VMAT radiation therapy treatment verification. *Phys. Med. Biol.* 60: 2217–2230.

Jirasek A, Carrick J and Hilts M (2012) An x-ray CT polymer gel dosimetry prototype: I. Remnant artefact removal. *Phys. Med. Biol.* 57: 3137–3153.

Jirasek A, Hilts M and McAuley KB (2010) Polymer gel dosimeters with enhanced sensitivity for use in x-ray CT polymer gel dosimetry. *Phys. Med. Biol.* 55: 5269–5281.

Johnston H et al. (2012) An x-ray CT polymer gel dosimetry prototype: II. Gel characterization and clinical application. *Phys. Med. Biol.* 57: 3155–3175.

Jordan K (2010) Review of recent advances in radiochromic materials for 3D dosimetry. *J. Phys. Conf. Ser.* 250: 012043.

Jordan K and Avvakumov N (2009) Radiochromic leuco dye micelle hydrogels: I. Initial investigation. *Phys. Med. Biol.* 54: 6773–6789.

Juang T (2015) Clinical and research applications in 3D dosimetry. PhD Thesis, Duke University, Durham, NC.

Juang T et al. (2013a) On the need for comprehensive validation of deformable image registration, investigated with a novel 3-dimensional deformable dosimeter. *Int. J. Radiat. Oncol. Biol. Phys.* 87: 414–421.

Juang T et al. (2014) On the feasibility of comprehensive high-resolution 3D remote dosimetry. *Med. Phys.* 41: 071706.

Juang T et al. (2013b) Customising PRESAGE® for diverse applications. *J. Phys. Conf. Ser.* 444: 012029.

Kakakhel MB et al. (2011) Improved image quality for x-ray CT imaging of gel dosimeters. *Med. Phys.* 38: 5130–5135.

Kelly RG, Jordan KJ and Battista JJ (1998) Optical CT reconstruction of 3D dose distributions using the ferrous-benzoic-xylenol (FBX) gel dosimeter. *Med. Phys.* 25: 1741–1750.

Kennan RP et al. (1996) The effects of cross-link density and chemical exchange on magnetization transfer in polyacrylamide gels. *J. Magn. Reson. B* 110: 267–277.

Koeva VI et al. (2009) Preliminary investigation of the NMR, optical and x-ray CT dose-response of polymer gel dosimeters incorporating cosolvents to improve dose sensitivity. *Phys. Med. Biol.* 54: 2779–2790.

Kroll F, Pawelke J and Karsch L (2013) Preliminary investigations on the determination of three-dimensional dose distributions using scintillator blocks and optical tomography. *Med. Phys.* 40: 082104.

Kron T, Metcalfe P and Pope JM (1993) Investigation of the tissue equivalence of gels used for NMR dosimetry. *Phys. Med. Biol.* 38: 139–150.

Krstajic N and Doran SJ (2006) Focusing optics of a parallel beam CCD optical tomography apparatus for 3D radiation gel dosimetry. *Phys. Med. Biol.* 51: 2055–2075.

Krstajic N and Doran SJ (2007a) Characterization of a parallel-beam CCD optical-CT apparatus for 3D radiation dosimetry. *Phys. Med. Biol.* 52: 3693–3713.

Krstajic N and Doran SJ (2007b) Fast laser scanning optical-CT apparatus for 3D radiation dosimetry. *Phys. Med. Biol.* 52: N257–N263.

Lepage M et al. (2001a) Dose resolution optimization of polymer gel dosimeters using different monomers. *Phys. Med. Biol.* 46: 2665–2680.

Lepage M et al. (2002) Magnetization transfer imaging for polymer gel dosimetry. *Phys. Med. Biol.* 47: 1881–1890.

Lepage M et al. (2001b) Modelling of post-irradiation events in polymer gel dosimeters. *Phys. Med. Biol.* 46: 2827–2839.

Lepage M et al. (2001c) The relationship between radiation-induced chemical processes and transverse relaxation times in polymer gel dosimeters. *Phys. Med. Biol.* 46: 1061–1074.

Letourneau D et al. (2005) Cone-beam-CT guided radiation therapy: Technical implementation. *Radiother. Oncol.* 75: 279–286.

Letourneau D et al. (2004) Evaluation of a 2D diode array for IMRT quality assurance. *Radiother. Oncol.* 70: 199–206.

Letourneau D et al. (2009) Novel dosimetric phantom for quality assurance of volumetric modulated arc therapy. *Med. Phys.* 36: 1813–1821.

Lopatiuk-Tirpak O et al. (2008) Performance evaluation of an improved optical computed tomography polymer gel dosimeter system for 3D dose verification of static and dynamic phantom deliveries. *Med. Phys.* 35: 3847–3859.

Lopatiuk-Tirpak O et al. (2012) Direct response to proton beam linear energy transfer (LET) in a novel polymer gel dosimeter formulation. *Technol. Cancer Res. Treat.* 11: 441–445.

Low DA and Dempsey JF (2003) Evaluation of the gamma dose distribution comparison method. *Med. Phys.* 30: 2455–2464.

Low DA et al. (2011) Dosimetry tools and techniques for IMRT. *Med. Phys.* 38: 1313–1338.

Mans A et al. (2010) Catching errors with in vivo EPID dosimetry. *Med. Phys.* 37: 2638–2644.

Maryanski MJ et al. (1994) Magnetic resonance imaging of radiation dose distributions using a polymer-gel dosimeter. *Phys. Med. Biol.* 39: 1437–1455.

Maryanski MJ et al. (1993) NMR relaxation enhancement in gels polymerized and cross-linked by ionizing radiation: A new approach to 3D dosimetry by MRI. *Magn. Reson. Imaging* 11: 253–258.

Maryanski MJ, Zastavker YZ and Gore JC (1996) Radiation dose distributions in three dimensions from tomographic optical density scanning of polymer gels: II. Optical properties of the BANG polymer gel. *Phys. Med. Biol.* 41: 2705–2717.

Mather ML and Baldock C (2003) Ultrasound tomography imaging of radiation dose distributions in polymer gel dosimeters: Preliminary study. *Med. Phys.* 30: 2140–2148.

McJury M et al. (1999) Experimental 3D dosimetry around a high-dose-rate clinical 192Ir source using a polyacrylamide gel (PAG) dosimeter. *Phys. Med. Biol.* 44: 2431–2444.

Mein S et al. (2015) Remote dosimetry with a novel PRESAGE formulation. *Med. Phys.* 42: 3492.

Mijnheer B et al. (2013) In vivo dosimetry in external beam radiotherapy. *Med. Phys.* 40: 070903.

Miles D (2016) Advanced applications of 3D dosimetry and 3D printing in radiation therapy. MSc thesis, Duke University, Durham, NC.

Molineu A et al. (2013) Credentialing results from IMRT irradiations of an anthropomorphic head and neck phantom. *Med. Phys.* 40: 022101.

Moran JM et al. (2011a) Safety considerations for IMRT: Executive summary. *Pract. Radiat. Oncol.* 1: 190–195.

Moran JM et al. (2011b) Safety considerations for IMRT: Executive summary. *Med. Phys.* 38: 5067–5072.

Nasr AT et al. (2013) Evaluation of the potential for diacetylenes as reporter molecules in 3D micelle gel dosimetry. *Phys. Med. Biol.* 58: 787–805.

Nelms BE et al. (2013) Evaluating IMRT and VMAT dose accuracy: Practical examples of failure to detect systematic errors when applying a commonly used metric and action levels. *Med. Phys.* 40: 111722.

Nelms BE, Zhen H and Tome WA (2011) Per-beam, planar IMRT QA passing rates do not predict clinically relevant patient dose errors. *Med. Phys.* 38: 1037–1044.

Newton J et al. (2011) Commissioning a small-field biological irradiator using point, 2D, and 3D dosimetry techniques. *Med. Phys.* 38: 6754–6762.

Niebanck MH (2012) A novel comprehensive verification method for multifocal rapidArc radiosurgery. MSc thesis, Duke University, Durham, NC.

Niebanck M et al. (2013) Investigating the reproducibility of a complex multifocal radiosurgery treatment. *J. Phys. Conf. Ser.* 444: 012072.

Niroomand-Rad A et al. (1998) Radiochromic film dosimetry: Recommendations of AAPM Radiation Therapy Committee Task Group 55. *Med. Phys.* 25: 2093–2115.

Niu CJ et al. (2012) A novel technique to enable experimental validation of deformable dose accumulation. *Med. Phys.* 39: 765–776.

Novotny J Jr et al. (2002) Application of polymer gel dosimetry in Gamma Knife radiosurgery. *J. Neurosurg.* 97(5 Suppl): 556–562.

Oldham M (2006) 3D dosimetry by optical-CT scanning. *J. Phys. Conf. Ser.* 56: 58–71.

Oldham M (2014) Methods and techniques for comprehensive 3D dosimetry. In: *Advances in Medical Physics*, Chap. 5. (Eds. Godfrey DJ. et al., Madison, WI: Medical Physics Publishing).
Oldham M et al. (1998a) An investigation into the dosimetry of a nine-field tomotherapy irradiation using BANG-gel dosimetry. *Phys. Med. Biol.* 43: 1113–1132.
Oldham M et al. (2005) Cone-beam-CT guided radiation therapy: A model for on-line application. *Radiother. Oncol.* 75: 271–278.
Oldham M. et al. (2010) Initial experience with optical-CT scanning of RadBall Dosimeters. *J. Phys. Conf. Ser.* 250: 012079.
Oldham M et al. (2012) A quality assurance method that utilizes 3D dosimetry and facilitates clinical interpretation. *Int. J. Radiat. Oncol. Biol. Phys.* 84: 540–546.
Oldham M and Kim L (2004) Optical-CT gel-dosimetry. II: Optical artifacts and geometrical distortion. *Med. Phys.* 31: 1093–1104.
Oldham M et al. (1998b) Improving calibration accuracy in gel dosimetry. *Phys. Med. Biol.* 43: 2709–2720.
Oldham M et al. (2013) How accurate is image guided radiation therapy (IGRT) delivered with a micro-irradiator? *J. Phys. Conf. Ser.* 444: 012070.
Oldham M et al. (2003) Optical-CT gel-dosimetry. I: Basic investigations. *Med. Phys.* 30: 623–634.
Oldham M et al. (2001) High resolution gel-dosimetry by optical-CT and MR scanning. *Med. Phys.* 28: 1436–1445.
Olding T and Schreiner LJ (2011) Cone-beam optical computed tomography for gel dosimetry II: Imaging protocols. *Phys. Med. Biol.* 56: 1259–1279.
Olding T et al. (2013) Stereotactic body radiation therapy delivery validation. *J. Phys. Conf. Ser.* 444: 012073.
Olding T, Holmes O and Schreiner LJ (2010) Cone beam optical computed tomography for gel dosimetry I: Scanner characterization. *Phys. Med. Biol.* 55: 2819–2840.
Olsson LE et al. (1992) Three-dimensional dose mapping from gamma knife treatment using a dosimeter gel and MR-imaging. *Radiother. Oncol.* 24: 82–86.
Olsson LE et al. (1990) MR imaging of absorbed dose distributions for radiotherapy using ferrous sulphate gels. *Phys. Med. Biol.* 35: 1623–1631.
Palmer AL et al. (2013) Comparison of methods for the measurement of radiation dose distributions in high dose rate (HDR) brachytherapy: Ge-doped optical fiber, EBT3 Gafchromic film, and PRESAGE(R) radiochromic plastic. *Med. Phys.* 40: 061707.
Pavoni JF et al. (2012) Tomotherapy dose distribution verification using MAGIC-f polymer gel dosimetry. *Med. Phys.* 39: 2877–2884.
Pierquet M et al. (2010) On the feasibility of verification of 3D dosimetry near brachytherapy sources using PRESAGE/optical-CT. *J. Phys. Conf. Ser.* 250: 012091.
Podgorsak MB and Schreiner LJ (1992) Nuclear magnetic relaxation characterization of irradiated Fricke solution. *Med. Phys.* 19: 87–95.
Qian X, Adamovics J and Wuu CS (2013) Performance of an improved first generation optical CT scanner for 3D dosimetry. *Phys. Med. Biol.* 58: N321–N331.
Ramm D et al. (2012) Optical CT scanner for in-air readout of gels for external radiation beam 3D dosimetry. *Phys. Med. Biol.* 57: 3853–3868.

Rankine LJ et al. (2013) Investigating end-to-end accuracy of image guided radiation treatment delivery using a micro-irradiator. *Phys. Med. Biol.* 58: 7791–7801.
Rankine L and Oldham M (2013) On the feasibility of optical-CT imaging in media of different refractive index. *Med. Phys.* 40: 051701.
Rintoul L, Lepage M and Baldock C (2003) Radiation dose distribution in polymer gels by Raman spectroscopy. *Appl. Spectr.* 57: 51–57.
Sakhalkar HS et al. (2009a) A comprehensive evaluation of the PRESAGE/optical-CT 3D dosimetry system. *Med. Phys.* 36: 71–82.
Sakhalkar HS and Oldham M (2008) Fast, high-resolution 3D dosimetry utilizing a novel optical-CT scanner incorporating tertiary telecentric collimation. *Med. Phys.* 35: 101–111.
Sakhalkar H et al. (2009b) Investigation of the feasibility of relative 3D dosimetry in the Radiologic Physics Center Head and Neck IMRT phantom using presage/optical-CT. *Med. Phys.* 36: 3371–3377.
Sandilos P et al. (2006) Mechanical and dose delivery accuracy evaluation in radiosurgery using polymer gels. *J. Appl. Clin. Med. Phys.* 7(4): 13–21.
Scharf K and Lee RM (1962) Investigation of spectrophotometric method of measuring ferric ion yield in ferrous sulfate dosimeter. *Radiat. Res.* 16: 115–124.
Schreiner LJ (2004) Review of Fricke gel dosimeters. *J. Phys. Conf. Ser.* 3: 9–21.
Schreiner LJ et al. (1994) Imaging of HDR brachytherapy dose distributions using NMR Fricke-gelatin dosimetry. *Magn. Reson. Imaging* 12: 901–907.
Sharpe J (2004) Optical projection tomography. *Ann. Rev. Biomed. Eng.* 6: 209–228.
Sharpe J et al. (2002) Optical projection tomography as a tool for 3D microscopy and gene expression studies. *Science* 296(5567): 541–545.
Stanley SJ et al. (2012) Locating, quantifying and characterising radiation hazards in contaminated nuclear facilities using a novel passive non-electrical polymer based radiation imaging device. *J. Radiol. Prot.* 32: 131–145.
Stein J et al. (1994) Dynamic x-ray compensation for conformal radiotherapy by means of multi-leaf collimation. *Radiother. Oncol.* 32: 163–173.
Thomas A et al. (2011a) Commissioning and benchmarking a 3D dosimetry system for clinical use. *Med. Phys.* 38: 4846–4857.
Thomas A, Newton J and Oldham M (2011b) A method to correct for stray light in telecentric optical-CT imaging of radiochromic dosimeters. *Phys. Med. Biol.* 56: 4433–4451.
Thomas A et al. (2013a) A comprehensive investigation of the accuracy and reproducibility of a multitarget single isocenter VMAT radiosurgery technique. *Med. Phys.* 40: 121725.
Thomas A et al. (2013b) Comprehensive quality assurance for base of skull IMRT. *J. Phys. Conf. Ser.* 444: 012050.
Thomas A and Oldham M (2010) Fast, large field-of-view, telecentric optical-CT scanning system for 3D radiochromic dosimetry. *J. Phys. Conf. Ser.* 250: 012007.
Thomas A et al. (2011c) A method to correct for spectral artifacts in optical-CT dosimetry. *Phys. Med. Biol.* 56: 3403–3416.
Thomas A et al. (2013c) The effect of motion on IMRT–looking at interplay with 3D measurements. *J. Phys. Conf. Ser.* 444: 012049.
Trapp JV et al. (2001) An experimental study of the dose response of polymer gel dosimeters imaged with x-ray computed tomography. *Phys. Med. Biol.* 46: 2939–2951.

Tseng YJ et al. (2002) The role of dose distribution gradient in the observed ferric ion diffusion time scale in MRI-Fricke-infused gel dosimetry. *Magn. Reson. Imaging* 20: 495–502.
Vandecasteele J and De Deene Y (2013a) On the validity of 3D polymer gel dosimetry: I. Reproducibility study. *Phys. Med. Biol.* 58: 19–42.
Vandecasteele J and De Deene Y (2013b) On the validity of 3D polymer gel dosimetry: II. Physico-chemical effects. *Phys. Med. Biol.* 58: 43–61.
Vandecasteele J and De Deene Y (2013c) On the validity of 3D polymer gel dosimetry: III. MRI-related error sources. *Phys. Med. Biol.* 58: 63–85.
Vandecasteele J and De Deene Y (2013d) Evaluation of radiochromic gel dosimetry and polymer gel dosimetry in a clinical dose verification. *Phys. Med. Biol.* 58: 6241–6262.
Vergote K (2003) Application of monomer/polymer gel dosimetry to study the effects of tissue inhomogeneities on intensity-modulated radiation therapy (IMRT) dose distributions. *Radiother. Oncol.* 67: 119–128.
Vergote K (2004) Validation and application of polymer gel dosimetry for the dose verification of an intensity-modulated arc therapy (IMAT) treatment. *Phys. Med. Biol.* 49: 287–305.
Vidovic AK et al. (2014) An investigation of a PRESAGE® in vivo dosimeter for brachytherapy. *Phys. Med. Biol.* 59: 3893–3905.
Wai P et al. (2009) Dosimetry of the microSelectron-HDR Ir-192 source using PRESAGE and optical CT. *Appl. Radiat. Isot.* 67: 419–422.
Wang Z et al. (2010) Dose verification of stereotactic radiosurgery treatment for trigeminal neuralgia with PRESAGE 3D dosimetry system. *J. Phys. Conf. Ser.* 250: 012058.
Webb S (1989) Optimisation of conformal radiotherapy dose distributions by simulated annealing. *Phys. Med. Biol.* 34: 1349–1370.
Winfree AT et al. (1996) Quantitative optical tomography of chemical waves and their organizing centers. *Chaos* 6: 617–626.
Wolodzko JG, Marsden C and Appleby A (1999) CCD imaging for optical tomography of gel radiation dosimeters. *Med. Phys.* 26: 2508–2513.
Wuu CS and Xu Y (2006) Three-dimensional dose verification for intensity modulated radiation therapy using optical CT based polymer gel dosimetry. *Med. Phys.* 33: 1412–1419.
Wuu C and Xu Y (2010) How to perform dosimetry with optical-CT. *J. Phys. Conf. Ser.* 250: 012044.
Xu Y, Wuu CS and Maryanski MJ (2004) Performance of a commercial optical CT scanner and polymer gel dosimeters for 3-D dose verification. *Med. Phys.* 31: 3024–3033.
Yates ES et al. (2011) Characterization of the optical properties and stability of Presage following irradiation with photons and carbon ions. *Acta Oncol.* 50: 829–834.
Yeo UJ et al. (2012) A novel methodology for 3D deformable dosimetry. *Med. Phys.* 39: 2203–2213.
Zhao L et al. (2012) Feasibility of using PRESAGE(R) for relative 3D dosimetry of small proton fields. *Phys. Med. Biol.* 57: N431–N443.

7

Electronic Portal Imaging Device Dosimetry

Boyd McCurdy, Peter Greer, and James Bedford

7.1 Introduction

Portal imaging was developed to improve the geometric accuracy of patient setup by generating radiographs of the patient just prior to or during treatment delivery. The portal image is formed by x-rays created by the megavoltage photon source in the linear accelerator (linac), and thus is shaped by the therapy collimation system or "portal." Historically, several different electronic technologies have been utilized to replace the role of radiographic films in capturing these images, including the scanning liquid ionization chamber array (commercialized as PortalVision by Varian Medical Systems, Palo Alto, CA), and camera-based systems viewing a phosphor scintillating screen. Camera-based systems used either analogue video cameras (e.g., Beamview Plus; Siemens Healthcare, Erlangen, Germany) or charge-coupled device (CCD) cameras (e.g., SRI-100; Elekta, Stockholm, Sweden). By modern standards, this first generation of electronic portal imaging devices (EPIDs) produced somewhat poor quality radiological images and possessed some promising but very limited dosimetric characteristics. Beginning in the early 2000s, the current generation of amorphous silicon (a-Si) flat-panel EPID technology eventually became commercially available from, at that time, three major linac manufacturers, and today they are a ubiquitous choice on newly purchased linacs. Current a-Si EPID technology not only generates better quality images but also demonstrates more useful dosimetric characteristics. Ultimately, the widespread availability and convenience of a-Si

EPIDs have made them the subject of strong research interest in anatomical and geometrical imaging and tracking, as well as a broad range of dosimetry applications including linac quality assurance (QA), pretreatment intensity-modulated radiation therapy (IMRT)/volumetric-modulated arc therapy (VMAT) QA, and real-time and *in vivo* patient treatment verification.

Radiation treatment delivery is continually becoming more complex, emphasized with the introduction of techniques such as dynamic IMRT and VMAT. VMAT is commercially available as either RapidArc™ (Varian Medical Systems, www.varian.com) or VMAT (Elekta, www.elekta.com). VMAT is especially complex as it delivers patient treatment while the gantry rotates and involves simultaneous modulation of radiation aperture, dose rate, gantry speed, and potential collimator rotation and couch motion. In addition to these new complex delivery technologies, there is increased clinical interest in aggressive treatment regimens that deliver the therapeutic dose in larger per-fraction doses and in fewer fractions compared to standard treatments. These techniques, including hypofractionation, stereotactic radiosurgery (SRS), and stereotactic body radiation therapy (SBRT), place greater importance on the dosimetric accuracy of each individual treatment fraction delivered. These developments emphasize the importance of recommendations of QA bodies throughout the world: patient treatment verification, preferably through *in vivo* patient dosimetry, is highly desirable for optimal patient safety during radiation treatment (Derreumaux et al., 2008; RCR, 2008; WHO, 2008; ICRU, 2010).

There are various ways to use EPIDs as dosimeters, ranging from daily output check devices to pretreatment verification systems to full three-dimensional (3D) *in vivo* dose estimation of the patient treatment from transmission images gathered during treatment delivery. These methods have been described in detail by van Elmpt et al. (2008) and are reviewed and updated in this chapter. Application examples for 3D patient dose estimation will be provided, since the rewards of daily 3D patient dose estimation are great: (1) mapping under- or overestimates of patient dose onto the anatomy, thus providing clinically relevant feedback; (2) availability of cumulative dose over the entire treatment course; (3) improved treatment through adaptive radiotherapy approaches; and (4) production of a medical/legal record of delivered patient dose. EPIDs provide a means for this type of verification as described in this chapter and in more detail in Chapter 18. Additional useful review materials on EPID dosimetry include Greer and Vial (2011), Greer (2013), McCurdy (2013), and Mijnheer et al. (2013).

7.2 a-Si EPID Systems and Dosimetric Characteristics

This section describes the image formation process of a-Si EPID systems and also discusses system characteristics with respect to measuring dose.

7.2.1 a-Si EPID Systems and Configuration

The modern a-Si flat-panel EPID is based on thin-film semiconductor technology. The current generation of commercially available EPIDs use a-Si deposited on a glass substrate to form arrays of photodiodes and field-effect transistors (FETs). An array of pixels is formed, with each pixel consisting of a photodiode and an FET (illustrated in Figure 7.1). The pixel arrays can detect radiation both directly and indirectly. Indirect detection, used on all current commercial systems, describes the configuration where the radiation energy deposited in

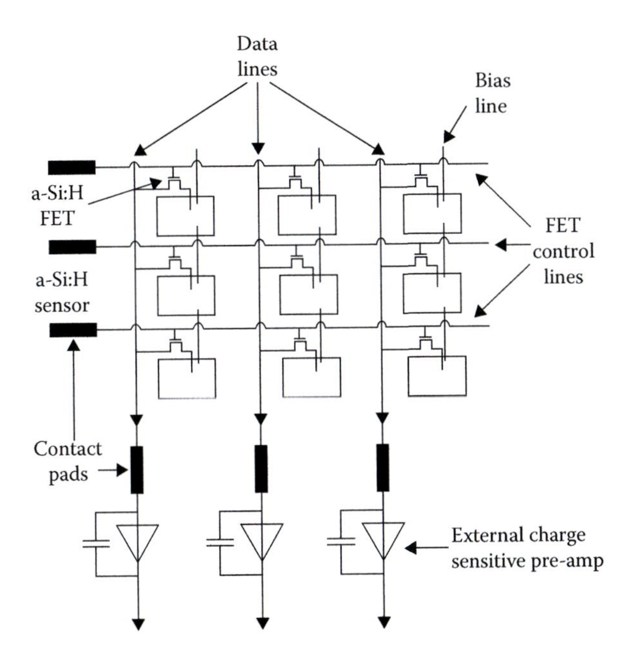

Figure 7.1

Schematic diagram from a portion of an amorphous silicon (a-Si) imaging array showing organization of photodiodes and field-effect transistors (FETs) as well as bias, FET control, and data lines. Charge sensitive preamplifiers are external to the imaging array. (Reprinted from Antonuk LE et al., *Med. Phys.*, 19, 1455–1466, 1992. With permission.)

the detector is converted into optical photons, which are then detected by the photodiodes. This conversion is typically accomplished through the use of a phosphor scintillating screen; the standard phosphor used in current commercial EPIDs is gadolinium oxysulfide doped with terbium, Gd_2O_2S:Tb. An advantage of the indirect configuration is the gain in detector signal of about a factor of 10 compared to a direct detection configuration (El-Mohri et al., 1999). To improve detector efficiency further, an additional thin sheet of metal is placed directly upstream of the scintillator to help convert more incident photons into electrons to increase dose deposited in the scintillator, therefore increasing signal in the photodiodes and improving sensitivity. The metal sheet also helps reduce low-energy patient scatter reaching the scintillator. The metal layer in current commercial EPIDs is typically 1 mm of copper (Varian and Elekta). Some users may place additional buildup on the detector, especially if performing dosimetry with high-energy therapy beams above 6 MV, to provide electron equilibrium in the phosphor layer, and further reduce low-energy photons reaching the phosphor layer. The direct detection configuration describes the situation where the incident radiation is not converted into optical photons and the photodiodes simply detect the charge directly deposited by the radiation energy. This configuration was examined as an approach to achieve a detector energy response closer to that of water (El-Mohri et al., 1999; Moran et al., 2005; Chen et al., 2007; Vial et al., 2008a, 2009; Sabet et al., 2010, 2012; Gustafsson et al., 2011), since the high-atomic-number phosphor screen used in the indirect configuration exhibits an energy response different than water (see Section 7.3.5). Even by simply physically blocking the optical photons from reaching the photodiodes, a nearly water-equivalent response is observed (Gustafsson et al., 2009). However, compared to indirect configurations, the direct detector configurations are much less sensitive to incident radiation because they do not experience the signal gain of the optical photon conversion step. Antonuk et al. (1992, 1995, 1996) helped pioneer the development of the modern a-Si EPID systems for radiotherapy, and publications by that group contain detailed discussions of the design and operation of the detector components for the interested reader.

7.2.2 EPID Acquisition Modes

Generally, there are two main acquisition modes of interest for dosimetric applications of EPIDs. The first is integrated mode, where all image frames during an acquisition are summed and a single "integrated" image is recorded and returned to the user. The second is cine mode (or "movie" mode), where images are recorded at a fixed time interval, thus creating a sequence or movie of images captured during an irradiation. Both major linac manufacturers have integrated acquisition mode options that are useful for dosimetry applications. However, it should be noted that currently available cine acquisition modes are not designed with dosimetric applications as the primary purpose. With this in mind, there are a few cine options available on commercial linacs that vary depending on the manufacturer and type of linac, and are described here. For the Varian Clinac series, the clinical acquisition software does not return a series of individual image frames, but rather a series of frame-averaged images. The number of image frames used to create each recorded image in the series is typically controlled by the user through a frame-averaging parameter. For the Varian TrueBeam series, the cine mode saves a normalized movie file that is not suitable for dosimetric applications; however, an "image-processing service" is provided that saves individual frames in DICOM (Digital Imaging and Communications in Medicine, National Electrical Manufacturers Association) format. For the Elekta EPID, Mans et al. (2010a) developed in-house acquisition software to save every detector frame during image acquisition, achieving about 2.5 frames per second. PerkinElmer, the supplier of Elekta EPIDs, provides software to run the imager in cine mode, but no gantry angle is provided.

When the current a-Si flat-panel EPID technology became commercially available in the early 2000s, research interests focused on characterizing the integrated acquisition mode, which is well suited to static gantry treatment delivery applications. However, with the clinical implementation of VMAT deliveries beginning around 2009, the integrated acquisition mode is no longer appropriate as it collapses all time-dependent (and gantry-dependent) dose information into a single image. Therefore, the movie acquisition mode became important due to its ability to capture the time-dependent nature of these deliveries.

7.2.3 EPID Imager Readout

The EPID imaging panels are not read out instantaneously. Rather, portions of the panel are read out at finite time intervals. This can become an important aspect of operation for time-resolved dosimetry applications, as the photodiodes will retain their charge information until they are read out. Therefore the dose pattern information contained in individual image frames will be effected by the timing of the readout and the spatial pattern of the readout process. The Varian panel is read out in a simple raster pattern, but with groups of rows being read out together instead of just a single line. Typically, around 15–30 lines form one group, and this value holds constant once it is set. This readout approach appears to be used for newer EPIDs on the Varian TrueBeam series, although this is not yet well described in the literature. Berger et al. (2006) described a method to adjust the number of lines read out between dose pulses to avoid detector saturation, and also described the timing relationship of the EPID row read out and the linac dose pulses. This timing is of interest as Varian linacs implement a pulse-drop servo control for VMAT delivery. This may result in pulse dropping artifacts in the EPID image frames acquired during VMAT. Since the different groups of rows in the imager are read out sequentially in time, a dropped pulse will result in a different output signal. The resulting image frame then will contain banding artifacts that appear as dark horizontal bands across the image (Yeo et al., 2013b). These are not true artifacts since the imager is accurately registering the dose pulses received, rather the visual disruption is due to the interaction between the time of the dose pulses and the timing and pattern of the image readout. Their impact on an image may be reduced through frame averaging, although this averages the dose delivery of an image over a larger gantry angle range. PerkinElmer (Waltham, MA) supplies EPID imagers for both Elekta and Siemens, and uses a more complex readout pattern. The panel is split into two halves separated by the in-plane axis through the panel center, each of which is further split into eight readout regions. These readout regions are read row by row from the outside edge of the panel to the center, and moving between adjacent readout regions in the gun-target direction on one half of the panel and in the target-gun direction on the other half of the panel. Podesta et al. (2012) described the panel readout in detail, and also demonstrated that the panel readout has a significant impact on absolute dosimetry applications at low monitor unit (MU) exposures but can be accounted for. A correction algorithm has also been proposed to reduce the effect of these artifacts in PerkinElmer EPID panels, although not applicable for small periods at beam startup or shutdown (Mooslechner et al., 2013).

For Varian EPID systems, achievable frame rates are dependent on the image acquisition parameters used, which impact the dose needed per image, and also on the dose rate of the beam. Typical frame rates are approximately 7.5–10 Hz (Varian aS1000), 9–12 Hz (Varian aS500). For Elekta EPID systems, the frame rate is currently fixed at 3.1 Hz (Winkler and Georg, 2006) but that is below the maximum frame rate of the PerkinElmer imaging panel used, so theoretically could be increased.

7.2.4 Dosimetric Characteristics of a-Si EPIDs

Desirable characteristics of any dosimeter include linearity (with dose and dose rate), reproducibility, high spatial resolution, no dead time, and real-time readout. Researchers have demonstrated many of these characteristics for a-Si EPIDs. For the PortalVision aS500/1000 (Varian) operated in integrating mode, many useful results are available (Greer and Popescu, 2003). Linearity of EPID response to dose and dose rate was established. A small dead-time effect was identified in that work and was subsequently removed by the manufacturer through a software upgrade in the acquisition computer (upgrading from IAS2 to IAS3). Kavuma et al. (2008) demonstrated the suitability of the Varian IAS3 software for dosimetry applications. Dosimetric properties of aS500/1000 EPIDs operated in movie mode were investigated by McCurdy and Greer (2009), where it was demonstrated that there was a small but nearly constant amount of missing signal in cine acquisition mode. This only had significant dosimetric effect for low MU irradiations (approximately <100 MU, much less than typical VMAT deliveries). However, it can be an important consideration when selecting the number of MUs for calibrating signal to dose. This effect was subsequently shown to be due to incomplete frame capture at the very beginning and very end of the acquisition session (Greer, 2013) and is discussed in more detail in Section 7.3.3. For the Elekta iViewGT EPID system, McDermott et al. (2004) and Winkler et al. (2005) documented many dosimetric performance studies of the system operated in integrating mode. McDermott et al. observed some nonlinearity in dose and dose rate; however, these could be corrected to within 1% of linearity through the use of a 5-mm copper buildup plate and a time-dependent ghosting correction factor (the ghosting effect is described in more detail in Section 7.3.4). Winkler et al. observed variations away from perfect-dose-rate linearity of almost 7%, attributing the variation to a dose-per-frame effect. However, these variations were reproducible and could be corrected with a custom (i.e., per EPID system) calibration to recover linearity with dose and dose rate and to improve dosimetric accuracy (Winkler et al., 2005).

Long-term reproducibility is a desirable feature of any dosimetric system, and a-Si flat-panel EPIDs have been shown to have this characteristic. Long-term reproducibility of the Varian EPID system has been demonstrated to be <1% (all pixels) over a 3-year period (King et al., 2011) and reproducibility of the Elekta EPID system was demonstrated to be <0.5% (all pixels) over nearly 2 years (Louwe et al., 2004). The pixel pitches of these systems (~0.4×0.4 mm^2) provide a high spatial resolution compared to most other available dosimeters, vastly superior to most ionization chambers, and also superior to patient computed tomography (CT) datasets where voxel sizes are typically about $1 \times 1 \times 2$ mm^3. EPID pixel pitches are currently 0.39×0.39 mm^2 (Varian) and 0.40×0.40 mm^2 (Elekta and Siemens), with varying active areas of 40×30 cm^2 (aS500 and aS1000; Varian), 40×40 cm^2 (dosimetry mode of aS1200; Varian), and 41×41 cm^2 (Elekta and Siemens), corresponding to 1024×768 pixels, 1024×1024 pixels, and 1024×1024 pixels, respectively.

Furthermore, the a-Si flat-panel imagers have been shown to be highly resistant to radiation damage (Boudry and Antonuk, 1994, 1996), which is another useful feature for imaging and dosimetry applications. However, if the EPID is to be used as a tool for routine clinical dosimetric applications (e.g., linac QA or *in vivo* dosimetry), it is necessary to implement a QA program on the EPID itself, which should be able to detect deteriorating image quality and dosimetric performance from accumulative radiation damage. For instance, corrections for a possible change in sensitivity of each EPID are determined weekly at the Netherlands Cancer Institute by irradiating a 20-cm-thick slab phantom under reference conditions with a 10 cm × 10 cm field and with two VMAT plans (Mijnheer et al., 2015).

7.3 Technical Challenges to Using EPIDs for Dosimetric Applications

There are many challenges to using EPIDs as dosimeters in clinical practice, as described in this section. Many of these issues have been addressed, although the vendors could make improvements to signal acquisition and clinical software to further improve the dosimetric performance of EPIDs.

7.3.1 Pixel Sensitivity Variation

The a-Si flat-panel imagers are known to exhibit pixel-to-pixel variations in response of approximately ±5%. These nondose-related variations may be due to intrinsic response differences of individual pixels, or due to differences in the response of the readout electronics between channels (Roberts et al., 2004). Ideally, the relative pixel sensitivities could be obtained by acquiring an image of a perfectly uniform incident energy fluence radiation distribution. In practice, such an ideal field is difficult to generate for megavoltage therapy beams. The approach commonly used by clinical EPID software is to make a correction using a flood-field image, defined as an open field large enough to just cover the imager sensitive area. However, this technique incorporates the incident beam fluence shape, as measured by the EPID, together with the pixel sensitivity variation. A more uniform intensity image at the EPID may be approximated by maximizing the source-to-imager distance and for beams using a flattening filter placing approximately 10 cm of equivalent water on top of the imager (Roberts et al., 2004; Siebers et al., 2004). The 10-cm-deep water is chosen because most linac manufacturers optimize their flattening filter design at this depth to achieve a flat beam profile, although this field will still exhibit off-axis changes due to spectral differences and inverse-square intensity differences. Another method to estimate the pixel sensitivity map is by using a single small, square field aligned with the central axis, and acquiring repeated EPID images with the imager offset laterally and longitudinally at several different physical positions (Parent et al., 2006). Use of the same, small (i.e., 10×10 cm^2) field ensures the energy spectrum remains reasonably constant across the field for all the EPID acquisitions. By stitching together the images, one can obtain the pixel sensitivity map for most of the active imager area. An efficient method where only three images are used to derive the pixel sensitivity variations has been proposed by Boriano et al. (2013). Greer used a small incident field and moved the EPID laterally to stitch together a pixel sensitivity profile, which was then divided by the flood-field profile to obtain an EPID response (Greer, 2005), as illustrated in Figure 7.2. This was assumed to be radially symmetric and used to generate a 2D EPID response image, which was then divided into the flood field to obtain the full 2D pixel sensitivity matrix. This approach has the drawback of removing any beam asymmetries from the corrected images. In the Netherlands Cancer Institute, an analogous representation of the sensitivity map is estimated using the ratio of the EPID flood field to a 2D profile measured using a small ionization chamber in a mini-phantom, because their back-projection model requires the primary dose map at the EPID level (Wendling et al. 2006).

7.3.2 Beam-Fluence Shape

As described in Section 7.3.1, the clinical image acquisition software for the various EPID systems uses a flood-field calibration image to normalize the returned acquired image in an effort to improve image quality for anatomy viewing by removing pixel sensitivity variations. However, this approach also "flattens" the image by removing the nonuniform, incident beam-fluence shape as well as the pixel-to-pixel response characteristics, both inherent in the raw EPID image (Greer, 2005). The incident beam fluence shape is a dosimetric characteristic of the beam and may be of interest in EPID dosimetric applications. It is important to remember that the incident beam fluence shape will be modified by the response of the device measuring it and that the measured signal is typically due

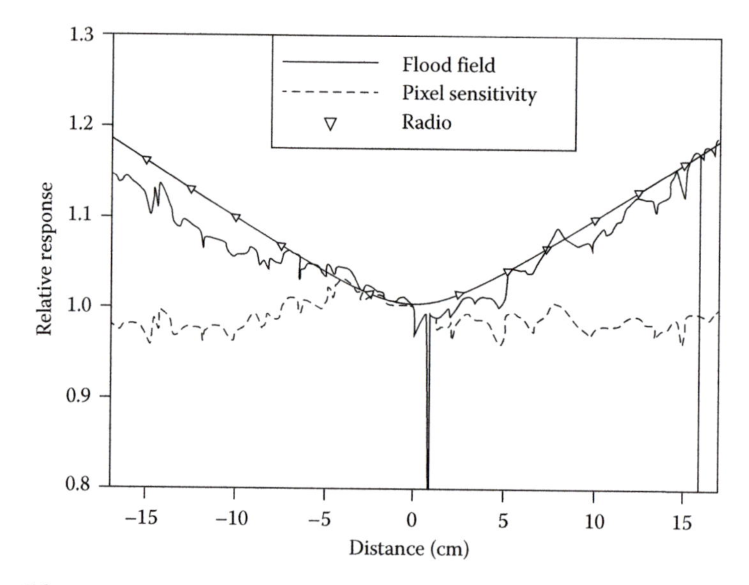

Figure 7.2

Cross section of the two-dimensional pixel sensitivity map for a 6-MV beam and an aS500 EPID (Varian Medical Systems). (Reprinted from Greer PB, *Med. Phys.*, 32, 3558–3568, 2005. With permission.)

to the incident fluence converted to dose deposited in the detector medium. This has important consequences for EPID dosimetric applications, since the beam profile shape as measured in an EPID, where phosphor is the detector medium, will be different from that measured using an ionization chamber in a scanning water tank.

Unfortunately in current clinical software, the image-processing step applying the flood-field normalization cannot be turned off by the user. If the user is interested in the incident beam fluence profile dosimetric information, it needs to be reintroduced to the acquired EPID images. This can be accomplished via multiplication of the stored flood-field calibrated image or by application of an independently estimated beam profile shape, combined with a separate correction of the pixel sensitivity variations. When reintroducing the beam profile information, the spatial position of the EPID relative to the profile position needs to be accounted for otherwise dosimetric errors will be introduced. Other applications may desire to convert the EPID response to a water-equivalent detector response as described in Section 7.3.5.

7.3.3 Incomplete Signal Acquisition

For cine acquisition mode in Varian a-Si EPIDs, it was shown that a small amount of image dose is missing from the total acquisition as compared to integrated acquisition mode (McCurdy and Greer, 2009). They demonstrated the time-resolved EPID signal agreed well with time-resolved ionization chamber measurements, but the acquisition mode underresponded at small dose amounts associated with short irradiation times. More recent work has shown this to be the result of the clinical software discarding two image frames at the beginning as well as at the end of an irradiation (Greer, 2013; McCowan et al., 2014). These images are discarded since they are considered “partial” frames, where the finite

EPID readout time results in image signal being present in only a portion of the full EPID image frame as mentioned in Section 7.2.4. In fact, since the photodiodes capture charge that is not cleared until readout occurs, the software may identify up to two initial frames as being only partial images. If the beginning of the EPID image frame readout (i.e., physically at the top of the imaging panel) happens to coincide closely in time with the start of irradiation, then only one frame is identified as a partial image and discarded. The same issue occurs at the end of irradiation. These partial frames are discarded, otherwise they may introduce visual artifacts into the EPID images; however, the partial frames do accurately represent the dose information delivered at the beginning and at the end of irradiation. The recent availability of separate frame-grabber systems that can capture any required frame sequence simultaneously with the clinical acquisition has shown that frames at the end of clinical integrated mode acquisition may not be captured with the vendor's software. This may be a significant factor in the observed EPID underresponse at low MU (Podesta et al., 2012) in the early literature. For example, comparisons with frame-grabber measurements for Varian systems in integrated acquisition mode have shown that the second partial frame after beam off was not being recorded (Greer, 2013). Recent software upgrades have corrected this discrepancy.

7.3.4 Ghosting and Lag

Flat-panel imagers using a-Si thin-film transistors and photodiodes are known to exhibit temporal artifacts (Siewerdsen and Jaffray, 1999; Overdick et al., 2001). These effects cause a small portion of the signal generated in the photodiodes to contribute to image formation at some time *after* the original signal was generated. These temporal artifacts are typically categorized as "image lag" and "image ghosting." Image lag is mainly attributed to trapped charge in the photodiode which, when read out in subsequent frames, results in that portion of EPID signal being offset in time. Smaller effects contributing to image lag may also include incomplete charge transfer into the readout electronics, and afterglow in the phosphor screen (Siewerdsen and Jaffray, 1999). Image ghosting refers to the change in individual pixel gains due to the trapped charge modifying the electric field strength in the photodiode, thus allowing prior irradiations to effect charge-collection efficiency and therefore effect EPID response. Many investigators have observed image lag and image ghosting effects with a-Si EPIDs (Siewerdsen and Jaffray, 1999; Partridge et al., 2002; McDermott et al., 2004, 2006; Winkler et al., 2005; Warkentin et al., 2012). These effects have been shown to result in a relative underresponse in the EPID of up to 11% compared to a perfect linear response, and this may have a direct impact on dosimetric applications of EPIDs. The largest magnitude effects of image lag and image ghosting occur with shorter irradiation times (i.e., low number of MUs), typically below those of routine clinical use, but including step and shoot IMRT delivery on Elekta linacs where each small MU segment is delivered with a beam on/off cycle. As discussed in Section 7.3.3, recent evidence of missing image frame data, even in integrated acquisition mode (Podesta et al., 2012; Greer 2013), may contribute significantly to the observed estimates of underresponse due to ghosting/lag effects in the early literature.

Several investigators have proposed techniques to nullify these effects, usually in the form of time-dependent signal correction. Time-dependent decay functions, typically using single or multiple exponential terms as a function of time or dose, can be used to model the temporal artifact effects and thus remove those effects (Hsieh et al., 2000; Roberts et al., 2004). Early strategies used a global temporal correction applied to the entire image (McDermott et al., 2004; Winkler et al., 2005) or a global dose-calibration strategy (Alshanqity et al., 2012; Deshpande et al., 2014). With the arrival of VMAT techniques, pixel-by-pixel strategies, accounting for the prior irradiation history of individual pixels, have been explored (Winkler et al., 2007; Warkentin et al., 2012; Podesta et al., 2014), and these approaches are also useful for static gantry dynamic IMRT delivery. The pixel-by-pixel correction methods have used exponential time decay functions (Podesta et al., 2014) but also an empirical correction strategy using measurement-estimated decay curves has also been demonstrated to account for these effects (Warkentin et al., 2012). Hardware modification could also reduce image lag and image ghosting effects, for example, by operating the photodiode at a higher bias voltage (Siewerdsen and Jaffray, 1999).

7.3.5 Nonwater-Equivalent Energy Response

The a-Si EPID is known to overrespond to low-energy photons (below about 0.5 MeV) as compared to a water-equivalent detector (Jaffray et al., 1994; McCurdy and Pistorius, 2000), as illustrated in Figure 7.3, due to the increased photoelectric effect occurring in the copper buildup plate and in the phosphor screen. The energy response of the EPID can be modeled explicitly via energy-dependent dose kernels (McCurdy et al., 2001; Chytyk-Praznik et al., 2013) or via direct Monte Carlo simulation (Siebers et al., 2004; Jarry and Verhaegen, 2007). Many investigators have been interested in converting the EPID signal, essentially dose-to-phosphor, into a water-equivalent detector signal through empirical means (Wendling et al., 2006, 2009; Nijsten et al., 2007; King et al., 2012; Zwan et al., 2014). A dose-to-water conversion approach offers the advantages of familiarity and simpler validation, i.e., through independent measurement with water-equivalent dosimeters, but brings the disadvantage of an additional processing step that will be associated with an inherent uncertainty. It is also more

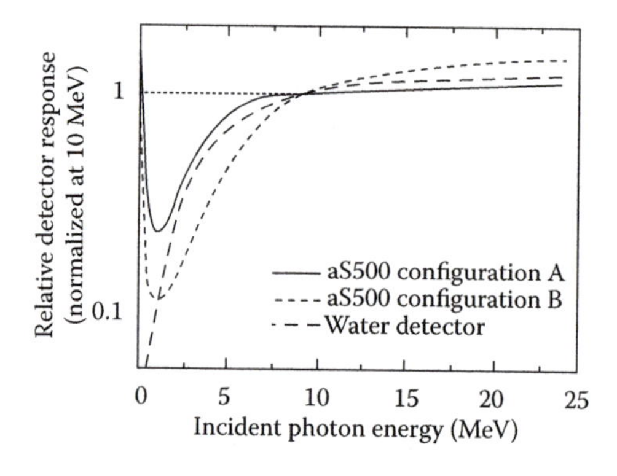

Figure 7.3

Energy sensitivity of an a-Si EPID in two configurations (A = no additional buildup, B = 3.0 cm of solid water buildup), normalized at 10 MeV, compared to a water-equivalent detector, illustrating the energy response differences, especially at lower energies below approximately 1 MeV. (Reprinted from McCurdy BM et al., *Med. Phys.*, 28, 911–924, 2001. With permission.)

difficult to account for the energy-dependent response of the EPID with these conversion models.

7.3.6 Self-scatter Signal

The a-Si EPID imaging units are fairly thin, cassette-like systems weighing a few kilograms, approximately 10–20 kg depending on the manufacturer, but this is enough material to cause a "self-scatter" signal. That is, photon scatter generated within the EPID itself, not just primary photons from the linac head and/or scattered photons originating in the patient, will contribute to image signal (McCurdy et al., 2001). The proportion of this source of EPID signal ranges from about 1–2% at high-energy photons up to 30% for low-energy (<1 MeV) photons. Dosimetric applications can account for this signal via numerous techniques, such as empirical methods (Wendling et al., 2006, 2009; Nijsten et al., 2007), analytical modeling (Chytyk-Praznik et al., 2013), or explicit modeling, i.e., Monte Carlo simulation (Siebers et al., 2004; Jarry and Verhaegen, 2007).

7.3.7 Optical Glare

Radiation energy deposited in the phosphor will result in the release of optical photons that are detected by the photodiodes and converted into an electrical signal. Since the phosphor used in commercial a-Si EPIDs is translucent, the optical photons will begin to spatially diffuse upon creation due to scattering processes. The diffusion can only take place over the thickness of the phosphor which is usually about 0.3 mm in the current commercial EPID systems. This results in a blurring of the deposited dose pattern and is termed an optical glare effect. However, due to the very short optical path length, this effect is much smaller in a-Si EPIDs as compared to the previous generation of camera-based systems which demonstrated extremely large amounts of glare signal. For example, for field-size response on the central axis, a-Si EPIDs show a 1–2% effect versus up to 25% for camera-based EPIDs (Munro et al., 1998). The cameras of the camera-based EPIDs could not be placed directly in the beam since they needed to be protected from radiation. Therefore they were mounted in a shielded location withdrawn into the main linac gantry, forcing the use of a mirror system to direct the optical photons emitted from the phosphor screen to the camera. The use of one (Siemen's Beamview Plus) or two (Elekta's SRI-100 or iView) mirrors and the long optical chain significantly increased the amount of scatter the optical photons suffered, making dosimetry applications using those systems extremely challenging.

The optical glare effect for the commercially available a-Si flat-panel EPIDs has been characterized and may be modeled via a convolution of dose pattern in the scintillator, with a simple exponential function representing the point spread function of the optical photons in the scintillator layer of the EPID. Kausch et al. (1999) used Monte Carlo simulation of radiation transport and optical photon transport to examine detective quantum efficiency (DQE) of metal/phosphor imaging systems and predicted significant improvement in moving from camera-based to a-Si EPIDs. Kirkby and Sloboda (2005) used Monte Carlo simulation to derive a point spread function for optical photons in Varian's aS500/1000 EPID imager, and validated this through independent measurement. More recently, Blake et al. (2013a) used GEANT4 to investigate the optical photon spread in the phosphor layer of the PerkinElmer EPID, again verifying the effect on field-size response and profile shape is small.

7.3.8 Patient Scatter

Photon scatter will be generated by the treatment beam interacting with the patient, creating an additional, complex source of mostly lower energy photons incident on the EPID. The physical nature of these scattered photons has been studied as a function of the field size, patient or phantom thickness, and the distance the imager is positioned behind the patient, usually termed "air gap" (Jaffray et al., 1994; Swindell and Evans, 1996; McCurdy and Pistorius, 2000). Many methods have been demonstrated to approximate or predict patient scatter at the EPID. Measurement-based approaches, where the patient or phantom scatter contribution is mapped over a range of clinically relevant situations, have been shown to be fast and reasonably accurate (Piermattei et al., 2006; Wendling et al., 2006; Peca and Brown, 2014). Pencil beam kernels representing either fluence or dose spread in the EPID due to patient-generated scatter may be applied in a superposition calculation framework. The kernels are usually generated via Monte Carlo techniques (Hansen et al., 1997; McCurdy and Pistorius, 2000; Spies et al., 2001; Chytyk-Praznik et al., 2013). Single-scattered photon fluence into the EPID can be exactly predicted analytically using Compton kinematics and Klein–Nishina cross-section functions (McCurdy and Pistorius, 1998), but no analogous exact analytical solution exists for multiple-scattered photons. However, the multiple-scattered component may be handled separately using a pencil beam kernel generated with Monte Carlo techniques (Spies et al., 2000). Full Monte Carlo simulations can calculate the patient scatter entering the EPID to arbitrary levels of accuracy at the cost of time (Jarry and Verhaegen, 2007). In general, empirical or correction-based techniques that have been used are attractive due to their simplicity, whereas model-based approaches trade-off an increase in complexity to achieve robustness over a wide range of operating conditions.

7.3.9 Backscatter from Mounting Arm

For the Varian aS500/1000 EPID, the imaging unit is mounted on a robotic arm that has been shown to contribute additional signal, up to 6% of maximum dose, to the image from increased backscattered photons. The backscatter signal contribution is known to be asymmetrical, field size dependent, and field location dependent (Rowshanfarzad et al., 2010a). This effect can be modeled using a simple backscatter kernel convolved with the portion of incident beam that impinges on the arm support components, which in turn is obtained by applying a simple binary mask representing the arm support shape to the incident beam shape at the EPID. This approach was further improved by King and Greer (2013), who optimized the estimated backscatter kernel and provided an iterative correction technique to estimate and remove the backscatter from the measured image. Hardware solutions also exist to mitigate this issue. It has been shown that placing additional backscatter material behind the EPID imaging cassette but upstream of the support arm components results in a uniform and symmetric backscatter response that can be corrected more simply. For example, Siebers et al. (2004) used 9.8 mm of water in Monte Carlo simulations, while others have explored a 2-mm lead sheet (Rowshanfarzad et al., 2010b, 2012a; King et al., 2012). The Varian aS1200 includes backscatter shielding material that was found to reduce the backscattered signal to less than 0.5% using a prototype imager with the same detector structure (King et al., 2012). This shielding layer adds additional uniform backscatter signal that changes the response compared to the previous imager models. The Elekta iView GT imager is not known to suffer from significant asymmetric backscattered photon signal.

7.3.10 Mechanical Flex

During clinical use, the EPID systems are extended outward from the main linac unit, into the treatment beam. When extended, the mounting systems are subject to gravitational forces that may cause mechanical flexion of the imager away from an ideal central-axis alignment. This introduces small shifts in the EPID image location (intended position versus real position) as a function of gantry angle, upward of ±1 mm in-plane and ±0.5 mm cross plane for Varian E-arm systems (Rowshanfarzad et al., 2012b) and ±2 mm for Elekta systems (Mans et al., 2010b; Rowshanfarzad et al., 2015), although another group shows up to ±4 mm shifts in the in-plane for Elekta EPIDs (Poludniowski et al., 2010). The older Varian R-arm systems demonstrated larger shifts approaching 10 mm (Gratton and McGarry, 2010).

The magnitude and direction of flex for individual linacs have been shown to be consistent and reproducible with gantry angle, and thus can be corrected once the relationship is mapped. Linac manufacturers may provide an option to correct for the EPID imager sag via directly measuring on the imaging system the position of a static object, for example, a ball bearing, aligned to isocenter. This feature is known as "IsoCal" on Varian linacs and can correct the imager arm sag to within 0.5 mm.

7.3.11 Gantry Angle Information

For many EPID dosimetry applications involving VMAT delivery, the gantry angle associated with each EPID image needs to be accurately known. It has been demonstrated (McCowan et al., 2014) that uncertainty in the gantry angle value in the header information associated with the individual acquired EPID images on Varian a-Si EPIDs typically exceeds the AAPM TG142 tolerance of 1° (Klein et al., 2009), with variations of up to ±3° observed. However, a simple method accounting for timing offsets and smoothing the sampling noise can correct these to an accuracy of within ±1°. The kilovoltage imaging system gantry angle signal, required for accurate cone-beam CT reconstruction, has also been used to obtain more accurate gantry angle information for the EPID (Woodruff et al., 2013). Independent gantry angle determination using inclinometers is another approach that has been explored (Rowshanfarzad et al., 2012b). Another group modified apertures in VMAT plans to allow imaging of a custom phantom that provided independent gantry angle information, to improve accuracy of an EPID-based pretreatment QA technique (Adamson and Wu, 2012). For Elekta linacs, an early investigation by Mans et al. (2010a) used an in-house developed prototype system for *in vivo* EPID dosimetry that recorded gantry angle data together with image frames via an iCom connection. Their work revealed an approximate 0.4 s time lag in the gantry angle data received by the acquisition software, which was accounted for in a back-projection model for VMAT verification. This small time lag was observed since the gantry angle stored for a given image frame corresponded to the average of those used by the VMAT arc between the start and the end of the frame.

7.4 Methods for EPID Dosimetry: A Literature Review

The extensive review of EPID dosimetry carried out by van Elmpt et al. (2008) in 2008 provides a detailed cataloging of the wide variety of methods found in the literature. The brief description that follows here is based on that work and includes an update of EPID dosimetry literature published since 2008.

EPID dosimetry can be categorized as either "pretreatment verification" where the EPID measures treatment beams without the patient present, or "treatment verification" where the EPID measures treatment beams during patient treatment. Further subcategorization can be applied depending on whether or not the treatment beams have passed through an attenuating medium before being measured by the EPID. If the treatment beams have not passed through an attenuating medium to reach the EPID, the method is referred to as "nontransmission dosimetry" (or "nontransit dosimetry"). Otherwise, the method is described as "transmission dosimetry" (or "transit dosimetry").

In general, dosimetry applications for EPIDs compare the dose estimated via an EPID measurement with an expected, independently calculated dose. Therefore, a third layer of subcategorization can be used, based on *where* the dose comparison is being made. There are two common locations where the comparisons can be made: First, at the detector level, i.e., comparing a measured image to an expected image, typically in 0D (point comparison) or 2D (image comparison), and second, at the patient/phantom level, i.e., inferring a dose in the patient/phantom from the measured EPID image and comparing it to an expected dose in the patient/phantom. This comparison is usually made in 0D (point comparison), 2D (plane comparison), or 3D (volumetric comparison).

The special case where patient dose (0D, 2D, or 3D) is estimated from EPID measurements during patient treatment, described by the earlier nomenclature as an "in-patient, transmission dosimetry, treatment-verification method," is simply described as "*in vivo* patient dosimetry." Since the clinical introduction of VMAT delivery techniques in 2009, after the van Elmpt review was published, a further distinction of "time-resolved" dosimetry can be made. This term is used when the application tracks dosimetry as a function of time during the beam irradiation. If the application uses 3D patient/phantom dosimetry combined with the time-resolved aspect, then it has been described as "4D" dosimetry.

Much work has been done investigating a-Si EPID dosimetric applications in all of these categories and the reader is referred to the review paper by van Elmpt et al. (2008) for a detailed summary of the EPID dosimetry literature up to 2008. The remaining discussion in this section focuses on updating the literature review and publication lists in Tables 2 and 3 of the paper by van Elmpt et al. to include publications between 2008 and mid-2016, numbering more than 100. For those interested, Chapter 18 contains a more detailed discussion of *in vivo* patient dosimetry applications of a-Si EPIDs.

7.4.1 Nontransmission EPID Dosimetry

7.4.1.1 QA of Treatment Machine

EPIDs continue to be used in QA applications for linacs. Multileaf collimator (MLC) positions can be identified with high accuracy (Clarke and Budgell, 2008;

Lee et al., 2008; Han-Oh et al., 2010; Richart et al., 2012; Rowshanfarzad et al., 2012d; Fuangrod et al., 2014; Mutic et al., 2016), as well as collimator jaw positions (Clews and Greer, 2009), and junction doses can also be examined (Madebo et al., 2010). Gantry angle may be monitored (Rowshanfarzad et al., 2012b) and isocenter accuracy may be investigated (Rowshanfarzad et al., 2011a,b; Sun et al., 2015) including related mechanical sag issues (Rowshanfarzad et al., 2012c,e, 2014). Several groups have used the EPID for more comprehensive linac QA (Nicolini et al., 2008b; Boylan et al., 2012; Lim et al., 2012; Rowshanfarzad et al., 2015). Some have explored using the EPID to monitor electron beam output (Beck et al., 2009; Wang et al., 2013), even though the commercial systems are designed for use with photon beams.

7.4.1.2 2D Verification at EPID

The commercial availability of 2D portal dosimetry software (Portal Dosimetry; Varian Medical Systems, Palo Alto, CA) produced a large number of published investigations since 2008 (Bailey et al., 2010, 2012; Iori et al., 2010; Roxby and Crosbie, 2010; Sharma et al., 2010; Varatharaj et al., 2010; Vinall et al., 2010; Adamson and Wu, 2012; Liu et al., 2013; Matsumoto et al., 2013; Van Esch et al., 2013; Zhu et al., 2013; Clemente et al., 2014; Min et al., 2014; Hobson and Davis, 2015; Merheb et al., 2015; Sjölin and Edmund, 2016). Some have proposed improvements to the Portal Dosimetry software (Vial et al., 2008b; Bailey et al., 2009, 2013; Berry et al., 2010; Pardo et al., 2016) or established action levels (Howell et al., 2008). Many groups have developed their own "in-house" algorithms or calibration techniques to predict dose or signal at the EPID level (Chytyk and McCurdy, 2009; Greer et al., 2009; Parker et al., 2009; Tyner et al., 2009; Wang et al., 2009; Fredh et al., 2010; Liebich et al., 2011; Conte et al., 2012; Steciw et al., 2013; Woodruff et al., 2013; Monville et al., 2014a,b; Podesta et al., 2014). Others have used existing treatment planning systems with some modification (Yohannes et al., 2015). As mentioned in Section 7.2.1, some have been interested in converting the EPID dose image to a water-equivalent dose image (King et al., 2012; King and Geer, 2013; Zwan et al., 2014; Camilleri et al., 2016), or using a direct-detection configured a-Si EPID for water-equivalent pretreatment QA measurements (Vial et al., 2008a, 2009; Sabet et al., 2010; Gustafsson et al., 2011). Related to this effort, some have examined replacing the metal plate/phosphor scintillating screen layers with segmented plastic scintillators to obtain a detector energy response much closer to water (Blake et al., 2013a,b). There has also been an investigation into the fluence error sensitivity of a-Si EPIDs used for pretreatment QA (Gordon et al., 2012).

7.4.1.3 2D Verification in Phantom or Planning CT

There have been several investigations of commercially available software to perform 2D treatment verification, including the GLAaS algorithm (Nicolini et al., 2008a,b, 2013; Fogliata et al., 2011; Huang et al., 2013), the EpiDose package (Nelms et al., 2010; Bailey et al., 2012), and the EpiQA package (Merheb et al., 2015). A Monte Carlo–based in-house algorithm is also developed (Lin et al., 2009). A simple approach to EPID dosimetry was also proposed where the EPID image is compared to the dose in water at a depth where the dosimetric properties are similar (Lee et al., 2009). Phantom-based reconstruction of open field EPID dose images is also explored for electron beams (Chatelain et al., 2013) and high dose–rate brachytherapy treatment (Smith et al., 2013).

7.4.1.4 3D Verification in Phantom or Planning CT

Commercially available solutions have been investigated (Nakaguchi et al., 2013; Narayanasamy et al., 2015). A Monte Carlo–based in-house algorithm was developed for 3D dose estimates in CT simulation or CBCT datasets (van Elmpt et al., 2008, 2010) and a GPU-accelerated collapsed-cone convolution technique was explored for dose reconstruction in phantom or in CT simulation datasets (Zhu et al., 2015). An ion chamber–based correction was investigated by Zhang et al. (2013). Sumida et al. (2016) explored deconvolving the EPID image to estimate incident fluence to compare to the treatment planning system fluence. 3D patient/phantom dose estimates for electron beams have also been examined using open-field EPID measurements (Ding et al., 2015).

7.4.2 Transmission EPID Dosimetry

7.4.2.1 Point Dose Verification

The research group in Rome, Italy, has been exploring the relationship between EPID transmission images and dose to a single reference point in a patient or phantom in great detail since 2008, including extensions to VMAT delivery, examining Elekta and Siemens EPIDs, and validation across many disease sites (Piermattei et al., 2008, 2009a,b, 2011; Cilla et al., 2010, 2011, 2014, 2016; Fidanzio et al., 2010, 2011a,b, 2014, 2015; Greco et al., 2013; Russo et al., 2015). A few other groups have investigated a similar approach since 2008 (Slosarek et al., 2010; Francois et al., 2011; Camilleri et al., 2014; Celia et al., 2016; Millin et al., 2016; Ricketts et al., 2016a,b).

7.4.2.2 2D Verification at EPID

Several groups have used full Monte Carlo simulation techniques to predict the EPID image with the patient in the beam path (Gardner et al., 2009; Juste et al., 2009, 2010; Cufflin et al., 2010, Yoon et al., 2016a), whereas others have used Monte Carlo–generated, EPID-specific dose kernels (Chytyk-Praznik et al., 2013) with the latter work also being applied to real-time error detection applications (Woodruff et al., 2015; Fuangrod et al., 2016). A measurement-based model has been developed for dynamic IMRT (Sabet et al., 2014). Berry et al. (2012, 2014) extended the in-air EPID image prediction work of Van Esch et al. (2004) to transit dosimetry through additional corrections for attenuation, patient scatter, and detector response. Another approach used the treatment planning system planar dose through isocenter and projected it to the EPID, with corrections applied (Bedford et al., 2014, 2016). Some have used the commercial treatment planning system and extended the patient calculation space to include a detector (Baek et al., 2014). The direct-detector a-Si EPID configuration has been shown to correspond to water-equivalent dose measurements for transit dosimetry (Sabet et al., 2012). Some investigators have related dose–volume histogram behavior to EPID transit dose images (Nijsten et al., 2009) and applied trend analysis on the differences in transit dose images through a treatment course (Persoon et al., 2012). Others have examined the effect of setup errors on transit dose images (Sukumar et al., 2012; Brouwers et al., 2015) and confirmed that integrated transit dose images are not very useful in detecting patient changes in lung cancer (Persoon et al., 2015). Schyns et al. (2016) also investigated the usefulness of time-resolved (4D) images in detecting anatomical changes during treatment.

7.4.2.3 2D Verification at Patient/Phantom

Estimates of the patient's radiological thickness (in 2D) using EPID transmission images have been explored. These approaches use a quadratic calibration method to relate imager response to radiological thickness of the patient or phantom in the beam path, as described by Morton et al. (1991), and have successfully demonstrated the method using a-Si EPIDs (Kairn et al., 2008; Kavuma et al., 2010, 2011; Tan et al., 2015). Matrix inversion techniques have been examined to take an EPID transmission image and reconstruct dose to a plane in the patient or phantom (Yeo et al., 2009), and these methods have been extended to 3D reconstruction as mentioned in Section 7.4.2.5. The single-point reconstruction technique of Piermattei et al. (2008) was extended to reconstruct a 2D dose distribution in a patient (Peca and Brown, 2014). A technique for using EPID transmission images to adjust a treatment plan fluence map to compensate for increased absorption in a metal hip prosthesis was proposed (Nielsen et al., 2008) and validated in a plane 10 cm behind the prosthesis. The technique to estimate 2D isocenter planar dose in the patient, originally developed by Wendling et al. (2006), was investigated to quantify performance in detecting several types of treatment error (Bojechko and Ford, 2015).

7.4.2.4 3D Verification at Patient/Phantom Using Geometric Features in the EPID Image

This section identifies work that extracts geometric features from EPID images in order to reconstruct the 3D dose distribution in the patient. One feature easily identified on the EPID is the position of the MLCs, which can be done as a function of time if cine imaging is used. This information may be utilized to reconstruct patient dose in 3D (Lee et al., 2008; Lin et al., 2012; Defoor et al., 2015). Furthermore, tumor motion may be tracked in the EPID allowing the intended beam fluence maps to be convolved with the recorded motion to achieve an estimate of the delivered 3D dose distribution to the patient. This has been demonstrated with direct lung tumor tracking (Aristophanous et al., 2011; Cai et al., 2015) and also using fiducial markers in the liver (Berbeco et al., 2008).

7.4.2.5 3D Verification at Patient/Phantom

Since 2008 there has been continued strong interest in full 3D dose verification in the patient. The group at the Netherlands Cancer Institute in the Netherlands extended their in-house 2D planar dose reconstruction method to a full 3D dose reconstruction method (Wendling et al., 2009) and made several enhancements to it. For example, extending to VMAT delivery (Mans et al., 2010a), utilizing CT data for attenuation calculations instead of a ratio of transmitted and measured open beams (Pecharromán-Gallego et al., 2011), extending to wedges (Spreeuw et al., 2015), online verification (Spreeuw et al., 2016), using a water-equivalent patient concept for lung treatment verification (Wendling et al., 2012), and automating the technique (Olaciregui-Ruiz et al., 2013). This research group has also been active in documenting error detection rates of a clinically implemented EPID dosimetry program (Mans et al., 2010b; Mijnheer et al., 2015), exploring methods to detect meaningful dose differences between treatment planning 3D doses and EPID-derived 3D reconstructed doses (Rozendaal et al., 2014), and streamlining the commissioning process (Hanson et al., 2014). Their technique has been shown to be applicable for flattening filter-free beams (Chuter et al., 2016) and to assess impact of on-treatment anatomy changes in head-and-neck cancer patients (Rozendaal et al., 2015). Another group in the Netherlands, at the Maastricht University Medical Centre, has developed a technique to estimate incident fluence from a measured EPID transmission image, followed by Monte Carlo techniques to estimate delivered 3D patient dose distributions (van Elmpt et al., 2009). The group has shown this approach to be clinically useful for detecting atelectasis on treatment (Persoon et al., 2013). Van Uytven et al. (2015) have developed a robust model-based algorithm that converts measured EPID transmission images into incident fluence and uses a collapsed-cone convolution calculation to estimate delivered patient dose. This approach has been validated for lung SBRT patients (McCowan et al., 2015) and the cine image acquisition has been optimized to reduce the introduction of sampling errors in the 3D patient dose reconstruction (McCowan and McCurdy, 2016). Others have used a matrix inversion approach to estimate delivered patient dose (Jung et al., 2012; Yeo et al., 2009; Yoon et al., 2016b). In yet another approach, the ratio of the transmission EPID image to open-field EPID images has been used to modify Monte Carlo–generated incident fluence maps, which may then be used to forward calculate dose in the patient and/or EPID (Lin et al., 2009). Furthermore, a commercial system has been investigated (Dosimetry Check) and found to be limited in terms of accuracy and integration with the record-and-verify system (Gimeno et al., 2014).

7.5 Summary

This chapter presents a technical overview of a-Si EPIDs in the context of dosimetry. Current, commercially available a-Si EPID systems were discussed, highlighting dosimetric characteristics and technical challenges to routine use. Dosimetry applications involving a-Si EPIDs that have been published since the 2008 review paper of van Elmpt et al. (2008) were briefly outlined.

Today's modern radiotherapy clinics are implementing complex treatments such as IMRT and VMAT, and specialized treatment techniques that incorporate large doses per fraction and short treatment courses such as hypofractionation, SRS, and stereotactic radiotherapy. These factors strongly motivate researchers to continue investigating EPIDs for patient treatment verification.

References

Adamson J and Wu Q (2012) Independent verification of gantry angle for pre-treatment VMAT QA using EPID. *Phys. Med. Biol.* 57: 6587–6600.

Alshanqity M, Duane S and Nisbet A (2012) A simple approach for EPID dosimetric calibration to overcome the effect of image-lag and ghosting. *Appl. Radiat. Isot.* 70: 1154–1157.

Antonuk LE et al. (1992) Demonstration of megavoltage and diagnostic x-ray imaging with hydrogenated amorphous silicon arrays. *Med. Phys.* 19: 1455–1466.

Antonuk LE et al. (1995) A real-time, flat-panel, amorphous silicon, digital x-ray imager. *Radiographics.* 15: 993–1000.

Antonuk LE et al. (1996) Megavoltage imaging with a large-area, flat-panel amorphous silicon imager. *Int. J. Radiat. Oncol. Biol. Phys.* 36: 661–672.

Aristophanous M et al. (2011) EPID-guided 3D dose verification of lung SBRT. *Med. Phys.* 38: 495–503.

Baek TS et al. (2014) Feasibility study on the verification of actual beam delivery in a treatment room using EPID transit dosimetry. *Radiat. Oncol.* 9: 273.

Bailey DW, Kumaraswamy L and Podgorsak MB (2009) An effective correction algorithm for off-axis portal dosimetry errors. *Med. Phys.* 36: 4089–4094.

Bailey DW, Kumaraswamy L and Podgorsak MB (2010) A fully electronic intensity-modulated radiation therapy quality assurance (IMRT QA) process implemented in a network comprised of independent treatment planning, record and verify, and delivery systems. *Radiol. Oncol.* 44: 124–130.

Bailey DW et al. (2012) EPID dosimetry for pretreatment quality assurance with two commercial systems. *J. Appl. Clin. Med. Phys.* 13(4): 82–99.

Bailey DW et al. (2013) A two-dimensional matrix correction for off-axis portal dose prediction errors. *Med. Phys.* 40: 051704.

Beck JA et al. (2009) Electron beam quality control using an amorphous silicon EPID. *Med. Phys.* 36: 1859–1866.

Bedford JL, Hanson IM and Hansen VN (2014) Portal dosimetry for VMAT using integrated images obtained during treatment. *Med. Phys.* 41: 021725.

Bedford JL et al. (2016) Quality of treatment plans and accuracy of in vivo portal dosimetry in hybrid intensity-modulated radiation therapy and volumetric modulated arc therapy for prostate cancer. *Radiother. Oncol.* 120: 320–326.

Berbeco RI et al. (2008) A novel method for estimating SBRT delivered dose with beam's-eye-view images. *Med. Phys.* 35: 3225–3231.

Berger L et al. (2006) Performance optimization of the Varian aS500 EPID system. *J. Appl. Clin. Med. Phys.* 7(1): 105–114.

Berry SL, Polvorosa CS and Wuu CS (2010) A field size specific backscatter correction algorithm for accurate EPID dosimetry. *Med. Phys.* 37: 2425–2434.

Berry SL et al. (2012) Implementation of EPID transit dosimetry based on a through-air dosimetry algorithm. *Med. Phys.* 39: 87–98.

Berry SL et al. (2014) Initial clinical experience performing patient treatment verification with an electronic portal imaging device transit dosimeter. *Int. J. Radiat. Oncol. Biol. Phys.* 88: 204–209.

Blake SJ et al. (2013a) Characterization of optical transport effects on EPID dosimetry using Geant4. *Med. Phys.* 40: 041708.

Blake SJ et al. (2013b) Characterization of a novel EPID designed for simultaneous imaging and dose verification in radiotherapy. *Med. Phys.* 40: 091902.

Bojechko C and Ford EC (2015) Quantifying the performance of in vivo portal dosimetry in detecting four types of treatment parameter variations. *Med. Phys.* 42: 6912–6918.

Boriano A et al. (2013) A new approach for the pixel map sensitivity (PMS) evaluation of an electronic portal imaging device (EPID). *J. Appl. Clin. Med. Phys.* 14(6): 234–250.

Boudry JM and Antonuk LE (1994) Radiation damage of amorphous silicon photodiode sensors. *IEEE Trans. Nucl. Sci.* 41: 703–707.

Boudry JM and Antonuk LE (1996) Radiation damage of amorphous silicon, thin-film, field-effect transistors. *Med. Phys.* 23: 743–754.

Boylan C et al. (2012) The impact of continuously-variable dose rate VMAT on beam stability, MLC positioning, and overall plan dosimetry. *J. Appl. Clin. Med. Phys.* 13(6): 254–266.

Brouwers PJ et al. (2015) Set-up verification and 2-dimensional electronic portal imaging device dosimetry during breath hold compared with free breathing in breast cancer radiation therapy. *Pract. Radiat. Oncol.* 5: 135–141.

Cai W et al. (2015) 3D delivered dose assessment using a 4DCT-based motion model. *Med. Phys.* 42: 2897–2907.

Camilleri J et al. (2014) Clinical results of an EPID-based in-vivo dosimetry method for pelvic cancers treated by intensity-modulated radiation therapy. *Phys. Med.* 30: 690–695.

Camilleri J et al. (2016) 2D EPID dose calibration for pretreatment quality control of conformal and IMRT fields: A simple and fast convolution approach. *Phys. Med.* 32: 133–140.

Celi et al. (2016) EPID based *in vivo* dosimetry system: Clinical experience and results. *J. Appl. Clin. Med. Phys.* 17(3): 262–276.

Chatelain C et al. (2013) Dosimetric properties of an amorphous silicon EPID for verification of modulated electron radiotherapy. *Med. Phys.* 40: 061710.

Chen Y et al. (2007) Performance of a direct-detection active matrix flat panel dosimeter (AMFPD) for IMRT measurements. *Med. Phys.* 34: 4911–4922.

Chuter RW et al. (2016) Feasibility of portal dosimetry for flattening filter-free radiotherapy. *J. Appl. Clin. Med. Phys.* 17(1): 112–120.

Chytyk K and McCurdy BM (2009) Comprehensive fluence model for absolute portal dose image prediction. *Med. Phys.* 36: 1389–1398.

Chytyk-Praznik K et al. (2013) Model-based prediction of portal dose images during patient treatment. *Med. Phys.* 40: 031713.

Cilla S et al. (2010) Correlation functions for Elekta aSi EPIDs used as transit dosimeter for open fields. *J. Appl. Clin. Med. Phys.* 12(1): 218–233.

Cilla S et al. (2011) Calibration of Elekta aSi EPIDs used as transit dosimeter. *Technol. Cancer Res. Treat.* 10: 39–48.

Cilla S et al. (2014) An in-vivo dosimetry procedure for Elekta step and shoot IMRT. *Phys. Med.* 30: 419–426.

Cilla S et al. (2016) Initial clinical experience with EPID-based in-vivo dosimetry for VMAT treatments of head-and-neck tumors. *Phys. Med.* 32: 52–58.

Clarke MF and Budgell GJ (2008) Use of an amorphous silicon EPID for measuring MLC calibration at varying gantry angle. *Phys. Med. Biol.* 53: 473–485.

Clemente S et al. (2014) To evaluate the accuracy of dynamic versus static IMRT delivery using portal dosimetry. *Clin. Transl. Oncol.* 16: 208–212.

Clews L and Greer PB (2009) An EPID based method for efficient and precise asymmetric jaw alignment quality assurance. *Med. Phys.* 36: 5488–5496.

Conte L et al. (2012) An empirical calibration method for an a-Si portal imaging device: Applications in pretreatment verification of IMRT. *Radiol. Med.* 117: 1044–1056.

Cufflin RS et al. (2010) An investigation of the accuracy of Monte Carlo portal dosimetry for verification of IMRT with extended fields. *Phys. Med. Biol.* 55: 4589–4600.

Defoor DL et al. (2015) Anatomy-based, patient-specific VMAT QA using EPID or MLC log files. *J. Appl. Clin. Med. Phys.* 16(3): 206–215.

Derreumaux S et al. (2008) Lessons from recent accidents in radiation therapy in France. *Radiat. Prot. Dosim.* 131: 130–135.

Deshpande S et al. (2014) Dose calibration of EPIDs for segmented IMRT dosimetry. *J. Appl. Clin. Med. Phys.* 15(6): 103–118.

Ding A, Xing L and Han B (2015) Development of an accurate EPID-based output measurement and dosimetric verification tool for electron beam therapy. *Med. Phys.* 42: 4190–4198.

El-Mohri et al. (1999) Relative dosimetry using active matrix flat-panel imager (AMFPI) technology. *Med. Phys.* 26: 1530–1541.
Fidanzio A et al. (2010) Breast in vivo dosimetry by EPID. *J. Appl. Clin. Med. Phys.* 11(4): 249–262.
Fidanzio A et al. (2011a) Generalized EPID calibration for in vivo transit dosimetry. *Phys. Med.* 27: 30–38.
Fidanzio A et al. (2011b) A generalized calibration procedure for in vivo transit dosimetry using Siemens electronic portal imaging devices. *Med. Biol. Eng. Comput.* 49: 373–383.
Fidanzio A et al. (2014) Quasi real time in vivo dosimetry for VMAT. *Med. Phys.* 41: 062103.
Fidanzio A et al. (2015) Routine EPID in-vivo dosimetry in a reference point for conformal radiotherapy treatments. *Phys. Med. Biol.* 60: N141–N150.
Fogliata A et al. (2011) Quality assurance of RapidArc in clinical practice using portal dosimetry. *Br. J. Radiol.* 84: 534–545.
Francois P et al. (2011) In vivo dose verification from back projection of a transit dose measurement on the central axis of photon beams. *Phys. Med.* 27: 1–10.
Fredh A, Korreman S and Munck af Rosenschöld P (2010) Automated analysis of images acquired with electronic portal imaging device during delivery of quality assurance plans for inversely optimized arc therapy. *Radiother. Oncol.* 94: 195–198.
Fuangrod T et al. (2014) A cine-EPID based method for jaw detection and quality assurance for tracking jaw in IMRT/VMAT treatments. *Phys. Med.* 31: 16–24.
Fuangrod T et al. (2016) Investigation of a real-time EPID-based patient dose monitoring safety system using site-specific control limits. *Radiat. Oncol.* 11: 106.
Gardner JK et al. (2009) Comparison of sources of exit fluence variation for IMRT. *Phys. Med. Biol.* 54: N451–N458.
Gimeno J et al. (2014) Commissioning and initial experience with a commercial software for in vivo volumetric dosimetry. *Phys. Med.* 30: 954–959.
Gordon JJ et al. (2012) Reliable detection of fluence anomalies in EPID-based IMRT pretreatment quality assurance using pixel intensity deviations. *Med. Phys.* 39: 4959–4975.
Gratton MW and McGarry CK (2010) Mechanical characterization of the Varian Exact-arm and R-arm support systems for eight aS500 electronic portal imaging devices. *Med. Phys.* 37: 1707–1713.
Greco F et al. (2013) aSi-EPID transit signal calibration for dynamic beams: A needful step for the IMRT in vivo dosimetry. *Med. Biol. Eng. Comput.* 51: 1137–1145.
Greer PB and Popescu CC (2003) Dosimetric properties of an amorphous silicon electronic portal imaging device for verification of dynamic intensity modulated radiation therapy. *Med. Phys.* 30: 1618–1627.
Greer PB (2005) Correction of pixel sensitivity variation and off-axis response for amorphous silicon EPID dosimetry. *Med. Phys.* 32: 3558–3568.
Greer PB et al. (2009) An energy fluence-convolution model for amorphous silicon EPID dose prediction. *Med. Phys.* 36: 547–555.
Greer PB and Vial P (2011) EPID dosimetry. In: *Concepts and Trends in Medical Radiation Dosimetry: Proceedings of the SSD Summer School*, pp. 129–144 (Eds. Rosenfeld A, Kron T and D'Errico F, Wollongong, Australia: AIP Conference Proceedings, Volume 1345, American Institute of Physics Publishing).

Greer PB (2013) 3D EPID based dosimetry for pre-treatment verification of VMAT—methods and challenges. *J. Phys. Conf. Ser.* 444: 012010.

Gustafsson H et al. (2009) EPID dosimetry: Effect of different layers of materials on absorbed dose response. *Med. Phys.* 36: 5665–5674.

Gustafsson H et al. (2011) Direct dose to water dosimetry for pretreatment IMRT verification using a modified EPID. *Med. Phys.* 38: 6257–6264.

Han-Oh S et al. (2010) Verification of MLC based real-time tumor tracking using an electronic portal imaging device. *Med. Phys.* 37: 2435–2440.

Hansen VN, Swindell W and Evans PM (1997) Extraction of primary signal from EPIDs using only forward convolution. *Med. Phys.* 24: 1477–1484.

Hanson IM et al. (2014) Clinical implementation and rapid commissioning of an EPID based in-vivo dosimetry system. *Phys. Med. Biol.* 59: N171–N179.

Hobson MA and Davis SD (2015) Comparison between an in-house 1D profile correction method and a 2D correction provided in Varian's PDPC Package for improving the accuracy of portal dosimetry images. *J. Appl. Clin. Med. Phys.* 16(2): 43–50.

Howell RM, Smith IP and Jarrio CS (2008) Establishing action levels for EPID-based QA for IMRT. *J. Appl. Clin. Med. Phys.* 9(3): 16–25.

Hsieh J, Gurmen OE and King KF (2000) A recursive correction algorithm for detector decay characteristics in CT. *Proc. SPIE* 3977: 298–305.

Huang YC et al. (2013) Clinical practice and evaluation of electronic portal imaging device for VMAT quality assurance. *Med. Dosim.* 38: 35–41.

ICRU (2010) Prescribing, recording, and reporting photon-beam intensity-modulated radiation therapy (IMRT). Report 83. (Oxford University Press, Oxford, UK: International Commission on Radiation Units and Measurements).

Iori M et al. (2010) Dosimetric verification of IMAT delivery with a conventional EPID system and a commercial portal dose image prediction tool. *Med. Phys.* 37: 377–390.

Jaffray DA et al. (1994) X-ray scatter in megavoltage transmission radiography: Physical characteristics and influence on image quality. *Med. Phys.* 21: 45–60.

Jarry G and Verhaegen F (2007) Patient-specific dosimetry of conventional and intensity modulated radiation therapy using a novel full Monte Carlo phase space reconstruction method from electronic portal images. *Phys. Med. Biol.* 52: 2277–2299.

Jung JW et al. (2012) Fast transit portal dosimetry using density-scaled layer modeling of aSi-based electronic portal imaging device and Monte Carlo method. *Med. Phys.* 39: 7593–7602.

Juste B et al. (2009) Dosimetric capabilities of the iView GT portal imager using MCNP5 Monte Carlo simulations. *Conf. Proc. IEEE Eng. Med. Biol. Soc.* 2009: 3743–3746.

Juste B et al. (2010) Monte Carlo simulation of the iView GT portal imager dosimetry. *Appl. Radiat. Isot.* 68: 922–925.

Kairn T et al. (2008) Radiotherapy treatment verification using radiological thickness measured with an amorphous silicon electronic portal imaging device: Monte Carlo simulation and experiment. *Phys. Med. Biol.* 53: 3903–3919.

Kausch C et al. (1999) Monte Carlo simulations of the imaging performance of metal plate/phosphor screens used in radiotherapy. *Med. Phys.* 26: 2113–2124.

Kavuma A et al. (2008) Assessment of dosimetrical performance in 11 Varian aS500 electronic portal imaging devices. *Phys. Med. Biol.* 53: 6893–6909.

Kavuma A et al. (2010) A novel method for patient exit and entrance dose prediction based on water equivalent path length measured with an amorphous silicon electronic portal imaging device. *Phys. Med. Biol.* 55: 435–452.

Kavuma A et al. (2011) Calculation of exit dose for conformal and dynamically-wedged fields, based on water-equivalent path length measured with an amorphous silicon electronic portal imaging device. *J. Appl. Clin. Med. Phys.* 12(3): 44–60.

King BW, Clews L and Greer PB (2011) Long-term two-dimensional pixel stability of EPIDs used for regular linear accelerator quality assurance. *Australas. Phys. Eng. Sci. Med.* 34: 459–466.

King BW, Morf D and Greer PB (2012) Development and testing of an improved dosimetry system using a backscatter shielded electronic portal imaging device. *Med. Phys.* 39: 2839–2847.

King BW and Greer PB (2013) A method for removing arm backscatter from EPID images. *Med. Phys.* 40: 071703.

Kirkby C and Sloboda R (2005) Comprehensive Monte Carlo calculation of the point spread function for a commercial a-Si EPID. *Med. Phys.* 32: 1115–1127.

Klein EE et al. (2009) Task Group 142 report: Quality assurance of medical accelerators. *Med. Phys.* 39: 4197–4212.

Lee C et al. (2009) A simple approach to using an amorphous silicon EPID to verify IMRT planar dose maps. *Med. Phys.* 36: 984–992.

Lee L, Mao W and Xing L (2008) The use of EPID-measured leaf sequence files for IMRT dose reconstruction in adaptive radiation therapy. *Med. Phys.* 35: 5019–5029.

Liebich J et al. (2011) Simple proposal for dosimetry with an Elekta iViewGT™ electronic portal imaging device (EPID) using commercial software modules. *Strahlenther. Onkol.* 187: 316–321.

Lim S et al. (2012) Development of a one-stop beam verification system using electronic portal imaging devices for routine quality assurance. *Med. Dosim.* 37: 296–304.

Lin MH et al. (2009) Measurement-based Monte Carlo dose calculation system for IMRT pretreatment and on-line transit dose verifications. *Med. Phys.* 36: 1167–1175.

Lin MH et al. (2012) 4D patient dose reconstruction using online measured EPID cine images for lung SBRT treatment validation. *Med. Phys.* 39: 5949–5958.

Liu B et al. (2013) A novel technique for VMAT QA with EPID in cine mode on a Varian TrueBeam linac. *Phys. Med. Biol.* 58: 6683–6700.

Louwe RJ et al. (2004) The long-term stability of amorphous silicon flat panel imaging devices for dosimetry purposes. *Med. Phys.* 31: 2989–2995.

Madebo M et al. (2010) Study of x-ray field junction dose using an a-Si electronic portal imaging device. *Australas. Phys. Eng. Sci. Med.* 33: 45–50.

Mans A et al. (2010a) 3D Dosimetric verification of volumetric-modulated arc therapy by portal dosimetry. *Radiother. Oncol.* 94: 181–187.

Mans A et al. (2010b) Catching errors with in vivo EPID dosimetry. *Med. Phys.* 37: 2638–2644.

Matsumoto K et al. (2013) Dosimetric properties and clinical application of an a-Si EPID for dynamic IMRT quality assurance. *Radiol. Phys. Technol.* 6: 210–218.

McCowan PM et al. (2014) Gantry angle correction methods for EPID images acquired in cine-mode. *J. Appl. Clin. Med. Phys.* 15: 187–201.

McCowan PM et al. (2015) An in vivo dose verification method for SBRT-VMAT delivery using the EPID. *Med. Phys.* 42: 6955–6963.

McCowan PM and McCurdy BMC (2016) Frame average optimization of cine-mode EPID images used for routine clinical in vivo patient dose verification of VMAT deliveries. *Med. Phys.* 43: 254.

McCurdy BMC and Pistorius S (1998) Analytical prediction of first scatter energy fluence into an electronic portal imaging device. In: Proceedings of the 5th International workshop on electronic portal imaging, pp. 104–105 (Phoenix, AZ).

McCurdy BMC and Pistorius S (2000) Photon scatter in portal images: Physical characteristics of pencil beam kernels generated using the EGS Monte Carlo code. *Med. Phys.* 27: 312–320.

McCurdy BM, Luchka K and Pistorius S (2001) Dosimetric investigation and portal dose image prediction using an amorphous silicon electronic portal imaging device. *Med. Phys.* 28: 911–924.

McCurdy BMC and Greer PB (2009) Dosimetric properties of an amorphous-silicon EPID used in continuous acquisition mode for application to dynamic and arc IMRT. *Med. Phys.* 36: 3028–3039.

McCurdy BMC (2013) Dosimetry in radiotherapy using a-Si EPIDs: Systems, methods, and applications focusing on 3D patient dose estimation. *J. Phys. Conf. Ser.* 444: 012002.

McDermott LN et al. (2004) Dose-response and ghosting effects of an amorphous silicon electronic portal imaging device. *Med. Phys.* 31: 285–295.

McDermott LN et al. (2006) Comparison of ghosting effects for three commercial a-Si EPIDs. *Med. Phys.* 33: 2448–2451.

Merheb C et al. (2015) Comparison between two different algorithms used for pretreatment QA via aSi portal images. *J. Appl. Clin. Med. Phys.* 16(3): 141–153.

Mijnheer et al. (2013) 3D EPID-based *in vivo* dosimetry for IMRT and VMAT. *J. Phys. Conf. Ser.* 444: 012011.

Mijnheer BJ et al. (2015) Overview of 3-year experience with large-scale electronic portal imaging device–based 3-dimensional transit dosimetry. *Pract. Radiat. Oncol.* 5: e679–e687.

Millin AE, Windle RS and Lewis DG (2016) A comparison of electronic portal dosimetry verification methods for use in stereotactic radiotherapy. *Phys. Med.* 32: 188–196.

Min S et al. (2014) Practical approach for pretreatment verification of IMRT with flattening filter free (FFF) beams using Varian Portal Dosimetry. *J. Appl. Clin. Med. Phys.* 16(1): 40–50.

Monville ME, Kuncic Z and Greer PB (2014a) An improved Monte-Carlo model of the Varian EPID separating support arm and rear-housing backscatter. *J. Phys. Conf. Ser.* 489: 012012.

Monville ME, Kuncic Z and Greer PB (2014b) Simulation of real-time EPID images during IMRT using Monte-Carlo. *Phys. Med.* 30: 326–330.

Mooslechner M et al. (2013) Analysis of a free-running synchronization artifact correction for MV-imaging with aSi:H flat panels. *Med. Phys.* 40: 031906.

Moran JM et al. (2005) An active matrix flat panel dosimeter (AMFPD) for in-phantom dosimetric measurements. *Med. Phys.* 32: 466–472.

Morton EJ et al. (1991) A linear array, scintillation crystal-photodiode detector for megavoltage imaging. *Med. Phys.* 18: 681–691.
Munro P et al. (1998) Glaring errors in transit dosimetry. *Med. Phys.* 25: 1097.
Mutic S, Pawlicki T and Orton CG (2016) Point-counterpoint: EPID-based daily quality assurance of linear accelerators will likely replace other methods within the next ten years. *Med. Phys.* 43: 2691–2693.
Nakaguchi Y et al. (2013) Development of multi-planar dose verification by use of a flat panel EPID for intensity-modulated radiation therapy. *Radiol. Phys. Technol.* 6: 226–232.
Narayanasamy G et al. (2015) Evaluation of Dosimetry Check software for IMRT patient-specific quality assurance. *J. Appl. Clin. Med. Phys.* 16(3): 329–338.
Nelms BE, Rasmussen KH and Tome WA (2010) Evaluation of a fast method of EPID-based dosimetry for intensity-modulated radiation therapy. *J. Appl. Clin. Med. Phys.* 11(2): 140–157.
Nicolini G et al. (2008a) Testing the GLAaS algorithm for dose measurements on low- and high-energy photon beams using an amorphous silicon portal imager. *Med. Phys.* 35: 464–472.
Nicolini G et al. (2008b) Testing the portal imager GLAaS algorithm for machine quality assurance. *Radiat. Oncol.* 3: 14.
Nicolini G et al. (2008c) The GLAaS algorithm for portal dosimetry and quality assurance of RapidArc, an intensity modulated rotational therapy. *Radiat. Oncol.* 3: 24.
Nicolini G et al. (2013) Evaluation of an aSi-EPID with flattening filter free beams: Applicability to the GLAaS algorithm for portal dosimetry and first experience for pretreatment QA of RapidArc. *Med. Phys.* 40: 111719.
Nielsen MS, Carl J and Nielsen J (2008) A phantom study of dose compensation behind hip prosthesis using portal dosimetry and dynamic MLC. *Radiother. Oncol.* 88: 277–284.
Nijsten SMJJG et al. (2007) A global calibration model for a-Si EPIDs used for transit dosimetry. *Med. Phys.* 34: 3872–3884.
Nijsten SM et al. (2009) Prediction of DVH parameter changes due to setup errors for breast cancer treatment based on 2D portal dosimetry. *Med. Phys.* 36: 83–94.
Olaciregui-Ruiz I et al. (2013) Automatic *in vivo* portal dosimetry of all treatments. *Phys. Med. Biol.* 58: 8253–8264.
Overdick M, Solf T and Wischmann H (2001) Temporal artifacts in flat dynamic x-ray detectors. *Proc. SPIE.* 4320: 47–58.
Pardo E et al. (2016) On flattening filter-free portal dosimetry. *J. Appl. Clin. Med. Phys.* 17(4): 132–145.
Parent L et al. (2006) Monte Carlo modelling of a-Si EPID response: The effect of spectral variations with field size and position. *Med. Phys.* 33: 4527–4540.
Parker BC et al. (2009) Pretreatment verification of IMSRT using electronic portal imaging and Monte Carlo calculations. *Technol. Cancer. Res. Treat.* 8: 413–423.
Partridge M, Hesse B-M and Müller L (2002) A performance comparison of direct- and indirect-detection flat-panel imagers. *Nucl. Instr. Meth. Phys. Res.* A 484: 351–363.
Peca S and Brown D (2014) Two-dimensional in vivo dose verification using portal imaging and correlation ratios. *J. Appl. Clin. Med. Phys.* 15(4): 117–128.

Pecharromán-Gallego R et al. (2011) Simplifying EPID dosimetry for IMRT treatment verification. *Med. Phys.* 38: 983–992.

Persoon LC et al. (2012) Interfractional trend analysis of dose differences based on 2D transit portal dosimetry. *Phys. Med. Biol.* 57: 6445–6458.

Persoon LC et al. (2013) First clinical results of adaptive radiotherapy based on 3D portal dosimetry for lung cancer patients with atelectasis treated with volumetric-modulated arc therapy (VMAT). *Acta Oncol.* 52: 1484–1489.

Persoon LC et al. (2015) Is integrated transit planar portal dosimetry able to detect geometric changes in lung cancer patients treated with volumetric modulated arc therapy? *Acta Oncol.* 54: 1501–1507.

Piermattei A et al. (2006) In vivo dosimetry by an aSi-based EPID. *Med. Phys.* 33: 4414–4422.

Piermattei A et al. (2008) Dynamic conformal arc therapy: Transmitted signal in vivo dosimetry. *Med. Phys.* 35: 1830–1839.

Piermattei A et al. (2009a) In patient dose reconstruction using a cine acquisition for dynamic arc radiation therapy. *Med. Biol. Eng. Comput.* 47: 425–433.

Piermattei A et al. (2009b) Integration between in vivo dosimetry and image guided radiotherapy for lung tumors. *Med. Phys.* 36: 2206–2214.

Piermattei A et al. (2011) Real-time dose reconstruction for wedged photon beams: A generalized procedure. *J. Appl. Clin. Med. Phys.* 12(4): 124–138.

Podesta M et al. (2012) Measured vs simulated portal images for low MU fields on three accelerator types: Possible consequences for 2D portal dosimetry. *Med. Phys.* 39: 7470–7479.

Podesta M et al. (2014) Time dependent pre-treatment EPID dosimetry for standard and FFF VMAT. *Phys. Med. Biol.* 59: 4749–4768.

Poludniowski G et al. (2010) CT reconstruction from portal images acquired during volumetric-modulated arc therapy. *Phys. Med. Biol.* 55: 5635–5651.

RCR (2008) Royal College of Radiologists, Society and College of Radiographers, Institute of Physics and Engineering in Medicine, National Patient Safety Agency, and the British Institute of Radiology Report: Towards safer radiotherapy. (London, UK: The Royal College of Radiologists).

Richart J et al. (2012) QA of dynamic MLC based on EPID portal dosimetry. *Phys. Med.* 28: 262–268.

Ricketts K et al. (2016a) Clinical experience and evaluation of patient treatment verification with a transit dosimeter. *Int. J. Radiat. Oncol. Biol. Phys.* 95: 1513–1519.

Ricketts K et al. (2016b) Implementation and evaluation of a transit dosimetry system for treatment verification. *Phys. Med.* 32: 671–680.

Roberts DA et al. (2004) Charge trapping at high doses in an active matrix flat panel dosimeter. *IEEE Trans. Nucl. Sci.* 51: 1427–1433.

Rowshanfarzad P et al. (2010a) Measurement and modeling of the effect of support arm backscatter on dosimetry with a Varian EPID. *Med. Phys.* 37: 2269–2278.

Rowshanfarzad P et al. (2010b) Reduction of the effect of non-uniform backscatter from an E-type support arm of a Varian a-Si EPID used for dosimetry. *Phys. Med. Biol.* 55: 6617–6632.

Rowshanfarzad P et al. (2011a) Verification of the linac isocenter for stereotactic radiosurgery using cine-EPID imaging and arc delivery. *Med. Phys.* 38: 3963–3970.

Rowshanfarzad P et al. (2011b) Isocenter verification for linac-based stereotactic radiation therapy: Review of principles and techniques. *J. Appl. Clin. Med. Phys.* 12(4): 185–195.

Rowshanfarzad P et al. (2012a) Improvement of Varian a-Si EPID dosimetry measurements using a lead-shielded support-arm. *Med. Dosim.* 37: 145–151.

Rowshanfarzad P et al. (2012b) Gantry angle determination during arc IMRT: Evaluation of a simple EPID-based technique and two commercial inclinometers. *J. Appl. Clin. Med. Phys.* 13(6): 203–214.

Rowshanfarzad P et al. (2012c) Detection and correction for EPID and gantry sag during arc delivery using cine EPID imaging. *Med. Phys.* 39: 623–635.

Rowshanfarzad P et al. (2012d) EPID-based verification of the MLC performance for dynamic IMRT and VMAT. *Med. Phys.* 39: 6192–6207.

Rowshanfarzad P et al. (2012e) Investigation of the sag in linac secondary collimator and MLC carriage during arc deliveries. *Phys. Med. Biol.* 57: N209–N224.

Rowshanfarzad P et al. (2014) An EPID-based method for comprehensive verification of gantry, EPID and the MLC carriage positional accuracy in Varian linacs during arc treatments. *Radiat. Oncol.* 9: 249–258.

Rowshanfarzad P et al. (2015) A comprehensive study of the mechanical performance of gantry, EPID and the MLC assembly in Elekta linacs during gantry rotation. *Br. J. Radiol.* 88: 20140581.

Roxby KJ and Crosbie JC (2010) Pre-treatment verification of intensity modulated radiation therapy plans using a commercial electronic portal dosimetry system. *Australas. Phys. Eng. Sci. Med.* 33: 51–57.

Rozendaal RA et al. (2014) *In vivo* portal dosimetry for head-and-neck VMAT and lung IMRT: Linking γ-analysis with differences in dose-volume histograms of the PTV. *Radiother. Oncol.* 112: 396–401.

Rozendaal RA et al. (2015) Impact of daily anatomical changes on EPID-based in vivo dosimetry of VMAT treatments of head-and-neck cancer. *Radiother. Oncol.* 116: 70–74.

Russo M et al. (2015) Step-and-shoot IMRT by Siemens beams: An EPID dosimetry verification during treatment. *Technol. Cancer Res. Treat.* 15: 535–545.

Sabet M, Menk FW and Greer PB (2010) Evaluation of an a-Si EPID in direct detection configuration as a water-equivalent dosimeter for transit dosimetry. *Med. Phys.* 37: 1459–1467.

Sabet M et al. (2012) Transit dosimetry in IMRT with an a-Si EPID in direct detection configuration. *Phys. Med. Biol.* 57: N295–N306.

Sabet M et al. (2014) Transit dosimetry in dynamic IMRT with an a-Si EPID. *Med. Biol. Eng. Comput.* 52: 579–588.

Schyns LE et al. (2016) Time-resolved versus time-integrated portal dosimetry: The role of an object's position with respect to the isocenter in volumetric modulated arc therapy. *Phys. Med. Biol.* 61: 3969–3984.

Sharma DS et al. (2010) Portal dosimetry for pretreatment verification of IMRT plan: A comparison with 2D ion chamber array. *J. Appl. Clin. Med. Phys.* 11(4): 238–248.

Siebers JV et al. (2004) Monte Carlo computation of dosimetric amorphous silicon electronic portal images. *Med. Phys.* 31: 2135–2146.

Siewerdsen JH and Jaffray DA (1999) A ghost story: Spatio-temporal response characteristics of an indirect-detection flat-panel imager. *Med. Phys.* 26: 1624–1641.

Sjölin M and Edmund JM (2016) Incorrect dosimetric leaf separation in IMRT and VMAT treatment planning: Clinical impact and correlation with pre-treatment quality assurance. *Phys. Med.* 32: 918–925.

Slosarek K et al. (2010) EPID in vivo dosimetry in RapidArc technique. *Rep. Pract. Oncol. Radiother.* 15: 8–14.

Smith RL et al. (2013) Source position verification and dosimetry in HDR brachytherapy using an EPID. *Med. Phys.* 40: 111706.

Spies L et al. (2000) Direct measurement and analytical modeling of scatter in portal imaging. *Med. Phys.* 27: 462–471.

Spies L et al. (2001) An iterative algorithm for reconstructing incident beam distributions from transmission measurements using electronic portal imaging. *Phys. Med. Biol.* 46: N203–N211.

Spreeuw H et al. (2015) Portal dosimetry in wedged beams. *J. Appl. Clin. Med. Phys.* 16(3): 244–257.

Spreeuw H et al. (2016) Online 3D EPID-based dose verification: Proof of concept. *Med. Phys.* 43(7): 3969–3974.

Steciw S, Rathee S and Warkentin B (2013) Modulation factors calculated with an EPID-derived MLC fluence model to streamline IMRT/VMAT second checks. *J. Appl. Clin. Med. Phys.* 14(6): 62–81.

Sukumar P et al. (2012) Exit fluence analysis using portal dosimetry in volumetric modulated arc therapy. *Rep. Pract. Oncol. Radiother.* 17: 324–331.

Sumida I et al. (2016) Intensity-modulated radiation therapy dose verification using fluence and portal imaging device. *J. Appl. Clin. Med. Phys.* 17(1): 259–271.

Sun B et al. (2015) Daily QA of linear accelerators using only EPID and OBI. *Med. Phys.* 42: 5584–5594.

Swindell W and Evans PM (1996) Scattered radiation in portal images: A Monte Carlo simulation and a simple physical model. *Med. Phys.* 23: 63–73.

Tan YI et al. (2015) A dual two dimensional electronic portal imaging device transit dosimetry model based on an empirical quadratic formalism. *Br. J. Radiol.* 88: 20140645.

Tyner E et al. (2009) Experimental investigation of the response of an a-Si EPID to an unflattened photon beam from an Elekta Precise linear accelerator. *Med. Phys.* 36: 1318–1329.

van Elmpt W et al. (2008) Literature review of electronic portal imaging for radiotherapy dosimetry. *Radiother. Oncol.* 88: 289–309.

van Elmpt W et al. (2009) 3D in vivo dosimetry using megavoltage cone-beam CT and EPID dosimetry. *Int. J. Radiat. Oncol. Biol. Phys.* 73: 1580–1587.

van Elmpt W et al. (2010) 3D dose delivery verification using repeated cone-beam imaging and EPID dosimetry for stereotactic body radiotherapy of non-small cell lung cancer. *Radiother. Oncol.* 94: 188–194.

Van Esch A, Depuydt T and Huyskens DP (2004) The use of an aSi-based EPID for routine absolute dosimetric pre-treatment verification of dynamic IMRT fields. *Radiother. Oncol.* 71(2): 223–234.

Van Esch A et al. (2013) Optimized Varian aSi portal dosimetry: Development of datasets for collective use. *J. Appl. Clin. Med. Phys.* 14(6): 82–99.

Van Uytven E et al. (2015) Validation of a method for in vivo 3D dose reconstruction for IMRT and VMAT treatments using on-treatment EPID images and a model-based forward-calculation algorithm. *Med. Phys.* 42: 6945–6954.

Varatharaj C et al. (2010) Implementation and validation of a commercial portal dosimetry software for intensity-modulated radiation therapy pre-treatment verification. *J. Med. Phys.* 35: 189–196.
Vial P et al. (2008a) Initial evaluation of a commercial EPID modified to a novel direct-detection configuration for radiotherapy dosimetry. *Med. Phys.* 35: 4362–4374.
Vial P et al. (2008b) The impact of MLC transmitted radiation on EPID dosimetry for dynamic MLC beams. *Med. Phys.* 35: 1267–1277.
Vial P et al. (2009) Direct-detection EPID dosimetry: Investigation of a potential clinical configuration for IMRT verification. *Phys. Med. Biol.* 54: 7151–7169.
Vinall AJ et al. (2010) Practical guidelines for routine intensity-modulated radiotherapy verification: Pre-treatment verification with portal dosimetry and treatment verification with in vivo dosimetry. *Br. J. Radiol.* 83: 949–957.
Wang S et al. (2009) Monte Carlo-based adaptive EPID dose kernel accounting for different field size responses of imagers. *Med. Phys.* 36: 3582–3595.
Wang Y et al. (2013) Quality assurance of electron beams using a Varian electronic portal imaging device. *Phys. Med. Biol.* 58: 5461–5475.
Warkentin B, Rathee S and Steciw S (2012) 2D lag and signal nonlinearity correction in an amorphous silicon EPID and their impact on pretreatment dosimetric verification. *Med. Phys.* 39: 6597–6608.
Wendling M et al. (2006) Accurate two-dimensional IMRT verification using a back-projection EPID dosimetry method. *Med. Phys.* 33: 259–273.
Wendling M et al. (2009) A simple backprojection algorithm for 3D in vivo EPID dosimetry of IMRT treatments. *Med. Phys.* 36: 3310–3321.
Wendling M et al. (2012) In aqua vivo EPID dosimetry. *Med. Phys.* 39: 367–377.
WHO (2008) Radiotherapy risk profile. (Geneva, Switzerland: World Health Organization).
Winkler P and Georg D. (2006) An intercomparison of 11 amorphous silicon EPIDs of the same type: implications for portal dosimetry. *Phys. Med. Biol.* 51(17): 4189–4200.
Winkler P, Hefner A and Georg D (2005) Dose-response characteristics of an amorphous silicon EPID. *Med. Phys.* 32: 3095–3105.
Winkler P, Hefner A and Georg D (2007) Implementation and validation of portal dosimetry with an amorphous silicon EPID in the energy range from 6 to 25 MV. *Phys. Med. Biol.* 52: N355–N365.
Woodruff HC et al. (2013) Gantry-angle resolved VMAT pretreatment verification using EPID image prediction. *Med. Phys.* 40: 081715.
Woodruff HC et al. (2015) First experience with real-time EPID-based delivery verification during IMRT and VMAT sessions. *Int. J. Radiat. Oncol. Biol. Phys.* 93: 516–522.
Yeo IJ et al. (2009) Dose reconstruction for intensity-modulated radiation therapy using a non-iterative method and portal dose image. *Phys. Med. Biol.* 54: 5223–5236.
Yeo IJ et al. (2013a) Feasibility study on inverse four-dimensional dose reconstruction using the continuous dose-image of EPID. *Med. Phys.* 40: 051702.
Yeo IJ et al. (2013b) Conditions for reliable time-resolved dosimetry of electronic portal imaging devices for fixed-gantry IMRT and VMAT. *Med. Phys.* 40: 072102.

Yohannes I, Prasetio H and Bert C (2015) Noncoplanar verification: A feasibility study using Philips' Pinnacle3 treatment planning system. *J. Appl. Clin. Med. Phys.* 16(6): 84–90.

Yoon J et al. (2016a) A Monte Carlo calculation model of electronic portal imaging device for transit dosimetry through heterogeneous media. *Med. Phys.* 43: 2242–2250.

Yoon J et al. (2016b) Four-dimensional dose reconstruction through in vivo phase matching of cine images of electronic portal imaging device. *Med. Phys.* 43: 4420–4430.

Zhang M et al. (2013) A clinical objective IMRT QA method based on portal dosimetry and electronic portal imager device (EPID) measurement. *Technol. Cancer Res. Treat.* 12: 145–150.

Zhu J et al. (2013) A comparison of VMAT dosimetric verifications between fixed and rotating gantry positions. *Phys. Med. Biol.* 58: 1315–1322.

Zhu J et al. (2015) Fast 3D dosimetric verifications based on an electronic portal imaging device using a GPU calculation engine. *Radiat. Oncol.* 10: 85.

Zwan BJ et al. (2014) Dose-to-water conversion for the backscatter-shielded EPID: A frame-based method to correct for EPID energy response to MLC transmitted radiation. *Med. Phys.* 41: 081716.

8

2D and Semi-3D Dosimetry Systems

Donald A. Roberts, Kelly C. Younge, and Jean M. Moran

8.1 Role of Multidimensional Systems

The adoption of intensity-modulated radiation therapy (IMRT) has resulted in a continuing clinical need for improved tools for commissioning, pretreatment measurements, and *in vivo* (during treatment) quality assurance (QA) methods. In the early days of IMRT, most medical physicists were restricted to the use of point detectors (Chapter 4) and radiographic film. As patient care continues to demonstrate the benefits of intensity-modulated treatment plans for target coverage and normal tissue sparing (De Neve et al., 2012), we continue to need methods to measure that the intended delivery is achieved. The need for efficient measurement devices (this chapter) and analysis tools (Chapter 14) is balanced with patient safety considerations.

Multidimensional systems have truly revolutionized dosimetry for applications such as IMRT pretreatment QA measurements. The primary benefit of radiographic film was high spatial resolution. However, it was challenging to use radiographic film for comprehensive measurements because of the need for adequate mixing of chemicals, time considerations (such as for extended dose range [EDR] film) (Pai et al., 2007; Low et al., 2011), and required supplemental measurements with an ion chamber. Multidimensional systems, whether in a two-dimensional (2D) or other array of detectors, provide real-time feedback at the time of measurements. This has resulted in significant efficiency improvements in the clinic. When superior spatial resolution is required, radiochromic film has become the preferred method because film processors continue to be removed from clinical radiation therapy environments due to the tremendous success of electronic portal imaging devices (EPIDs) for patient imaging.

Multidimensional systems have been designed to be positioned in the treatment field in place of the patient. However, another pressing need from a patient safety perspective is to ensure that the correct dose is delivered for each treatment fraction. A number of devices have been developed that perturb the treatment beam minimally and measure the fluence. These have the potential to be used to hold off a treatment beam during treatment delivery. Such tools are complementary to the Quality Assurance Plan Veto (QAPV) check proposed by the Integrating the Healthcare Enterprise in Radiation Oncology (IHE-RO) effort (Noel et al., 2014). When coupled with more sophisticated computational tools, transmission systems may enable a more efficient workflow for on-treatment patient QA.

Novel dosimetry systems have been developed for different applications as discussed in Section 8.4. For example, detectors have been developed to measure Cherenkov radiation during delivery, whereas newly developed plastic scintillator systems can be placed in phantoms and have the advantage of being tissue equivalent. (Gel dosimeters, radiochromic three-dimensional [3D] detectors, and EPIDs are discussed in Chapters 5, 6, and 7, respectively.)

8.2 Planar Systems

8.2.1 Radiochromic Film

Radiochromic film has become increasingly popular for dosimetric measurements in radiation therapy departments, because it is self-developing and therefore is ideal for processor-less environments. These films also offer significantly better energy independence compared to radiographic films (Muench et al., 1991), are relatively light insensitive, and provide a submillimeter resolution (Lynch et al., 2006; Devic et al., 2012; Lewis et al., 2012). Currently, the most frequently used radiochromic film products for dose verification in external beam therapy are EBT2 and EBT3, originally manufactured by International Specialty Products (ISP, Wayne, NJ), which is now part of Ashland (Bridgewater, NJ).

The active component of radiochromic film is a radiation-sensitive monomer, which when subject to radiation polymerizes to form a colored dye. EBT2 film has an asymmetric composition that causes the scanner response to vary depending on which side of the film faces the scanner glass. EBT3 films were redesigned to eliminate this asymmetry, and additionally have a special surface coating to prevent the formation of Newton rings during scanning. Both EBT2 and EBT3 films have a rotational sensitivity. The suggested film orientation for scanning is landscape mode because of the higher film sensitivity along this direction. When working with radiochromic film, it should be marked to ensure the orientation remains constant throughout calibration, irradiation, and readout.

Radiochromic film can be used in the dose range of approximately 1 cGy up to 40 Gy and produces a colored image when exposed to radiation. The strongest response is in the red-colored channel for doses up to 8 Gy, and then the green-colored channel becomes more sensitive. The blue-colored channel is relatively insensitive to radiation and can be used as a measure of the thickness of the film's active layer as part of a nonuniformity correction. Several different methods have been published for calibration of radiochromic film (Micke et al., 2011; Mayer et al., 2012; Perez Azorin et al., 2014).

Depending on the level of accuracy needed, the film can be calibrated either once per batch or for each irradiation. Special techniques can also be used to linearize the response of the film for relative measurements (Devic et al., 2012). The typical calibration method is to cut a single piece of 20.3×25.4 cm^2 film into multiple pieces and irradiate each piece to a different known dose over the range of doses needed for the given set of tests. The film can be calibrated in the red channel alone or for all three color channels as described earlier. Because the film is self-developing, its response changes over time; film scanned immediately after irradiation will show a different level of darkening compared to film scanned 24 hours after exposure. To overcome this limitation, all films, both calibration and measurement, are typically scanned at least 24 hours after irradiation when the film response change is much slower. Special protocols have also been developed to allow earlier readout of the film (Lewis et al., 2012).

8.2.2 Electronic Arrays (Positioned on a Tabletop)

There are many available options for 2D and semi-3D arrays for external beam dosimetry. These devices are popular because of their ease of use, fast readout, and analysis. The two major classes are diode and ionization chamber arrays. These types of arrays are designed to be placed inside a solid phantom for dose measurement and verification. Advantages and disadvantages exist for each type of array, which is why multiple measurement types (such as an array measurement plus ion-chamber point measurement) are often useful. Diodes typically act as point measurements, whereas ion chambers measure over the volume within the chamber. The volume-averaging effect of ion chambers leads to a blurring of the measured dose, yet also means that more of the radiation fluence pattern is sampled during the measurement.

An important consideration of all measurement arrays is the change in detector response as a function of irradiation angle. Essentially all of these devices exhibit some sort of angular dependence. For 2D systems, the detector array can either solely be irradiated perpendicularly, i.e., with the detector on the couchtop or mounted to the gantry, or an angular correction can be applied if the measurement results can be correlated with gantry angle. In any case, the need for an angular correction factor should always be evaluated if the measurement geometry will involve beams with nonperpendicular incidence (Li et al., 2010; Wolfsberger et al., 2010; Shimohigashi et al., 2012).

8.2.2.1 Diode Arrays

The Sun Nuclear MapCHECK 2 (www.sunnuclear.com) is a 2D diode array made up of 1527 diode detectors with uniform 0.707 cm spacing (along the diagonal) across the 32×26 cm^2 array. This model is an upgrade of the original MapCHECK with a larger measurement plane and more finely spaced detectors. The MapCHECK 2 has been shown to have excellent linearity over a wide range of doses as well as very good reproducibility with limited temperature dependence, making it well suited for IMRT QA (Jursinic and Nelms, 2003; Letourneau et al., 2004). The MapCHECK software SNC Patient includes an "arc" correction factor that can be applied for the angular dependence of the diodes when appropriate. A potential drawback of diode arrays is their tendency to exhibit radiation damage and therefore these arrays should be checked regularly to make sure that their calibration has not changed (Low et al., 2011).

An additional feature available for use with MapCHECK is the 3DVH software that is used to estimate how measured dose deviations affect doses to target volumes and organs at risk (OARs). Careful commissioning of this feature is required before the results can be meaningfully interpreted (Song et al., 2014).

8.2.2.2 Ionization Chamber Arrays

The MatriXX Evolution detector from IBA Dosimetry (www.iba-dosimetry.com) contains 1020 vented ionization chambers over an active area of 24 × 24 cm^2 (Table 8.1). The chamber size is 0.45 cm and the center-to-center spacing is 0.76 cm. The MatriXX is supplied with a gantry angle sensor to correlate the delivery with gantry angle. The sensor is used to compensate for angular dependence or to view QA results as a function of gantry angle. The MatriXX has been widely used for both IMRT and more recently volumetric-modulated arc therapy (VMAT) QA because of its excellent linearity and dose-rate independence, along with the ion chamber's innate energy independence (Herzen et al., 2007; Saminathan et al., 2010).

The PTW seven29 (www.ptw.de) array has been replaced with the OCTAVIUS 729. This array has 729 vented ionization chambers over an active area of 27 × 27 cm^2 (Table 8.1). The detectors are cubic shaped and spaced in 1 cm increments. The VeriSoft software allows the user to shift the array to increase the effective resolution of the measurement, as well as to measure over a larger field of view, and the dose measurements can then be composited. In addition, PTW manufactures the OCTAVIUS 1000 SRS (Markovic et al., 2014) and the OCTAVIUS 1500 (Stelljes et al., 2015). The SRS version has 977 liquid-filled ionization chambers over a 10 × 10 cm^2 grid, with a spacing of 0.25 cm in the central 5 × 5 cm^2. The 1500 version has the same dimensions as the 729 but includes 1405 vented ionization chambers for a resolution about half that of the 729. This array was found to have increased instantaneous measurement stability but more pronounced directional dependence compared to its 729 counterparts (Van Esch et al., 2014).

All of these models have been shown to be reliable and accurate for patient-specific QA of radiotherapy treatment plans with sufficient commissioning, calibration, and routine quality control (Markovic et al., 2014; Van Esch et al., 2014; Stelljes et al., 2015). Although ion chambers do not exhibit the same radiation damage as diodes, verification of array calibration should be a regular part of the array QA (Low et al., 2011).

Table 8.1 Basic Characteristics of Ionization Chamber Arrays

Manufacturer and Model	Number and Type of Detector	Active Area (cm^2)	Detector Features and Spacing
IBA Dosimetry: MatriXX Evolution	1020 vented ion chambers	24 × 24	Chamber volume: 0.08 cm^3 Center-to-center: 0.76 cm
PTW: OCTAVIUS 729	729 vented ion chambers	27 × 27	Chamber volume: 0.125 cm^3 Center-to-center: 1 cm
PTW: OCTAVIUS 1000 SRS	977 liquid-filled ion chambers	10 × 10	Chamber volume: 0.003 cm^3 0.25-cm spacing in central: 5 × 5 cm^2
PTW: OCTAVIUS 1500	1405 vented ion chambers	27 × 27	Chamber volume: 0.06 cm^3 Center-to-center: 0.71 cm

8.2.3 Transmission Systems for Online Treatment Monitoring

Online treatment monitoring of IMRT delivery is now available through a few different commercial systems. These systems are unique in that they are placed in the accessory slot of the head of the linear accelerator and remain in place during patient dose delivery. They offer wireless communication and real-time display of results during dose delivery. Transmission systems attached to the head of the treatment machine reduce the amount of clearance available between the patient and the head of the treatment machine. Clearance should always be carefully evaluated on a per-patient basis.

8.2.3.1 PTW DAVID

The PTW DAVID (device for the advanced verification of IMRT deliveries; www.ptw.de) is a transparent, vented, multiwire transmission-type ionization chamber that is placed in the accessory tray of the linear accelerator head (Poppe et al., 2010). The wires are arranged such that they align to the midline of each multileaf collimator (MLC; necessitating that the system be designed for the specific MLC being used). Figure 8.1a shows a cross section of the device with the collection wires shown in air between two polymethyl methacrylate (PMMA) layers (Johnson et al., 2014). Each wire measures ionization across the length of a single MLC leaf pair, thus the signal from the wire represents the line integral of the ionization in that region, as well as some spillover from the neighboring regions. A correction for the lateral response function of the DAVID detector is not required as long as a reference fluence measurement exists but can be performed to increase sensitivity to MLC positioning errors (Looe et al., 2010).

The measured fluence for each treatment fraction is compared to a reference fluence obtained either during the first fraction or during QA of the treatment plan on a verification phantom. The deviation of online measured values from reference values is then required to be within specific tolerance limits.

An important aspect of this type of system is the attenuation of the radiation beam caused by the presence of the detector. This attenuation depends on field size and energy and for the PTW DAVID is approximately 5% (Poppe et al., 2010). This attenuation must be accounted for during planning, similar to the way a tray factor is applied, or could even be included in the commissioning of the planning system. Additionally, the presence of the detector can increase surface dose, the magnitude of which depends on the distance of the detector from the patient surface. This dose enhancement should be quantified and understood before use.

8.2.3.2 IBA Dosimetry Dolphin

The IBA Dosimetry Dolphin (www.iba-dosimetry.com) is a high-resolution, high-sensitivity ion-chamber array made up of 1513 air-vented, plane-parallel ionization chambers. The chambers have a 0.5 cm center-to-center spacing in the central 15×15 cm^2 portion with a full width of 24.3×24.3 cm^2 and can be used to measure fields up to 40×40 cm^2. The system attaches to the head of the treatment field (Figure 8.1b) and is used to measure the fluence prior to treatment or during a patient's treatment. Data are transferred wirelessly. The detector response is characterized by Monte Carlo modeling. It is coupled with software that can be used to support an evaluation of a patient's treatment plan.

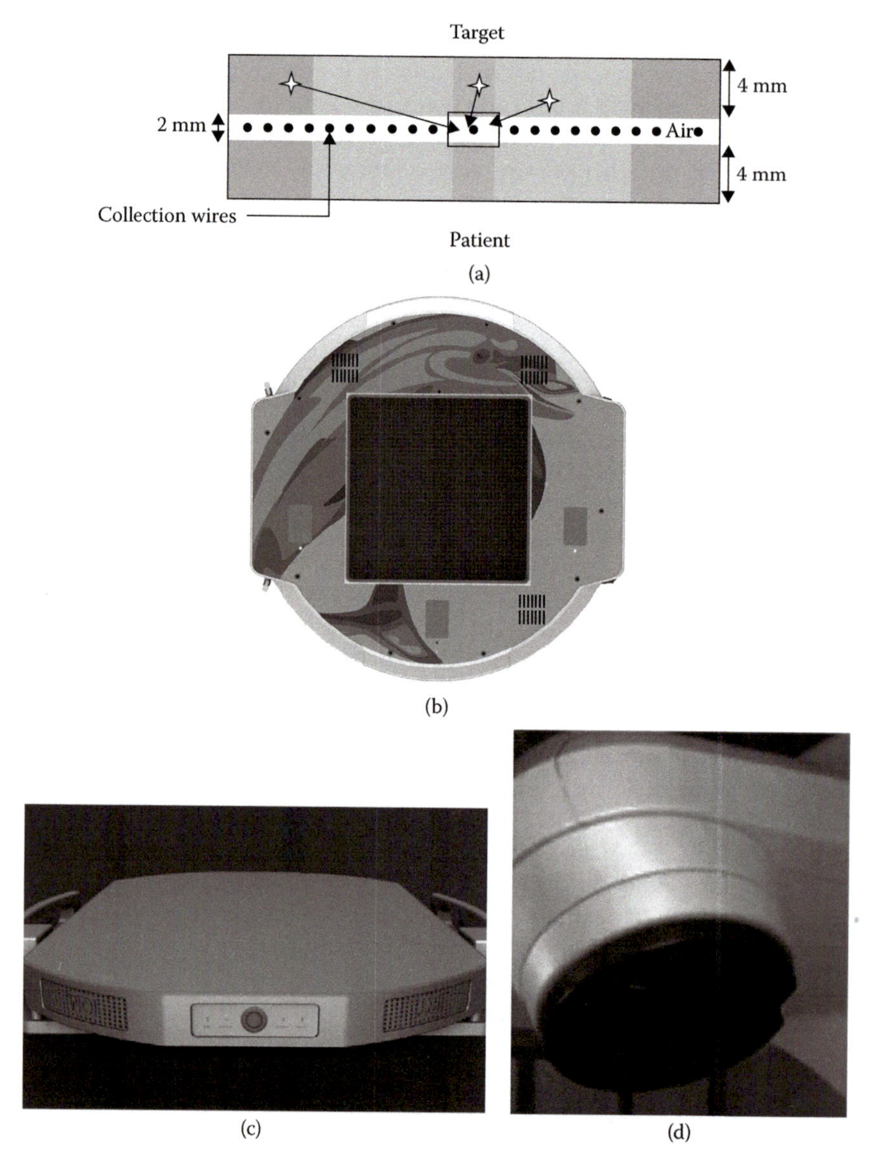

Figure 8.1

Transmission detectors mounted to the exit of the collimator head. (a) Cross section of the PTW DAVID. (Reproduced from Johnson D et al., *J. Appl. Clin. Med. Phys.*, 15, 4, 2014. With permission.) (b) IBA Dolphin system with 1513 ionization chambers. (c) iRT IQM system consisting of a single ionization chamber. (d) Delta[4] Discover system consisting of 4040 disc-shaped diodes.

8.2.3.3 iRT Systems Integral Quality Monitor

The iRT Systems Integral Quality Monitor (IQM; www.i-rt.de) is based on a prototype developed by Islam et al. (2009). The commercial system is battery powered and consists of a single air-vented ion chamber with an active size of 26.5 × 26.5 cm^2 placed at the exit of the collimator, which can be used to measure fields up to 40 × 40 cm^2 (Figure 8.1c). The system includes a built-in inclinometer and data are transferred wirelessly. The commercial version is undergoing characterization by a number of investigators. The system has been characterized for a range of photon conditions (Hoffman et al., 2017). Transmission of the system was determined to be a function of energy and ranged from 5.4% for 6 MV to 4.2% for 15 MV. In-depth dosimetric evaluation was performed for 1×1 cm^2 and 10×10 cm^2 fields. The IQM was found to be reproducible and when errors were introduced it was sensitive to IMRT and VMAT errors. The system includes an inclinometer, accurate to within 0.3° when compared to a spirit level and plumb bob, and a barometer, accurate to within 2.3 mmHg compared to a mercury barometer.

8.2.3.4 ScandiDos Delta^4 Discover

The Delta^4 Discover (www.delta4family.com) is a high-resolution, ultrathin detector array consisting of 4040 disc-shaped p-Si diodes with a diameter of 0.1 cm. The system is battery operated and the diodes have a spacing of 0.25 cm along and 0.5 cm transverse to the MLC trajectories covering an area of 19.5×25.0 cm^2 at the isocenter. The device is disc shaped, has a diameter of 79.0 cm, has a total thickness of 5.5 cm, and covers the whole collimator area (Figure 8.1d). The attenuation and the additional skin dose are both less than 1% according to the manufacturer's website.

8.3 Multidimensional Arrays (Non-gel)

Between 2D planar and full-3D detectors, semi-3D detectors (non-gel) sample a portion of a 3D volume. These detectors are sometimes referred to as four-dimensional (4D) detectors because they are able to synchronize with pulses from the linear accelerator and thus report dose as a function of time/gantry angle, and such systems are sometimes used on motion platforms. However, these systems should not be confused with detector systems (such as gels and radiochromic detectors) that are truly 3D (Chapters 5 and 6).

8.3.1 Diode Systems

8.3.1.1 ArcCHECK

The Sun Nuclear ArcCHECK (www.sunnuclear.com) is a cylindrical phantom with 1386 diode detectors arranged in a helical manner about the phantom over a total length of 21 cm. The detectors are spaced 1 cm apart and have a size of 0.08×0.08 cm^2. The spiral design effectively doubles the resolution of the detectors along the beam's-eye-view direction. The phantom measures both entrance and exit dose and correlates the two to determine the gantry angle during each part of the delivery. A 10×10 cm^2 area contains 221 detectors (including entrance and exit), which is equivalent to the resolution of the MapCHECK 2. The ArcCHECK diameter is 26.6 cm with a 15-cm-diameter hollow central cavity that allows for specialized inserts that can be used for measuring point doses at various locations or for film (Letourneau et al., 2009; Li et al., 2013).

The diode response of the ArcCHECK is similar to MapCHECK. The ArcCHECK exhibits excellent reproducibility and linearity but has a field size and angular dependence as well as dose rate and dose per pulse dependence (Letourneau et al., 2009; Kozelka et al., 2011; Li et al., 2013). The ArcCHECK software can automatically correct for its angular dependence during the dose reconstruction.

8.3.1.2 Delta[4]

The ScandiDos Delta[4] (www.scandidos.com) consists of two perpendicular diode arrays constructed inside a cylindrical PMMA phantom (Bedford et al., 2009). The 1069 diodes are spaced 0.5 cm apart in the central 6×6 cm^2 area of each plane, and 1 cm apart extending to a total area of 20×20 cm^2. The phantom itself has a diameter of 22 cm and length of 40 cm. The Delta[4] system includes an inclinometer for correlating measurements with gantry angle. It also synchronizes with pulses from the linear accelerator in order to pair measured dose with specific control points, thus allowing the Delta[4] software to apply a gantry angle–specific correction factor for angular dependence. The Delta[4] has been shown to have very good uniformity, angular response, and linearity, and is well suited for IMRT and VMAT QA measurements (Bedford et al., 2009; Sadagopan et al., 2009).

8.3.2 Ionization Chamber Systems

The PTW OCTAVIUS 1500 and 729 detector arrays described in Section 8.2.2.2 can both be placed inside the OCTAVIUS 2D phantom, a hexagonal cylindrically symmetric polystyrene phantom with a length and diameter of 32 cm and weight of 24 kg. This phantom can be rotated into eight different orientations to increase the number of measurement points within the volume of the phantom.

A more sophisticated version of the OCTAVIUS 2D is the OCTAVIUS 4D. This system is a cylindrical polystyrene phantom with a slot to hold any of the three PTW 2D arrays described earlier (OCTAVIUS 729, 1500, 1000 SRS). The phantom has a diameter of 32 cm and a length of 34.3 cm, and weighs 29 kg. This phantom has a wireless inclinometer and rotates with the gantry, such that the detector remains perpendicular to the radiation beam at all times during the delivery, thus avoiding the angular dependence issues discussed earlier. This measurement geometry is used to reconstruct a 3D dose distribution in the phantom. The reconstruction algorithm uses percent depth dose curves for field sizes between 4×4 cm^2 and 26×26 cm^2, which are established at the time of commissioning. The software determines the effective field size during the irradiation by analyzing which detectors received dose above a given threshold (Stathakis et al., 2013). Measurements with the OCTAVIUS 4D have been found to be consistent with radiochromic film with a gantry angle accuracy within 0.4° (McGarry et al., 2013).

8.4 Novel Dosimetry Systems

8.4.1 Cherenkov Radiation Detectors

Cherenkov light is emitted when a charged particle passes through a polarizable (dielectric) medium at a velocity that is greater than the phase velocity of light within that medium (Cherenkov, 1934). For many years, experimental nuclear physics, nuclear astrophysics, and particle physics fields have often used the detection of Cherenkov radiation to identify the velocity of individual particles by imaging and parameterizing the characteristic light rings generated by the particles as they pass through a material. Cherenkov light peaks in the blue and ultraviolet range and has a cutoff in the soft x-ray range determined by the index of refraction of the dielectric medium. The light can be easily imaged with existing technologies (Andreozzi et al., 2015). However, due to the relatively low yield of Cherenkov photons, a wavelength shifter is often added to a detector system to increase the signal strength to better match the spectral shape to the response of the detector system. Typically, Cherenkov light will produce ~1% of the light produced by the same particle in a scintillator (Knoll, 1989).

In comparison, Cherenkov detection in medical physics is used to measure the light generation of a large ensemble of particles passing through a material. Many electrons scattered by photon therapy beams result in a large number of scattered electrons such that the Cherenkov light can be imaged. Cherenkov detectors have been used to measure superficial dose (Zhang et al., 2013a,b), to image photon beams (Glaser et al., 2013a,b), and to perform optical dosimetry for photon-treatment plans (Glaser et al., 2014a,b). These detectors have also been used for portal dosimetry (Mei et al., 2006).

When scintillating materials are used to detect the light produced by ionizing radiation, there can be Cherenkov background light in those materials (Beddar et al., 2004; Archambault et al., 2006). Although subtraction of this Cherenkov background may be challenging, it is required to produce high-quality results with these scintillators. Non-scintillating materials, such as glass, crystal, or plastic fibers, can also be used (Yamamoto et al., 2014; Somlai-Schweiger and Ziegler, 2015) for Cherenkov imaging, or Cherenkov light can be directly imaged with a camera system.

8.4.2 Plastic Scintillators

Plastic scintillators have been developed to detect the light produced when ionizing radiation goes through a material. These scintillators typically respond within nanoseconds. Generally, plastic scintillators will have little or no long phosphorescence components making them ideal for high counting rate situations. The best characteristics of a scintillator for dosimetry include a high transparency to the emitted light, a high scintillation efficiency, high linearity along with a density, and atomic composition similar to water (Beddar et al., 1992a). The scintillation efficiency, which is the fraction of deposited particle energy converted into light, is typically low with the majority of the energy dissipated either as phonons in the plastic lattice or as heat (Beddar et al., 1992a). As a result of the low yield of light in the medium, scintillators are generally coupled to high-gain detector devices such as photomultiplier tubes (PMTs) although charge-coupled devices (CCDs) have been used as well. The scintillator is generally coupled to the detector by an intermediate "light pipe," which serves to transmit the light to the detector but does not scintillate. However, these coupling devices may produce light inside a radiation field from other processes, such as Cherenkov light production as mentioned in Section 8.4.1. Therefore, scintillation-based dosimeters incorporate a method to estimate the light produced in the light pipe. For example, a scintillation detector was created consisting of two PMTs, two optical fibers, and a small plastic scintillator embedded in a polystyrene capsule (Beddar et al., 1992a,b). The first fiber-optic cable was coupled between the PMT and the scintillation detector. Light detected from this fiber has a combination of light produced in the scintillator and light produced in the fiber-optic cable. The second fiber was terminated in the polystyrene capsule but was not attached to the scintillator; the light detected in this fiber was assumed to be the same as the background light produced in the first fiber. As a result, the true scintillation could be determined by subtraction of the two signals. This dosimetry system was linear in energy and dose rate, had excellent spatial resolution, and was reproducible and stable (Beddar et al., 1992a). Since 1992 many other such systems have been developed for 1D, 2D, and 3D scintillator-based detectors, but the basis of each of these detectors remains the same with evolutionary improvements in this design (Fluhs et al., 1996; Ranade et al., 2006; Lacroix et al., 2010; Goulet et al., 2013). A recent publication provides a comprehensive introduction to plastic scintillation dosimetry and its use in the field of radiation dosimetry (Beddar and Beaulieu, 2016).

Video-based scintillation devices have also been reported that measure dose using a scintillator plate, mirror, and video system (Zeidan et al., 2004; Ranade et al., 2006; Collomb-Patton et al., 2009), but none are commercially developed for medical physics applications at this time. For each of these systems, the scintillator plate is placed normal to the incident beam with a 45° mirror, which reflects the light produced by scintillation to the video camera. The entire system is sealed in a light-tight container so that no ambient light reaches the video system, which typically operates at approximately 10 frames per second. A frame grabber–type system is used to capture and integrate the video images. The raw data are corrected on a frame-by-frame basis and summed together to generate the measured fluence for IMRT plans. These systems can be made of water-equivalent materials such as polystyrene, can be corrected for Cherenkov contamination, and have high spatial resolution.

8.5 Considerations Regarding Use

Many of the 2D and semi-3D systems have been designed to work with geometric phantoms. The PTW OCTAVIUS and the IBA Dosimetry MatriXX systems can be used with geometric phantoms purchased from the manufacturer or with other materials such as solid water. The ScandiDos Delta[4] consists of panels and is designed to sample multiple planes simultaneously in a standard geometry. The Sun Nuclear ArcCHECK system cannot be placed in a phantom but has the option for an insert to be placed at the center of its cylindrical design. All of these systems have been designed for intensity-modulated fields.

8.5.1 Leveraging Diode Systems through the Addition of a Reference Chamber

Because diodes are known to have their responses drift over time, a QA program needs to incorporate a check of the response of all diodes. When using a diode system, users should consider introducing an ionization chamber to their standard phantom geometry. It can be helpful to calculate the dose to the ion chamber in addition to the diode array(s) to ensure a reasonably stable point of measurement for the ion chamber. Such a measurement is also valuable because of the geometry of the multiplanar diode array systems may be less intuitive to interpret, especially for VMAT delivery. When combined with a standard measurement field as a constancy check for each measurement session, the physicist can monitor the behavior of the overall system and the results are enhanced by tracking the dose to the ionization chamber array.

8.5.2 *En Face* vs Composite Deliveries: Advantages and Disadvantages of Each Approach

When commissioning an IMRT program, it is essential to make measurements with the detector system *en face*, i.e., perpendicular to the beam, and in a composite geometry. The *en face* setup is most appropriate during commissioning to understand the limits of the system since composite measurements may mask differences if there are competing under- and overdose areas. When measuring the delivery accuracy for pretreatment measurements, it is valuable to use the patient's beam angles. Planar systems can be used mounted to the gantry if their angular dependence is too large to be used for reliable composite measurements. When questions are raised from a composite delivery, *en face* measurements can be instrumental in identifying the cause of any discrepancies. It is important to note that QA failures may occur due to the use of suboptimal treatment plans that can be a combination of inaccurate modeling of the beam and limitations of the delivery system (Younge et al., 2012). Planar versus composite approaches of IMRT QA are also discussed in Chapter 17.

8.5.3 Resolution Considerations

The resolution of the system should be considered with respect to the limits placed in the treatment planning system for the minimum gap between opposing leaves and the allowed leaf-travel speed for VMAT delivery. Commissioning of an IMRT program should be done with ion chambers and high-resolution systems (such as radiochromic film). Any new detector array needs to be adequately characterized. There are certain commissioning tests that are much more feasible when performed with detector arrays. When selecting a system, it is also important to note the chamber-to-chamber distance and the volume averaging that is present for ion chamber systems. Once a program is commissioned, spatial resolution is less of a factor unless treatment plans contain small fields such as those used for modulated stereotactic radiotherapy delivery.

8.5.4 Phantom Considerations

Multiplanar arc systems have a design that prevents them from being placed within a phantom. However, alterations to the geometry can be made, such as adding an ion chamber measurement to the setup (see Section 8.5.1). 2D planar systems can be placed in solid water phantoms or combined with other materials such as lung- or bone-equivalent density materials. The versatility of these systems is very desirable during the commissioning process. The transmission detector systems are designed to be used without an additional phantom geometry. Plastic scintillators can be used in custom phantoms.

8.5.5 The Relationship between Measurements and Calculations

When interpreting measurements, it is critical to understand any corrections that may be applied to the measurements. As noted in Chapter 14, measurements and calculations should be reviewed simultaneously. The physicist should inspect the calculated and measured dose distribution to locate areas of high modulation. Tools such as dose difference displays should be used to assess areas of disagreement between the measurements and calculations. The physicist must also understand the limitations of the measurements when interpreting any regions of disagreement. For instance, one might expect a larger dose difference in a high-gradient region. Furthermore, the physicist must also be alert for excess dose in regions outside the treatment field, even if they are of low value, since it may indicate poor modeling of transmission in the treatment-planning system or a drift in the performance of a detector system (Ezzell et al., 2009).

8.6 QA of Planar and Multidimensional Arrays

8.6.1 Commissioning Considerations

Multidimensional arrays, either 2D or semi-3D, require considerable commissioning efforts prior to routine use for clinical or research purposes. As noted in Table 8.2, the commissioning process should investigate factors such as the response during warm up, background, linearity of dose response, linearity with dose rate, long- and short-term reproducibility, stability, calibration, geometric accuracy, temperature and pressure response if needed, dead-time corrections,

Table 8.2 Major Categories When Commissioning a Multidimensional Detector Array and Sample Tests or Activities for Each Topic Area

Topic	Sample Tests (or Activities)
Basic response	• Warm-up response • Background • Sensitivity • Effective point of measurement • Energy • Temperature • Angular • Background • Dead-time corrections (where appropriate) • Density effects
Homogeneity of response	• Evaluate response of individual detectors • Perform a flip test for 2D detectors • Rotate phantom geometry where appropriate for multidimensional systems • Sensitivity to position: introduce known errors such as 1 cm position offset
Geometry	• Physical characterization (confirm center-to-center detector distances, chambers similar sizes) • Spatial resolution • Field-size dependence (e.g., full-width half-maximum over a range of field sizes) • Sensitivity to translational and/or rotational errors
Dose	• Compare to other previously characterized 2D or semi-3D systems (such as film or other array system) • Short-term stability (measurements within days) • Long-term stability (measurements over >1 week) • Dose rate • Reproducibility (within a measurement session and over time) • Dose response (low to high doses) • Complex distributions (modulation) • Central axis and off-axis • Standard test suite (such as AAPM Task Group 119)
Calibration	• Against a primary standard traceable detector with a standard geometry and dose where appropriate • Use a standard field for each setup and monitor response over time
Software	• Verify software calculations as part of commissioning • Review documentation and release notes • Understand any software corrections applied to measured data
Training	• Setup of detector system • Safe use of detector system • Use and limitations of analysis software • Functionality and limitations of accessories purchased with the detector system

effective points of measurement, field-size dependence, angular response of both the longitudinal angle and the angle in the axial plane, homogeneity tests of detectors ("flip" and "rotisserie" tests), and density effects (Spezi et al., 2005; Herzen et al., 2007; Feygelman et al., 2009, 2011; Stambaugh et al., 2014).

The spatial resolution of a system can be examined using delivery of fields with different size beamlets. For example, measurements made with 2 × 2 cm^2 and then 0.5 × 0.5 cm^2 beamlets demonstrate the impact of volume averaging when using a MatriXX system that consists of ionization chambers. 2D measurements are shown for the checkerboard test with radiographic film measurement (Kodak) on the left, MatriXX in the middle, and an extracted profile denoted on each measurement (Figure 8.2). The position of each detector is superimposed on the 2D figures. Because of volume averaging, the peaks and valleys of the measurement are no longer present in the MatriXX measurement for the stringent example of 0.5 × 0.5 cm^2 beamlets.

The physicist is responsible for thoroughly commissioning dosimetric devices, documenting the results, and providing training to other users (Low et al., 2011). The physicist should also identify tests that will be repeated as part of the regular QA of the system and at the time of any software upgrades. Figure 8.3 depicts sample commissioning tests for a multidimensional diode system (ArcCHECK) demonstrating the dependence of the system on warmup, background, dose, dose-rate, field-size dependence, and the sensitivity of response when the system is tilted.

Example fields should preferably also be measured with multiple systems. When the same plan is calculated on different phantom configurations and detectors, the distributions will look substantially different. This is shown in Figure 8.4 for one of the fields of an example IMRT plan (Figure 8.4a) measured

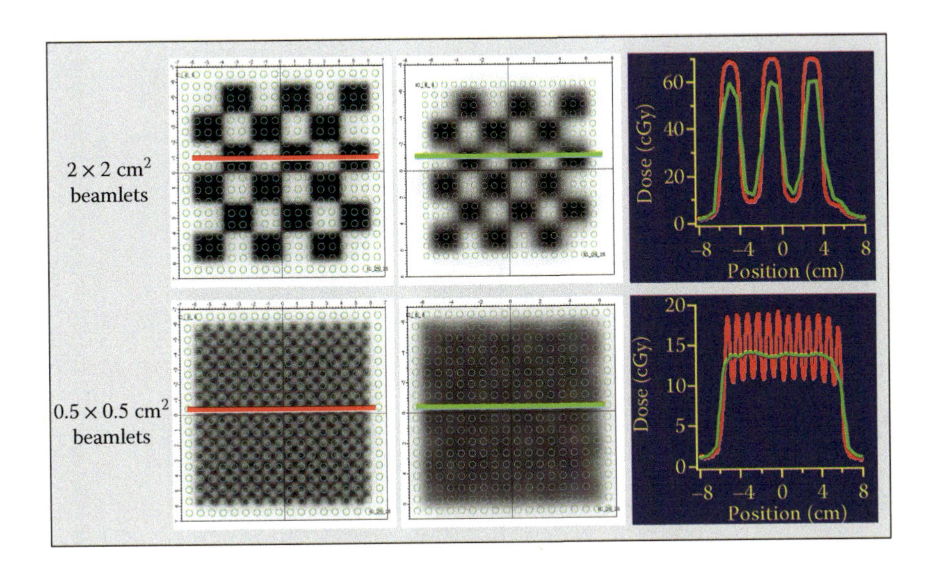

Figure 8.2

Example of a test to evaluate volume averaging of an ionization chamber array (IBA MatriXX) versus film for a strict test of a checkerboard with 2 × 2 cm^2 and then 0.5 × 0.5 cm^2 beamlets. (Courtesy of Natan Shtraus, Dale Litzenberg and Jean Moran.)

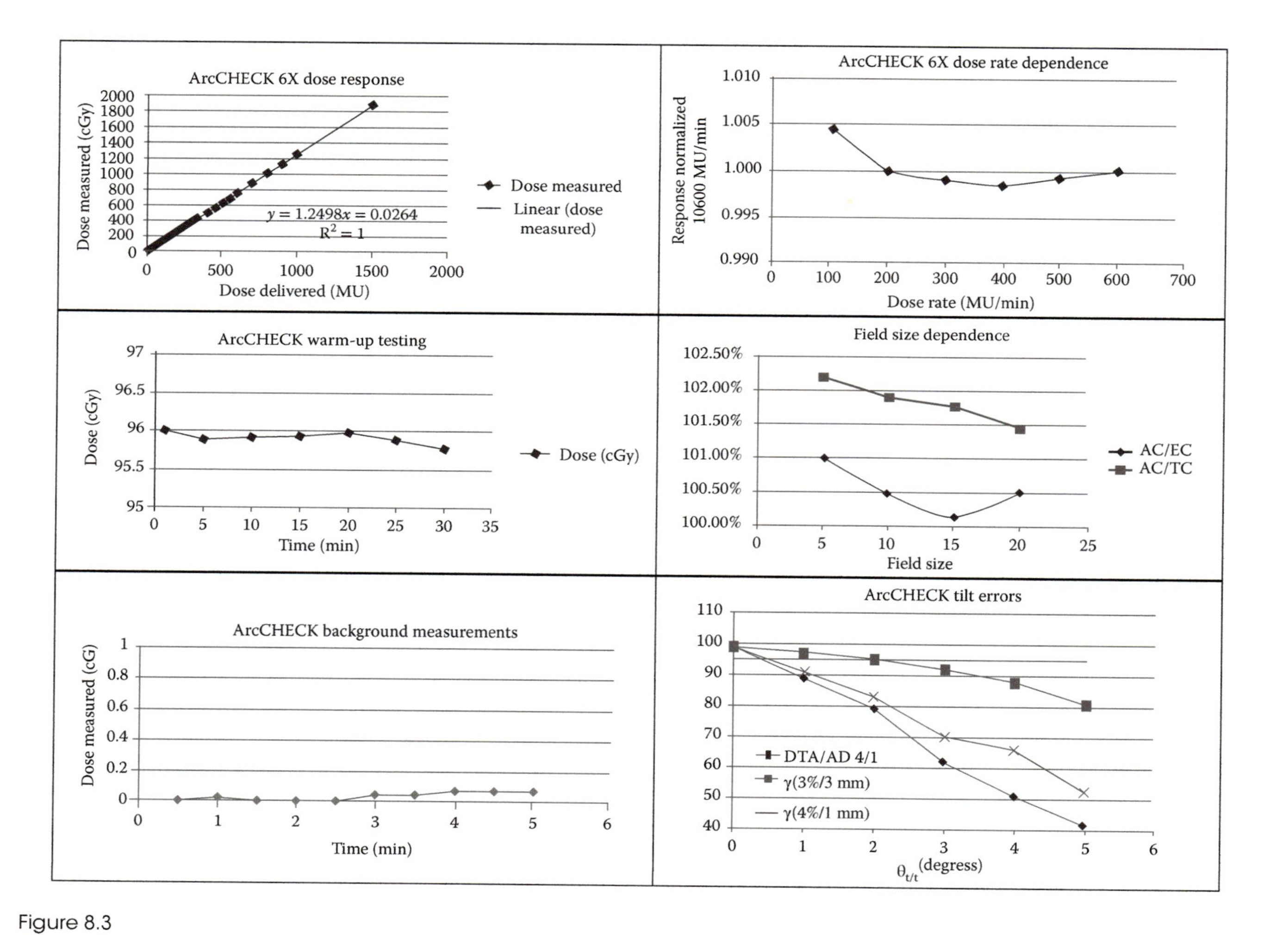

Figure 8.3

Example test results for a multidimensional diode system (Sun Nuclear ArcCHECK) when commissioning a device for static gantry intensity-modulated radiation therapy (IMRT) and volumetric-modulated arc therapy (VMAT) delivery. (Courtesy of Don Roberts.)

with an ionization chamber array system (MatriXX) in Figure 8.4b, and a diode array system (ArcCHECK) in Figure 8.4c.

8.6.2 Quality Checks for Each Measurement Set

A standard set of field irradiations should be established during commissioning to normalize detector responses to known dose distributions. These field irradiations should be repeated as part of each QA measurement session to check the normalization of the detector array(s). In addition to accounting for the variations in the daily output of the linear accelerators, the results of the premeasurement irradiations should be confirmed to assure that the array is working correctly and to establish a global normalization factor if needed (Low et al., 2011). These irradiations act as a constancy check for the systems.

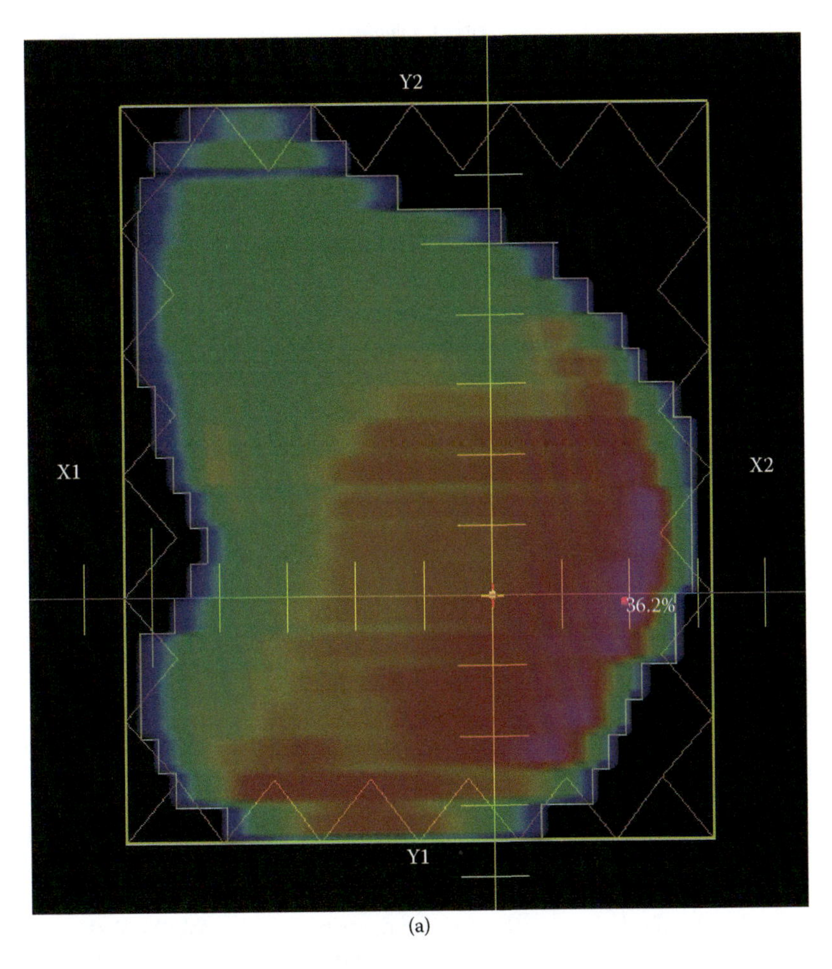

(a)

Figure 8.4

Example of measurements of an IMRT plan: (a) fluence of one field.

(Continued)

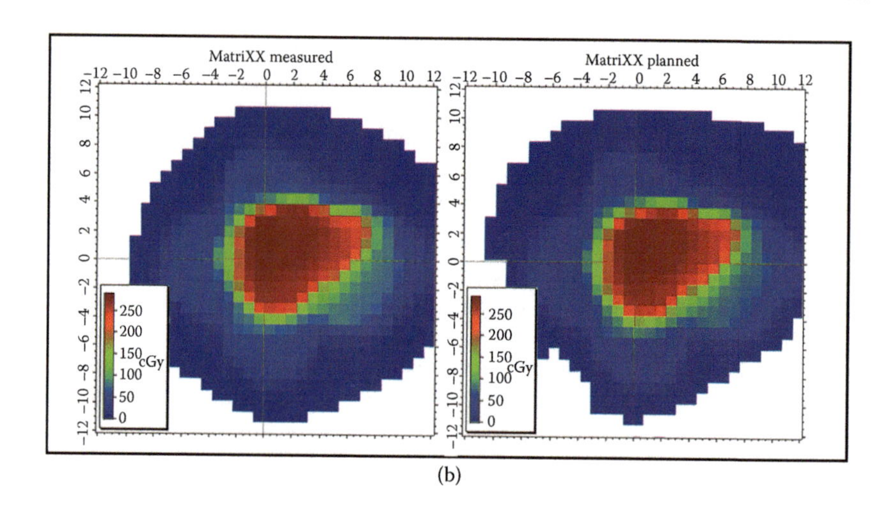

(b)

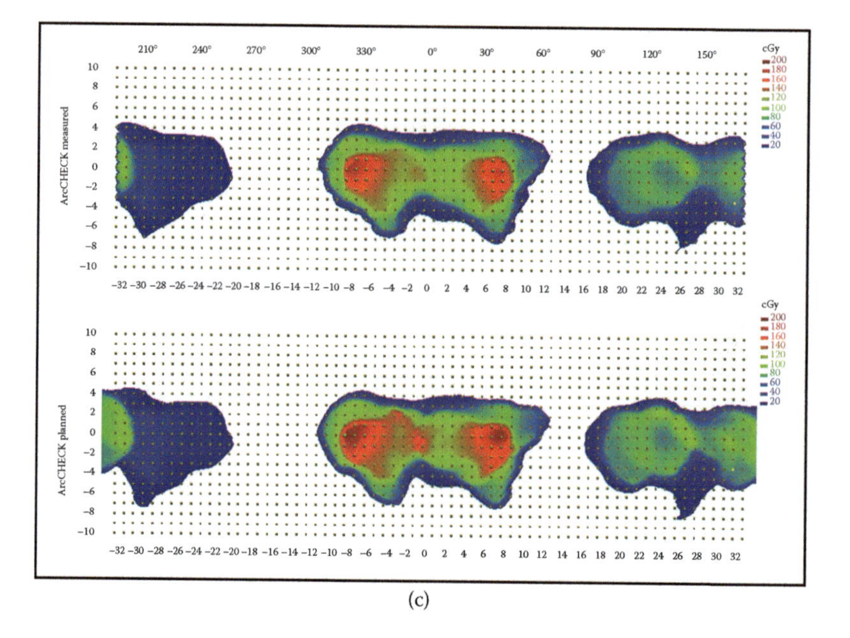

(c)

Figure 8.4 (Continued)

Example of measurements of an IMRT plan: (b) comparison with calculations on an ion chamber array (IBA MatriXX) and (c) with a multidimensional diode array (Sun Nuclear ArcCHECK).

A set of plans should be selected as a regular part of the QA program to verify the reproducibility of measurements on a particular system. These plans may be taken from an anonymized clinical set used for commissioning or from a set of community standard plans such as those described in the AAPM Task Group 119 report (Ezzell et al., 2009). The plans should be delivered repeatedly during commissioning to assess the stability of the response and at regular intervals determined by each clinic to confirm the continued proper functioning of the array. These tests are also helpful as an additional QA check to monitor delivery of a complete plan as a sensitive constancy check for departments with multiple accelerators.

8.7 Software Analysis

Several software tools have been developed to aid in interpreting the differences between measurements and calculations, especially based on an individual patient's anatomy. Since the early days of IMRT, one of the challenges for physicists has been to assess areas of disagreements and to use that information to consult with a physician on the implications of those disagreements to their specific patient. Tools that recalculate the dose on the patient's anatomy have been developed in response to that conundrum. These tools are still under investigation. If such tools are used in a clinical environment, it is important for the physicist and other personnel to note that the measurement itself might be based on incomplete information. For instance, the spatial resolution of the detector may result in volume averaging of an ion chamber or in a large sensitivity to positioning of a diode. Also, the geometry does not represent the patient; often only homogeneous phantoms are used whereas the detectors themselves have different compositions in material and density. Dose calculations may also use less accurate methods such as pencil beam methods (Narayanasamy et al., 2015; Park et al., 2015). The recalculation is done in the patient anatomy, which may or may not represent that patient's anatomy at the time of treatment. There may be additional approximations in the recalculation that affect the accuracy of the information provided to the physicist from the calculation. Such tools continue to be investigated and provide complementary information to that available by a direct comparison of calculations and measurements. It is strongly recommended that such tools are not used independent of a review of measurements and calculations in the same phantom geometry as the measurement.

8.8 Treatment Planning Considerations

8.8.1 Sensitivity

Multidimensional detectors consist of arrays of detectors, either ion chambers or diodes. For ion chambers, the sensitivity of the detector (defined as collected charge per unit of deposited dose) varies from ~0.1 to 2 nC/Gy and is correlated with the volume of the detector and the materials used for its construction. For diodes, the sensitivity typically varies from 6 to 35 nC/Gy. From these numbers, it is apparent that diode sensitivities are one to two orders of magnitude greater than those for typical ion chambers (Low et al., 2011). Ion chambers have volumes of 10^{-3} cc and greater, whereas diode volumes are typically in the 10^{-5} cc range. For ion-chamber measurements, volume-averaging effects and resolution of the dose calculations must be carefully considered and accounted for to obtain reasonable results from measurements. Additionally, care must be taken when positioning ion chambers near steep dose gradients or in regions where the dose distribution is convex (Low et al., 2011). Consequently, ion chambers are generally positioned in areas of high-dose homogeneity such that the maximum dose variation across the volume of the ion chamber is less than 10% (Low et al., 2011). Although diodes are also subject to geometric-positioning errors, the size of the region requiring target dose homogeneity is smaller.

8.8.2 Absolute versus Relative Dose

Although dose measurements with detector arrays may not be directly traceable to a Primary Standards Laboratory via an Accredited Dosimetry Calibration

Laboratory (ADCL) calibration, it is relevant if measurements and calculations are normalized in percent such as to the highest dose value, or if the results are reported in dose units (cGy or Gy) based on how the device is calibrated. As noted in Section 8.6.2, a sanity check of the detector response should be performed at each measurement session. In order to calculate a reliable dose from a measurement, the detector response should be relatively flat across the energy spectrum being measured, the response should be isotropic, and the detector must exhibit long-term stability that is independent of the amount of irradiation the detector may experience (Greene, 1962; Greene et al., 1962). If a detector meets these requirements, absolute dose can be calculated when the detector used can be traced back to a primary standard without renormalization (Low et al., 2011) using a dosimetry protocol such as the TG-51 protocol, or by cross calibration as defined by that protocol (Almond et al., 1999). Diodes can have a large energy dependence, particularly for low energies due to the Z dependence of the photoelectric effect. Attempts to moderate the low-energy overresponse of diodes by installing shielding, which produces a "compensated" diode, usually create asymmetries in the angular response. Additionally, diodes may have long-term stability issues associated with damage to the diode by radiation. For these reasons, diodes are not absolute dosimeters and the dose calculated from their response is relative, meaning that normalization is needed. It should be noted that some ion chambers, particularly those containing high Z material, may also have some of the same issues and may also not be suitable for absolute dose measurements (Low et al., 2011). Often small ion chambers will have high Z collection electrodes in order to increase the signal in the chamber, but this modification will then often lead to low-energy overresponse of the chamber similar to diodes, and a corresponding loss of absolute normalization. For these reasons, it is important to understand the construction of the detectors that make up a multidimensional array to determine whether measurements made by the device can be converted to absolute measurements or whether these measurements are relative. When measurements are relative, the dose calculations generally require normalization to some value. It is recommended that absolute dose measurements be done whenever possible.

8.8.3 Normalization Settings

For detectors used to make relative measurements, a normalization procedure is often used where the response of the detectors is calibrated against an absolute dosimeter, such as an ion chamber having a calibration traceable to a Primary Standards Laboratory (Feygelman et al., 2011), and then the dose measurements are referred to as "absolute dose." The validity of this approach must be verified during commissioning as mentioned earlier. The user should be cognizant that the vendor may have modified detectors significantly in order to achieve this goal. For instance, diodes may be shielded to compensate for low-energy photons, which may lead to asymmetries in the angular response of the diodes, i.e., some under- or overresponse, which may then be corrected in the manufacturer's software. However, when software corrections are made to the raw measurement signals to compensate for challenges in the hardware construction of the detector system, there may be a possibility that the corrections are incorrectly applied for some situations. At a minimum, the user should be aware that the presence of such corrections affects the legitimacy of an absolute dose measurement from the system.

For these reasons, users may choose to normalize the measured data at some point in the measured dose distribution. For example, the global dose maximum can be used where all errors are determined relative to that value (Van Dyk et al., 1993). Generally the analysis will exclude regions of low dose, typically <10% of the global maximum. Particularly for diodes, users should exclude regions of low-energy scatter where the detectors may overrespond. Another normalization choice is to pick a specific point, such as the intersection of the central axis with the detector system at a specified gantry angle as a normalization point.

8.8.4 Spatial Resolution

The detector system should be adequately represented in the dose calculation. The calculated dose distribution will depend on the distribution of the voxels in the dose grid. The user must determine what dose grid and what structure resolution are adequate to ensure that the dose estimates for the detector arrays are accurate.

Even with an adequately fine spatial resolution in the calculation grid, the response of ionization chamber detectors, and therefore the array of detectors, may have large measurement errors associated with the perturbation of the dose distributions due to the presence of the chambers (Bouchard and Seuntjens, 2004) in non-homogeneous areas of the dose distribution. The user should be aware of the possible errors and the possible consequences of any correction procedures that may be used to account for them. The computed-tomography (CT) scans or models of the measurement phantom also must be of adequate resolution to allow accurate dose calculations. Some manufacturers may provide a CT model to the user to represent the detector system in a phantom. The user is responsible for confirming the accuracy of that model for both static and modulated treatment fields.

8.9 Summary

It is essential that multidimensional detectors are adequately characterized for clinical use. Ionization chamber and diode array systems have been explicitly designed for use in pretreatment measurement systems and many can be used as part of a geometric phantom. Novel dosimetry systems include the measurement of Cherenkov radiation, whether in a phantom or a patient, and plastic scintillators that can be set up in arrays more similar to other detector systems. Interesting developments have been made for systems that provide real-time feedback of a patient's delivery based on the exit fluence from the collimator. These systems are mounted to the collimator head and can be used with or without the patient present. When used with a patient present, it is crucial that the gantry is still able to rotate around the patient without the risk of collision. These systems are relatively new to the radiotherapy market but provide additional QA checks.

References

Almond PR et al. (1999) AAPM's TG-51 protocol for clinical reference dosimetry of high-energy photon and electron beams. *Med. Phys.* 26: 1847–1870.

Andreozzi JM et al. (2015) Camera selection for real-time in vivo radiation treatment verification systems using Cherenkov imaging. *Med. Phys.* 42: 994–1004.

Archambault L et al. (2006) Measurement accuracy and Cerenkov removal for high performance, high spatial resolution scintillation dosimetry. *Med. Phys.* 33: 128–135.
Beddar AS, Mackie TR and Attix FH (1992a) Water-equivalent plastic scintillation detectors for high-energy beam dosimetry: I. Physical characteristics and theoretical consideration. *Phys. Med. Biol.* 37: 1883–1900.
Beddar AS, Mackie TR and Attix FH (1992b) Water-equivalent plastic scintillation detectors for high-energy beam dosimetry: II. Properties and measurements. *Phys. Med. Biol.* 37: 1901–1913.
Beddar AS, Suchowerska N and Law SH (2004) Plastic scintillation dosimetry for radiation therapy: Minimizing capture of Cerenkov radiation noise. *Phys. Med. Biol.* 49: 783–790.
Beddar S and Beaulieu L (2016) Scintillation dosimetry. Boca Raton, FL: CRC Press, Taylor & Francis Group.
Bedford JL et al. (2009) Evaluation of the Delta4 phantom for IMRT and VMAT verification. *Phys. Med. Biol.* 54: N167–N176.
Bouchard H and Seuntjens J (2004) Ionization chamber-based reference dosimetry of intensity modulated radiation beams. *Med. Phys.* 31: 2454–2465.
Cherenkov PA (1934) Visible emission of clean liquids by action of gamma-radiation. *Dokl. Akad. Nauk. SSSR.* 2: 451–452.
Collomb-Patton V et al. (2009) The DOSIMAP, a high spatial resolution tissue equivalent 2D dosimeter for LINAC QA and IMRT verification. *Med. Phys.* 36: 317–328.
De Neve W, De Gersem W and Madani I (2012) Rational use of intensity-modulated radiation therapy: The importance of clinical outcome. *Semin. Radiat. Oncol.* 22: 40–49.
Devic S et al. (2012) Linearization of dose-response curve of the radiochromic film dosimetry system. *Med. Phys.* 39: 4850–4857.
Ezzell GA et al. (2009) IMRT commissioning: Multiple institution planning and dosimetry comparisons, a report from AAPM Task Group 119. *Med. Phys.* 36: 5359–5373.
Feygelman V et al. (2009) Evaluation of a biplanar diode array dosimeter for quality assurance of step-and-shoot IMRT. *J. Appl. Clin. Med. Phys.* 10(4): 64–78.
Feygelman V et al. (2011) Evaluation of a new VMAT QA device, or the "X" and "O" array geometries. *J. Appl. Clin. Med. Phys.* 12(2): 146–168.
Fluhs D et al. (1996) Direct reading measurement of absorbed dose with plastic scintillators—The general concept and applications to ophthalmic plaque dosimetry. *Med. Phys.* 23: 427–434.
Glaser AK et al. (2013a) Projection imaging of photon beams by the Cerenkov effect. *Med. Phys.* 40: 012101.
Glaser AK et al. (2013b) Projection imaging of photon beams using Cerenkov-excited fluorescence. *Phys. Med. Biol.* 58: 601–619.
Glaser AK et al. (2014a) Video-rate optical dosimetry and dynamic visualization of IMRT and VMAT treatment plans in water using Cherenkov radiation. *Med. Phys.* 41: 062102.
Glaser AK et al. (2014b) Optical dosimetry of radiotherapy beams using Cherenkov radiation: The relationship between light emission and dose. *Phys. Med. Biol.* 59: 3789–3811.
Goulet M et al. (2013) 3D tomodosimetry using long scintillating fibers: A feasibility study. *Med. Phys.* 40: 101703.

Greene D (1962) The use of an ethylene-filled polythene chamber for dosimetry of megavoltage x-rays. *Phys. Med. Biol.* 7: 213–224.

Greene D, Massey JB and Meredith WJ (1962) Exposure dose measurements in megavoltage therapy. *Phys. Med. Biol.* 6: 551–560.

Herzen J et al. (2007) Dosimetric evaluation of a 2D pixel ionization chamber for implementation in clinical routine. *Phys. Med. Biol.* 52: 1197–1208.

Hoffman D et al. (2017) Characterization and evaluation of an integrated quality monitoring system for online quality assurance of external beam radiation therapy. *J. Appl. Clin. Med. Phys.* 18: 40–48.

Islam MK et al. (2009) An integral quality monitoring system for real-time verification of intensity modulated radiation therapy. *Med. Phys.* 36: 5420–5428.

Johnson D et al. (2014) A simple model for predicting the signal for a head-mounted transmission chamber system, allowing IMRT in-vivo dosimetry without pretreatment linac time. *J. Appl. Clin. Med. Phys.* 15(4): 270–279.

Jursinic PA and Nelms BE (2003) A 2-D diode array and analysis software for verification of intensity modulated radiation therapy delivery. *Med. Phys.* 30: 870–879.

Knoll G (1989) Radiation detection and measurement. New York, NY: Wiley.

Kozelka J et al. (2011) Optimizing the accuracy of a helical diode array dosimeter: A comprehensive calibration methodology coupled with a novel virtual inclinometer. *Med. Phys.* 38: 5021–5032.

Lacroix F et al. (2010) Simulation of the precision limits of plastic scintillation detectors using optimal component selection. *Med. Phys.* 37: 412–418.

Letourneau D et al. (2004) Evaluation of a 2D diode array for IMRT quality assurance. *Radiother. Oncol.* 70: 199–206.

Letourneau D et al. (2009) Novel dosimetric phantom for quality assurance of volumetric modulated arc therapy. *Med. Phys.* 36: 1813–1821.

Lewis D et al. (2012) An efficient protocol for radiochromic film dosimetry combining calibration and measurement in a single scan. *Med. Phys.* 39: 6339–6350.

Li QL et al. (2010) The angular dependence of a 2-dimensional diode array and the feasibility of its application in verifying the composite dose distribution of intensity-modulated radiation therapy. *Chin. J. Cancer* 29: 617– 620.

Li G et al. (2013) Evaluation of the ArcCHECK QA system for IMRT and VMAT verification. *Phys. Med.* 29: 295–303.

Looe HK et al. (2010) Enhanced accuracy of the permanent surveillance of IMRT deliveries by iterative deconvolution of DAVID chamber signal profiles. *Phys. Med. Biol.* 55: 3981–3992.

Low DA et al. (2011) Dosimetry tools and techniques for IMRT. *Med. Phys.* 38: 1313–1338.

Lynch BD et al. (2006) Important considerations for radiochromic film dosimetry with flatbed CCD scanners and EBT GAFCHROMIC film. *Med. Phys.* 33: 4551–4556.

Markovic M et al. (2014) Characterization of a two-dimensional liquid-filled ion chamber detector array used for verification of the treatments in radiotherapy. *Med. Phys.* 41: 051704.

Mayer RR et al. (2012) Enhanced dosimetry procedures and assessment for EBT2 radiochromic film. *Med. Phys.* 39: 2147–2155.

McGarry CK et al. (2013) Octavius 4D characterization for flattened and flattening filter free rotational deliveries. *Med. Phys.* 40: 091707.

Mei X, Rowlands JA and Pang G (2006) Electronic portal imaging based on Cerenkov radiation: A new approach and its feasibility. *Med. Phys.* 33: 4258–4270.

Micke A, Lewis DF and Yu X (2011) Multichannel film dosimetry with nonuniformity correction. *Med. Phys.* 38: 2523–2534.

Muench PJ et al. (1991) Photon energy dependence of the sensitivity of radiochromic film and comparison with silver halide film and LiF TLDs used for brachytherapy dosimetry. *Med. Phys.* 18: 769–775.

Narayanasamy G et al. (2015) Evaluation of Dosimetry Check software for IMRT patient-specific quality assurance. *J. Appl. Clin. Med. Phys.* 16(3): 329–338.

Noel CE et al. (2014) Quality assurance with plan veto: Reincarnation of a record and verify system and its potential value. *Int. J. Radiat. Oncol. Biol. Phys.* 88: 1161–1166.

Pai S et al. (2007) TG-69: Radiographic film for megavoltage beam dosimetry. *Med. Phys.* 34: 2228–2258.

Park JC et al. (2015) Adaptive beamlet-based finite-size pencil beam dose calculation for independent verification of IMRT and VMAT. *Med. Phys.* 42: 1836–1850.

Perez Azorin JF, Ramos Garcia LI and Marti-Climent JM (2014) A method for multichannel dosimetry with EBT3 radiochromic films. *Med. Phys.* 41: 062101.

Poppe B et al. (2010) Clinical performance of a transmission detector array for the permanent supervision of IMRT deliveries. *Radiother. Oncol.* 95: 158–165.

Ranade MK et al. (2006) A high-speed scintillation based electronic portal imaging device to quantitatively characterize IMRT delivery. *Med. Phys.* 33: 106–110.

Sadagopan R et al. (2009) Characterization and clinical evaluation of a novel IMRT quality assurance system. *J. Appl. Clin. Med. Phys.* 10(2): 104–119.

Saminathan S et al. (2010) Dosimetric study of 2D ion chamber array matrix for the modern radiotherapy treatment verification. *J. Appl. Clin. Med. Phys.* 11(2): 116–127.

Shimohigashi Y et al. (2012) Angular dependence correction of MatriXX and its application to composite dose verification. *J. Appl. Clin. Med. Phys.* 13(5): 198–214.

Somlai-Schweiger I and Ziegler SI (2015) CHERENCUBE: Concept definition and implementation challenges of a Cherenkov-based detector block for PET. *Med. Phys.* 42: 1825–1835.

Song JY et al. (2014) Dosimetric evaluation of MapCHECK 2 and 3DVH in the IMRT delivery quality assurance process. *Med. Dosim.* 39: 134–138.

Spezi E et al. (2005) Characterization of a 2D ion chamber array for the verification of radiotherapy treatments. *Phys. Med. Biol.* 50: 3361–3373.

Stambaugh C et al. (2014) Evaluation of semiempirical VMAT dose reconstruction on a patient dataset based on biplanar diode array measurements. *J. Appl. Clin. Med. Phys.* 15(2): 169–180.

Stathakis S et al. (2013) Characterization of a novel 2D array dosimeter for patient-specific quality assurance with volumetric arc therapy. *Med. Phys.* 40: 071731.

Stelljes TS et al. (2015) Dosimetric characteristics of the novel 2D ionization chamber array OCTAVIUS Detector 1500. *Med. Phys.* 42: 1528–1537.

Van Dyk J et al. (1993) Commissioning and quality assurance of treatment planning computers. *Int. J. Radiat. Oncol. Biol. Phys.* 26: 261–273.
Van Esch A et al. (2014) The Octavius1500 2D ion chamber array and its associated phantoms: Dosimetric characterization of a new prototype. *Med. Phys.* 41: 091708.
Wolfsberger LD et al. (2010) Angular dose dependence of Matrixx TM and its calibration. *J. Appl. Clin. Med. Phys.* 11(1): 241–251.
Yamamoto S et al. (2014) Development of a PET/Cerenkov-light hybrid imaging system. *Med. Phys.* 41: 092504.
Younge KC et al. (2012) Penalization of aperture complexity in inversely planned volumetric modulated arc therapy. *Med. Phys.* 39: 7160–7170.
Zeidan OA et al. (2004) Verification of step-and-shoot IMRT delivery using a fast video-based electronic portal imaging device. *Med. Phys.* 31: 463–476.
Zhang R et al. (2013a) Superficial dosimetry imaging of Cerenkov emission in electron beam radiotherapy of phantoms. *Phys. Med. Biol.* 58: 5477–5493.
Zhang R et al. (2013b) Superficial dosimetry imaging based on Cerenkov emission for external beam radiotherapy with megavoltage x-ray beams. *Med. Phys.* 40: 101914.

SECTION III

Measurement and Computation

9

Small-Field Dosimetry in Photon Beams

Indra J. Das and Paolo Francescon

9.1 Introduction

The evolution of advanced technologies for the management of cancer patients with radiation, such as the implementation of new treatment techniques, for example, intensity-modulated radiation therapy (IMRT), volumetric-modulated arc therapy (VMAT), stereotactic radiosurgery (SRS), stereotactic radiotherapy (SRT), and stereotactic body radiotherapy (SBRT), as well as of new treatment machines such as Gamma Knife, CyberKnife, and TomoTherapy units, has changed the paradigm on the limit of radiation fields. Most of these modalities use subcentimeter field dimensions for patient treatment. Traditionally, national and international codes of practice for dosimetry (Almond et al., 1999; IAEA, 2000) provided guidelines based on dosimetry in a reference field, usually with dimensions of 10×10 cm^2. In large fields, dosimetric parameters are well defined and can be accurately measured. However, with shrinking field-size lateral electron equilibrium cannot be established and traditional reference dosimetry cannot be utilized. Manufacturers have tried to provide many types of detectors ranging from micro, mini, and standard detectors (Das et al., 2008a) whose characteristics in small fields are not well known. This has led to many radiation accidents with significant over-dosage and harm to patients (e.g., Bogdanich and Ruiz, 2010). The complexity of small-field dosimetry has also been compounded by the radiation-source size that plays an important role in dosimetry.

Small-field dosimetry is performed ad-hoc based on comparison with many detectors using a daisy chain or other intercomparison approaches without understanding radiation transport and knowing the characteristics of the detector in nonequilibrium conditions (Dieterich and Sherouse, 2011). In SRS/SRT, the impact of using small-field sizes is eluded in a review article indicating significant variation in dose (Taylor et al., 2011). Das et al. (2000) provided a comparison of SRS data obtained with many detectors used in various institutions. Figure 9.1 provides a look on the variability of data among detectors and institutions indicating ±14% and ±12% change for small fields, respectively. Fan et al. (2009a) showed that even for a single Varian Trilogy SRS machine, the output variation

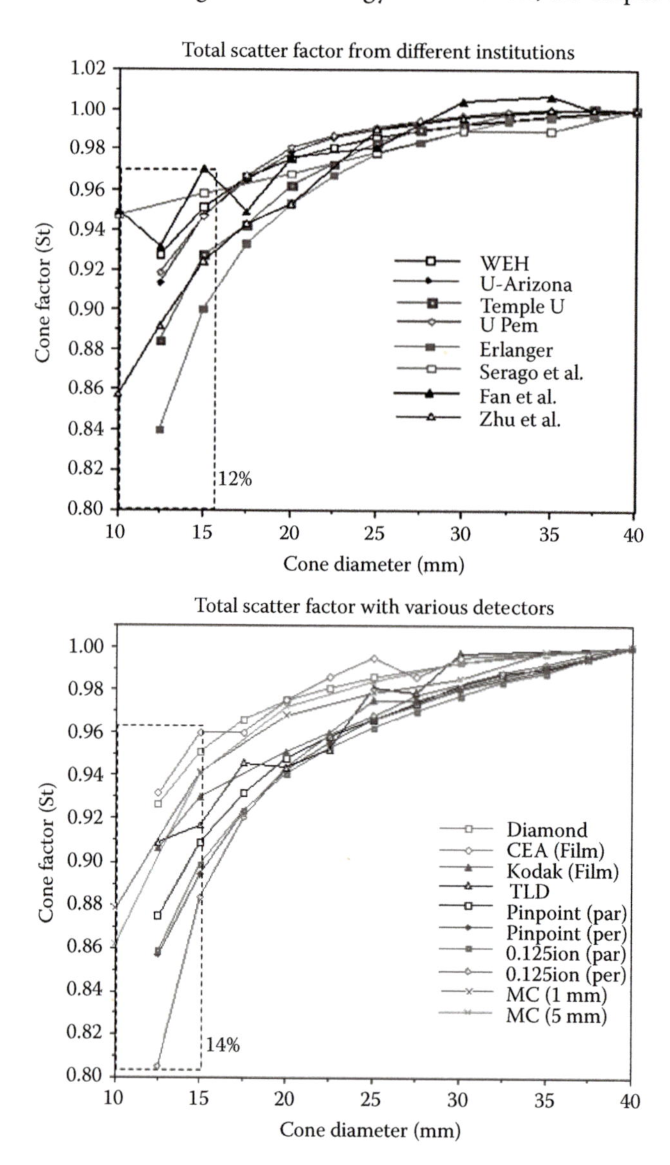

Figure 9.1

Variation in output factor from stereotactic radiosurgery cones for a 6 MV beam. various institutions and detectors (Adapted from Das IJ et al., *J. Radiosurg.*, 3, 177–185, 2000.)

in small cones was 10% among ion chambers compared to Monte Carlo (MC) calculation. Similarly, Capote et al. (2004) showed 9% difference in dose using ion chambers and material-density perturbation. Variation is even more pronounced (42%) for output factors (OFs) when using commercial diodes and ion chambers as shown by Dieterich and Sherouse (2011). Of course, such variability should be reduced to approximately the same ±2% dosimetric accuracy limit for OFs of larger fields (Almond et al., 1999). Understanding such variability was not realized till recently, which is the main subject of this chapter.

The International Atomic Energy Agency (IAEA) and American Association of Physicists in Medicine (AAPM) realized the importance of small-field dosimetry and formed a task group to tackle this issue (Alfonso et al., 2008; Das et al., 2017). This chapter provides detailed information to better understand the complexity of using small fields in radiotherapy and its implications for dosimetry purposes. Some recommendations based on our current knowledge of the science of small-field dosimetry are also provided.

9.2 Small-Field Definition

The definition of a small field is rather subjective and is dependent on the photon-beam energy. However, scientifically, three physical conditions should be fulfilled for a photon beam that can be designated as small: (1) loss of lateral charged-particle equilibrium (LCPE), (2) partial occlusion of the primary photon source by the collimating devices, and (3) the size of the detector being large compared to the beam dimensions. Insttitute of Physics and Engineering in Medicine (IPEM) Report No. 103 (Aspradakis et al., 2010) provides an overview of dosimetry issues related to small-field dosimetry. The loss of LCPE occurs in photon beams if the beam half-width is smaller than the maximum range of secondary electrons r_{LEE}. This condition has been quantified by Li et al. (1995) who evaluated the minimal radius of a circular photon field for which collision kerma in water and absorbed dose to water are equal:

$$r_{LEE}(g/cm^2) = 5.973\ (TPR_{10}^{20}) - 2.688 \tag{9.1}$$

where TPR_{10}^{20} is the beam quality index, which can be either measured or computed from a formula by Kalach and Rogers (2003) based on the depth dose beam quality specifier $\%dd(10)_x$ (Almond et al., 1999).

The initial Bragg–Gray cavity theory and its modifications with cutoff energy (Spencer and Attix, 1955), and density effect (Fano, 1954), provide the concept of the flow of secondary electrons and cavity size. However, this size limitation often becomes unacceptably restrictive. For small fields, the lack of lateral electron equilibrium is mainly due to the lack of electron fluence, which significantly decreases with increasing distance from the central axis of the beam, and in particular decreases within the volume occupied by the detector (Figure 9.2). In these conditions, we can neither apply the Fano theorem nor assume that the cavity is sufficiently small. Therefore, we expect that the electron fluence perturbation will be greater if the size of the cavity is larger with respect to the size of the field in the plane perpendicular to the beam axis.

The breakdown of LCPE reflects an imbalance between electron fluence entering and exiting a region of interest. A detailed description of cavity theory in small fields has been discussed by many investigators (Bouchard, 2012;

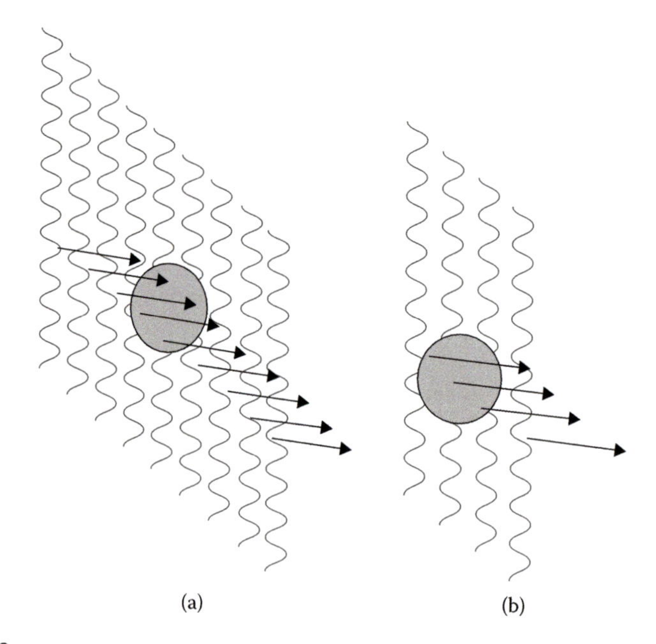

Figure 9.2

Schematic view of photons (curly lines), Compton electrons (arrows) and a cavity, indicating (a) large-field and (b) small-field conditions. Note that for the large-field situation, the same number of electrons is entering and leaving the cavity, thus a condition for electron equilibrium is fulfilled, whereas for the small-field situation there is no equilibrium.

Fenwick et al., 2013) as well as in Chapter 3. Figure 9.2 shows a schematic and very simplistic view of photon interactions producing secondary electrons in a homogeneous medium. Large and small fields are shown along with a small volume cavity that is traversed by secondary electrons. Note that in the large-field situation there is electron equilibrium whereas with the small field it is perturbed. Additionally, the introduction of a cavity with a different density to the surrounding medium substantially perturbs the electron fluence.

The second condition is related to the finite size of the primary photon-beam source. When the field size is reduced, it obstructs the source size thus limiting the photon fluence. This primary source occlusion effect becomes important when the field diameter is comparable to or smaller than the size of the primary photon source. For modern linear accelerators where the primary photon source size is not larger than 1 mm, direct source occlusion usually occurs at field sizes smaller than those where lateral electron disequilibrium starts to be seen. As the field size is reduced the penumbras from opposing jaws overlap, and there is a drop in dose at the center of the field. As a result, the full width at half maximum (FWHM) of the dose profile is no longer equal to the collimator setting. The actual field size becomes broader than the field size defined by the projected collimator settings, an effect called as apparent widening of the field. For a given source to detector distance, this effect is dependent on the source–collimator distance as shown in Figure 9.3.

The loss of LCPE and the primary photon source occlusion effect are both responsible for a sharp drop in beam output with decreasing field size (Nyholm et al., 2006). Additionally, Charles et al. (2014) have introduced the concept of a "very

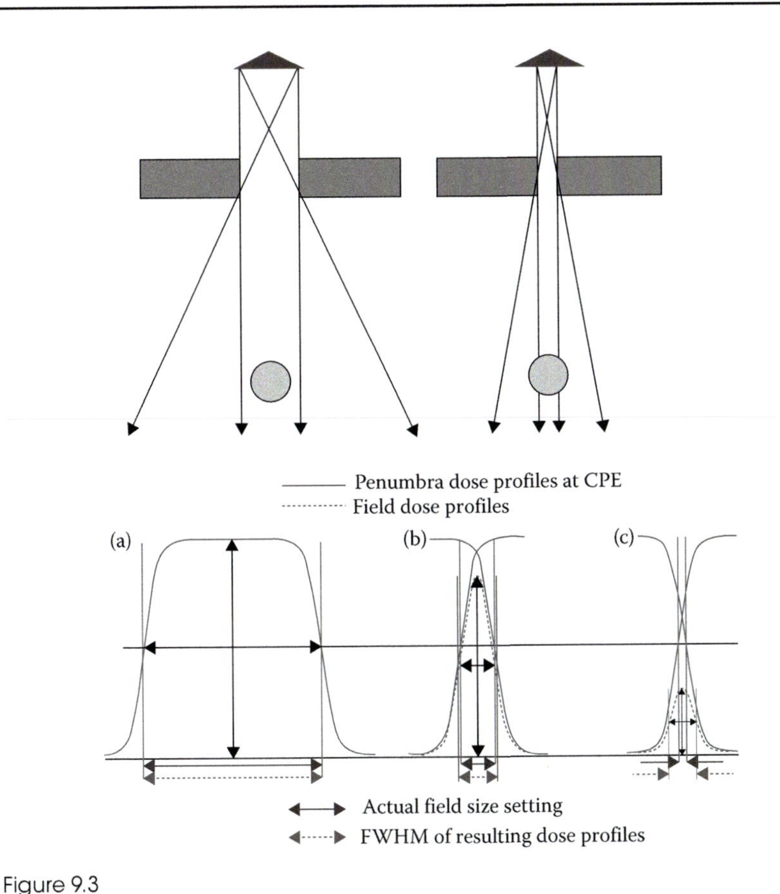

Figure 9.3

Pictorial view of large and small fields blocking the source (upper panel), and the associated impact on the dose distribution (lower panel). (a) Large field, (b) intermediate field, and (c) small field. (Adapted from Das IJ et al., *Med. Phys.*, 35, 206–215, 2008b).

small" field size, below which the OF falls by 1% or more per millimeter decrease in field width. For a 6-MV beam in water, such fields are narrower than 15 mm.

The third feature that influences the characterization of a small field is the size of the detector relative to the size of the radiation field. In small fields, detector readings are affected both by volume averaging and by the densities of the detector-sensitive volume and surrounding components (Fan et al., 2009a,b; Pantelis et al., 2010; Tyler et al., 2013; Underwood et al., 2013; Morales et al., 2014; Papaconstadopoulos et al., 2014). Its atomic number also affects detector readings to a lesser extent, via differences between photon spectra in broad and narrow fields. In the presence of large dose gradients and in the absence of LCPE conditions, fluence perturbations become large and difficult to model. Corrections for volume averaging will also have a larger uncertainty. For these reasons, small-field conditions can be assumed to exist when the external edge of the detector volume is at a distance from the field edge smaller than the LCPE range.

The combination of the above three factors dictates the definition of small fields. In this context, Kamio and Bouchard (2014) provided an elegant method describing the limit of small fields. This method tried to provide conditions where detectors can be treated as being in a correction-less condition in a small field, which a clinical physicist would like to know.

9.3 IAEA/AAPM Framework

The IAEA undertook a project to define and provide a unified approach to dosimetry in small fields. Alfonso et al. (2008) provided a conceptual view to derive the dose as below:

$$D_{w,Q_{msr}}^{f_{msr}} = M_{Q_{msr}}^{f_{msr}} N_{D,w,Q_o} k_{Q,Qo} k_{Q_{msr},Q}^{f_{msr},f_{ref}} \tag{9.2}$$

where the $D_{w,Q_{msr}}^{f_{msr}}$ is the absorbed dose at a reference depth in water in the absence of a detector at its point of measurement in a field size, machine-specific reference (MSR) specified by f_{msr} and beam quality Q_{msr}. f_{msr} is a variable depending upon the machine, i.e., different for conventional linac, Gamma Knife, CyberKnife, or TomoTherapy unit. The reading measured by the detector (corrected for variations in environmental conditions, polarity, leakage, stem correction, and ion recombination corrections) is denoted by M. The f_{ref} represents the conventional reference field in standard dosimetry protocols (typically 10×10 cm^2), for which the calibration coefficient of an ionization chamber in terms of absorbed dose to water is provided by a standard dosimetry laboratory, Q is the beam quality in the f_{ref}, and Q_o is the energy of the beam under reference conditions, traditionally a ^{60}Co beam. $N_{D,w}$ is the chamber's specific calibration coefficient in terms of absorbed dose to water for ^{60}Co. The last factor in Equation 9.2, $k_{Q_{msr},Q}^{f_{msr},f_{ref}}$, is a correction factor accounting for the detector's difference in dose response between the conditions of field size, geometry, phantom material, and beam quality of the conventional reference field and the actual f_{msr} field. The $k_{Q_{msr},Q}^{f_{msr},f_{ref}}$ values for selected detectors are works in progress for over 7–8 years and are reported by many investigators (Francescon et al., 2009, 2011, 2012, 2014b; Rosser and Bedford, 2009; Chung et al., 2010; Pantelis et al., 2010; Sterpin et al., 2010). Even though the correction factors are published in various journals, authenticity of these values has not always been established. In collaboration with the IAEA, Azangwe et al. (2014) published data for a large set of detectors normalized to f_{ref} of 3×3 cm^2. Additionally, a comprehensive set of data has recently been published in a report by the IAEA-AAPM on small-field dosimetry (IAEA, 2017).

For relative dosimetry of parameters such as percent depth dose (PDD), tissue phantom ratio/tissue maximum ratio (TPR/TMR), off-axis ratio (OAR), and total OF, TG-155 (Das et al., 2017) provides additional guidelines. Following the IAEA notation (Alfonso et al., 2008), the absorbed dose to water in composite fields used in treatments with a Gamma Knife, or with IMRT and VMAT with a conventional linac, at a point in a phantom is given by

$$D_{w,Q\,clin}^{f_{clin}} = D_{w,Q_{msr}}^{f_{msr}} \Omega_{Q_{clin}Q_{msr}}^{f_{clin},f_{msr}} \tag{9.3}$$

where $\Omega_{Q_{clin}Q_{msr}}^{f_{clin}f_{msr}}$ is a field factor that converts the absorbed dose to water per monitor unit (MU) for the machine-specific reference field to the absorbed dose to water in the clinical field. It can be determined as the ratio of a detector reading multiplied by a detector correction factor, $k_{Q_{clin},Q_{msr}}^{f_{clin},f_{msr}}$ as shown in Equation 9.2.

$$\frac{D_{w,Q_{clin}}^{f_{clin}}}{D_{w,Q_{msr}}^{f_{msr}}} = \left[\frac{M_{Q_{clin}}^{f_{clin}}}{M_{Q_{msr}}^{f_{msr}}}\right] k_{Q_{clin},Q_{msr}}^{f_{clin},f_{msr}} \tag{9.4}$$

$$\Omega_{Q_{\text{clin}}Q_{\text{msr}}}^{f_{\text{clin}}f_{\text{msr}}} = \frac{M_{Q_{\text{clin}}}^{f_{\text{clin}}}}{M_{Q_{\text{msr}}}^{f_{\text{msr}}}}\left[\frac{\left(D_{w,Q_{\text{clin}}}^{f_{\text{clin}}}\right)/\left(M_{Q_{\text{clin}}}^{f_{\text{clin}}}\right)}{\left(D_{w,Q_{\text{msr}}}^{f_{\text{msr}}}\right)/\left(M_{Q_{\text{msr}}}^{f_{\text{msr}}}\right)}\right] = \frac{M_{Q_{\text{clin}}}^{f_{\text{clin}}}}{M_{Q_{\text{msr}}}^{f_{\text{msr}}}} k_{Q_{\text{clin}},Q_{\text{msr}}}^{f_{\text{clin}},f_{\text{msr}}} \tag{9.5}$$

Thus, the standard practice of equating the ratio of readings to the ratio of doses (PDD, TMR, OAR, OF) is inaccurate in small fields since in Equations 9.4 and 9.5 the values of $k_{Q_{\text{clin}},Q_{\text{msr}}}^{f_{\text{clin}},f_{\text{msr}}}$ and $\Omega_{Q_{\text{clin}}Q_{\text{msr}}}^{f_{\text{clin}}f_{\text{msr}}}$ can be different from unity depending on the detector and machine.

Several investigators (Caprile and Hartmann, 2009a; Rosser and Bedford, 2009; Chung et al., 2010; Pantelis et al., 2010; Sterpin et al., 2010; Cranmer-Sargison et al., 2011a,b; Francescon et al., 2011; Charles et al., 2013; Czarnecki and Zink, 2013; Gago-Arias et al., 2013; Lechner et al., 2013; Tyler et al., 2013; Benmakhlouf et al., 2014) have recently calculated $k_{Q_{\text{clin}},Q_{\text{msr}}}^{f_{\text{clin}},f_{\text{msr}}}$ values for several detectors in small fields, which was reviewed by Azangwe et al. (2014).

It is now common understanding that small volume, air-filled ionization chambers provide a large perturbation and $k_{Q_{\text{clin}},Q_{\text{msr}}}^{f_{\text{clin}},f_{\text{msr}}}$ can be expressed as the ratio of the product of field-size-dependent stopping-power ratios, $[L/\rho]_{\text{air}}^{w}$, and the overall chamber perturbation correction factor, p, for the f_{clin} and f_{msr} field size, respectively, as shown in Equation 9.6 (Bouchard et al., 2009).

$$k_{Q_{\text{clin}},Q_{\text{msr}}}^{f_{\text{clin}},f_{\text{msr}}} = \frac{\left[\left(\frac{\bar{L}}{\rho}\right)_{\text{air}}^{w} \cdot P_{\text{fl}} \cdot P_{\text{grad}} \cdot P_{\text{stem}} \cdot P_{\text{cell}} \cdot P_{\text{wall}}\right]_{f_{\text{clin}}}}{\left[\left(\frac{\bar{L}}{\rho}\right)_{\text{air}}^{w} \cdot P_{\text{fl}} \cdot P_{\text{grad}} \cdot P_{\text{stem}} \cdot P_{\text{cell}} \cdot P_{\text{wall}}\right]_{f_{\text{msr}}}} \tag{9.6}$$

It is noted that $P_{\text{grad}} \equiv P_d \cdot P_{\text{vol}}$ and accounts for the perturbation due to variation in density and volume (Bouchard et al., 2009). The mean values of $k_{Q_{\text{clin}},Q_{\text{msr}}}^{f_{\text{clin}},f_{\text{msr}}}$ for Siemens and Elekta machines averaged over FWHM and energy for various microdetectors were published by Francescon et al. (2011, 2012) and for Varian machines by Benmakhlouf et al. (2014).

9.4 Dosimetric Parameters

Dosimetric parameters can be divided into three sets: PDD, TPR/TMR, and OAR measurements. These parameters are ratios of doses and hence require exploration in small fields as was performed in the past for conventional field sizes (Das et al., 2008a). For small fields, Francescon et al. (2014a) provided a detailed analysis, which is discussed in the following sections.

9.4.1 Percent Depth Dose

From the perspective of measurement of PDDs in small beams, Francescon et al. (2014a) provided MC as well as experimental data using various detectors. It is rather difficult to find an ideal detector that does not have a perturbation in a small field; however, a set or class of detectors can be found that is energy independent, linear in dose and dose rate, and has minimum angular dependence. In small fields, the photon spectrum becomes harder with depth. This is different from larger fields where the beam-hardening effect is offset by an increasing amount of scattered radiation, which, depending on the field size, may lead to the effective softening of photon spectrum with depth.

The types of errors that one may encounter during measurements of depth dose curves in small fields are discussed below.

9.4.1.1 Effective Point of Measurement

For a cylindrical ion chamber, the effective point of measurement is dictated by the secondary electron gradients that seem to move it closer to the surface (Attix, 1986). The choice of the effective point of measurement is not trivial as shown by Tessier and Kawrakov (2010). A simple method to determine the effective point of measurement is by comparison of MC simulation of PDD curves obtained with the actual detector with the MC calculated depth dose distribution in water. This leads to shifts, expressed as a fraction of the chamber radius, that depend on the details of the chamber materials and construction. It is important to underline that the effective point of measurement changes with field size, because the fluence perturbation of the beam within the sensitive volume of the detector changes with field size; the smaller the field, the larger the perturbation of the detector response. Therefore, the position of the effective point of measurement is fundamentally dependent on the field size as discussed in many publications (e.g., Bouchard et al., 2009, 2011).

9.4.1.2 Positioning Error

The mounting of the detector, i.e., the alignment of the detector axis with the central axis of the beam, must be kept proper when the dosimeter is measuring at different depths during a vertical scan. The detector should be mounted with its axis parallel to the beam axis (vertical mounting) in order to keep the same amount of volume of the detector with depth. The perturbation of the field with depth is also constant in this way. A misalignment of the detector axis with the beam central axis may result in a change in the PDD by a few percent (Cheng et al., 2007). A variation in misalignment of the detector along the scan depth can result in an even higher percentage difference in PDD (Li and Zhu, 2006; Cheng et al., 2007). Methods to minimize positioning errors should be applied, as shown in the literature (Dieterich and Sherouse, 2011).

9.4.1.3 Effect of Collimator Jaw Setting on PDD

The effect of collimator jaw setting on the conversion of percent depth ionization into PDD was analyzed by Cheng et al. (2007). The effect of ±2 mm collimator jaw setting uncertainty was found to have a negligible influence on the stopping-power ratio of water-to-air, consistent with an only modest dependence of that quantity as a function of field size. The Cheng et al. (2007) study also reported a variation in the extrapolated zero-field PDD of a maximum of 2% for a field size change of +2 mm.

The PDD has been traditionally taken to be a ratio of ionization readings, which is generally correct when $k_{Q_{clin},Q_{msr}}^{f_{clin},f_{msr}}$ is 1.0 for large (>3 × 3 cm^2) fields. However, for small fields $k_{Q_{clin},Q_{msr}}^{f_{clin},f_{msr}} \neq 1.0$, and must be accounted for in the measurements. Therefore, one should carefully determine whether $k_{Q_{clin},Q_{msr}}^{f_{clin},f_{msr}}$ for a detector remains constant at all measuring depths, since field size increases with depth during PDD measurements. The correction factor may thus decrease since it is field-size dependent. The unshielded stereotactic diodes, except the Sun Nuclear EDGE diode, which has a layer of copper below the sensitive volume, reproduced the PDD and TMR in water to within 2% at all depths beyond the buildup region. Near the surface, i.e., at a depth of 0.2 cm, the diodes exhibit an electron fluence perturbation in the sensitive volume due to the materials around the sensitive volume and atomic composition of the silicon of the sensitive volume. Microchambers show a PDD response that increases with increasing depths. This effect is greater if the stem axis is perpendicular to the beam axis due to the area occupied by the chamber in the plane perpendicular to the beam axis which is greater. This behavior mainly depends on the perturbation due to the presence of a material of very low density (air) in the cavity, and on the dimensions of the chamber compared to the varying field dimensions with depth.

9.4.2 TPR Measurements

Some treatment-planning systems (TPSs) require the measurement of a TPR (or TMR) instead of PDD. TPR data are also often used for independent MU checks. The motivation for measuring TPR instead of PDD measurements is that a TPR measurement is potentially more accurate than a PDD measurement because the detector positioning of the beam axis is performed only once and the requirement for an absolutely perfect alignment between the beam axis and the scanning path is not needed. In addition, because the field size does not change, corrections due to volume-averaging effect may cancel out. Despite this, TPR measurements are seldom performed as regular scanning tanks do not have a TPR data acquisition mode, or the accurate determination of the water level is challenging (McEwen et al., 2008), which may adversely affect the accuracy of the measurements in the buildup region. It is worth noting that a TPR measurement does not avoid issues related to the change in beam spectrum as the amount of attenuating material in front of the detector changes. In clinical applications, users thus end up measuring PDD and convert them to TPRs, as discussed in detail in Chapter 15. This process introduces uncertainties as well (Li et al., 2004; Cheng et al., 2007; Thomas et al., 2014).

In TMR or TPR measurements, by definition, the field size remains constant for every depth and thus the correction factor remains unchanged related to the change of field dimensions. Therefore, even microchambers can be used without applying a correction factor for measurements of TPR for all the field dimensions as long as the detector is comparatively small.

9.5 Beam Profiles and Penumbra

The commissioning for small fields typically involves the acquisition of profiles in both directions (gun-target and left-right) at a variety of depths, for a variety of small fields down to 0.5×0.5 cm^2. Use of a small volume detector is extremely important for profile measurements to avoid significant penumbra blurring as the active volume of the detector moves through the steep lateral penumbra of the profile. Some of the common criteria for profiles were discussed in the AAPM TG-106 report (Das et al., 2008a) that should be used in small fields too. Deconvolution-based spatial response functions (Sibata et al., 1991; Higgins et al., 1995; Charland et al., 1998; Bednarz et al., 2002; Herrup et al., 2005) for correcting profiles were suggested; however, direct measurements with high-resolution detectors (e.g., diodes, liquid ionization chambers, and diamond detectors) are often preferred. In such situations, users should verify whether or not the reading depends on dose rate changes provoked by changes with the distance from the central axis. If detectors with a directional asymmetry (e.g., ionization chambers) are used, they can be mounted vertically in order to minimize the magnitude of penumbra blurring in both lateral and in-plane profiles. In general, the detector should be used in the orientation that optimizes its spatial resolution. However, the user should first verify the absence of any significant stem effect or polarity effect, which may occur as a result of asymmetric scanning.

Alternatively, radiochromic films can be used to measure dose profiles, provided an accurate film data-processing protocol is developed and validated by comparison with conventional techniques in large fields (Devic, 2011). Other precautions associated with film, particularly the potential penumbra blurring associated with the film scanner, may also be needed to be taken into account. There are, however, many contradicting publications in favor and against radiochromic films for small-field characterization (Tyler et al., 2013; Garcia-Garduno et al., 2014; Gonzalez-Lopez et al., 2015; Larraga-Gutierrez et al., 2015; Underwood et al., 2015).

Neither stereotactic diodes nor microchambers correctly reproduce the dose profiles in water. Only the Exradin W1 plastic scintillator detector (PSD) can be considered as water equivalent and is suitable for these measurements (see Section 9.6.2.5). Unfortunately, this detector requires point-by-point measurements and cannot be used as a scanning detector. Diodes reproduce the OAR with an acceptable accuracy in water up to the penumbra region, while in the tail regions the diodes significantly underestimate the OAR in water. Microchambers overestimate the OAR in water in the penumbra region, but in the tail region the overestimation is almost constant. As the collimator diameter increases, the overestimation decreases.

Exradin W1 PSD is the only detector that can reproduce the PDD and OAR in water with remarkable accuracy compared to the MC simulation data. However, it is not possible to use this dosimeter for scanning data measurements of OAR and PDD. Therefore, its use in clinical practice remains difficult until the manufacturer makes such a detector available for scanning. The manufacturer is aware of this limitation and improvements of its design to make it suitable for scanning are underway. In the future, near water–equivalent dosimeters, such as those fabricated with synthetic microdiamonds, could be used with minimum correction. Another best choice is to utilize a stereotactic diode that achieves PDDs that mimic those in water with a systematic error of less than 2%. The stereotactic diode correctly reproduced the OAR in water up to the penumbra zone but significantly underestimated the value of OAR in the tail region. One could argue that this systematic error has little clinical importance, as it is associated with a region of very low dose. However, the fact that CyberKnife and IMRT treatments use many fields, and that the low doses occupy relatively large volumes should prompt one to reflect carefully on the possible underestimation of the long-term effects. Moreover, microchambers should not be used for OAR measurements of small fields.

Figures 9.4 and 9.5 provide measured PDD and OAR data for a CyberKnife unit corrected for $k_{Q_{clin},Q_{msr}}^{f_{clin},f_{msr}}$ factors. A difference plot is also shown in each panel. Similar data using MC simulation are provided for various source sizes and cones indicating that source size plays a role only for very small fields (Sham et al., 2008).

9.6 Output Factors

As shown earlier, OFs can be determined as the ratio of detector readings multiplied by a detector correction factor, $k_{Q_{clin},Q_{msr}}^{f_{clin},f_{msr}}$. Figure 9.6 shows $k_{Q_{clin},Q_{msr}}^{f_{clin},f_{msr}}$ values of

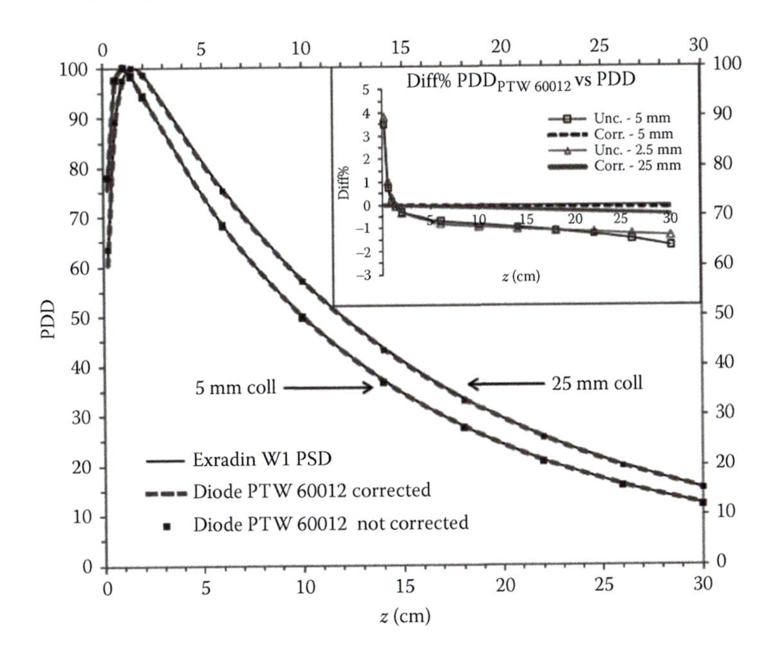

Figure 9.4

PDD for two CyberKnife beams measured with two detectors: PTW diode compared with a plastic scintillator detector. (Adapted from Francescon P et al., *Med. Phys.*, 41, 101708, 2014a).

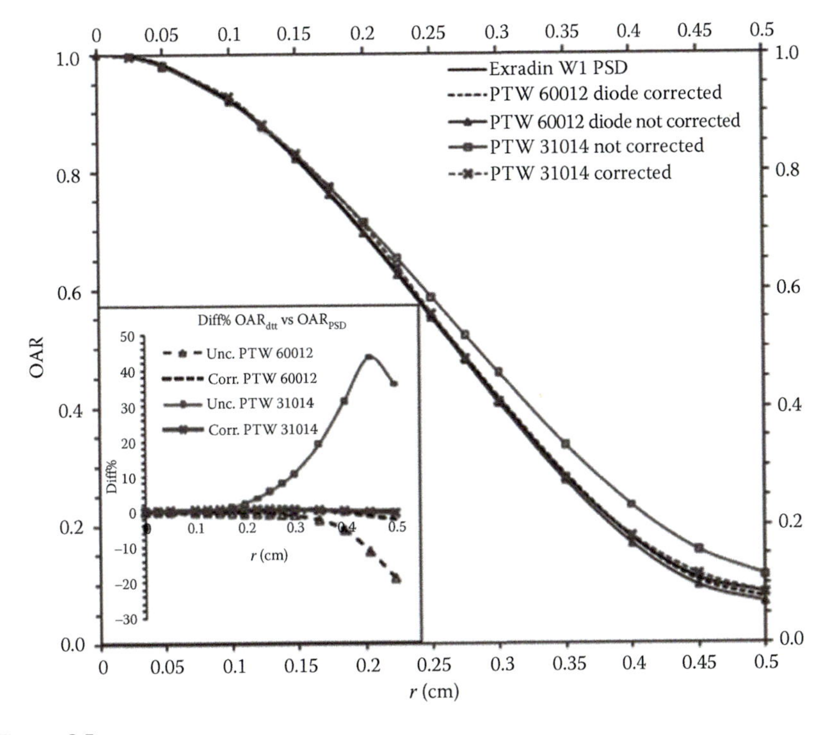

Figure 9.5

OAR for a 5 mm CyberKnife beam with various detectors. (Adapted from Francescon P et al., *Med. Phys.*, 41, 101708, 2014a).

various detectors for Siemens and Elekta machines. Several investigators have also provided data for various detectors to be used with a CyberKnife unit with similar accuracy (Francescon et al., 2012, 2014b; Gago-Arias et al., 2013; Chalkley and Heyes, 2014). For the same nominal energy, the $k_{Q_{clin},Q_{msr}}^{f_{clin},f_{msr}}$ factor mostly depends on the field size and the type of detector. It is less dependent on the linac model, the radial FWHM and energy of the beam, and the distance between the exit window and the target. Therefore, a mean value of $k_{Q_{clin},Q_{msr}}^{f_{clin},f_{msr}}$ could be used with an acceptable uncertainty (Francescon et al., 2011, 2012; Benmakhlouf et al., 2014). It is important to mention that the values of $k_{Q_{clin},Q_{msr}}^{f_{clin},f_{msr}}$ reported in Table 9.1(a) and (b) refer to field dimensions obtained by MC-simulated profiles in water. Thus, to properly use these correction factors, it is necessary to measure the field dimensions using a detector which does not introduce a significant distortion to the shape of the profile compared to the "true" profile in water. It must be emphasized that the numerical values of $k_{Q_{clin},Q_{msr}}^{f_{clin},f_{msr}}$ reported in Table 9.2 cannot be applied to other types of detectors and machines of differing nominal energies. For nonstandard machines such as the Gamma Knife, CyberKnife, and TomoTherapy unit, users must refer to specific data collection using other types of detectors (Kawachi et al., 2008; Sterpin et al., 2008, 2010, 2012; Francescon et al., 2012; Thomas et al., 2014).

The values for OFs have been evolving since the original data provided by Francescon et al. (2008). The actual magnitude of detector correction factors is detector-, machine-, and focal spot-dependent. Figure 9.6 shows a trend of such data. Note that the value of $k_{Q_{clin},Q_{msr}}^{f_{clin},f_{msr}}$ approaches unity after a 2 cm diameter field for a 6-MV beam, but is critical if fields smaller than 2 cm are used.

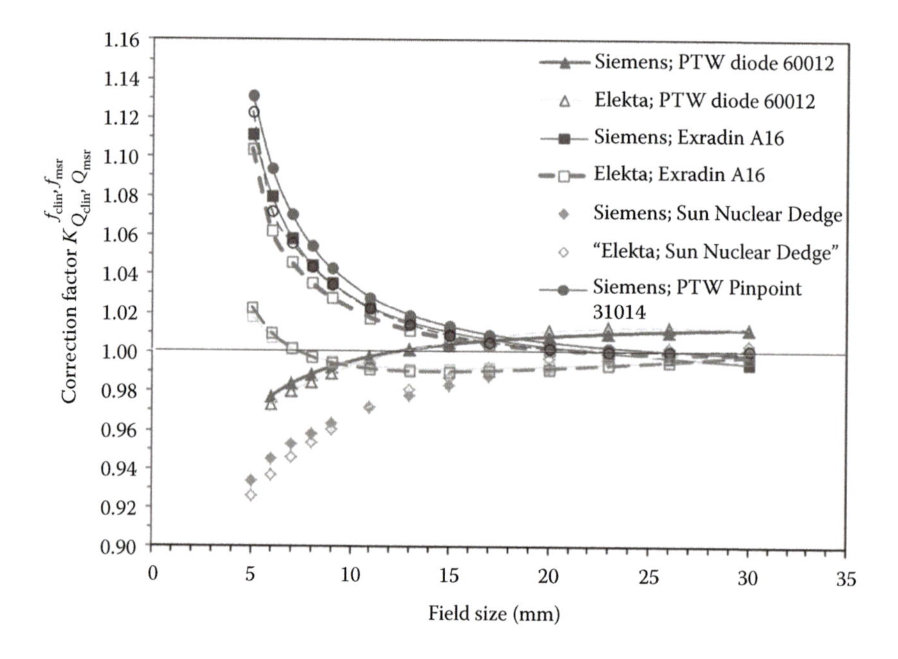

Figure 9.6

k Factors for 6 MV beams for different detectors for Siemens and Elekta machines. Note that the *k* values can be either smaller or larger than 1.0. All machine and detector values approach unity for field sizes larger than 20 mm.

Table 9.1 Average Detector Correction Factors, $k_{Q_{clin},Q_{msr}}^{f_{clin},f_{msr}}$ and Estimated Combined Standard Uncertainties (within Parenthesis) for 6 MV Radiation Therapy Photon Beams, Obtained from Published Data

(a)

Vendor	Model	Type	4 × 4 cm² (%)	2 × 2 cm² (%)	1 × 1 cm² (%)	0.5 × 0.5 cm² (%)
PTW	T60016	Diode (photon/shielded)	0.997 (0.11)	0.994 (0.26)	0.957 (0.29)	0.913 (0.56)
PTW	T60017	Diode (electron/ unshielded)	1.013 (0.21)	1.014 (0.24)	0.991 (0.48)	0.951 (0.30)
PTW	T31016	Ion chamber PinPoint 3D parallel	1.004 (0.15)	1.003 (0.01)	1.002 (1.08)	1.103 (0.45)
PTW	T31018	Micro-liquid ion chamber (LIC)	1.003 (0.12)	1.003 (0.24)	0.992 (0.03)	1.011 (0.33)
IBA	PFD	Diode (photon/shielded)	0.992 (0.16)	0.983 (0.22)	0.952 (0.33)	0.947 (0.53)
IBA	EFD	Diode (electron/ unshielded)	1.015 (0.17)	1.020 (0.35)	1.003 (0.02)	0.991 (0.20)
IBA	SFD	Diode (stereotactic/ unshielded)	1.021 (0.24)	1.025 (0.29)	1.016 (0.30)	0.979 (0.95)
IBA	CC01	Ion chamber (stereotactic/IMRT)	1.003 (0.14)	1.004 (0.22)	1.006 (0.48)	1.071 (0.55)

(b)

Vendor	Model	Field Size (cm²) 0.5 × 0.5	0.75 × 0.75	1.0 × 1.0	1.25 × 1.25	1.5 × 1.5	3.0 × 3.0
PTW	60012 diode	0.968 ± 0.003	0.984 ± 0.002	0.995 ± 0.001	1.001 ± 0.001	1.006 ± 0.001	1.013 ± 0.001
Sun Nuclear	EDGE diode	0.932 ± 0.003	0.951 ± 0.002	0.967 ± 0.001	0.978 ± 0.003	0.986 ± 0.002	1.001 ± 0.002
PTW	PinPoint 31014	1.128 ± 0.018	1.053 ± 0.007	1.024 ± 0.002	1.010 ± 0.001	1.005 ± 0.001	1.000 ± 0.001
Standard Imaging	Exradin-A16	1.112 ± 0.018	1.044 ± 0.007	1.020 ± 0.001	1.007 ± 0.001	1.002 ± 0.001	0.999 ± 0.001

Source: Data shown in (a) are from (Benmakhlouf H et al., *Med. Phys.*, 41, 041711, 2014) and in (b) from (Francescon P et al., *Med. Phys.*, 38, 6513-6527, 2011) for various detectors and field sizes.

Note: The corrections are normalized to a 10 × 10 cm² field size. The long axis of the detector is parallel to the beam axis.

9.6.1 MC Approach

For most of the detectors, $k_{Q_{clin},Q_{msr}}^{f_{clin},f_{msr}}$ is usually quantified by means of MC simulations (Francescon et al., 2009, 2011, 2012, 2014b; Charles et al., 2013; Czarnecki and Zink, 2013; Fenwick et al., 2013; Gago-Arias et al., 2013; Benmakhlouf et al., 2014). For microion chambers, *k* values can be computed based on Equation 9.6; however, still some parameters require MC simulation since most factors are not available for detectors in use.

Table 9.2 Values of $k_{Q_{clin},Q_{msr}}^{f_{clin},f_{msr}}$ (80 cm) Calculated by Monte Carlo Simulation of Six Small Field Sizes on the CyberKnife System

Detector	Detector Type	Orientation	Field Diameter (cm)					
			0.5	0.75	10	1.25	1.5	2.5
PTW diode 60008	Diode	Parallel	0.946	0.962	0.975	0.986	0.992	1.004
PTW diode 60012	Diode	Parallel	0.964	0.975	0.984	0.991	0.996	1.002
PTW diode 60017	Diode	Parallel	0.958	0.971	0.981	0.990	0.996	1.003
Sun Nuclear Edge	Diode	Parallel	0.950	0.959	0.973	0.979	0.986	1.000
IBA SFD	Diode	Parallel	0.967	0.991	0.999	1.003	1.003	1.003
Exradin D1V	Diode	Parallel	0.975	0.983	0.988	0.993	0.998	1.002
PTW 31018	Micro-liquid ion chamber (μLion)	Parallel	1.026	1.000	0.996	0.994	0.996	0.998
Exradin W1	Scintillator	Parallel	1.003	0.999	1.002	0.999	0.999	0.999
Exradin A16	Ion chamber	Parallel	1.097	1.033	1.012	1.006	1.003	1.002
Exradin A16	Ion chamber	Perpendicular	1.173	1.051	1.026	1.012	1.007	0.999
PTW 31014	Pinpoint ion chamber	Parallel	1.102	1.037	1.014	1.007	1.004	0.998
PTW 31014	Pinpoint ion chamber	Perpendicular	1.350	1.119	1.064	1.032	1.019	1.001
IBA CC01	Ion chamber	Parallel	1.074	1.018	1.010	1.007	1.005	1.005
IBA CC01	Ion chamber	Perpendicular	1.159	1.036	1.017	1.008	1.005	1.003

Source: Francescon P et al., *Phys. Med. Biol.*, 57, 3741–3758, 2012; *Phys. Med. Biol.*, 59, N11–17, 2014b.

Note: These values are the average of values calculated for a dose rate of 400–600 and 800 MU/min with fixed and mechanically variable (IRIS) collimators. Note that the detector orientation is indicated for each detector, with perpendicular meaning the long axis of the detector perpendicular to the beam axis, and parallel that beam and detector axes are parallel as shown elsewhere.

9.6.2 Experimental Approach

Chapter 4 discusses the detector response in reference fields that can also be extended to small fields. The factor, $k_{Q_{clin},Q_{msr}}^{f_{clin},f_{msr}}$ as shown in the above equations cannot be experimentally derived in general since we do not have gold standard data for small fields. However, they could be derived using the daisy chain method as described by several investigators (e.g., Dieterich and Sherouse, 2011), and by intercomparison (Das et al., 2014) based on suitable and confident data such as recently shown for values obtained with microdiamonds (Benmakhlouf et al., 2014; Chalkley and Heyes, 2014; Morales et al., 2014; Papaconstadopoulos et al., 2014; Larraga-Gutierrez et al., 2015).

The measurement of OFs should be done using a water phantom which allows to center the detector by two orthogonal scans, one along the x-axis and the other along the y-axis. The two dose profiles obtained from these scans allow determining the shift of the detector, initially set at the origin of the coordinate system of the water phantom, with respect to the point of maximum dose that coincides with the center of the field. To achieve an accurate positioning, a very small field should be used, such as a 5 × 5 mm^2. Potential scanning system hysteresis effects must be verified. The detector must be placed with the same orientation with respect to the beam axis that is used to calculate its correction factor $k_{Q_{clin},Q_{msr}}^{f_{clin},f_{msr}}$. Typically, for very small fields it is preferred to put the dosimeter with the longest axis parallel to the beam axis. For the microchambers this setup reduces the dimensions of the active volume in the plane perpendicular to the beam axis. To avoid leakage, particular attention must be paid to reduce parts of the cable within the radiation field as much as possible. Particular attention must also be paid to the choice of the effective point of measurement of the detector. In fact, this depends on the active volume, wall material, energy, and dimensions of the beam. As it is difficult to take into account all these factors, the best solution is to put the dosimeter in the same geometric conditions used to calculate its correction factor $k_{Q_{clin},Q_{msr}}^{f_{clin},f_{msr}}$. In this way, the choice of the effective point of measurement is directly incorporated into the factor $k_{Q_{clin},Q_{msr}}^{f_{clin},f_{msr}}$.

9.6.2.1 Choice of the Detector

The magnitude of r_{LEE} to achieve LCPE greatly restricts the physical dimensions of the detectors that can be used for the experimental determination of small-field dosimetric parameters, such as OFs, beam profiles, PDD, TMR, or TPR. When used for measurements in small fields, perturbation of particle fluence caused by the physical size, density of active volume, and nonwater equivalence of the dosimeter must be accounted for. Small volume ionization chambers experience low signal-to-noise ratios and possibly high polarity effects. Solid-state dosimeters such as diodes and metal-oxide-semiconductor field-effect transistors (MOSFETs) have smaller sensitive volumes. However, these dosimeters exhibit energy dependence as a function of field size. Thus, their response with field size and the possible effects of beam hardening at the measurement depth must be considered (Yin et al., 2002; Francescon et al., 2008, 2009) Also, as shown by Scott et al. (2008, 2012) for small fields, the dose absorbed by the detector-sensitive volume depends upon its density; high-density detectors overrespond while low-density detectors underrespond.

The dosimeters most commonly applied in small fields will be discussed in the following sections. A more comprehensive description of various types of commercially available point detectors used in radiotherapy, along with their particular dosimetric characteristics and clinical applications, is given in Chapter 4.

9.6.2.2 Air-Filled Ionization Chambers

Air-filled ionization chambers are most commonly used for dosimetric measurements because of their high sensitivity, long-term stability, reproducibility, robustness, and traceability to calibration protocols. The small cavity in small ionization chambers results in decreased sensitivity in comparison to larger chambers. Despite this, small volume ionization chambers are successfully used for dosimetric measurements with proper correction factors (Rice et al., 1987; Bjärngard et al., 1990; Martens et al., 2000; Stasi et al., 2004; Tyler et al., 2013) in fields down to 1 × 1 cm^2.

If microchambers are used, all ionization chamber measurements should be performed at both polarities since polarity effects for several types of commercial chambers were found to depend upon field size and chamber type (Martens et al., 2000; Stasi et al., 2004).

Irradiation of stem and cable of small volume cylindrical chambers can contribute to the already weak signal from the small volume chamber (Spokas and Meeker, 1980; Lee et al., 2002; Das et al., 2008a). It has been reported that the stem and cable effect can result in an erroneous signal increase as high as 2.5%. It is therefore important to irradiate the stem and cable as little as possible. If possible, the same length of the cable should be kept within the field. This is because the radiation-induced current increases with the length of irradiated cable and this increases with increasing field size. This is especially true for radial scans where the irradiated cable length changes with detector position. It is possible to evaluate the cable effects by irradiating known lengths of cable and extrapolating to zero length.

9.6.2.3 Silicon Diodes

The high sensitivity of diodes enable designs with very small dimensions which make them very promising for use in small-field dosimetry. The active volume of a diode is determined by the diffusion length, which is usually in the range between 20 and 80 μm depending on the model.

Shielded diodes are energy compensated, to absorb some of the low-energy scattered photons, and contain high-density material, for example, tungsten. However, the presence of tungsten increases the fluence of secondary electrons in silicon due to the higher mass-energy absorption coefficient of tungsten, for low-energy photon beams. This causes overresponse of a diode, and therefore the use of unshielded diodes is recommended. Diodes have characteristics which need additional corrections such as those arising from dose rate dependence (Shi et al., 2003; Saini and Zhu, 2004), variation of response with accumulated dose (up to 10%), and temperature dependence (~0.3%/°C).

9.6.2.4 Synthetic Single Crystal Microdiamonds

The characteristics of diamond detectors were studied by many investigators and have been widely reported in the literature (Planskoy, 1980; Heydarian et al., 1993; Vatnitsky and Järvinen, 1993; Hoban et al., 1994; Rustgi, 1995; Laub et al., 1999; De Angelis et al., 2002; Das, 2009). A diamond detector has the advantage of being nearly tissue equivalent due to its atomic number ($Z = 6$), which is close to that of water ($Z = 7.4$). Their relatively high spatial resolution and high sensitivity make them suitable for dosimetry. However, manufacturing natural diamond detectors is generally costly due to labor involvement in producing an individual detector. Additionally, there is significant response variability which requires specimen-dependent correction factors. For these reasons, natural diamond detectors did not become popular.

With the advancement in crystal design, single crystal diamonds (known as microdiamonds) became commercially available. Due to their tissue equivalence and very small size, this detector has been shown to measure dose without any perturbation for even very small-field sizes. It is recently shown by many investigators that $k_{Q_{clin}Q_{msr}}^{f_{clin}f_{msr}}$ for synthetic diamonds can be treated as unity in small fields within the limit of experimental uncertainty (Lechner et al., 2013; Marsolat et al., 2013; Benmakhlouf et al., 2014; Chalkley and Heyes, 2014; Morales et al., 2014; Papaconstadopoulos et al., 2014; Larraga-Gutierrez et al., 2015).

9.6.2.5 Plastic Scintillator Detectors

The use of PSDs is a relatively new development in radiotherapy dosimetry. The light generated in the scintillator during its irradiation is carried away by an optical fiber to a photomultiplier tube located outside the irradiation room. Scintillator response is generally linear in the absorbed dose to water range of therapeutic interest. The various studies have indicated that perturbation correction factors in small fields are close to unity (see Table 9.2 and Figure 9.7). Plastic scintillators are almost water equivalent in terms of electron density and atomic composition. Typically, they match the water mass stopping power and mass-energy absorption coefficient to within ±2% for the range of beam energies in clinical use including the keV region. Scintillators are nearly energy independent and can be used directly for relative absorbed dose determination. Plastic scintillation dosimeters can be made very small (about 1 mm^3 or less) and yet give adequate sensitivity for clinical dosimetry applications. Due to their high spatial resolution, flat energy dependence, and small size, plastic scintillators can be adequately used for small beam dosimetry applications. The only commercially available PSD is the Exradin W1. The measurements must be corrected for Cerenkov emission. The method for calibrating this detector for small-field measurements is different from the method described by the manufacturer, which applies to large-field measurements. In this case, the measurements must be performed in water with the scintillator axis oriented parallel to the beam axis following the procedure described by Morin et al. (2013). Additionally, various group have shown that PSDs provide suitable data in small fields (Cho et al., 2005; Klein et al., 2010; Morin et al., 2013; Wang and Beddar, 2011; Gagnon et al., 2012; Carrasco et al., 2015).

9.7 Verification of TPSs

Dose calculation algorithms in TPSs have evolved from algorithms based on actual beam data, obtained from water tank measurements, to model based. Modern TPSs are all model based and do not require a comprehensive set of measured data. The model-based TPSs provide sophisticated algorithms that are superior in dose calculation. However, most of them are modeled only for conventional fields. For small fields, these models need to be properly tested. For example, the MC-derived algorithm for a CyberKnife unit provides surpassed accuracy in lung-dose calculation compared to other algorithms as shown by Sharma et al. (2007). Similar observations are also noted by other investigators with model-based TPSs dealing with tissue heterogeneities (Ahnesjö, 1989; Alaei et al., 2000; Nisbet et al., 2004; Fogliata et al., 2006, 2011, 2012; Van Esch et al., 2006; Morgan et al., 2008; Garcia-Garduno et al., 2014; Ojala et al., 2014). If a TPS is used for small fields, even including IMRT, this should be validated by in-phantom measurements with small fields. The measurement must include the appropriate correction values as discussed above for comparison. MC-based TPSs

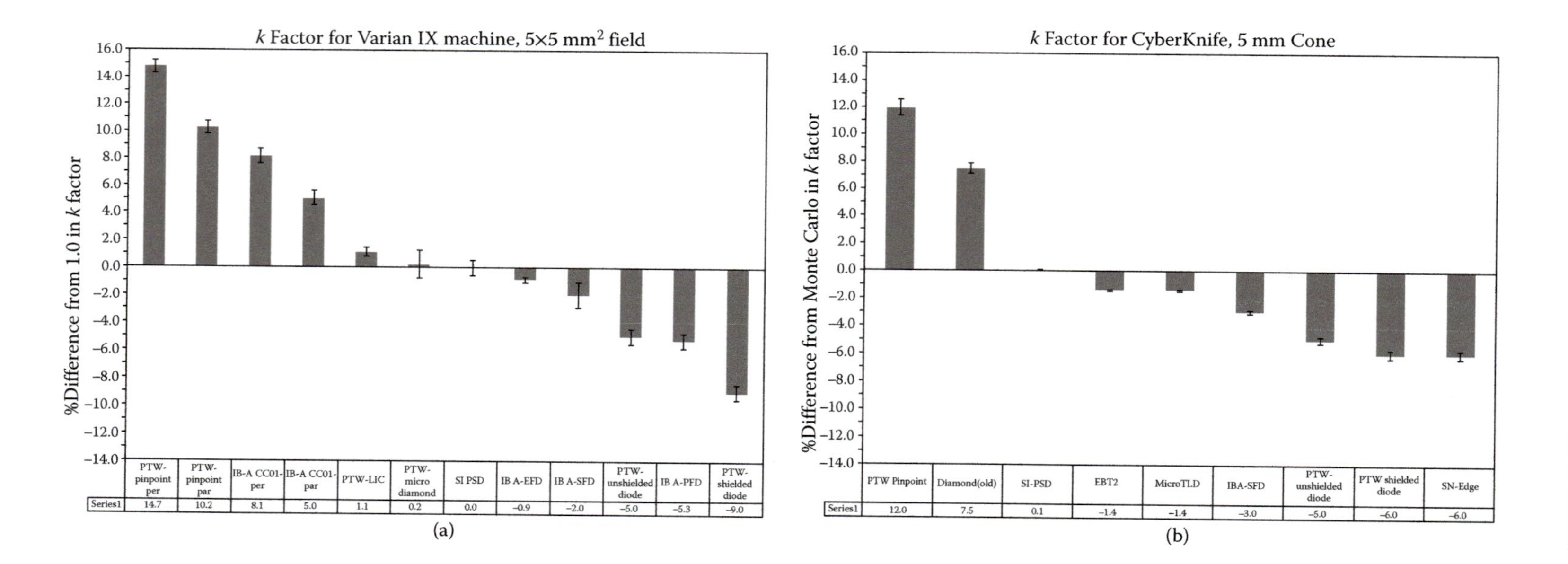

Figure 9.7

Comparison of various detectors where *per* stands for perpendicular and *par* for parallel geometry in (a) a Varian iX machine for a 5 × 5 mm^2 field and (b) a CyberKnife unit with a 5 mm cone. These data are compared against data provided by Benmakhlouf H et al., *Med. Phys.*, 41: 041711, 2014 in (a) and Moignier C et al., *Med. Phys.*, 41, 071702, 2014 in (b). This intercomparison clearly points out to detectors that should not be used in small field dosimetry. The Standard Imaging PSD (SI-PSD) data are adapted from Morin J, *Med. Phys.*, 40, 011719, 2013 and Underwood TS et al., *Phys. Med. Biol.*, 60: 6669–6683, 2015 for the CyberKnife unit and Varian machine, respectively. The PSD data are compiled from Wang LL and Beddar S, *Med. Phys.*, 38, 1596–1599, 2011, Morin J, *Med. Phys.*, 40, 011719, 2013, and Underwood TS et al., *Phys. Med., Biol.*, 60: 6669–6683, 2015.

are evolving for small-field treatment planning (Craig et al., 2008; Caprile and Hartmann, 2009b; Gete et al., 2013) as discussed in Chapter 13.

9.8 Verification of Clinically Used IMRT and VMAT Fields

IMRT and VMAT fields are by definition irregular and a combination of small fields. These fields are split into segments (fields that have the same number of MUs) for delivery purpose. Delivery of small fields can be reduced by choosing a minimum field size and the associated number of MUs. Typically 2–4 cm^2 are the size limits of such fields. A frequency distribution of the area of these fields shows that these very small fields are only seen in highly modulated fields as used for instance for head-and-neck treatments (Wu et al., 2010), and the overall contribution of such fields to the total dose is generally very small in step-and-shoot IMRT. A weighted sum of dose contributions indicates that small-field issues get diluted as long as the small-field modeling is clearly followed in IMRT. As VMAT consists of the dynamic delivery of small fields, similar considerations are valid for VMAT. There are several other publications indicating that as long as IMRT/VMAT fields are large enough, small segments do not pose a significant problem (Bouchard et al., 2009; Azimi et al., 2012). This is due to the temporal invariance of complex IMRT and VMAT fields. Therefore, composite IMRT/VMAT fields should be looked at rather than the individual beamlets. Currently, the most common practice is to verify IMRT/VMAT fields by direct measurement in composite fields using various devices (see Chapters 8 and 17) while individual or small fields are not frequently tested in clinical situations. Followill et al. (2012, 2014) presented a consistent data set for small-field OFs that can be used as a redundant quality assurance (QA) check of a treatment-TPS dosimetry data for small-field treatments. An analysis by Pulliam et al. (2014) showed that for a set of single institutional QA data of 13,000 patients only a fraction of the patients were reevaluated who failed the QA process. Thus, it is prudent to evaluate clinical QA data with respect to the TPS, and if needed small-field modeling in the TPS should be iteratively examined to pass the QA criteria.

9.9 Practical Guidelines for Accurate 3D Dosimetry in Small Fields

Figure 9.7 shows data for *k* given either as a deviation from unity, or compared to MC simulation values, for linear accelerator and CyberKnife fields. It clearly shows classes of detectors having large correction factors that should not be encouraged for use in small fields. The detectors that are best suited are liquid ion chamber (LIC), microdiamond, PSD, external beam therapy (EBT) film, and electron diodes. These detectors have favorable characteristics such as water equivalence, small volume, and minimum field perturbation.

Some further recommendations with respect to small-field dosimetry are as follows:

- Understand the limitations of small-field dosimetry in view of current publications (Das et al., 2008b; Aspradakis et al., 2010; Charles et al., 2014).
- Variability of focal spot in older machines ranges from 1 to 10 mm (Munro et al., 1988; Jaffray et al., 1993) and would make small-field dosimetry more complex due to its dependence on source size; this is not

compromised in modern machines with <1 mm focal spot (Czarnecki et al., 2012; Papaconstadopoulos et al., 2014).

- Detectors that are water equivalent such as MicroLion chambers, microdiamonds, and plastic scintillators are best suited for small-field dosimetry (Lechner et al., 2013; Marsolat et al., 2013; Benmakhlouf et al., 2014; Chalkley and Heyes, 2014; Morales et al., 2014; Papaconstadopoulos et al., 2014; Larraga-Gutierrez et al., 2015; Underwood et al., 2015). Gafchromic films can also be used, however, extreme care and corrections are needed as discussed in the literature (Wilcox and Daskalov, 2007; Tyler et al., 2013; Garcia-Garduno et al., 2014; Gonzalez-Lopez et al., 2015; Larraga-Gutierrez et al., 2015; Underwood et al., 2015), as well as in Chapter 8.
- Use proper correction factors to correct detector response for the dose from various types of machines (Francescon et al., 2008, 2009, 2011, 2012; Benmakhlouf et al., 2014). Fortunately, for modern machines $k_{Q_{clin}Q_{msr}}^{f_{clin}f_{msr}}$ for the same energy can be used interchangeably as shown by Liu et al. (2014). Conditions for correction-less small-field dosimetry as discussed by Kamio and Bouchard (2014) should be explored for the minimum field size.
- Recently published guidelines as the IAEA report TRS-483 (IAEA, 2017) and AAPM TG-155 report (Das et al., 2017) should be followed.

9.10 Summary

In this chapter, a definition of small fields is provided for fields where lateral electron equilibrium is not maintained. Consequently, the range of small-field sizes is dependent on the beam energy, but for simplicity we may consider every treatment field smaller than 3×3 cm^2 as a small field. IAEA is publishing a code of practice to provide data and methods for calibration of nonstandard fields. With the introduction of correction factors $k_{Q_{clin}Q_{msr}}^{f_{clin}f_{msr}}$ depending on detector type, field size, beam energy, and type of machine (focal spot), one could accurately measure parameters for small fields. The $k_{Q_{clin}Q_{msr}}^{f_{clin}f_{msr}}$ factor must be used in the ratio of detector readings to get the ratio of dose values in deriving dosimetric parameters such as PDD, TMR, OAR, and OF.

References

Ahnesjö A (1989) Collapsed cone convolution of radiant energy for photon dose calculation in heterogeneous media. *Med. Phys.* 16: 577–592.

Alaei P, Gerbi BJ and Geise R (2000) Evaluation of a model-based treatment planning system for dose computations in the kilovoltage energy range. *Med. Phys.* 27: 2821–2826.

Alfonso R et al. (2008) A new formalism for reference dosimetry of small and nonstandard fields. *Med. Phys.* 35: 5179–5186.

Almond PR et al. (1999) AAPM's TG-51 protocol for clinical reference dosimetry of high-energy photon and electron beams. *Med. Phys.* 26: 1847–1870.

Aspradakis MM et al. (2010) IPEM Report No 103: Small field MV dosimetry. York, UK: Institute of Physics and Engineering in Medicine.

Attix FH (1986) *Introduction to Radiological Physics and Radiation Dosimetry.* New York, NY: John Wiley & Sons.

Azangwe G et al. (2014) Detector to detector corrections: A comprehensive study of detector specific correction factors for beam output measurements for small radiotherapy beams. *Med. Phys.* 41: 072103.
Azimi R, Alaei P and Higgins P (2012) The effect of small field output factor measurements on IMRT dosimetry. *Med. Phys.* 39: 4691–4704.
Bednarz G, Huq S and Rosenow U (2002) Deconvolution of detector size effect for output factor measurement for narrow Gamma Knife radiosurgery beams. *Phys. Med. Biol.* 47: 3643–3649.
Benmakhlouf H, Sempau J and Andreo P (2014) Output correction factors for nine small field detectors in 6 MV radiation therapy photon beams: A PENELOPE Monte Carlo study. *Med. Phys.* 41: 041711.
Bjärngard BE, Tsai J-S and Rice RK (1990) Doses on the central axes of narrow 6-MV x-ray beams. *Med. Phys.* 17: 794–799.
Bogdanich W and Ruiz RR (2010) Radiation errors reported in Missouri. *The New York Times*, February 24.
Bouchard H (2012) A theoretical re-examination of Spencer-Attix cavity theory. *Phys. Med. Biol.* 57: 3333–3358.
Bouchard H, Seuntjens J and Kawrakow I (2011) A Monte Carlo method to evaluate the impact of positioning errors on detector response and quality correction factors in nonstandard beams. *Phys. Med. Biol.* 56: 2617–2634.
Bouchard H et al. (2009) Ionization chamber gradient effects in nonstandard beam configurations. *Med. Phys.* 36: 4654–4663.
Capote R et al. (2004) An EGSnrc Monte Carlo study of the microionization chamber for reference dosimetry of narrow irregular IMRT beamlets. *Med. Phys.* 31: 2416–2422.
Caprile P and Hartmann GH (2009) Development and validation of a beam model applicable to small fields. *Phys. Med. Biol.* 54: 3257–3268.
Carrasco P et al. (2015) Characterization of the Exradin W1 scintillator for use in radiotherapy. *Med. Phys.* 42: 297–304.
Chalkley A and Heyes G (2014) Evaluation of a synthetic single-crystal diamond detector for relative dosimetry measurements on a CyberKnife. *Br. J. Radiol.* 87: 20130768.
Charland P, el-Khatib E and Wolters J (1998) The use of deconvolution and total least squares in recovering a radiation detector line spread function. *Med. Phys.* 25: 152–160.
Charles PH et al. (2013) Monte Carlo-based diode design for correction-less small field dosimetry. *Phys. Med. Biol.* 58: 4501–4512.
Charles PH et al. (2014) A practical and theoretical definition of very small field size for radiotherapy output factor measurements. *Med. Phys.* 41: 041707.
Cheng CW et al. (2007) Determination of zero field size percent depth doses and tissue maximum ratios for stereotactic radiosurgery and IMRT dosimetry: Comparison between experimental measurements and Monte Carlo simulation. *Med. Phys.* 34: 3149–3157.
Cho SH et al. (2005) Reference photon dosimetry data and reference phase space data for the 6 MV photon beam from Varian Clinac 2100 series linear accelerators. *Med. Phys.* 32: 137–148.
Chung E, Bouchard H and Seuntjens J (2010) Investigation of three radiation detectors for accurate measurement of absorbed dose in nonstandard fields. *Med. Phys.* 37: 2404–2413.

Craig J et al. (2008) Commissioning a fast Monte Carlo dose calculation algorithm for lung cancer treatment planning. *J. Appl. Clin. Med. Phys.* 9(2): 83–97.

Cranmer-Sargison G et al. (2011a) Implementing a newly proposed Monte Carlo based small field dosimetry formalism for a comprehensive set of diode detectors. *Med. Phys.* 38: 6592–6602.

Cranmer-Sargison G et al. (2011b) Experimental small field 6MV output ratio analysis for various diode detector and accelerator combinations. *Radiother. Oncol.* 100: 429–435.

Czarnecki D, Wulff J and Zink K (2012) The influence of linac spot size on scatter factors. *Metrologia* 49 S215–S218.

Czarnecki D and Zink K (2013) Monte Carlo calculated correction factors for diodes and ion chambers in small photon fields. *Phys. Med. Biol.* 58: 2431–2444.

Das IJ et al. (2000) Choice of radiation detector in dosimetry of stereotactic radio-surgery-radiotherapy. J. *Radiosurg.* 3: 177–185.

Das IJ et al. (2008a) Accelerator beam data commissioning equipment and procedures: Report of the TG-106 of the therapy physics committee of the AAPM. *Med. Phys.* 35: 4186–4215.

Das IJ, Ding GX and Ahnesjö A (2008b) Small fields: Non-equilibrium radiation dosimetry. *Med. Phys.* 35: 206–215.

Das IJ (2009) Diamond detector. In *Clinical Dosimetry Measurements in Radiotherapy*, pp 891–912. (Eds. Rogers DWO and Cygler JE), Madison, WI: Medical Physics Publishing.

Das I, Akino Y and Francescon P (2014) Experimental determination of k factor in small field dosimetry. *Med. Phys.* 41: 374.

Das IJ et al. (2017) Small fields and non-equilibrium condition photon beam dosimetry: AAPM Task Group 155 Report. *Med. Phys.* (in press)

De Angelis C et al. (2002) An investigation of the operating characteristics of two PTW diamond detectors in photon and electron beams. *Med. Phys.* 29: 248–254.

Devic S (2011) Radiochromic film dosimetry: Past, present, and future. Phys. Med. 27: 122–134.

Dieterich S and Sherouse GW (2011) Experimental comparison of seven commercial dosimetry diodes for measurement of stereotactic radiosurgery cone factors. *Med. Phys.* 38: 4166–4173.

Fan J et al. (2009) Determination of output factors for stereotactic radiosurgery beams. *Med. Phys.* 36: 5292–5300.

Fano U (1954) Inelastic collisions and the Moliere theory of multiple scattering. *Phys. Rev.* 93: 117–120.

Fenwick JD et al. (2013) Using cavity theory to describe the dependence on detector density of dosimeter response in non-equilibrium small fields. *Phys. Med. Biol.* 58: 2901–2923.

Fogliata A et al. (2006) Dosimetric validation of the anisotropic analytical algorithm for photon dose calculation: Fundamental characterization in water. *Phys. Med. Biol.* 51: 1421–1438.

Fogliata A et al. (2011) Dosimetric evaluation of Acuros XB advanced dose calculation algorithm in heterogeneous media. *Radiat. Oncol.* 6: 82.

Fogliata A et al. (2012) Critical appraisal of Acuros XB and Anisotropic Analytic Algorithm dose calculation in advanced non-small-cell lung cancer treatments. *Int. J. Radiat. Oncol. Biol. Phys.* 83: 1587–1595.

Followill DS et al. (2012) The Radiological Physics Center's standard dataset for small field size output factors *J. Appl. Clin. Med. Phys.* 13(5): 282–289.

Followill DS (2014) Erratum: The Radiological Physics Center's standard dataset for small field size output factors. *J. Appl. Clin. Med. Phys.* 15(2): 356–357.

Francescon P, Cora S and Cavedon C (2008) Total scatter factors of small beams: A multidetector and Monte Carlo study. *Med. Phys.* 35: 504–513.

Francescon P et al. (2009) Application of a Monte Carlo-based method for total scatter factors of small beams to new solid state micro-detectors. *J. Appl. Clin. Med. Phys.* 10(1): 147–152.

Francescon P, Cora S and Satariano N (2011) Calculation of k(Q(clin),Q(msr)) (f(clin),f(msr)) for several small detectors and for two linear accelerators using Monte Carlo simulations. *Med. Phys.* 38: 6513–6527.

Francescon P, Kilby W, Satariano N and Cora S (2012) Monte Carlo simulated correction factors for machines specific reference field dose calibration and output factor measurement using fixed and iris collimators on the CyberKnife system. *Phys. Med. Biol.* 57: 3741–3758.

Francescon P et al. (2014a) Variation of k(fclin,fmsr, Qclin, Qmsr) for the small-field dosimetric parameters percentage depth dose, tissue-maximum ratio, and off-axis ratio. *Med. Phys.* 41: 101708.

Francescon P, Kilby W and Satariano N (2014b) Monte Carlo simulated correction factors or output factor measurement with the CyberKnife system-results for new detectors and correction factor dependence on measurement distance and detector orientation. *Phys. Med. Biol.* 59: N11–N17.

Gagnon JC et al. (2012) Dosimetric performance and array assessment of plastic scintillation detectors for stereotactic radiosurgery quality assurance. *Med. Phys.* 39: 429–436.

Gago-Arias A (2013) Correction factors for ionization chamber dosimetry in CyberKnife: Machine-specific, plan-class, and clinical fields. *Med. Phys.* 40: 011721.

Garcia-Garduno OA (2014) Effect of dosimeter type for commissioning small photon beams on calculated dose distribution in stereotactic radiosurgery. *Med. Phys.* 41: 092101.

Gete E (2013) A Monte Carlo approach to validation of FFF VMAT treatment plans for the TrueBeam linac. *Med. Phys.* 40: 021707.

Gonzalez-Lopez A, Vera-Sanchez JA and Lago-Martin JD (2015) Small fields measurements with radiochromic films. J. *Med. Phys.* 40: 61–67.

Herrup D et al. (2005) Determination of penumbral widths from ion chamber measurements. *Med. Phys.* 32: 3636–3640.

Heydarian M et al. (1993) Evaluation of a PTW diamond detector for electron beam measurements. *Phys. Med. Biol.* 38: 1035–1042.

Higgins PD et al. (1995) Deconvolution of detector size effect for small field measurement. *Med. Phys.* 22: 1663–1666.

Hoban PW et al. (1994) Dose rate dependence of a PTW diamond detector in the dosimetry of a 6 MV photon beam. *Phys. Med. Biol.* 39: 1219–1229.

IAEA (2000) Absorbed dose determination in external beam radiotherapy: An international code of practice for dosimetry based on standards of absorbed dose to water. Technical Reports Series No. 398. Vienna, Austria: International Atomic Energy Agency.

IAEA (2017) Dosimetry of small static fields used in external beam radiotherapy: An IAEA-AAPM international code of practice for reference and relative

dose determination. Technical Reports Series No. 483. Vienna, Austria: International Atomic Energy Agency.
Jaffray DA et al. (1993) X-ray sources of medical linear accelerators: Focal and extra-focal radiation. *Med. Phys.* 20: 1417–1427.
Kalach NI and Rogers DWO (2003) Which accelerator photon beams are "clinic-like" for reference dosimetry purposes? *Med. Phys.* 30: 1546–1555.
Kamio Y and Bouchard H (2014) Correction-less dosimetry of nonstandard photon fields: A new criterion to determine the usability of radiation detectors. *Phys. Med. Biol.* 59: 4973–5002.
Kawachi T et al. (2008) Reference dosimetry condition and beam quality correction factor for CyberKnife beam. *Med. Phys.* 35: 4591–4598.
Klein DM et al. (2010) Measuring output factors of small fields formed by collimator jaws and multileaf collimator using plastic scintillation detectors. *Med. Phys.* 37: 5541–5549.
Larraga-Gutierrez JM et al. (2015) Properties of a commercial PTW- 60019 synthetic diamond detector for the dosimetry of small radiotherapy beams. *Phys. Med. Biol.* 60: 905–924.
Laub WU, Kaulich TW and Nusslin F (1999) A diamond detector in the dosimetry of high-energy electron and photon beams. *Phys. Med. Biol.* 44: 2183–2192.
Lechner W et al. (2013) Detector comparison for small field output factor measurements in flattening filter free photon beams. *Radiother. Oncol.* 109: 356–360.
Lee H-R, Pankuch M and Chu J (2002) Evaluation and characterization of parallel plate microchamber's functionalities in small beam dosimetry. *Med. Phys.* 29: 2489–2496.
Li J and Zhu TC (2006) Measurement of in-air output ratios using different mini-phantom materials. *Phys. Med. Biol.* 51: 3819–3834.
Li XA et al. (1995) Lateral electron equilibrium and electron contamination in measurements of head-scatter factors using miniphantoms and brass caps. *Med. Phys.* 22: 1167–1170.
Liu PZ, Suchowerska N and McKenzie DR (2014) Can small field diode correction factors be applied universally? *Radiother. Oncol.* 112: 442–446.
Marsolat F et al. (2013) A new single crystal diamond dosimeter for small beam: Comparison with different commercial active detectors. *Phys. Med. Biol.* 58: 7647–7660.
Martens C, De Wagter C and De Neve W (2000) The value of the PinPoint ion chamber for characterization of small field segments used in intensity-modulated radiotherapy. *Phys. Med. Biol.* 45: 2519–2530.
Moignier C, Huet C and Makovicka L (2014) Determination of the KQclinfclin,Qmsr fmsr correction factors for detectors used with an 800 MU/min CyberKnife((R)) system equipped with fixed collimators and a study of detector response to small photon beams using a Monte Carlo method. *Med. Phys.* 41: 071702.
Morales JE et al. (2014) Dosimetry of cone-defined stereotactic radiosurgery fields with a commercial synthetic diamond detector. *Med. Phys.* 41: 111702.
Morgan AM et al. (2008) Clinical implications of the implementation of advanced treatment planning algorithms for thoracic treatments. *Radiother. Oncol.* 86: 48–54.

Morin J (2013) A comparative study of small field total scatter factors and dose profiles using plastic scintillation detectors and other stereotactic dosimeters: The case of the CyberKnife. *Med. Phys.* 40: 011719.

Munro P, Rawlinson JA and Fenster A (1988) Therapy imaging: Source sizes of radiotherapy beams. *Med. Phys.* 15: 517–524.

Nisbet A et al. (2004) Dosimetric verification of a commercial collapsed cone algorithm in simulated clinical situations. *Radiother. Oncol.* 73: 79–88.

Nyholm T et al. (2006) Modeling lateral beam quality variations in pencil kernel based photon dose calculations. *Phys. Med. Biol.* 51: 4111–4118.

Ojala JJ et al. (2014) Performance of dose calculation algorithms from three generations in lung SBRT: Comparison with full Monte Carlo-based dose distributions. *J. Appl. Clin. Med. Phys.* 15(2): 4–18.

Pantelis E et al. (2010) On the implementation of a recently proposed dosimetric formalism to a robotic radiosurgery system. *Med. Phys.* 37: 2369–2379.

Papaconstadopoulos P, Tessier F and Seuntjens J (2014) On the correction, perturbation and modification of small field detectors in relative dosimetry. *Phys. Med. Biol.* 59: 5937–5952.

Planskoy B (1980) Evaluation of diamond radiation dosemeters. *Phys. Med. Biol.* 25: 519–532.

Pulliam KB et al. (2014) A six-year review of more than 13,000 patient-specific IMRT QA results from 13 different treatment sites. *J. Appl. Clin. Med. Phys.* 15(5): 196–206.

Rice RK et al. (1987) Measurements of dose distributions in small beams of 6 MV x-rays. *Phys. Med. Biol.* 32: 1087–1099.

Rosser KE and Bedford JL (2009) Application of a new dosimetry formalism to volumetric modulated arc therapy (VMAT). *Phys. Med. Biol.* 54: 7045–7061.

Rustgi SN (1995) Evaluation of the dosimetric characteristics of a diamond detector for photon beam measurements. *Med. Phys.* 22: 567–570.

Saini AS and Zhu TC (2004) Dose rate and SDD dependence of commercially available diode detectors. *Med. Phys.* 31: 914–924.

Scott AJ et al. (2012) Characterizing the influence of detector density on dosimeter response in non-equilibrium small photon fields. *Phys. Med. Biol.* 57: 4461–4476.

Scott AJ, Nahum AE and Fenwick JD (2008) Using a Monte Carlo model to predict dosimetric properties of small radiotherapy photon fields. *Med. Phys.* 35: 4671–4684.

Sham E et al. (2008) Influence of focal spot on characteristics of very small diameter radiosurgical beams. *Med. Phys.* 35: 3317–3330.

Sharma SC et al. (2007) Commissioning and acceptance testing of a CyberKnife linear accelerator. *J. Appl. Clin. Med. Phys.* 8(3): 119–125.

Shi J, Simon WE and Zhu TC (2003) Modeling the instantaneous dose rate dependence of radiation diode detectors. *Med. Phys.* 30: 2509–2519.

Sibata CH et al. (1991) Influence of detector size in photon beam profile measurements. *Phys. Med. Biol.* 36: 621–631.

Spencer LV and Attix FH (1955) A theory of cavity ionization. *Radiat. Res.* 3: 239–254.

Spokas JJ and Meeker RD (1980) Investigation of cables for ionization chambers. *Med. Phys.* 7: 135–140.

Stasi M et al. (2004) The behavior of several microionization chambers in small intensity modulated radiotherapy fields. *Med. Phys.* 31: 2792–2795.
Sterpin E et al. (2010) Monte Carlo-based analytical model for small and variable fields delivered by TomoTherapy. *Radiother. Oncol.* 94: 229–234.
Sterpin E, Mackie TR and Vynckier S (2012) Monte Carlo computed machine-specific correction factors for reference dosimetry of TomoTherapy static beam for several ion chambers. *Med. Phys.* 39: 4066–4072.
Sterpin E et al. (2008) Monte Carlo simulation of helical tomotherapy with PENELOPE. *Phys. Med. Biol.* 53: 2161–2180.
Taylor ML, Kron T and Franich RD (2011) A contemporary review of stereotactic radiotherapy: Inherent dosimetric complexities and the potential for detriment. *Acta Oncol.* 50: 483–508.
Tessier F and Kawrakow I (2010) Effective point of measurement of thimble ion chambers in megavoltage photon beams. *Med. Phys.* 37: 96–107.
Thomas SJ (2014) Reference dosimetry on TomoTherapy: An addendum to the 1990 UK MV dosimetry code of practice. *Phys. Med. Biol.* 59: 1339–1352.
Tyler M et al. (2013) Characterization of small-field stereotactic radiosurgery beams with modern detectors. *Phys. Med. Biol.* 58: 7595–7608.
Underwood TS et al. (2013) Mass-density compensation can improve the performance of a range of different detectors under non-equilibrium conditions. *Phys. Med. Biol.* 58: 8295–8310.
Underwood TS et al. (2015) Application of the Exradin W1 scintillator to determine Ediode 60017 and microDiamond 60019 correction factors for relative dosimetry within small MV and FFF fields. *Phys. Med. Biol.* 60: 6669–6683.
Van Esch A et al. (2006) Testing of the analytical anisotropic algorithm for photon dose calculation. *Med. Phys.* 33: 4130–4148.
Vatnitsky S and Järvinen H (1993) Application of natural diamond detector for the measurement of relative dose distributions in radiotherapy. *Phys. Med. Biol.* 38: 173–184.
Wang LL and Beddar S (2011) Study of the response of plastic scintillation detectors in small-field 6 MV photon beams by Monte Carlo simulations. *Med. Phys.* 38: 1596–1599.
Wilcox EE and Daskalov GM (2007) Evaluation of GAFCHROMIC EBT film for CyberKnife dosimetry. *Med. Phys.* 34: 1967–1974.
Wu H et al. (2010) Impacting parameter analysis for IMRT quality. *Med. Phys.* 37: 3148.
Yin FF et al. (2002) Dosimetric characteristics of Novalis shaped beam surgery unit. *Med. Phys.* 29: 1729–1738.

10

Special Delivery Techniques

Dedicated to Michael Sharpe

Tomas Kron

10.1 Introduction and Background: Why More Than a "Linac" on a C-Gantry

Radiotherapy techniques and technology are linked. Based on the definition of the Royal Australian and New Zealand College of Radiologists (RANZCR), we define a technique as a method for accomplishing a desired radiation therapy dose distribution, while technology describes a method used to facilitate the delivery of a radiation therapy treatment technique (FRO, 2014). While the chapter is focused on techniques, these are often intrinsically linked to technologies, and techniques will be often discussed in the context of technologies. As radiation oncology is a relatively small and technology-focused discipline, these technologies are often also intrinsically associated with one manufacturer's product.

This chapter is focused on external beam radiotherapy while brachytherapy will be discussed in Chapter 20. It is also important to note that many modern radiotherapy delivery techniques rely on imaging for target localization and treatment planning (compare Chapter 22). In this chapter, image guidance is considered in the context of its impact on dosimetry. Finally, while this chapter is entitled Special Delivery Techniques, it has the overall aim of the book in mind and emphasis is given to the interrelationship of delivery techniques and dosimetry.

Radiotherapy with C-type gantries as shown in Figure 10.1a has a long tradition. The design allows for easy access to the patient while facilitating a large number of possible beam directions that can include noncoplanar deliveries if

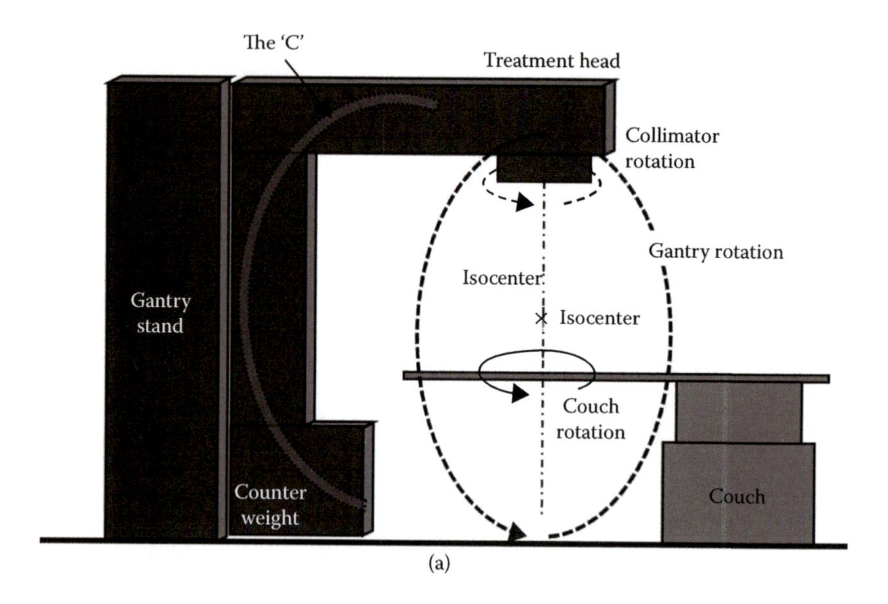

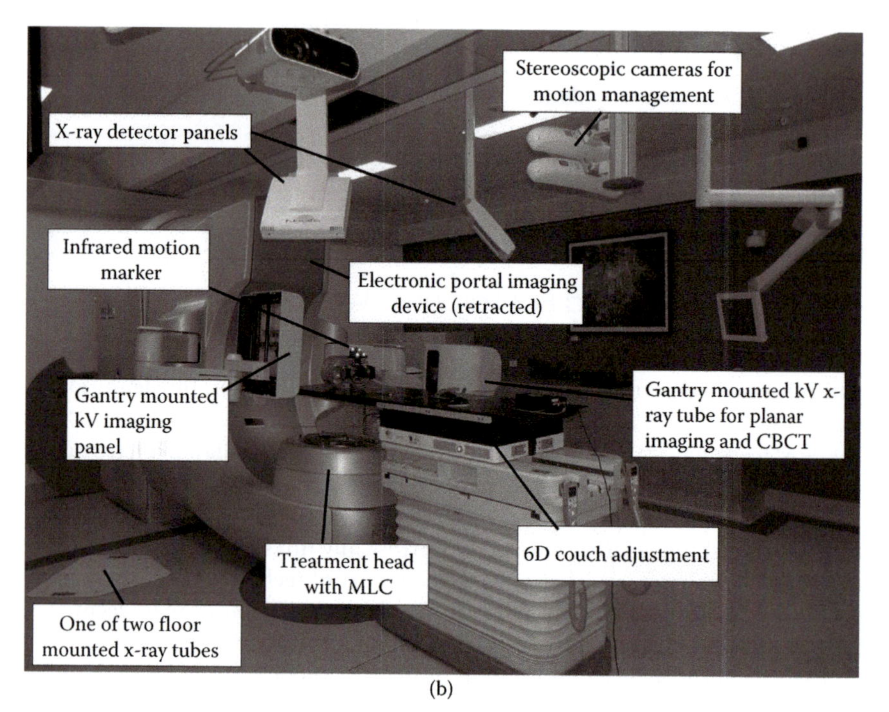

Figure 10.1

A Varian Truebeam STX (Varian Medical Systems) linear accelerator mounted on a "C"-shaped gantry. (a) Schematic drawing of a conventional linear accelerator. (b) Photograph of a modern linear accelerator with several "add-ons" to improve dose delivery and image guidance. Gantry is rotated by 180 degrees.

the couch is rotated. This leads to the 3D dose distributions that are desirable for radiotherapy. However, there are several disadvantages to the C-shaped design including

- Large weight of gantry leads to sag.
- Rotations of the gantry limited to one rotation per minute due to risk of injuries at faster speed.
- Rotation of the gantry is at maximum 180° in each direction—no continuous rotation is possible.
- Isocenter height typically too high for ergonomic patient setup.
- Additional collimation devices often added externally reducing clearance around patients.
- Field size limited usually to 40×40 cm^2 or less.
- Image guidance tools need to be added externally.

Figure 10.1b shows a modern linear accelerator with many of the desirable imaging options added. It clearly has lost some of its simplicity, and despite the fact that most of the accessories are retractable, access to the patient is more restricted.

As radiation oncologists, physicists, radiation therapists, and engineers consider how to improve the design of radiotherapy treatment units, a number of specific objectives provide design goals as follows:

1. Improvement of dose delivery including very small and/or very large fields
2. Improvement of target visualization
3. Motion management
4. Cost-effectiveness
5. Consideration of different radiation types

Improvements related to any of these considerations can be incorporated into the C-arm designs; however, new designs offer further potential improvements and are often driven most strongly by one of the aspects listed above.

Over the years, improvement of dose distribution has been the most important imperative for the development of treatment techniques and technologies as can be seen in Table 10.1. Since more than 50 years these developments have contributed to radiotherapy becoming faster, more penetrating, and stronger ("*citius, altius, fortius*" so to speak), thus improving cancer treatment (Moran et al., 2005). Image guidance (Dawson and Sharpe, 2006) and motion management (Keall, 2006) ensure that these advances are delivered to the correct target.

Cost-effectiveness is not a particular emphasis of this chapter even if it has become an integral part of modern medicine. However, it is necessary to remember that while designing equipment it is important to keep in mind that economic thinking does not affect the ability to perform accurate dosimetry.

The consideration of different radiation types, point "e" in the list above, is beyond the scope of this chapter. Protons and carbon ions are covered in Chapters 12 and 23. Possibly the pinnacle in dosimetric challenge is the use of microbeams (Slatkin et al., 1992; Brauer-Krisch et al., 2015). This refers to kV x-rays typically from a synchrotron with very small divergence and extremely high dose rates (>100 Gy/s). These beams are collimated to "microbeams" of 20–50 μm width

Table 10.1 Key Developments in Radiotherapy and the Associated Delivery Technologies

Treatment Technique	Time	Key Advantage	Enabling Technologies	Treatment Technology	Comments
Fixed SSD treatment	1950s	Easy setup		Orthovoltage, Co-60	Appropriate for single fields—still used for electrons and kV
Isocentric treatment	1970s	Better setup, no patient movement	High output and MV to facilitate long enough SAD	Linear accelerator	More complex planning: fixed SSD replaced by isocentric treatments
3D conformal treatment	1980s	Better targeting	CT scanner for planning; computerized treatment planning		Relies on virtual simulation using a 3D model of a patient
Intensity modulation	1990s	Better sparing of normal structures close to the target	Multileaf collimator; computer optimization	Helical tomotherapy, IMRT; VMAT followed in the 2010s	Computer optimization required
Image guidance	2000s	Better and more reproducible targeting	Imaging, particularly volumetric, available on treatment unit	CBCT Helical tomotherapy, VERO, MRI-based systems	Many options and customization
"Stereotactic" treatments	1960s (Intracranial) and 2010s	High precision, patient convenience (less fractions), retreatment options	Improved mechanical performance of treatment units, motion management	Gamma Knife, CyberKnife, VERO	Improvements in dosimetry; relies on image guidance and motion management

Note: The selection is neither meant to be complete nor exhaustive and represents the personal view of the author.
CBCT, cone beam CT; IMRT, intensity-modulated radiation therapy; MRI, magnetic resonance imaging; VMAT, volumetric modulated arc therapy.

separated by gaps of some 200 μm. There is preclinical evidence that these "spatially" intensity-modulated beams provide normal tissues with a significant radiobiological advantage (Dilmanian et al., 2001). Dosimetry in these beams is difficult due to the use of kV beams, the high spatial resolution required, and the extremely high dose rate. Medical physics issues related to microbeam radiotherapy are covered in more detail in Chapter 25.

The present chapter explicitly considers the points (1)–(3) in the list above, followed by comments on some specific aspects on dosimetry relevant to special techniques. The final conclusion includes some comments on health economics and a general outlook.

10.2 Techniques and Technology Developments to Improve Dose Delivery

Dose delivery in radiotherapy can be improved through several measures on the delivery side:

- Subdivision of radiation fields in smaller segments with different weightings increases the flexibility of delivery. Intensity-modulated radiation therapy (IMRT) has become the umbrella term for these approaches (Webb, 2005; Ezzell et al., 2009) which also include volumetric modulated arc therapy (VMAT) (Otto, 2008) and helical tomotherapy (Mackie, 2006).
- Faster treatment can be facilitated through rotational deliveries (Teoh et al., 2011) and increased dose rate, for example, by omitting the flattening filter (Georg et al., 2011).
- Considering more beam directions through noncoplanar delivery provides more access points to radiation beams and as such often more conformity and the ability to avoid critical structures more effectively (Nguyen, 2014).

It must be noted that all these improvements are largely facilitated by computer usage. This has implications for dosimetry as treatment plans are often not intuitively verifiable and patient-specific measurements are commonly considered necessary. For obvious reasons, many of the improvements in dose delivery make dosimetry more difficult. Small fields (Chapter 9) and dynamic deliveries (Chapter 11) highlight this aspect in particular.

10.2.1 Very Small and/or Very Large Fields

Small fields are not only used for stereotactic applications but also an integral part of IMRT. The dosimetric challenges are a result of charged particle disequilibrium (Das et al., 2008; IPEM, 2010) which applies to electron as well as photon beams (Kron et al., 2013). A lot of work and many publications are concerned with small-field dosimetry, and Chapter 9 summarizes this well. On the other hand, large fields have seen less attention but can equally be dosimetrically challenging.

Large fields are obviously important for half or total body irradiation (TBI) (Quast, 1987; Van Dyk, 1987). However, large fields are also used in other circumstances, such as craniospinal irradiation, and if more than one lesion shall be treated in a single plan "large" fields are often required. Historically, these large treatment fields were achieved using extended distances or field junctions.

More recently, dynamic treatments with or without patient movement have also become common (Hui et al., 2005; Mancosu et al., 2015).

From a dosimetric perspective, large radiation fields have several complicating factors:

- Scatter is different from standard conditions, within the patient as well as from walls and floor (Sanchez-Nieto et al., 1993).
- Standard scanning water phantom measurements are not possible.
- Most array detectors, including electronic portal imaging (EPI) devices, are too small.
- Many treatment-planning systems are not designed to do dose calculations at extended focus–surface distance (FSD) and special consideration must be given to treatment planning (Lavallee et al., 2009).
- Dose rate is often lower leading to signal-to-noise and leakage problems in detection.

In addition to these factors, a lot of relevant literature is relatively old, and when calculations are difficult, dose measurements including *in vivo* dosimetry (Kron et al., 1993) continue to play an essential role.

Not surprisingly there is a lot of interest in dynamic deliveries of large treatment fields where either the beam is moved across the patient (Hugtenburg et al., 1994) or the patient is moved through the beam (Hussain et al., 2011). The latter is a standard process in helical tomotherapy which makes this approach an interesting option for TBI (Hui et al., 2005) and craniospinal irradiation (Bauman et al., 2005). In the case of TBI, this technique also allows to target only bone marrow rather than the whole body (Corvo et al., 2011), a technique which is now also available using VMAT (Han et al., 2012; Mancosu et al., 2015). Given the complexity of the target volume, which needs to be covered completely, and the significant risk of toxicity, actual dose measurements, typically in anthropomorphic phantoms, are an essential requirement (Wilkie et al., 2008).

10.2.2 High Dose Rate Options: Flattening Filter Free

Increasing complexity of delivery often comes at the expense of more monitor units delivered in smaller fields or field segments. Therefore, it is attractive to increase dose rate to ensure that patient treatment times are not prolonged. This is also beneficial for gated deliveries and hypofractionated stereotactic ablative radiotherapy (SABR), which has become increasingly popular (Timmerman et al., 2007). The most common method for increasing dose rate is the omission of a flattening filter, which is used in conventional linear accelerators to produce a homogeneous dose profile over a 40 cm wide beam at 100 cm distance from the target (Georg et al., 2011; Fogliata et al., 2012). Depending on the radiation quality, flattening filter-free (FFF) radiation beams can increase the dose at the central axis of the beam by a factor of 2.5 (6 MV x-rays) to 4 (10 MV x-rays). Even higher ratios could in principle be achieved with higher x-ray energies, but most manufacturers limit their FFF beams to 10 MV.

FFF beam configuration is an inherent part of helical tomotherapy (Mackie, 2006) and CyberKnife (Kilby et al., 2010) design as discussed in the next section. Figure 10.2 shows a helical tomotherapy unit. A profile for the FFF 6 MV fan-beam is shown in the inset. From a dosimetric perspective, FFF beams produce two major challenges:

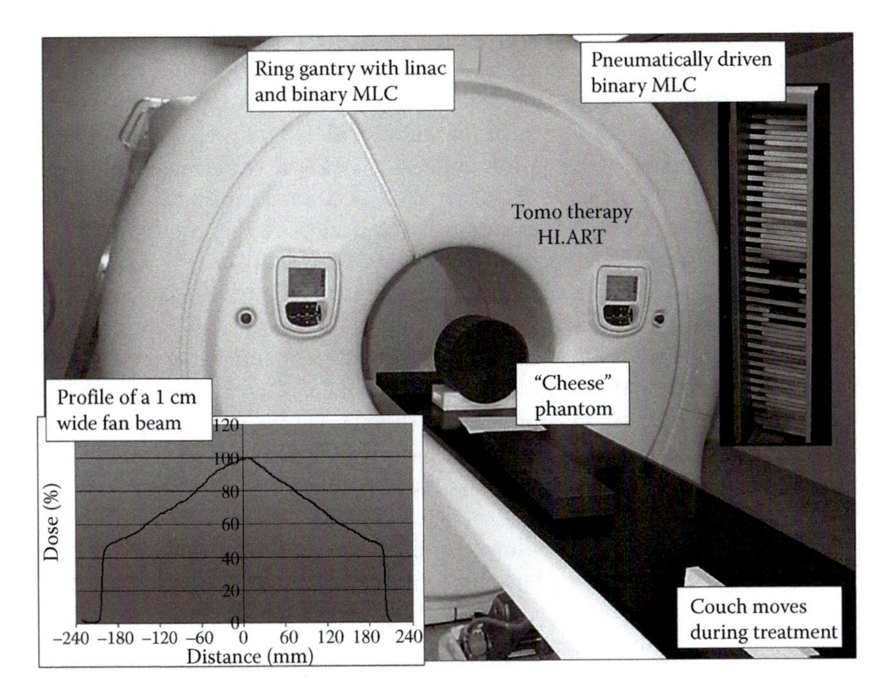

Figure 10.2

Photograph of a helical tomotherapy unit (Accuray). The inserts show the binary MLC (right upper corner) and the lateral profile of a 1 cm wide fan beam with all 64 leaves over the maximum field width of 40 cm open (lower left corner). Not visible is the exit detector located on the opposite side of the linac which was adapted from a diagnostic CT scanner.

- The high-dose rate will affect the performance of some detectors: ionization chambers and diodes experience larger recombination effects (Kry et al., 2012; Wang et al., 2012). This has most recently also been demonstrated for diamond detectors (Brualla-Gonzalez et al., 2016).
- The dose distribution is not uniform which may affect large detectors that average dose. It also leads to significant dose variations with gantry angle in arc treatments of lesions that are not at central axis. More homogeneous dose distributions can be achieved through the use of intensity-modulated delivery.

It is noteworthy that rotational deliveries also tend to speed up treatment as the linac is permanently on through an arc, while there is always time between static gantry positions when no dose is delivered (Wolff et al., 2009; Popescu et al., 2010; Nguyen et al., 2012). This is particularly noticeable when moving from IMRT to VMAT-type deliveries that often also feature less monitor units for similar plan quality.

10.2.3 Noncoplanar Deliveries

Radiation must be delivered to the target using distinct pathways. In many clinical scenarios, it is advantageous to use many pathways to keep the dose to normal structures as low as possible in individual paths. This is particularly relevant for "stereotactic" treatments (be it intra- or extracranial) where the target dose per fraction is usually very high. Stereotactic treatments also usually feature small targets where overlap of adjacent beam trajectories is typically small. To keep the dose in each delivery path as low as 2 Gy or less (where radiobiology is well known) for a target dose of some 20 Gy, it is necessary to utilize 10 or more beams. To make this feasible often noncoplanar beams are needed.

For C-shaped gantries, noncoplanar deliveries are facilitated by couch rotation. This is not necessarily the best approach as it requires movement of the patient, and even the best immobilization aid does not prevent subtle patient movement when she/he experiences the motion of their support assembly. This problem is overcome in the VERO (Brainlab/Mitsubishi, Figure 10.3), CyberKnife (Accuray, Figure 10.4), and Gamma Knife (Elekta, Figure 10.5) systems, which will be described in the next section. Each of them employs a unique method to minimize restrictions in terms of beam access to the patient:

- VERO: the ring gantry allows movement of the linac around the patient while the ring itself rotates up to ±60° around the couch which itself remains static.
- The CyberKnife consists of an X-band linac (higher frequency, smaller accelerating structure, smaller field size) mounted on an industrial robot

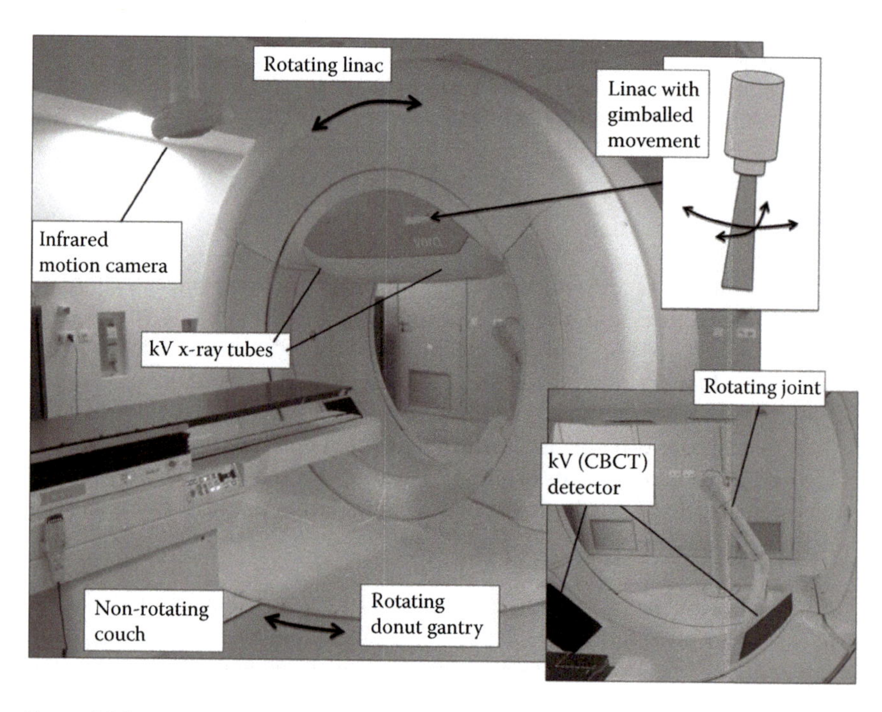

Figure 10.3

Photograph of a VERO radiotherapy unit (Mitsubishi). This system was designed from scratch to include many features of modern radiotherapy: linear accelerator mounted on a ring gantry which can rotate continuously due to a rotating joint; two kV x-ray units "on board" which can generate cone beam CTs (CBCTs) when the gantry rotates; ceiling mounted infrared motion detector; and a rotating mount for the gantry that allows rotation of the gantry around the patient without moving the couch.

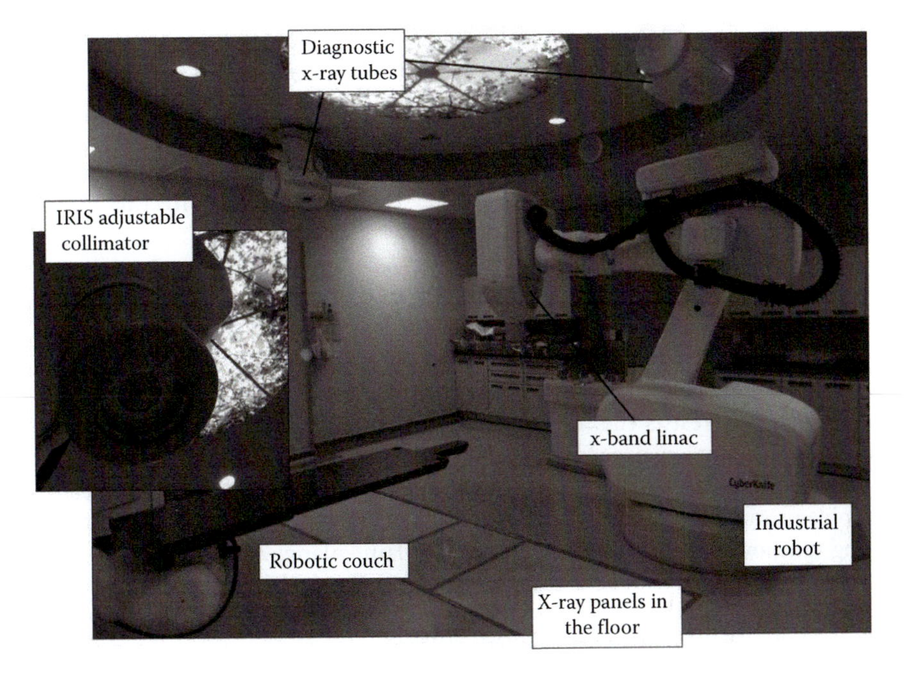

Figure 10.4

Photograph of a CyberKnife (Accuray). The insert shows the "IRIS" collimator that facilitates the delivery of different diameter circular radiation fields. Also shown are the two ceiling mounted x-ray units that allow for acquisition of orthogonal x-ray images at a user-defined timing prior to and during treatment delivery.

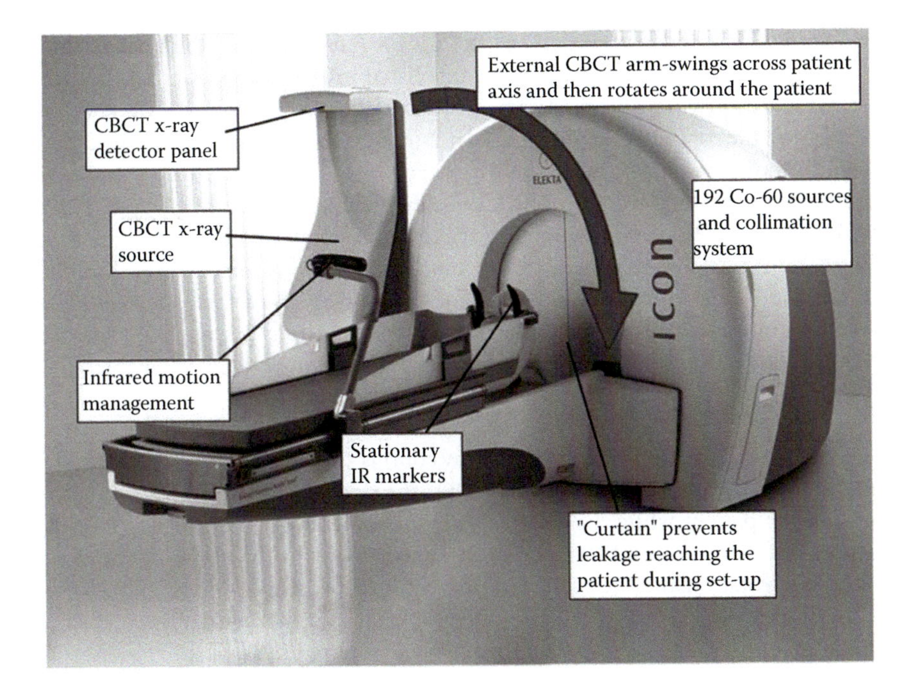

Figure 10.5

Photograph of the Gamma Knife Icon (Elekta), which includes a CBCT and infrared (IR) motion management system for image guidance.

similar to the ones used in car assembly. In principle, it allows any beam direction with respect to the patient, but mechanical restrictions limit the practical range of beam directions.
- The Gamma Knife consists of about 200 radioactive Co-60 sources mounted in a hemispherical arrangement. By opening and blocking the sources, dose can selectively be delivered from any direction in this hemisphere.

While noncoplanar deliveries are inherently 3D, they also pose specific challenges for dosimetry. The design of phantoms suitable for assessment of noncoplanar deliveries requires more attention, particularly if weight and need for computed tomography (CT) acquisition for planning are considered. Also, of dosimetric interest is the integral dose (Nguyen, 2014). Noncoplanar delivery allows spreading low dose further while reducing dose in each delivery path.

In addition, any variation of detector response with beam direction will cause problems in noncoplanar delivery if all beams are to be assessed in a single setting (Suchowerska et al., 2001; Roshau and Hintenlang, 2003; Lehmann et al., 2014). This problem can be overcome by judiciously placing the detectors.

10.2.4 Technological Developments for Improved Dose Delivery

The following section describes a few developments designed specifically to facilitate improvements in dose delivery. This is not meant to be an exhaustive or complete list but aims to highlight features that were deemed to be significant enough to warrant redesign of existing technology. Similar summaries will be given for two other drivers of technology development in radiotherapy (image guidance and motion management). As the different fields of improvement often link with each other, also several technologies will feature more than once.

Most of the technological developments listed in this section were developed for more accuracy and flexibility in dose delivery. This is matched by a need to perform dosimetry with a higher spatial accuracy and often assess steep dose gradients. More problematic is the fact that some of the units described cannot be calibrated using standard protocols for reference dosimetry (Almond et al., 1999; IAEA, 2000). These protocols require reference dosimetry to be performed in a 10×10 cm^2 field at a distance close to 100 cm, conditions neither helical tomotherapy nor CyberKnife and Gamma Knife can achieve. Therefore, a new process involving intermediate calibration fields is proposed (Alfonso et al., 2008) and the required correction factors are derived, as discussed in more detail in Chapter 9.

10.2.4.1 Helical Tomotherapy

A helical tomotherapy unit (www.accuray.com) is shown in Figure 10.2 (Mackie et al., 1993; Jeraj et al., 2004; Mackie, 2006). The ring gantry rotates continuously around the patient while the couch slowly moves through the gantry resulting in a helical delivery. A 6 MV x-ray linac is mounted with the target 85 cm from the axis of rotation and the beam profile is collimated to fanbeam geometry with a width of 40 cm and no flattening filter as shown in the figure. The fanbeam of variable thickness is modulated using a binary multileaf collimator (MLC) where each of the 64 leaves can be either open or shut with transition occurring in 20–40 ms. This allows creation of different fluence patterns from the fanbeam shown in the insert as the MLC pattern changes up to 51 times per full gantry rotation.

10.2.4.2 VERO

The VERO system was one of the most recent developments in collaboration between Mitsubishi Heavy Industries and Brainlab (Burghelea et al., 2014; Solberg et al., 2014). Unfortunately this unit is no longer marketed internationally, but the machine is still being built by Mitsubishi, called MHI-TM2000 (https://www.mhi-global.com/products/category/radiotherapy_unit.html). Several units are used clinically due to its interesting design features. It combines many desirable features for beam delivery, image guidance, and motion management as seen in Figure 10.3. From a delivery perspective, the continuous gantry rotation and noncoplanar delivery options are of interest. The maximum field size of 15 × 15 cm^2 and the excellent isocenter definition afforded by the ring gantry and gimballed linac (better than 0.5 mm diameter) indicate that the unit is primarily designed for stereotactic applications. Important for these applications is also the fact that the couch does not need to be moved for noncoplanar deliveries as the ring itself rotates to ±60°.

10.2.4.3 CyberKnife

The Accuray CyberKnife shown in Figure 10.4 consists of a linear accelerator that is kept small through use of a higher radiofrequency (X-band) accelerator that is mounted on an industrial robot which can position the linac in any desired direction in relation to the patient, thus facilitating noncoplanar delivery with maximum degrees of freedom (Calcerrada Diaz-Santos et al., 2008; Kilby et al., 2010). A disadvantage is that some posterior beams are not possible due to difficulties to position the linac underneath the couch. Figure 10.4 shows an interesting additional development to achieve different field sizes: the IRIS collimator. More recently, an MLC has also become available.

10.2.4.4 Gamma Knife

The Elekta Gamma Knife was originally designed in 1968. It consists of 201 (in the recently released Gamma Knife Icon 192) Co-60 sources in five bands from 6° to 36° from the axial slice (Wu et al., 1990; Yu et al., 2000; Drzymala et al., 2008). All sources are collimated in a way to focus the radiation to a point in space. Four different collimation sizes are available (4, 8, 14, and 18 mm) for each source. Over the years, several improvements have been introduced to automate the collimation selection and move the patient into the correct location (Gamma Knife Perfexion) (Cho et al., 2010; Ruschin et al., 2010). The most recent generation 6, the Gamma Knife Icon, is shown in Figure 10.5. It includes also a cone beam CT (CBCT) and infrared motion management system which allows combining the highly accurate delivery with isocentricity better than a radius of 0.2 mm with image guidance and motion management.

Another development utilizing radioactive sources is the “Rotating Gamma Knife” popular in the Chinese market (Goetsch et al., 1999; Kubo and Araki, 2002). This unit consists of 30 Co-60 sources that rotate around the patient yielding a similar dose distribution and spatial accuracy as the Elekta Gamma Knife (Cheung and Yu, 2006).

10.2.5 Intraoperative Radiation Delivery

Intraoperative radiotherapy refers to the delivery of radiation to patients on the operating table. This has the obvious advantage of being able to direct the radiation to the target or suspected residual disease while exposed. Disadvantages include the need to work in sterile conditions and under considerable time pressure. In addition to the radiation protection issues, emergency procedures can be tricky as a shielded theater may be needed and the patient cannot be left without monitoring.

At present, there are three common variants: brachytherapy (Tan et al., 2013), miniaturized kV x-ray systems (Vaidya et al., 2010; Jones et al., 2014; Eaton, 2015), and specialized cones for megavoltage electron delivery (Orecchia and Veronesi, 2005; Veronesi et al., 2013). Figure 10.6 shows the INTRABEAM System (www.zeiss.com) as an example for a miniaturized 35 or 50 kVp x-ray source at the end of a needle that is inserted into a spherical applicator. The applicator size is selected to fit the surgical cavity after lumpectomy for early stage breast cancer. At the end of surgery, the applicator is placed in the cavity and dose delivered to the surface of the cavity.

It is not difficult to appreciate the dosimetric challenges in this approach as given below:

- kV x-ray attenuation and absorption depends on the atomic composition of the tissue.
- The dose fall off close to the source is very rapid.
- Dose is difficult to determine at the surface of the applicator.

In many circumstances, dosimetry for intraoperative radiotherapy is similar to brachytherapy where irradiation geometry and direct relation between source and target play the predominant role.

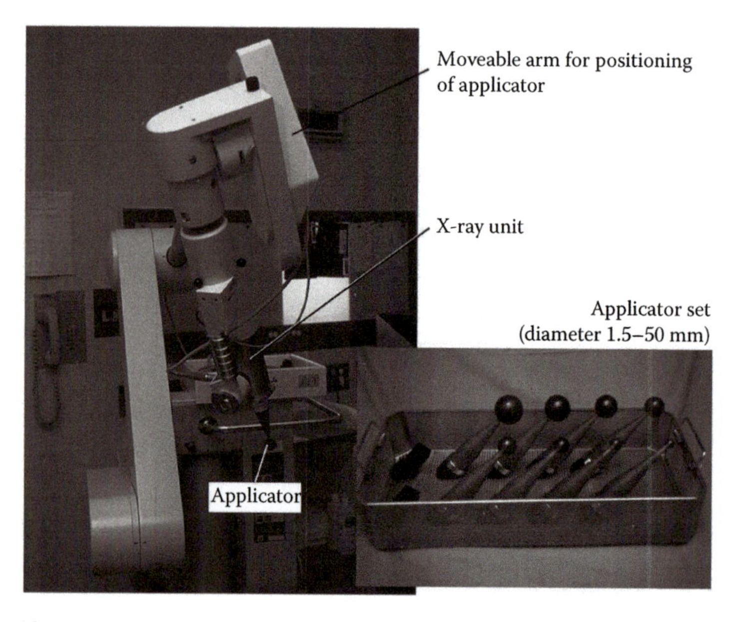

Figure 10.6

Photograph of the INTRABEAM System for intraoperative radiotherapy (Zeiss). The unit allows for flexible positioning of spherical applicators shown in the insert. An applicator of appropriate size for the excision cavity after lumpectomy for breast cancer is placed in the cavity during the operation.

10.3 IGRT-Driven Developments of Techniques and Technology

Image guidance has made a significant difference to radiotherapy practice and outcomes (Dawson and Sharpe, 2006; Simpson et al., 2010). Not surprisingly, a large number of image guidance tools are available as part of conventional linear accelerators as seen in Figure 10.1b. While EPI is easy to integrate (Herman, 2005), most other imaging modalities, in particular if they are volumetric, are often tricky to realize and do not necessarily provide the best possible image quality for a given modality. This is easily appreciable when comparing diagnostic CT with CBCT images acquired on a linear accelerator (Bissonnette et al., 2008). From a dosimetric perspective, three key questions arise as follows:

1. How does image guidance influence the dose distribution delivered in the first place, for example, by reducing margins (Skarsgard et al., 2010) or adapting to the image information (Ghilezan et al., 2010; Wu et al., 2011; Kron et al., 2012)?
2. Does imaging contribute to the radiation dose received by the patient and how is it accounted for (Murphy et al., 2007; Ding and Munro, 2013)?
3. Does the use of image guidance tools, such as fiducial markers, affect the dose distribution in the patient?

In any case, the need for optimal image guidance leads to the development of treatment units that are specifically designed around imaging modalities such as ring gantry systems (helical tomotherapy or VERO) and integrated magnetic resonance imaging (MRI) and MV treatment units (Lagendijk et al., 2008; Wooten et al., 2015).

10.3.1 Technologies Developed to Integrate X-Ray-Based Image Guidance

10.3.1.1 Orthogonal kV X-Ray Imaging (Linacs and CyberKnife)

Traditionally, imaging of patients was performed using the treatment beam of the linear accelerator with portal films or more recently with EPI. However, the image quality of EPI is rather limited due to the MV radiation source, and kV onboard imaging was therefore introduced in the 1990s to overcome this limitation. The onboard imaging system seen in Figure 10.1b in horizontal projection is linked to the linear accelerator, and cannot be typically utilized while the treatment beam is on. The decoupling of x-ray imaging from the linac is achieved by using two x-ray sources in the floor or the ceiling projecting orthogonally across the isocenter (Brainlab ExacTrac). This system is realized in the linac shown in Figure 10.1b as well as in the CyberKnife unit shown in Figure 10.4. In the VERO unit, the two x-ray tubes and detectors are embedded in the rotating gantry as can be seen in Figure 10.3. The advantage of this setup is that imaging is independent of delivery and can be performed while the beam is on, a prerequisite for motion management. The images can also be used to update other nonionizing radiation-based monitoring systems such as infrared markers shown in Figures 10.1b and 10.3. This is important to reduce the imaging dose: if treatment takes many minutes even modest exposure settings can yield high doses to the patient.

An interesting issue from a dosimetric perspective is that many image guidance approaches with planar x-ray imaging rely on fiducial markers. They may affect the dose distribution due to their high density and high atomic number (Vassiliev et al., 2012), and could also cause artifacts in the planning images, which lead to an additional dose uncertainty depending on their size.

10.3.1.2 Cone Beam CT

More useful than planar imaging is often volumetric imaging as the same image type is used for planning and verification. CBCT has become the standard for this type of imaging as images for a whole volume can be acquired in a single rotation as is essential for the slow-rotating gantries in the standard C-shape gantries. CBCT was originally introduced by Jaffray and Siewerdsen (2000) and has proven to be so useful that even systems that do not require a moveable gantry, such as the Gamma Knife, now include one as in Figure 10.5. Here an external arm is attached to the unit. When a CT scan is to be taken, the arm moves over the patient and can then be rotated around the patient. Like in many CBCT applications, only a half rotation is required for image reconstruction.

While not the primary focus of this book, the difficulty of measuring dose in CBCT is worth mentioning (e.g., Alaei and Spezi, 2015), which is further discussed in Chapter 22. Two issues make CBCT dosimetry difficult in particular when aiming to account for the imaging dose in the treatment plan. First, kV imaging is used for which no good, widely available dose calculation algorithms exist; Monte Carlo calculations provide the best estimates of dose in patients (Ding and Coffey, 2009, 2010). This also leads to dose distributions with higher dose in bone. As the system is used to perform both planar imaging and CBCT, it is not fully optimized for any of them in regards of filtration and detector design including the use of antiscatter grids. Second, the width of the beam exceeds the CT dose index (CTDI) phantoms used for CT dosimetry. This makes conventional reporting of imaging dose in terms of CTDI typically an underestimation of dose as scatter exceeds what is considered in phantoms of limited size. As such it is common to account for CBCT dose with a single dose value in Gy that is uniformly applied to the whole imaged volume (Kron et al., 2010).

10.3.1.3 Helical Tomotherapy (MVCT)

A helical tomotherapy unit as shown in Figure 10.2 is not only outwardly similar to a diagnostic system but also employs methods similar to diagnostic CT scanning to radiotherapy delivery. While the original proposal included a diagnostic CT acquisition system below 90° from the delivery beam (Mackie et al., 1993), this was dropped from the commercial unit. It turned out that the Xe-filled ionization chamber array adapted from the original GE CT scanner produced acceptable images when detuning the linear accelerator used for treatment delivery to a lower energy and collimating the fan beam to a narrower profile (Ruchala et al., 1999). The advantages of using MV x-rays for imaging are that the dose distribution resembles the one delivered using treatment beams much closer, and the imaging dose can be relatively easily included in the treatment plan.

An interesting observation is that the variations in dose rate of the MV source during imaging can actually affect the image quality sufficiently to cause differences in dose calculation based on the MVCT images (Duchateau et al., 2010).

10.3.1.4 VERO

The VERO concept includes two kV imaging systems as well as EPI. The kV imaging systems shown in Figure 10.3 are integrated in the rotating gantry unlike in the linac and CyberKnife shown in Figures 10.1 and 10.4, respectively. This allows their use for kV imaging during treatment as well as for CBCT. To make matters even more interesting, the two x-ray systems are independent of each other and as such can in principle be used with two different energies.

10.3.2 Technologies Using Nonionizing Image Guidance

Image guidance using nonionizing radiation has several immediate advantages as follows:

- No dose has to be considered in the treatment plan.
- Imaging can be repeated as often as needed without any detriment.
- The imaging modality is entirely independent of the beam delivery.

Nonionizing image guidance includes ultrasound, optical, and MRI-based systems (Tome et al., 2002; Cury et al., 2006; Raaymakers et al., 2009; Mutic and Dempsey, 2014). From a dosimetric perspective, MRI is the most complex and requires special consideration. All systems share the problem of access to the patient which is particularly relevant for ultrasound which must be in contact with the patient. Optical imaging as yet another alternative is very fast but limited to surface structures. Like ultrasound it allows real-time imaging which is important in motion management.

10.3.2.1 MRI Linacs and Cobalt Units

MRI in particular would be of considerable interest as it provides excellent soft tissue contrast. As it uses a method completely independent of the treatment delivery, MRI can also, at least in principle, be used in real time to monitor motion and changes due to treatment. As such, it is not surprising that several groups are currently working on prototype units despite the formidable challenges of combining strong magnetic fields with the electromagnetic components of a linac (Fallone et al., 2009; Constantin et al., 2011; Lechner et al., 2013). A method to overcome this problem is to replace the linac with one (or more) Co-60 sources (Kron et al., 2006; Mutic and Dempsey, 2014). The concept proved to be faster to realize than the MRI–linac combinations, and the Viewray MRIdian system has become functional treating patients since early in 2015 (Hu et al., 2015; Li et al., 2015; Wooten et al., 2015). The system consists of three equally spaced Co-60 sources each of which features its own MLC. Treatment planning optimizes the delivery of all sources combined, which results in dose rate comparable to linear accelerators. The magnetic field strength of 0.35 T is low compared to a diagnostic MRI scanner. However, the image quality and acquisition time are compatible with the purpose of providing high-quality image guidance.

From a dosimetric perspective, a magnetic field poses problems as dosimetric equipment needs to be hardened against magnetic fields, and secondary electrons are forced on a spiral pathway. These problems will be briefly discussed in this section, while 3D dosimetry in magnetic fields will be discussed in more detail in Chapter 26.

The group at the University Medical Center, Utrecht, the Netherlands, has tested many pieces of dosimetric equipment in magnetic fields up to 1.5 T (Meijsing et al., 2009; Smit et al., 2013, 2014). Provided equipment is carefully selected or customized for use in magnetic fields, both absolute and relative dosimetries can be performed. This was also confirmed by other groups

(Gargett et al., 2015). In general, the dosimetric issues increase with magnetic field strength (Raaijmakers et al., 2008; Kirkby et al., 2010). This is unfortunate as higher field strength provides faster imaging, a direct link to MR simulators usually operating at higher field strength, and the possibility to generate contrast from functional processes. Many publications describe the effects of magnetic fields on dose distribution. A particularly interesting setup was designed by Keall and coworkers in Sydney, Australia. As shown schematically in Figure 10.7, the group is exploring dosimetric (and other) differences as a function of the direction of the magnetic field with respect to the treatment beam (Constantin et al., 2011).

The main dosimetric issues are a result of secondary electrons being forced on a curved pathway in the presence of a magnetic field. While the typical field strength of MRI scanners appears to be not high enough to affect biological effectiveness, for example, by having a curved electron path traversing the same part of the DNA several times (Nettelbeck et al., 2008), the dose distribution is affected. Not surprisingly, the most important differences occur in small fields and at interfaces (Raaijmakers et al., 2005; Kirkby et al., 2010).

In the presence of a transversal field (the more common scenario shown in Figure 10.7a), changes in depth dose, build-up region, and particularly at the exit side have been observed (Raaijmakers et al., 2005). The so called "electron return effect" increases dose at interfaces from high- to low-density material, such as at the exit side of the patient. Most of the studies on the effects of magnetic fields on dose distribution were done using Monte Carlo calculations; however, in the various prototype systems they have also been experimentally confirmed (Raaijmakers et al., 2007).

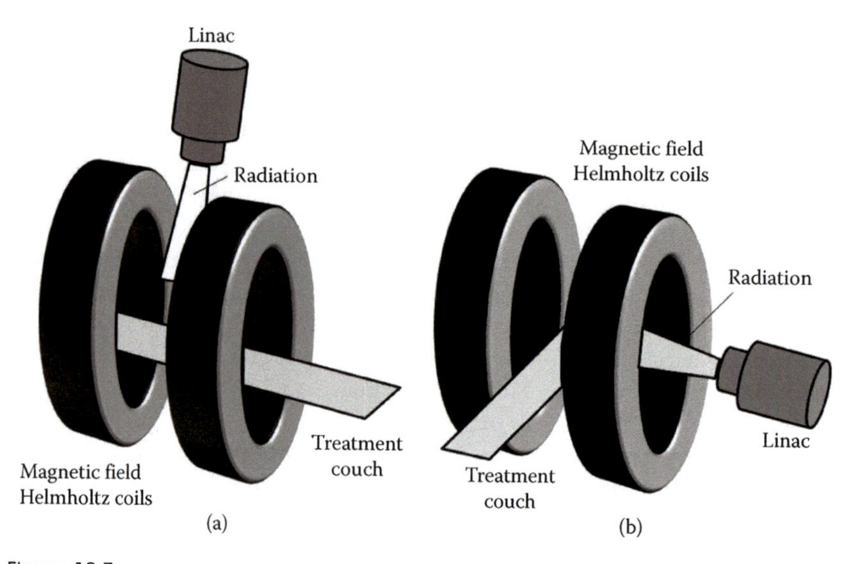

Figure 10.7

Schematic drawing of the MRI linear accelerator combination designed by Keall et al. As discussed in the text, the dose distribution delivered by a linear accelerator is affected by the direction and magnitude of a magnetic field present during irradiation. Therefore, two designs are explored as illustrated in the figure: (a) orthogonal (transversal, Figure 10.7a) and (b) in line (longitudinal) with the magnetic field.

In longitudinal magnetic fields, on the other hand, the effect can actually be beneficial by confining the secondary electrons to the forward direction, which possibly improves the dose distribution in low-density organs such as the lung (Kirkby et al., 2010).

The impact of image guidance in general on radiation dosimetry is difficult to assess. Clearly, reduced margins due to image guidance provide opportunities for tighter dose conformality and potential dose escalation. Both will increase, at least in principle, the requirement for accuracy in dose determination. An interesting side issue is that the availability of image guidance tools also allows to verify the location (and possibly even the integrity) of the dosimeter used. This can prove very useful in performing dose measurements.

10.4 Motion Management

Motion management in the context of radiotherapy is intrinsically linked to image guidance (Korreman et al., 2008; Keall, 2011; Korreman, 2015), and some of the comments in the previous section apply to it. It typically refers to consideration of intrafraction breathing motion (Keall, 2006). While other types of motions (swallowing, peristalsis, heart beat) also require attention, it is the regularity and often large amplitude motion due to breathing that is currently of most concern. Like most of the improvements in treatment delivery, motion management is linked to appropriate treatment-planning processes. Diagnostic and planning images acquired with consideration of motion are essential to enable motion-managed radiation delivery. In principle, three approaches are possible (Dawson and Balter, 2004).

1. An internal target volume (ITV) approach includes all possible locations of the target in the breathing cycle into the treatment volume (ICRU, 2000). A related method, the midventilation approach was proposed by the NKI group in Amsterdam, the Netherlands, where the probability of finding the target in a certain location is used to reduce the required margins (Wolthaus et al., 2006; Peulen et al., 2014). From a dosimetric perspective, an ITV can be problematic if the treatment is delivered using dynamic delivery techniques, such as sliding window IMRT, VMAT, or helical tomotherapy. In this case, "interplay" effects, where target and delivery move in synchrony, can occur (Riley et al., 2014; Tudor et al., 2014). In general, this leads to dose blurring in the target, but in multifraction treatments the overall effect is typically small (Duan et al., 2006). Care must be taken in single-fraction treatments, such as stereotactic approaches, and in circumstances where critical structures are close to the target (Ong et al., 2013; Zou et al., 2014) as discussed below.
2. A gated or breath-hold approach where radiation is only delivered when the target is in a prespecified location (Hanley et al., 1999). Deep inspiration breath-hold (DIBH) in particular has become a widely used technique to reduce heart dose when treating patients with left-side breast cancer (Korreman et al., 2006; Hjelstuen et al., 2012; Bartlett et al., 2013). In both cases, an indication of the breathing status of the patient typically via a surrogate marker is required. Gated and breath-hold target positions have some residual uncertainty but in general result in smaller margins in particular for targets with larger motion (Engelsman et al., 2005).

It also results in potentially, significantly increased treatment times. From a dosimetric perspective, the link of delivery with the breathing cycle is important and either integrative dosimeters such as radiochromic film, or detectors that can determine dose in real time, are helpful.

3. The third approach is called motion adaptive where the treatment field follows the target motion (Keall et al., 2001). This can be facilitated by moving the linac or using MLC motion. Another suggested approach is the use of treatment couch movement to maintain the target in the radiation field.

While the last approach is optimal both in terms of limiting margins and delivering the treatment quickly, it is most complex and as such requires more quality control, including dosimetric tools that can verify delivery. In principle, two approaches to motion adaptation are possible: patient or treatment field movement to compensate for target motion.

10.4.1 Moving Fields

From a patient perspective, movement of radiation fields appears to be a less invasive method. Three systems of this nature are clinically in use as follows:

- The CyberKnife shown in Figure 10.4 is mounted on a robot which allows following the target motion. Patients wear a reflective vest which allows a camera system to determine the phase of the breathing cycle of the patient. This can be updated with internal motion information at regular intervals using the ceiling mounted orthogonal x-ray units. The CyberKnife was the first real-time adaptation system.
- VERO allows a similar workflow for motion detection as the CyberKnife. However, as indicated in Figure 10.3, motion is accounted for by a gimballed linac that can follow the target motion.
- MLC tracking is in principle possible for all linacs with MLC. The concept was introduced in 2001 (Keall et al., 2001) and is recently realized (Colvill et al., 2014, 2015).

10.4.2 Moving Patient

Any internal motion can also be compensated for by moving the whole patient (Menten et al., 2012). In some radiotherapy applications employing radiation, beams cannot be moved due to the size of the generating apparatus and moving the patient is the only option. Negative pions (Skarsgard et al., 1980) and synchrotron x-rays (Brauer-Krisch et al., 2015) are examples of this. However, most treatment couches on recent radiotherapy equipment have at least three and often six degrees of motion (Hyde et al., 2012; Schmidhalter et al., 2013). Provided the patient is well immobilized (and informed), the couch can move to ensure the tumor is always in the radiation beam path (Buzurovic et al., 2012; Menten et al., 2012).

From a conceptual point, moving the patient has advantages in particular for dynamic treatment deliveries as the compensation of motion combined with a dynamic movement of MLC and gantry can lead to highly complex deliveries.

Dosimetry for moving fields or moving patients requires accounting for a pattern of dose delivery with time. Not only absolute dose and dose distribution need to be verified, but it is essential to also establish that the motion compensation is in synchronicity with the target movement. As an added complexity, patient breathing is rarely regular and dosimetry must ensure appropriate mechanisms to deal with such irregularities.

All motion management systems also have some degree of latency between the detection of motion and the ability of the system to compensate for it. This is typically of the order of a few hundred milliseconds (Hong et al., 2011; Rottmann and Berbeco, 2014), and one of the more important aspects of dosimetry is to determine and verify this lag. In any case, predictive models are used to predict on planning data and the latest information from the patient where a target would be at a particular point in time (Vedam et al., 2004; Hong et al., 2011; Poulsen et al., 2012; Rottmann and Berbeco, 2014). More details on 4D dosimetry issues required to assess motion-managed radiotherapy is provided in Chapter 11.

10.5 Dosimetric Tools for Special Techniques and Technologies

As mentioned in the introduction, many techniques are intrinsically linked to a single manufacturer's technology. As such, close collaboration with the manufacturer is required, which also pertains to dosimetry. For example, many of the modern delivery technologies are based on ring gantries. They do not accommodate the "classical" scanning water phantom, and other smaller devices have to be designed for relative and reference dosimetry. They often come customized for the purpose as the "cheese phantom" shown in connection with helical tomotherapy in Figure 10.2 (Fenwick et al., 2004; Balog and Soisson, 2008). This phantom with its cylindrical design is ideally suited for the delivery process in helical tomotherapy and has become the major tool for dosimetry. It is now even adapted for independent dose audits (Schiefer et al., 2015).

This machine-specific thinking also applies to reference dosimetry and modern approaches for reference dosimetry are often based on "machine-specific" reference fields (Alfonso et al., 2008). The required correction factors for this formalism are typically derived using Monte Carlo calculations (Cranmer-Sargison et al., 2011; Francescon et al., 2012).

For relative dosimetry, verification for special techniques and technologies often involves the assessment of treatment aspects that are otherwise ignored. Automatic couch movements during treatment for motion management, presence of magnetic fields, or delivery of large fields are good examples for this.

Customized phantoms and the use of EPI devices also have the potential of automating and streamlining dosimetric quality assurance. An example of the Truebeam linacs of Varian Medical Systems is the machine performance check (MPC) which automatically determines and records a large number of parameters that define machine performance in a few minutes (Clivio et al., 2015). It will be interesting to see how more of these methods are incorporated into future developments of technology and how operators can maintain a good understanding of the working of the machine if both delivery and verification are automatic.

Two additional tools must be mentioned. First, *in vivo* dosimetry, which is covered in Chapter 18, plays a particularly important role in the context of special techniques and technologies (Essers and Mijnheer, 1999; Mijnheer, 2008; Mijnheer et al., 2013). By their very nature, these approaches are not widely available and documentation such as protocols and guidelines are rare. In addition to this, many special techniques are complex as many components must work in synchronicity typically under computer control. All this increases the residual uncertainty for delivery, and *in vivo* dosimetry is one of the most direct tools to verify dose delivery from large fields (Bloemen-van Gurp et al., 2007) to dynamic treatments (Engstrom et al., 2005; Van Uytven et al., 2015) and intraoperative radiotherapy (Fogg et al., 2010; Eaton et al., 2012).

Second, Monte Carlo simulations play a fundamental role for dosimetry in general and for special techniques and technologies in particular (Seco and Verhaegen, 2013), and is the topic of Chapter 13. They are essential in characterizing the detectors as demonstrated in correction factors derived for reference dosimetry using ionization chambers (Francescon et al., 2012). Furthermore, Monte Carlo calculations allow determination of the anticipated radiation quality (e.g., x-ray spectrum) and angular distribution at a point of interest which would be helpful in interpreting dosimetric results. They are useful in designing dosimeters for specific purposes (Charles et al., 2013) and are also the method of choice for dose calculations in the presence of tissue heterogeneities (Disher et al., 2012; Zhuang et al., 2013) or magnetic fields (Oborn et al., 2009; Gargett et al., 2015). Both of these are often difficult to do with physical detectors and as such Monte Carlo simulations complement and enrich dose measurements.

10.6 Outlook

Radiotherapy without dosimetry and dose measurement is unthinkable as dose is the therapeutic agent and can be the cause of significant side effects. However, dose measurement has not been one of the major drivers for technology development. The only exception where dosimetry is actively integrated in the treatment unit is EPI (Herman, 2005; van Elmpt et al., 2008). While originally designed for imaging, EPI has become an essential tool for dosimetry. Even systems that have far superior imaging modalities, such as VERO (Solberg et al., 2014) or MRI linacs (Raaymakers et al., 2011), employ EPI. More information about EPI dosimetry is given in Chapter 7.

The lack of integrated, independent dosimetric systems reflects at least in part the overall excellent reliability of dose delivery methods, mostly based on well-established and controlled linear accelerator technology. However, the technologies discussed in this chapter and the development sketched in Table 10.1 introduce additional complexity in the way linacs are used, which would increase the risk of error (Klein et al., 2005). Dose measurement for commissioning and dose verification during at least some treatments remains an essential task for medical physicists.

If one considers the dose distributions achievable with modern radiotherapy equipment, there appear to be only a few remaining technical goals that would make significant difference to the delivery (both actual delivery and image guidance for targeting). Motion management and adaptation as discussed earlier are probably the most important future technical challenges. However, one can expect that other challenges will increase in relevance as follows:

1. **Access:** While the availability of quality radiotherapy equipment in the world increases, there is still a large gap between rich and poor countries (Kron et al., 2015).
2. **Cost:** On the other hand, health-care costs are also rapidly increasing in the developed world and radiotherapy professionals are increasingly confronted with the need to justify all expenditures. The cost of equipment

will be a major consideration in the future; this includes maintenance and associated costs (such as commissioning and quality assurance), and cheap equipment will not necessarily be the most cost effective. Medical physicists need to contribute to this discussion and wherever possible ask for the most efficient tools to perform tasks such as dosimetry.

3. **Computerization and automation:** Computerization and automation have been shown to make radiotherapy in general safer (Fraass, 2012). They also have the potential to reduce cost and increase complexity.
4. **Optimization:** Not all features are relevant for all patients and it can be expected that equipment will be customizable for need based on a robust standard base model.

Considering these changes, there is no doubt that the nature of dosimetry, i.e., the detector types, evaluation process, and data management will also change. Statistical process control and risk management thinking will affect the way dosimetry interfaces with radiotherapy delivery, particularly in the case of special techniques for which there may not be a recipe for the best dose measurement approach. However, dose measurement can be expected to stay.

References

Alaei P and Spezi E (2015) Imaging dose from cone beam computed tomographyin radiation therapy. *Phys. Med.* 31: 647–658.

Almond PR et al. (1999) AAPM's TG-51 protocol for clinical reference dosimetryof high-energy photon and electron beams. *Med. Phys.* 26: 1847–1870.

Alfonso R et al. (2008) A new formalism for reference dosimetry of small andnonstandard fields. *Med. Phys.* 35: 5179–5186.

Balog J and Soisson E (2008) Helical tomotherapy quality assurance. *Int. J. Radiat. Oncol. Biol. Phys.* 71: S113–S117.

Bartlett FR et al. (2013) The UK HeartSpare Study: Randomised evaluation of voluntary deep-inspiratory breath-hold in women undergoing breast radiotherapy. *Radiother. Oncol.* 108: 242–247.

Bauman G et al. (2005) Helical tomotherapy for craniospinal radiation. *Br. J. Radiol.* 78: 548–552.

Bissonnette JP, Moseley DJ and Jaffray DA (2008) A quality assurance program for image quality of cone-beam CT guidance in radiation therapy. *Med. Phys.* 35: 1807–1815.

Bloemen-van Gurp EJ et al. (2007) Total body irradiation, toward optimal individual delivery: Dose evaluation with metal oxide field effect transistors, thermoluminescence detectors, and a treatment planning system. *Int. J. Radiat. Oncol. Biol. Phys.* 69: 1297–1304.

Brauer-Krisch E et al. (2015) Medical physics aspects of the synchrotron radiation therapies: Microbeam radiation therapy (MRT) and synchrotron stereotactic radiotherapy (SSRT). *Phys. Med.* 31: 568–583.

Brualla-Gonzalez L et al. (2016) Dose rate dependence of the PTW 60019 micro-Diamond detector in high dose-per- pulse pulsed beams. *Phys. Med. Biol.* 61: N11–N19.

Burghelea M et al. (2014) Feasibility of using the Vero SBRT system for intracranial SRS. *J. Appl. Clin. Med. Phys.* 15(1): 90–99.

Buzurovic I et al. (2012) Implementation and experimental results of 4D tumor tracking using robotic couch. *Med. Phys.* 39: 6957–6967.

Calcerrada Diaz-Santos N et al. (2008) The safety and efficacy of robotic image-guided radiosurgery system treatment for intra- and extracranial lesions: A systematic review of the literature. *Radiother. Oncol.* 89: 245–253.

Charles PH et al. (2013) Monte Carlo-based diode design for correction-less small field dosimetry. *Phys. Med. Biol.*58: 4501–4512.

Cheung JY and Yu KN (2006) Rotating and static sources for gamma knife radiosurgery systems: Monte Carlo studies. *Med. Phys.* 33: 2500–2505.

Cho YB et al. (2010) Verification of source and collimator configuration for Gamma Knife Perfexion using panoramic imaging. *Med. Phys.* 37: 1325–1331.

Clivio A et al. (2015) Evaluation of the machine performance check application for TrueBeam linac. *Radiat. Oncol.* 10: 97.

Colvill E et al. (2014) DMLC tracking and gating can improve dose coverage for prostate VMAT. *Med. Phys.* 41: 091705.

Colvill E et al. (2015) Multileaf collimator tracking improves dose delivery for prostate cancer radiation therapy: Results of the first clinical trial. *Int. J. Radiat. Oncol. Biol. Phys.* 92: 1141–1147.

Constantin DE, Fahrig R and Keall PJ (2011) A study of the effect of in-line and perpendicular magnetic fields on beam characteristics of electron guns in medical linear accelerators. *Med. Phys.* 38: 4174–4185.

Corvo R et al. (2011) Helical tomotherapy targeting total bone marrow after total body irradiation for patients with relapsed acute leukemia undergoing an allogeneic stem cell transplant. *Radiother. Oncol.* 98: 382–386.

Cranmer-Sargison G et al. (2011) Implementing a newly proposed Monte Carlo based small field dosimetry formalism for a comprehensive set of diode detectors. *Med. Phys.* 38: 6592–6602.

Cury FL et al. (2006) Ultrasound-based image guided radiotherapy for prostate cancer: Comparison of cross-modality and intramodality methods for daily localization during external beam radiotherapy. *Int. J. Radiat. Oncol. Biol. Phys.* 66: 1562–1567.

Das IJ, Ding GX and Ahnesjo A (2008) Small fields: Nonequilibrium radiation dosimetry. *Med. Phys.* 35: 206–215.

Dawson LA and Balter JM (2004) Interventions to reduce organ motion effects in radiation delivery. *Semin. Radiat. Oncol.* 14: 76–80.

Dawson LA and Sharpe MB (2006) Image-guided radiotherapy: Rationale, benefits, and limitations. *Lancet Oncol.* 7: 848–858.

Dilmanian FA et al. (2001) Response of avian embryonic brain to spatially segmented x-ray microbeams. *Cell Mol. Biol.* 47: 485–493.

Ding GX and Coffey CW (2009) Radiation dose from kilovoltage cone beam computed tomography in an image-guided radiotherapy procedure. *Int. J. Radiat. Oncol. Biol. Phys.* 73: 610–617.

Ding GX and Coffey CW (2010) Beam characteristics and radiation output of a kilovoltage cone-beam CT. *Phys. Med. Biol.* 55: 5231–5248.

Ding GX and Munro P (2013) Radiation exposure to patients from image guidance procedures and techniques to reduce the imaging dose. *Radiother. Oncol.* 108: 91–98.

Disher B et al. (2012) An in-depth Monte Carlo study of lateral electron disequilibrium for small fields in ultra-low-density lung: Implications for modern radiation therapy. *Phys. Med. Biol.* 57: 1543–1559.

Drzymala RE, Wood RC and Levy J (2008) Calibration of the Gamma Knife using a new phantom following the AAPM TG51 and TG21 protocols. *Med. Phys.* 35: 514–521.

Duan J et al. (2006) Dosimetric and radiobiological impact of dose fractionation on respiratory motion induced IMRT delivery errors: A volumetric dose measurement study. *Med. Phys.* 33: 1380–1387.

Duchateau M et al. (2010) The effect of tomotherapy imaging beam output instabilities on dose calculation. *Phys. Med. Biol.* 55: N329–N336.

Eaton DJ (2015) Electronic brachytherapy—Current status and future directions. *Br. J. Radiol.* 88: 20150002.

Eaton DJ et al. (2012) In vivo dosimetry for single-fraction targeted intraoperative radiotherapy (TARGIT) for breast cancer. *Int. J. Radiat. Oncol. Biol. Phys.* 82: e819–e824.

Engelsman M et al. (2005) How much margin reduction is possible through gating or breath hold? *Phys. Med. Biol.* 50: 477–490.

Engstrom PE et al. (2005) In vivo dose verification of IMRT treated head and neck cancer patients. *Acta Oncol.* 44: 572–578.

Essers M and Mijnheer BJ (1999) In vivo dosimetry during external photon beam radiotherapy. *Int. J. Radiat. Oncol. Biol. Phys.* 43: 245–259.

Ezzell GA et al. (2009) IMRT commissioning: Multiple institution planning and dosimetry comparisons, a report from AAPM Task Group 119. *Med. Phys.* 36: 5359–5373.

Fallone BG (2009) First MR images obtained during megavoltage photon irradiation from a prototype integrated linac-MR system. *Med. Phys.* 36: 2084–2088.

Fenwick JD et al. (2004) Quality assurance of a helical tomotherapy machine. *Phys. Med. Biol.* 49: 2933–2953.

Fogg P et al. (2010) Thermoluminescence dosimetry for skin dose assessment during intraoperative radiotherapy for early breast cancer. *Australas. Phys. Eng. Sci. Med.* 33: 211–214.

Fogliata A et al. (2012) Definition of parameters for quality assurance of flattening filter free (FFF) photon beams in radiation therapy. *Med. Phys.* 39: 6455–6464.

Fraass BA (2012) Impact of complexity and computer control on errors in radiation therapy. *Ann. ICRP* 41: 188–196.

Francescon P et al. (2012) Monte Carlo simulated correction factors for machine specific reference field dose calibration and output factor measurement using fixed and iris collimators on the CyberKnife system. *Phys. Med. Biol.* 57: 3741–3758.

FRO (2014) Faculty of Radiation Oncology position paper: Techniques and technologies in radiation oncology. 2013 Horizon scan Australia. (Sydney: The Royal Australian and New Zealand College of Radiologists).

Gargett M et al. (2015) Monte Carlo simulation of the dose response of a novel 2D silicon diode array for use in hybrid MRILINAC systems. *Med. Phys.* 42: 856–865.

Georg D, Knoos T and McClean B (2011) Current status and future perspective of flattening filter free photon beams. *Med. Phys.* 38: 1280–1293.

Ghilezan M, Yan D and Martinez A (2010) Adaptive radiation therapy for prostate cancer. *Semin. Radiat. Oncol.* 20: 130–137.

Goetsch SJ et al. (1999) Physics of rotating gamma systems for stereotactic radiosurgery *Int. J. Radiat. Oncol. Biol. Phys.* 43: 689–696.
Han C, Schultheisss TE and Wong JY (2012) Dosimetric study of volumetric modulated arc therapy fields for total marrow irradiation. *Radiother. Oncol.* 102: 315–320.
Hanley J et al. (1999) Deep inspiration breath-hold technique for lung tumors: The potential value of target immobilization and reduced lung density in dose escalation. *Int. J. Radiat. Oncol. Biol. Phys.* 45: 603–611.
Herman MG (2005) Clinical use of electronic portal imaging. *Semin. Radiat. Oncol.* 15: 157–167.
Hjelstuen MH et al. (2012) Radiation during deep inspiration allows locoregional treatment of left breast and axillary-, supraclavicular- and internal mammary lymph nodes without compromising target coverage or dose restrictions to organs at risk. *Acta. Oncol.* 51: 333–344.
Hong SM, Jung BH and Ruan D (2011) Real-time prediction of respiratory motion based on a local dynamic model in an augmented space. *Phys. Med. Biol.* 56: 1775–1789.
Hu Y et al. (2015) Characterization of the onboard imaging unit for the first clinicalmagnetic resonance image guided radiation therapy system. *Med. Phys.* 42: 5828–5837.
Hugtenburg RP et al. (1994) Total-body irradiation on an isocentric linear accelerator: A radiation output compensation technique. *Phys. Med. Biol.* 39: 783–793.
Hui SK et al. (2005) Feasibility study of helical tomotherapy for total body or total marrow irradiation. *Med. Phys.* 32: 3214–3224.
Hussain A et al. (2011) Aperture modulated, translating bed total body irradiation. *Med. Phys.* 38: 932–941.
Hyde D et al. (2012) Spine stereotactic body radiotherapy utilizing cone-beam CT image-guidance with a robotic couch: Intrafraction motion analysis accounting for all six degrees of freedom. *Int. J. Radiat. Oncol. Biol. Phys.* 82: e555–e562.
IAEA (2000) Absorbed dose determination in external beam radiotherapy. Technical Report Series Report TRS 398. (Vienna, Austria: International Atomic Energy Agency).
ICRU (2000) Prescribing, recording, and reporting photon beam therapy (Supplement to ICRU Report 50), Report 62. (Bethesda, MD: International Commission on Radiological Units and Measurements).
IPEM (2010) Small field MV photon dosimetry. Report 103. (York, UK: Institute of Physics and Engineering in Medicine).
Jaffray DA and Siewerdsen JH (2000) Cone-beam computed tomography with a flat-panel imager: Initial performance characterization. *Med. Phys.* 27: 1311–1323.
Jeraj R et al. (2004) Radiation characteristics of helical tomotherapy. *Med. Phys.* 31: 396–404.
Jones R et al. (2014) Dosimetric comparison of (192)Ir high-dose-rate brachytherapy vs. 50 kV x-rays as techniques for breast intraoperative radiation therapy: Conceptual development of image-guided intraoperative brachytherapy using a multilumen balloon applicator and in-room CT imaging. *Brachytherapy* 13: 502–507.

Keall P (2011) Locating and targeting moving tumors with radiation beams. *Front. Radiat. Ther. Oncol.* 43: 118–131.

Keall PJ (2006) The management of respiratory motion in radiation oncology report of AAPM Task Group 76. *Med. Phys.* 33: 3874–3900.

Keall PJ, Kini VR, Vedam SS and Mohan R (2001) Motion adaptive x-ray therapy: A feasibility study *Phys. Med. Biol.* 46: 1–10.

Kilby W et al. (2010) The CyberKnife robotic radiosurgery system in 2010. *Technol. Cancer Res. Treat.* 9: 433–452.

Kirkby C et al. (2010) Lung dosimetry in a linac- MRI radiotherapy unit with a longitudinal magnetic field. *Med. Phys.* 37: 4722–4732.

Klein EE et al. (2005) Errors in radiation oncology: A study in pathways and dosimetric impact. *J. Appl. Clin. Med. Phys.* 6(3): 81–94.

Korreman SS et al. (2006) Reduction of cardiac and pulmonary complication probabilities after breathing adapted radiotherapy for breast cancer. *Int. J. Radiat. Oncol. Biol. Phys.* 65: 1375–1380.

Korreman SS et al. (2008) The role of image guidance in respiratory gated radiotherapy. *Acta Oncol.* 47: 1390–1396.

Korreman SS (2015) Image-guided radiotherapy and motion management in lung cancer. *Br. J. Radiol.* 88: 20150100.

Kron T et al. (1993) Clinical thermoluminescence dosimetry: How do expectations and results compare? *Radiother. Oncol.* 26: 151–161.

Kron T et al. (2006) Magnetic resonance imaging for adaptive cobalt tomotherapy: A proposal. *J. Med. Phys.* 31: 242–254.

Kron T et al. (2010) Adaptive radiotherapy for bladder cancer reduces integral dose despite daily volumetric imaging. *Radiother. Oncol.* 97: 485–487.

Kron T et al. (2012) Credentialing of radiotherapy centres for a clinical trial of adaptive radiotherapy for bladder cancer (TROG 10.01). *Radiother. Oncol.* 103: 293–298.

Kron T, Taylor ML and Thwaites D (2013) Small field dosimetry. *In: Modern Technology of Radiation Oncology,* Vol 3, pp. 245–299. (Ed. Van Dyk J, Madison, WI: Medical Physics Publishing).

Kron T et al. (2015) Medical physics aspects of cancer care in the Asia Pacific region: 2014 survey results. *Australas. Phys. Eng. Sci. Med.* 38: 493–501.

Kry SF et al. (2012) Ion recombination correction factors (P(ion)) for Varian TrueBeam high-dose-rate therapy beams. *J. Appl. Clin. Med. Phys.* 13(6): 318–325.

Kubo HD and Araki F (2002) Dosimetry and mechanical accuracy of the first rotating gamma system installed in North America. *Med. Phys.* 29: 2497–2505.

Lagendijk JJ et al. (2008) MRI/linac integration. *Radiother. Oncol.* 86: 25–29.

Lavallee MC et al. (2009) Commissioning and evaluation of an extended SSD photon model for PINNACLE3: An application to total body irradiation. *Med. Phys.* 36: 3844–3855.

Lechner W et al. (2013) Detector comparison for small field output factor measurements in flattening filter free photon beams. *Radiother. Oncol.* 109: 356–360.

Lehmann J et al. (2014) Angular dependence of the response of the nanoDot OSLD system for measurements at depth in clinical megavoltage beams. *Med. Phys.* 41: 061712.

Li HH et al. (2015) Patient-specific quality assurance for the delivery of (60)Co intensity modulated radiation therapy subject to a 0.35-T lateral magnetic field. *Int. J. Radiat. Oncol. Biol. Phys.* 91: 65–72.

Mackie TR (2006) History of tomotherapy. *Phys. Med. Biol.* 51: R427–R453.

Mackie TR et al. (1993) Tomotherapy: A new concept for the delivery of dynamic conformal radiotherapy. *Med. Phys.* 20: 1709–1719.

Mancosu P et al. (2015) Plan robustness in field junction region from arcs with different patient orientation in total marrow irradiation with VMAT. *Phys. Med.* 31: 677–682.

Meijsing I et al. (2009) Dosimetry for the MRI accelerator: The impact of a magnetic field on the response of a Farmer NE2571 ionization chamber. *Phys. Med. Biol.* 54: 2993–3002.

Menten MJ et al. (2012) Comparison of a multileaf collimator tracking system and a robotic treatment couch tracking system for organ motion compensation during radiotherapy. *Med. Phys.* 39: 7032–7041.

Mijnheer B (2008) State of the art of *in vivo* dosimetry. *Radiat. Prot. Dos.* 131: 117–122.

Mijnheer B et al. (2013) *In vivo* dosimetry in external beam radiotherapy. *Med. Phys.* 40: 070903.

Moran JM, Elshaikh MA and Lawrence TS (2005) Radiotherapy: What can be achieved by technical improvements in dose delivery? *Lancet Oncol.* 6: 51–58.

Murphy MJ et al. (2007) The management of imaging dose during image-guided radiotherapy: Report of the AAPM Task Group 75. *Med. Phys.* 34: 4041–4063.

Mutic S and Dempsey JF (2014) The ViewRay system: Magnetic resonance-guided and controlled radiotherapy. *Semin. Radiat. Oncol.* 24: 196–199.

Nettelbeck H, Takacs GJ and Rosenfeld AB (2008) Effect of transverse magnetic fields on dose distribution and RBE of photon beams: Comparing PENELOPE and EGS4 Monte Carlo codes. *Phys. Med. Biol.* 53: 5123–5137.

Nguyen BT et al. (2012) Optimising the dosimetric quality and efficiency of post-prostatectomy radiotherapy: A planning study comparing the performance of volumetric-modulated arc therapy (VMAT) with an optimised seven-field intensity-modulated radiotherapy (IMRT) technique. *J. Med. Imaging Radiat. Oncol.* 56: 211–219.

Nguyen D (2014) Integral dose investigation of non-coplanar treatment beam geometries in radiotherapy. *Med. Phys.* 41: 011905.

Oborn BM et al. (2009) High resolution entry and exit Monte Carlo dose calculations from a linear accelerator 6 MV beam under the influence of transverse magnetic fields. *Med. Phys.* 36: 3549–3559.

Ong CL et al. (2013) Dosimetric impact of intrafraction motion during RapidArc stereotactic vertebral radiation therapy using flattened and flattening filter-free beams. *Int. J. Radiat. Oncol. Biol. Phys.* 86: 420–425.

Orecchia R and Veronesi U (2005) Intraoperative electrons. *Semin. Radiat. Oncol.* 15: 76–83.

Otto K (2008) Volumetric modulated arc therapy: IMRT in a single gantry arc. *Med. Phys.* 35: 310–317.

Peulen H et al. (2014) Mid-ventilation based PTV margins in stereotactic body radiotherapy (SBRT): A clinical evaluation. *Radiother. Oncol.* 110: 511–516.

Popescu CC et al. (2010) Volumetric modulated arc therapy improves dosimetry and reduces treatment time compared to conventional intensity-modulated radiotherapy for locoregional radiotherapy of left-sided breast cancer and internal mammary nodes. *Int. J. Radiat. Oncol. Biol. Phys.* 76: 287–295.

Poulsen PR et al. (2012) Image-based dynamic multileaf collimator tracking of moving targets during intensity-modulated arc therapy. *Int. J. Radiat. Oncol. Biol. Phys.* 83: e265–e271.

Quast U (1987) Total body irradiation—Review of treatment techniques in Europe. *Radiother. Oncol.* 9: 91–106.

Raaijmakers AJ, Raaymakers BW and Lagendijk JJ (2005) Integrating a MRI scanner with a 6 MV radiotherapy accelerator: Dose increase at tissue-air interfaces in a lateral magnetic field due to returning electrons. *Phys. Med. Biol.* 50: 1363–1376.

Raaijmakers AJ, Raaymakers BW and Lagendijk JJ (2007) Experimental verification of magnetic field dose effects for the MRI-accelerator. *Phys. Med. Biol.* 52: 4283–4291.

Raaijmakers AJ, Raaymakers BW and Lagendijk JJ (2008) Magnetic-field-induced dose effects in MR-guided radiotherapy systems: Dependence on the magnetic field strength. *Phys. Med. Biol.* 53: 909–923.

Raaymakers BW et al. (2009) Integrating a 1.5 T MRI scanner with a 6 MV accelerator: Proof of concept. *Phys. Med. Biol.* 54: N229–N237.

Raaymakers BW et al. (2011) Integrated megavoltage portal imaging with a 1.5 T MRI linac. *Phys. Med. Biol.* 56: N207–N214.

Riley C et al. (2014) Dosimetric evaluation of the interplay effect in respiratory-gated RapidArc radiation therapy. *Med. Phys.* 41: 011715.

Roshau JN and Hintenlang DE (2003) Characterization of the angular response of an "isotropic" MOSFET dosimeter. *Health Phys.* 84: 376–379.

Rottmann J and Berbeco R (2014) Using an external surrogate for predictor model training in real-time motion management of lung tumors. *Med. Phys.* 41: 121706.

Ruchala KJ et al. (1999) Megavoltage CT on a tomotherapy system. *Phys. Med.* Biol. 44: 2597–2621.

Ruschin M et al. (2010) Performance of a novel repositioning head frame for Gamma Knife Perfexion and image-guided linac-based intracranial stereotactic radiotherapy. *Int. J. Radiat. Oncol. Biol. Phys.* 78: 306–313.

Sanchez-Nieto B et al. (1993) Backscatter correction algorithm for TBI treatment conditions. *Med. Dosim.* 18: 107–111.

Schiefer H et al. (2015) Design and implementation of a "cheese" phantom-based Tomotherapy TLD dose intercomparison. *Strahlenther. Onkol.* 191: 855–861.

Schmidhalter D et al. (2013) Evaluation of a new six degrees of freedom couch for radiation therapy. *Med. Phys.* 40: 111710.

Seco J and Verhaegen F (2013) *Monte Carlo Techniques in Radiation Therapy.* (Boca Raton, FL: CRC Press).

Simpson DR et al. (2010) A survey of image-guided radiation therapy use in the United States. *Cancer* 116: 3953–3960.

Skarsgard LD, Henkelman RM and Eaves CJ (1980) Pions for radiotherapy at TRIUMF. *J. Can. Assoc. Radiol.* 31: 3–12.

Skarsgard D et al. (2010) Planning target volume margins for prostate radiotherapy using daily electronic portal imaging and implanted fiducial markers. *Radiat. Oncol.* 5: 52.

Slatkin DN et al. (1992) Microbeam radiation therapy. *Med. Phys.* 19: 1395–1400.

Smit K et al. (2013) Towards reference dosimetry for the MR-linac: Magnetic field correctionof the ionization chamber reading. *Phys. Med.* Biol. 58: 5945–5957.

Smit K et al. (2014) Relative dosimetry in a 1.5 T magnetic field: An MR-linac compatible prototype scanning water phantom. *Phys. Med. Biol.* 59: 4099–4109.

Solberg TD et al. (2014) Commissioning and initial stereotactic ablative radiotherapy experience with Vero. *J. Appl. Clin. Med. Phys.* 15(2): 205–225.

Suchowerska N et al. (2001) Directional dependence in film dosimetry: Radiographic and radiochromic film. *Phys. Med.* Biol. 46: 1391–1397.

Tan J et al. (2013) Prospective single-arm study of intraoperative radiotherapy for locally advanced or recurrent rectal cancer. *J. Med. Imaging Radiat. Oncol.* 57: 617–625.

Teoh M et al. (2011) Volumetric modulated arc therapy: A review of current literature and clinical use in practice. *Br. J. Radiol.* 84: 967–996.

Timmerman RD et al. (2007) Stereotactic body radiation therapy in multiple organ sites. *J. Clin. Oncol.* 25: 947–952.

Tome WA et al. (2002) Commissioning and quality assurance of an optically guided three-dimensional ultrasound target localization system for radiotherapy. *Med. Phys.* 29: 1781–1788.

Tudor GS, Harden SV and Thomas SJ (2014) Three-dimensional analysis of the respiratory interplay effect in helical tomotherapy: Baseline variations cause the greater part of dose inhomogeneities seen. *Med. Phys.* 41: 031704.

Vaidya J, Joseph D and Tobias J (2010) Targeted intraoperative radiotherapy versus whole breast radiotherapy for breast cancer (TARGIT-A trial): An international, prospective, randomised, non-inferiority phase 3 trial. *Lancet* 376(9735): 91–102.

Van Dyk J (1987) Dosimetry for total body irradiation. *Radiother. Oncol.* 9: 107–118.

van Elmpt W et al. (2008) A literature review of electronic portal imaging for radiotherapy dosimetry. *Radiother. Oncol.* 88: 289–309.

Van Uytven E et al. (2015) Validation of a method for in vivo 3D dose reconstruction for IMRT and VMAT treatments using on-treatment EPID images and a model-based forward-calculation algorithm. *Med. Phys.* 42: 6945–6954.

Vassiliev ON et al. (2012) Dosimetric impact of fiducial markers in patients undergoing photon beam radiation therapy. *Phys. Med.* 28: 240–244.

Vedam SS et al. (2004) Predicting respiratory motion for four-dimensional radiotherapy. *Med. Phys.* 31: 2274–2283.

Veronesi U et al. (2013) Intraoperative radiotherapy versus external radiotherapy for early breast cancer (ELIOT): A randomised controlled equivalence trial. *Lancet Oncol.* 14: 1269–1277.

Wang Y, Easterling SB and Ting JY (2012) Ion recombination corrections of ionization chambers in flattening filter-free photon radiation. *J. Appl. Clin. Med. Phys.* 13(5): 262–268.

Webb S (2005) Intensity-modulated radiation therapy (IMRT): A clinical reality for cancer treatment, "any fool can understand this". *Br. J. Radiol.* 78(Spec No 2): S64–S72.

Wilkie JR et al. (2008) Feasibility study for linac-based intensity modulated total marrow irradiation. *Med. Phys.* 35: 5609–5618.

Wolff D et al. (2009) Volumetric modulated arc therapy (VMAT) vs. serial tomotherapy, step-and-shoot IMRT and 3D-conformal RT for treatment of prostate cancer. *Radiother. Oncol.* 93: 226–233.

Wolthaus JW et al. (2006) Mid-ventilation CT scan construction from fourdimensional respiration-correlated CT scans for radiotherapy planning of lung cancer patients. *Int. J. Radiat. Oncol. Biol. Phys.* 65: 1560–1571.

Wooten HO et al. (2015) Quality of intensity modulated radiation therapy treatment plans using a (60)Co magnetic resonance image guidance radiation therapy system. *Int. J. Radiat. Oncol. Biol. Phys.* 92: 771–778.

Wu A et al. (1990) Physics of Gamma Knife approach on convergent beams in stereotactic radiosurgery. *Int. J. Radiat. Oncol. Biol. Phys.* 18: 941–949.

Wu QJ, Li T, Wu Q and Yin FF (2011) Adaptive radiation therapy: Technical components and clinical applications. *Cancer J.* 17: 182–189.

Yu C, Petrovich Z and Luxton G (2000) Quality assurance of beam accuracy for Leksell Gamma Unit *J. Appl. Clin. Med. Phys.* 1(1): 28–31.

Zhuang T et al. (2013) Dose calculation differences between Monte Carlo and pencil beam depend on the tumor locations and volumes for lung stereotactic body radiation therapy. *J. Appl. Clin. Med. Phys.* 14(2): 38–51.

Zou W et al. (2014) Dynamic simulation of motion effects in IMAT lung SBRT. *Radiat. Oncol.* 9: 225.

4D Dosimetry

Emma Colvill, Jeremy T. Booth, and Paul Keall

11.1 What Is 4D Dosimetry?

Four-dimensional (4D) dosimetry is the measurement of dose in three dimensions while the detector positions are changing with time during the dose delivery. The rationale for performing 4D dosimetry is that human anatomy, and specifically that of cancer patients, changes with time. These temporal changes in the three-dimensional (3D) anatomy have driven the development of 4D dosimetry.

The ideal detector for 4D dosimetry would be an anthropomorphic phantom with tissue densities to match that of a patient. The phantom would be deformable (both target and surrounding normal tissues) with programmable target motion (both translation and rotation), while external motion such as the chest used for surrogate measurements would also be programmable. The detector system itself would be 3D and have high spatial and dosimetric accuracy, and would supply information for both the target volume and the normal tissues. Under this definition, the requirement for 4D dosimetry includes a volumetric (3D) detector, where the detector positions are mechanically driven in a manner that mimics human cancer and/or normal tissue motion during radiotherapy treatments. All current 4D detector systems are an approximation of this.

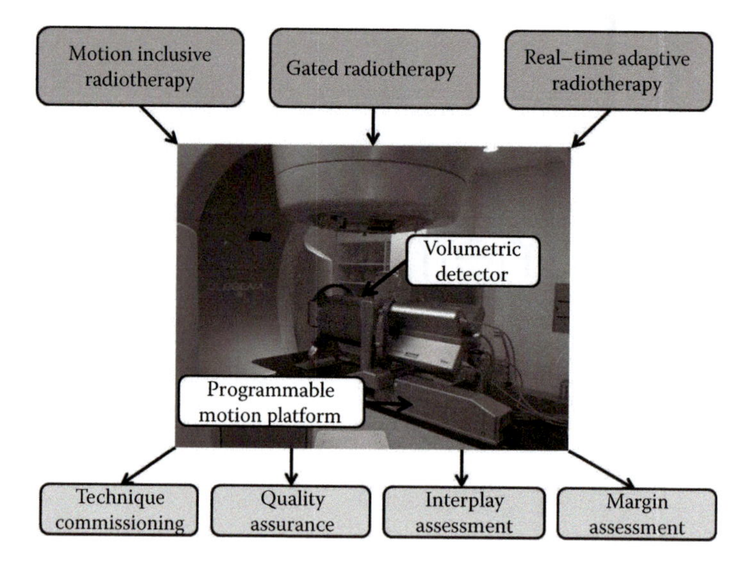

Figure 11.1

Illustration of the use of 4D dosimetry for three types of radiotherapy techniques that take temporal changes in the 3D anatomy into account.

The scope of this chapter includes photon-beam delivery methods in which anatomic motion is implicit and covered in margins (motion inclusive), as well as delivery methods in which anatomic motion is explicitly accounted for, i.e., gating and real-time adaptive radiotherapy on conventional, robotic, and gimballed linear accelerators. Excluded from this chapter are 4D dosimetry approaches for proton and carbon ion beams (see Chapters 12 and 23), and abdominal compression and breath-hold techniques that reduce motion.

The three types of radiotherapy techniques for which 4D dosimetry is used are motion-inclusive (i.e., using margins), gated, and real-time adaptive radiotherapy as indicated in Figure 11.1. 4D dosimetry can be used for the commissioning and quality assurance of these techniques, and the assessment of interplay effects and margins.

11.2 Why Is 4D Dosimetry Needed?

4D dosimetry is needed because patient motion creates one of the largest uncertainties in the treatment chain, along with contouring uncertainties (Groenendaal et al., 2010; Louie et al., 2010; Khoo et al., 2012). Since the routine application of image-guided radiation therapy (IGRT) began, including marker or soft-tissue matching, setup and interfraction motion uncertainties can be minimized (Grills et al., 2008; Guckenberger et al., 2008; Barney et al., 2011). Intrafraction organ motion during radiotherapy is often due to respiratory motion, affecting thoracic and abdominal treatments (Keall et al., 2006c; Bissonnette et al., 2009; Case et al., 2010; Whitfield et al., 2012). Treatments in the pelvic region, such as for the prostate, are often affected by bladder or rectal changes throughout treatment delivery (Langen et al., 2008; Adamson and Wu, 2010).

The geometric changes caused by intrafraction motion translate to dosimetric uncertainties (Li et al., 2008a; Seco et al., 2008; Langen et al., 2012), which can, along with residual interfraction motion, be accounted for in a number of ways. The most common method is via the use of treatment margins, i.e., motion-inclusive treatments (van Herk et al., 2000; Litzenberg et al., 2006; Li et al., 2011; Whitfield et al., 2012), which means that additional dose is delivered to the surrounding normal tissues to ensure dosimetric coverage of the target. Gating of the treatment beam when the observed motion of the target (or a surrogate, such as abdominal motion) exceeds a certain threshold, based on either displacement or phase (Hugo et al., 2002; Keall et al., 2006a; Li et al., 2006), allows for a potential reduction in the treatment margins as any larger motions result in a beam-hold. Another way of accounting for motion uncertainties is through real-time adaptive radiotherapy techniques that require the continual observation of and adaptation to the target motion throughout treatment beam delivery. Methods for real-time adaptation were discussed in more detail in Chapter 10 and include the following:

- Robotic tracking by the CyberKnife system, which involves the manipulation of the linear accelerator (Koong et al., 2004; Dieterich and Pawlicki, 2008; Hoogeman et al., 2009; Malinowski et al., 2012; Poels et al., 2014).
- Gimballed tracking using the Vero system, where the linear accelerator is moving in a specific way (Depuydt et al., 2011, 2014; Mukumoto et al., 2012; Poels et al., 2013).
- Multileaf collimator (MLC) tracking, which shifts the beam (Keall et al., 2006b, 2014; Falk et al., 2010; Krauss et al., 2011; Fast et al., 2014).
- Couch tracking, where the patient position is adjusted (D'Souza et al., 2005; Menten et al., 2012; Wilbert et al., 2013; Lang et al., 2014).

4D dosimetry can be applied in the commissioning and quality assurance for all of these techniques that account for interfraction and intrafraction motions. It is important to establish that the methods ensure dosimetric coverage of the treatment target (Keall et al., 2006c; Benedict et al., 2010). 4D dosimetry is used for technique commissioning and quality assurance through end-to-end testing and can also be used to assess the adequacy of margins. The effects of interplay between the treatment beam and the target caused by the relative motions of the machine and target (Bortfeld et al., 2004; Court et al., 2010a; Thomas et al., 2013) can also be assessed using 4D dosimetry (see Figure 11.2). Understanding the interplay effect on dose delivery is important as it cannot be accounted for through the addition of treatment margins. The use of 4D dosimetry to validate dose reconstruction methods is also valuable, as dose reconstruction allows for augmentation of the 4D dosimetry process. Figure 11.2 shows schematically the interplay between organ motion and leaf motion for the delivery of intensity-modulated radiation therapy (IMRT) with an MLC. The leaves move from left to right. The star symbolizes a point in an organ that moves up and down. The two different versions of the star represent two different phases of the motion (let us say that the filled star represents exhalation and the open star represents inhalation). Depending on the phase relative to the leaf motion, the point can receive very different dose values. In the phase shown by the filled star, the point does not receive any primary dose between time t_1 and t_2, and it may in fact receive no primary dose at all. In the phase symbolized by the open star, in which the point

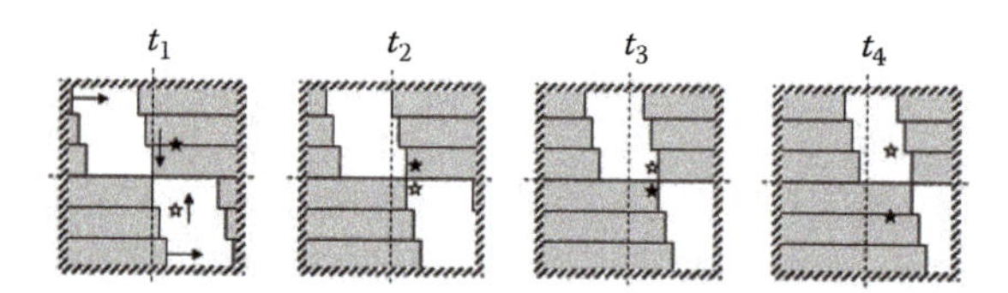

Figure 11.2

Illustration of the interplay between organ motion and leaf motion for the delivery of intensity-modulated radiation therapy (IMRT) with a multileaf collimator (MLC). (From Bortfeld T et al., *Phys. Med. Biol.*, 47, 2303–2320, 2002. With permission.)

moves up between t_1 and t_2, it is treated with the full primary dose at all times. This example, of course, represents the extreme limit of possible dose variation and such extremes are certainly relatively improbable.

11.3 How Is 4D Dosimetry Performed?

4D dosimetry requires the addition of motion into the dosimetry chain. To perform 4D dosimetry, you need a detector and a programmable motion phantom. Treatments are delivered to a moving detector and the measured doses are compared to either the planned dose distribution or a static dose measurement. Different applications of 4D dosimetry have different equipment requirements including detector dimensions and accuracy and motion requirements.

11.3.1 Volumetric Detectors with Programmable Phantoms

4D dosimetry is commonly performed for both experimental and clinical reasons using volumetric detectors moved by programmable phantoms. The volumetric detectors used for 4D dosimetry are most often diode or chamber detector arrays including the Delta^4, ArcCHECK, and OCTAVIUS (see Figure 11.3 and Chapter 8). These detectors interpolate from a single 2D rotating, curved or multiplane array dose measurement to 3D dose distributions, and allow for comparison between the measured 4D dose distributions and the planned or static measured dose distributions. Stacked or orthogonal film dosimeters are also often used for volumetric dose measurements (Dieterich et al., 2011; Chan et al., 2013; Garibaldi et al., 2015).

Programmable phantoms used for 4D dosimetry are designed for data input of 3D target motion; some also have a programmable fourth external motion stage to mimic surrogate motion used in radiotherapy treatment delivery of upper abdominal or thoracic sites such as the chest wall (Zhou et al., 2004; Park et al., 2013). Phantoms with a greater number of degrees of freedom are being developed, such as the HexaMotion system (ScandiDos, Uppsala, Sweden), which has programmable rotation about tilt and roll, as well as the three translation target dimensions. Most commercially available programmable platforms are designed for use with specific detector systems. The HexaMotion is designed for use with the Delta^4 detector (Pommer et al., 2013; Colvill et al., 2014) but can be used with other detectors (Jönsson et al., 2014; Anneli et al., 2015). Other programmable platforms (mostly custom-made) have been designed to be more versatile (Malinowski et al., 2007a; Nakayama et al., 2008; Davies et al., 2013).

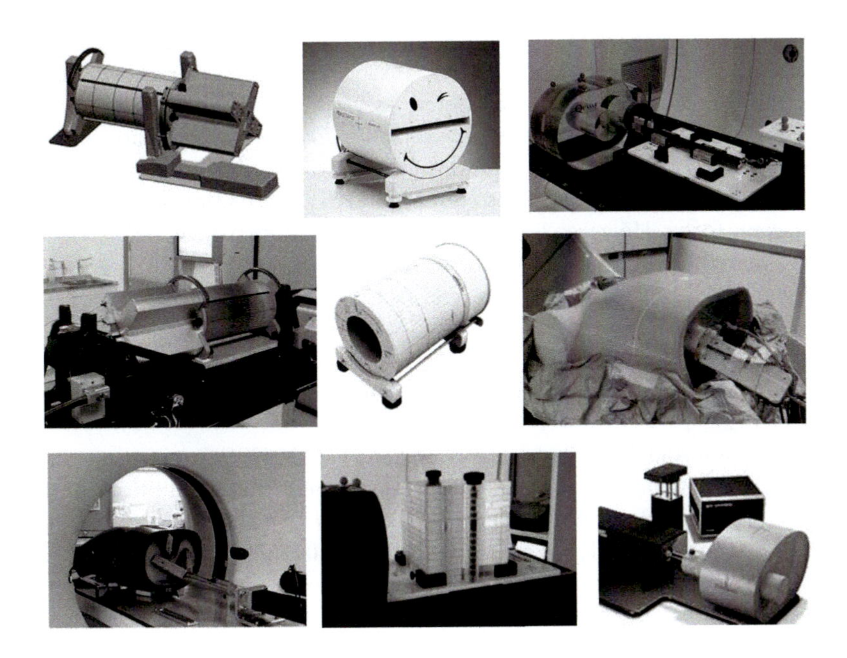

Figure 11.3

Volumetric detectors and motion platforms used for 4D dosimetry.

11.3.2 Programmable Anthropomorphic Phantoms

Anthropomorphic programmable phantoms, which are heterogeneous, are often used for treatments such as lung where there are large differences in tissue density through which the radiation travels. Deformable anthropomorphic programmable thorax phantoms commonly work by moving the detector/tumor insert itself, or via moving an artificial diaphragm, or inflating and deflating the lungs of the phantom with a programmed (1D) motion (Kashani et al., 2007; Court et al., 2010b; Cherpak et al., 2011; Steidl et al., 2012; Mayer et al., 2015). This motion in turn moves the chest wall and lungs. Nondeformable thoracic phantoms are most commonly fixed solid thorax body with varied density for lung, bone, and soft tissue, with a programmable insert that moves within the phantom. This input motion can result in either a 1D motion of the detector within the phantom or, depending on the design, a 2D or 3D motion (Chan et al., 2012, 2013).

The detectors used within heterogeneous or anthropomorphic phantoms are most often multiple films, though gel dosimetry systems have also been developed (Ceberg et al., 2013; De Deene et al., 2015). A limitation of these phantoms is that the detectors, such as the film, are often in the target volume alone so that any dose delivered outside of the target is not measured. Also, due to the detectors most often being film, the measurements obtained yield integral dose only. A collection of volumetric detectors and motion platforms used for 4D dosimetry is shown in Figure 11.3.

11.3.3 Other Motion-Inclusive Dosimetry Systems

Any physical detector that can be mounted on a motion platform (electronic portal imaging devices [EPIDs] or transmission type of detectors are not suitable) has the potential to be used for motion-inclusive dosimetry purposes. This includes 2D arrays (Sawant et al., 2008; Court et al., 2010a) and point detectors such as ionization chambers and diodes. These systems, however, move further away from the ideal 4D dosimetry system as fewer physical dimensions of the detectors could mean that potentially valuable dosimetric information is not measured. An exception to this is gel dosimetry (Ceberg et al., 2013; De Deene et al., 2015), which has high spatial resolution, a full-3D distribution, and is deformable in some forms. However, gel dosimetry has limitations, which means it is less utilized than other dosimetry systems to date (see Chapters 5 and 6).

Many "4D dosimetry" studies have also used motion platforms using fewer than three dimensions of motion, including the following:

- 1D programmable platforms (Smith and Becker, 2009; Abdellatif et al., 2012)
- 1D sinusoidal motion platforms (Chen et al., 2009; Court et al., 2010a)
- Combinations of 1D sinusoidal platforms resulting in elliptical motion (Sawant et al., 2008)
- 2D programmable platforms (Hugo et al., 2002; Park et al., 2009)

These systems are further from the ideal 4D dosimetry system. However all systems are approximations, and generally selecting the largest dimension of motion, often the superior–inferior direction, can be a good approximation for respiratory motion such as that seen in the thoracic treatments.

11.3.4 Data for 4D Dosimetry

The data required for 4D dosimetry include treatment plans and motion traces as shown in Figure 11.4. Treatment plans either can be created on patient computed tomography (CT) and structure sets and a verification plan transferred to that of the detector or dosimetry phantom, or can be planned directly on the detector/phantom CT sets. Once the treatment plans are created and ready to deliver to the dosimeter and phantom, motion traces are required as input into the motion platform to mechanically drive the detector to mimic organ or target motion. Ideally, the motion traces are direct patient measurements for the cancer site under investigation, for example, lung and prostate; however, the complexity of anatomic motion is often simplified due to the limitations of the motion stage.

11.4 Applications of 4D Dosimetry for Motion-Inclusive, Gated, and Real-Time Adaptive Radiotherapy

11.4.1 4D Dosimetry for Motion-Inclusive Radiotherapy

The most common method of accounting for intrafraction motion is the use of motion-inclusive treatments, which involves planning on 3D image sets and delivering that 3D plan to the patient. 4D image sets can be used for motion estimation and margin creation for treatment sites with respiratory motion such as lung, liver, and pancreas. 4D dosimetry can be used to assess motion-inclusive radiotherapy through assessing internal target volume (ITV), maximum intensity projection (MIP) (Huang et al., 2010), and planning target volume (PTV) margins. 4D dosimetry can also be used to assess the effect of interplay on treatments (Nguyen et al., 2013), which cannot be corrected for by margins. 4D dosimetry for assessing the adequacy of margins for target coverage, the interplay effect, along with commissioning and quality assurance of motion-inclusive

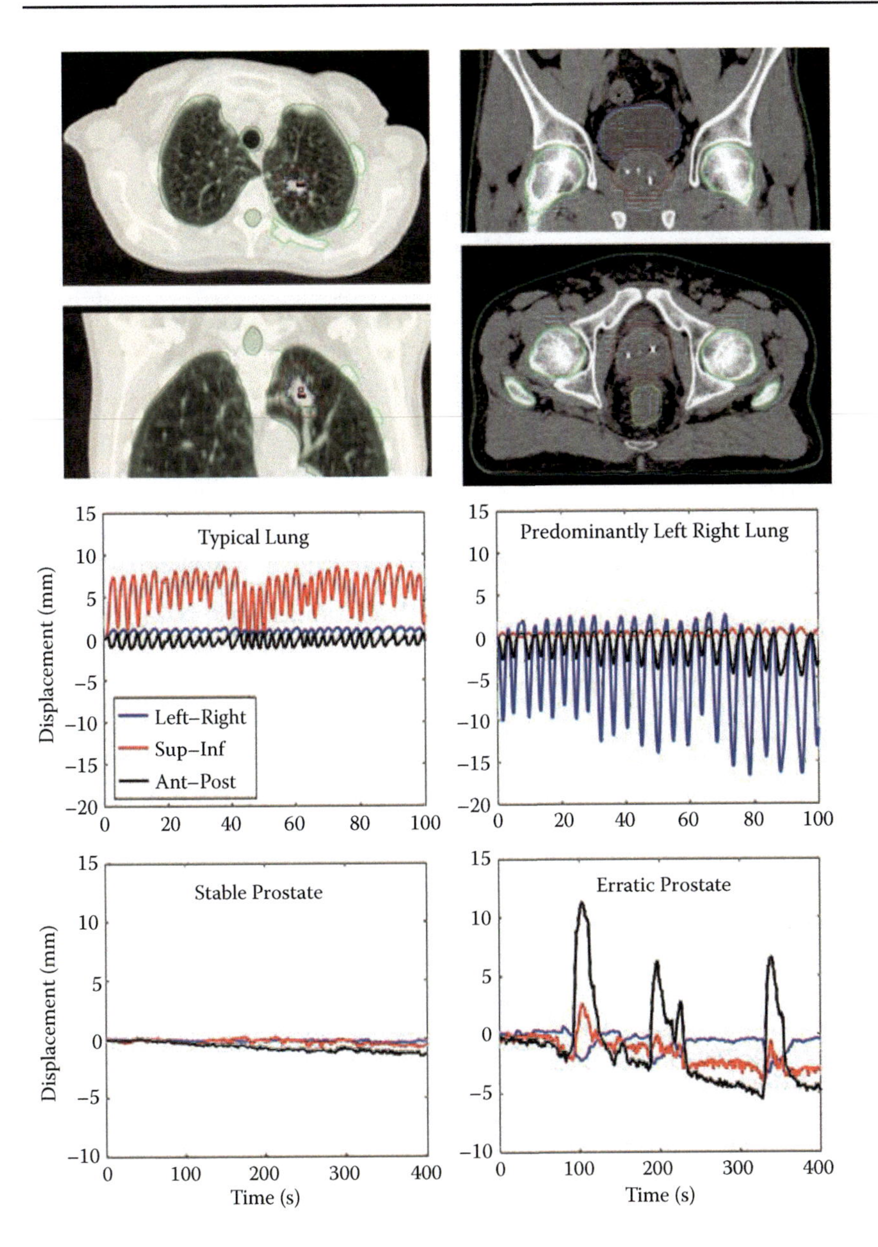

Figure 11.4

Example of data required for 4D dosimetry, which includes plans created using CT and structure sets (top) and patient-measured motion traces (bottom).

radiotherapy techniques, requires a detector with programmed 3D target motion, either a volumetric detector with a programmable motion platform or a programmable anthropomorphic phantom.

Many 4D dosimetry studies have been performed for motion-inclusive radiotherapy treatment techniques, including 3D conformal radiation therapy (3D-CRT) (Court et al., 2010b; Huang et al., 2010), IMRT (Thomas et al., 2013), and volumetric-modulated arc therapy (VMAT) (Court et al., 2010a; Ong et al., 2011; Nguyen et al., 2013). These studies used 4D dosimetry to commission

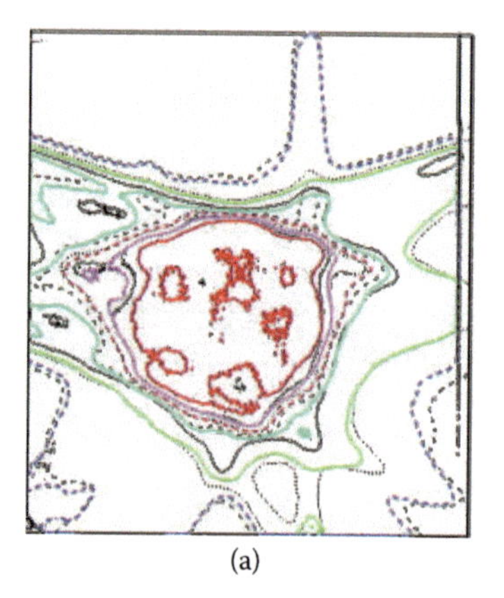

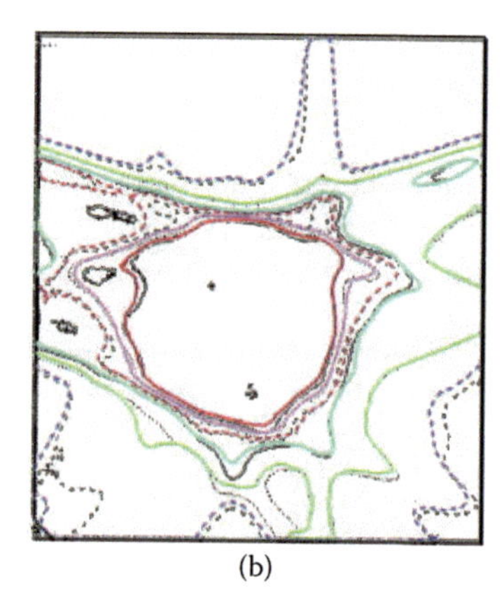

Figure 11.5

Isodose lines on the central axial plane of the moving phantom (color lines) for (a) one and (b) five fractions overlaying with those of the stationary phantom (black lines). Different solid and dashed lines represent different isodose values. (From Duan J et al., *Med. Phys.*, 33, 1380–1387, 2006. With permission.)

treatment-delivery techniques, assess margins, and evaluate the effects of interplay for treatment sites such as lung and liver. The results show that the effect of interplay can be important for individual treatment fractions with large motion and can be reduced with increased number of beams or arcs in each treatment and with increased fractionation (see Figure 11.5).

11.4.2 4D Dosimetry for Gated Radiotherapy

Gated radiotherapy is a treatment-delivery technique during which beam-holds are applied when the observed motion of the target (or a surrogate, such as abdominal motion) exceeds a predetermined threshold (Hugo et al., 2002; Keall et al., 2006a; Li et al., 2006, 2008a; Langen et al., 2012). Gated radiotherapy potentially allows for the reduction of margins. Commissioning and quality assurance should be performed where changes to standard margins and delivery are applied when using gating. 4D dosimetry is used to assess the effect of beam-holds on treatment dose delivery and the adequacy of margins, as well as the interplay effect, which is reduced due to less motion when treatment beam is on.

Gating of the treatment beam can be applied either via a direct measurement of the target position or phase (Smith et al., 2009), or via a surrogate motion (Dietrich et al., 2005) measurement such as the external chest wall for an internal lung lesion (Tenn et al., 2005). These two different gating configurations have different 4D dosimetry system requirements to assess the interplay and margins of such treatments, and for commissioning and quality assurance. Gating via direct measurement requires a detector with programmed 3D target motion, either a volumetric detector with a programmable motion platform or a programmable anthropomorphic phantom. Gating using a surrogate motion requires a 4D dosimetry system with both the programmed 3D target motion for the detector and a fourth external surrogate motion (Park et al., 2013) using either a volumetric detector with programmable motion platform or a deformable anthropomorphic programmable phantom with external motion.

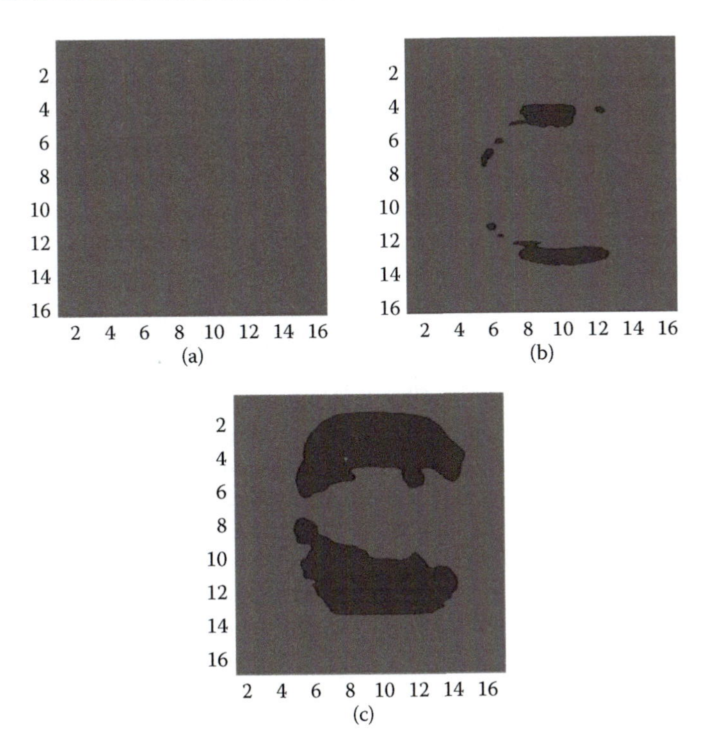

Figure 11.6

Results of interplay effect for three gating windows shown in gamma pass-fail maps. The passing pixels are shown in light gray and the failing pixels are in dark gray. (a) Gating window 29%–68% phases; (b) gating window 23%–77% phases; (c) gating window 0%–100% phases. (Adapted from Chen H et al., *Med. Phys.*, 36, 893–903, 2009. With permission.)

Various 4D dosimetry studies have been performed for commissioning, margin assessment, and assessment of the interplay effect for gated radiotherapy techniques for lung, liver, and pancreas. These studies showed that the beam-holds resulted in minimizing the maximum motion amplitude while the treatment beam was on. Consequently, the interplay effect was reduced (Keall et al., 2006a; Viel et al., 2015). The size of the gating window (either phase or amplitude derived) used determines the size of the interplay effect reduction (see Figure 11.6).

11.4.3 4D Dosimetry for Real-time Adaptive Radiotherapy

Real-time adaptive radiotherapy involves the adaptation of the beam to account for intrafraction motion during treatment delivery. This includes CyberKnife (robotic tracking) and Vero (gimballed tracking) (see Chapter 10), the manipulation of the radiation beam, i.e., MLC tracking, or couch tracking where the patient position is adjusted.

4D dosimetry is used for assessing the interplay effect and margins, and for commissioning and quality assurance of real-time adaptive radiotherapy systems through end-to-end testing. The 4D dosimetry system must be compatible with and encompass not only the delivery system but also the real-time motion-detection system. CyberKnife (Zhou et al., 2004; Dieterich and Pawlicki, 2008;

Dieterich et al., 2011) and Vero (Mukumoto et al., 2012; Depuydt et al., 2014) use two simultaneous motion-detection systems for real-time adaptation of respiratory motion. That information is used to build a prediction and correlation model between the external surrogate motion of the chest wall (measured using optical imaging) and the internal target motion (measured using low-frequency, orthogonal kilovoltage [kV] imaging). For this reason, the 4D dosimetry systems required to assess the interplay and margins of such treatments, and for commissioning and quality assurance, may require a programmable phantom to have both the 3D target motion to move the detectors and the fourth external surrogate motion. This can be two separate phantoms (Garibaldi et al., 2015) or a single phantom (Malinowski et al., 2007b).

MLC tracking and couch tracking are real-time adaptive radiotherapy techniques that are compatible with standard linear accelerators. The motion is generally detected directly, such as when using an electromagnetic transponder (Menten et al., 2012; Wilbert et al., 2013; Keall et al., 2014), or kV–MV (Liu et al., 2008; Cho et al., 2009) tracking, rather than through an external surrogate. For this approach, a simpler 3D programmable phantom is required for the 4D dosimetry process. 4D dosimetry is used in the commissioning and quality assurance process as an end-to-end test to encompass all components of the system, from detection to correction and delivery (Sawant et al., 2010).

An increasing number of 4D dosimetry studies are performed for real-time adaptive radiotherapy treatment techniques. As the treatments account for motion in real-time during treatment, it is important to incorporate motion into commissioning and treatment assessment. The various 4D dosimetry studies for CyberKnife (Nioutsikou et al., 2008; Chan et al., 2013), Vero (Ono et al., 2014; Garibaldi et al., 2015), MLC tracking (Krauss et al., 2011; Davies et al., 2013; Pommer et al., 2013), and couch tracking (Menten et al., 2012; Wilbert et al., 2013) show that there is improved precision in dose delivery (Colvill et al., 2016), with far less interplay effect, with the implementation of real-time adaptive treatments when compared to static treatment deliveries (see Figure 11.7).

11.5 Using 4D Dosimetry to Validate Dose Reconstruction Methods

Dose reconstruction, also called dose accumulation, is a process that allows for the retrospective estimation of delivered dose as shown schematically in Figure 11.8. The use of dose reconstruction methods augments 4D dosimetry, allows for the exploration of a larger solutions space, allows for a possible reduction in the frequency of 4D dosimetry, and can potentially be used for adaptation of treatment and doses. The dose reconstruction process begins with a patient treatment plan along with the original planned 3D dose distribution. The motion traces of the tumor, and possibly the treatment beam log file, obtained during treatment delivery, are then combined with the original treatment plan to create a reconstructed dose distribution for comparison with the original treatment plan.

Dose reconstruction can be performed for any treatment during which the intrafraction motion of the target was observed, such as with motion-inclusive and gated radiotherapy techniques. These methods, which are not necessarily

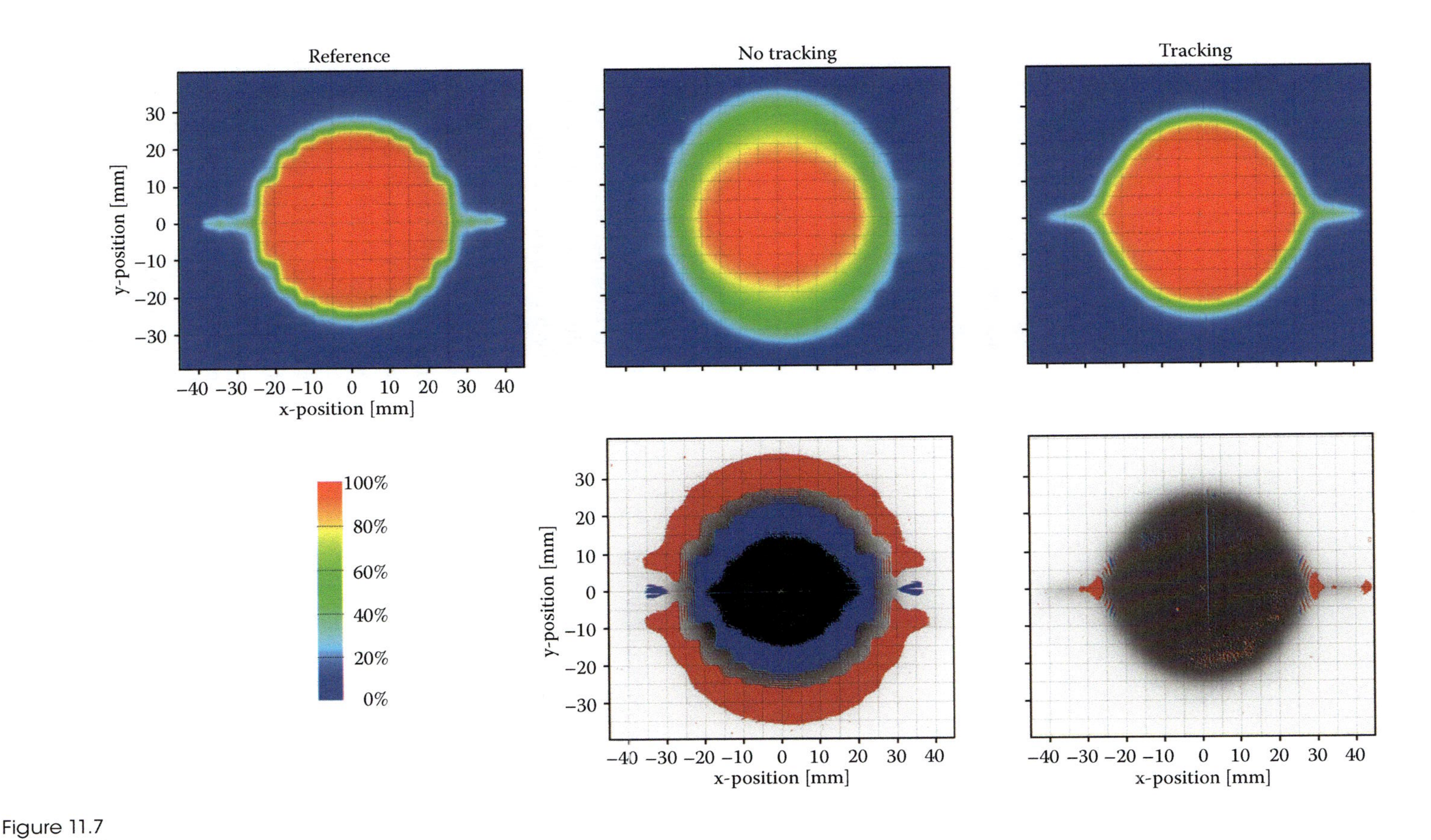

Figure 11.7

Dose distributions of a circular field applied to a target moving in two directions (upper row). Gamma test results displayed as grayscale dose distribution with gamma indexes > 1 marked in red for overdosage and blue for underdosage (lower row). (From Krauss A et al., *Int. J. Radiat. Oncol. Biol. Phys.*, 79, 579–587, 2011. With permission.)

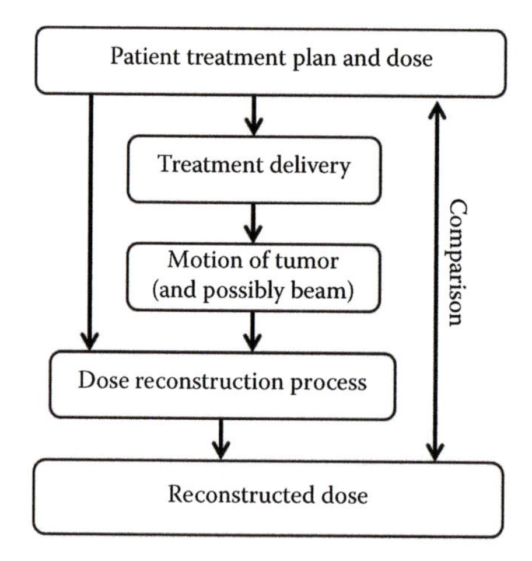

Figure 11.8

Schematic depiction of the dose reconstruction process.

time resolved, can also potentially be used for couch tracking real-time radiotherapy treatments. Various methods have been developed to incorporate translational motion, including the following:

- Convolution (Li et al., 2008b; Adamson et al., 2011; Poulsen et al., 2011; Langen et al., 2012)
- Perturbation (Feygelman et al., 2013a; Nelms and Feygelman, 2013)
- Isocenter shifts (Poulsen et al., 2012; Colvill et al., 2014)

These approaches provide an estimated 3D dose distribution delivered to the patient and can be calculated on planning (Poulsen et al., 2012) or cone beam (Adamson et al., 2011) CT sets. The 4D dosimetry data for the validation of dose reconstruction methods are more viable for translation and possibly rotation than for deformation due to restrictions in detector systems available to date.

Dose reconstruction methods for MLC tracking real-time adaptive radiotherapy need to be time-resolved so that the adaptation of the beam and motion of the target are modeled correctly. Several methods have been developed that can be used for retrospective reconstruction of the dose delivered during these treatments using data obtained during treatment, including motion trajectories, machine delivery log files, and measurements (Lin et al., 2012; Poulsen et al., 2012; Feygelman et al., 2013b; Nelms et al., 2014). CyberKnife and Vero treatment methods both have the potential for dose reconstruction with time-resolved methods as the motion of the treatment delivery and of organs are in principle known throughout treatment.

All dose reconstruction methods are approximations, however, they can be considered a form of patient-specific quality assurance, as shown in Figure 11.9. Once these methods are commissioned, ongoing 4D dosimetry may not be routinely needed on a per-patient basis for treatment plan quality assurance in the presence of motion.

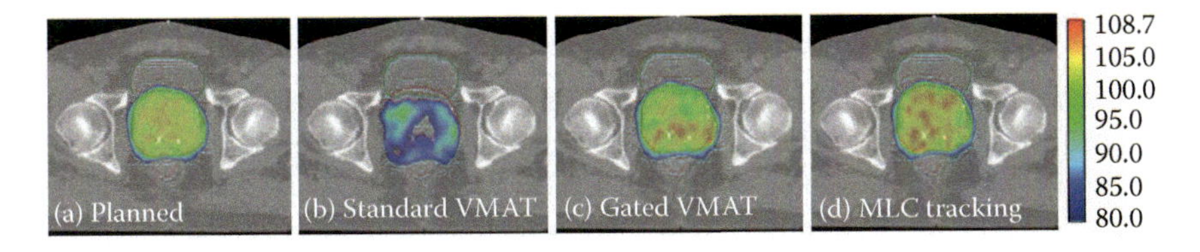

Figure 11.9

Dose reconstruction for a volumetric-modulated arc therapy (VMAT) prostate plan with the 80% isodose lines surrounding the planning target volume (PTV) (red), rectum (pink), and bladder (green). (a) Planned dose; (b) standard motion-inclusive VMAT; (c) gated VMAT with gating threshold of >3 mm for >5 s; and (d) multileaf collimator (MLC) tracking real-time adaptive VMAT.

11.6 4D Dosimetry Challenges and Future Directions

4D dosimetry is forming an increasingly important role in the emerging fields of real-time adaptive radiotherapy and the assessment of other existing treatment techniques. There are however various limitations with the current 4D dosimetry systems. The use of volumetric detectors with programmable motion platforms provides the user with relatively large 3D dose distributions. However, because the entire detector is programmed with the target motion, it provides a challenge in that the moving target (gross tumor volume [GTV] and clinical target volume [CTV]) dose is the only useful portion of the measurement whereas the nonmoving (ITV/PTV) structures and critical structures portion of the measurement are poorly modeled. Volumetric detectors are also largely homogeneous, and for this reason, the dose measured is not an accurate model of that delivered during patient treatment, particularly for thoracic and head and neck cancers, where significant density heterogeneities are present in patients. The programmable anthropomorphic phantoms that do have heterogeneous materials and model differential motion of the target and the nonmoving structures, similarly have detectors only in the target volume, so also are missing valuable information about the dose delivered to adjacent structures.

To date most 4D dosimetry has been performed with translation of the target volume; however, rotation (Plathow et al., 2006; Huang et al., 2015) and deformation (von Siebenthal et al., 2007; Kyriakou and McKenzie, 2012) of the target volume are likely to be just as significant factors effecting target dose. As real-time adaptive radiotherapy techniques develop to encompass rotation and deformation (Wu et al., 2012; Ge et al., 2014), 4D dosimetry will have to evolve too. Accounting for rotation will be the simpler of these, with a programmable motion platform already accounting for two of the three rotation axes (Pommer et al., 2013). Accounting for both differential deformation, such as the case of a lung tumor and a node moving in different trajectories, and target deformation will potentially be harder. However, each addition of complexity to the system moves them closer to an ideal 4D dosimetry system.

4D dosimetry could lead in the development of improved real-time adaptive radiotherapy techniques, with higher spatial and temporal resolution detectors allowing for better understanding of dosimetric uncertainties.

11.7 Summary

4D dosimetry is a fast-emerging and important field of treatment verification to account for the rapid technological advances in our understanding of, and ability to account for, patient motion during radiotherapy. All 4D dosimetry systems are approximations of treatment situations but these approximations allow for assessment, commissioning, and quality assurance of motion-inclusive, gated, and real-time adaptive radiotherapy techniques. 4D dosimetry allows us to better understand the 3D dose distribution that is actually delivered during radiotherapy and could lead the way to overcoming approximations of existing radiotherapy treatment-planning and delivery systems.

References

Abdellatif A et al. (2012) Experimental assessments of intrafractional prostate motion on sequential and simultaneous boost to a dominant intraprostatic lesion. *Med. Phys.* 39: 1505–1517.

Adamson J and Wu Q (2010) Prostate intrafraction motion assessed by simultaneous kilovoltage fluoroscopy at megavoltage delivery. I. Clinical observations and pattern analysis. *Int. J. Radiat. Oncol. Biol. Phys.* 78: 1563–1570.

Adamson J, Wu Q and Yan D (2011) Dosimetric effect of intrafraction motion and residual setup error for hypofractionated prostate intensity-modulated radiotherapy with online cone beam computed tomography image guidance. *Int. J. Radiat. Oncol. Biol.* Phys. 80: 453–461.

Anneli E et al. (2015) Verification of motion induced thread effect during tomotherapy using gel dosimetry. *J. Phys. Conf. Ser.* 573: 012048.

Barney BM et al. (2011) Image-guided radiotherapy (IGRT) for prostate cancer comparing kV imaging of fiducial markers with cone beam computed tomography (CBCT). *Int. J. Radiat. Oncol. Biol. Phys.* 80: 301–305.

Benedict SH et al. (2010) Stereotactic body radiation therapy: The report of AAPM Task Group 101. *Med. Phys.* 37: 4078–4101.

Bissonnette J-P et al. (2009) Quantifying interfraction and intrafraction tumor motion in lung stereotactic body radiotherapy using respiration-correlated cone beam computed tomography. *Int. J. Radiat. Oncol. Biol. Phys.* 75: 688–695.

Bortfeld T et al. (2002) Effects of intrafraction motion on IMRT dose delivery: Statistical analysis and simulation. *Phys. Med. Biol.* 47: 2303–2320.

Bortfeld T, Jiang SB and Rietzel E (2004) Effects of motion on the total dose distribution. *Sem. Radiat. Oncol.* 14: 41–51.

Case RB et al. (2010) Interfraction and intrafraction changes in amplitude of breathing motion in stereotactic liver radiotherapy. *Int. J. Radiat. Oncol. Biol. Phys.* 77: 918–925.

Ceberg S et al. (2013) Evaluation of breathing interplay effects during VMAT by using 3D gel measurements. *J. Phys. Conf. Ser.* 444: 012098.

Chan MKH et al. (2012) Accuracy and sensitivity of four-dimensional dose calculation to systematic motion variability in stereotactic body radiotherapy (SBRT) for lung cancer. *J. Appl. Clin. Med. Phys.* 13(6): 303–317.

Chan MKH et al. (2013) Experimental evaluations of the accuracy of 3D and 4D planning in robotic tracking stereotactic body radiotherapy for lung cancers. *Med. Phys.* 40: 041712.

Chen H et al. (2009) Dosimetric evaluations of the interplay effect in respiratory-gated intensity-modulated radiation therapy. *Med. Phys.* 36: 893–903.

Cherpak A et al. (2011) 4D dose-position verification in radiation therapy using the RADPOS system in a deformable lung phantom. *Med. Phys.* 38: 179–187.

Cho B et al. (2009) First demonstration of combined kV/MV image-guided real-time dynamic multileaf-collimator target tracking. *Int. J. Radiat. Oncol. Biol. Phys.* 74: 859–867.

Colvill E et al. (2014) DMLC tracking and gating can improve dose coverage for prostate VMAT. *Med. Phys.* 41: 091705.

Colvill E et al. (2016) A dosimetric comparison of real-time adaptive and non-adaptive radiotherapy: A multi-institutional study encompassing robotic, gimbaled, multileaf collimator and couch tracking. *Radiother. Oncol.* 119: 159–165.

Court L et al. (2010a) Evaluation of the interplay effect when using RapidArc to treat targets moving in the craniocaudal or right-left direction. *Med. Phys.* 37: 4–11.

Court LE et al. (2010b) Use of a realistic breathing lung phantom to evaluate dose delivery errors. *Med. Phys.* 37: 5850–5857.

D'Souza WD et al. (2005) Real-time intra-fraction-motion tracking using the treatment couch: A feasibility study. *Phys. Med. Biol.* 50: 4021–4033.

Davies GA et al. (2013) An experimental evaluation of the Agility MLC for motion-compensated VMAT delivery. *Phys. Med. Biol.* 58: 4643–4657.

De Deene Y et al. (2015) FlexyDos3D: A deformable anthropomorphic 3D radiation dosimeter: Radiation properties. *Phys. Med. Biol.* 60: 1543–1563.

Depuydt T et al. (2011) Geometric accuracy of a novel gimbals based radiation therapy tumor tracking system. *Radiother. Oncol.* 98: 365–372.

Depuydt T et al. (2014) Treating patients with real-time tumor tracking using the Vero gimbaled linac system: Implementation and first review. *Radiother. Oncol.* 112: 343–351.

Dietrich L et al. (2005) Compensation for respiratory motion by gated radiotherapy: An experimental study. *Phys. Med. Biol.* 50: 2405–2414.

Dieterich S and Pawlicki T (2008) CyberKnife image-guided delivery and quality assurance. *Int. J. Radiat. Oncol. Biol. Phys.* 71: S126–S130.

Dieterich S et al. (2011) Report of AAPM TG 135: Quality assurance for robotic radiosurgery. *Med. Phys.* 38: 2914–2936.

Duan J et al. (2006) Dosimetric and radiobiological impact of dose fractionation on respiratory motion induced IMRT delivery errors: A volumetric dose measurement study. *Med. Phys.* 33: 1380–1387.

Falk M et al. (2010) Real-time dynamic MLC tracking for inversely optimized arc radiotherapy. *Radiother. Oncol.* 94: 218–223.

Fast MF et al. (2014) Dynamic tumor tracking using the Elekta Agility MLC. *Med. Phys.* 41: 111719.

Feygelman V et al. (2013a) Experimental verification of the planned dose perturbation algorithm in an anthropomorphic phantom. *J. Phys. Conf. Ser.* 444: 012047.

Feygelman V et al. (2013b) Motion as a perturbation: Measurement-guided dose estimates to moving patient voxels during modulated arc deliveries. *Med. Phys.* 40: 021708.

Garibaldi C et al. (2015) Geometric and dosimetric accuracy and imaging dose of the real-time tumour tracking system of a gimbal mounted linac. *Phys. Med.* 31: 501–509.

Ge Y et al. (2014) Toward the development of intrafraction tumor deformation tracking using a dynamic multileaf collimator. *Med. Phys.* 41: 061703.

Grills IS et al. (2008) Image-guided radiotherapy via daily online cone-beam CT substantially reduces margin requirements for stereotactic lung radiotherapy. *Int. J. Radiat. Oncol. Biol. Phys.* 70: 1045–1056.

Groenendaal G et al. (2010) Validation of functional imaging with pathology for tumor delineation in the prostate. *Radiother. Oncol.* 94: 145–150.

Guckenberger M et al. (2008) Image-guided radiotherapy for liver cancer using respiratory-correlated computed tomography and cone-beam computed tomography. *Int. J. Radiat. Oncol. Biol. Phys.* 71: 297–304.

Hoogeman M et al. (2009) Clinical accuracy of the respiratory tumor tracking system of the CyberKnife: Assessment by analysis of log files. *Int. J. Radiat. Oncol. Biol. Phys.* 74: 297–303.

Huang L et al. (2010) A study on the dosimetric accuracy of treatment planning for stereotactic body radiation therapy of lung cancer using average and maximum intensity projection images. *Radiother. Oncol.* 96: 48–54.

Huang C-Y et al. (2015) Six degrees-of-freedom prostate and lung tumor motion measurements using kilovoltage intrafraction monitoring. *Int. J. Radiat. Oncol. Biol. Phys.* 91: 368–375.

Hugo GD, Agazaryan N and Solberg TD (2002) An evaluation of gating window size, delivery method, and composite field dosimetry of respiratory-gated IMRT. Med. Phys. 29: 2517–2525.

Jönsson M et al. (2014) Technical evaluation of a laser-based optical surface scanning system for prospective and retrospective breathing adapted computed tomography. *Acta Oncol.* 54: 261–265.

Kashani R et al. (2007) Technical note: A deformable phantom for dynamic modeling in radiation therapy. *Med. Phys.* 34: 199–201.

Keall P et al. (2006a) The clinical implementation of respiratory-gated intensity-modulated radiotherapy. *Med. Dosim.* 31: 152–162.

Keall PJ et al. (2006b) Geometric accuracy of a real-time target tracking system with dynamic multileaf collimator tracking system. *Int. J. Radiat. Oncol. Biol. Phys.* 65: 1579–1584.

Keall PJ et al. (2006c) The management of respiratory motion in radiation oncology report of AAPM Task Group 76. *Med. Phys.* 33: 3874–3900.

Keall PJ et al. (2014) The first clinical implementation of electromagnetic transponder-guided MLC tracking. *Med. Phys.* 41: 020702.

Khoo ELH et al. (2012) Prostate contouring variation: Can it be fixed? *Int. J. Radiat. Oncol. Biol. Phys.* 82: 1923–1929.

Koong AC et al. (2004) Phase I study of stereotactic radiosurgery in patients with locally advanced pancreatic cancer. *Int. J. Radiat. Oncol. Biol. Phys.* 58: 1017–1021.

Krauss A et al. (2011) Electromagnetic real-time tumor position monitoring and dynamic multileaf collimator tracking using a Siemens 160 MLC: Geometric and dosimetric accuracy of an integrated system. *Int. J. Radiat. Oncol. Biol. Phys.* 79: 579–587.

Kyriakou E and McKenzie DR (2012) Changes in lung tumor shape during respiration. *Phys. Med. Biol.* 57: 919–935.

Lang S et al. (2014) Development and evaluation of a prototype tracking system using the treatment couch. *Med. Phys.* 41: 021720.

Langen KM et al. (2008) Observations on real-time prostate gland motion using electromagnetic tracking. *Int. J. Radiat. Oncol. Biol. Phys.* 71: 1084–1090.
Langen KM et al. (2012) The dosimetric effect of intrafraction prostate motion on step-and-shoot intensity modulated radiation therapy plans: Magnitude, correlation with motion parameters, and comparison with helical tomotherapy plans. *Int. J. Radiat. Oncol. Biol. Phys.* 84: 1220–1225.
Li XA, Stepaniak C and Gore E (2006) Technical and dosimetric aspects of respiratory gating using a pressure-sensor motion monitoring system. *Med. Phys.* 33: 145–154.
Li HS et al. (2008a) Dosimetric consequences of intrafraction prostate motion. *Int. J. Radiat. Oncol. Biol. Phys.* 71: 801–812.
Li HS, Chetty IJ and Solberg TD (2008b) Quantifying the interplay effect in prostate IMRT delivery using a convolution-based method. *Med. Phys.* 35: 1703–1710.
Li FX et al. (2011) Comparison of the planning target volume based on three dimensional CT and four-dimensional CT images of non-small-cell lung cancer. *Radiother. Oncol.* 99: 176–180.
Lin M-H et al. (2012) 4D patient dose reconstruction using online measured EPID cine images for lung SBRT treatment validation. *Med. Phys.* 39: 5949–5958.
Litzenberg DW et al. (2006) Influence of intrafraction motion on margins for prostate radiotherapy. *Int. J. Radiat. Oncol. Biol. Phys.* 65: 548–553.
Liu W et al. (2008) Real-time 3D internal marker tracking during arc radiotherapy by the use of combined MV–kV imaging. *Phys. Med. Biol.* 53: 7197–7213.
Louie AV et al. (2010) Inter-observer and intra-observer reliability for lung cancer target volume delineation in the 4D-CT era. *Radiother. Oncol.* 95: 166–171.
Malinowski K et al. (2007a) Use of the 4D phantom to test real-time targeted radiation therapy device accuracy. *Med. Phys.* 34: 2611.
Malinowski K et al. (2007b) Development of the 4D phantom for patient-specific, end-to-end radiation therapy QA. In: Proceeding of SPIE Vol. 6510, pp. 65100E-65101–65100E-65109 (Bellingham, Washington: SPIE-The International Society of Optical Engineering).
Malinowski K et al. (2012) Incidence of changes in respiration-induced tumor motion and its relationship with respiratory surrogates during individual treatment fractions. *Int. J. Radiat. Oncol. Biol. Phys.* 82: 1665–1673.
Mayer R et al. (2015) 3D printer generated thorax phantom with mobile tumor for radiation dosimetry. *Rev. Sci. Instr.* 86: 074301.
Menten MJ et al. (2012) Comparison of a multileaf collimator tracking system and a robotic treatment couch tracking system for organ motion compensation during radiotherapy. *Med. Phys.* 39: 7032–7041.
Mukumoto N et al. (2012) Positional accuracy of novel x-ray-image-based dynamic tumor-tracking irradiation using a gimbaled MV x-ray head of a Vero4DRT (MHI-TM2000). *Med. Phys.* 39: 6287–6296.
Nakayama H et al. (2008) Development of a three-dimensionally movable phantom system for dosimetric verifications. *Med. Phys.* 35: 1643–1650.
Nelms B and Feygelman V (2013) A novel method for 4D measurement-guided planned dose perturbation to estimate patient dose/DVH changes due to interplay. *J. Phys. Conf. Ser.* 444: 012097.
Nelms BE et al. (2014) Motion as perturbation. II. Development of the method for dosimetric analysis of motion effects with fixed-gantry IMRT. *Med. Phys.* 41: 061704.

Nguyen D et al. (2013) Dosimetric impact of interplay effect on a volumetric modulated arc therapy (VMAT) stereotactic lung (SBRT) delivery: Validation using a 6D motion platform and 3D dosimeter. *Phys. Med.* 29, Suppl 1: e29–e30.

Nioutsikou E et al. (2008) Dosimetric investigation of lung tumor motion compensation with a robotic respiratory tracking system: An experimental study. *Med. Phys.* 35: 1232–1240.

Ong C et al. (2011) Dosimetric impact of interplay effect on RapidArc lung stereotactic treatment delivery. *Int. J. Radiat. Oncol. Biol. Phys.* 79: 305–311.

Ono T et al. (2014) Geometric and dosimetric accuracy of dynamic tumortracking conformal arc irradiation with a gimbaled x-ray head. *Med. Phys.* 41: 031705.

Park K et al. (2009) Do maximum intensity projection images truly capture tumor motion? *Int. J. Radiat. Oncol. Biol. Phys.* 73: 618–625.

Park Y-K et al. (2013) Development of real-time motion verification system using in-room optical images for respiratory-gated radiotherapy. *J. Appl. Clin. Med. Phys.* 14(5): 25–42.

Plathow C et al. (2006) Quantification of lung tumor volume and rotation at 3D dynamic parallel MR imaging with view sharing: Preliminary results. *Radiol.* 240: 537–545.

Poels K et al. (2013) A complementary dual-modality verification for tumor tracking on a gimbaled linac system. *Radiother. Oncol.* 109: 469–474.

Poels K et al. (2014) Improving the intra-fraction update efficiency of a correlation model used for internal motion estimation during real-time tumor tracking for SBRT patients: Fast update or no update? *Radiother. Oncol.* 112: 352–359.

Pommer T et al. (2013) Dosimetric benefit of DMLC tracking for conventional and sub-volume boosted prostate intensity-modulated arc radiotherapy. *Phys. Med. Biol.* 58: 2349–2361.

Poulsen PR et al. (2012) A method of dose reconstruction for moving targets compatible with dynamic treatments. *Med. Phys.* 39: 6237–6246.

Poulsen PR et al. (2011) Delivered target dose reconstruction in actual and simulated SBRT liver treatments with MV image guided dynamic MLC tracking. *Int. J. Radiat. Oncol. Biol. Phys.* 81: S772.

Sawant A et al. (2010) Failure mode and effect analysis-based quality assurance for dynamic MLC tracking systems. *Med. Phys.* 37: 6466–6479.

Sawant A et al. (2008) Management of three-dimensional intrafraction motion through real-time DMLC tracking. *Med. Phys.* 35: 2050–2061.

Seco J et al. (2008) Dosimetric impact of motion in free-breathing and gated lung radiotherapy: A 4D Monte Carlo study of intrafraction and interfraction effects. *Med. Phys.* 35: 356–366.

Smith RL et al. (2009) Evaluation of linear accelerator gating with real-time electromagnetic tracking. *Int. J. Radiat. Oncol. Biol. Phys.* 74: 920–927.

Smith WL and Becker N (2009) Time delays in gated radiotherapy. *J. Appl. Clin. Med. Phys.* 10(3): 140–154.

Steidl P et al. (2012) A breathing thorax phantom with independently programmable 6D tumour motion for dosimetric measurements in radiation therapy. *Phys. Med. Biol.* 57: 2235–2250.

Tenn SE, Solberg TD and Medin PM (2005) Targeting accuracy of an image guided gating system for stereotactic body radiotherapy. *Phys. Med. Biol.* 50: 5443–5462.

Thomas A et al. (2013) The effect of motion on IMRT—Looking at interplay with 3D measurements. *J. Phys. Conf. Ser.* 444: 012049.

van Herk M et al. (2000) The probability of correct target dosage: Dose-population histograms for deriving treatment margins in radiotherapy. *Int. J. Radiat. Oncol. Biol. Phys.* 47: 1121–1135.

Viel F et al. (2015) Amplitude gating for a coached breathing approach in respiratory gated 10 MV flattening filter-free VMAT delivery. *J. Appl. Clin. Med. Phys.* 16(4): 78–90.

von Siebenthal M et al. (2007) Systematic errors in respiratory gating due to intrafraction deformations of the liver. *Med. Phys.* 34: 3620–3629.

Whitfield G et al. (2012) Quantifying motion for pancreatic radiotherapy margin calculation. *Radiother. Oncol.* 103: 360–366.

Wilbert J et al. (2013) Accuracy of real-time couch tracking during 3-dimensional conformal radiation therapy, intensity modulated radiation therapy, and volumetric modulated arc therapy for prostate cancer. *Int. J. Radiat. Oncol. Biol. Phys.* 85: 237–242.

Wu J et al. (2012) Electromagnetic detection and real-time DMLC adaptation to target rotation during radiotherapy. *Int. J. Radiat. Oncol. Biol. Phys.* 82: e545–e553.

Zhou T et al. (2004) A robotic 3-D motion simulator for enhanced accuracy in CyberKnife stereotactic radiosurgery. *Int. Congress Ser.* 1268: 323–328.

12

Light-Ion Beam Dosimetry

Hugo Palmans

12.1 Introduction

An introduction to the nature, use, and implementation of proton and carbon ion beams in radiotherapy is given in Chapter 23. The need for relative dosimetry, range measurement, and the accurate determination of spot positions is also discussed. As for all forms of radiotherapy, beam commissioning of proton and ion beams requires the determination of a full 3D dose map delivered by the treatment, either as a combination of all treatment fields or for each field individually. For scanned beams, the most practical approach is to start from 3D dose distributions delivered by individual pencil beams.

In this chapter, the different detector technologies to determine the quantity of interest, absorbed dose to water, in proton and carbon ion beams are described and categorized followed by a discussion of their application to 3D dosimetry. The main issues in their application to 3D dosimetry are the resolution in each spatial direction and the energy dependence of the detector response along the beam direction. Similarly as in intensity-modulated radiation therapy (IMRT), for dosimetry in scanned beams traditional scanning tanks with a point detector or linear array are not an option due to timing constraints, and 2D- or 3D-array or continuous detector systems need to be used. The practical aspects related to this as well as to the problem of detector resolution are discussed in Chapter 23. This chapter is restricted to a discussion of the energy dependence of the detector response and its effect on the measurement of depth-dose distributions.

12.2 Interactions of Light-Ions with Matter

To understand the operation of detectors for dosimetry of light-ions, their interaction with matter needs to be understood (Palmans, 2015) and a brief overview is given here. Light-ions interact with matter by two interaction mechanisms: electromagnetically with atomic electrons and target nuclei and by (strong force) nuclear interactions with target nuclei which can result in target fragmentation and, in the case of ions heavier than protons, also in projectile fragmentation.

12.2.1 Electromagnetic Interactions

Electromagnetic interactions with atomic electrons are categorized as a function of the classical impact parameter, b, defined as the closest distance between the initial trajectory of the incident light-ion and the nucleus, by comparing b with the atomic radius, a.

When $b \gg a$, the incident ion interacts with the atom as a whole, and only a small amount of energy is transferred from the incident ion to the atom. These interactions are often called soft collisions. The energy transfer can result in atomic excitation (raising an orbital electron to a higher allowed orbital state) or ionization.

When $b \approx a$, the incident ion can interact with a single orbital electron, resulting in a large energy transfer to that electron, termed as "knock-on electron." These interactions are often called hard collisions. The knock-on electrons ejected from the atom are also termed as "γ-rays." Since the electron binding energy is usually small compared to the energy transferred to the electrons, the collision can be approximated by the Rutherford cross section based on classical mechanics or the Bhabha cross section based on relativistic mechanics.

When $b \ll a$, the incident ion can interact with the nucleus either via elastic and inelastic coulomb interactions or via nonelastic interactions described in the next section. Both elastic and inelastic scattering can result in large angular deflections, in the case of protons usually with limited energy transferred to the target nucleus due to the small ratio of the mass of the incident proton to the target nucleus.

The stopping power is the macroscopic quantity that represents the mean energy loss per unit of path length for a large number of light-ions due to the combination of the different electromagnetic interaction mechanisms (ICRU, 1993, 2005). Since, for individual particles, the energy loss deviates from that described by the stopping power due to the stochastic nature of the interaction mechanisms, the beam undergoes energy straggling resulting in a gradual broadening of the energy spectrum with increasing penetration depth. The scattering power is the macroscopic quantity that represents the mean angular deflection per unit of path length for a large number of protons or ions due to the combination of the different electromagnetic interaction mechanisms. Of course, the effect of the electromagnetic interactions with electrons is also that ionization is produced, which forms the basis of many measurement methods, either by measuring the amount of ionization directly or by quantifying the effect of ionization on physical, chemical, or biomolecular systems.

12.2.2 Strong-Force Nuclear Interactions

Nuclear interactions resulting from the strong nuclear force can be categorized as inelastic or nonelastic (ICRU, 2000). In an inelastic nuclear interaction, the kinetic energy is not conserved (as opposed to an elastic nuclear interaction) but the target nucleus remains unchanged. The latter can remain excited after the interaction and/or emit a γ-ray. In a nonelastic nuclear interaction, kinetic energy is not conserved and in addition the target nucleus undergoes break-up and/or a particle transfer reaction occurs. Note that this categorization differs slightly from the one used in ICRU Report 63 (ICRU, 2000) where nonelastic nuclear interactions refer to all nuclear interactions in which kinetic energy is not conserved; the reason for this is to avoid the use of the expression "nonelastic interactions which are not inelastic."

Inelastic nuclear interactions can be treated in a similar way as inelastic coulomb interactions in the sense that the primary light-ion continues to be transported and γ-rays can be emitted. Nonelastic nuclear interactions result in an attenuation process in which the projectile ions are removed from the primary beam either by absorption or by projectile fragmentation.

In case the projectile is absorbed, its nucleons penetrate the nucleus and an intranuclear cascade takes place. This is essentially a collision avalanche with individual nucleons leading to ejection of forward-directed protons, neutrons, and/or even pions in case the projectile's kinetic energy is high enough for pion formation. The energy not transported away by escaping particles leaves the remaining nucleus in a highly excited state and leads to potential emission of nucleons and light fragments both during the redistribution of this excitation energy (the precompound stage) and after the energy distribution over the nucleons has reached an equilibrium state (compound stage). This compound nucleus is characterized only by its mass, charge, and excitation energy, so the history of the collision and cascade is "forgotten" at that moment, and any further emission of particles is isotropic. The compound nucleus can lose energy by evaporation emitting protons, neutrons, or light fragments (mainly alpha particles), and by emitting γ-rays. Eventually, a stable nucleus is reached. The kinetic energy that the emitted particles carry away at all stages can contribute to the local energy deposition, which explains the contribution of the reduction in rest mass to the definition of energy imparted and, thus, to the absorbed dose. The charged particles emitted in the intranuclear cascade stage usually have high energy, of the same order of magnitude as the kinetic energy per nucleon of the projectile light-ion, and can thus carry away the energy from the projectile absorption point. Those generated in the evaporation stage, on the other hand, have low energy and will deposit all their energy close to the generation point.

The nonelastic interaction has rather the character of projectile fragmentation in the case of grazing incidence on the target nucleus due to which the projectile is stripped of one or a few nucleons resulting in lighter fragments with approximately the same energy per nucleon as the incident ion. Due to their lower stopping power, these will have a longer range and form a fragmentation tail beyond the distal edge of the Bragg peak.

Neutrons and γ-rays emitted in nonelastic nuclear interactions make a negligible contribution to the local energy deposition but need to be considered in shielding, radioprotection, and estimation of secondary cancer risks. They are also important to consider in prompt gamma imaging to reconstruct the distribution of the primary proton absorption point by locating the origin of the γ-rays. The production of radionuclides leads to the possibility of treatment plan verification using positron emission tomography (PET). Prompt gamma imaging and PET will be discussed in Chapter 23.

12.3 Specific Detector Requirements for Light-Ion Beams

12.3.1 Reference Dosimetry

Calorimeters are the primary instruments to measure absorbed dose to medium and make use of the assumption that all the energy deposited by the radiation will appear as heat under the condition that there is no other form of internal energy storage or release, i.e., no change in chemical, phase, or lattice energy takes place (nuclear changes are allowed since they are accounted for in the definition of absorbed dose). The energy deposition in calorimeters is measured by quantifying the electrical energy dissipation needed to realize the same temperature rise in the medium as the radiation does. If the specific heat capacity, c_{med}, of the medium is known, absorbed dose to the medium can also be derived from a measurement of the temperature rise ΔT as $D_{\text{med}} = c_{\text{med}}\Delta T$. The measurement can be performed with thermistors in a continuous medium if the thermal diffusivity is low, such as in water, or in a sample of the medium, called a core, which is thermally isolated from the phantom if the thermal diffusivity is high, such as in graphite calorimetry. Calorimeters were only realized as point detectors till now and while in principle they could be used for step-by-step profile measurements, their low sensitivity and the resulting long acquisition times involved make them impractical for that purpose. The minimal size of calorimeter cores is also limited to a few mm so that the spatial resolution is restricted.

The more common instrument to perform reference dosimetry in a particle therapy is the air-filled ionization chamber. Dose to air can in principle be derived directly from the ionization in the cavity via the air volume from which ionization is collected, and the mean energy required to produce an ion pair in air, W_{air}. Dose to water or dose to the medium can be derived via application of Bragg–Gray cavity theory with inclusion of correction terms for fluence perturbations. In practice, the volume of commercial ionization chambers is not known with sufficient accuracy and they require calibrations against primary standards in calibration beams.

12.3.2 Macroscopic 3D Dosimetry

A wide range of detectors is used for relative dosimetry in light-ion beams (Karger et al., 2010). To a large extent, similar detector systems are used as for photon and electron beams. Both active and passive detector systems are used that require no detector calibration, but just verification of the response linearity, water equivalence, energy independence, and sufficient spatial resolution within the dynamic range encountered in the required measurements. A suitable detector choice has to consider the time structure and delivery method of the beam. In passively scattered beams that contain only static beam shaping devices or a very fast spinning modulator wheel, the dose distribution is static or quasi-static and a plotting tank with a point detector can be used to register the entire 3D dose distribution of interest, although multidimensional detector systems play an important role in reducing the time needed for the measurement. In scanned light-ion beams, on the other hand, the use of multidimensional detector systems is essential since the dose at a single point in the radiation field can constitute of contributions from multiple spots delivered over the course of the entire field delivery (Karger et al., 2010; Vatnitsky and Palmans, 2015). Using a point detector, the entire dose distribution should be delivered for each dose point making the measurement times impractically large. This situation is very similar as in complex IMRT delivery schemes.

For plotting systems, point detectors that are commonly used are small-volume ionization chambers, diodes, and diamond detectors. For multiple point-detector arrays, ionization chambers and diodes are typically used. Some passive detectors can also be used as an array or a 3D arrangement of point detectors such as alanine, thermoluminescent dosimeters (TLDs), and optically stimulated luminescence dosimeters (OSLDs). Continuous 2D detectors that are used for light-ions are radiographic or radiochromic films and scintillating screens. 3D gels and plastic detectors are used for direct acquisition of the 3D dose distribution.

Some detectors have been developed to address specific needs for dosimetry of scanned light-ion beams. One of the beam commissioning requirements is the measurement of the 3D dose distribution of single spots in rather narrow fields. These are sometimes difficult to characterize by moving a point detector in a plotting tank given the difficulty to align the detector with the central axis of the beam spot. To overcome this, an alternative approach is to determine the laterally integrated dose (dose–area–product) as a function of depth, and for this purpose dedicated scanning systems with large-area ionization chambers were developed, for example, the PTW Peak Finder (www.ptw.de), as well as multilayer ionization chamber detectors, for example, the IBA Giraffe system (www.iba-dosimetry.com).

In later subsections, the issues of spatial resolution, water equivalence, and energy dependence of detectors, including those used in multidimensional detector systems, are discussed.

12.3.3 Track Structure, Microdosimetry, and Nanodosimetry

It is assumed that the distribution of energy deposition on the microscale affects the indirect DNA damage inflicted by radiation via diffusion of radiation-induced reactive species. Microdosimetry concerns the determination of the spatial and temporal distribution of interactions of ionizing radiation within micrometer-sized volumes of matter. The energy dependence of the radiobiological effectiveness (RBE) of light-ions can be related to the variation of the microdosimetric properties. When ionization clustering in the vicinity or within the DNA becomes very high, the diffusion and long-term chemistry (on a time scale longer than 10^{-7} s) of reactive species becomes less important, and substantial DNA damage will be more directly correlated with the cluster density distribution. This is especially important for heavier ions while for protons and very light-ions the occurrence of such high cluster densities is mainly restricted to the distal edge of the Bragg peak. The measurement or simulation of the clustering distributions within the track structure on the nanoscale is the subject of nanodosimetry. The characterization of track structure is based on the stochastic quantity, called ionization cluster size, and its frequency distribution (ionization cluster size distribution [ICSD]).

12.3.4 Spatial Resolution—Volume Averaging

Requirements on spatial resolution depend always strongly on the quantity or beam characteristics of interest and are in many respects similar as for photon and electron beams. For reference dose measurements at a single point, usually a small detector volume is important while for measurements of depth-dose characteristics a small dimension in the scanning direction may be a sufficient requirement. A higher resolution will often be required in the Bragg peak region than in the plateau region of the depth-dose curve. For the measurement of lateral beam profiles in broad scattered beams, a high lateral resolution in the scanning direction is only required in the penumbrae, while for the characterization of a single spot for an actively scanning beam system a high resolution is required for the entire profile. These requirements for measuring profiles are discussed in detail in Chapter 23. In this section only the volume averaging for a point detector is discussed in depth, the considerations apply equally to individual detectors in arrays, and a distinction is made between the cavity of an air-filled ionization chamber with a density three orders of magnitude smaller than water and solid detectors with densities different by a factor up to four from water.

12.3.5 Water Equivalence

An ideal detector for multidimensional dosimetry is water equivalent, which means all its physical properties, i.e., density, stopping power, scattering power, and nuclear interaction cross sections, are the same as those for water. In reality, these conditions are only met in a water calorimeter and the water equivalence of detectors refers rather to the study to what extent the deviations from those ideal conditions compromise the performance of a detector in determining absorbed dose to water or a spatial distribution of dose to water. If a detector is perfectly water equivalent apart from its mass density then we speak of a Fano-cavity; indeed, the theorem of Fano (1954) says that "in a medium of a given composition exposed to a uniform flux of radiation the flux of secondary radiation is also uniform and independent of the density of the medium as well as of the density variations from point to point."

A second aspect of water-equivalence is related to the medium in which measurements are performed, if that medium is not water. In this section, the various aspects of water equivalence in this sense are discussed. These include scattering properties, electromagnetic interactions, and nuclear interactions. The emphasis is put on the dosimetric relevance of the water equivalence of both detector systems and phantom materials.

12.3.5.1 Stopping Power and Range

The first aspects that are usually mentioned concerning water equivalence of a material for ion beams are stopping power and range. The water equivalent thickness of a sample of nonwater medium is defined as the thickness of a water layer that, when traversed, results in the same reduction of the particle range as the sample of the medium. A different way of expressing this is to say that the mean energy losses of the beam over the equivalent water layer and the sample are the same. It means that water equivalence in terms of range is very closely related to equivalence in terms of energy loss, i.e., stopping power. Indeed, if the stopping power of two materials is the same then the range in both materials will be the same, and if the mass stopping power of two materials is the same then the range, expressed in $g{\cdot}cm^{-2}$, in both materials will also be the same. For the design of anthropomorphic phantoms, it is important that the materials are water or tissue equivalent in terms of range while for dosimetric phantoms it is often sufficient to be able to scale depths adequately to water equivalent depths. For detector materials, it can be assumed that water equivalence in terms of mass stopping powers is the ideal to minimize fluence perturbations for a given amount of energy loss via the application of cavity theory. In particular, the constancy of the mass stopping-power ratios of water to medium is, in the absence of perturbations, a prerequisite for the relative dose in a detector to be proportional to the relative dose in water.

The full Bethe formula with additional terms for the shell correction, density effect correction, Bloch correction, and Barkas correction is discussed in detail in ICRU Report 49 (ICRU, 1993) and forms the basis of stopping-power calculations based on experimental or theoretical values of the mean excitation energy, I. For the calculation of stopping-power ratios as a function of depth in ion beams, Lühr et al. (2011) proposed and validated the following approximation:

$$s_{w,\text{med}}(z_w) = \frac{\left(\frac{Z}{A}\right)_w \ln\left(\frac{E_0}{I_W}\right) + 0.58824 \ln\left(1 - \frac{z_w}{R_p}\right) - 6.1291}{\left(\frac{Z}{A}\right)_{\text{med}} \ln\left(\frac{E_0}{I_{\text{med}}}\right) + 0.58824 \ln\left(1 - \frac{z_w}{R_p}\right) - 6.1291} \tag{12.1}$$

where z_w is the depth in water, Z/A is the ratio of the atomic number and atomic weight of the material, E_0 is the incident ion beam energy, and R_p is the practical range in water.

Stopping-power ratios for a number of relevant dosimetric materials calculated using Equation 12.1 are shown in Figure 12.1. It is clear that dosimetric materials such as air, alanine, diamond, and the sensitive layer of external beam therapy (EBT)-type radiochromic films perform very well as water equivalent materials in this respect, while lithium fluoride, silicon, aluminum oxide, and photographic emulsions exhibit a considerable depth dependence, which would result in considerable differences between relative distributions of the detector signal and absorbed dose to water. Note that the lower stopping power of these materials will contribute to the signal quenching in the Bragg peak observed for many detectors, discussed in Section 12.4.

The continuous slowing-down approximation (CSDA) range is the distance a light-ion of energy E_0 travels along its path if it continuously loses energy according to the mass stopping power, (S/ρ), and is calculated as follows:

$$r_0(E_0) = \int_{E_0}^{0} \left(-S/\rho\right)^{-1} \mathrm{d}E \tag{12.2}$$

In practical dosimetry aplications, a matter of interest is the depth particles reach on average, i.e., the depth reached by 50% of the protons that have not undergone nuclear interaction. The ratio between the range and the CSDA range is expressed as a detour factor in ICRU Report 49 (ICRU, 1993) and is very close to unity (within 0.1%) for clinical proton energies. As a rule of thumb, the CSDA range corresponds with the depth, z_{80}, on the distal edge of the Bragg peak where the dose is reduced to 80% of the maximum dose (Moyers et al., 2007).

Figure 12.2 illustrates the issue of water equivalence in terms of range depending on the application for a number of water equivalent or tissue equivalent plastics. As mentioned earlier, in the construction of anthropomorphic phantoms, the equivalence of the range in cm is more important while for dosimetric equivalence the range in g·cm^{-2} is more relevant. One can see for example that

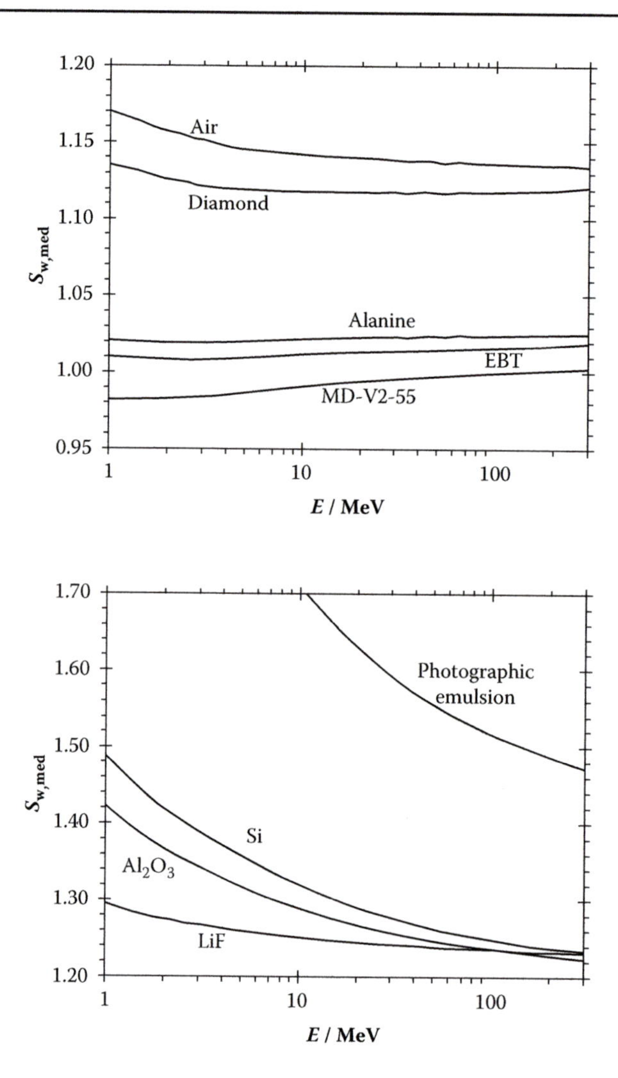

Figure 12.1

Water-to-medium mass stopping-power ratios for a number of relevant dosimetric materials as a function of proton kinetic energy. (From ICRU, Stopping powers and ranges for protons and alpha particles. ICRU Report 49. (Bethesda, MD: International Commission on Radiation Units and Measurements), 1993.)

polymethylmethacrylate (PMMA) performs worse than polyethylene in the first sense, but the opposite is true in terms of dosimetric equivalence.

12.3.5.2 Scattering Properties

As light-ions pass through a medium, they are deflected by many small-angle coulomb-scattering events with the nuclei. The combined effect of multiple events is called multiple coulomb scattering. The single-scattering events are described by the Rutherford cross section (Rutherford, 1911) with corrections for screening. Theories to model multiple scattering generally assume that the number of individual scattering events is large so that the average angular deflection can be derived using a statistical approach. The angular distribution of an initially parallel pencil beam of ions after passing through an absorber of thickness t can be,

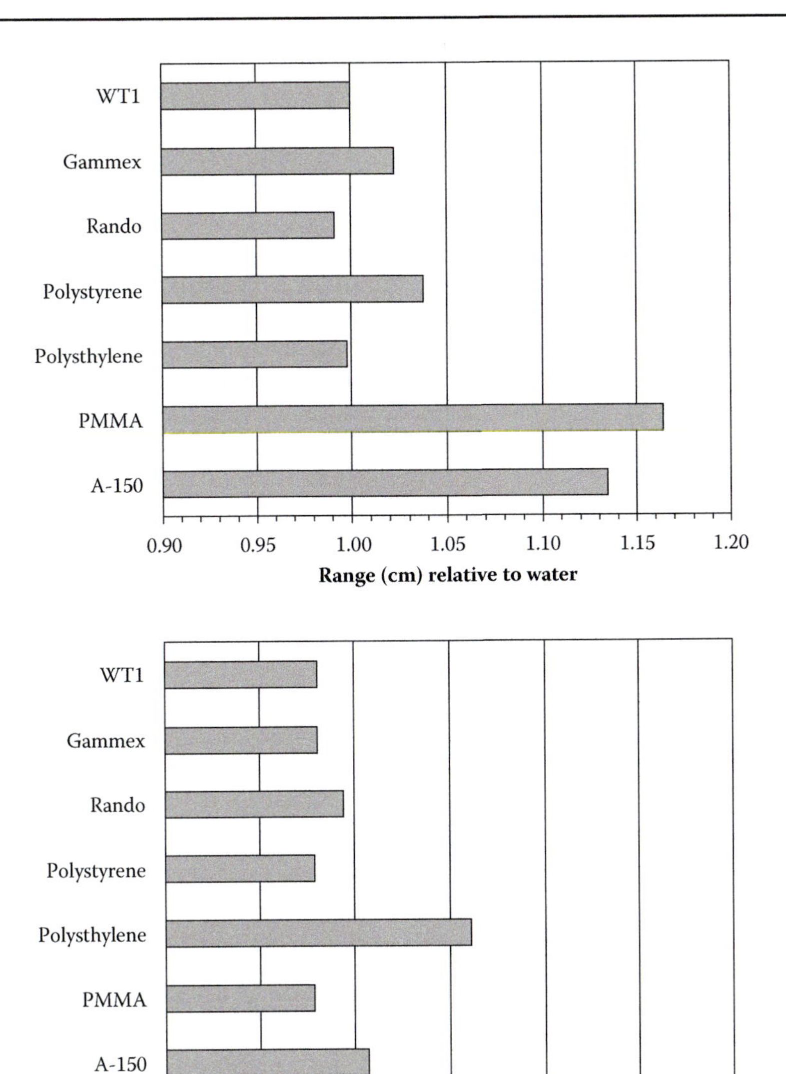

Figure 12.2

Ratio of ranges in some water equivalent and tissue equivalent phantom materials relative to water expressed in cm (upper graph) and expressed in g·cm^{-2} (lower graph). (From Lourenço A, PhD thesis: Water-equivalence of phantom materials in proton and carbon-ion dosimetry. University College London, 2016.)

to first order, approximated by a Gaussian distribution with a root mean square angle of $\theta_0(t) = \sqrt{\overline{\theta^2}(t)}$, using the small-angle approximation ($\sin\theta \approx \theta$):

$$f(\theta,t)\theta\mathbf{d}\theta = \frac{2\theta}{[\theta_0(t)]^2} e^{\frac{\theta^2}{[\theta_0(t)]^2}\mathbf{d}\theta} \quad (12.3)$$

An adequate multiple scattering model for light-ions that is widely used was formulated by Molière (1948), and a good approximation to the root mean square angle of the Molière distribution is given by the formula of Highland (Highland, 1975; Gottschalk, 2010) as follows:

$$\theta_0(t) = \frac{20\,\mathrm{MeV}}{pv} z \sqrt{\frac{t}{L_R}} \left[1 + \frac{1}{9} \log_{10}\left(\frac{t}{L_R} \right) \right] \tag{12.4}$$

where pv and z are the kinetic energy and the charge number of the incident ion, respectively, and L_R is the radiation length which is a material-specific property. For completeness we mention that Goudsmit and Saunderson (1940) and Lewis (1950) developed more exact theories that include higher order statistical moments than the Molière theory and correctly model the large-angle scattering tails.

Figure 12.3 represents the water equivalence in terms of scattering angle as a function of Z by plotting the ratio $\sqrt{L_{R,\text{water}} / L_{R,Z}}$, denoted the Rossi ratio. Rossi (1952) proposed a simplified version of the Highland formula without the factor in square brackets in Equation 12.4, where the scattering lengths $L_{R,Z}$ for elemental materials were taken from the Particle Data Group and $L_{R,\text{water}} = 35.9\,\mathrm{g\,cm^{-2}}$. The black full line is a fit to the data for $Z > 2$. A fit to the ratio, given below,

$$\sqrt{\frac{L_{R,\text{water}}}{L_{R,Z}}} \left[\frac{1 + \frac{1}{9} \log_{10}\left(\frac{t}{L_{R,Z}} \right)}{1 + \frac{1}{9} \log_{10}\left(\frac{t}{L_{R,\text{water}}} \right)} \right],$$

denoted the Highland ratio, is shown as well for $t = 1\ \mathrm{g\,cm^{-2}}$. The values of these ratios vary only very little with t. Figure 12.3 shows that for high-Z materials the root mean square (rms) scattering angles are up to 2.5 times higher than that for water. This means that per unit of mass thickness, high-Z materials are more effective in scattering. For $Z < 20$, there is still a considerable variation from about 0.6 times less to 1.5 times more scattering per unit mass thickness as in water. This becomes even more pronounced when considering that the mass stopping power decreases with Z. Figure 12.3 also shows the ratio given below:

$$\sqrt{\frac{L_{R,\text{water}}}{L_{R,Z} \cdot s_{Z,w}}} \left[\frac{1 + \frac{1}{9} \log_{10}\left(\frac{t}{L_{R,Z} \cdot s_{Z,w}} \right)}{1 + \frac{1}{9} \log_{10}\left(\frac{t}{L_{R,\text{water}}} \right)} \right],$$

denoted the scaled Highland ratio, i.e., where a mass thickness for the elemental material is considered over which the ions lose the same amount of energy as over the mass thickness of water, t. The higher effectiveness of scattering for the same energy loss explains why high-Z foils are normally used in double-scattered ion beams.

Differences in scattering properties have an influence on detector perturbations although given the magnitude of typical scattering angles this influence can be assumed to be very small. Another application where scattering plays a role is

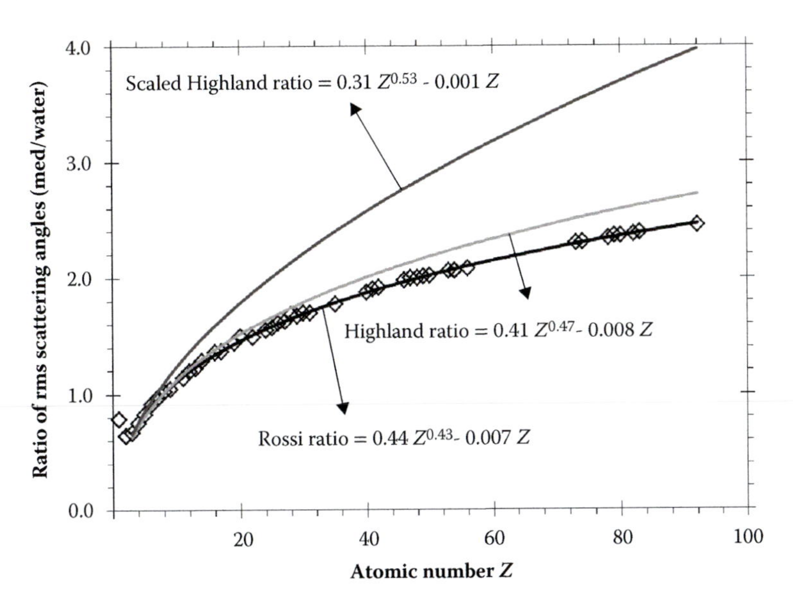

Figure 12.3

The rms scattering angle over slabs of 1 g·cm^{-2} of elemental materials relative to the rms scattering angle over a slab of 1 g·cm^{-2} of water as a function of the atomic number, calculated according to the Rossi-formula and the Highland formula. Also shown is a "scaled Highland ratio" in which the scattering angles are also calculated using the Highland formula but the slab thickness of the elemental material is such that the energy loss over it is the same as over a slab of 1 g·cm^{-2} of water. (From Tsai YS, *Rev. Mod. Phys.*, 46, 815–851, 1974.)

the field size or the size of a large-area plane-parallel chamber needed for integral depth-dose measurements. This can especially be considerable for protons.

12.3.5.3 Nonelastic Nuclear Interactions

Nonelastic nuclear interactions contribute to the total absorbed dose via the secondary charged particles that are emitted. For proton projectiles, roughly 60% of the energy of the absorbed particle is transferred to secondary charged particles, and thus will contribute to the dose (Palmans, 2015). The total nonelastic nuclear interaction cross section is almost independent of energy (the resonance at low energy accounts only for the last 2 cm of the range), so apart from a secondary proton buildup at the entrance, the contribution, in absolute terms, of nuclear interactions to the dose is almost constant as a function of depth. This represents, however, a varying fraction of the total dose, given the variation of the mass electronic stopping powers as a function of depth/energy. For example, for 230 MeV protons the dose contribution in the plateau region is about 20% of the total dose, while for 60 MeV protons this is only about 3%. For carbon ions, the secondary particle spectrum consists of a mix of projectile and target fragments. The latter are short-range particles that deposit all their energy close to the nuclear interaction point, while the projectile fragments have long ranges in the direction of the beam (in the case of the lighter fragments much longer than the primary particles). For example, in a 420 MeV/u carbon ion beam fragments contribute an increasing fraction of the dose with depth reaching about 50% of the total dose shortly before the Bragg peak. Especially at higher energies it is clear that nonelastic nuclear interactions play an important role in dosimetry and it has been shown that they are an important factor in evaluating the water equivalence of phantom materials. As is the case for electron beams, the charged particle fluence at equivalent depths in two different materials (scaled by the particle range) is different, which is entirely due to the nonelastic nuclear interactions. When performing dosimetry in a plastic phantom, as a substitute for water, this requires a fluence scaling factor as defined in IAEA Report TRS-398 (IAEA, 2000), and values of such factors as a function of depth in proton and carbon ion beams for some typical water substitutes are shown in Figure 12.4.

12.3.5.4 Detector Perturbation

Due to the ballistic properties of light-ion beams, it is generally assumed that the charged particle fluence is minimally altered by the presence of a small dosimeter or detector. Potential identified sources of perturbation are related to short-range charged particles such as secondary electrons and low-energy fragments from nonelastic nuclear interactions.

While the secondary electron spectrum is dominated by very low-energy particles, the ranges of the highest energy electrons are sufficient to cross an air cavity in an ionization chamber. In proton beams, Medin and Andreo (1997) showed that these contributions affect the Spencer–Attix water-to-air stopping-power ratio by about 0.5% and Verhaegen and Palmans (2001) and Palmans (2011a) showed that the secondary electron spectra are altered by changing the composition of the wall material leading to wall perturbation factors of similar order of magnitude (0.5%) for some wall materials.

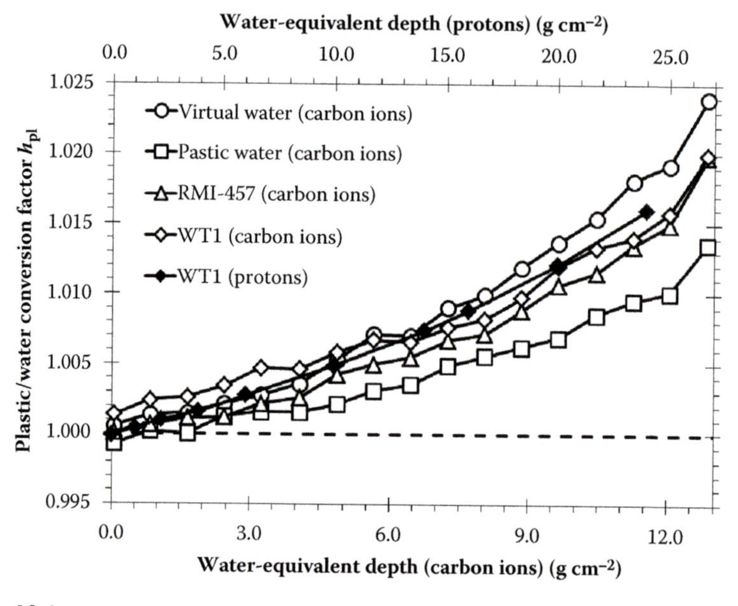

Figure 12.4

Plastic-to-water fluence scaling factors for some common water equivalent phantom materials for protons and carbon ions as a function of water equivalent depth. (From Lourenço A, PhD thesis: Water-equivalence of phantom materials in proton and carbon-ion dosimetry. University College London, 2016.)

The secondaries from nonelastic nuclear interactions affect detector response compared to water in a different way as the stopping power. This can be understood from the following example: protons incident on graphite produce about three times more alpha particles per unit of mass of medium than protons incident on water (ICRU, 2000), which is very different from the stopping-power ratio, which differs only by about 12%. Given that alpha particles contribute about 0.5% of the absorbed dose to water in a 200 MeV proton beam, this perturbation can be substantial in a carbon-rich dosimeter material. Despite this, it remains an open topic that has not yet been investigated in the literature.

12.3.5.5 Microdosimetry

The basic quantities determined in regional microdosimetry are

1. The specific energy, z, in a volume V, defined as $z = \varepsilon / \rho V$, where ε is the energy imparted in the volume and ρ the mass density of the medium, and its probability distribution $f(z)$.
2. The lineal energy y, defined as $y = \varepsilon_1 / \bar{l}$, where ε_1 is the energy imparted in the volume in a single event and $\bar{l}$ is the mean chord length of the volume, and its probability distribution $f(y)$, usually presented as $y \cdot f(y)$ or $y^2 \cdot f(y)$ plotted against $\log(y)$.

Common microdosimeters are tissue-equivalent proportional counters (TEPCs) filled with a tissue equivalent gas mixture and silicon microdosimeters. It is an on question if the relevant medium in which the microdosimetric spectrum, i.e., the distribution of lineal energy, has to be known is water, tissue, cytoplasm, or intranuclear composition. This will mainly depend on the biological endpoint one is interested in (Palmans, 2015). The issue of water equivalence discussed here can be translated into similar issues such as tissue equivalence and cytoplasm equivalence. The mass of the detector is adapted, in a TEPC via the gas pressure and in a silicon dosimeter via the geometry, such that the mean imparted energies in the site of interest in water, $\bar{\varepsilon}_w$, and in the detector, $\bar{\varepsilon}_{det}$, are equal. This can also be written as follows:

$$\left(\frac{S}{\rho}\right)_w \cdot \rho_w \cdot \bar{l}_w = \left(\frac{S}{\rho}\right)_{det} \cdot \rho_{det} \cdot \bar{l}_{det} \tag{12.5}$$

where S/ρ are the mass stopping powers, ρ the mass densities, and $\bar{l}$ the mean chord lengths of the sites. It is thus clear that water equivalence of a microdosimeter, apart from its mass density, also implies the stopping power of the detector medium to be equal. Nevertheless, it is not so important that the amplitude of microdosimetric spectra is accurately known because spectra are usually normalized. It is more important that the normalized distributions in both media are the same, for which it is assumed that the ratio of stopping powers between water and the detector medium being constant is a sufficient condition.

Water equivalence in nanodosimetry is not determined by energy deposition but by the distribution of ionization and ionization clusters (ICSDs) at the nanoscale. While distributions of energy transfer could be expected to be very different from one low-Z material to another because the distributions of cross sections are heavily determined by the molecular structure of the medium, it has been shown that ICSDs are much less influenced by the medium, and that, for example, ICSDs in a low-Z gas are representative for those in water (Grosswendt, 2002).

12.4 Energy Dependence

The energy dependence of the detector response for a given absorbed dose is a major, and incompletely resolved, issue in relative dosimetry for light-ion beams. Most, if not all, detectors exhibit energy dependence because of the signal per unit of energy deposition, either directly or indirectly, being dependent on the linear energy transfer (LET) for a given light-ion type. Even calorimeters that are considered the most direct absolute instruments for the measurement of absorbed dose do not escape from this LET dependence. Most other detectors are based on the direct or indirect measurement of ionization, and their energy dependence is in the first place determined by the energy dependence of the amount of ionization produced per unit of energy deposited, the initial recombination of ionization formed, and the dependence of any subsequent physicochemical processes on the initial ionization density. In general, all these mechanisms that are causing an underresponse at low particle energies are denoted quenching mechanisms. In the following sections, the energy dependence of a number of representative detector systems is discussed.

12.4.1 Calorimeters

Given that calorimetry is generally considered the most direct way of measuring absorbed dose according to its definition, i.e., the amount of energy deposited by ionizing radiation per unit of mass, a calorimeter would in principle be the ultimate reference for 3D dosimetry, albeit an impractical instrument for this purpose given its low sensitivity and, consequently, long measurement acquisition time. While for photon and electron beams, there is sufficient evidence that the absence of physicochemical state changes can be guaranteed for water and graphite calorimeters, there is far more limited evidence this is also the case in proton beams and even less so in heavier light-ion beams.

The main issue with water calorimetry is the chemical heat defect; the radiolysis of the water results in the presence of free reactive species initiating a chain of chemical reactions that can be endothermic or exothermic, resulting overall in a net chemical heat defect (h) and a correction factor $1/(1-h)$ for the dose determination. This correction is generally regarded as the main source of uncertainty for absolute dosimetry using water calorimeters. The primary chemical yields of the initial species, i.e., those formed by about 10^{-7} s after passage of the ionizing particles being the timescale before substantial diffusion of aqueous specifies takes place, is dependent on the LET. It can thus be expected that the chemical heat defect that results from the reactions by these species in bulk with each other, with equilibrium species in the water and with impurities, will also be LET dependent. Limited information is available about the chemical heat defect in light-ion beams; the following paragraph summarizes the theoretical and experimental evidence that has been reported in the literature.

For high-energy (low-LET) protons, pure water saturated with a chemically inert gas, such as argon or nitrogen, exhibits a small (subpercent) initial chemical heat defect and a steady state is reached after preirradiation (Palmans et al., 1996). For high-LET ions, simulations of the entire chemical reaction chain involving

38 reactions show a steady increase in the chemical energy in pure water systems due to a higher production of hydrogen peroxide than what is decomposed, resulting in a nonzero endothermic heat defect (Sassowsky and Pedroni, 2005). Experimental work comparing the heat defect of water with that of aluminum in a dual-component water/aluminum absorber confirmed this result (Brede et al., 1997). The LET dependence of the chemical heat defect of pure water derived from this work can be represented as follows:

$$h = (0.041 \pm 0.004)(e^{-(0.035\pm0.010)LET} - (1.000 \pm 0.001)) \tag{12.6}$$

where the LET is expressed in keV.μm^{-1} and the uncertainties are standard (1σ) uncertainties (Palmans, 2011b). For pure water saturated with hydrogen gas, simulations and relative experimental determination by comparison with other water systems of the chemical heat defect indicate that it is zero over the entire LET range. The theoretical explanation for this is an enhanced decomposition of hydrogen peroxide compared to the nitrogen system. In the presence of unavoidable initial oxygen concentrations, however, the hydrogen system exhibits an initial exothermic heat defect, which increases until depletion of oxygen, after which the heat defect drops abruptly to zero. This forms an effective indication to monitor when the steady-state, zero-heat defect condition is reached. Pure water to which a known quantity of sodium formate is added, serving as a deliberately added organic impurity, and which is then saturated with oxygen, is shown to exhibit an exothermic chemical heat defect in modulated proton beams that is only about half of that in a ^{60}Co beam with the same dose rate. This can be explained by a combination of the lower chemical yields for certain species by the high-LET component in the Bragg peak and the time structure of the formation of chemical species due to the beam modulation (Palmans et al., 1996). A water calorimeter has the advantage of measuring the quantity of interest, absorbed dose to water, directly and, although further confirmation is needed, the hydrogen saturated pure water system would appear to be a potential candidate to serve as an absolute reference for 3D dose measurements. Its low sensitivity, however, makes that it would need a considerable amount of time to measure distributions with sufficient resolution, for example, to resolve in detail the LET dependence of other detectors. Another issue to mention is that conductive heat transfer can be difficult to manage or to correct for in the vicinity of steep gradients. Hence, absolute measurements in the Bragg peak region, near the distal edge in modulated beams or near the penumbrae may have large uncertainty unless very high dose rates are available.

While the use of graphite calorimetry introduces a complication because it does not measure the quantity of interest, it could offer some advantages. Because of the three orders of magnitude larger thermal diffusivity of graphite compared to water, a sample, called the core, must be insulated from the rest of the medium, which is usually done by creating vacuum gaps. This makes it less prone to influence from the environment and thus to heat transfers resulting from steep dose gradients. Graphite calorimeters can also be operated in isothermal mode, in which all components are kept at a constant temperature elevated above the environmental temperature, further minimizing the effect of the environment by keeping any radiative heat transfers over the vacuum gap (or gaps) constant. When deriving a dose distribution in water from that in graphite in a quasi-homogeneous calorimeter, the difference in fluence due

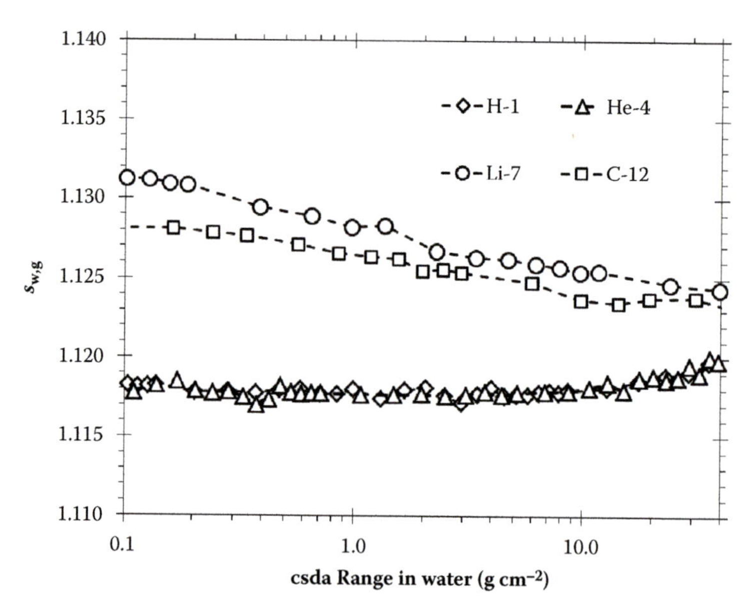

Figure 12.5

Water-to-graphite mass stopping-power ratios used in the conversion of dose-to-graphite obtained by a graphite calorimeter to dose-to-water for four types of light-ions as a function of the csda range in water. (From ICRU, Stopping powers and ranges for protons and alpha particles. ICRU Report 49. (Bethesda, MD: International Commission on Radiation Units and Measurements), 1993; ICRU, Stopping of ions heavier than helium. ICRU Report 73. (Bethesda, MD: International Commission on Radiation Units and Measurements), 2005.)

to differences in nuclear interaction cross sections is an issue leading to depth-dependent correction factors (Lühr et al., 2011; Lourenço, 2016). This could be solved by using a small probe-calorimeter (Renaud et al., 2013) or, if one is only interested in depth-dose distributions, a thin pancake-like graphite calorimeter (Picard et al., 2009) in water. In that case, assuming the variation of other fluence perturbations can be minimized, only accurate values of the variation of graphite-to-water stopping-power ratios are needed, which are for light-ions almost constant as a function of depth. To illustrate this, in Figure 12.5 water-to-graphite stopping-power ratios for monoenergetic light-ions as a function of the CSDA range in water are shown.

12.4.2 Ionization Chambers

Ionization chambers are considered the gold standard in relative dosimetry for radiotherapy, and also in light-ion beams (ICRU, 1998), either as point-like detectors in a scanning system or as 1D, 2D, and even 3D arrays. In general, the relative response of ionization chambers is considered almost constant for clinical light-ion beams, mainly supported by the modest variation of the water-to-air stopping-power ratio. Few have considered taking into account the variation of stopping-power ratios in light-ion beams, which can amount to a difference in peak-to-plateau ratio of about 0.5%, but variations of the mean energy required to produce an ion pair and perturbation correction factors are normally ignored.

The relation between dose to water and the charge reading, M_Q, of the ionization chamber, corrected for influence quantities such as temperature, pressure, humidity, polarity effect, and charge recombination, in the light-ion beam with beam quality Q, can be written as follows:

$$D_{w,Q} = \left[M_Q\right]\left[\frac{1}{\rho_{\text{air}} V_{\text{cav}}}\right]\left[\left(W_{\text{air}}/e\right)_Q \left(s_{w,\text{air}}\right)_Q p_Q\right]$$

where W_{air} is the mean energy required to produce 1 C of charge of each sign in air, e is the elementary charge, $s_{w,\text{air}}$ is the Spencer–Attix water-to-air stopping-power ratio for the charged particle spectrum at the measurement point in water, and p is the correction factor to account for any deviation from the conditions under which Bragg–Gray cavity theory is valid, such as the perturbation of the fluence by the cavity's presence, the wall and the central electrode (Palmans, 2011b; Palmans and Vatnitksy, 2015). ρ_{air} is the mass density of air and V_{cav} is the constant volume of the cavity such that the only factors that determine the energy dependence of ionization chambers in light-ion beams are $\left(s_{w,\text{air}}\right)_Q$, $\left(W_{\text{air}}/e\right)_Q$ and p_Q.

Dennis (1973) and ICRU Report 31 (ICRU, 1979) proposed a simple semi - empirical model for the energy dependence of $\left(W_{\text{air}}/e\right)_Q$ for particle energies per nucleon, E/A, higher than 1 MeV. The model is based on the assumption that in the high-energy limit of relativistic protons the value should be the same as that for high-energy electrons, for which the current best estimate is $\left(W_{\text{air}}/e\right)_e = 33.97$ J·C^{-1}, and is given by

$$\left(W_{\text{air}}/e\right)_Q = \left(W_{\text{air}}/e\right)_e \frac{(E/A)}{(E/A)-k}$$

where k is a constant dependent on the ion type and the gas. Values of k for various light-ions in air as well as the resulting variation in $\left(W_{\text{air}}/e\right)_e$ values for particle energies per atomic weight unit, E/A, between 1 and 300 MeV are given in Table 12.1. Figure 12.6 compares these values with the ones from other theoretical and experimental investigations. These variations are substantial but mainly significant for $E/A < 10$ MeV, which corresponds with the last millimeters of the ion range.

The influence of perturbation factors on ionization chamber response in light-ion beams is not necessary negligible, as discussed in Section 12.3.5.4,

Table 12.1 k Values for the Dennis Model for the Mean Energy Required to Produce 1 C of Charge of Each Sign in Air as Derived by Verhey and Lyman (1992) and the Variation Δ of the Value between E/A = 1 and 300 MeV

Ion	k (V&L)	Δ (in %)	k (G&B)	Δ (in %)
^{1}H	0.08513	9.3	0.05264	5.5
^{4}He	0.05921	6.3	—	—
^{12}C	0.04762	5.1	—	—
^{16}O	0.05218	5.6	—	—

Source: Verhey LJ and Lyman JT, *Med. Phys.*, 19, 151–153, 1992.
Note: The last two columns provide for protons the values obtained from fitting the Dennis model to the theoretical data of Grosswendt and Baek (1998).

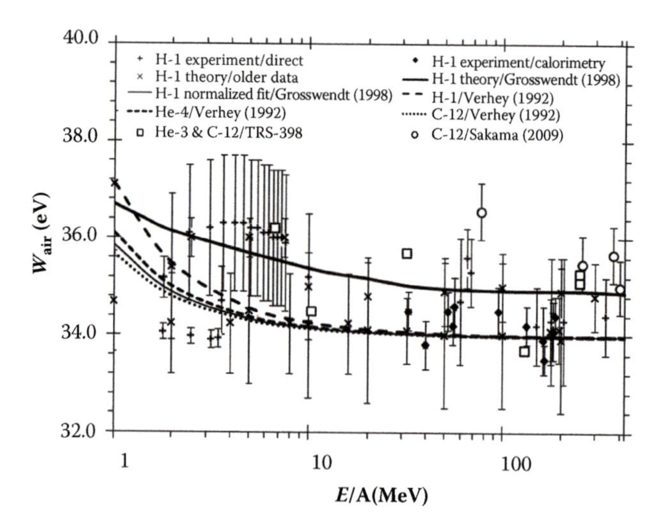

Figure 12.6

Mean energy required to produce one ion pair in air as a function of kinetic particle energy for different types of light-ions obtained from experiments (symbols). (From the compilation by Palmans H., In Das IJ and Paganetti H [eds.], *Principles and Practice of Proton Beam Therapy*, Madison, Medical Physics Publishing, 2015; by models (lines) from Verhey LJ and Lyman JT, *Med. Phys.*, 19, 151–153, 1992; and Grosswendt B and Baek WY., *Phys. Med. Biol.*, 43, 325–337, 1998).

but observed variations are generally well below 1%. Also the variation of the water-to-air stopping-power ratio was discussed previously. Overall, for ionization chambers it can be concluded that in the clinical energy range there are small, and only partially investigated, variations of their response as a function of energy. These are well within the uncertainty of the variation of the RBE with energy. In current proton therapy practice, RBE is assigned a constant value of 1.1, while variations of 10% and more were reported (Paganetti et al., 2002). For heavier ions, the RBE variation has to be taken into account in clinical planning but also their uncertainties are large (Friedrich et al., 2013). Ionization chambers can thus be safely considered as the gold standard for relative dosimetry in light-ion beams, but for studies aiming at more detailed and accurate quantifications of RBE it is nevertheless worth considering the variation of the ionization chamber response with energy.

12.4.3 Alanine

The amino acid L-alpha-alanine (chemical formula $\mathbf{CH_3CH(NH_2)COOH}$) is an attractive reference or relative dosimeter because of its stable signal after irradiation (apart from short-term fading effects), nondestructive readout, its near tissue equivalent composition and radiological properties, its linear response over a wide range of dose levels and dose rates, and its very low dependence on temperature. It is used in crystalline powder form either in a small container, embedded in film or compressed with a binding agent, such as paraffin wax, in pellets. The free radicals formed by ionizing radiation, the main radical detected is $\mathbf{CH_3C^{\cdot}HCOOH}$, remain stable in the crystal and can be quantified by electron spin resonance (ESR). Given the low sensitivity of the technique the detectors cannot be made very small for radiotherapy level doses but are nevertheless well suited for point measurements distributed in a phantom, for example, for audit purposes or end-to-end test procedures. Assuming that alanine is a one-hit detector because each molecule can only form a free radical once, the energy dependence of alanine for light-ions can be modeled by amorphous track structure models as has been described by Hansen and Olsen (1985). The relative effectiveness of alanine exposed to a mixed particle spectrum in a light-ion beam is defined as the ratio of absorbed dose to alanine in a ^{60}Co beam and the absorbed dose to alanine at the measurement point in the light-ion beam that results in the same alanine ESR signal. It can be calculated as a dose-weighted average of the energy-dependent relative effectiveness of alanine for monoenergetic particles over the particle spectrum at the point of measurement given as follows:

$$\bar{\eta}_{A1} = \frac{\sum_{t=1}^{n_{\text{proj}}} \int_{0}^{E_{\max i}} (S/\rho)_{\text{A1}} \eta_{\text{A1},t}(E) \Phi_{E,t} \mathrm{d}E}{\sum_{t=1}^{n_{\text{proj}}} \int_{0}^{E_{\max i}} (S/\rho)_{\text{A1}} \Phi_{E,t} \mathrm{d}E}, \quad (12.7)$$

where the summation is over all charged projectile types, i, and $\phi_{E,t}$ is the fluence differential in energy for charged particle type i (Bassler et al., 2008). The relative effectiveness as a function of the kinetic energy of various low-Z ions and an example illustration that the underresponse in a mixed particle field can be well modeled are shown in Figure 12.7.

For the measurement of depth-dose curves using a stack of pellets in a plastic phantom, one must consider that dose tails due to primary ions, which are normally stopped around the distal edge of the Bragg peak, can occur beyond the Bragg peak due to in-scatter from the surrounding phantom material if the pellet density is higher than that of the phantom materials. For protons this tail signal would be obvious, but for heavier ions this could complicate the interpretation and quantification of the fragmentation tail. Tunneling of ions within air gaps between pellets and phantom could also result in a tail signal, which can be avoided by orienting the stack under a small angle with respect to the beam axis.

12.4.4 Chemical Dosimeters

The chemistry that happens in an aqueous or other environment after the formation of primary reactive species by ionizing radiation offers a range of sensitive methods to quantify the amount of energy locally deposited (Spinks and Woods, 1964). These methods are usually based on chemical probes that are altered due to a chain of chemical reactions and that can be detected optically or by magnetic resonance methods. Common to all these methods is that they exhibit energy dependence due to the LET dependence of the chemical yield, the number of molecules formed per 100 eV of energy imparted, of initial reactive species formed. In addition, the energy dependence can be affected by other factors such as recombination or termination reactions that depend on the ionization density. The energy dependence of a few chemical dosimeters is discussed in the following sections.

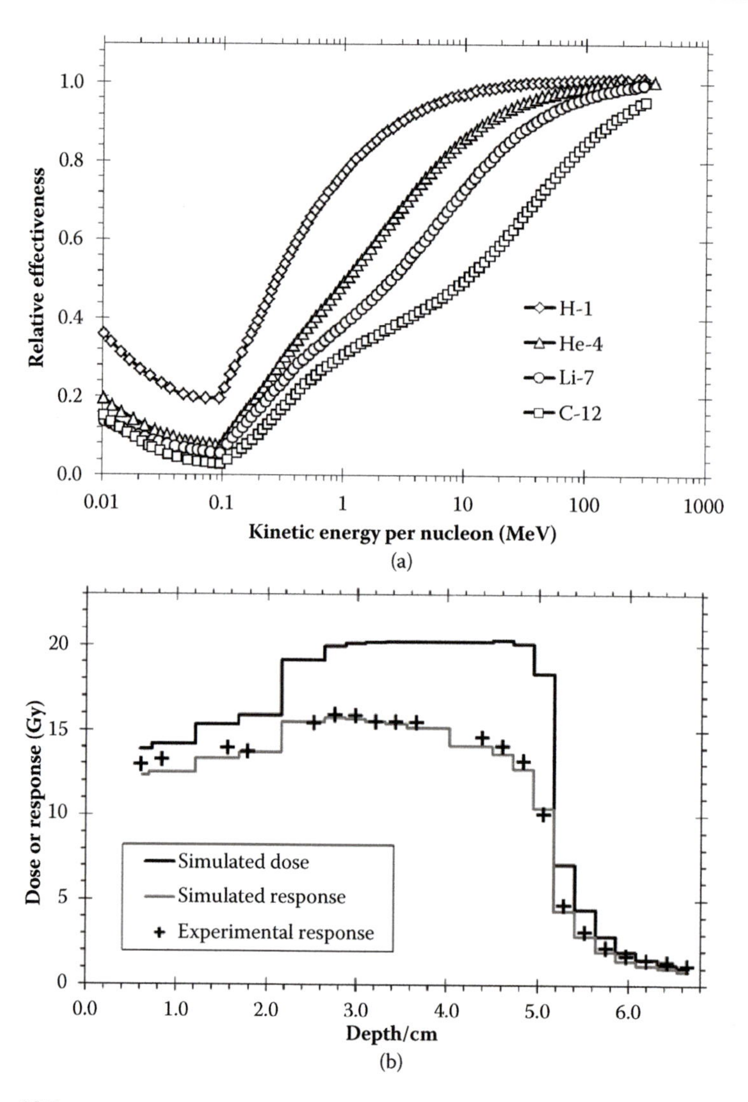

Figure 12.7

(a) Relative effectiveness of alanine for four light-ion types as a function of particle energy. (From Herrmann R, PhD thesis: Prediction of the response behaviour of one-hit detectors in particle beams. Aarhus University, Denmark, 2012.) (b) Depth-dose curve in a modulated scanned carbon ion beam obtained from ionization chamber measurements and the corresponding measured alanine response as well as the predicted alanine response based on Equation 12.7. (Reproduced from Herrmann R et al., *Med. Phys.*, 38, 1859–1866, 2011. With permission.)

12.4.4.1 Fricke dosimeter

The ferrous sulfate or Fricke dosimeter is a well-known chemical system consisting of an aerated acidic ferrous sulfate solution in which reactive species cause ferrous ions to be converted to ferric ions, mainly via reactions with hydroxyl radicals, hydroperoxyl radicals, and hydrogen peroxide. Thanks to the strongly peaked absorption of light with wavelengths of 224 and 303 nm by ferric ions, the amount of converted ions can be quantified via

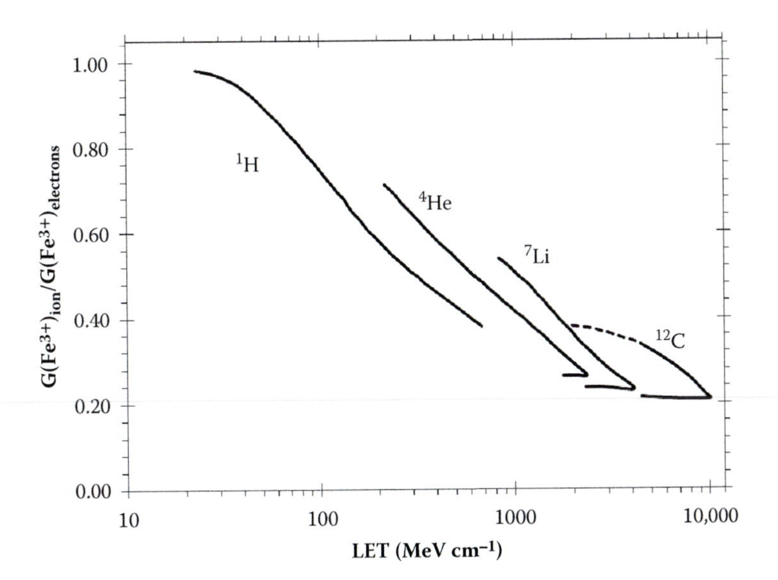

Figure 12.8

Chemical yields of ferric ions in the Fricke dosimeter relative to those of high-energy electron beams for four light-ion types as a function of LET. (From LaVerne JA and Schuler RH, *J. Phys. Chem.*, 91, 5770–5776, 1987.)

spectrophotometry. Given the dependence of the response on the concentration of primary species it is dependent on LET. Figure 12.8 shows the chemical yields of ferric ions in an aerated standard Fricke solution for different ions relative to those for relativistic electrons as a function of LET experimentally determined by LaVerne and Shuler (1987). It is clear that the yields depend on LET but for a given LET also on the type of ion. This indicates that a picture of the primary yields as, for example, given by Ross and Klassen (1996) is too simplistic, although sufficiently accurate for many applications. The application of the Fricke dosimeter to relative dosimetry of ion beams has been restricted mainly to its use in 3D gels.

12.4.4.2 Radiochromic dosimeters

These are based on a change of color as a result of radiation-induced chemical reactions. Examples are aminotriphenyl-methane dyes, leuco dyes, and pentacosa-10,12-diyonic acid (PCDA) or its lithium salt LiPCDA. Radiochromic films, such as MD-55 and EBT-types, are attractive 2D detectors that are widely used in radiotherapy since they require no development process and are nearly tissue equivalent. They consist of one or multiple sensitive layers embedded in polyester films. The sensitive layers contain PCDA or LiPCDA in crystalline format which polymerize after irradiation forming an intense coloration (Rink, 2008). This dosimeter exhibits a similar underresponse as a Fricke dosimeter and response quenching of up to 60% is observed in clinical carbon ion beams (Martišíková and Jäkel, 2010). Another dosimeter that has received considerable attention recently is the 3D PRESAGE dosimeter, consisting of a free radical initiator and leuco dye locked in a polyurethane matrix, as discussed in detail in Chapter 6. This dosimeter exhibits similar response quenching in light-ion beams as other radiochromic detectors.

12.4.4.3 Polymer gels

Another example of a chemical detector exhibiting response quenching in light-ion beams is a polymer gel in which radicals (generic symbol $R^{\bullet}$), such as hydroxyl, react with monomers that are locked up in the gel matrix in a suitable aqueous environment (Baldock et al. 2010): $R^{\bullet} + M \rightarrow R + M^{\bullet}$. These radicalized monomers initiate propagation reactions resulting in polymerization ($\dot{M}_m + M_n \rightarrow \dot{M}_{m+n}$) until a termination reaction takes place ($\dot{M}_m + \dot{M}_n \rightarrow M_{m+n}$ or $\dot{R} + \dot{M}_n \rightarrow RM_n$). The resulting polymers form opaque clusters in the gel matrix which can be quantified in 3D by optical tomography or, since the polymers also have a different magnetic susceptibility than the original solution with the monomers, by magnetic resonance imaging, as discussed in detail in Chapter 5. In such polymerizing detectors, the energy dependence can be understood as that of a one-hit, two-component detector, i.e., the underresponse is not only due to the reduced amount of radicals produced per unit of energy imparted at higher ionization density, but also by the increased probability of termination reactions. This explains in general a more complex behavior as a function of energy, absorbed dose, and dose rate than for simple one-hit detectors such as alanine.

12.4.5 Solid-State Devices: Diodes and Luminescence Detectors

The operation of a considerable number of dosimetric devices is based on solid-state physics. Two main operational principles can be distinguished as will be discussed in the following sections.

12.4.5.1 Diodes

In diode detectors, the energy levels limiting the band gap are different in two regions in the device, either by a different doping (p–n junction) or by a metallic contact (Schottky diode). This results in the negative and positive charge carriers (electrons and holes) in an intermediate region drifting away in opposite directions resulting in a depletion zone, where no charge carriers are present and over which an internal electric field exists. Radiation-induced electrons and holes will then be collected by the electric field and a measureable current is induced in the device. In silicon diode detectors, p–n junctions are most commonly established by implanting electron donor and acceptor atoms on different sides. In natural diamond detectors, the depletion layer is created by applying an external bias voltage while for artificial, chemical vapor deposition (CVD) diamond detectors, generally a Schottky diode, is created. The energy dependence of the mean energy required to produce an ion pair in silicon diode detectors was found to be extremely small (Scholze et al., 1998) and the same can be assumed for diamond detectors. The main mechanism for energy dependence is then initial recombination. This can explain why in a very thin commercial CVD diamond (thus having a very high electric field across the small depletion layer) the LET dependence is found to be extremely minimal (Marinelli et al., 2015) and only in very high-LET beams a small underresponse is observed (Rossomme et al., 2016).

12.4.5.2 TLDs and OSLDs

Luminescence detectors are crystalline solid-state devices in which impurities or interstitial atoms exhibit energy levels in the bandgap forming trapping centers for electrons or holes. These traps can be populated by radiation-induced electrons or holes and subsequently released by thermal or optical stimulation.

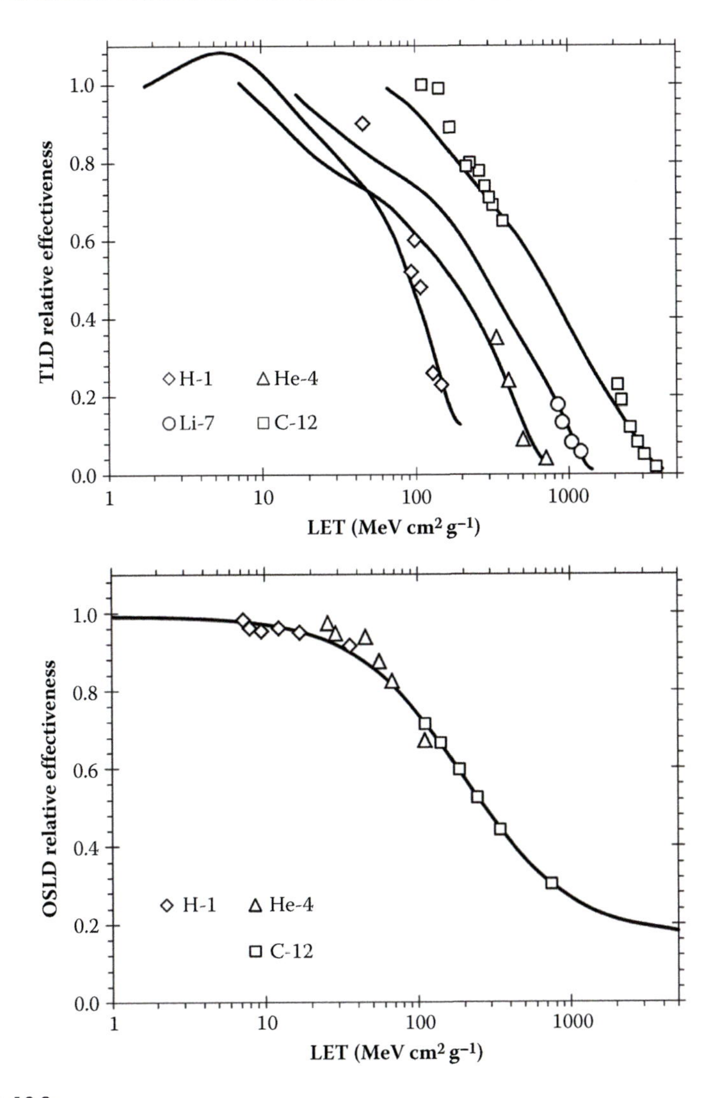

Figure 12.9

Relative effectiveness of a TLD-100 and an Al_2O_3-based OSL detector for four light-ion types as a function of LET. (Upper graph: From Geiss OB et al., *Nucl. Instrum. Methods Phys. Res. B* 142: 592–598, 1998; lower graph: From Yukihara EG et al., *Phys. Med. Biol.*, 60, 6613–6638, 2015.)

The behavior of such detectors can be complicated due to the distribution of the energy levels of shallow electron and hole traps in thermal equilibrium with the conduction and valence bands, respectively, the distribution of energy levels of luminescence centers and the different mobilities of electrons and holes. It is therefore difficult to model their behavior as a function of ionization cluster distributions or LET, and the complexity can be substantially dependent on the type of crystal used, the distribution of luminescence centers, and the level of control on the purity and energy levels of trapping centers that can be exerted during the production process. For example, Figure 12.9 shows that for TLD-100 (LiF-based) both theoretical models and experimental data indicate that the dependence of the relative response as a function of LET is complex and dependent on the ion type, while an Al_2O_3-based optically stimulated luminescence detector exhibits a more unique dependence on LET, regardless of the ion type. In Chapter 4, it is shown that analysis of TLD and OSLD curves offer the potential to provide not only dose measurements but also information on the LET of the particle beam, particularly in and beyond the spread-out Bragg peak.

12.4.5.3 Scintillators

Scintillators form another class of luminescence detectors that emit light immediately following the absorption of ionizing radiation by fluorescence or phosphorescence and are used in small-volume detectors combined with optical fibers and photodetectors, 2D screens combined with charge coupled device (CCD) cameras, and also 3D configurations embedded in organic plastics or liquids combined with multiple CCD cameras. An interesting approach to the energy dependence was developed by Safai et al. (2004) after observing that two different phosphors embedded in a polyethylene matrix exhibited an opposite energy dependence; Gd_2O_2S:Tb showed a decreasing response with increasing LET, while the response of (Zn,Cd)S:Ag increased. This allowed preparing a mixture (in a ratio 4:1) of both phosphors yielding a scintillating system that exhibited no quenching.

12.4.6 Micro- and Nanodosimeters

Energy dependence in micro- and nanodosimeters has to be considered rather differently since, in principle, for a given kinetic energy of an ion the lineal energy or ionization cluster size can have any value. For microdosimetric devices, it is of importance that the lineal energy scale is adequately calibrated and that the signal is proportional to the lineal energy. The calibration is usually performed based on characteristic edges in the $f(y)$ distribution when the detector is exposed to a reference radioactive source (often an alpha-emitter). The requirement of proportionality often limits the lineal energy range for which a detector can be used. A detector threshold may also be determined by electronic noise becoming dominant at lower signal values. Similar considerations apply to nanodosimeters.

12.5 Conclusion

In this chapter, aspects of the interaction of light-ions with matter that make light-ion beams distinct from photon and electron beams are discussed, followed by an overview of different detector technologies to determine the absorbed dose to water at a point or in 3D, in light-ion beams. The main issues with respect to 3D dosimetry that are discussed are the water equivalence and energy dependence of detectors while other issues such as spatial resolution are treated in Chapter 23. Ionization chambers can in general be considered the gold standard for relative dosimetry in light-ion beams (ICRU, 1998), but their response nevertheless exhibits a small energy dependence that is worth considering when aiming for more detailed and better quantitative data on experimental RBE values. Other relative detectors can exhibit substantial response quenching as a function of energy and should therefore always be benchmarked against ionization chambers prior to clinical use.

References

Baldock C et al. (2010) Polymer gel dosimetry. *Phys. Med. Biol.* 55: R1–R63.

Bassler N et al. (2008) The antiproton depth dose curve measured with alanine detectors. *Nucl. Instrum. Meth. Phys. Res. B* 266: 929–936.

Brede HJ, Hecker O and Hollnagel R (1997) Measurement of the heat defect in water and A-150 plastic for high-energy protons, deuterons and α-particles. *Radiat. Prot. Dosim.* 70: 505–508.

Dennis JA (1973) Computed ionization and kerma values in neutron irradiated gases. *Phys. Med. Biol.* 18: 379–395.

Fano U (1954) Note on the Bragg-Gray cavity principle for measuring energy dissipation. *Radiat. Res.* 1: 237–240.

Friedrich T et al. (2013) Systematic analysis of RBE and related quantities using a database of cell survival experiments with ion beam irradiation. *J. Radiat. Res.* 54: 494–514.

Geiss OB, Krämer M and Kraft G (1998) Efficiency of thermoluminescent detectors to heavy charged particles. *Nucl. Instrum. Meth. Phys. Res. B* 142: 592–598.

Gottschalk B (2010) On the scattering power of radiotherapy protons. *Med. Phys.* 37: 352–367.

Goudsmit S and Saunderson JL (1940) Multiple scattering of electrons. *Phys. Rev.* 57: 24–29.

Grosswendt B (2002) Formation of ionization clusters in nanometric structures of propane-based tissue-equivalent gas or liquid water by electrons and α-particles. *Radiat. Environ. Biophys.* 41: 103–112.

Grosswendt B and Baek WY (1998) W values and radial dose distributions for protons in TE-gas and air at energies up to 500 MeV. *Phys. Med. Biol.* 43: 325–337.

Hansen JW and Olsen KJ (1985) Theoretical and experimental radiation effectiveness of the free radical dosimeter alanine to irradiation with heavy charged particles. *Radiat. Res.* 104: 15–27.

Herrmann R et al. (2011) Dose response of alanine detectors irradiated with carbon ion beams. *Med. Phys.* 38: 1859–1866.

Herrmann R (2012) PhD thesis: Prediction of the response behaviour of one-hit detectors in particle beams. Aarhus University, Denmark.

Highland VL (1975) Some practical remarks on multiple scattering. *Nucl. Instr. Meth.* 129: 497–499.

IAEA (2000) Absorbed dose determination in external beam radiotherapy: An international code of practice for dosimetry based on standards of absorbed dose to water. Technical Reports Series No. 398. (Vienna, Austria: International Atomic Energy Agency).

ICRU (1979) Average energy required to produce an ion pair. ICRU Report 31. (Washington, DC: International Commission on Radiation Units and Measurements).

ICRU (1993) Stopping powers and ranges for protons and alpha particles. ICRU Report 49. (Bethesda, MD: International Commission on Radiation Units and Measurements).

ICRU (1998) Clinical proton dosimetry: Part I. Beam production, beam delivery and measurement of absorbed dose. ICRU Report No 59. (Bethesda, MD: International Commission on Radiation Units and Measurements).

ICRU (2000) Nuclear data for neutron and proton radiotherapy and for radiation protection dose. ICRU Report 63. (Bethesda, MD: International Commission on Radiation Units and Measurements).

ICRU (2005) Stopping of ions heavier than helium. ICRU Report 73. (Bethesda, MD: International Commission on Radiation Units and Measurements).

Karger CP et al. (2010) Dosimetry for ion beam radiotherapy. *Phys. Med. Biol.* 55: R193–R234.

LaVerne JA and Schuler RH (1987) Radiation chemical studies with heavy ions: Oxidation of ferrous ion in the Fricke dosimeter. *J. Phys. Chem.* 91: 5770–5776.

Lewis HW (1950) Multiple scattering in infinite medium. *Phys. Rev.* 78: 526–529.

Lourenço A (2016) PhD thesis: Water-equivalence of phantom materials in proton and carbon-ion dosimetry. University College London, UK.

Lühr A et al. (2011) Fluence correction factors and stopping power ratios for clinical ion beams. *Acta Oncol.* 50: 797–805.

Marinelli M et al. (2015) Dosimetric characterization of a microDiamond detector in clinical scanned carbon ion beams. *Med. Phys.* 42: 2085–2093.

Martišíková M and Jäkel O (2010) Dosimetric properties of Gafchromic EBT films in monoenergetic medical ion beams. *Phys. Med. Biol.* 55: 3741–3751.

Medin J and Andreo P (1997) Monte Carlo calculated stopping-power ratios water/air for clinical proton dosimetry (50–250 MeV). *Phys. Med. Biol.* 42: 89–105.

Molière G (1948) Theorie der Streuung schneller geladener Teilchen II—Mehrfach- und Vielfachstreuung. *Z. Naturforsch.* 3a: 78–97.

Moyers MF et al. (2007) Calibration of a proton beam energy monitor. *Med. Phys.* 34: 1952–1966.

Paganetti H et al. (2002) Relative biological effectiveness (RBE) values for proton beam therapy. *Int. J. Radiat. Oncol. Biol. Phys.* 53: 407–421.

Palmans H et al. (1996) Water calorimetry and ionization chamber dosimetry in an 85-MeV clinical proton beam. *Med. Phys.* 23: 643–650.

Palmans H and Vynckier S (2002) Reference dosimetry for clinical proton beams. In: *Recent Developments in Accurate Radiation Dosimetry*, pp. 157–194. (Eds. Seuntjens JP and Mobit PN, Madison, WI: Medical Physics Publishing).

Palmans H (2011a) Secondary electron perturbations in Farmer type ion chambers for clinical proton beams. In: Standards, Applications and Quality Assurance in Medical Radiation Dosimetry—Proceedings of an International Symposium, Vienna, 9–12 November 2010, Vol. 1, pp. 309–317. (Vienna, Austria: International Atomic Energy Agency).

Palmans H (2011b) Dosimetry. In: *Proton Therapy Physics*, pp. 191–219. (Ed. Paganetti H, London: CRC Press—Taylor & Francis).

Palmans H (2015) Proton beam interactions. In: *Principles and Practice of Proton Beam Therapy*, pp. 43–80. (Eds. Das IJ and Paganetti H, Madison, WI: Medical Physics Publishing).

Palmans H and Vatnitsky SM (2015) Dosimetry and beam calibration. In: *Principles and Practice of Proton Beam Therapy*, pp. 317–351. (Eds. Das IJ and Paganetti H, Madison, WI: Medical Physics Publishing).

Palmans H et al. (2015) Future development of biologically relevant dosimetry. *Br. J. Radiol.* 87: 20140392.

Picard S, Burns DT and Roger P (2009) Construction of an absorbed-dose graphite calorimeter. Rapport BIPM-2009/01. (Sèvres, France: Bureau International des Poids et Mesures).

Renaud J et al. (2013) Development of a graphite probe calorimeter for absolute clinical dosimetry. *Med. Phys.* 40: 020701.

Rink A (2008) PhD thesis: Point-based ionizing radiation dosimetry using radiochromic materials and a fibreoptic readout system. University of Toronto, ON, Canada.

Ross CK and Klassen NV (1996) Water calorimetry for radiation dosimetry. *Phys. Med. Biol.* 41: 1–29.

Rossi B (1952) *High-Energy Particles.* (New York, NY: Prentice-Hall).

Rossomme S et al. (2016) Under response of a PTW-60019 microDiamond detector in the Bragg peak of a 62 MeV/n carbon ion beam. *Phys. Med. Biol.* 61: 4551–4563.

Rutherford E (1911) The scattering of α and β particles by matter and the structure of the atom. *Philos. Mag.* 21: 669–688.

Safai S, Lin S and Pedroni E (2004) Development of an inorganic scintillating mixture for proton beam verification dosimetry. *Phys. Med. Biol.* 49: 4637–4655.

Sassowsky M and Pedroni E (2005) On the feasibility of water calorimetry with scanned proton radiation. *Phys. Med. Biol.* 50: 5381–5400.

Scholze F, Rabus H and Ulm G (1998) Mean energy required to produce an electron-hole pair in silicon for photons of energies between 50 and 1500 eV. *J. Appl. Phys.* 84: 2926–2939.

Spinks JWT and Woods RJ (1964) *An Introduction to Radiation Chemistry.* (New York, NY: Wiley).

Tsai YS (1974) Pair production and bremsstrahlung of charged leptons. *Rev. Mod. Phys.* 46: 815–851.

Vatnitsky SM and Palmans H (2015) Detector systems. In: *Principles and Practice of Proton Beam Therapy*, pp. 275–316. (Eds. Das IJ and Paganetti H, Madison, WI: Medical Physics Publishing).

Verhaegen F and Palmans H (2001) A systematic Monte Carlo study of secondary electron fluence perturbation in clinical proton beams (70–250 MeV) for cylindrical and spherical ion chambers. *Med. Phys.* 28: 2088–2095.

Verhey LJ and Lyman JT (1992) Some considerations regarding w values for heavy charged-particle radiotherapy. *Med. Phys.* 19: 151–153.

Yukihara EG et al. (2015) Time-resolved optically stimulated luminescence of Al2O3:C for ion beam therapy dosimetry. *Phys. Med. Biol.* 60: 6613–6638.

13

Monte Carlo Applications in Clinical Three-Dimensional Dosimetry

Indrin J. Chetty and Joanna E. Cygler

13.1 Introduction

13.1.1 Rationale for Monte Carlo Methods for Photon and Electron Dose Calculations

When high (megavoltage [MV]) energy x-rays impinge on patient tissues, they interact with electrons primarily via the Compton effect resulting in scattered photons and electrons. These electrons interact within the tumor and normal patient tissues, depositing dose locally. One could then imagine that the most accurate dose algorithm for modeling such dose deposition would be one that mimics the actual patient treatment. The Monte Carlo (MC) method performs calculation of photon and electron tracks within the linear accelerator (linac) and patient tissues, and as such imitates how radiation is physically delivered to the patient. Historically, patient dose calculation algorithms were developed using empirical approaches assuming the patient was composed of water-equivalent tissues. Algorithms evolved over time to account for physical interactions in the linac treatment head and the patient, for example, the superposition–convolution (SC) method. Although such algorithms are accurate for most clinical situations, there exist treatments, for example, irradiations with external electron beams, and of small "island-like" lung lesions (surrounded by lung) with photon beams, in which the use of MC-based algorithms may be warranted (Benedict et al., 2010). Interested readers are encouraged to review other relevant articles on MC methods (e.g., Chetty et al., 2007; Reynaert et al., 2007; Verhaegen and Seco, 2013) and dose calculation algorithms in general for additional details (e.g., Ahnesjo and Aspradakis, 1999; Papanikolaou et al., 2004).

The impact of accurate dose distributions on patient clinical outcomes is of ultimate importance. Evidence exists that dose differences on the order of 7% are clinically detectable (Dutreix, 1984; Papanikolaou et al., 2004). It has also been demonstrated that 5% changes in dose can result in 10%–20% changes in tumor control probability (TCP) or up to 20%–30% changes in normal tissue complication probabilities (NTCPs) if the prescribed dose falls along the steepest region of the dose–response curves (Papanikolaou et al., 2004). A comprehensive overview of radiobiological and clinical aspects related to dose–response curves observed in radiation therapy is given in Chapter 2. Although more data are necessary to better understand dose–volume–effect relationships, it is clear that the accuracy of dose distributions is of significant concern in radiation therapy. It is therefore crucial that highly accurate dose algorithms are used in routine treatment planning to span the range of clinically observed situations as discussed in detail in Chapter 2.

13.2 Review of the Monte Carlo Method

13.2.1 Overview of Monte Carlo-Based Photon and Electron Transport

Most generally, the MC method is a numerical technique in which repeated random sampling is performed to simulate a stochastic process. In the limit of a large number of events, the MC method estimates the aggregate behavior of a system based on the behavior of individual events. In the context of simulation of radiation transport, Rogers and Bielajew (1990) have provided the following eloquent description: "The Monte Carlo technique for the simulation of the transport of electrons and photons through bulk media consists of using knowledge of the probability distributions governing the individual interactions of electrons and photons in materials to simulate the random trajectories of individual particles. One keeps track of physical quantities of interest for a large number of histories to provide the required information about the average quantities."

Because of the low probability of photon interactions in patient-like tissues, photon transport can be handled by simulating individual interaction events (analog transport), without a loss in calculation efficiency. For instance, a 2-MeV photon interacting in water has a mean free path of approximately 20 cm. Electrons, however, being charged particles experience numerous interactions. For instance, a 2-MeV electron has a range of just 1 cm in water, and in so doing will undergo over a million collisional-type interactions. Therefore, simulation of electrons using interaction-by-interaction transport would be prohibitively long, requiring days or weeks, even using currently available computational technology. In 1963, Berger proposed the condensed history technique (CHT) for electron transport, with the realization that most electron interactions lead to very small changes in the electron energy and/or direction (Berger, 1963). Many such "small-effect" interactions can therefore be grouped into relatively few condensed history "steps" and their cumulative effect taken into account by sampling energy, direction, and position changes from appropriate distributions of grouped single interactions (Berger, 1963). The CHT is considered the most influential development in the field of MC transport of photon/electron transport in matter; without it, MC-based patient calculations in radiation therapy would not be practically feasible. Detailed reviews of MC-based transport of photons and electrons are provided in the following references: Rogers and Bielajew (1990), Faddegon and Cygler (2006), Ma and Sheikh-Bagheri (2006), Chetty et al. (2007), and Verhaegen and Seco (2013).

A large number of general purpose codes have been developed for simulating transport of photons and electrons, including the EGS (Ford and Nelson, 1978; Nelson et al., 1985; Kawrakow, 2000a,b; Kawrakow and Rogers, 2000), ITS (Halbleib and Melhorn, 1984; Halbleib, 1988), MCNP systems (Briesmeister, 1993; Brown, 2003), PENELOPE (Baro et al., 1995), and GEANT4 (Agostinelli et al., 2003) code systems. Many of these codes have incorporated specialized features for simulation of interactions in the linac treatment head. For instance, the BEAM code system (Rogers et al., 1995, 2004; Walters and Rogers, 2004) includes a number of variance reduction techniques to enhance the efficiency of the simulation (Kawrakow et al., 2004). Comprehensive reviews of MC simulation of radiotherapy beams from linacs are available elsewhere (Ma and Jiang, 1999; Verhaegen and Seuntjens, 2003).

A novel class of MC-based codes optimized for photon and electron beams in patient-specific geometries has invigorated interest in the use of MC-based dose calculations for radiotherapy treatment planning. These codes, for example, Macro Monte Carlo (MMC), Voxel Monte Carlo (VMC++), X-Voxel Monte Carlo (XVMC), Dose Planning Method (DPM), have made it possible to perform MC-based photon beam dose calculations within minutes even on a single processor (Chetty et al., 2007; Gardner et al., 2007; Hasenbalg et al., 2008; Fragoso et al., 2009). The main advantage of these "second-generation codes" over "first-generation codes," such as EGS4, MCNP, GEANT4, and Penelope, is that the transport mechanics and boundary crossing implementations are optimized for radiation transport in the therapeutic energy ranges, and over a range of material atomic numbers and densities characteristic of human tissues. Consequently, they converge faster, i.e., fewer condensed history steps are required for the same precision relative to first-generation codes. Commercially available, Food and Drug Administration (FDA)-approved, MC-based treatment planning systems for photon and electron beams, including Nucletron (Oncentra), Elekta (XiO), Elekta (Monaco), Varian (Eclipse), BrainLab (iPlan), and Accuracy (Multiplan) utilize "second-generation" code systems implemented on multiple CPU processors, typically four to eight, running in parallel on a single workstation.

13.2.2 General Techniques for Simulating linacs

The goal of MC simulation of interactions in the linac treatment head is to produce the phase space, which is defined, nominally, by the following parameters for each particle: x, y, z (position); u, v, w (direction); E (energy); and Q (charge). Detailed simulation of photon and electron interactions in the linac treatment head can provide a wealth of information about the spatial, energy, and angular fluence distributions of particles interacting in different components of the treatment head. Sensitivity studies have been performed to understand how the geometric design and material construction of the various treatment head structures affect the fluence distributions

(Chetty et al., 2007). For photon beams, several important trends of the effects on dose distributions (and output factors) of the incident electron on target spatial, energy and angular distributions, and the geometric and material constructions of the primary collimator and flattening filter have been reported (Sheikh-Bagheri and Rogers, 2002; Tzedakis et al., 2004). In the context of electron beams, sensitivity analysis of dose to source and geometry details has been performed using large electron fields (Huang et al., 2005; Schreiber and Faddegon, 2005). Both symmetrical effects (beam energy and spot size, width of the peak in the energy spectrum, distance between the scattering foils, and thickness of the foils) and asymmetrical effects (lateral shifts of the secondary scattering foil and outer wall of the monitor chamber) were analyzed (Huang et al., 2005; Schreiber and Faddegon, 2005).

A beam model, derived from the phase space, is subsequently used for the patient-specific dose calculations during treatment planning. The AAPM Task Group Report No. 105 (Chetty et al., 2007) defines three possible routes for constructing a beam model: (1) Direct use of phase-space information from the linac treatment head simulation, which provides details on the physical interactions within the treatment head, but may not be practical for routine clinical application (Ma and Rogers, 1995a; Ma et al., 1997, 1999; Schach von Wittenau et al., 1999; Fix et al., 2001). This approach requires detailed knowledge of simulation parameters and the geometric and material compositions of the linac components, which may be subject to inaccuracies, and proprietary concerns. (2) Creation of virtual, multiple-source models derived from the original simulated phase-space data, in the form of correlated histogram distributions that approximately retain correlation of the particle's position, energy, and direction (Ma and Rogers, 1995a; Ma et al., 1997; Faddegon et al., 1998; Ma, 1998; Schach von Wittenau et al., 1999; Chetty et al., 2000; Deng et al., 2000; Fix et al., 2000, 2001). Multiple-source models are also reliant on simulation parameters and geometric and material specifications of the component structures, however, may be "adjusted" (without redoing the simulations), based on measurements to optimize agreement between calculations and measured data. (3) Development of models derived from a standard set of measurements (measurement-driven models), which have the advantage of being independent of details of the linac treatment head. Fluence distributions may be developed using analytical models with parameters optimized by minimizing differences between calculations and measurements (Faddegon and Blevis, 2000; Deng et al., 2001a; Janssen et al., 2001; Yang et al., 2004; Siljamaki et al., 2005; Ulmer et al., 2005; Aljarrah et al., 2006). Measurement-driven models are similar to those of conventional dose algorithms and do not require expertise with MC linac simulation. However, care must be exercised during measurements to minimize systematic errors, which will be propagated during development of the model.

13.3 Application of Monte Carlo in Three-Dimensional Photon and Electron Dosimetry

13.3.1 Monte Carlo–Based Simulation for Three-Dimensional and Intensity-Modulated Radiotherapy Planning

13.3.1.1 Methods of Modeling of the Jaws, Multileaf Collimator, and Other Patient-Specific Beam Modifiers

MC-based transport through the patient-specific components (e.g., the field-defining jaws or multileaf collimator [MLC]) can generally be classified into explicit and approximate schemes (Chetty et al., 2007). In an explicit scheme, the detailed structure geometry is included in the simulation and all particles (with appropriate energy cutoff values) are transported through these components (Ma et al., 1999; Heath and Seuntjens, 2003; Fippel, 2004). Approximate transport methods are utilized to increase simulation efficiency (Liu et al., 2001; Aaronson et al., 2002; Siebers et al., 2002; Tyagi et al., 2007). Examples include adapted transport simulations through the beam modifying devices, such as the tracking of first Compton-scattered photons only (Siebers et al., 2002), or photon-only transport (Tyagi et al., 2007). Other approximate methods involve beam "sub-source" fluence distributions derived for structures, such as the jaws (Fix et al., 2000, 2001), electron applicators (Ma et al., 1997; Fix et al., 2013), and the MLC (Chetty et al., 2000). Appropriate benchmarking must be performed by developers and vendors to evaluate the trade-offs between speed and accuracy related to the use of various beam model approaches in MC-based modeling of patient-specific beam modifiers.

13.3.1.2 Absolute Dose and Independent Dose/Monitor Unit Calculation

MC typically calculates dose per initial source particle (primary history) in the entire accelerator model in units of Gy/initial particle. This quantity has then to be linked to the clinically used absolute dose/monitor unit (MU), in units Gy/MU. Therefore a calibration coefficient is required for conversion of Gy/initial particle to absolute dose/MU. In order to determine this calibration coefficient, a simulation under reference calibration conditions, using large enough number of particles (histories) to achieve low statistical uncertainty (below 1%), must be performed (Ma et al., 2004; Popescu et al., 2005). The calibration factor derived this way is the ratio of the calibration dose at the reference point to the MC-calculated dose in Gy/initial particle at the same point. This calibration can be then used to determine the dose in Gy/MU delivered in each voxel in the patient/phantom. MC algorithms have been shown to accurately predict MU values needed to deliver prescribed dose (Zhang et al., 1999; Cygler et al., 2004; Ding et al., 2006; Edimo et al., 2009, 2014; Zhang et al., 2013; Vandervoort et al., 2014; Zhao et al., 2014). However, it is always prudent to perform independent MU verification as part of routine quality assurance (QA) of a patient's plan. One has to bear in mind that traditional methods of MU calculations are based on water tank collected beam data and geometries, and do not account for arbitrary angle of incident beam, irregularities of anatomical contours, or heterogeneities present in a patient's body, all of which can change the value of the isodose surface at the prescription depth. Therefore, differences, especially for electron beams, between MC and hand-calculated MUs may occur (Zhang et al., 2013).

13.3.1.3 Intensity-Modulated Radiotherapy Optimization

Beamlet calculations for intensity-modulated radiotherapy (IMRT) optimization are often still computed using pencil-beam (PB) algorithms, primarily because of the large calculation time required to perform the many MC-based beamlet calculations required for inverse planning. With MC simulation during optimization one can incorporate tissue heterogeneity effects, as well as MLC leakage and scattered radiation. Inaccurate dose calculation algorithms used during optimization have been shown to produce convergence errors in which the optimized fluence pattern differs from that corresponding to the optimal dose distribution (Jeraj et al., 2002). Example studies in the area of MC calculations for IMRT optimization have focused on the areas of evaluation of cost functions and convergence (Jeraj et al., 2002), MC-based beamlet dose matrices for planning using direct aperture optimization (Bergman et al., 2006), and efficient approaches for MC-based beamlet calculations (Zhong and Chetty, 2012). Detailed reviews of other research related to MC simulation of IMRT and inverse planning can be found elsewhere (Verhaegen and Seuntjens, 2003; Siebers and Ma, 2006).

13.3.1.4 Statistical Fluctuations and Impact on Dose Distributions

Given the stochastic nature of radiation dose deposition, it is not possible to achieve absolute precision with MC dose calculation, unless an infinite number of histories are simulated. Dose is computed from the average energy of individual particle tracks deposited within each voxel. It has been shown that the standard error in the mean dose is proportional to the square root of the mean dose deposited (Keall et al., 2000; Sempau and Bielajew, 2000; Kawrakow, 2004). A useful metric for evaluation of the statistical uncertainty of clinical dose distributions is the fractional uncertainty (or the standard error in mean dose divided by the mean dose), which following from the previous argument is inversely proportional to the square root of the mean dose (Chetty et al., 2007). This suggests that the fractional uncertainty in a voxel decreases as the dose increases, implying that the relative uncertainty in the dose in high-dose regions will be smaller than in low-dose regions, even though the absolute uncertainty is usually larger.

The AAPM Task Group Report No. 105 recommends that specification of uncertainties to single voxels (e.g., maximum dose point) be avoided because of the significantly large statistical fluctuation in the dose to individual voxels (Chetty et al., 2007). Instead, specification of statistical uncertainties over a volume, such as the planning target volume (PTV) or a dose–volume, such as the volume receiving greater than X% of the treatment dose, will provide a much more reliable estimate of the dose uncertainty (Rogers and Mohan 2000; Chetty et al., 2007).

In general, integrated dose quantities, such as dose–volume histograms (DVHs) are less sensitive to statistical uncertainty (Buffa and Nahum, 2000; Jiang et al., 2000; Keall et al., 2000; Sempau and Bielajew, 2000; Kawrakow, 2004; Siebers et al., 2005; Chetty et al., 2006). Therefore, for "parallel" organs (e.g., the lung, liver), where the dose–volume–effect relationships are driven by integrated metrics, such as the mean dose, large statistical fluctuations will likely not have a significant impact on such dose indices. However, for "serial" organs, such as the spinal cord, where the maximum dose is of clinical relevance, the statistical uncertainty to the near-maximum dose voxels (e.g., D1% or D2%) will be of significance (Chetty et al., 2006). Statistical noise in dose distributions can be mitigated as follows: (1) by increasing the number of histories simulated, which can require significantly larger computational time; (2) by deconvolving the uncertainty from the DVH (Jiang et al., 2000; Sempau and Bielajew, 2000), which enables substantial reduction in the required number of histories simulated, depending on the complexity of the DVH; and (3) by denoising the dose distributions (see Section 13.3.1.5).

Decisions about the acceptable amount of statistical jitter in MC isodose distributions should be made with input from the clinical team (physicians and planners) and should include consideration of the uncertainty in the three-dimensional (3D) dose distribution in the target as well as surrounding normal organs, and their dose–volume–effect dependencies (serial or parallel). In this regard, tools for viewing statistical uncertainties in combination with planned dose distributions will be of significant benefit. For instance, Figure 13.1a through d shows maps of the statistical uncertainties mapped onto the isodose distributions, as a function of the number of histories simulated (Chetty et al., 2006). Such information will enable the planning team to determine the acceptable statistical uncertainty level by evaluating the uncertainties within the PTV and normal organs at all levels of the 3D dose distribution. Figure 13.2a through d shows DVHs along with uncertainty-volume histograms (UVHs) for the gross tumor volume for an example patient dose distribution (Chetty et al., 2006).

13.3.1.5 Denoising of Dose Distributions

The idea of denoising distributions was first proposed for electron beam dose distributions using several digital filtering techniques to remove statistical noise from isodose lines (IDLs) (Deasy, 2000). Algorithms for smoothing include wavelet threshold denoising (WTD) (Deasy et al., 2002), locally adaptive Savitzky–Golay (LASG) filter (Kawrakow, 2002), anisotropic diffusion (AD) denoising (Miao et al., 2003), and the iterative reduction of noise (IRON) method (Fippel and Nusslin, 2003). Calculation efficiency increases of 2–20 have been observed with these algorithms, while maintaining the general features of the dose distributions.

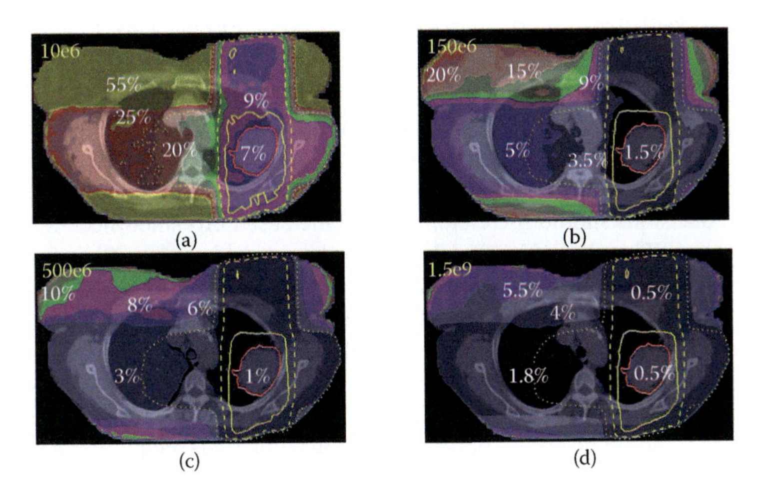

Figure 13.1

Color-wash displays of the statistical uncertainties in dose, shown in the axial plane for an example lung cancer patient, as a function of the total number of simulated histories: (a) 10E6, (b) 150E6, (c) 500E6, and (d) 1.5E9. Values on the plots indicate the percentage relative uncertainties averaged over the dose voxels in the local region. The 90%, 50%, and 10% isodose lines are demarcated in the yellow solid, dashed, and dotted lines, respectively. The gross tumor volume is illustrated in the solid red line. (Reproduced from Chetty IJ et al., *Int. J. Radiat. Oncol. Biol. Phys.*, 65, 1249–1259, 2006. With permission.)

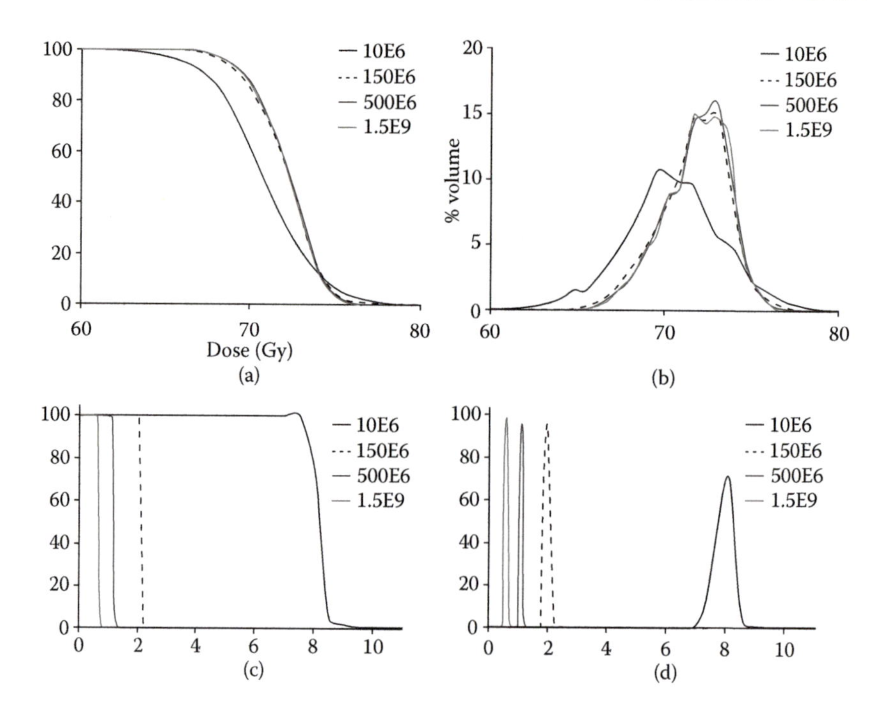

Figure 13.2

Dose–volume histograms for the gross tumor volume (GTV), shown in (a) cumulative and (b) differential forms, for an example lung cancer patient. Uncertainty-volume histograms for the GTV corresponding to the dose distribution in (a) and (b) plotted in (c) cumulative and (d) differential forms. Curves are shown for plans computed with 10E6, 150E6, 500E6, and 1.5E9 simulated histories. (Reproduced from Chetty IJ et al., *Int. J. Radiat. Oncol. Biol. Phys.*, 65, 1249–1259, 2006. With permission.)

Smoothing a noisy dose distribution may introduce bias into the final dose values essentially by converting the statistical uncertainty of the dose distribution into a systematic deviation of the dose value (Fippel and Nusslin, 2003). Clinical decisions should therefore not be made based on "smoothed" MC dose distributions (Fippel and Nusslin, 2003). Similar results have been demonstrated by testing smoothing functions in a commercial electron beam MC treatment planning system, and emphasize that careless application of smoothing methods can introduce systematic errors or artifacts (Ding et al., 2006). Figure 13.3 shows an example of how inappropriate use of smoothing can alter dose distributions systematically (Ding et al., 2006). Careful benchmarking against good experimental data should be always performed in the clinic to develop criteria for correct use of smoothing techniques. Comprehensive comparisons between various denoising algorithms can be found in other review articles (El Naqa et al., 2005; Kawrakow and Bielajew, 2006).

13.3.1.6 Computed Tomography to Material Conversions

MC algorithms utilize the material density and the material atomic composition when performing particle transport to account for dependencies of particle interactions on the materials, which can lead to notable discrepancies in high-atomic-number materials (Siebers et al., 2000). Material compositions cannot be determined solely from a single energy computed-tomography (CT) scan,

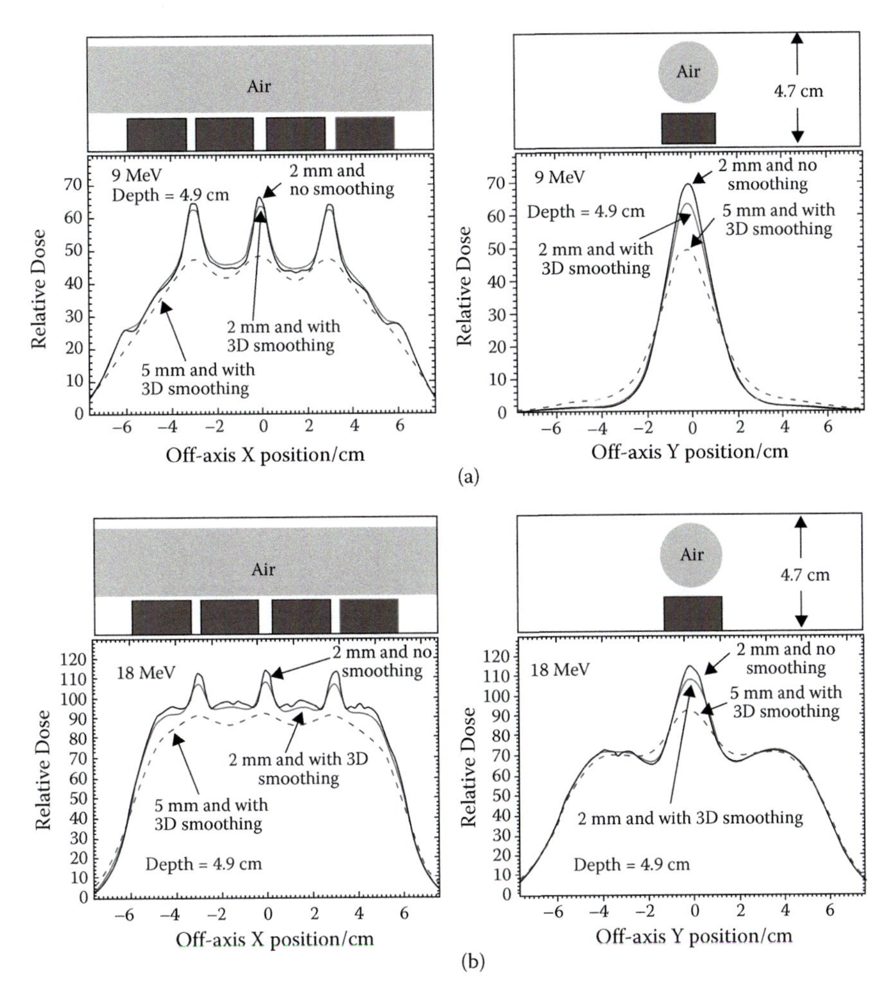

Figure 13.3

Effect of using different calculation voxel sizes and smoothing techniques on the dose distributions for (a) a 9-MeV and (b) an 18-MeV electron beam incident on the trachea and spine phantom, 10 × 10 cm^2 cone and source-to-surface distance (SSD) = 110 cm. (Reproduced from Ding GX et al., *Phys. Med. Biol.*, 51, 2781–2799, 2006. With permission.)

(Continued)

they can be (1) indirectly approximated by estimating the mass density from the electron density followed by assigning a material to each voxel (Ma and Rogers, 1995b; Kawrakow et al., 1996; DeMarco et al., 1998; du Plessis et al., 1998; Siebers et al., 2000); or (2) directly estimated by relating the CT numbers in Hounsfield units (HUs) to material interaction coefficients, based on parameterization of materials representative of the patient (Kawrakow et al., 1996). Although the importance of exact material specification has been established in other studies (Verhaegen and Devic, 2005), more work in this area of research for clinical treatment planning is warranted. CT number artifacts caused by issues such as beam hardening in the CT scanning process or by high-density structures, such as hip prostheses or dental fillings, are potentially important in MC dose calculation

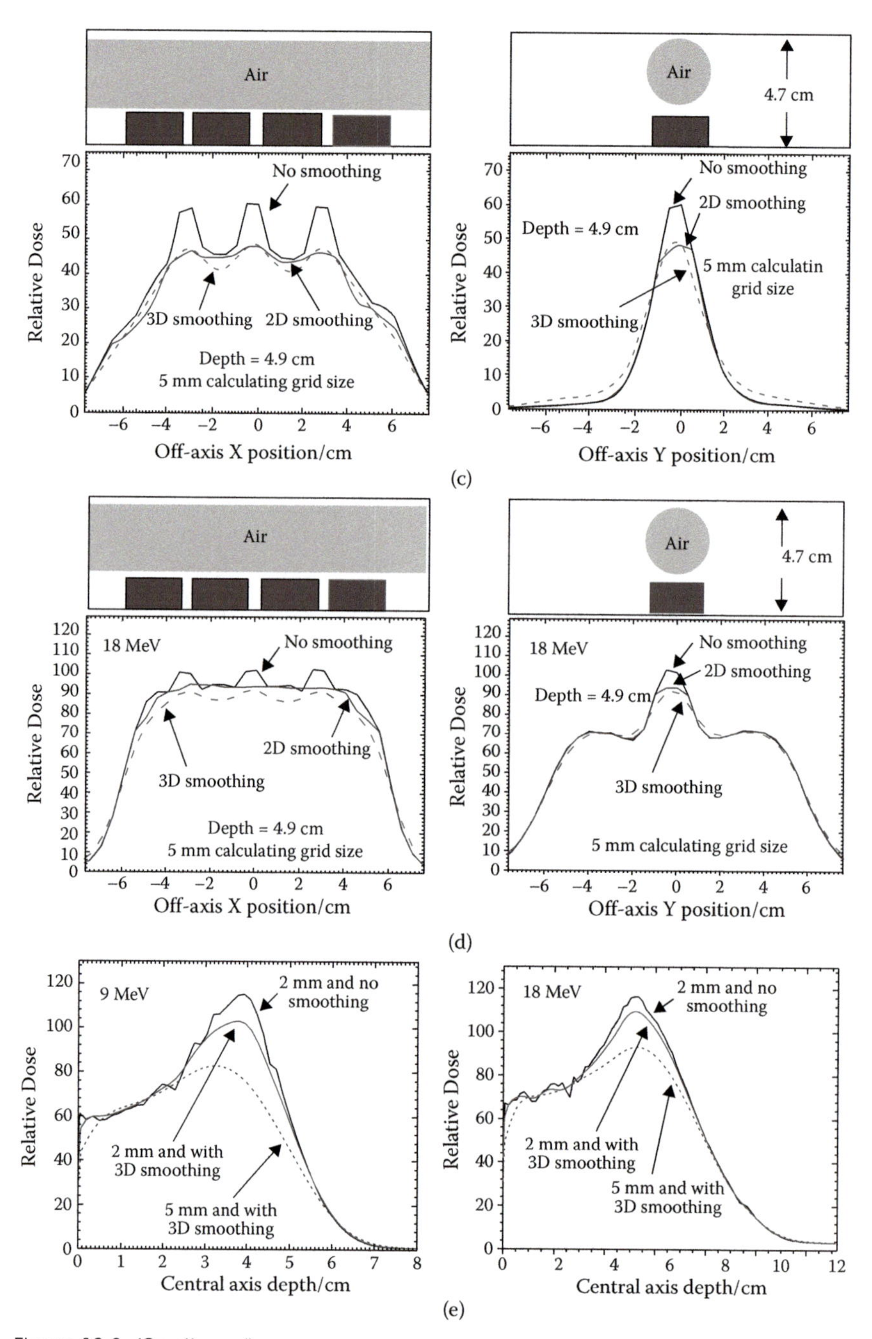

Figure 13.3 (Continued)

Effect of using different smoothing techniques with the same calculation voxel sizes on the dose distributions for (c) a 9-MeV and (d) an 18-MeV beam incident on the trachea and spine phantom, 10 × 10 cm² cone and SSD = 110 cm. (e) Effect of using different calculation voxel sizes and smoothing techniques on the depth-dose distributions for a 9-MeV and an 18-MeV beam incident on the trachea and spine phantom, 10 × 10 cm² cone and SSD = 110 cm. (Reproduced from Ding GX et al., *Phys. Med. Biol.*, 51, 2781–2799, 2006. With permission.)

(Chetty et al., 2007). As with any dose calculation algorithm, testing should be performed to estimate the influence of streaking artifacts from high-density materials on dose distributions computed with the MC method (Reft et al., 2003; Bazalova et al., 2007).

13.3.1.7 Dose-to-Water and Dose-to-Medium

MC simulations model radiation transport and energy deposition in realistic patient/treatment geometries. To account for heterogeneities present in patient anatomy, the particle transport is carried out in patient-specific media, using voxel-by-voxel properties of tissues derived from CT numbers. The dose can be locally scored either in medium denoted as $D_{m,m}$ or in water denoted as $D_{w,m}$. The second m in the subscript indicates that transport is performed in medium (as opposed to water) and is frequently omitted resulting in notation D_m or D_w for scoring done in voxels consisting of a given medium or of water. MC dose calculation algorithms generally compute and report the absorbed dose to the medium, contained in the dose voxel, D_m. If so required, D_m can be converted to D_w using the Bragg–Gray relationship, which states that the ratio of D_w to D_m is equal to the unrestricted mass collision stopping power ratio of water to medium, averaged over the electron energy spectrum at the point of interest.

Conventional treatment planning systems for photon and electron beams calculate D_w. There can be significant differences between D_m and D_w calculations if the water/medium stopping power ratio differs from unity (Siebers et al., 2000; Ding et al., 2006; Dogan et al., 2006; Gardner et al., 2007). There have been a number of papers discussing differences between D_m and D_w calculations for photon and electron beams. For instance, in the case of electron beams, it has been shown that the difference for hard bone depends on beam energy and depth in tissue and can be of the order of 12% (Ding et al., 2006). For photon beams, systematic errors between D_m and D_w in MC-calculated IMRT treatment plans have been shown to be up to 6% for head-and-neck, and 8% for prostate cases when the hard bone–containing structures such as femoral heads were present (Dogan et al., 2006). Prospective studies of treatment outcomes will provide definitive answers about which approach (D_m or D_w) correlates better with biological effects and clinical outcomes in radiotherapy (Liu and Keall, 2002). Until then, it is reasonable to request that vendors of treatment planning systems state explicitly to which material dose is reported and allow for conversion between D_w and D_m (Chetty et al., 2007).

13.3.2 Patient-Specific Quality Assurance for Photon and Electron Beams

13.3.2.1 Methods for Monte Carlo-Based Simulation of Patient-Specific Quality Assurance

A key reason for consideration of MC as a tool in the IMRT quality control (QC) process is the fact that one is able to perform detailed simulation of complex delivery techniques. This section highlights studies in the literature related to simulation of complex delivery techniques, such as IMRT, toward the goal of establishing the feasibility of MC in the treatment and delivery QA process.

The BEAM code system (Rogers et al., 1995) has been instrumental in enabling detailed simulations of the treatment head structures (e.g., the MLC), and the role of various effects on IMRT planning and delivery, such as inter- and intraleaf transmission in the MLC, the tongue-and-groove effect, and transmission through the leaf tips and scattering, on IMRT planning and delivery. As an example, a component module entailing a detailed model of the Varian Millennium 120-leaf MLC (Varian Medical Systems, Palo Alto, CA), incorporating details of the geometry, has been developed (Heath and Seuntjens, 2003). The MCDOSE (Ma et al., 2000a) code system developed to model the details of IMRT delivery showed differences up to 10% in the maximum dose between calculations performed with and without inclusion of the tongue-and-groove effect (Deng et al., 2001b). The modulated electron beam radiation therapy (MERT) system was developed using MC simulations, and a specially designed MLC was proposed to deliver optimized electron beam distributions (Ma et al., 2000b, 2003). Dosimetric characterization of the MLC included analysis of the required leaf widths to deliver complex field shapes, analysis of scatter distributions using rounded versus straight leaf ends, and analysis of electron intra- and interleaf leakage (Ma et al., 2000b). A model of the MLC applicable to both dynamic and segmental IMRT beam delivery was incorporated within the MCV code system to perform patient IMRT dose calculations as well as IMRT treatment delivery verification (Siebers et al., 2002). Accounting for details in the MLC delivery with MC resulted in better agreement of fluence prediction on flat phantom measurements in comparison with the SC algorithm. Differences greater than 5% between SC and MC in the patient geometries (in some cases) were attributed to improved fluence modulation prediction with MC (Sakthi et al., 2006). The DPM MC code system was adapted to incorporate details of the Varian Millennium 120-leaf MLC leaf geometry for IMRT dose verification calculations (Tyagi et al., 2007). Full photon transport through the detailed Millennium MLC (including air spaces between leaves) was performed with electrons depositing energy locally (Tyagi et al., 2007). Figure 13.4 shows excellent agreement between MC calculations and film measurements for a complex IMRT head-and-neck split field case (Tyagi et al., 2007). More comprehensive reviews of the literature substantiating the role of MC in verification calculations of complex delivery techniques, such as IMRT, are provided in other articles (Chetty et al., 2007) and book chapters (Siebers and Ma, 2006; Chetty, 2011; Li and Ma, 2013).

13.3.2.2 Monte Carlo Simulation Incorporating Detector and Dosimetry Systems

The MC method can be used to perform radiation transport simulation within ion chambers (Ma and Nahum, 1991, 1995; Kawrakow, 2000b; Wulff et al., 2008) and other detectors (Ma and Nahum, 1993; Mainegra-Hing et al., 2008). Careful MC-based simulation of different ionization chamber geometries has led to improved understanding of the effective point of measurement correction in ion chamber measurements (Kawrakow, 2006; McEwen et al., 2008). It is conceivable that responses and associated perturbation factors for various ion chambers may be computed at calibration laboratories or research centers, for example, National Research Council of Canada (NRCC), and applied broadly for use in IMRT QA (Bouchard et al., 2009) as elucidated in Chapter 3. Investigators have also demonstrated that MC may be a useful tool for simulation of electronic portal image dosimetry (EPID)-based systems (see Chapter 7). In one such study, an EGS4-based simulation was performed of a Varian aS500 (Varian Medical

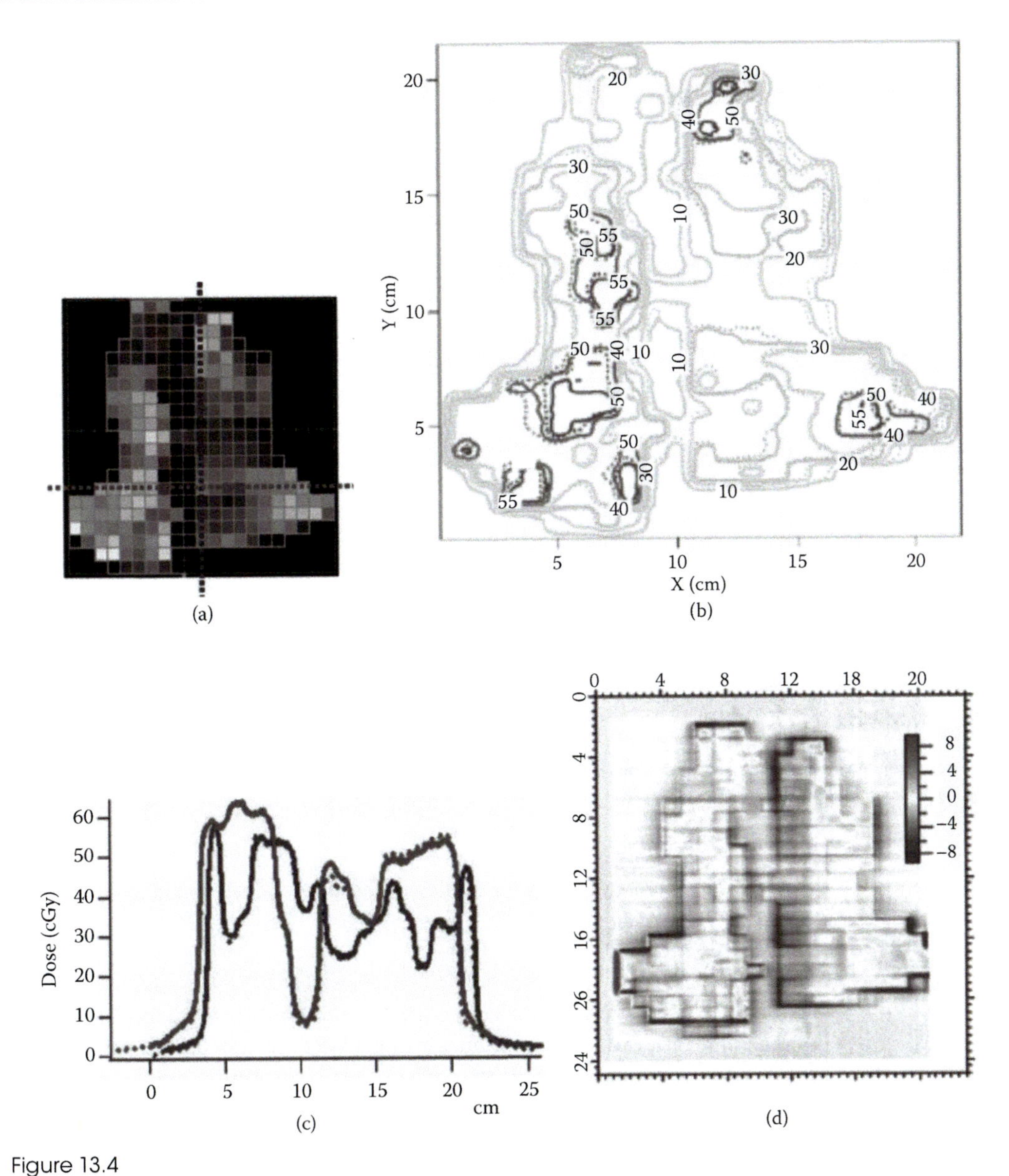

Figure 13.4

Head-and-neck intensity-modulated radiotherapy beam (split field) simulated using SMLC with (119 + 73) MUs and (290 + 190) segments. (a) Beam intensity map. (b) Isodose display for film measurement and DPM calculation; film is shown in solid and DPM in dashed lines. (c) One-dimensional profile comparisons between film measurement and DPM calculation; film is shown in solid and DPM in dashed lines. (d) Dose-difference map in cGy: (DPM-film). (Reproduced from Tyagi N et al., Med. Phys., 34, 651–663, 2007. With permission.)

(Continued)

Systems, Palo Alto, CA) flat-panel imager, incorporating the details of the imager geometry, to investigate various dosimetric characteristics of the imager, which showed good agreement with measurements (Siebers et al., 2004).

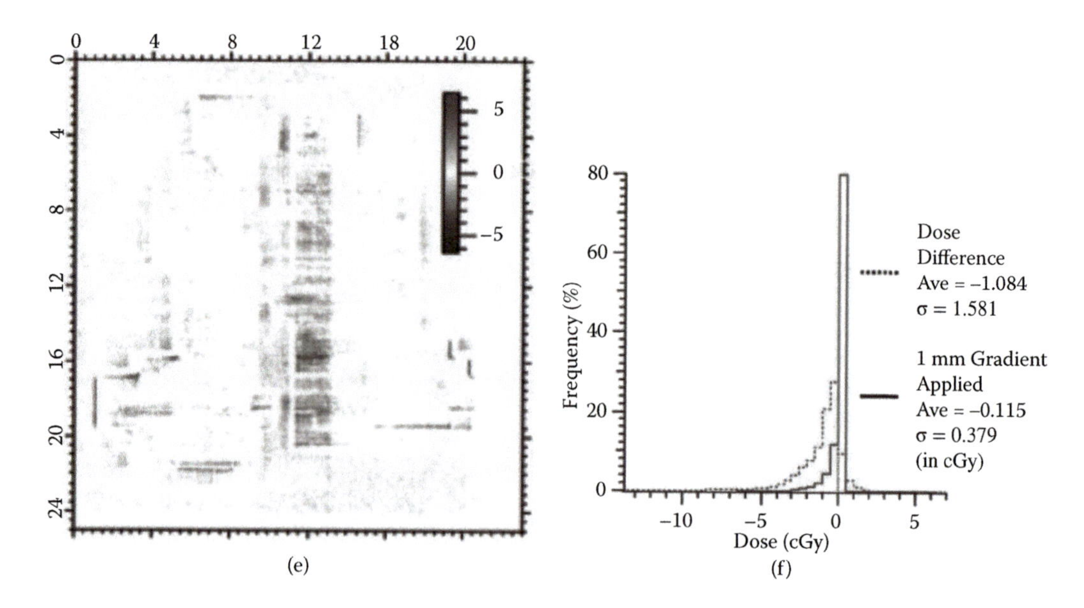

Figure 13.4 (Continued)

Head-and-neck intensity-modulated radiotherapy beam (split field) simulated using SMLC with (119 + 73) MUs and (290 + 190) segments. (e) Dose-difference map in cGy (generated by applying a 1-mm gradient compensation). (f) Dose-difference histogram of the dose-difference map (dotted line) and the gradient compensated dose-difference map (solid line). (Reproduced from Tyagi N et al., *Med. Phys.*, 34, 651–663, 2007. With permission.)

13.4 Role of Monte Carlo Calculations for Small-Field and Non-equilibrium Dosimetry

13.4.1 Role of Monte Carlo in Small-Field Dosimetry

Small radiation fields (less than 3×3 cm^2) are commonly used in modern radiation therapy in treatment techniques using IMRT or stereotactic approaches on linacs or specialized machines such as the CyberKnife. Small-field measurements required for commissioning of these procedures are technically difficult and require a careful choice of detectors and methodology as discussed in detail in Chapter 9. MC simulations play an important role in measurement dosimetry in general and for small non-standard radiation fields in particular. They provide correction factors for various detectors in non-standard clinical beams (Francescon et al., 2008, 2009, 2014; Crop et al., 2009; Scott et al., 2012; Sterpin et al., 2012; Underwood et al., 2013). Dosimetric measurements in small fields require suitable, small-volume radiation detectors to avoid signal averaging over their volume and special attention to detector positioning (Araki, 2006; Francescon et al., 2008; Scott et al., 2008, 2012; Sterpin et al., 2012). In addition, detector material is of great importance because it influences the detector response (Scott et al., 2012; Underwood et al., 2013). An ideal detector should be made of a water-equivalent material and have close to zero active volume. Because an ideal detector (water-equivalent with an infinitely small active volume) does not exist, various correction factors have to be applied to real detector readings to arrive at the quantity of interest, namely the dose-to-water value at the measurement position. In addition to detector characteristics, special dosimetric conditions, such as lack of lateral charged-particle equilibrium (CPE) present in small radiation fields and partial occlusion of the beam source as viewed from the detector position, introduce additional perturbations and uncertainties to the measurements (Das et al., 2008; Aspradakis et al., 2010). A methodology was proposed allowing one to measure accurately dose-to-water in small and composite fields (such as IMRT) through measurements and application of a series of correction factors, accounting for detector volume and material, as well as difference in field size, beam quality, source obscuration, lack of lateral CPE, and phantom geometry between reference and clinical fields (Alfonso et al., 2008). Various MC codes such as EGSnrc, BEAMnrc, egs_chamber, GEANT, and PENELOPE have been used to model detectors and derive correction factors (Seuntjens, 2006). MC correction factors have been computed for static cones and collimator (Iris)-defined CyberKnife fields, equipment shown in Chapter 10, and for a number of detectors such as ionization chambers, diodes, and MOSFETs in a variety of measurement conditions (Francescon et al., 2008, 2014). Examples of these correction factors are provided in Chapter 9. One can also clearly see the effect of the detector material on the raw measurements: diodes overrespond by about 5%, while microion chambers underrespond by as much as 10%. The latter finding has been confirmed by other investigators (Scott et al., 2012; Underwood et al., 2013), who found that at the smallest field sizes overresponse of high-density detectors and underresponse of low-density detectors correlate with the mass density of the detector material relative to that of water. It can be concluded that MC-calculated correction factors play an important role when using specific types of detectors in small-field dosimetry such as diodes or microion chambers.

13.4.2 Clinical Examples of Monte Carlo and Other Algorithms for Treatment Planning

The examples presented here are provided to demonstrate principles and are in no way meant to be comprehensive. Detailed discussions of the role of MC in treatment planning are available in the following review articles, among others (Cygler et al., 2006; Chetty et al., 2007; Reynaert et al., 2007; Verhaegen and Seco, 2013). PB or equivalent-path-length-type algorithms do not account for electron transport in an accurate way and therefore are severely limited in accuracy for electron beam, and small-field photon beam calculations, especially in low-density (e.g., lung equivalent) tissues (Chetty et al., 2007; Benedict et al., 2010). Figure 13.5 shows an example dose distribution for a patient with neck cancer. The treatment plan consists of a boost to the tumor region using 9 MeV electrons (15×15 cm^2 electron insert with an individualized cutout, 100 cm source-to-surface distance [SSD], at a gantry angle of 16°). IDLs at the 70%, 50%, and 30% level, calculated by the PB and electron Monte Carlo (eMC) algorithms, are shown in Figure 13.5a and b, respectively. Significant outward bowing of the MC-based dose calculation is observed in the trachea region consistent with the scattering of electrons in the low-density air, which is not accurately accounted for with the PB algorithm. The differences in the 50% and 30% IDLs are up to 10 mm between eMC and PB algorithms in the trachea.

Figure 13.6 shows an example comparison between PB and MC algorithms for a lung cancer patient treated with stereotactic body radiation therapy (SBRT), using 6 MV photons (Fragoso et al., 2010). The PB-computed dose coverage of the 95% and 80% IDLs are shown in the lower left view (Figure 13.6a). The

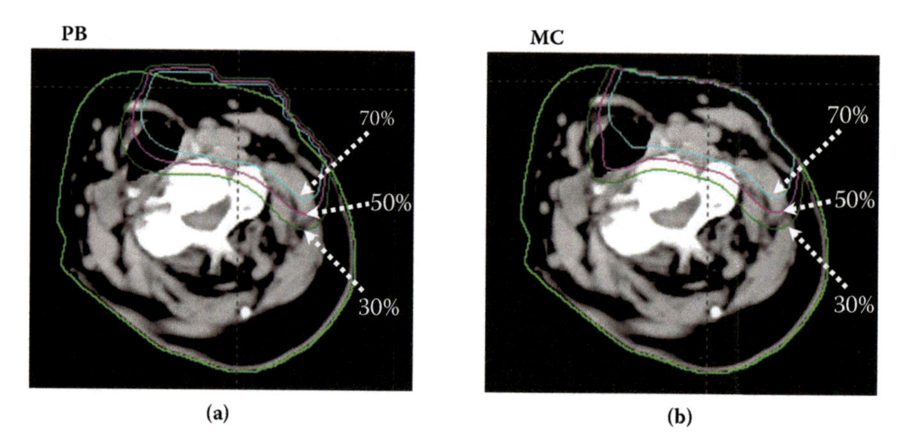

Figure 13.5

Comparison of isodose lines (IDLs) in the transverse plane for an example patient calculated with (a) pencil beam (PB) and (b) electron Monte Carlo (eMC) dose algorithms. Shown are the 70%, 50%, and 30% IDLs, respectively in each figure. The reduction and "dipping" of the MC IDLs, especially at the 50% and 30% levels, are clearly observable on the eMC calculation in (b). This is due to the outward scattering of electrons in the air cavity, an effect not accurately handled by the PB algorithm.

corresponding MC dose calculation is shown in the lower right view, where a severe reduction in dose to the PTV is noted. Figure 13.6b shows the DVHs computed with the MC and PB algorithms for the PTV (left) and normal lung tissue (right). The minimum PTV doses (D99) are 97% and 58% of the prescription dose in the PB- and MC-based DVHs, respectively. This example demonstrates the significant problem associated with the use of PB-type dose algorithms in small-field lung cancer treatment planning (Fragoso et al., 2010). There are two issues at hand: (1) small-field effects, especially loss of lateral CPE, which occurs when the tumor size (field size) is so small that the lateral ranges of the secondary electrons become comparable to (or greater than) the field size (Das et al., 2008); and (2) increased range of scattered electrons in the forward and lateral directions, in low density tissues. When electrons scattered within lung tissue impinge on the tumor, the much higher density of the (water-equivalent) tumor will cause electrons to stop, depositing their energy in the tumor. However, the electrons have a range; therefore, dose builds up along the electron range at the edge of the tumor, causing a dose reduction preferentially at the edges of the tumor. These electron and photon scattering effects are not properly handled with PB dose algorithms (Chetty et al., 2013). The preferential underdosing at the tumor edge and the resulting heterogeneous coverage of the PTV is demonstrated clearly in Figure 13.6b (left). DVHs of the normal lung tissue, shown in Figure 13.6b (right), demonstrate that larger volumes of lung receive lower doses in the MC DVH, which is consistent with greater range of lateral electrons scattered in lower density lung tissue. Other studies demonstrating limitations of the PB algorithm and showing significant underdosage of the target with MC (for plans based on PB) have been performed (Knöös et al., 1995; Rassiah-Szegedi et al., 2006; van Elmpt et al., 2008; van der Voort van Zyp et al., 2010; Wilcox et al., 2010). Dose overestimation with the PB calculation for studies of lung SBRT patients showed a strong dependence on tumor size and location in the lung

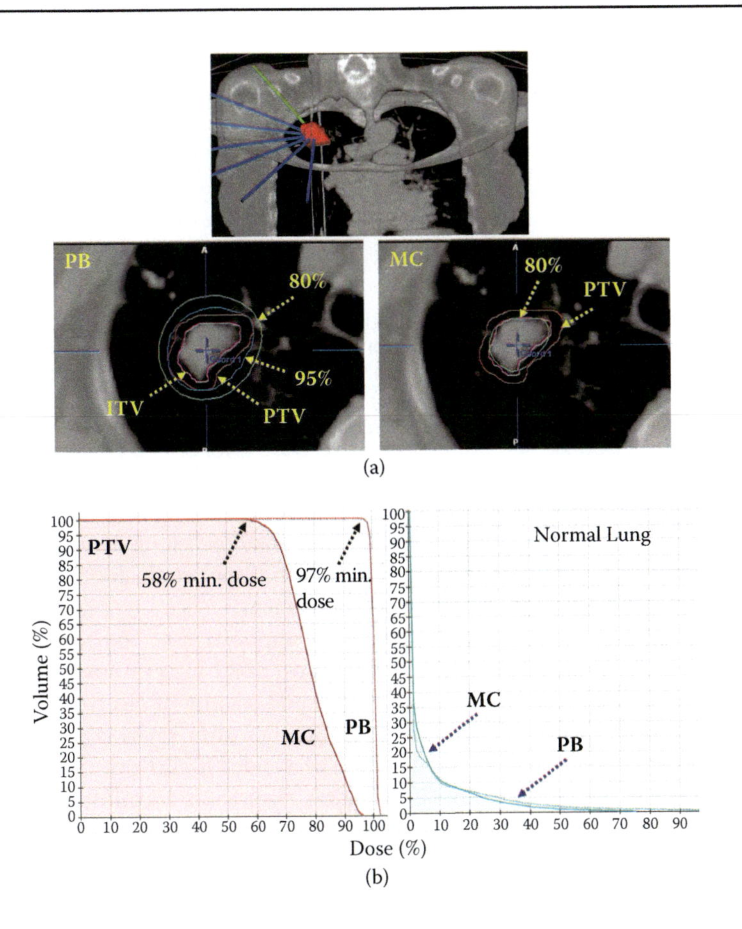

Figure 13.6

(a) Plan comparison for a lung cancer patient treated with stereotactic body radiation therapy. The tumor volume was 14.6 cm^3 and the maximum planning target volume (PTV) width was 3.2 cm. Top: Beam arrangements for the treatment plan. Bottom left: Pencil-beam (PB) dose distribution in the axial view showing the internal target volume (ITV), PTV, and the 95% and 80% isodose lines (IDLs). Note that the 95% IDL encompasses the PTV. Bottom right: Dose distribution recomputed with Monte Carlo (MC), showing the 80% IDL in relation to the PTV. Underdosage of the PTV is so severe that even the 80% IDL fails to cover the PTV. (b) Dose–volume histograms (DVHs) for the PTV (left) and normal lung tissue (right) for the patient plan in Figure 13.6a. The minimum PTV dose (D99) is 97% in the PB calculation but only 58% of the prescription dose in the MC-computed plan. The MC DVH shows increased heterogeneity, since a larger dose reduction occurs at the tumor periphery relative to the center, which produces a differential dose difference on the tumor relative to the PB algorithm. For the normal lung tissue, the MC calculation shows that a larger volume of lung receives increased low dose relative to the PB algorithm. This is consistent with increased lateral scattering of electrons in the lung tissue, an effect not properly accounted for with the PB algorithm. (Reproduced from Fragoso M et al., *Phys. Med. Biol.*, 55, 4445–4464, 2010. With permission.)

substantiating a recommendation that a different prescription dose should be selected for lung SBRT depending on the tumor size and location (van der Voort van Zyp et al., 2010; Chetty et al., 2013).

Even more sophisticated algorithms, such as convolution methods (based on invariant kernels), may be limited for lung cancer planning. In one such study, plans calculated with MC showed significant underdosage of the PTV as

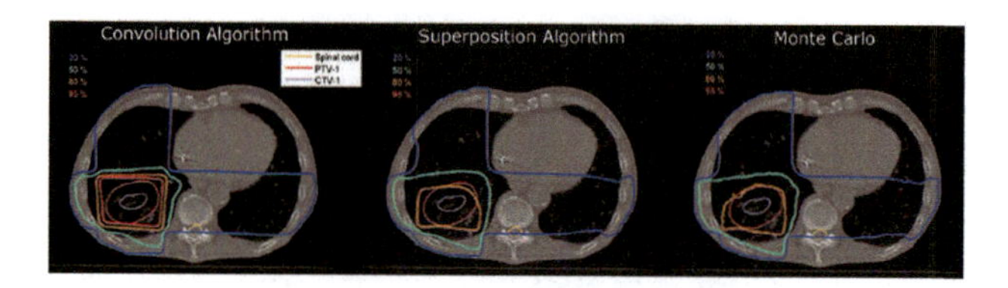

Figure 13.7

Dose distributions for an example lung cancer patient using the convolution, superposition-convolution (superposition algorithm [SA]), and Monte Carlo (MC) dose algorithms. For each calculation, the same beam setup is applied with identical number of monitor units (MUs). Note that the MC or SA algorithm does not even reach coverage of the 80% isodose around the planning target volume (PTV); the 100% isodose line corresponds to a dose level of 79.2 Gy. (Reproduced from van Elmpt W et al., *Radiother. Oncol.*, 88, 326–334, 2008. With permission.)

compared to a convolution algorithm for non-small cell lung cancer (NSCLC) tumors planned with 10 MV photon beams (van Elmpt et al., 2008) (see Figure 13.7). The SC and MC algorithms were in agreement.

The aforementioned studies are generally supportive of the recommendations of the AAPM Task Group 101, that for most SBRT lung tumors, algorithms accounting for 3D scatter (including photon and scaling of electron scattering kernels in the relevant medium), such as the SC algorithm, provide accurate dose distributions and that PB algorithms should be avoided for lung cancer treatment planning (Benedict et al., 2010). "Calculation algorithms accounting for better photon and electron transport such as Monte Carlo would be ideal for the most demanding circumstances, such as a small lesion entirely surrounded by a low-density medium" (Benedict et al., 2010).

13.5 Monte Carlo-Based Calculation Incorporating Motion and Time Dependence

Fluence convolution, in which the radiation fluence distribution is convolved with a Gaussian function representing random setup errors, was proposed to account for patient random setup errors using MC-based dose calculation (Beckham et al., 2002). A "fluence translation" approach, where the fluence was "translated" as opposed to "convolved," was used to account for both random and systematic setup errors using MC-based dose calculation (Chetty et al., 2003). This approach was applied to account for respiratory-induced motion of thoracic tumors by translating the fluence according to a patient respiratory-induced motion function (Chetty et al., 2003, 2004). A major advantage of the MC method over analytic algorithms is that the fluence translation is performed in a single dose calculation because individual particle histories are translated after sampling respective motion distributions. This is not the case with analytic algorithms, where independent calculations are warranted each time the incident fluence is translated. The MC algorithm has been proposed for dose computations for four-dimensional (4D) treatment planning (Keall et al., 2004) as discussed in Chapter 11. It was shown that MC computations on each of the N, 3D CT datasets, required approximately $1/N$ fewer particles than that necessary for a full 3D plan, implying similar calculation times for the 3D and 4D methods at roughly the same statistical uncertainty (Keall et al., 2004). Novel techniques for dose accumulation have been developed in the setting of 4D, MC-based dose calculation (Rosu et al., 2005; Heath and Seuntjens, 2006; Siebers and Zhong, 2008).

13.6 Summary

As supported by Das et al. (2008), it is expected that the MC techniques will increasingly be used in assessing the accuracy, verification, and calculation of dose, and they will aid in perturbation calculations of detectors used in small and highly conformal radiation beams.

Research evaluating dose–volume–effect relationships, as performed previously (Lindsay et al., 2007; Stroian et al., 2008; Chetty et al., 2013; Liu et al., 2013; Latifi et al., 2014), will help elucidate the benefit of the more accurate dose distributions afforded by the MC method on observed clinical outcomes. More studies on this topic are warranted.

References

Aaronson RF et al. (2002) A Monte Carlo based phase space model for quality assurance of intensity modulated radiotherapy incorporating leaf specific characteristics. *Med. Phys.* 29: 2952–2958.

Agostinelli S et al. (2003) GEANT4—a simulation toolkit. *Nucl. Instrum. Methods Phys. Res. A* 506: 250–303.

Ahnesjo A and Aspradakis MM (1999) Dose calculations for external photon beams in radiotherapy. *Phys. Med. Biol.* 44: R99–R155.

Alfonso R et al. (2008) A new formalism for reference dosimetry of small and nonstandard fields. *Med. Phys.* 35: 5179–5186.

Aljarrah K et al. (2006) Determination of the initial beam parameters in Monte Carlo linac simulation. *Med. Phys.* 33: 850–858.

Araki F (2006) Monte Carlo study of a Cyberknife stereotactic radiosurgery system. *Med. Phys.* 33: 2955–2963.

Aspradakis MM et al. (2010) Small field MV photon dosimetry. IPEM Report No. 103. (York, UK: Institute of Physics and Engineering in Medicine).

Baro J et al. (1995) PENELOPE—an algorithm for Monte-Carlo simulation of the penetration and energy-loss of electrons and positrons in matter. *Nucl. Instrum. Methods Phys. Res. A* 100: 31–46.

Bazalova M et al. (2007) Correction of CT artifacts and its influence on Monte Carlo dose calculations. *Med. Phys.* 34: 2119–2132.

Beckham WA, Keall PJ and Siebers JV (2002) A fluence-convolution method to calculate radiation therapy dose distributions that incorporate random set-up error. *Phys. Med. Biol.* 47: 3465–3473.

Benedict SH et al. (2010) Stereotactic body radiation therapy: The report of AAPM Task Group 101. *Med. Phys.* 37: 4078–4101.

Bergman AM et al. (2006) Direct aperture optimization for IMRT using Monte Carlo generated beamlets. *Med. Phys.* 33: 3666–3679.

Berger MJ (1963) Monte Carlo calculations of the penetration and diffusion of fast charged particles. In: *Methods in Computational Physics*, Vol. 1. (Eds. Fernbach S, Alder B and Rothenberg M, New York, NY: Academic Press). 135–215.

Bouchard H et al. (2009) Ionization chamber gradient effects in nonstandard beam configurations. *Med. Phys.* 36: 4654–4663.

Bergman AM, Bush K, Milette, MP, Popescu IA, Otto K and Duzenli C (2006) Direct aperture optimization for IMRT using Monte Carlo generated beamlets. *Med. Phys.* 33: 3666-3679.

Briesmeister JF (1993) MCNP—A general Monte Carlo n-particle transport code, version 4A. Report LA-12625-M. (Los Alamos, NM: Los Alamos National Laboratory).

Brown FB (2003) MCNP—A general Monte Carlo n-particle transport code, version 5. Report LA-UR-03 1987. (Los Alamos, NM: Los Alamos National Laboratory).

Buffa FM and Nahum AE (2000) Monte Carlo dose calculations and radiobiological modelling: Analysis of the effect of the statistical noise of the dose distribution on the probability of tumour control. *Phys. Med. Biol.* 45: 3009–3023.

Chetty IJ (2011) The role of Monte Carlo in patient-specific quality control. In: *Quality and Safety in Radiotherapy*, pp. 517–526. (Eds. Pawlicki T, Dunscombe P, Mundt AJ and Scalliet P, Boca Raton, FL: CRC Press).

Chetty IJ et al. (2003) A fluence convolution method to account for respiratory motion in three-dimensional dose calculations of the liver: A Monte Carlo study. *Med. Phys.* 30: 1776–1780.

Chetty IJ et al. (2004) Accounting for center-of-mass target motion using convolution methods in Monte Carlo-based dose calculations of the lung. *Med. Phys.* 31: 925–932.

Chetty IJ et al. (2006) Reporting and analyzing statistical uncertainties in Monte Carlo-based treatment planning. *Int. J. Radiat. Oncol. Biol. Phys.* 65: 1249–1259.

Chetty IJ et al. (2007) Report of the AAPM Task Group No. 105: Issues associated with clinical implementation of Monte Carlo-based photon and electron external beam treatment planning. *Med. Phys.* 34: 4818–4853.

Chetty IJ et al. (2013) Correlation of dose computed using different algorithms with local control following stereotactic ablative radiotherapy (SABR)-based treatment of non-small-cell lung cancer. *Radiother. Oncol.* 109: 498–504.

Chetty IJ, DeMarco JJ and Solberg TD (2000) A virtual source model for Monte Carlo modeling of arbitrary intensity distributions. *Med. Phys.* 27: 166–172.

Crop F et al. (2009) The influence of small field sizes, penumbra, spot size and measurement depth on perturbation factors for microionization chambers. *Phys. Med. Biol.* 54: 2951–2969.

Cygler JE et al. (2004) Evaluation of the first commercial Monte Carlo dose calculation engine for electron beam treatment planning. *Med. Phys.* 31: 142–153.

Cygler JE et al. (2006) Monte Carlo systems in pre-clinical and clinical planning: Pitfalls and triumphs. In: *Integrating New Technologies into the Clinic: Monte Carlo and Image-Guided Radiation Therapy*, pp. 199–231. (Eds. Curran B, Balter JM and Chetty IJ, Madison, WI: Medical Physics Publishing).

Das IJ, Ding GX and Ahnesjo A (2008) Small fields: Nonequilibrium radiation dosimetry. *Med. Phys.* 35: 206–215.

Deasy JO (2000) Denoising of electron beam Monte Carlo dose distributions using digital filtering techniques. *Phys. Med. Biol.* 45: 1765–1779.

Deasy JO, Wickerhauser MV and Picard M (2002) Accelerating Monte Carlo simulations of radiation therapy dose distributions using wavelet threshold de-noising. *Med. Phys.* 29: 2366–2373.

DeMarco JJ, Solberg TD and Smathers JB (1998) A CT-based Monte Carlo simulation tool for dosimetry planning and analysis. *Med. Phys.* 25: 1–11.

Deng J et al. (2000) Photon beam characterization and modelling for Monte Carlo treatment planning. *Phys. Med. Biol.* 45: 411–427.

Deng J et al. (2001a) Derivation of electron and photon energy spectra from electron beam central axis depth dose curves. *Phys. Med. Biol.* 46: 1429–1449.

Deng J et al. (2001b) The MLC tongue-and-groove effect on IMRT dose distributions. *Phys. Med. Biol.* 46: 1039–1060.

Ding GX et al. (2006) First macro Monte Carlo based commercial dose calculation module for electron beam treatment planning—new issues for clinical consideration. *Phys. Med. Biol.* 51: 2781–2799.

Dogan N, Siebers JV and Keall PJ (2006) Clinical comparison of head and neck and prostate IMRT plans using absorbed dose to medium and absorbed dose to water. *Phys. Med. Biol.* 51: 4967–4980.

du Plessis FC et al. (1998) The indirect use of CT numbers to establish material properties needed for Monte Carlo calculation of dose distributions in patients. *Med. Phys.* 25: 1195–1201.

Dutreix A (1984) When and how can we improve precision in radiotherapy? *Radiother. Oncol.* 2: 275–292.

Edimo P et al. (2009) Evaluation of a commercial VMC++ Monte Carlo based treatment planning system for electron beams using EGSnrc/BEAMnrc simulations and measurements. *Phys. Med.* 25: 111–121.

El Naqa I et al. (2005) A comparison of Monte Carlo dose calculation denoising techniques. *Phys. Med. Biol.* 50: 909–922.

Faddegon B et al. (1998) Clinical considerations of Monte Carlo for electron radiotherapy treatment planning. *Radiat. Phys. Chem.* 53: 217–227.

Faddegon BA and Blevis I (2000) Electron spectra derived from depth dose distributions. *Med. Phys.* 27: 514–526.

Faddegon BA and Cygler JE (2006) Use of the Monte Carlo method in accelerator head simulation and modeling for electron beams. In: *Integrating New Technologies into the Clinic: Monte Carlo and Image-Guided Radiation Therapy*, pp. 51–69. (Eds. Curran B, Balter JM and Chetty IJ, Madison, WI: Medical Physics Publishing).

Fippel M (2004) Efficient particle transport simulation through beam modulating devices for Monte Carlo treatment planning. *Med. Phys.* 31: 1235–1242.

Fippel M and Nusslin F (2003) Smoothing Monte Carlo calculated dose distributions by iterative reduction of noise. *Phys. Med. Biol.* 48: 1289–1304.

Fix MK et al. (2000) Simple beam models for Monte Carlo photon beam dose calculations in radiotherapy. *Med. Phys.* 27: 2739–2747.

Fix MK et al. (2001) A multiple source model for 6 MV photon beam dose calculations using Monte Carlo. *Phys. Med. Biol.* 46: 1407–1427.

Fix MK et al. (2013) Generalized eMC implementation for Monte Carlo dose calculation of electron beams from different machine types. *Phys. Med. Biol.* 58: 2841–2859.

Ford RL and Nelson WR (1978) The EGS code system–Version 3. Report SLAC-210. (Stanford, CA: SLAC National Accelerator Laboratory).

Fragoso M et al. (2009) Fast, accurate photon beam accelerator modeling using BEAMnrc: A systematic investigation of efficiency enhancing methods and cross-section data. *Med. Phys.* 36: 5451–5466.

Fragoso M et al. (2010) Dosimetric verification and clinical evaluation of a new commercially available Monte Carlo-based dose algorithm for application in stereotactic body radiation therapy (SBRT) treatment planning. *Phys. Med. Biol.* 55: 4445–4464.

Francescon P, Cora S and Cavedon C (2008) Total scatter factors of small beams: A multidetector and Monte Carlo study. *Med. Phys.* 35: 504–513.

Francescon P et al. (2009) Application of a Monte Carlo-based method for total scatter factors of small beams to new solid state micro-detectors. *J. Appl. Clin. Med. Phys.* 10(1): 147–152.

Francescon P, Kilby W and Satariano N (2014) Monte Carlo simulated correction factors for output factor measurement with the CyberKnife system-results for new detectors and correction factor dependence on measurement distance and detector orientation. *Phys. Med. Biol.* 59: N11–N17.

Gardner J, Siebers J and Kawrakow I (2007) Dose calculation validation of Vmc++ for photon beams. *Med. Phys.* 34: 1809–1818.

Halbleib JA (1988) Structure and operation of the ITS code system. In: *Monte Carlo Transport of Electrons and Photons*, pp. 249–262. (Eds. Nelson WR, Jenkins TM, Rindi A, Nahum AE and Rogers DWO, New York, NY: Plenum Press).

Halbleib JA and Melhorn TA (1984) ITS: The integrated TIGER series of coupled electron/photon Monte Carlo transport codes. Sandia Report SAND84-0573. (Albuquerque, NM: Sandia National Laboratory).

Hasenbalg F et al. (2008) VMC++ versus BEAMnrc: A comparison of simulated linear accelerator heads for photon beams. *Med. Phys.* 35: 1521–1531.

Heath E and Seuntjens J (2003) Development and validation of a BEAMnrc component module for accurate Monte Carlo modelling of the Varian dynamic Millennium multileaf collimator. *Phys. Med. Biol.* 48: 4045–4063.

Heath E and Seuntjens J (2006) A direct voxel tracking method for four-dimensional Monte Carlo dose calculations in deforming anatomy. *Med. Phys.* 33: 434–445.

Huang VW et al. (2005) Experimental determination of electron source parameters for accurate Monte Carlo calculation of large field electron therapy. *Phys. Med. Biol.* 50: 779–786.

Janssen JJ et al. (2001) A model to determine the initial phase space of a clinical electron beam from measured beam data. *Phys. Med. Biol.* 46: 269–286.

Jeraj R, Keall PJ and Siebers JV (2002) The effect of dose calculation accuracy on inverse treatment planning. *Phys. Med. Biol.* 47: 391–407.

Jiang SB, Pawlicki T and Ma CM (2000) Removing the effect of statistical uncertainty on dose-volume histograms from Monte Carlo dose calculations. *Phys. Med. Biol.* 45: 2151–2161.

Kawrakow I (2000a) Accurate condensed history Monte Carlo simulation of electron transport. I. EGSnrc, the new EGS4 version. *Med. Phys.* 27: 485–498.

Kawrakow I (2000b) Accurate condensed history Monte Carlo simulation of electron transport. II. Application to ion chamber response simulations. *Med. Phys.* 27: 499–513.

Kawrakow I (2002) On the de-noising of Monte Carlo calculated dose distributions. *Phys. Med. Biol.* 47: 3087–3103.

Kawrakow I (2004) The effect of Monte Carlo statistical uncertainties on the evaluation of dose distributions in radiation treatment planning. *Phys. Med. Biol.* 49: 1549–1556.
Kawrakow I (2006) On the effective point of measurement in megavoltage photon beams. *Med. Phys.* 33: 1829–1839.
Kawrakow I and Bielajew AF (2006) Monte Carlo treatment planning: Interpretation of noisy dose distributions and review of denoising methods. In: *Integrating New Technologies into the Clinic: Monte Carlo and Image-Guided Radiation Therapy, AAPM Medical Physics Monograph No. 32*, pp. 179–197. (Eds. Curran BH, Balter JM and Chetty IJ, Madison, WI: Medical Physics Publishing).
Kawrakow I, Fippel M and Friedrich K (1996) 3D electron dose calculation using a Voxel based Monte Carlo algorithm (VMC). *Med. Phys.* 23: 445–457.
Kawrakow I and Rogers DWO (2000) The EGSnrc code system: Monte Carlo simulation of electron and photon transport. Technical Report PIRS-701. (Ottawa, Ontario, Canada: National Research Council of Canada).
Kawrakow I, Rogers DWO and Walters BRB (2004) Large efficiency improvements in BEAMnrc using directional bremsstrahlung splitting. *Med. Phys.* 31: 2883–2898.
Keall PJ et al. (2000) The effect of dose calculation uncertainty on the evaluation of radiotherapy plans. *Med. Phys.* 27: 478–484.
Keall PJ et al. (2004) Monte Carlo as a four-dimensional radiotherapy treatment-planning tool to account for respiratory motion. *Phys. Med. Biol.* 49: 3639–3648.
Knöös T et al. (1995) Limitations of a pencil beam approach to photon dose calculations in lung tissue. *Phys. Med. Biol.* 40: 1411–1420.
Latifi K et al. (2014) Study of 201 non-small cell lung cancer patients given stereotactic ablative radiation therapy shows local control dependence on dose calculation algorithm. *Int. J. Radiat. Oncol. Biol. Phys.* 88: 1108–1113.
Li JS and Ma CM (2013) Monte Carlo as a QA tool for advanced radiation therapy. In: *Monte Carlo Techniques in Radiation Therapy*, pp. 145–154. (Eds. Verhaegen F and Seco J, Boca Raton, FL: CRC Press).
Lindsay PE et al. (2007) Retrospective Monte Carlo dose calculations with limited beam weight information. *Med. Phys.* 34: 334–346.
Liu MB et al. (2013) Clinical impact of dose overestimation by effective path length calculation in stereotactic ablative radiation therapy of lung tumors. *Pract. Radiat. Oncol.* 3: 294–300.
Liu HH and Keall PJ (2002) D_m rather than D_w should be used in Monte Carlo treatment planning. *Med. Phys.* 29: 922–924.
Liu HH, Verhaegen F and Dong L (2001) A method of simulating dynamic multileaf collimators using Monte Carlo techniques for intensity-modulated radiation therapy. *Phys. Med. Biol.* 46: 2283–2298.
Ma CM (1998) Characterization of computer simulated radiotherapy beams for Monte-Carlo treatment planning. *Radiat. Phys. Chem.* 53: 329–344.
Ma CM et al. (1997) Accurate characterization of Monte Carlo calculated electron beams for radiotherapy. *Med. Phys.* 24: 401–416.
Ma CM et al. (1999) Clinical implementation of a Monte Carlo treatment planning system. *Med. Phys.* 26: 2133–2143.
Ma CM et al. (2000a) MCDOSE—A Monte Carlo dose calculation tool for radiation therapy treatment planning. In: Proceedings of the 13th International

Conference on the Use of Computer in Radiation Therapy (ICCR), pp. 123–125. (Eds. Bortfeld T and Schlegel W, Heidelberg, Germany: Springer-Verlag).
Ma CM et al. (2000b) Energy- and intensity-modulated electron beams for radiotherapy. *Phys. Med. Biol.* 45: 2293–2311.
Ma CM et al. (2003) A comparative dosimetric study on tangential photon beams, intensity-modulated radiation therapy (IMRT) and modulated electron radiotherapy (MERT) for breast cancer treatment. *Phys. Med. Biol.* 48: 909–924.
Ma CM et al. (2004) Monitor unit calculation for Monte Carlo treatment planning. *Phys. Med. Biol.* 49: 1671–1687.
Ma CM and Jiang SB (1999) Monte Carlo modelling of electron beams from medical accelerators. *Phys. Med. Biol.* 44: R157–R189.
Ma CM and Nahum AE (1991) Bragg-Gray theory and ion chamber dosimetry for photon beams. *Phys. Med. Biol.* 36: 413–428.
Ma CM and Nahum AE (1993) Dose conversion and wall correction factors for Fricke dosimetry in high-energy photon beams: Analytical model and Monte Carlo calculations. *Phys. Med. Biol.* 38: 93–114.
Ma CM and Nahum AE (1995) Monte Carlo calculated stem effect correction for NE2561 and NE2571 chambers in medium-energy x-ray beams. *Phys. Med. Biol.* 40: 63–72.
Ma CM and Rogers DWO (1995a) BEAM Characterization: A multiple-source model. NRC Report PIRS-0509(C). (Ottawa, Ontario, Canada: National Research Council of Canada).
Ma CM and Rogers DWO (1995b) BEAMDP user's manual. NRC Report PIRS-0509(D). (Ottawa, Ontario, Canada: National Research Council of Canada).
Ma CM and Sheikh-Bagheri D (2006) Monte Carlo methods for accelerator simulation and photon beam modeling. In: *Integrating New Technologies into the Clinic: Monte Carlo and Image-Guided Radiation Therapy*, pp. 21–49. (Eds. Curran B, Balter JM and Chetty IJ, Madison, WI: Medical Physics Publishing).
Mainegra-Hing E, Reynaert N and Kawrakow I (2008) Novel approach for the Monte Carlo calculation of free-air chamber correction factors. *Med. Phys.* 35: 3650–3660.
McEwen MR, Kawrakow I and Ross CK (2008) The effective point of measurement of ionization chambers and the build-up anomaly in MV x-ray beams. *Med. Phys.* 35: 950–958.
Miao B et al. (2003) Adaptive anisotropic diffusion filtering of Monte Carlo dose distributions. *Phys. Med. Biol.* 48: 2767–2781.
Nelson WR, Hirayama H and Rogers DWO (1985) The EGS4 code system. Report SLAC-265. (Stanford, CA: SLAC National Accelerator Laboratory).
Papanikolaou N et al. (2004) AAPM Report No. 85: Tissue inhomogeneity corrections for megavoltage photon beams. (Madison, WI: Medical Physics Publishing).
Popescu IA et al. (2005) Absolute dose calculations for Monte Carlo simulations of radiotherapy beams. *Phys. Med. Biol.* 50: 3375–3392.
Rassiah-Szegedi P et al. (2006) Monte Carlo characterization of target doses in stereotactic body radiation therapy (SBRT). *Acta Oncol.* 45: 989–994.
Reft C et al. (2003) Dosimetric considerations for patients with hip prostheses undergoing pelvic irradiation. Report of the AAPM Radiation Therapy Committee Task Group 63. *Med. Phys.* 30: 1162–1182.

Reynaert N et al. (2007) Monte Carlo treatment planning for photon and electron beams. *Rad. Phys. Chem.* 76: 643–686.

Rogers DWO et al. (1995) BEAM: A Monte Carlo code to simulate radiotherapy treatment units. *Med. Phys.* 22: 503–524.

Rogers DWO and Bielajew AF (1990) Monte Carlo techniques of electron and photon transport for radiation dosimetry. In: *The Dosimetry of Ionizing Radiation*, pp. 427–539. (Eds. Bjärngard B, Kase K and Attix F, New York, NY: Academic Press).

Rogers DWO and Mohan R (2000) Questions for comparisons of clinical Monte Carlo codes. In: Proceedings of the 13th International Conference on the Use of Computer in Radiation Therapy (ICCR), pp. 120–122. (Eds. Bortfeld T and Schlegel W, Heidelberg, Germany: Springer-Verlag).

Rogers DWO, Walters B and Kawrakow I (2004) BEAMnrc user's manual: NRC Report PIRS 509(a)revH. (Ottawa, Ontario, Canada: National Research Council of Canada).

Rosu M et al. (2005) Dose reconstruction in deforming lung anatomy: Dose grid size effects and clinical implications. *Med. Phys.* 32: 2487–2495.

Sakthi N et al. (2006) Monte Carlo-based dosimetry of head-and-neck patients treated with SIB-IMRT. *Int. J. Radiat. Oncol. Biol. Phys.* 64: 968–977.

Schach von Wittenau AE et al. (1999) Correlated histogram representation of Monte Carlo derived medical accelerator photon-output phase space. *Med. Phys.* 26: 1196–1211.

Schreiber E and Faddegon BA (2005) Sensitivity of large-field electron beams to variations in a Monte Carlo accelerator model. *Phys. Med. Biol.* 50: 769–778.

Scott AJ, Nahum AE and Fenwick JD (2008) Using a Monte Carlo model to predict dosimetric properties of small radiotherapy photon fields. *Med. Phys.* 35: 4671–4684.

Scott AJ et al. (2012) Characterizing the influence of detector density on dosimeter response in non-equilibrium small photon fields. *Phys. Med. Biol.* 57: 4461–4476.

Sempau J and Bielajew AF (2000) Towards the elimination of Monte Carlo statistical fluctuation from dose volume histograms for radiotherapy treatment planning. *Phys. Med. Biol.* 45: 131–157.

Seuntjens J (2006) Measurement issues in commissioning and benchmarking of Monte Carlo treatment planning systems. In: *Integrating New Technologies into the Clinic: Monte Carlo and Image-Guided Radiation Therapy*, pp. 117–144. (Eds. Curran B, Balter JM and Chetty IJ, Madison, WI: Medical Physics Publishing).

Sheikh-Bagheri D and Rogers DWO (2002) Sensitivity of megavoltage photon beam Monte Carlo simulations to electron beam and other parameters. *Med. Phys.* 29: 379–390.

Siebers JV et al. (2000) Converting absorbed dose to medium to absorbed dose to water for Monte Carlo based photon beam dose calculations. *Phys. Med. Biol.* 45: 983–995.

Siebers JV et al. (2002) A method for photon beam Monte Carlo multileaf collimator particle transport. *Phys. Med. Biol.* 47: 3225–3249.

Siebers JV et al. (2004) Monte Carlo computation of dosimetric amorphous silicon electronic portal images. *Med. Phys.* 31: 2135–2146.

Siebers JV and Ma CM (2006) Monte Carlo applications in IMRT planning and quality assurance. In: *Integrating New Technologies into the*

Clinic: Monte Carlo and Image-Guided Radiation Therapy, AAPM Medical Physics Monograph No. 32, pp. 145–177. (Eds. Curran BH, Balter JM and Chetty IJ, Madison, WI: Medical Physics Publishing).

Siebers JV and Zhong H (2008) An energy transfer method for 4D Monte Carlo dose calculation. *Med. Phys.* 35: 4096–4105.

Siebers JV, Keall PJ and Kawrakow I (2005) Monte Carlo dose calculations for external beam radiation therapy. In: *The Modern Technology of Radiation Oncology*, Vol. 2, pp. 91–130. (Ed. Van Dyk J, Madison, WI: Medical Physics Publishing).

Siljamaki S et al. (2005) Determining parameters for a multiple-source model of a linear accelerator using optimization techniques. *Med. Phys.* 32: 2113–2114.

Sterpin E, Mackie TR and Vynckier S (2012) Monte Carlo computed machine-specific correction factors for reference dosimetry of TomoTherapy static beam for several ion chambers. *Med. Phys.* 39: 4066–4072.

Stroian G et al. (2008) Local correlation between Monte-Carlo dose and radiation-induced fibrosis in lung cancer patients. *Int. J. Radiat. Oncol. Biol. Phys.* 70: 921–930.

Tyagi N et al. (2007) Experimental verification of a Monte Carlo-based MLC simulation model for IMRT dose calculation. *Med. Phys.* 34: 651–663.

Tzedakis A et al. (2004) Influence of initial electron beam parameters on Monte Carlo calculated absorbed dose distributions for radiotherapy photon beams. *Med. Phys.* 31: 907–913.

Ulmer W, Pyyry J and Kaissl W (2005) A 3D photon superposition/convolution algorithm and its foundation on results of Monte Carlo calculations. *Phys. Med. Biol.* 50: 1767–1790.

Underwood TS et al. (2013) Detector density and small field dosimetry: Integral versus point dose measurement schemes. *Med. Phys.* 40: 082102.

Vandervoort EJ et al. and Cygler JE (2014) Evaluation of a new commercial Monte Carlo dose calculation algorithm for electron beams. *Med. Phys.* 41: 21711.

van der Voort van Zyp NC et al. (2010) Clinical introduction of Monte Carlo treatment planning: A different prescription dose for non-small cell lung cancer according to tumor location and size. *Radiother. Oncol.* 96: 55–60.

van Elmpt W et al. (2008) Transition from a simple to a more advanced dose calculation algorithm for radiotherapy of non-small cell lung cancer (NSCLC): Implications for clinical implementation in an individualized dose-escalation protocol. *Radiother. Oncol.* 88: 326–334.

Verhaegen F and Devic S (2005) Sensitivity study for CT image use in Monte Carlo treatment planning. *Phys. Med. Biol.* 50: 937–946.

Verhaegen F and Seco J (2013) (Eds). *Monte Carlo Techniques in Radiation Therapy*. (Boca Raton, FL: CRC Press).

Verhaegen F and Seuntjens J (2003) Monte Carlo modelling of external radiotherapy photon beams. *Phys. Med. Biol.* 48: R107–R164.

Walters BRB and Rogers DWO (2004) DOSXYZnrc user's manual, NRC Report PIRS 794 (rev. B). (Ottawa, Ontario, Canada: National Research Council of Canada).

Wilcox EE et al. (2010) Comparison of planned dose distributions calculated by Monte Carlo and Ray-Trace algorithms for the treatment of lung tumors with CyberKnife: A preliminary study in 33 patients. *Int. J. Radiat. Oncol. Biol. Phys.* 77: 277–284.

Wulff J, Zink K and Kawrakow I (2008) Efficiency improvements for ion chamber calculations in high energy photon beams. *Med. Phys.* 35: 1328–1336.
Yang J et al. (2004) Modelling of electron contamination in clinical photon beams for Monte Carlo dose calculation. *Phys. Med. Biol.* 49: 2657–2673.
Zhang GG et al. (1999) Monte Carlo investigation of electron beam output factors versus size of square cutout. *Med. Phys.* 26: 743–750.
Zhang A et al. (2013) Comprehensive evaluation and clinical implementation of commercially available Monte Carlo dose calculation algorithm. *J. Appl. Clin. Med. Phys.* 14(2): 4062.
Zhao Y et al. (2014) A clinical study of lung cancer dose calculation accuracy with Monte Carlo simulation. *Radiat. Oncol.* 16: 287.
Zhong H and Chetty IJ (2012) Generation of a novel phase-space-based cylindrical dose kernel for IMRT optimization. *Med. Phys.* 39: 2518–2523.

14

Quantifying Differences in Dose Distributions

David Westerly and Moyed Miften

14.1 Introduction

An important component of radiation therapy is the creation of a patient's treatment plan, including calculation of the dose distribution to be deposited. As the computing power available for treatment planning and delivery has increased, so has the complexity of the plans being generated. Due to the large number of variables used in modern planning systems, it is often necessary to compare dose distributions among multiple plans created for the same patient using different parameters and/or calculation algorithms. Additionally, with the advent of intensity-modulated radiation therapy (IMRT) in the early 1990s, it has become standard practice to verify the delivery of IMRT plans using either direct measurement of the radiation dose distribution in a phantom (Palta et al., 2003; Nelms and Simon, 2007; Mijnheer and Georg, 2008; Hartford et al., 2009) or indirect methods that involve measuring various machine parameters during the treatment delivery and plugging these measured values back into the dose calculation algorithm to compute a delivered dose distribution (Pawlicki et al., 2008; Siochi et al., 2009; Sun et al., 2012). These verification processes also necessitate the comparison of different dose distributions.

Such comparisons are the subject of this chapter. Although at first, comparing dose distributions may seem to be a trivial task, there are a number of subtleties associated with the comparison process that make extraction of clinically meaningful results difficult. In the sections that follow, we discuss challenges associated with making these comparisons and describe both qualitative and quantitative approaches that have shown to be useful in clinical applications. We describe the strengths and weaknesses of these methods and discuss various parameters that can impact comparison results. We also look at some practical considerations that come into play when making these comparisons. Finally, we discuss the interpretation of comparison results and look at various problems and limitations that arise using the different methods.

14.2 Challenges in Comparing Dose Distributions

The problem of comparing calculated and/or measured dose distributions scales in complexity with the dimensionality of the distributions being evaluated. As an example, consider the zero-dimensional (0D) case where two point doses are compared. In this relatively simple case, a mathematical difference can be used as a metric for comparison. The difference can be expressed in absolute terms (e.g., Gy) or as a percentage of the summands depending on the situation. Interpretation is straightforward since the difference lies on a continuous number line and a single dose tolerance value can be used to gauge whether the calculated difference is clinically relevant.

Now consider the one-dimensional (1D) case, comparing dose profiles. In this case, things are more complicated. The added difficulty is not mathematical in nature, a difference can still be computed between corresponding points in the two registered profiles, rather the problem is one of interpretation. Consider the dose profiles in Figure 14.1. This figure shows the transverse dose profile for a 6 MV unflattened photon beam along with the same profile shifted by 5 mm. Dose differences between the two are small (<3%) in regions of uniform dose but become large (>10%) in high-gradient regions where the profiles do not perfectly align. Dose differences in high-gradient regions may or may not be clinically meaningful depending on where they occur within the patient anatomy. This type of subjectivity precludes the use of a single threshold value to determine whether observed differences are significant and confounds simple interpretation of comparison results.

The problem previously described is exacerbated when comparing 2D planar or 3D volumetric dose distributions, since the dimensionality of the local dose gradient is increased. Computational methods that account for these difficulties typically require a search algorithm and are also more complex due to the increased degrees of freedom in the search space. Add to this the complications that arise when dealing with choice of dose normalization and/or differences in spatial resolution between distributions, and one can see that comparing dose distributions in a clinically meaningful way can be quite challenging. In the sections that follow, we describe some of the more common approaches to handling these problems.

14.3 Qualitative Comparison Methods

One of the most commonly used methods of comparison is visual inspection of the dose distributions. Though qualitative in nature, visual comparison allows for fast identification of gross differences in the dose distributions, which

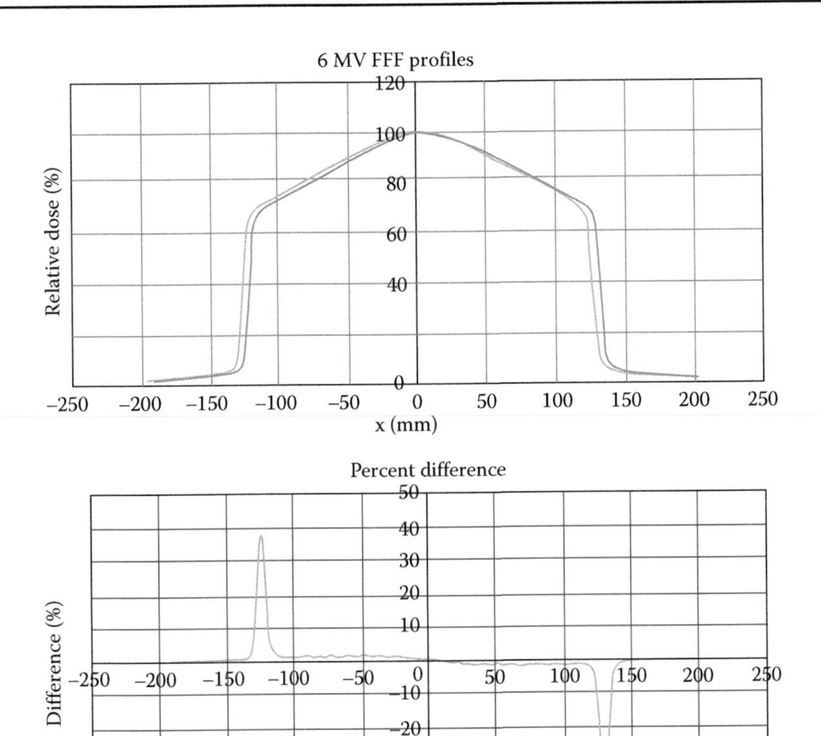

Figure 14.1

Dose profiles for a 6 MV, flattening filter-free (FFF) photon beam and the same profile shifted by 5 mm (upper). Percent difference (relative to maximum dose) computed between the two dose profiles (lower). Dose differences are small in low-gradient regions but become increasingly large in high-gradient regions where the profiles do not align.

is useful for both treatment planning and quality assurance (QA) tasks. The effectiveness of this method depends on the manner in which the dose distributions are displayed. Many software applications use a linked side-by-side display that shows different dose distributions overlaid on the same planning image in adjacent viewing panels. The dose can usually be displayed either in color-wash or as discrete isodose lines. In the case of 3D planning images, the user can scroll through different slices of the image to compare the dose distributions in different regions of the patient's anatomy. Figure 14.2 shows an example of this type of comparison implemented in Varian's Eclipse treatment-planning system (version 13.6; Varian Medical Systems, Palo Alto, CA). The main advantage of this display is that it provides an easy way to discern any gross differences in the dose distributions while still allowing for a complete display of each plan; i.e., the dose distribution can be displayed using a continuous color-wash scale.

An alternative to the side-by-side display is to overlay discrete isodose contours from two different plans on a single planning image. An example of this type of comparison is shown in Figure 14.3 and comes from TomoTherapy's Planned

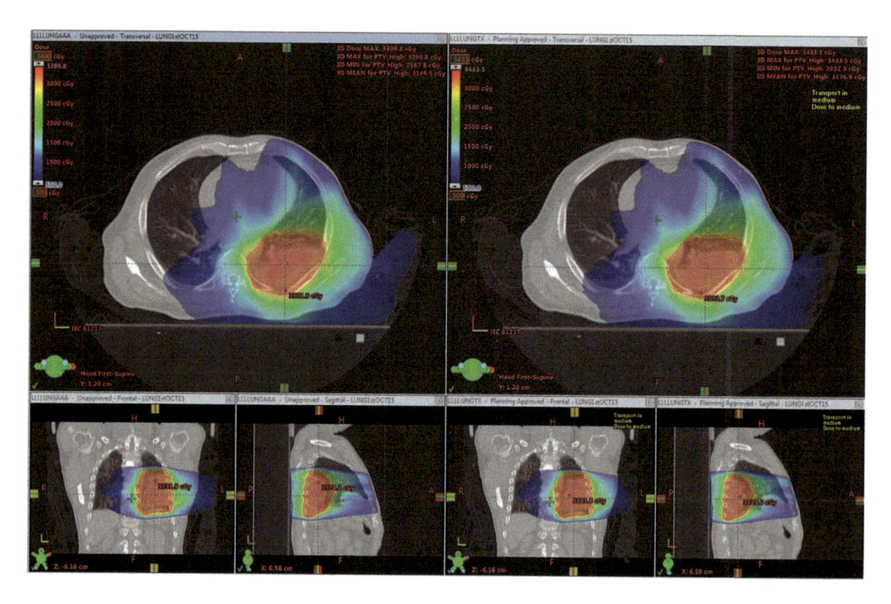

Figure 14.2

Illustration of Varian's Eclipse plan evaluation tool. Two dose distributions computed on the same computed-tomography (CT) set can be viewed side by side in adjacent panels. The two displays are linked so that the user can scroll through different slices of the plans simultaneously. This tool allows for both dose distributions to be displayed using a continuous color-wash scale.

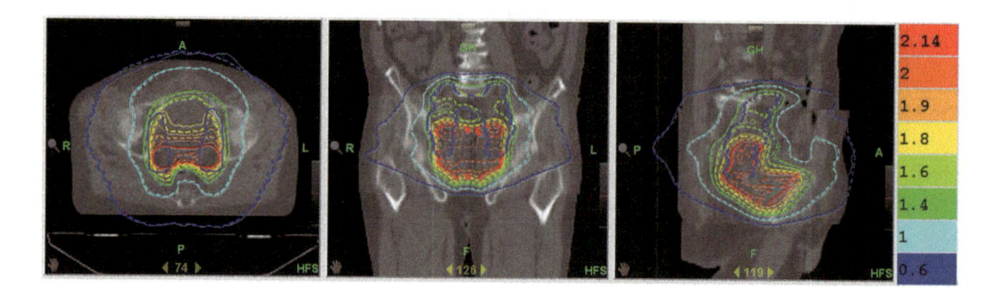

Figure 14.3

Isodose distributions for a TomoTherapy prostate bed and pelvic lymph node plan. The image is a fusion of the original planning computed therapy (CT) and a daily megavoltage CT (MVCT) image acquired to verify patient setup. Solid isodose lines are for the original plan, whereas the dashed lines are the dose values that result from recalculation of the plan on the daily MVCT. The dose color scale is in Gy.

Adaptive software (version 4.2; Accuray, Sunnyvale, CA). In this figure, the solid and dashed lines represent isodose contours for the original treatment plan and the treatment plan recalculated on the daily megavoltage computed-tomography (CT) image, respectively. Areas where the isodose lines diverge indicate differences in the dose distributions. The advantage of this type of display over the side-by-side comparison is that it allows for quick identification of regions where the dose distributions differ. This can sometimes be difficult to see in the side-by-side comparison unless the differences are sufficiently large. The real drawback to this method, however, is that one is limited to the isodose line display, which provides only a sampling of the dose distribution at specified levels. Regardless of display format, both of these methods lack the ability to discern small differences in the dose distributions on the order of 3–5%. For this reason, one requires more quantitative methods that are discussed in the following sections.

14.4 Quantitative Comparison Methods

A more rigorous assessment of differences between two dose distributions requires a quantitative approach. At this point, it is worthwhile to make clear certain definitions and assumptions. In particular, it is useful to have a standardized way of referring to the dose distributions being compared. To be consistent with the existing literature on this topic, we have chosen to follow the nomenclature used by Low and Dempsey (2003), where the two dose distributions are referred to as the evaluated and the reference distribution. The evaluated distribution is a dose distribution being compared to a reference dose distribution, which is the "standard" of comparison. In all cases, it is assumed that both dose distributions are coregistered in a common reference frame; however, no constraint is placed on the dimensionality or spatial resolution of the dose distributions being compared.

14.4.1 Dose Difference

One of the most straightforward means of quantitatively comparing two dose distributions is by computing the dose difference. The difference δ between the evaluated and the reference dose distributions is calculated as follows:

$$\delta(\vec{r}) = D_{\text{eval}}(\vec{r}) - D_{\text{ref}}(\vec{r}) \tag{14.1}$$

where $\vec{r}$ is a vector that extends from a common origin to a given point of interest. Practically speaking, this difference is calculated over the intersection of the evaluated and the reference dose distributions for every point in the higher resolution dose grid, interpolating the lower resolution grid as necessary.

The dose difference provides useful information in regions of uniform dose where both dose distributions are insensitive to small spatial perturbations. However, in high-dose-gradient regions, small spatial misalignments and/or differences in the local dose gradients can result in large dose differences. Such misalignments may or may not be clinically relevant depending on their origin. In cases where one or both distributions are generated via measurement, uncertainties associated with positioning the measurement device can result in small misalignments that lead to large dose differences. These misalignments are a result of the measurement process and in no way reflect the accuracy of treatment delivery. On the other hand, if the misalignment is due to a mechanical error with the treatment machine, then the dose differences observed would be clinically relevant and actions should be taken to correct the underlying problem.

When comparing calculated dose distributions, spatial misalignments are typically not an issue since the calculations often share the same geometry. More often in these cases, differences in local dose gradients produce large dose discrepancies. For local gradients near the edge of a target volume, it is unlikely that small differences in the dose falloff will be clinically significant unless they occur at the border between the target and a dose-limiting organ at risk. On the other hand, if the gradients appear inside a target region for an IMRT plan, it is unclear whether such differences will be clinically significant without further analysis.

Such analysis might include monitoring to see if the discrepancy results in an underdosing of the target volume, and if it does, to what volume and by how much. The answer to these questions can have a significant impact on tumor control probability and must be assessed on a case-by-case basis.

14.4.2 Distance-to-Agreement

The concept of distance-to-agreement (DTA) was developed as a means for comparing dose distributions in regions characterized by large dose gradients. DTA measures the closest Euclidean distance between a point in the evaluated dose distribution and a corresponding point in the reference distribution having the same dose value. Several groups used DTA to validate treatment-planning system calculations in high-gradient regions during commissioning (e.g., Van Dyk et al., 1993). Later, Harms et al. (1998) incorporated DTA into an algorithm that allowed for comparison of 2D dose distributions and provided pass/fail results based on whether the DTA was below a certain distance threshold. This algorithm used a search of the evaluated dose distribution to find the set of all points with doses equal to each point in the reference distribution. The distances between all matching points were computed and the shortest distance was designated the DTA for each reference dose point. In cases where the DTA exceeded the preset search distance, the DTA for that point would be assigned the search distance (in the original paper, the search distance was 1.0 cm).

The advantage of DTA over dose difference is that it is insensitive to dose differences in high-gradient regions. However, DTA has the problem that it is overly sensitive to differences that occur in regions of relatively uniform dose. An additional problem with DTA is that it also depends on the assignment of the evaluated and reference distributions. Dose difference on the other hand is invariant to this assignment to within a sign change. To see this dependency, consider the case where the maximum dose in the evaluated distribution is greater than all dose points in the reference distribution. When a search of the reference distribution is performed for the maximum evaluated dose point, no matching value will be found and the DTA will be assigned the search distance. Now reverse the dose distribution assignments. Since the dose at this point is now less than the maximum dose in the reference distribution, a match will be found, and it is possible that this match is at a distance less than the search distance, resulting in a different DTA.

14.4.3 Composite Test of Dose Difference and Distance-to-Agreement

Given the complementary nature of the dose difference and DTA metrics, it makes sense that these two should be combined into a composite test. This was done by the software tool developed by Harms et al. (1998) and a variation has since been adopted by a commercial vendor (Jursinic and Nelms, 2003). With the composite test, both dose difference and DTA are computed for each point in the reference distribution; again interpolating the lower resolution grid if necessary. User-defined dose difference and DTA tolerances are then used as a basis of comparison. If a point in the reference distribution has a dose difference *or* DTA less than their respective tolerance values, the point is deemed passing the composite test. If, however, the dose difference and DTA are both greater than their respective tolerance values, then the point is deemed to fail the composite test.

The advantage of the composite test is that it allows for a quantitative comparison of two dose distributions that is not overly sensitive to regions of high- or low-dose gradient. A major disadvantage, however, is that the composite test gives only pass/fail results; there is no numerical answer that can be evaluated on a continuous scale. Another disadvantage is that the composite test considers dose difference and DTA separately. As such, for a dose difference tolerance of 3% of the local reference point dose and DTA tolerance of 3 mm, a point in the evaluated distribution could be deemed failing even if its nearest neighbor located 2 mm away is within the 3% dose tolerance.

14.4.4 Gamma Analysis

In the late 1990s, Low noted that the main difficulty with comparing dose difference and DTA simultaneously was that the two quantities have different units. To overcome this problem, he normalized the dose and distance metrics by their respective tolerance values. This allowed for computation of the Euclidean distance between two dose points in the reference and evaluated distributions in normalized dose difference–distance space. This quantity was denoted Γ and can be written as follows:

$$\Gamma(\vec{r}_{\text{eval}} \cdot \vec{r}_{\text{ref}}) = \sqrt{\left(\frac{D_{\text{eval}}(\vec{r}_{\text{eval}}) - D_{\text{ref}}(\vec{r}_{\text{ref}})}{D_{\text{tol}}}\right)^2 + \left(\frac{|\vec{r}_{\text{eval}} - \vec{r}_{\text{ref}}|}{r_{\text{tol}}}\right)^2} \quad (14.2)$$

where $\vec{r}_{\text{eval}}$ and $\vec{r}_{\text{ref}}$ are vectors extending from a common origin to points in the evaluated and reference dose distributions, respectively. D_{tol} is the dose difference tolerance and r_{tol} is the distance tolerance. The Γ function is defined for each set of points $\vec{r}_{\text{eval}}$ and $\vec{r}_{\text{ref}}$. By performing a search of the evaluated dose distribution for each point in the reference distribution (similar to the DTA calculation), the minimum value of Γ is determined for each point in the reference distribution. This minimum value is commonly denoted as lowercase gamma (γ) (Low et al., 1998) and can be expressed mathematically as follows:

$$\gamma(\vec{r}_{\text{ref}}) = min\{\Gamma(\vec{r}_{\text{eval}}, \vec{r}_{\text{ref}})\} \forall \{\vec{r}_{\text{eval}}\} \quad (14.3)$$

Gamma is the minimum Euclidean distance between points in the evaluated and reference dose distributions calculated in the normalized dose difference–distance space. The meaning of gamma can be understood by recognizing that Equation 14.3 describes a unit circle, sphere, or hypersphere in dose difference–distance space, depending on the dimensionality of the dose distributions. Points that lie on or within this contour, surface, or hypersurface (points for which $\gamma \leq 1$) represent dose and spatial difference combinations with magnitudes less than or equal to that produced by a dose difference D_{tol} or spatial misalignment r_{tol} alone. These points are deemed passing with respect to the tolerance values (Low et al., 1998; Low and Dempsey, 2003; Low, 2010).

There are several advantages of using the gamma metric approach compared to other methods. First and foremost, gamma allows for a quantitative comparison of dose distributions that considers, but is not overly sensitive to, differences occurring in both high- and low-dose-gradient regions. In regions of low-dose gradient, gamma behaves like a dose difference test, choosing points that minimize spatial separation, whereas in regions of high-dose gradient, it behaves like the DTA test, allowing for small spatial differences that result in a smaller dose difference.

Another advantage of gamma is that it yields a numerical value in addition to a binary pass/fail result. This value can be used to assess by how much a given point fails the set criteria. For instance, $\gamma = 1.1$ implies a point failed by 10% relative to the set criteria. For a D_{tol} and r_{tol} of 3% and 3 mm, respectively, this implies a discrepancy of 3.3% or 3.3 mm, which may still be considered acceptable. On the other hand, $\gamma = 2$ indicates a failure by a factor of 2, which corresponds to a difference of 6% or 6 mm (Low, 2010).

Since its introduction, gamma has become one of the more commonly used tools to quantitatively assess differences in dose distributions. This is especially true when comparing measured versus calculated dose distributions, as is commonly done for IMRT QA. Figure 14.4 shows an example gamma calculation for a TomoTherapy plan delivered to a cylindrical solid water phantom containing a sheet of radiographic film. In this case, the reference dose was taken to be the dose measured by the film, whereas the evaluated distribution is the dose distribution resulting from calculation of the treatment plan on the phantom geometry. Values for D_{tol} and r_{tol} are 3% of the maximum reference dose and 3 mm, respectively. The color map indicates regions with different values of gamma. Regions that are not colored fall below the lowest color band ($\gamma = 0.33$).

14.4.5 Region-of-Interest Analysis

Another common methodology used to assess differences in dose distributions is region-of-interest (ROI) analysis. ROI analysis seeks to extract relevant information by focusing on regions of the dose distributions that are contextually relevant to the comparison being made. Typically, ROIs are anatomic structures, though regions corresponding to specific isodose levels can also be used. All of the methods previously described can be used in conjunction with ROI analysis. In addition, a number of methods specific to ROI analysis have been developed, including descriptive statistics, dose–volume histogram analysis, and various indices that quantify dose conformity or homogeneity.

Descriptive statistics offer one of the most straightforward ways to compare dose distributions within an ROI. Commonly used statistics include maximum dose, minimum dose, median dose, and mean dose, whereas the standard deviation is also sometimes used. The advantage of descriptive statistics is that they simplify the comparison of multidimensional dose distributions by providing

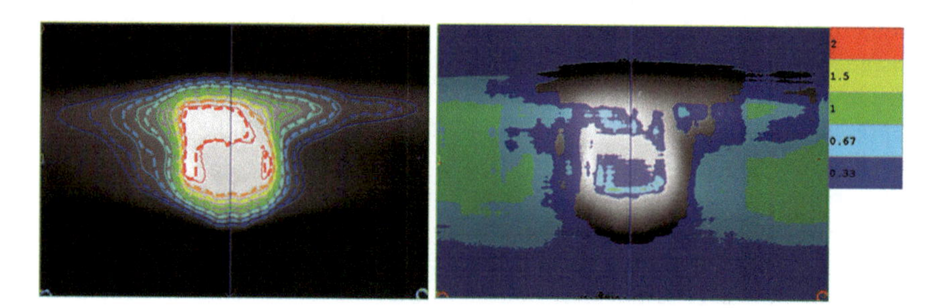

Figure 14.4

Isodose lines overlaid on a coronal film measurement of a TomoTherapy treatment plan (left). Gamma calculation overlaid on the same film performed using dose difference and distance-to-agreement (DTA) tolerance criteria of 3% and 3 mm, respectively (right). Gamma values ≤ 1 are considered passing the stated criteria. Regions not colored have $\gamma < 0.33$.

a single-valued numeric measure for each distribution. These numbers are usually easy to compare and have a clear interpretation. In addition, because the calculated statistics are specific to the ROI, they often contain more useful information than if the statistics were computed for the dose distribution as a whole. The main drawback of descriptive statistics is that they provide no information about the spatial distribution of dose within the ROI, which may be important to the analysis.

A more advanced form of ROI analysis splits the volume elements (voxels) within an ROI into bins based on the dose that each voxel receives. This process produces a differential dose–volume histogram (dDVH), which is displayed graphically in Figure 14.5a. A plot of a dDVH shows the total ROI volume receiving each dose level. An alternative way of displaying these data is to compute a

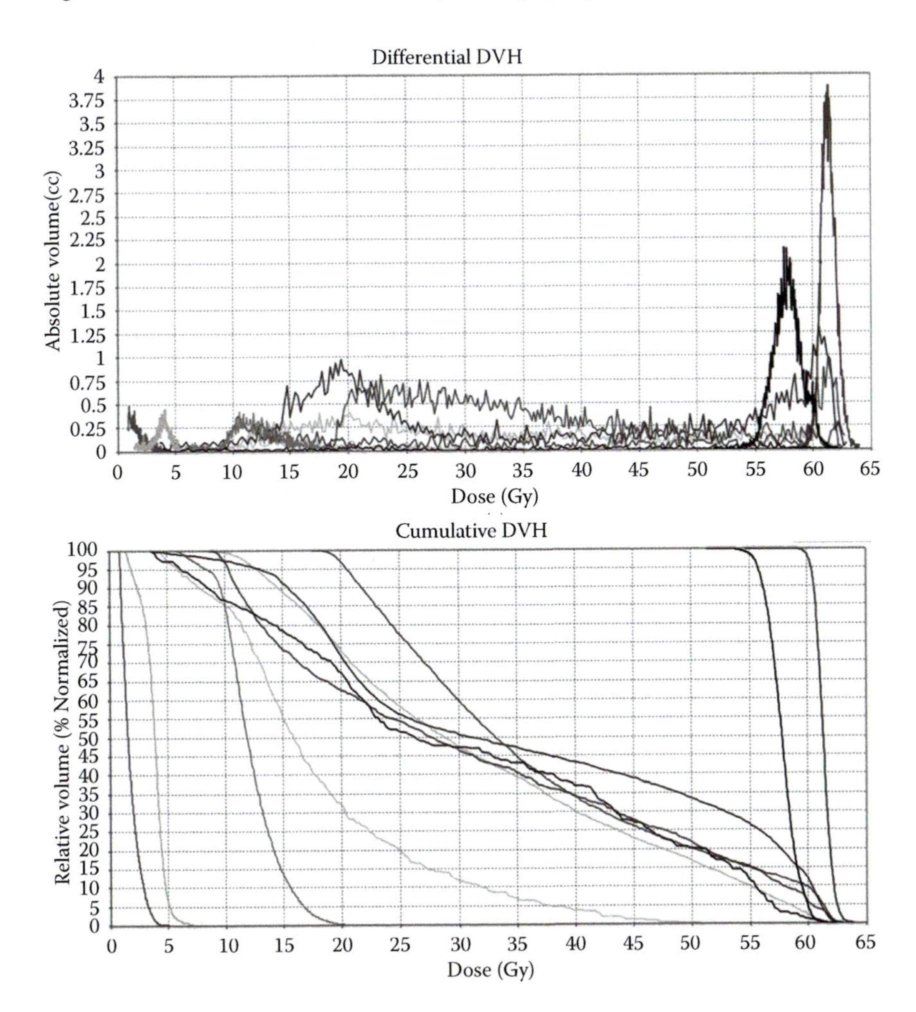

Figure 14.5

(a) Differential dose–volume histograms (DVHs) for regions of interest (ROIs) in a head-and-neck intensity-modulated radiation therapy (IMRT) plan. (b) Cumulative DVHs for the same plan and ROIs. Target ROIs in the differential DVH appear as peaks spanning a relatively narrow dose range. This corresponds to the nearly vertical lines in the cumulative DVH. The dose range for the normal tissue ROIs is more broadly distributed in both DVH representations.

cumulative DVH (cDVH). With a cDVH, the volume assigned to a particular dose bin is the total ROI volume receiving that bin dose or lower. Mathematically, this can be written as the volume summation taken over the dDVH:

$$V_k^c(D_k) = \sum_{i=0}^{k} V_i^d(D_i) \tag{14.4}$$

where D_i is the dose assigned to the ith dose bin, and V_i^d and V_i^c are the volume associated with the ith dose bin for the dDVH and the cDVH, respectively. An example of a cDVH is shown in Figure 14.5b. Other common variants of the DVH involve normalizing the volume axis to the ROI volume, and/or the dose axis to either the maximum dose contained by the ROI or the prescription dose used for the treatment plan(s) being analyzed.

The advantage of DVHs compared to descriptive statistics is that they provide information pertaining to the relationship between dose and volume within the ROI. This allows for analysis of specific dose–volume points, for example, $V_{20\text{ Gy}}$, which is the volume of the ROI (either fractional or absolute) receiving at least 20 Gy. These dose–volume metrics are often used to correlate the dose received by a particular anatomical structure with side effects observed after radiation exposure. Calculating these values for different dose distributions can provide a comparison that is clinically relevant to the outcomes experienced by the patient. The main drawback of DVHs in general is that, like descriptive statistics, they lack any information about the spatial distribution of dose within the ROI. This is a result of separating the ROI voxels into dose bins without considering the spatial location of the volume elements.

The last group of ROI metrics discussed in this chapter uses indices calculated from the area or volume of the ROI and/or information about the dose distribution encompassed by the ROI. These indices are typically used to evaluate how well the shape of a particular isodose level corresponds to the shape of a given ROI (conformity), or how uniform the dose distribution is within the ROI (homogeneity). While these indices allow for comparison of specific characteristics of different dose distributions as they relate to a given ROI, they are inherently singular in nature and only offer information about one particular aspect of the distributions being compared. Nonetheless, they can be a useful tool, particularly when trying to develop a plan that optimizes one of these parameters. Some of the more commonly used indices are listed later. Note that the descriptions below are made in reference to 3D dose distributions and ROIs. However, the methods apply equally well in 2D; simply replace references to ROI volumes and isodose surfaces with ROI areas and isodose lines.

14.4.5.1 Homogeneity Index

The homogeneity index (HI) measures how uniform the dose distribution is within the specified ROI. The International Commission on Radiation Units and Measurements (ICRU) has recommended that the HI should be computed as follows (ICRU, 2010):

$$\text{HI} = \frac{D_{2\%} - D_{98\%}}{D_{\text{median}}} \tag{14.5}$$

where $D_{2\%}$, $D_{98\%}$, and D_{median} are the dose to 2%, 98%, and 50% of the ROIs, respectively. An HI value of 0 indicates uniform dose throughout the ROI volume.

14.4.5.2 Target Coverage

Target coverage (TC) is a useful metric in radiation therapy treatment planning because it indicates the percent volume of a target ROI that receives prescription dose. Mathematically, TC can be written as follows:

$$\mathrm{TC} = \frac{V_{\mathrm{target}} \cap V_{\mathrm{pres}}}{V_{\mathrm{target}}} \times 100 \tag{14.6}$$

where V_{target} is the volume of the target ROI, V_{pres} is the volume enclosed by the prescription isodose surface, and $V_{\mathrm{target}} \cap V_{\mathrm{pres}}$ represents the volume of the intersection. A TC value of 100 indicates perfect target coverage.

14.4.5.3 Prescription Isodose-to-Target Volume Ratio

The prescription isodose-to-target volume (PITV) ratio is defined as the ratio of the volume enclosed by the prescription isodose surface to the volume of the target ROI (Shaw et al., 1993). Mathematically, this can be written as follows:

$$\mathrm{PITV} = \frac{V_{\mathrm{pres}}}{V_{\mathrm{target}}} \tag{14.7}$$

A PITV ratio of 1 indicates that the volume enclosed by the prescription isodose surface and the target ROI volume are equal. Note, however, that this ratio contains no information about how these two volumes overlap.

14.4.5.4 Conformation Number

The conformation number (CN) proposed by van't Riet et al. (1997) measures dose conformity with respect to a target ROI. Mathematically, it is defined as follows:

$$\mathrm{CN} = \frac{(V_{\mathrm{target}} \cap V_{\mathrm{pres}})^2}{V_{\mathrm{target}} \cdot V_{\mathrm{pres}}} \tag{14.8}$$

The CN considers both the TC and the conformity of the dose distribution by including a multiplicative factor that is the ratio of volume inside the target ROI covered by the prescription isodose to the total volume enclosed by the prescription isodose. This index varies between 0 and 1 with a CN of 1 indicating perfect conformity; i.e., the prescription isodose surface overlaps the target ROI exactly.

14.4.5.5 Dice Coincidence Index

Another index that provides information about the conformity of the prescription isodose surface is the dice coincidence index (DCI) (Dice, 1945). The DCI is defined as follows:

$$\mathrm{DCI} = \frac{2 \cdot (V_{\mathrm{target}} \cap V_{\mathrm{pres}})}{V_{\mathrm{target}} + V_{\mathrm{pres}}} \tag{14.9}$$

It looks at how closely the prescription isodose surface overlaps the target ROI. However, whereas the CN uses the intersection of the target ROI volume and the volume enclosed by the prescription isodose surface to scale the target ROI coverage, the DCI normalizes this intersection by the sum of the volumes. The DCI also varies between 0 and 1 with 1 indicating perfect conformity.

14.4.6 Additional Methods

In addition to the methods described thus far, a number of other methods have been proposed or put into practice, though for various reasons these methods have not been as widely adopted. Bakai et al. (2003) simplified the gamma calculation by normalizing local dose differences to a spatially varying dose difference tolerance. This local tolerance value incorporates spatial uncertainties by adding in quadrature the reference dose gradient multiplied by a distance tolerance value and traditional dose difference criteria used in gamma calculations. This approach to deriving the local dose acceptance criteria can be viewed with respect to error propagation since it effectively combines the dose difference uncertainties inherent to the measurement (or calculation) with those arising from spatial uncertainties, and eliminates the need for an exhaustive search of the evaluated dose distribution for every point in the reference distribution. With this method, points with dose differences less than the acceptance criteria (normalized values <1) are considered to pass, whereas those greater than the local criteria are deemed failing and require further analysis.

Moran et al. (2005) developed the gradient compensation method. Similar to the method described by Bakai et al. (2003), this method compares the local dose difference to a spatially varying tolerance value. However, with the gradient compensation method, this tolerance value is just the product of the local reference dose gradient and a distance parameter that accounts for spatial uncertainties. Points with dose differences less than the local tolerance value are deemed passing, whereas points with dose differences larger than the local tolerance value are assumed to have dose errors not attributable to spatial uncertainties and therefore require further investigation.

The normalized agreement test (NAT) index introduced by Childress and Rosen (2003) represents the average deviation from the percent dose difference and DTA tolerances for every pixel/voxel calculated, ignoring regions with deviations less than a set criterion. The NAT index is calculated from NAT values, which are determined in the following way: comparison points that fall within the stated dose difference or DTA tolerances have a NAT value of 0 as do reference points with doses below 75% of the maximum reference dose. Otherwise the NAT value is calculated as $D_{scale}(\delta - 1)$, where δ is the lesser of the ratio of dose difference (absolute value) to the dose difference tolerance, or the ratio of DTA to the DTA tolerance, and D_{scale} is the greater of the reference or evaluated dose at the comparison point normalized to the maximum evaluated dose.

Stock et al. (2005) developed a novel algorithm to improve the speed of gamma calculations and also incorporated new tools into their gamma analysis. In particular, they looked at gamma angle calculations, which provide information about whether the gamma value is dominated more by dose difference or DTA. They also used gamma-value histograms as a way to easily visualize the proportion of points that are failing the gamma test.

Jiang et al. (2006) introduced the concepts of equivalent dose tolerance, maximum allowed dose difference (MADD), and normalized dose difference (NDD). These quantities are derived by transferring spatial tolerances into the dose domain. Specifically, the equivalent dose tolerance at a given point is calculated by shifting the evaluated dose distribution along the dose axis until the DTA for that point is equal to preset DTA criteria. The equivalent dose tolerance then is the dose difference between the evaluated and the reference dose distributions at this point. MADD is a more general concept than the equivalent dose tolerance. MADD defines an acceptance region in dose space that takes into account both the traditional dose difference criteria and the equivalent dose tolerance, which represents spatial uncertainties. If at a given point, the dose difference is less than MADD, then the point is considered passing. To further simplify the analysis, the authors use the NDD, which is just the dose difference normalized by the ratio of MADD to the traditional dose difference criteria. Thus, the NDD allows the dose difference comparison with MADD to be expressed in terms of the traditional dose difference criteria.

14.5 Practical Considerations

While the methods described in this chapter provide a suite of useful tools for quantitatively analyzing dose distributions, it is important to consider certain practical aspects when making dose comparisons. In particular, spatial resolution, choice of dose normalization, and choice of dose threshold for comparison are important to consider when making dose comparisons, since these parameters can have a dramatic impact on the comparison results. Each of these topics is discussed in more detail in the following sections.

14.5.1 Spatial Resolution

Spatial resolution is a basic property of any sampled representation of a continuous function. In the case of dose distributions, spatial resolution specifies the distance between known dose values. This might reflect the physical spacing of measurement points in a detector array or it might correspond to the calculation grid used by a treatment-planning system. For most clinical dose distributions, the spatial resolution is constant, with values of 1–3 mm often used to accommodate steep local dose gradients that may exceed a few percent per millimeter.

Spatial resolution is an important parameter to consider when comparing dose distributions. In many cases, it is convenient, if not necessary, for the dose distributions being compared to have equal spatial resolution since it makes computing dose differences straightforward. Additionally, it is important that the evaluated distribution has a fine enough spatial resolution so that distance calculations (including search operations performed for DTA and gamma calculations) do not introduce artifacts that can result from the spacing between dose points being a significant fraction of the distance tolerance. In cases where this is an issue, interpolation can be used to ensure a fine enough resolution. A general rule of thumb suggested by Low (2010) is to ensure the spatial resolution of the evaluated dose grid is at most one-third of the distance tolerance criterion. For most clinical applications, this amounts to a 1-mm dose grid.

Spatial resolution also plays a role when performing ROI analysis on small volumes/areas. Since the use of a coarse dose grid tends to average out higher frequency dose variations, using a coarse dose grid to perform ROI analysis can reduce the accuracy of certain metric calculations by skewing the results toward the answer that would be observed if all of the dose points were equal to the ROI mean dose. In addition, because small structures tend to encompass only a few dose points in a coarse dose grid, calculation results are more sensitive to small perturbations in the ROI delineation. To ensure calculation accuracy, it is important to make sure that any ROIs being used for analysis are dimensionally large compared to the spatial resolution of the dose distributions.

14.5.2 Normalization

Another practical consideration for comparing dose distributions is dose normalization. Dose normalization can apply either to the scaling of individual dose distributions or to the scaling of dose differences. In the case of absolute plan normalization, it is typical to scale (or normalize) a treatment plan in order to achieve a given TC, for example, 95% of the planning target volume receives 100% of the prescription dose, or to ensure that a critical organ dose is not exceeded, for example, maximum spinal cord dose <45 Gy. When comparing dose distributions, it is necessary that both plans are normalized in order to meet the same planning constraint(s). In this way, the plans are being compared on an equal footing.

Normalization of dose differences is more challenging. In this case, one is looking to compute a percent difference at a point in the dose distribution; however, the question is, with respect to what? Some would argue that percent difference should be a locally varying quantity that is computed with respect to the local reference dose. This however has the undesirable side effect of magnifying differences in regions of low dose which are less likely to be clinically significant. Another option is to normalize the dose difference to the global maximum dose. This downplays differences occurring in low-dose regions; however, it also may be overly forgiving if the clinical dose distribution is characterized by large-dose heterogeneity.

A third option that represents the middle ground is to normalize dose differences by the prescription dose since this provides a clinically relevant reference point. This solution also offers some difficulty though when comparing measured dose distributions with calculations of patient treatment plans in a measurement phantom. Differences between the patient anatomy and the phantom geometry can result in dose levels in target regions that are significantly different from the prescription dose in the patient. In these cases, one can compare the maximum dose in the clinical patient plan to the prescription dose and scale the phantom dose calculation by an equal percentage. Regardless of the normalization method used, it is important for the user to have a good understanding of the implications this scaling will have on the comparison results.

14.5.3 Dose Threshold for Comparison

A third consideration for comparing dose distributions is selection of a minimum dose threshold below which dose comparisons will not be made. This is usually done to avoid biasing aggregate comparison results (e.g., the percentage of points that pass a given test) with dose differences occurring in low-dose regions that are not clinically relevant. Use of a nonzero-dose threshold can also have the added benefit of speeding up comparison metric calculations substantially.

A drawback to using a nonzero-dose threshold is that improper selection of this threshold can impact aggregate comparison metrics and may give misleading results. For this reason, it is recommended that dose threshold levels be chosen in a manner that is both consistent and clinically relevant. An example would be setting the dose threshold to 10% of the maximum reference point dose based on the physician's statement that they are not concerned with dose discrepancies occurring at dose levels below this value. It is also recommended that dose comparisons be evaluated in an absolute manner in addition to considering any aggregate comparison metrics.

14.6 Interpretation of Dose Difference Results: Limitations, Problems, and Issues

One of the major challenges with comparing dose distributions lies not in the computational methods, but rather in the interpretation of results. Dose comparisons are only meaningful in the context of a specific goal or endpoint; for example, does plan A achieve better or worse TC than plan B. Without appropriate tolerances or quality criteria, interpretation of dose distribution comparisons is not possible. In clinical practice, dose distributions are compared for a variety of reasons and the quality endpoints that are used when interpreting comparison results are often specific to the task. For this discussion, we consider three broad categories of dose distribution comparisons: (1) comparison of calculated dose distributions created using the same patient or phantom study set but different plan parameters; (2) comparison of measured and calculated dose distributions, typically performed as part of patient specific QA; and (3) comparison of dose distributions calculated for the same treatment plan applied to different study sets. An example of this would be the recalculation of a patient's treatment plan on a subsequent CT scan acquired after the patient had been observed to lose a significant amount of weight.

When comparing dose distributions calculated using the same patient or phantom dataset, one is often interested in the differential effects produced by using different beam parameters, for example, beam energy, orientation, and fluence modulation, or in the effects of different dose calculation algorithms. Comparisons are usually made on the basis of clinically relevant endpoints such as TC and normal tissue dose–volume constraints with the goal of producing the optimal treatment plan for the patient. While interpretation of common ROI-based metrics is straightforward, such metrics are limited in their ability to assess the sensitivity of different plan configurations to the uncertainties inherent to radiation therapy. In particular, increasing the modulation of an IMRT plan may improve certain ROI metrics; however, this may result in a plan that is more difficult for the machine to deliver accurately and is also more sensitive to daily setup variations. These uncertainties can negate any advantages initially achieved during planning. Information pertaining to these types of uncertainties is not contained within the dose distribution.

Comparing measured versus calculated dose distributions falls within the purview of patient-specific QA performed for IMRT treatments, though it also finds a role in systems commissioning with end-to-end testing (Nelms and Simon, 2007; Ezzell et al., 2009). These topics are discussed in more detail in Chapters 17 and 19. Typically, for measurement-based QA, a clinical treatment plan is computed on a phantom geometry and a dose measurement with the phantom is acquired. Comparisons are usually interpreted with respect to dose difference and DTA tolerances, which are built into comparison algorithms such as the gamma calculation and composite tests. Interpretation of comparison results in this category are especially challenging because the phantom geometry often differs significantly from the patient geometry used to create the plan. Also, the relationship between dose discrepancies and clinical outcomes for modest size errors is uncorrelated (Kruse, 2010; Nelms et al., 2011; Zhen et al., 2011). Finally, reports from Imaging and Radiation Oncology Core (IROC) Houston have found that the ability to distinguish dosimetrically acceptable versus unacceptable treatment plans depends on the QA device being used (Kry et al., 2014; McKenzie et al., 2014).

Ideally, one would compare the planned dose distribution to the actual dose distribution delivered to the patient. This would allow the same metrics used to evaluate the original treatment plan to be applied when assessing the treatment delivery. As discussed in a more comprehensive way in Chapter 8, various methods have been developed, or are being investigated, that allow for this type of comparison. One vendor solution uses phantom dose measurements to perturb the planned dose distribution calculated in the patient in order to simulate the delivered dose distribution (Nelms et al., 2012), whereas another uses a head-mounted ion-chamber array to measure the photon fluence as it exits the machine. This measured fluence is then used to reconstruct the dose in the patient's planning CT (Boggula et al., 2011; Korevaar et al., 2011). Other methods currently under investigation involve back-projecting fluence measurements made with an electronic portal imaging device (see Chapter 7) to estimate the delivered fluence, which can then be used by the treatment-planning system to calculate the delivered dose distribution (Renner et al., 2005; van Zijtveld et al., 2007; van Elmpt et al., 2008).

While the methods described earlier will go a long way toward improving the relevance of phantom-based IMRT QA measurements, as of this writing, these methods are not widely used by the majority of radiation therapy clinics. Instead, most centers rely on dose difference, DTA, and gamma calculations performed on phantom dose distributions or detector fluence patterns to evaluate the accuracy of dose delivery. Interpretation of these comparisons can be made on an absolute basis, where numerical values calculated with the different tests are compared directly to dose difference and/or DTA criteria for each measurement point, or on a binary, pass/fail basis with each point either meeting the stated criteria or not (Depuydt et al., 2002). In the latter case, it is common for a percentage threshold to be used as an aid to determining whether a dose distribution is clinically acceptable. For example, a user might require that 95% of measured points have a gamma value less than one when using local dose difference and DTA criteria of 3% and 3 mm, respectively.

One of the major problems with this type of analysis is that it removes all information about the magnitude of differences observed in points that fail the dose difference and/or DTA criteria. To remedy this problem, a hybrid approach can be used where a percentage pass rate for measurement points is specified in conjunction with an absolute criterion, for example, no point can have a gamma value >1.5. Combining these types of tests with ROI methods may also help focus the analysis and yield results that are more clinically meaningful.

The last category of dose distribution comparisons deals with comparing dose distributions from the same treatment plan calculated on different data sets. There are a variety of scenarios where this might come into play and the comparison metrics used will vary accordingly. In addition to the example given previously where a patient's plan is recalculated on a new CT scan after the patient has experienced weight loss, another example would be to recalculate a treatment plan on CT scans of the same patient acquired with and without contrast to assess the effects of contrast on the dose calculation. The former comparison would likely rely more on ROI-based metrics to ascertain whether the original treatment plan is still acceptable in terms of meeting the stated clinical goals. The latter would use metrics similar to those utilized for IMRT QA since in this case it is of more interest to determine the magnitude and location of differences in the dose distribution.

14.7 Conclusion

An important component of a radiation treatment plan is the dose distribution. Given the complexity and large numbers of variables involved in modern radiation therapy treatment planning, it is often necessary to compare calculated and/or measured dose distributions. A number of quantitative methods have been developed to make these comparisons, including dose difference and DTA methods, gamma evaluation, and ROI analysis metrics. Each of these methods has advantages and disadvantages when applied in different situations.

One of the major challenges with quantitatively comparing dose distributions is how to interpret the comparison results. In part this difficulty arises from the spatially varying nature of most clinical dose distributions, which can result in large dose differences in regions with local dose gradients. Many of the comparison algorithms discussed in this chapter have been designed to take into account dose differences occurring in low-gradient regions as well as those arising from spatial misalignments in high-gradient regions. Despite these more advanced quantitative methods, problems of interpretation persist, especially when comparing measured dose distributions acquired in a phantom geometry. This difficulty often stems from a desire to extract clinical meaning from physical data when such a translation is not entirely clear. Another issue stems from the need to assign physical dose difference and distance tolerances as well as thresholds for the number of points in a comparison that pass a given test. Historically, it has been left to individual institutions to determine appropriate values for these parameters, though some guidance may be found in the literature (Ezzell et al., 2009).

In general, all dose comparisons are made with respect to some tolerance limits or quality endpoints. Ideally, these tolerances are tied to clinical outcomes so that a clear relationship between comparison metrics and clinical goals can be established. When this type of relationship cannot clearly be defined, as is often the case when verifying measurements of patient treatment plans in a phantom, one must recognize that such comparisons can only be interpreted as verifying the delivery accuracy of a treatment plan with respect to stated tolerance values, and that information about the impact of differences, whether large or small, are left to the judgment of the experienced clinical practitioner.

References

Bakai A, Alber M and Nusslin F (2003) A revision of the gamma-evaluation concept for the comparison of dose distributions. *Phys. Med. Biol.* 48: 3543–3553.

Boggula R et al. (2011) Patient-specific 3D pretreatment and potential 3D online dose verification of Monte Carlo–calculated IMRT prostate treatment plans. *Int. J. Radiat. Oncol. Biol. Phys.* 81: 1168–1175.

Childress NL and Rosen II (2003) The design and testing of novel clinical parameters for dose comparison. *Int. J. Radiat. Oncol. Biol. Phys.* 56: 1464–1479.

Depuydt T, Van Esch A and Huyskens DP (2002) A quantitative evaluation of IMRT dose distributions: Refinement and clinical assessment of the gamma evaluation. *Radiother. Oncol.* 62: 309–319.

Dice LR (1945) Measures of the amount of ecologic association between species. *Ecology* 26: 297–302.

Ezzell GA et al. (2009) IMRT commissioning: Multiple institution planning and dosimetry comparisons, a report from AAPM Task Group 119. *Med. Phys.* 36: 5359–5373.

Harms WB et al. (1998) A software tool for the quantitative evaluation of 3D dose calculation algorithms. *Med. Phys.* 25: 1830–1836.

Hartford AC et al. (2009) American Society for Therapeutic Radiology and Oncology (ASTRO) and American College of Radiology (ACR) practice guidelines for intensity-modulated radiation therapy (IMRT). *Int. J. Radiat. Oncol. Biol. Phys.* 73: 9–14.

ICRU (2010) Prescribing, recording, and reporting intensity-modulated photon-beam therapy (IMRT). Report No.83. *J. ICRU* 10. (Bethesda, MD: International Commission on Radiological Units and Measurements).

Jiang SB et al. (2006) On dose distribution comparison. *Phys. Med. Biol.* 51: 759–776.

Jursinic PA and Nelms BE (2003) A 2-D diode array and analysis software for verification of intensity modulated radiation therapy delivery. *Med. Phys.* 30: 870–879.

Korevaar EW et al. (2011) Clinical introduction of a linac head-mounted 2D detector array based quality assurance system in head and neck IMRT. *Radiother. Oncol.* 100: 446–452.

Kruse JJ (2010) On the insensitivity of single field planar dosimetry to IMRT inaccuracies. *Med. Phys.* 37: 2516–2524.

Kry SF et al. (2014) Institutional patient-specific IMRT QA does not predict unacceptable plan delivery. *Int. J. Radiat. Oncol. Biol. Phys.* 90: 1195–1201.

Low DA et al. (1998) A technique for the quantitative evaluation of dose distributions. *Med. Phys.* 25: 656–661.

Low DA and Dempsey JF (2003) Evaluation of the gamma dose distribution comparison method. *Med. Phys.* 30: 2455–2464.

Low DA (2010) Gamma dose distribution evaluation tool. *J. Phys. Conf. Ser.* 250: 012071.

McKenzie EM et al. (2014) Toward optimizing patient-specific IMRT QA techniques in the accurate detection of dosimetrically acceptable and unacceptable patient plans. *Med. Phys.* 41: 121702.

Mijnheer B and Georg D (Eds.) (2008). *Guidelines for the Verification of IMRT. ESTRO Booklet No. 9* (Brussels, Belgium: European Society for Radiotherapy and Oncology).

Moran JM, Radawski J and Fraass BA (2005). A dose gradient analysis tool for IMRT QA. *J. Appl. Clin. Med. Phys.* 6(2): 62–73.

Nelms BE and Simon JA (2007) A survey on IMRT QA analysis. *J. Appl. Clin. Med. Phys.* 8(3): 76–90.

Nelms BE, Zhen H and Tome WA (2011) Per-beam, planar IMRT QA passing rates do not predict clinically relevant patient dose errors. *Med. Phys.* 38: 1037–1044.

Nelms BE et al. (2012) VMAT QA: Measurement-guided 4D dose reconstruction on a patient. *Med. Phys.* 39: 4228–4238.

Palta JR, Mackie TR and Chen Z (2003) Intensity-modulated radiation therapy—The state of the art. *Med. Phys.* 30: 3265.

Pawlicki T et al. (2008) Moving from IMRT QA measurements toward independent computer calculations using control charts. *Radiother. Oncol.* 89: 330–337.

Renner WD, Norton KJ and Holmes TW (2005) A method for deconvolution of integrated electronic portal images to obtain fluence for dose reconstruction. *J. Appl. Clin. Med. Phys.* 6(4): 22–39.

Shaw E et al. (1993) Radiation therapy oncology group: Radiosurgery quality assurance guidelines. *Int. J. Radiat. Oncol. Biol. Phys.* 27: 1231–1239.

Siochi RAC et al. (2009) Radiation therapy plan checks in a paperless clinic. *J. Appl. Clin. Med. Phys.* 10(1): 43–62.

Stock M, Kroupa B and Georg D (2005) Interpretation and evaluation of the γ index and the γ index angle for the verification of IMRT hybrid plans. *Phys. Med. Biol.* 50: 399–411.

Sun B et al. (2012) Evaluation of the efficiency and effectiveness of independent dose calculation followed by machine log file analysis against conventional measurement based IMRT QA. *J. Appl. Clin. Med. Phys.* 13(5): 140–155.

Van Dyk J et al. (1993) Commissioning and quality assurance of treatment planning computers. *Int. J. Radiat. Oncol. Biol. Phys.* 26: 261–273.

van Elmpt W et al. (2008) The next step in patient-specific QA: 3D dose verification of conformal and intensitymodulated RT based on EPID dosimetry and Monte Carlo dose calculations. *Radiother. Oncol.* 86: 86–92.

van't Riet A et al. (1997) A conformation number to quantify the degree of conformality in brachytherapy and external beam irradiation: Application to the prostate. *Int. J. Radiat. Oncol. Biol. Phys.* 37: 731–736.

van Zijtveld M et al. (2007) 3D dose reconstruction for clinical evaluation of IMRT pretreatment verification with an EPID. *Radiother. Oncol.* 82: 201–207.

Zhen H, Nelms BE and Tome WA (2011) Moving from gamma passing rates to patient DVH-based QA metrics in pretreatment dose QA. *Med. Phys.* 38: 5477–5489.

SECTION IV
Clinical Applications

15

Acceptance Testing, Commissioning, and Quality Assurance of Linear Accelerators

Michael Altman and Eric E. Klein

15.1 Introduction

Linear accelerators (linacs) are one of the most common machines in the delivery of radiation therapy. Following installation, proper preparation of these machines for clinical use is pivotal, as this will ensure the ability to treat patients with the utmost safety, accuracy, and precision. Furthermore, establishing well-known values of accelerator and beam characteristics, including depth dose and profile information, is imperative to delivering a planned dose distribution accurately. This process of evaluation and beam characterization is composed of two parts: acceptance testing and commissioning. After this is completed, linacs are regularly evaluated to establish that the machine is correctly functioning and that it, among other things, is still delivering dose within some tolerance level of the characterization determined during acceptance testing and commissioning. This regular quality assurance (QA) process comprises of different tests and/or levels of rigor depending on several factors including the frequency with which a given test is performed: annually, monthly, daily, and so on.

There are a number of different approaches and techniques for acceptance testing, commissioning, and regular linac QA. For example, some groups espouse the benchmarking of their beam and dosimetric data against manufacturer provided "idealized" machine data, sometimes called "golden beam" data (Murray et al., 2006; Stern et al., 2011). This process can include adjusting the machine's beam and dosimetric characteristics to match the golden beam data. Others caution against the reliance on these datasets (Das et al., 2008). In terms of regular QA, many of the current techniques and protocols for the QA of linacs are driven by the assigning of a fixed set of tests and tolerances along with specified frequencies to achieve a specific outcome, such as 5% overall dosimetric error in treatment (Kutcher et al., 1994; IAEA, 2008). However, an increasing number of groups and studies espouse a risk assessment approach in which the treatment and dosimetric processes are analyzed to alter the QA techniques, tolerances, and frequencies to address the determined failure points (Fraass, 2008; Huq et al., 2008; Ford et al., 2015). As these debates continue, and there are differences even within those that fall on either side, this chapter is not intending to act as a complete review of all equipment available, methodologies, test tolerances, and so on for linac acceptance testing, commissioning, and routine QA. Rather, the intent here is to review a basic framework of resources, equipment, and techniques central to dosimetric acceptance testing and commissioning, as well as how a protocol for regular dosimetric linac QA could be constructed.

As mentioned above, a number of protocols abound for each of these techniques. Frequently, clinics will choose to follow the recommendations of international organizations, national organizations, and/or professional societies, many of which have made specific recommendations as to the content, frequency, and tolerances for acceptance, commissioning, and QA tests for linacs used in the delivery of radiation therapy. For example, the International Atomic Energy Agency (IAEA) has published works such as TECDOC-989, among others, guidelines for setting up radiation therapy programs, including descriptions of linac commissioning aspects, recommended QA regiments and tolerances, and equipment (IAEA, 1997, 2008). Similar documents have been produced by other groups, including, but not limited to, Report 94 of the Institute of Physics and Engineering in Medicine (IPEM) (Kirby et al., 2006) and Reports 60976 and 60977 of the International Electrotechnical Commission (IEC) (IEC, 2007, 2008). In the United States, the American Association of Physicists in Medicine (AAPM) has published a number of task group (TG) reports, which include the report of TG 106 addressing the recommended equipment and techniques involved in linac beam data acceptance testing and commissioning (Das et al., 2008), and the reports of TGs 40 and 142 discussing recommended linac QA frequencies and tolerances (Kutcher et al., 1994; Klein et al., 2009). Ultimately, it is up to the end user to select which of the various protocols makes the greatest sense to apply in his or her own clinic, or to follow one of the espoused risk assessment-based techniques (Fraass, 2008; Huq et al., 2008; Williamson et al., 2008).

Acceptance testing and commissioning are intrinsically linked, but are different concepts. When a linac is purchased, the buyer and the vendor agree that the machine installed will have certain capabilities and precision. The acceptance testing is then a set of tests the vendor performs along with the physicist to demonstrate the machine installed has the abilities advertised during purchase. Commissioning is the set of tests performed by the end user, the "buyer" as defined above, to prepare the linac for its full use in the clinic, including the collection of data which is needed as input into a treatment planning system (TPS).

15.1.1 Effort and Personnel

Before beginning the acceptance testing and commissioning process, it is important to allocate the proper resources in terms of personnel and time to complete the tests. Some of the publications provide guidance for this: the IAEA, for example, includes a table outlining the times necessary for an individual to complete each aspect of the acceptance and commissioning processes (IAEA, 2008). Instead of providing an explicit breakdown by personnel required, the AAPM details a formula to estimate the total time required for commissioning based on the number of energies/modalities and scanning datasets required (Das et al., 2008). Guidance is then given about the blocks of time needed to inform data analysis and point measurements, with an ultimate conclusion that "the typical time allotted for commissioning is 4–6 weeks."

The differences in these approaches illustrate there is no definitive methodology for determining how many people and how much time is needed to perform a given acceptance and commissioning project. As the acceptance testing and commissioning of linacs dictates the accuracy, efficacy, and safety of radiation therapy treatments and treatment planning, it is imperative to not shortchange the process in terms of time or personnel available. It is also important to note that new accessory devices or treatment modes which require commissioning will add additional time to this process. Furthermore, as technology develops and/or the complexity of linacs increase, the guidance provided by published sources may not be sufficient to estimate the time/personnel needed to properly address those items.

15.2 Equipment and Setup

An array of equipment is needed for acceptance testing and commissioning of a linac. Equipment for acceptance testing is typically very similar compared to commissioning. The difference tends to be that acceptance testing equipment is owned and provided by the vendor, while commissioning equipment is provided by the team performing the work and should include devices (or similar devices) that will be on hand for the periodic QA: annually, monthly, daily, and so on. Commissioning equipment need not be the same specific devices as that used for acceptance testing but must be capable of acquiring the data with sufficient accuracy and precision necessary for creating beam models of interest in the TPS, as well as setting baselines for periodic QA.

15.2.1 Scanning Water Tanks

For beam scanning, a three-dimensional (3D) scanning water tank is typically the instrument of choice (Figure 15.1). These devices have a stage upon which a radiation measurement instrument, ionization chamber, diode detector, and so on, can be mounted and moved throughout an incident radiation field in all three dimensions. These tanks can have different geometries, for example: cubic, rectangular cubic, or cylindrical. Some scanning water tanks, including those used by certain vendors for acceptance testing, may only have stages that can move in two directions, along one axis perpendicular and one axis parallel to the radiation field. Those who use these "two-dimensional (2D)" scanning tanks may assume the radiation field is cylindrically symmetric, thus only one profile of the beam is needed to characterize the field. Typically for most types of scanning water tanks, the tank itself and/or the scanning stage must be leveled in each direction.

Different sizes of scanning water tanks are available commercially. Recommendations are made in the literature (Das et al., 2008) to have a tank large enough to include all of the fields of interest. This is generally limited by

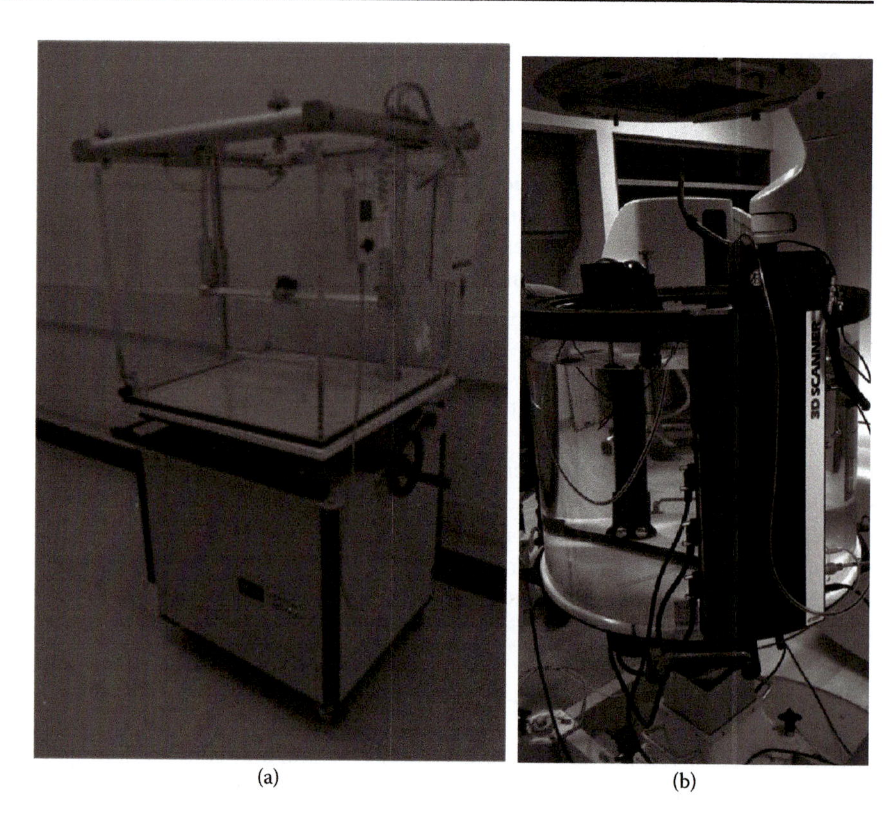

(a) (b)

Figure 15.1

Scanning water systems: (a) rectangular cubic (Blue Phantom², IBA, Schwarzenbruck, Germany); (b) cylindrical (3D SCANNER, Sun Nuclear, Melbourne, FL).

the size of field data the end user will need to fully commission the TPS. Bear in mind that, for most linacs, field size is defined at the isocenter, typically 100 cm source-to-axis distance (SAD). Scanning data needed for TPSs generally require profile data for setups with source-to-detector distance (SDD) greater than the SAD, resulting in effective field sizes at those SDDs that are greater than the nominal SAD field size. For example, if a 20 cm × 20 cm field is required to be scanned at 100 cm source-to-surface distance (SSD) and 30 cm depth, the field size at that depth is 26 cm × 26 cm. Furthermore, nominal field size is typically defined for flattened fields at the 50% falloff of the penumbra. In order to capture data encompassing the full edge of the penumbra, which is generally good practice to generate high-quality field models in TPSs, an even greater distance is required. Typically, 5 cm added to the nominal at-depth field size on either side can frequently give more than sufficient data. TPSs are notoriously limited in their ability to model the peripheral dose from radiation field, as discussed in detail in Chapter 21, thus measuring further outside the radiation field is generally unnecessary. Measurements outside of the field may be performed during the acceptance testing; however, alternative detector systems would generally be employed. In the example above, this would require a tank of at least 36 cm wide to accommodate the desired field.

With that said, tanks which can accommodate scanning 40 cm × 40 cm fields at appropriate SDDs are prohibitively large and heavy to be widely produced and/or used. If large fields are required, one solution is to offset the water tank from being centered with the beam and scan one or more half-beam profiles; i.e., scanning from central axis (CAX) past the penumbra. Most commercially available scanning water tanks also come with a platform which allows them to be moved and leveled as the heft and size of the filled tanks may be difficult for patient support systems to stably hold and maneuver, and may exceed limits. Most modern water tanks come with computer software interface systems, which allow the tanks to be driven remotely and include integrated data collection modules. Other accessories/features are available with some scanning water tanks, such as mobile water storage and pumping systems, integrated pump/movement systems to facilitate tissue-phantom ratio (TPR)/tissue-maximum ratio (TMR) measurements, and automated leveling systems.

15.2.2 Scanning System Setup

Proper setup of the water scanning system requires many steps. What follows here is a simplified review and recommendations. More complete information about setup and QA for scanning systems is found in the literature (Mellenberg et al., 1990; Purdy et al., 2006). The first step for proper scanning water tank setup, and really for most beam measurements, is to ensure that the gantry head is leveled at 0°. After this, etchings or markings on most water tanks, combined room laser, linac crosshairs, and/or other reference tools are used to align the water scanning system with isocenter. Axial and coronal room lasers can be especially useful for setting up the scanning system at specified SDDs for in-air measurements required by some Monte Carlo-based treatment planning algorithms. The scanning device, for example, ionization chamber or diode, and its corresponding mount can be attached at this time, ensuring that there is sufficient slack with the cables to allow for movement among the full translational limits of the tank.

For in-water measurements, the tank is then filled with water. Distilled water is generally preferred, although other water sources generally do not perturb the results severely provided the water and tank are fairly clean. The water should be of sufficient height in the tank such that (1) full scatter conditions are achieved, (2) all of the required depths are accessible by the scanning apparatus, and (3) there is sufficient clearance of the tank walls by the linac head such that any accessories such as trays, or electron applicators, can be attached. For example, with the tank in place, electron applicators can require tilting of the gantry head slightly to attach, but if the water is too shallow, even this is impossible.

Once filled, the scanning apparatus is leveled by tilting either the entire tank or the scanning apparatus itself, with the idea that chamber movement when scanning beam profiles will be parallel to the water surface. It may be that the tank is somewhat tilted to offset couch sag. If scanning is to be done in both the cross-plane (left-to-right) and in-plane (gun-to-target) direction of the field, it should be leveled in both dimensions. There are different methods to accomplish this. A common one is to move the scanning stage to one side of its movement range, set the chamber or some visible reference at a known point relative to the surface, and move the stage through the entire range of travel in that dimension to see if that reference changes relative to the water surface. An example of this is shown in Figure 15.2 for one vendor who provides a cap for their scanning system ionization chambers with an etched "X" on it that can be used for this process. As a double check of this process, and to ensure that the scanning system is fairly leveled relative to the gantry, the detector can be placed at or near center

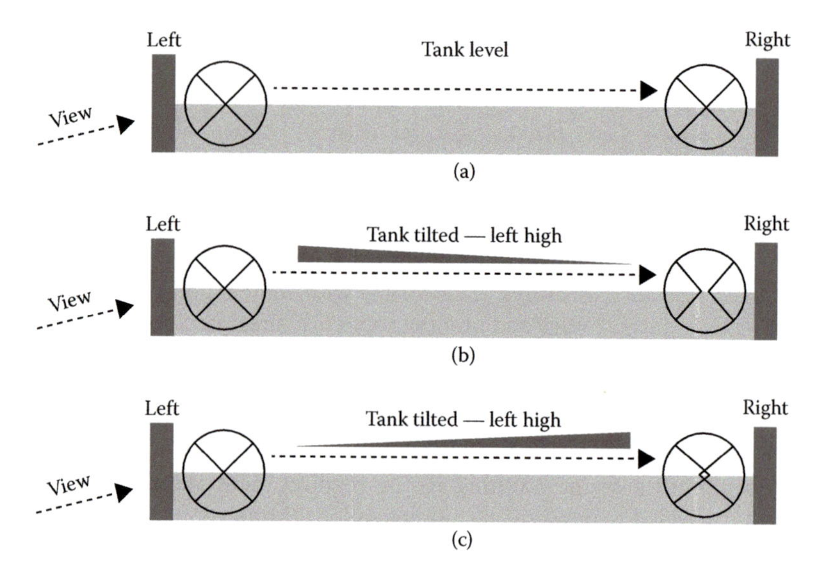

Figure 15.2

Example of leveling of a scanning water system from a front view, where the chamber is leveled at one end of the tank and moved to the other side. (a) If the tank is leveled, the reference is unchanged. If the reference changes, the way it changes can reflect how the tank is skewed, either (b) tilted with the left side of the tank high or (c) the right side of the tank high. The earlier mentioned example is for a commercially available ionization chamber cap with an etched "X" mark on the face.

and dropped down into the tank to ensure that the chamber shadow tracks well with the crosshairs.

Once the scanning apparatus has been leveled, the user should determine the surface level to set as a "zero" position for the detector. One common way of doing this is through observation of the reflection of the detector on the water surface from just below the surface; for cylindrical detectors, this is called the "perfect circle" test, while for flat-topped detectors such as parallel-plate chambers or diodes, this is the point where the detector and its reflection just meet (Figure 15.3). The surface of the water should then be set to the preferred SSD with the end user's preferred method. Note that the scanning apparatus can be bulky, so minimize any errors due to water displacement by submerging any moving parts of the assembly before the SSD is set. The SSD should be routinely checked if the tank is sitting for extended periods of time, i.e., a few hours or more, during or between measurements, as the SSD can change due to evaporation and/or settling of the heavy tank into the floor.

The detectors can then be positioned. Due to the desire to minimize the impact of instantaneous fluctuations or drifts in the beam, two detectors are typically recommended for beam scanning: a "field detector," which moves within the tank, and a stationary "reference detector." These are frequently, but do not need to be, the same model and type of detector. The reference detector is usually placed out of the water and "upstream" of the field detector and should not occlude the field detector (Figure 15.4). If it is not possible to leave the field detector unblocked such as for small fields, the reference chamber can be placed between the upper and lower jaws on some linacs, or the field detector can be slowly rastered through the field, integrating the measurement at each point.

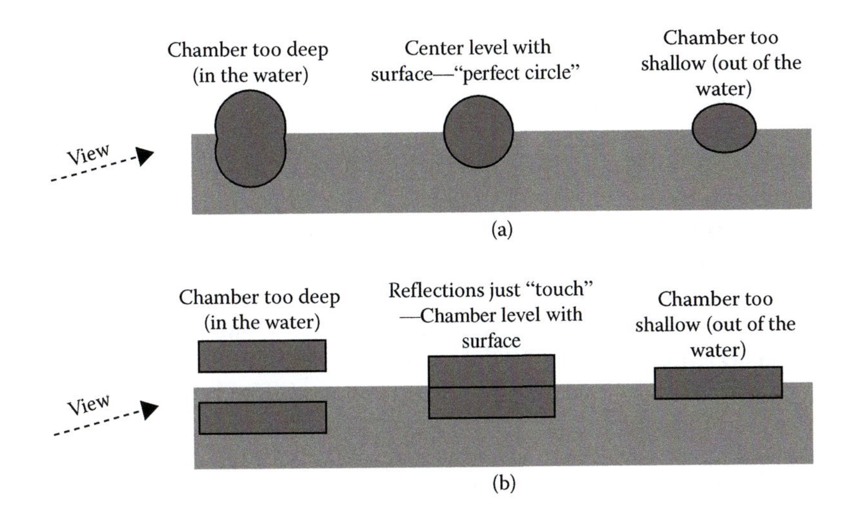

Figure 15.3

Example of setting a detector to the surface of a water tank for (a) a cylindrical chamber and (b) a flat topped detector such as a parallel plate chamber. In (a), the surface is found, by looking from just below the surface, when the bottom half of the chamber and its reflection form a circle (the "perfect circle" test). In (b), the surface is found when the chamber and its reflection just meet.

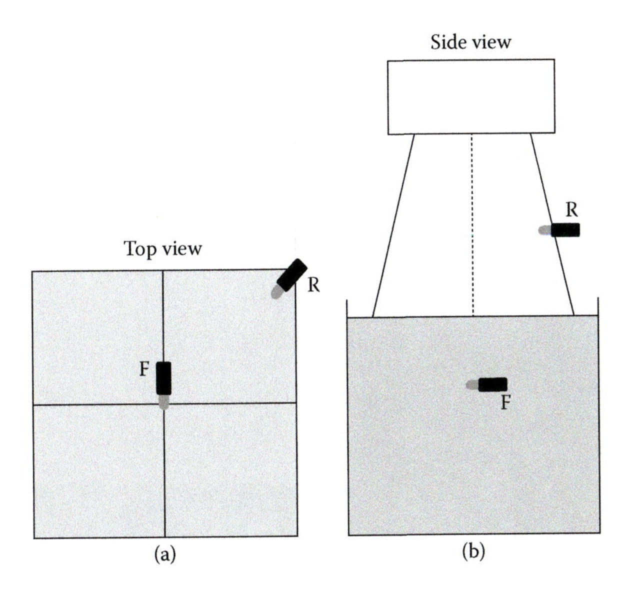

Figure 15.4

(a) Top view and (b) side view of the placement of field (F) and reference (R) chamber in a scanning water tank.

The chambers and movement arms are then connected to an electrometer/control unit. In modern systems, this unit will interface with a computer and allow remote control and monitoring. One important aspect is to ensure that the correct cables and connectors are used to attach the chambers to the control unit. For instance, trying to run a biaxial requiring diode detector with a coaxial cable can damage the diode. The electrometer is used to set both detectors to the correct voltage. Some water tanks allow for an initial scan to verify and/or correct for small offsets in centering and angulation of the chamber relative to the beam. After this, and any other testing of proper functionality the user wishes to perform, scanning is ready to commence.

15.2.3 Other Water Systems and Phantoms

Absolute dose calibration of radiation beams can be performed in a scanning water tank, but does not need to be. A smaller water tank, which can provide full scatter conditions for calibration-sized fields, which are typically between 10 cm × 10 cm and 20 cm × 20 cm, which allows the detector to be affixed and moved only parallel to the radiation field may be called a "one-dimensional (1D)" water tank (Figure 15.5a). The AAPM, for example, recommends a water tank of minimum dimension 30 cm × 30 cm × 30 cm (Almond et al., 1999). These smaller water tanks can be used on the patient support system and their relative ease in storage and setup compared to scanning tanks makes them a common choice for dose calibrations and other types of QA.

Relative point dosimetric measurements, such as the output factors (OFs) or wedge factors (WFs), can be performed in either a scanning or 1D water tank, or in block (slab) phantoms comprised of one or more rectangular cuboid blocks of water equivalent plastic as shown in Figure 15.5b. For block phantoms, one or more slabs are drilled to accommodate a detector and are flushed with the solid slabs. Ultimately, any of these phantoms should just have sufficient size to provide full scatter conditions for the field sizes of interest. More information about water and water equivalent phantoms can be found in Chapter 3.

15.2.4 Detectors and Devices

As described above, scanning water systems can utilize two detectors: field and reference. The choice of which types of detectors are employed is up to the end user; ionization chambers (cylindrical or parallel plate), diodes, diamond detectors, and liquid-filled chambers are all possibilities. A more in-depth discussion

(a)

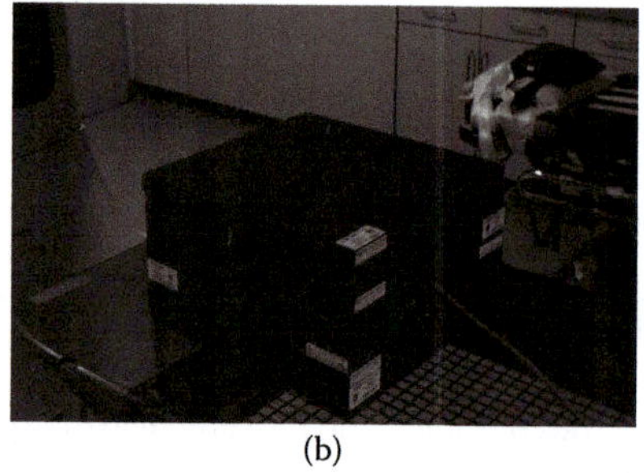

(b)

Figure 15.5

(a) One-dimensional water tank used for absolute dose calibration measurements. The chamber can move only up and down in the tank. (b) Water equivalent slab phantom for use in point measurements with an ionization chamber inserted.

of the physics of these detectors is found in Chapters 3 and 4. However, the choice is informed by ensuring the collecting volume is small enough that the fields can be sufficiently well sampled to generate good data for modeling the TPS. At the same time, one should not sacrifice sensitivity resulting in data that is too noisy. Any detector employed should also be compatible with the type of radiation measured; electron versus photon diode, for example. Air-filled ionization chambers are a common choice due to their relatively low cost and flexibility of use across radiation types and energies. As the detectors are to be used in water, they can themselves be waterproof or will require some type of waterproof sleeve. The waterproof sleeve should be sufficiently thin that it does not perturb the incident radiation field. Verify from documentation that a chamber is waterproof before submerging it in water to prevent damage. Small fields require special consideration in terms of the choice of chamber so that the beam profiles are not undersampled and to avoid volume averaging effects. This is discussed in more detail in Chapter 9. For scanning systems, the electrometers are frequently integrated into the control unit of the system, but it is important to ensure that the electrometer is compatible with, and will run the desired voltages for the chambers used.

Absolute dosimetry measurements require a traceable calibrated detector. In the United States, this traceability is obtained from calibration at the National Institute of Standards and Technology (NIST) or an Accredited Dosimetry Calibration Laboratory (ADCL). According to AAPM's TG-51, a cylindrical ionization chamber, such as a 0.6-cc Farmer-type, is allowed for all photon energies and electron energies >6 MeV (Almond et al., 1999). Parallel plate ionization chambers are required for electron beams of energy ≤6 MeV. Electrometers used for absolute calibration can either be calibrated with the calibrated chamber or separately.

Other devices can be used for dosimetric acceptance testing/commissioning, although each has its own limitation. Detector arrays, having 1D, 2D, or 3D measurement geometries, can be used for an array of tasks including the centering of the radiation beam ("beam steering"). The use of these devices is discussed in greater detail in Chapter 8. However, for fully characterizing the radiation field, these devices can suffer from issues related to water nonequivalence, field size limitations, and field undersampling due to the wide spacing of detectors. Film can be used for acceptance/commissioning beam characterization due to its high spatial resolution and, in the case of radiochromic films, possessing near-water equivalence. Film is not reusable, however, and uncertainties due to the development/scanning process must be carefully considered. Other options, including gels and 3D radiochromic detectors (see Chapters 5 and 6) and electronic portal imaging devices (see Chapter 7) can be used as well. Ultimately, there are many usable options, so selecting a device which appropriately addresses each task is at hand.

15.2.5 Monthly and Daily Verification Equipment

Monthly and daily verification of the beam output and other parameters is suggested by many professional societies and considered a standard part of good practice. Both of these verification procedures need equipment that can take many forms, from vendor-provided devices to slab phantoms among others. Typically both of these are much less complex than commissioning or acceptance equipment, or, use that equipment, such as a 1D water tank, in a simplified way. These more routine dosimetric checks tend to consist of only single, or few, point samplings under a fixed and easily repeatable geometry. After the commissioning is completed, monthly and daily verification tests with their associated equipment should be performed as these devices are used to determine any variation from commissioning levels. Monthly verification typically employs traceable detectors and electrometers, calibrated by a standards laboratory, or secondary calibrated against a local standard. Cross calibration of systems should occur minimally annually. Daily verification can be performed using similar equipment, or by using one of a number of commercial devices with integrated detectors. Commercial devices can contain multiple detectors to assess various properties of the beam.

15.3 Acceptance Testing

Acceptance testing verifies that the machine the vendor provided and installed has the capabilities as described when purchased. Successful acceptance testing is thus, essentially, the completion of a purchase contract between the vendor and the purchaser of the linac. As such, the components, specifications, and equipment used for acceptance testing are provided for and designed by the vendor. With the specific tests and levels of acceptability varying from company to company, and potentially from model to model, an exhaustive review of these tests and equipment will not be provided. With that said, dosimetric acceptance testing for linacs has several common items which are being tested, and those are reviewed in this section.

15.3.1 Photon Beam Output

For acceptance testing, photon beam output tests, including relative OFs, ensure that the beam is delivering an appropriate amount of dose under a vendor-specified setup with a specified number of monitor units (MUs). This would typically be performed in a vendor-provided water tank or slab phantom with their detector. Output at multiple depths may be collected and compared, essentially providing percent depth dose (PDD) or TMR/TPR values at specific depths, which also act as a gauge that the energy spectrum of the beam is within specifications. Outputs will be tested for each photon energy mode available. Note that this should not be regarded as the absolute calibration of the beam for clinical use, even if the vendor follows the protocol you wish to use, which they may not. Absolute calibration should be performed with the on-site chambers and equipment used for commissioning, according to a national protocol, for example, AAPM's TG-51 (Almond et al., 1999; McEwen et al., 2014), or international protocol, for example, IAEA's TRS-398 (IAEA, 2000).

15.3.2 Photon Beam Flatness and Symmetry

For photon beams, the shape of the radiation field is fundamentally defined by the angulation and centering of the incident electron beam on the target and subsequent photon beam on the flattening filter, if used. Each energy/mode of the machine will have its own target and flattening filter combination. The position/angle of the incident beam on the target and flattening filter can be adjusted, a process called "beam steering," which is typically performed by linac engineers. Ideally, the incident beams would be centered and perfectly perpendicular on both. In practice, alignment should be achieved within some acceptability criteria as defined by the vendor. To do this, the alignment is determined by characterizing the shape of the radiation field through two parameters: flatness and symmetry. Flatness is a parameter which is limited by the shape and composition of the flattening filter, or lack thereof, itself and characterizes how close the beam is to a perfectly flat field under some specific setup (typically defined at a specific depth). Symmetry, as the name implies, is a parameter that shows how consistent the field profile is on either side off-center to the left, right, up, or down. It can be defined point-by-point, or by comparing areas under each half of the beam profile, among other ways. Flatness and symmetry are typically measured from beam profiles acquired with some kind of scanning tank or an array device.

Definitions vary from vendor to vendor and from device to device, although an example can be illustrative. For linacs made by Varian Medical Systems (Palo Alto, CA), the full-beam width is defined as the beam profile between the points on the penumbrae where the beam falls off to 50% of the CAX dose. Flatness and symmetry are defined only on the central 80% of this full width, known as the "flattened area" or FA. Flatness is determined using the maximum and minimum doses measured within the FA, D_{max}, and D_{min}, respectively, and is given by

$$\text{Flatness} = \frac{|D_{max} - D_{min}|}{(D_{max} + D_{min})} \times 100 \tag{15.1}$$

Symmetry is defined as the maximum difference in dose between two points equidistant to, but in opposite directions from, the CAX along the beam profile, D_{CAX+x}, and D_{CAX-x}, respectively:

$$\text{Symmetry} = \text{Max}\left(|D_{CAX+x} - D_{CAX-x}|\right) \tag{15.2}$$

Note that definitions which rely on these percentages of falloff were developed before the advent of clinically employed unflattened linac photon beams. In these instances, the same definitions can be retained for specifications analyzed during acceptance testing, although there is debate in the literature regarding how flatness, especially, and symmetry should be defined for such beams (Hrbacek et al., 2011).

15.3.3 Electron Beam Output, Flatness, and Symmetry

If available and activated, most modern linacs are equipped to deliver multiple distinct electron energies. The broad electron fields are achieved by spreading a pencil electron beam through a scattering foil. These foils can be single or dual foil systems, the latter of which may provide some functionality as a flattening filter. Each electron energy mode of a linac may have its own individual scattering foil.

Output, flatness, and symmetry for electron fields are assessed similarly during acceptance testing to their photon counterparts. One main difference is that different detectors may be used; parallel plate ionization chambers or diodes for electron beams compared to cylindrical ionization chambers for photon beams, due to the sharp falloff of electron beams, especially at low energies. Another difference is that each parameter, output/flatness/symmetry, may be defined at a different depth, such as the depth of maximum depth dose, d_{max}, for each individual electron beam energy. It is more likely that these parameters are defined at the same depth for all photon energies/modes.

15.3.4 Field Modification Devices

Different linacs are available with an array of beam altering devices including wedges, physical or electronic, and block trays among others. Some acceptance tests will simply verify that these devices are functional and/or that the linac will recognize when a specific device is in place and, especially for wedges, what orientation it is in. Additional tests will use similar setup to the output measurements described above to take point measurements under the devices. By comparing the output with the modification device in place to the output without, these tests will determine if the modification device is properly centered and/or attenuating the beam to within agreed upon acceptability criteria.

15.3.5 Output Stability and Linearity

It is incumbent for the vendor to show during acceptance testing that the machine is stable with the alteration of a number of different parameters which can be adjusted during the delivery of a field. The most common ones to be tested for most linacs is stability of each mode with changing dose rate, linearity with changing MU, and output constancy with changing gantry angle. The first two can be tested with a similar point dose setup and detectors to the output tests described above. Dose rate constancy is tested for each available energy/mode of the machine by taking a series of measurements with varying dose rate through some range of available rates to determine if the output is constant across all rates to within some level of acceptability. The MU linearity is established for each available energy/mode by varying to total number of MUs delivered and ensuring that the measured output scales similarly; for instance, 200 MU should produce two times the output of 100 MU.

Output constancy with gantry angle must be typically analyzed with a different setup: a cylindrically symmetric phantom with the detector at the center, or some holder for the detector used with a cylindrically symmetric buildup cap, for example. Most setups employ a cylindrically symmetric detector placed at isocenter in a phantom, if used, which provides the same buildup no matter which gantry angle is used for irradiation. A common variation of the test is to take output readings with this setup at the four cardinal gantry angles (0°, 90°, 270°, and 180°) and to verify that the output is consistent across all four angles to within some acceptance level.

15.4 Commissioning

Commissioning is the process of preparing the linac for clinical use in delivering radiation treatments to patients. As such, commissioning tests are performed to determine that the linac can perform the full breadth of functionality required by its use in the clinic, and these tests tend to be much broader in scope than acceptance tests. At the same time, commissioning is the process where point and scanning data are collected to enter into the TPS to inform the beam models used for dose calculation. The accuracy of these data will thus determine the quality of treatment plans. The data should also encompass data needed for secondary software to check dose calculations performed by the TPS, and/or for databooks which can be used for secondary checks as well as for hand calculations for on-treatment-defined treatment plans ("clinical setups"). Finally, during commissioning, baselines are set with equipment used for regular annual, monthly, or daily QA.

15.4.1 Photon Beams

Most commercially available linacs produce one or more photon beams with different energies: typically one to three flattened beams with a low (4–6 MV), medium (8–12 MV), and/or high (15–25 MV) energy beams. Some modern machines offer flattening filter-free (FFF) versions of one or more of these energies. Each of these beams desired for clinical use must be fully commissioned. In general, there are two types of data required: continuous data and point data. The most typical methodology for acquiring the "continuous" data is with a scanning tank, and in this chapter those items are described as "scanning" data. The alternative dosimeters described above could be used for these measurements as well. For some data, scanning tanks are less useful and an alternative tool such as an array device may be preferable.

15.4.1.1 Scanning Data Measurements

The two types of common scanning data needed for beam commissioning are depth doses, either PDDs or TMR/TPRs, and beam profiles. For each photon beam energy/mode desired to be used clinically, an array of scans will be needed with each corresponding to different setup conditions, such as PDDs at different field sizes. The total number of scans, depth doses, and profiles needed is dependent on various factors such as the following:

- If a brand new beam model is being commissioned in the TPS, the minimum number of scans is determined by the requirements of the treatment planning software.
- Only a smaller subset of scans is required if the beam is being "matched" to preexisting beam data from a previous commissioning dataset, or matched to vendor-provided ideal data; sometimes called "golden beam" data (Beyer, 2013).
- Scans may be added if required to commission the secondary check software, or to flesh out the databook if the requisite data are not covered by the previous two options.

Note that for the second option, it is incumbent on the physicist to determine the array of data sufficient to determine a matched machine.

Most TPSs require in-water PDD/TPR/TMR scans for an array of different field sizes along the CAX. If physical wedges are to be used, depth dose scans along central axes with those devices in place may be required. For some Monte Carlo-based calculation packages, longitudinal in-air CAX scans will be required as well, although these are no longer PDD/TMR/TPRs, but rather a measure of the change in scatter with distance, as well as inverse square-based falloff. All three of these quantities are, in general, defined as follows:

$$\frac{D_{\text{depth}}}{D_{\text{Ref}}} = \text{PDD or TMR or TPR} \tag{15.3}$$

where D_{depth} is the dose at some depth and D_{ref} is the dose at some reference depth. PDD is a fixed-SSD quantity where the reference depth is the depth of maximum dose (D_{max}). TMR and TPR are both fixed-SAD quantities where, for TMR, $D_{\text{ref}} = \text{D}_{\text{max}}$, while for TPR, D_{ref} is some user-defined depth that need not be the depth of D_{max}. PDDs are typically easier to measure than TMR/TPRs, although most photon-based treatments are fixed SAD. As a result, many software packages allow PDDs as an input and/or convert PDDs into TMR/TPRs; conversion relations are also provided in a number of publications, for example, in Khan and Gibbons (2014).

A "clean" family of depth–dose curves is shown in Figures 15.6a and b. Note that as energy increases, and field size decreases, the depth of D_{max} increases. Meanwhile, beyond D_{max}, PDD/TMR/TPR should decrease as depth increases, and increase with increasing field size and energy. If these features are not evident, this could indicate an issue with the beam or scanning system and should be investigated. Depth dose scans should always be performed with the chamber starting at depth and moving toward the surface. This is to avoid the so-called "meniscus effect," where a chamber descending can, due to surface tension, pull the water surface down below the set "surface" level (Figure 15.6c). This can result in a "hooked" appearance of the depth dose at the surface of the scan (Figure 15.6d). Note the hooked appearance can also occur from an improperly set surface level, which should be checked periodically, every few hours depending on the humidity, due to evaporation. Although some noise in the raw measured data is expected, an extreme amount can imply issues with the detectors or improper gain settings on the field and/or reference detectors.

Most TPSs require in-plane and/or cross-plane beam profiles over a range of field sizes and depths at a given field size. A clean set of profiles at a specific field size but over a range of depths is seen in Figure 15.7. From these data, flatness and symmetry can be tabulated, and should be within some user-defined acceptable level. If they are not, the beam can be steered with the aid of machine engineers. The flatness and symmetry determined from

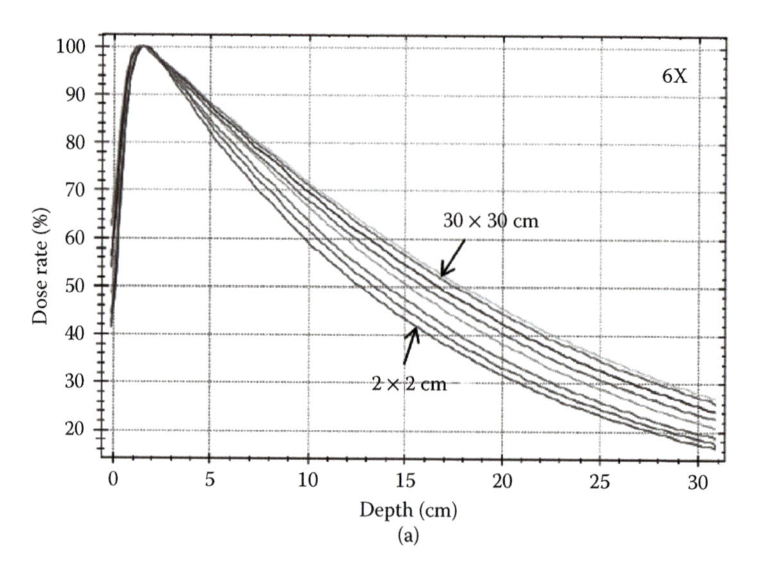

Figure 15.6

(a) Percent depth dose (PDD) curves for a 6-MV photon beam for field sizes from 2 cm × 2 cm to 30 cm × 30 cm. Note the PDDs increase with increasing field size past the point of maximum dose.

(Continued)

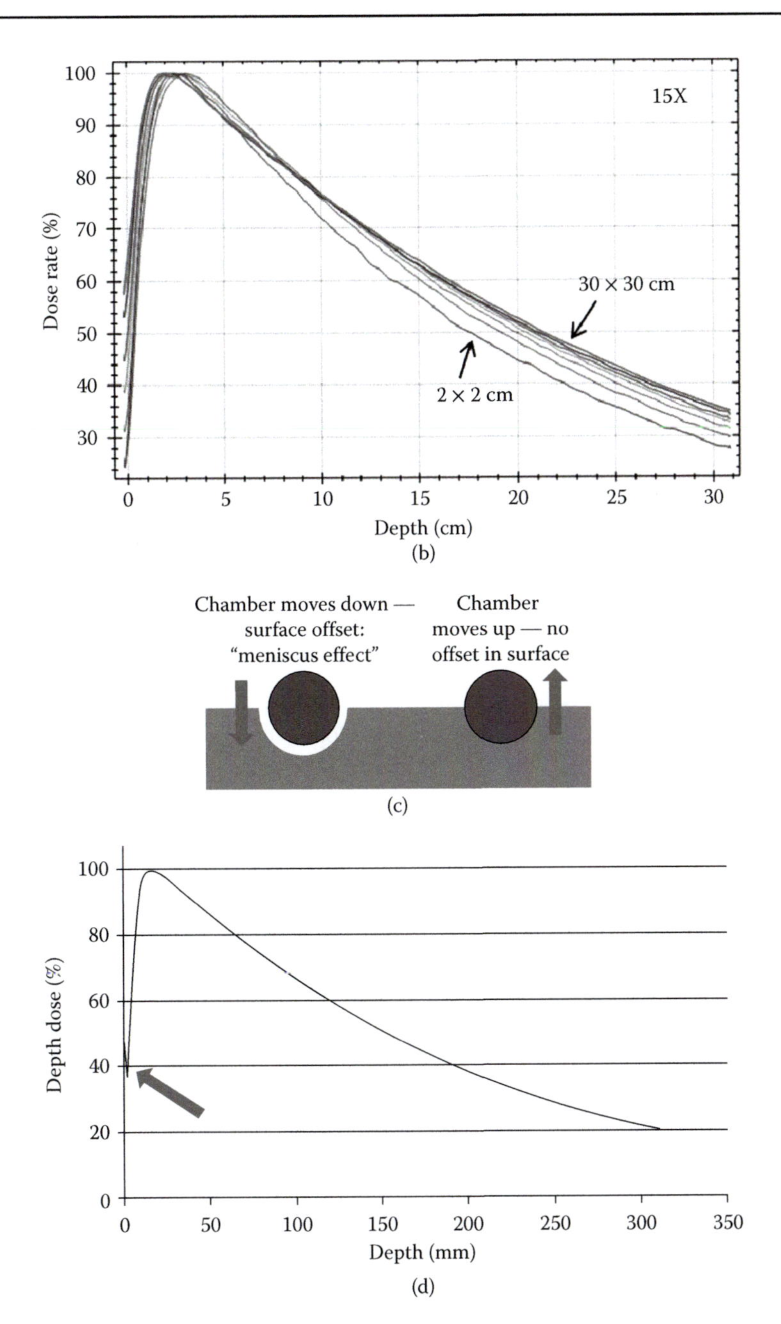

Figure 15.6 (Continued)

(b) Same as (a) for a 15-MV photon beam. (c) Schematic depiction of the "meniscus effect" whereby scanning a water tank down for a PDD pulls the water surface down with the chamber causing an offset. (d) PDD with the impact of the meniscus effect where the PDD is "hooked" and pulled down slightly just at the surface.

commissioning measurements under some defined conditions for each beam energy/mode can then serve as baselines for future regular QA. If physical wedges are to be used, a similar array of profiles with the wedges in place may also be required.

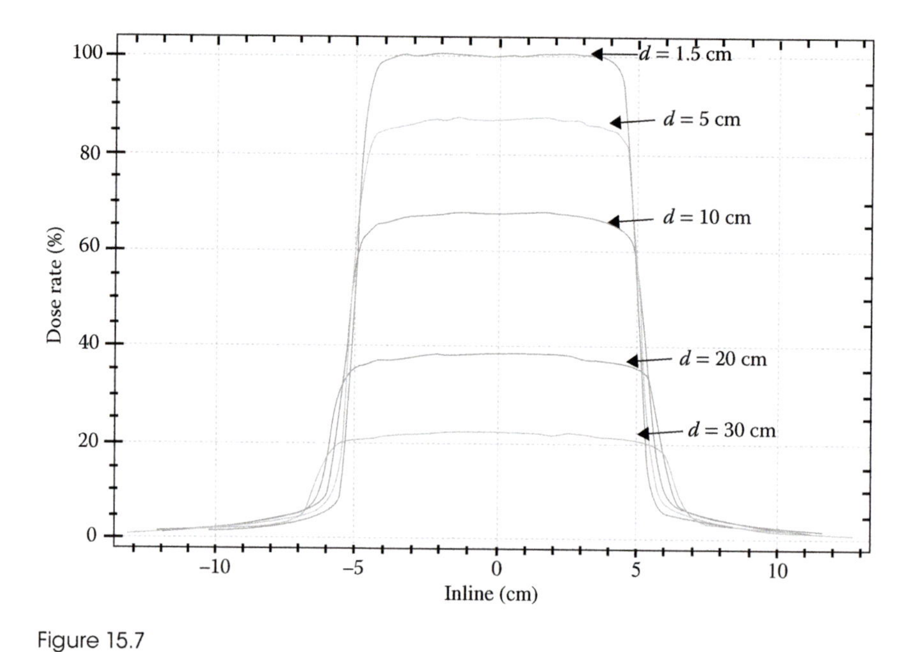

Figure 15.7

Inline 6-MV photon beam profiles for 10 cm × 10 cm fields at depths 1.5, 5, 10, 20, and 30 cm. Shallower depths are closer to the top of the plot due to the gain being set at 1.5 cm depth and the subsequent reduction of signal with depth.

Beam profiles should be relatively flat, symmetric, and well centered. Small centering offsets, ≤1 mm, are tolerable and can be removed in postprocessing. Larger offsets should be investigated as an issue with detector and/or beam centering. Tilted or nonsymmetric beams can typically be recognized by eye and may be a sign of beam steering. However, before immediately proceeding to beam steering, it can be useful to make measurements with other beam energies/modes first. As each energy/mode employs different steering/flattening filters, similar tilts or asymmetries across different energies can imply issues with setup, for example, gantry/chamber tilt or other issues. If a machine has only a single photon energy mode, electron energies can be tested, if available, or physical verification of setup efficacy should be performed. Asymmetries in the width of the field, for example, to the 50% falloff point for a flattened field, could imply issues with collimator position/symmetry.

A "wavy" or sinusoidal appearance along the profile can imply that the scanning speed of the chamber is too fast as the chamber is perturbing the water sufficiently to create ripples in the surface when scanning. This can especially be an issue for shallow beam profiles. Similar to the "open" part of the field, asymmetry in the size/shape of the beam penumbrae for a given scan should also be investigated, and could be related to issues with beam steering or collimation. Beam penumbra size should increase with field size and increasing depth, the former due to increasing scatter contribution, and the latter due to both this and the increasing magnification of the source size which increases the geometric penumbra. The appropriateness of the penumbra size can be best assessed at/near the SAD, where the geometric penumbra should be about the size of the source, ~3 mm for a typical linac, and the additional penumbra components will only add ~0.5–1 mm more. As with the depth–dose curves, overly noisy profiles, especially in the center of the beam, can imply issues with the chamber or gain settings. For in-air in-line profiles, asymmetries in the buildup cap/chamber in the direction of the beam can cause issues with penumbra asymmetry; scanning tanks may need to be rotated 90° to acquire scans while obviating this issue.

15.4.1.2 Point Dose Measurements

Along with continuous/scanning measurements, most TPSs, secondary check systems, and/or databooks require an array of point measurements for each energy/mode being commissioned. The most widely used of these is the OF, also called the total scatter factor (S_{cp}), and defined for a given energy/mode as follows:

$$\frac{D(d,\mathrm{FS}_x)}{D(d,\,\mathrm{FS}_{\mathrm{ref}})} = \mathrm{OF} = S_{\mathrm{cp}} \tag{15.4}$$

where $D(d,\mathrm{FS}_x)$ is the CAX dose/output at a depth d with a field size x, and $D(d,\mathrm{FS}_{\mathrm{ref}})$ is the CAX dose/output at depth d with a reference field size. Practically, OF is measured by acquiring output at some fixed depth (commonly $d_{\max}$, 5 cm, or 10 cm) first with a predetermined reference field size (10 cm × 10 cm is common), then with an array of varying field sizes. The field sizes can either be symmetric (square) or asymmetric (rectangular). As the jaws used to define the fields in both the lateral and gun-target directions are not in the same plane, it cannot be assumed that asymmetric OFs are the same regardless of orientation; i.e., the OF for 5 cm × 10 cm is not the same as that for 10 cm × 5 cm, provided the collimator orientation is unchanged.

Although the chamber is static during these measurements, some statistical variability is expected. Typically, this implies averaging over a minimum of three measurements at each setup condition to determine the dose/output for that condition. It can also be considered good practice to ensure that the measurements at a given setup condition are stable, i.e., do not keep increasing or decreasing with each subsequent measurement. To determine if any measurements were outliers or obvious errors, it can be useful to plot the OFs in various configurations and review the reasonableness of the trend or distribution. Some systems will require that S_{cp} be split into two factors, a collimator scatter factor (S_c) and a phantom scatter factor (S_p). S_c relates to the scatter dose due to all elements except the patient/phantom, and is measured in air using a set of buildup caps and/or "mini-phantoms" which provide a reasonable dose measurement in the detector while introducing minimal additional scatter. Greater detail about the measurement of S_c can be found, for instance, in AAPM's TG 74 report (Zhu et al., 2009). S_p is then tabulated from the other two factors by the relationship $S_{cp} = S_c \times S_p$.

If any accessories are used such as wedges or trays, each of them will add some attenuation to the beam and will require individual transmission factors. An accessory OF is, in general, defined as the ratio of CAX outputs with the accessory in place to that with it removed at a given depth or field size. WFs vary by the wedge angle and energy/mode used for both physical and electronic wedges. The WFs for physical wedges will vary with field size, given that different amounts of the wedge are exposed for different field sizes, resulting in different scatter contributions. Physical and electronic wedges also are not necessarily perfectly centered with the beam, thus it is common practice to take a WF measurement with all available orientations of the wedge, without moving the collimator, and averaging them. WFs for electronic wedges derived from jaw motion do not differ by field size, but can differ by which jaw is used to create the wedged field; so WFs for each orientation derived from a different jaw should be collected.

Tray factors (TFs) are not considered to vary substantially by field size, thus they are typically measured only at a single field size and depth. They do vary by energy/mode, and should therefore be collected for each commissioned energy/mode. A TF should be measured for each available tray configuration, for example, if one tray has slots and the other has holes. For trays with long slots or large, irregular perforations, it can be useful to move the chamber around relative to the tray, by moving the patient support table with the chamber in a 1D water tank, for example, and averaging the readings to get a TF. Any additional accessories which may be inserted and attenuate the field will also need measured factors.

15.4.1.3 Array/Film Measurements

As described earlier, arrays or films (or other options) can be used to acquire the continuous data with careful consideration as to the limitations of these detectors. Water scanning systems are still considered the standard for most of these, however. With that said, array devices/films/others are more useful than water scanning systems for certain tasks. When the beam is tilted and/or asymmetric and needs to be steered, the process of iteratively steering and rescanning in a water tank can be tedious. Most array devices (especially) have sufficient sampling to determine tilting/asymmetry and can offer real-time feedback when beam steering. After the beam appears sufficiently leveled and/or symmetric on the array device, water tank scanning (if desired) can commence to derive flatness and symmetry baselines.

Dynamic wedge profiles, either jaw-motion generated, or by motion of a physical wedge within the treatment head (Klein et al., 2009), can be measured in a scanning water system using an integrate mode, where the chamber dwells at certain positions for a period of time. However, due to both the motion of the collimation when creating the field as well as the fact that different wedge angles are commonly created as a weighted sum of open and wedged fields, such measurements can be very inefficient with scanning water systems. Array devices and film can perform this task well and should be used if required for TPSs, secondary check systems, databooks, or as baselines for regular QA.

15.4.1.4 Small Field Considerations

For both continuous and point measurements, the largest issue when moving to small fields, typically $\leq$3 cm $\times$ 3 cm, is the correct choice of detector. Too large of a detector will result in volume averaging which can result in overly low point measurements and distorted continuous data. Different smaller detectors have advantages and disadvantages that should be fully vetted before selecting. In addition, for scanning systems, the continuous-motion scanning may move too quickly to capture small fields accurately. Many modern scanning systems also offer an "integrate-type" mode, where the chamber dwells at each position. Distances between each position may be user determined, for a user-defined amount of time, which can be preferable for small fields. For point measurements such as OFs, some detectors, such as diodes, may respond correctly at small fields but can have response issues at larger fields, such as the reference OF field size, as well as at depth. OFs acquired with these fields are often bootstrapped to those for larger fields by comparing values at some intermediate field size, for example, 4 cm × 4 cm. These, and other, issues related to measurements in small fields are discussed in greater detail in Chapter 9.

15.4.2 Electron Beams

In addition to photon beams, most commercially available linacs produce multiple electron beams at different energies: typically four to six beams somewhat evenly distributed between 4 and 25 MeV. Each of these beams desired for clinical use must be fully commissioned. Most of the commissioning for electron beams is similar to photon beams: a set of scanning and point data needs to be collected, with the amount of data depending on the implementation. Typically, a lesser range of data is needed to commission electron beams compared to photon beams, among other differences, which will be elucidated in the next sections.

15.4.2.1 Scanning Data Measurements

Electron treatments are typically delivered at fixed SSDs, and therefore many applications and clinics will only require PDDs. A family of "clean" electron beam depth dose scans is shown in Figure 15.8a. Electron collimators, called applicators or cones, typically are accessories added to the treatment head and end usually 5 cm from the patient or phantom surface. These applicators are generally only offered in an array of fixed field sizes, and a typical allotment for a given clinic is 3–6 sizes ranging from 6 cm × 6 cm to 25 cm × 25 cm. Many TPSs or secondary check systems will require PDDs for each energy at each available applicator size. As electron beams do not follow the inverse square law, PDDs at different SSDs may also be required.

The good practices and issues described for photon depth dose scanning apply also to electron depth dose scanning. Due to the rapid buildup near the surface, shallow peaks, and rapid falloff, smaller detectors and/or parallel plate ionization chambers may be preferable for electron PDDs, as well as scanning at slower speeds or in integrate modes. This may be especially true for low-energy electron beams. As the practical range of clinical electron beams is rather shallow, the temptation may be to not begin depth dose scans nearly as deep as for photon beams. Good quality models from the TPS will frequently need data from the Bremsstrahlung tail region, so scans should extend 10 cm or more past the practical range.

Electron beam profiles have similar properties to photon beam profiles as shown in Figure 15.8b. Due to their rapid falloff, profiles at only 1–3 depths may be required for each electron energy and each applicator size. As the amount of flattening achievable for an electron beam is influenced by the presence of a single or dual foil system as well as the collimation geometry and phantom scattering, electron beams tend to be far less "flat" than photon beams, with flatness getting worse at lower electron energies. A quality beam should still maintain a high degree of symmetry. Penumbra size is affected by a similarly large number of factors and appropriateness is not as easily assessed as for photon beams. Similar to photon beams, tilted profiles can imply issues in either leveling or beam alignment with the scattering foil. For electron beams, this can also result from tilted or damaged applicators. This can be identified by physical/visual inspection of

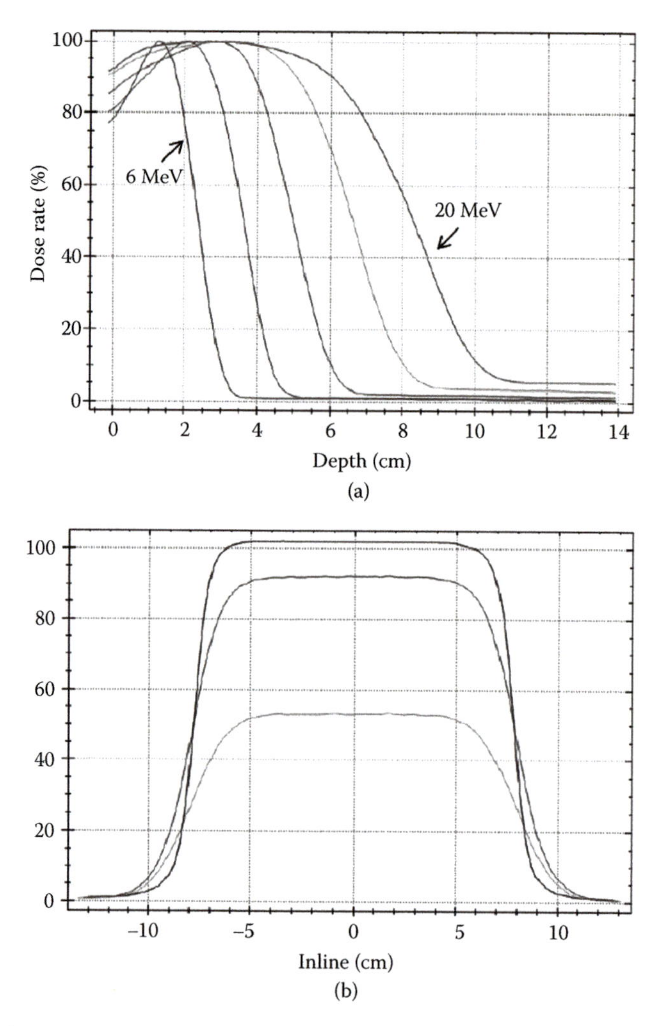

Figure 15.8

(a) Electron beam percent depth-dose curves for a 15 cm × 15 cm collimator (applicator). Energies are 6, 8, 12, 16, and 20 MeV from the left to the right. (b) Electron inline beam profiles for the 16-MeV electron beam shown in (a) at the depths of d_{max}, 90% depth dose, and 50% depth dose.

the applicator and/or scanning the same setup/energy with different applicators to see if beam asymmetry issues resolve. Due to the significant impact of ripples, electron beam profiles generally take longer by slowing the scan speed.

15.4.2.2 Point Dose Measurements

Applicator or cone factors (CFs) are defined similar to OFs between different applicators: for a given electron beam energy, it is the ratio of the output of a given applicator to a reference applicator at a depth, such as d_{max}. 10 cm × 10 cm or 15 cm × 15 cm are common choices for the reference applicator. To define a field of smaller size and different shape, aperture blocks or cutouts are fashioned which can be inserted into the applicators. Typically the block portion is made from a low melting point lead alloy such as Cerrobend. For commissioning, a series of cutouts should be created which are arrays of standard square, rectangular, and/or circular sizes. Cutout factors (CutFs) can then be measured similarly

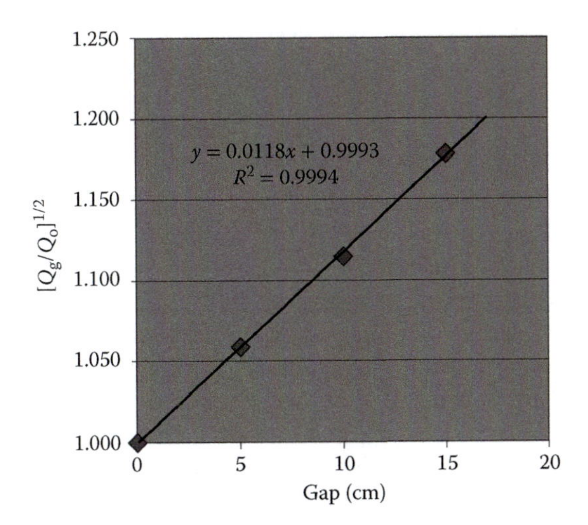

Figure 15.9

Plot of the gap between the electron applicator and the surface versus the ratio of the central axis readings with a gap of length *g* and a gap of 0. The slope of this plot can be used to calculate the effective source-to-surface distance by means of Equation 15.5.

to OFs or CFs with the reference for a given applicator size being the nominal or "open" size, for example, the 10 cm × 10 cm field for the applicator of that size. The array of needed CutFs is guided by the requirements of the TPS, secondary check system, and/or databook, although several TPSs do not require CutFs. OFs for a given applicator size and cutout can then be tabulated from the relationship OF = CF × CutF; some clinics will alternatively simply measure and define OFs as, for a given energy, the ratio of the output for a given applicator and cutout to the output at the same depth for a reference applicator and cutout.

As the electron beams do not obey inverse square law, there are two common solutions for determining OFs at varying SSDs. One is to simply remeasure the table of OFs at a few SSDs covering the clinical range of used electron SSDs, such as every 5 cm from 100 to 115 cm. OFs at intermediate SSDs are then found through interpolation. Another option is using the concepts of virtual SSD and effective SSD. Virtual SSDs are, essentially, back projections of the divergence of the radiation field to an effective point source upstream of the scattering foil. Inverse square is then assessed from this point; however, a set of correction factors is still needed to get the calculations to function correctly. Due to the need for correction factors, virtual SSDs are not as widely used as effective SSDs (SSD_{Eff}), which are SSD values for which the electron beams effectively obey inverse square law over a range of clinically useful SSDs. This is found by creating a plot of $[Q_o/Q_g]^{1/2}$ versus *g*, where *g* is the "gap" between the collimator (applicator) and the surface, Q_o is the detector reading with zero gap, and Q_g is the detector reading with gap *g*. Gap zero is the nominal distance between the detector and the surface; if a vendor designs the applicator to be 5 cm from the surface at 100 SSD, then that is the $g = 0$ distance. The resultant curve should be a straight line (Figure 15.9); the slope of this line is then related to the SSD_{Eff} by

$$SSD_{Eff} = \frac{1}{\text{slope}} - d \tag{15.5}$$

where d is the depth of measurement. SSD_{Eff} needs to be calculated for all beam energies and applicator sizes. The process of determining virtual SSDs, effective SSDs, and correction factors is described in AAPM's TG 25 report (Khan et al., 1991).

15.4.3 Data Processing and Analysis

After the dosimetric and beam data have been collected, and before it can be employed in a TPS, secondary calculation software, or databook, the data needs to be processed and/or reviewed. All scanning data need processing to reduce noise, center, and/or symmetrize them to optimize the performance of the TPS. The amount of processing depends on a number of factors including scan mode (continuous or point-by-point), characteristics of the equipment used (chambers, cables, etc.), and accuracy of the setup, among others. The amount and type of processing that needs to be employed is ultimately left to the end user.

If depth–dose curves are acquired with cylindrical ionization chambers, the first step is to shift the curves to correct for the effective point of measurement (0.6 × the radius of the cavity, r_{cav}, for photon beams, and 0.5 × r_{cav} for electron beams). The curves can then be smoothed to limit noise. Most modern scanning beam software packages include a number of different smoothing algorithms and iteration options among other factors. There is no consensus on the optimal smoothing algorithm or number of iterations to employ; the decision of approach is left to the end user. The primary caution is that some smoothing is useful, but oversmoothing can distort the curves, resulting in errors in treatment planning. One way to gauge this is to plot both the smoothed and unsmoothed data and compare the two. Finally, for electron PDDs, differences in restricted stopping power ratios between water and air, and the rapid change of these ratios with depth, require that a conversion is needed to take the measured data, technically percent depth ionization values, and convert them to a PDD. Many modern software packages provide an array of built-in conversion protocols from which the end user can choose, although some analytic methods are also published in the literature (Gerbi et al., 2009).

For beam profiles, centering can be used to correct for minor offsets in the setup of the tank, chamber, and/or radiation field. Symmetrization, in the form of averaging both sides or mirroring one side over the center line, among other operations, is used to account for small discrepancies between the two sides of the profiles along CAX. Following symmetrization, the profiles can be smoothed. Oversmoothing of profiles is most evident in the penumbrae and should be carefully analyzed.

Point measurements typically require minimal processing outside of that described above. However, as mentioned earlier, it can be useful to plot OFs as curves or surfaces to easily visualize any outliers. If these exist, remeasurement may be needed.

15.4.4 Absolute Dose Calibration

At the conclusion of beam scanning and steering, the photon and electron beams of the linac should be in their optimal condition. Thus, absolute dose calibration of all beam energies/modes should commence as soon as possible after this. Different calibration protocols exist, for instance, the IAEA Report TRS-398 (IAEA, 2000), and in the United States (and elsewhere), a common one is the AAPM's TG-51 and its associated addendum (Almond et al., 1999; McEwen et al., 2014). Absolute dose calibration is performed with a traceable calibrated ionization chamber and electrometer. It can be performed in the scanning water system, although the smaller 1D water tanks described earlier are generally preferred due to their ease of setup if incidental beam calibrations are required when scanning systems are not being used, and their ability to accommodate larger ionization chambers such as Farmer-type chambers which are typical local standards. Greater detail about the formalisms for absolute dose calibration is provided in Chapter 3.

15.5 Verification and Documentation

Following beam scanning and point dose measurements, as well as absolute dose calibration, internal and external verification of the commissioning data is important to ensure the quality of the collected data. Internal calibration can be achieved through different methods. If similar machines (vendor and model) have been calibrated in the clinic or system previously, the collected data can be benchmarked against that previous data. Any deviations between the two can be further investigated. This may have essentially been done previously if the machines were "matched" during the acceptance and commissioning process. Machines which a clinic has never used before can potentially be benchmarked against data acquired from other clinics with similar machines, from publications in the literature, or from vendor factory data, if any of these are available. Finally, after the TPS is commissioned, end-to-end tests with an array of phantoms and dosimeters can also provide verification. Commissioning and QA of TPSs are discussed in detail in Chapter 16.

External verification can be used as well to provide an unbiased and outside verification. Groups such as the Imaging and Radiation Oncology Core (IROC) Houston in the United States (http://rpc.mdanderson.org/RPC/home.htm) provide phantoms and dosimeters which can perform dosimetric verification of linacs. Point dosimeters to verify absolute dose calibration as well as an array of other task-specific phantoms with dosimeters can be ordered from these groups and irradiated under predetermined conditions. After returning the irradiated phantoms, potentially with some additional information regarding the measured characteristics of the radiation beam, reports will be furnished with the external group's measurements which can be benchmarked against the local measurements. Discrepancies above determined threshold should be analyzed and resolved. Audits using end-to-end tests are discussed in detail in Chapter 19.

The final step in an acceptance testing and commissioning process of a linac is to compile and document all of the used data into one or more reports. This is the data that all subsequent regular QA measurements will be benchmarked against, and high-quality documentation can be very useful in diagnosing issues with the machine. Furthermore, this data could be used to benchmark additional machines in the clinic or in the community.

15.6 Conclusions

Acceptance testing, commissioning, and QA of a linac are pivotal steps in engineering high-quality radiation treatments. Although this chapter has sought to provide a basis in the general thought process and most commonly used equipment and technique, the wide variety of implemented techniques, as well as the increasing variety of treatment machines and modalities, means that the full scope of these tasks will be left to those who will be responsible for the linacs. As such, before a linac purchase and installation is completed, it is imperative that users understand the necessary steps in both of these processes for their specific device and, especially for commissioning, the equipment and manpower that will need to be on hand to perform these processes correctly and efficiently. Guidance for acceptance testing will largely be provided by the vendor, while there is largely more freedom in the types and quantities of measurements needed for commissioning. As discussed herein, the various professional societies have provided some guidance for commissioning, although little to none may be available for more newly developed technologies such as magnetic resonance (MR)-guided linacs (Whelan et al., 2016). In these instances, physicists will need to assess how previous methods may be adapted to these new techniques, while any available literature, as well as input from those in the broader community who have previously worked with these technologies can also be extremely useful.

References

Almond PR et al. (1999) AAPM's TG-51 protocol for clinical reference dosimetry of high-energy photon and electron beams. *Med. Phys.* 26: 1847–1870.

Beyer GP (2013) Commissioning measurements for photon beam data on three TrueBeam linear accelerators, and comparison with Trilogy and Clinac 2100 linear accelerators. *J. Appl. Clin. Med. Phys.* 14(1): 273–288.

Das IJ et al. (2008) Accelerator beam data commissioning equipment and procedures: Report of the TG-106 of the Therapy Physics Committee of the AAPM. *Med. Phys.* 35: 4186–4215.

Ford E et al. (2015) Patterns of practice for safety-critical processes in radiation oncology in the United States from the AAPM safety profile assessment survey. *Pract. Radiat. Oncol.* 5: e423–e429.

Fraass BA (2008) Errors in radiotherapy: Motivation for development of new radiotherapy quality assurance paradigms. *Int. J. Radiat. Oncol. Biol. Phys.* 71: S162–S165.

Gerbi BJ et al. (2009) Recommendations for clinical electron beam dosimetry: Supplement to the recommendations of Task Group 25. *Med. Phys.* 36: 3239–3279.

Hrbacek J, Lang S and Klöck S (2011) Commissioning of photon beams of a flattening filter-free linear accelerator and the accuracy of beam modeling using an anisotropic analytical algorithm. *Int. J. Radiat. Oncol. Biol. Phys.* 80: 1228–1237.

Huq MS et al. (2008) A method for evaluating quality assurance needs in radiation therapy. *Int. J. Radiat. Oncol. Biol. Phys.* 71: S170–S173.

IAEA (1997) Quality assurance in radiotherapy. TECDOC-989. In *Proceedings of the Working Meeting on National Programmes: Design, Harmonisation and Structures* (Vienna, Austria: International Atomic Energy Agency).

IAEA (2000) Absorbed dose determination in external beam radiotherapy: An international code of practice for dosimetry based on standards of absorbed dose to water. Technical Reports Series No. 398. (Vienna, Austria: International Atomic Energy Agency).

IAEA (2008) Setting up a radiotherapy programme: Clinical, medical physics, radiation protection and safety aspects. Report 1296. (Vienna, Austria: International Atomic Energy Agency).

IEC (2007) Medical electrical equipment—Medical electron accelerators—Guidelines for functional performance characteristics. Report 60976. (Geneva, Switzerland: International Electrotechnical Commission). www.iec.ch.

IEC (2008) Medical electrical equipment—Medical electron accelerators—Guidelines for functional performance characteristics. Report 60977. (Geneva, Switzerland: International Electrotechnical Commission). www.iec.ch

Khan FM et al. (1991) Clinical electron-beam dosimetry: Report of AAPM Radiation Therapy Committee Task Group No. 25. *Med. Phys.* 18: 73–109.

Khan FM and Gibbons JP (2014) *The Physics of Radiation Therapy*, ed. 5. (Philadelphia, PA: Lippincott, Williams & Wilkins).

Kirby M, Ryde S, Hall C and Beavis A (2006) Acceptance testing and commissioning of linear accelerators. IPEM Report 94. (York, England: Institute of Physics and Engineering in Medicine).

Klein EE et al. (2009) Task Group 142 Report: Quality assurance of medical accelerators. *Med. Phys.* 36: 4197–4212.

Kutcher GJ et al. (1994) Comprehensive QA for radiation oncology: Report of AAPM Radiation Therapy Committee Task Group 40. *Med. Phys.* 21: 581–618.

McEwen M et al. (2014) Addendum to the AAPM's TG-51 protocol for clinical reference dosimetry of high-energy photon beams. *Med. Phys.* 41: 041501.

Mellenberg DE, Dahl RA and Blackwell CR (1990) Acceptance testing of an automated scanning water phantom. *Med. Phys.* 17: 311–314.

Murray B et al. (2006) Experimental validation of the Eclipse AAA algorithm. *Med. Phys.* 33: 2661.

Levitt SH, Puirdy JA, Perez CA, and Vijayakumar S (Eds.) (2006) Quality assurance in radiation oncology. In: *Technical Basis of Radiation Therapy.* (Heidelberg, Germany: Springer Verlag).

Stern RL et al. (2011). Verification of monitor unit calculations for non-IMRT clinical radiotherapy: Report of AAPM Task Group 114. *Med. Phys.* 38: 504–530.

Whelan B et al. (2016) A novel electron accelerator for MRI-Linac radiotherapy. *Med. Phys.* 43: 1285–1294.

Williamson JF et al. (2008) Quality assurance needs for modern image-based radiotherapy: Recommendations from 2007 Int. Symp. Quality assurance of radiation therapy: Challenges of advanced technology. *Int. J. Radiat. Oncol. Biol. Phys.* 71: S2–S12.

Zhu TC et al. (2009) Report of AAPM Therapy Physics Committee Task Group 74: In-air output ratio, Sc, for megavoltage photon beams. *Med. Phys.* 36: 5261–5291.

16

Commissioning and Quality Assurance of Treatment Planning Systems

Nesrin Dogan, Ivaylo B. Mihaylov, and Matthew T. Studenski

16.1 Introduction

During the last decade, the capabilities of image-based commercial treatment planning systems (TPSs) have been enhanced due to significant developments in treatment imaging, treatment planning, treatment delivery, and computer technology. These developments included the dynamic multileaf collimator (DMLC) (Chui et al., 1994; Losasso, 2008; Ezzell et al., 2009), intensity-modulated radiation therapy (IMRT) (Fraass et al., 1999; Webb, 2000), volumetric-modulated arc therapy (VMAT) (Otto, 2008; Song et al., 2009), four-dimensional radiation therapy (4DRT) (Keall, 2004; Keall et al., 2006; Suh et al., 2008), stereotactic radiosurgery (SRS) (Niranjan et al., 2003), stereotactic body radiotherapy (SBRT) (Grau et al., 2006; Dahele and McLaren, 2015; Franks et al., 2015; Henderson et al., 2015), and kilovoltage (kV) onboard imaging systems such as kV cone beam computed tomography (CBCT) (Oelfke et al., 2006; Boda-Heggemann et al., 2011).

IMRT and VMAT are complex treatment techniques that have imposed high demands on the TPSs. The accurate delivery of IMRT and VMAT treatments strongly depends on the proper commissioning and testing of TPSs (Saw et al., 2001; Venencia and Besa, 2004; Aspradakis et al., 2005; Ezzell et al., 2009; Wen et al., 2014). The integrated imaging, planning, and delivery systems such as TomoTherapy (Langen et al., 2010; Kupelian and Langen, 2011), CyberKnife (Sharma et al., 2007; Pantelis et al., 2008; Van Dyk, 2008), and ViewRay (Mutic and Dempsey, 2014; Wooten et al., 2015) come with their own TPSs and require additional special quality assurance (QA) procedures to be done. The addition of these new technologies has allowed for delivery of highly conformal dose distributions with sharp dose gradients which in turn allows escalated doses to be given to the tumor while better sparing critical structures. At the same time, this had required the TPS to provide more accurate dose calculations and sophisticated automated optimization tools in conjunction with inverse planning optimization. All of this imposes more stringent and new QA requirements due to the complex and less intuitive nature of beam modulation (Palta et al., 2008). Furthermore, the availability of integrated onboard imaging systems and development of commercial deformable image registration tools have become more readily available as a part of the radiation treatment planning and delivery process (Kumarasiri et al., 2014; Garcia-Molla et al., 2015). The use of such technology and tools has allowed the automatic generation of deformed contours and the calculation of actual dose delivered on patient's daily image (Andersen et al., 2012; Thor et al., 2014). It is very important to commission and perform QA of both imaging and dose deformation and accumulation software to make sure that the warped contours and accumulated doses produce accurate and clinically sensible results (Van Dyk, 2008).

With all of this in mind, it is critical to implement a comprehensive TPS QA program, which can incorporate all requirements of these rapidly evolving new technologies. Such a QA program is expected to significantly reduce the magnitude and frequency of errors associated with the treatment planning process. The purpose of this chapter is to review the commissioning and QA procedures for a modern TPS. Such procedures consist of series of tests which should be used prior to initial clinical use of a TPS, as well as the periodic testing of an existing TPS.

16.1.1 Goals and Scope

Commissioning and QA of a TPS is one of the most important aspects of a QA program in a radiotherapy department. A comprehensive set of QA tests need to be performed not only to understand the capabilities and limitations of the TPS, but also to assure that it will satisfy a set of predetermined requirements prior to its use for patient treatments (Olszewska et al., 2003; Mijnheer et al., 2004; Able and Thomas, 2005; Camargo et al., 2007; Jamema et al., 2008; Smilowitz et al., 2015). The purpose of such tests is to ensure that each patient will receive an optimal treatment, as planned and without any error. The comprehensive commissioning of a TPS includes both nondosimetric and dosimetric components as described in many reports (Fraass et al., 1998; Olszewska et al., 2003; Mijnheer et al., 2004). The commissioning of the nondosimetric aspects of a TPS is not discussed here. Commissioning and QA of a TPS for treatments with small fields and for brachytherapy are covered in Chapters 9 and 20, respectively.

This chapter summarizes the measurements, testing, and validation of dosimetric aspects of a modern TPS. It includes a description of experimental techniques to assure a comprehensive and accurate beam model, the commissioning process, the validation of state-of-the-art 3D dose calculation algorithms, the dosimetric effects of immobilization devices and couch, and the guidelines for performing routine QA tests and end-to-end tests for all types of external beam irradiations including IMRT, VMAT, and stereotactic treatments.

16.1.2 Clinical Utilization of TPSs

Although the current commercially available TPSs offer a variety of hardware and software features depending on the specific treatment delivery techniques utilized, the fundamental components of TPSs are generally very similar. Treatment planning is a well-known and complex process which requires multiple steps (e.g., Jamema et al., 2008). The treatment planning process starts with 3D imaging studies generally from computed tomography (CT), but often supplemented with magnetic resonance (MR), positron emission tomography (PET), and/or other functional imaging studies. These different image datasets can be fused using image registration techniques to assess, define, and delineate the tumor and relevant normal tissues. The next step in the treatment planning process is to design treatment beams, which will provide the desired dose distribution based on the prescription doses to the tumor and normal structures. Forward or inverse planning techniques may be used to achieve the desired dose distribution (e.g., Hunt and Burman, 2003). Once the dose distribution is calculated for the volume of interest in the patient, the treatment planner and/or radiation oncologist evaluates the treatment plan by analyzing the dose distributions and dose–volume histograms (DVHs) of tumor and normal tissues. Further improvements can be achieved by adjusting the beam parameters such as individual beam angles (or arc angles), beam energy, weighting, shaping, and beam modifiers. For 3D treatment planning (forward planning), the adjustment of these parameters is manual, and is an iterative process that may take a considerable amount of time depending on the skills of the treatment planner.

For IMRT/VMAT treatment planning utilizing inverse optimization (Bortfeld et al., 1994; Webb, 1994, 1998), a computer algorithm automatically adjusts the parameters to achieve the desired dose distributions defined by the treatment planner. The inverse planning process is fundamentally different than the forward planning process where the computer calculates the resulting dose distribution, not necessarily the desired, based on the beam parameters selected by the treatment planner. The key to the inverse planning optimization process is a definition of an objective (or cost) function which provides a score for the goodness of a treatment plan. A computer optimization algorithm tries to minimize this objective function as it adjusts the beam intensity at each iteration. Most inverse TPSs are based on dose–volume type of objective functions (e.g., Hunt and Burman, 2003), although some utilize biological-based objective functions (e.g., the Elekta Monaco system). The evaluation of plans based on both forward (three-dimensional conformal radiotherapy [3DCRT]) and inverse plannings rely generally on the review and analysis of the dose distributions and DVHs. Although the radiobiological evaluation tools are also available in some TPSs, caution must be taken using such tools due to the limitations of the biological models, uncertainties in model parameters, and data which are derived from incomplete clinical data (Van Dyk, 2005).

16.1.3 Brief Overview of Existing Reports on Commissioning and QA of TPSs

Many reports and papers on QA of TPSs have been published since their initial clinical utilization. The majority of published papers cover the commissioning of dose calculation algorithms (Van Dyk et al., 1993; Kuchnir et al., 2000; Scielzo et al., 2002; Bedford et al., 2003; Borca et al., 2005; Vanderstraeten et al., 2006; Wang et al., 2006; Breitman et al., 2007; Hu et al., 2008; Li and Zhang, 2008; Morgan et al., 2008; Van Dyk, 2008; Moradi et al., 2012). There are studies as early as in the 1980s (Dahlin et al., 1983; McCullough and

Holme, 1985) on the performance evaluation of computerized TPSs. ICRU Report 42, published in 1987, described the details of a state-of-the-art TPS without going into the details of the QA aspect (ICRU, 1987). During the 1990s and 2000s, many reports by both national and international organizations have been published as a result of increased attention paid to the QA of TPSs. One report published by Van Dyk and colleagues in 1993 described a systematic way of dosimetric commissioning and QA of a treatment planning computer (Van Dyk et al., 1993). A report from the United Kingdom, published in 1996, focused on the possible errors and issues related to both hardware and software of a TPS (IPEMB, 1996). However, not much detail on how to perform the QA tests was given in this report. In 1997, the Swiss Society for Radiobiology and Medical Physics (SSRPM) published a report which included a detailed description of a number of recommended tests for commissioning and QA of a TPS (SSRMP, 1997). The American Association of Physicists in Medicine (AAPM) Task Group (TG) Report 53: Quality assurance for clinical radiotherapy treatment planning by Fraass et al., published in 1998, provided a comprehensive general guidance on commissioning and QA of all aspects of the 3D treatment planning process (Fraass et al., 1998). This report described a wide range of tests for both dosimetric and nondosimetric aspects of TPS commissioning, including both external beam radiotherapy and brachytherapy. Another report from the United Kingdom by IPEM (Mayles et al., 1999) described the QA of certain aspects of the treatment planning process. However, not much detail is provided on how the tests should be performed. A 1999 publication by Van Dyk et al. provided a TG-53-like report for evaluation of nondosimetric aspects of a TPS, and also gives detailed descriptions of tests for dose calculation algorithm validation (Van Dyk et al., 1999). In 2004, both IAEA (2004) and the European Society for Radiotherapy and Oncology (ESTRO) (Mijnheer et al., 2004) published reports on the commissioning and QA of a TPS. IAEA Report TRS-430 is an extension of the AAPM TG 53 report and describes the comprehensive process of treatment planning, including the capabilities of a modern TPS, and provides a generic guideline for the commissioning of TPSs for both external beam therapy and brachytherapy. The AAPM TG 65 report, published in 2004, provided a detailed description of dose calculation algorithms, focusing on inhomogeneity corrections utilized in TPSs (Papanikolaou et al., 2004). In 2005, a report published by the Netherlands Commission on Radiation Dosimetry (NCS) described the QA of TPSs (Bruinvis et al., 2005). This report is complementary to the AAPM TG 53 report and provided examples of practical tests for TPS QA. In 2007, the IAEA published an additional report on the specification and acceptance testing of TPSs (IAEA, 2007). This report gave specific details related to the acceptance testing of a TPS as compared to the 2004 IAEA report. Another IAEA report supplied specific commissioning tests for the verification of dose calculations (Gershkevitsh et al., 2008).

It should be noted that none these reports did include the commissioning and QA related to IMRT, stereotactic RT, or other special techniques. These advanced techniques were implemented by many institutions around the world during the last decade. Although general guidelines for QA of such sophisticated techniques have been described in a report of the AAPM IMRT Subcommittee (Ezzell et al., 2003), and in the AAPM TG 119 report (Ezzell et al., 2009), they did not provide comprehensive commissioning and QA procedures for both IMRT and VMAT. The AAPM TG 148 report, published in 2010, provided a comprehensive QA for helical tomotherapy, including QA of the TomoTherapy TPS (Langen et al., 2010). This report addressed the unique QA needs of helical tomotherapy which was not included in the other reports. The recent NCS Reports 22 and 23 (NCS, 2013, 2015), focused on the comprehensive QA of IMRT and VMAT, respectively, which was not included in the older NCS report. These two reports utilized the previous NCS Report 15 (Bruinvis et al., 2005) as foundation and focused on the additional commissioning and QA demands imposed by the more stringent requirements for IMRT and VMAT applications. These reports included machine QA, treatment planning, and TPS commissioning, and reviewed QA methods and tests published in the literature. The report of AAPM TG 166, published in 2012, provides guidelines on the acceptance testing, as well as commissioning and periodic QA tests for a biologically based TPS (Li et al., 2012). This report provided additional QA tests for the verification of biological metrics, including TPS-specific recommendations and precautions of clinical use of biologically based models in treatment plan optimization.

16.1.4 Common Sources of Error

Many potential sources of error are present in the treatment planning process. The identification, reduction, or prevention of such errors will ensure the accurate and precise delivery of the dose to the patient, which plays an important role in the outcome of radiation therapy treatment (Gershkevitsh et al., 2008; Buzdar et al., 2013). Publications by the IAEA (1998, 2000) and International Commission on Radiological Protection (ICRP) (ICRP 2000; Lopez et al., 2009) reported on the accidental exposures of patients undergoing radiation therapy due to errors associated with the different components of the radiation therapy process. These reports emphasized the importance of the design of an overall quality and safety program to prevent such events. The complete commissioning of a TPS is a very important part of the QA program and needs to be done prior to its first clinical use (IAEA, 2004; Mijnheer et al., 2004; Jamema et al., 2008; Lopez-Tarjuelo et al., 2014). To develop a TPS QA program, the potential sources of errors when using the TPS system need to be identified, and a failure mode and effects analysis is recommended (Ford et al., 2009; IAEA, 2012). Previously reported radiation therapy accidents showed that the major factors that contribute to errors in treatment planning are due to misunderstanding of the input, output, and dose calculations of the TPS, improper commissioning of the TPS, lack of independent verification of the basic data entered into the system, use of TPS in a different way than instructed by the manufacturer without properly validating and testing the modified use, and lack of independent calculation checks of absorbed doses to selected points or verification measurements on both homogenous and anthropomorphic phantom geometries as well as *in vivo* dosimetry (IAEA, 2004). As summarized by Van Dyk et al. (1993) and in IAEA Report TRS-430 (Jamema et al., 2008) the clear understanding of the capabilities and limitations of the TPS system is extremely important. This can be achieved by thorough commissioning of the TPS system and adequate training of the personnel, both by the vendor and by the institution on the basic understanding of the capabilities of the TPS system (e.g., dose normalization, monitor unit (MU) calculation, and inhomogeneity corrections). Good communication, among the personnel involved in the treatment planning process, as well as the clear documentation of the procedures is key to avoid errors (Van Dyk, 2005).

16.1.5 Accuracy Requirements and Acceptability Criteria for Dose Calculations

Determination of the required accuracy and acceptability criteria for a TPS is a difficult task since uncertainties from a variety of sources (e.g., dosimetric, imaging, patient related, beam measurement, and dose calculation) contribute to the total uncertainty in the treatment planning process (Van Dyk, 2008). The dose calculation algorithm accuracy is one of the main contributors to overall uncertainty in the final dose delivered to the patient, and therefore it is important to perform a variety of validation tests to recognize the limitations of dose calculation algorithms under different clinical scenerios. ICRU Report 24 (ICRU, 1976) recommended that the overall accuracy in dose delivered to the patient, i.e., the maximum difference between planned and delivered dose, be 5%, which requires a 3% accuracy in the dose calculation. Various groups (e.g., Mijnheer et al., 1987; Van Dyk, 2005) argued that the dose calculation accuracy should be defined in terms of confidence intervals (e.g., one standard deviation), and the criteria can be tightened as the dose calculation algorithms are improved. Van Dyk recommended the 67% confidence level (one standard deviation) acceptability criteria with 2% accuracy for dose calculation algorithms except in high-dose gradient areas. The ESTRO report on QA of TPSs (Mijnheer et al., 2004) also suggested that the accuracy of dose calculations should be around 2% inside the central part of the beam and 2 mm in the regions of high-dose gradient (Venselaar et al., 2001). These and other issues related to physical and clinical aspects of accuracy requirements for 3D dosimetry in advanced radiotherapy are discussed in more detail in Chapter 2.

16.2 Dosimetric Commissioning of TPSs

Modern TPSs perform complex dose calculations and handle many different operations such as image processing and contouring. Ensuring that the TPS functions properly and safely is an involved process and requires more than simply accepting the software from the vendor and turning it on. A well-designed plan for commissioning in addition to ongoing monitoring of the performance of the TPS through periodic QA mitigates the risks and errors that can reach the patient from this complex piece of software. Commissioning and periodic QA require testing of both nondosimetric (contouring, image processing, etc.) and dosimetric components of the TPS. Although nondosimetric tests and periodic QA are vital to the performance of a TPS, this chapter focuses on the dosimetric aspects of TPS commissioning as stated earlier.

16.2.1 Dose Calculation Algorithms

Dosimetric accuracy of a TPS strongly depends on the beam data input quality and dose calculation algorithms (Bruinvis et al., 2005). The dose calculation algorithm is the core of the TPS and it computes the 3D dose distribution deposited by a beam or beams of ionizing radiation in a 3D image. There are two types of dose calculation algorithms: measurement-based and model-based. Early dose calculation algorithms were measurement-based as they required much less computational power and could easily be verified with simple checks, but they tend to be inaccurate, especially in regions with tissue heterogeneity. Modern computers are powerful, and model-based algorithms are now standard in almost all commercially available TPSs. Model-based algorithms rely on precomputed dose kernels that can more accurately account for the effects of scatter and attenuation created in heterogeneous tissues (Bedford et al., 2003; Vanderstraeten et al., 2006; Li and Zhang, 2008). A benefit of a model-based algorithm is that many of the lengthy computations can be done before the actual dose calculation in the TPS so that accuracy can be maintained without a significant increase in computing time. The precomputed dose kernels are typically calculated using the Monte Carlo method, which uses millions of histories and physical material properties to calculate the dose distribution with a high degree of accuracy (Ma et al., 2002; Paganetti et al., 2008; Zhou et al., 2010). The dose calculation algorithm simply needs to calculate the photon fluence in the patient, and the kernels can be superimposed over this fluence. More details about the use of the Monte Carlo method for dose calculations can be found in Chapter 13.

An algorithm must provide an accurate depiction of the dose in the patient because the treating physician will use the calculated dose distribution to decide if this is the plan to be delivered to the patient or if modifications are required. The dose calculation algorithm is the link between the final dose delivered at the linear accelerator and the final treatment plan visualized by the physician in the TPS. A robust dose calculation algorithm should be able to accurately and quickly compute the delivered dose for any treatment scenario that is used clinically. The algorithm should be stable if the treatment plan consists either of a single photon field or of a multiple arc SBRT plan with intensity modulation.

During a course of radiation therapy, the goal is to mitigate random and systematic errors that result in a deviation from the dose distribution approved by the physician in the TPS. Random errors typically result from variation in patient setup or internal anatomical motion. The effect of these errors can be reduced by using treatment margins around the target and through the use of image guidance (van Herk, 2004). On the other hand, poor commissioning of a TPS can result in systematic errors that propagate through the entire treatment delivery for all patients. Although these systematic errors can be more severe than random errors and affect more patients, they can be reduced by understanding the limitations of the TPS and the dose calculation algorithm. By generating a comprehensive set of commissioning and periodic QA tests, the accuracy of the algorithm can be assessed. Independent verification of the dose calculation prior to treatment can also serve to ensure individuals have proper training.

Following installation of a TPS, there is an acceptance test between the user and the vendor, which is legally binding and confirms that the TPS is operating within the specifications provided by the vendor (Khan, 2010). Acceptance tests can be developed by the vendor but the user can request any additional tests to demonstrate that the TPS is functioning properly. Acceptance testing can be extensive but usually does not cover all aspects of the dose calculation algorithm; therefore, the first step after accepting the TPS from the vendor is to thoroughly commission the TPS. The main priority of TPS commissioning is to ensure that the dose distribution displayed on the screen represents what will be delivered at the treatment machine (Gershkevitsh et al., 2008).

16.2.2 Data Requirements for a Dose Calculation Algorithm

To reduce the probability of a systematic error reaching the patient due to an error in the dose calculation algorithm, it is not only important to understand the dose calculation algorithm, but also extensive testing must be done as well to verify the algorithm, and the results must be documented for future reference.

Additionally, all personnel involved in treatment planning should be aware of the limitations and uncertainties in the algorithm, and the consequences these limitations could have on patient treatments.

16.2.2.1 Measurement of Basic Input Data (Vendor)

The process of creating a beam that mimics the treatment beam in a model-based dose calculation algorithm is called beam modeling (Verhaegen and Seuntjens, 2003). The vendor is responsible for providing the user with a comprehensive list of data that is required to model a beam in the TPS. The amount and type of data will vary from one TPS to another. The vendor also has the responsibility of testing the robustness of the algorithm as in many cases the code is proprietary and can be considered as a "black box." It is important for the user to discuss with the vendor to ensure that the robustness of the algorithm has been validated (Olszewska et al., 2003). For model-based algorithms, the vendor typically has precalculated all of the beam kernels and has also modeled the treatment head for typical linear accelerators. This helps to alleviate some of the work required, although the user must verify all of these parameters during commissioning.

16.2.2.2 Data Required for a Dose Calculation Algorithm (User)

The user is responsible for measuring all of the data required for the specific TPS beam modeling. A typical dataset consists of percent depth dose (PDD) curves, profiles, output factors (OFs), wedge factors, wedge profiles, and multileaf collimator (MLC) parameters (Fraass et al., 1998). The vendor may also provide "golden beam" data, which is a compilation of datasets from many linacs. This data can be used as a reference when acquiring the dataset but it is not recommended to use only "golden beam" data to model the beam. Even if linacs are "matched" so that the parameters are similar to a certain degree, there can be individual variations that cannot be detected unless a full dataset is obtained from each specific linac (NCS, 2015). This can lead to problems at a later time during periodic QA if an accurate baseline is not established. Before the commissioning and measurements begin, it is a good idea to have a plan laid out as to what data must be collected and how the data is going to be collected. The entire dataset should be consistent even if multiple users are collecting data (Fraass et al., 1998).

Special tools must be used to acquire the required data. One of the most important tools is the computer-controlled scanning water tank. The dimensions of this tank are larger than the largest field size and usually have a possible depth of around 40 cm to ensure proper scatter conditions. Ion chambers, diodes, or other detectors can be positioned and moved in the tank to acquire profiles and PDDs. The choice of detector is very important depending on the data being acquired, as discussed in the previous chapter. For example, if treatments will use intensity-modulated beams, apertures are very small and irregularly shaped (<3 cm $\times$ 3 cm); therefore, the user needs to select a detector that is small enough to avoid volume averaging which can result in a smoothed penumbra region in the scan (Low et al., 2011). It can also be difficult to measure the OFs of small fields; therefore, the use of multiple detectors with overlapping field sizes is a good idea to compare results. Issues related to the specific problems encountered during the dosimetry of small fields are discussed in detail in Chapter 9. Using inappropriate data will compromise the quality of the beam model in the TPS and the quality of the delivered plan. Film is also a good option as the spatial resolution is much higher than that of an ion chamber or diode. There are other factors to consider when collecting data for commissioning such as scanning speed, scanning direction, and tank setup that can be found in more detail in AAPM TG Report 106 (Das et al., 2008). Acceptance testing, commissioning, and QA of linear accelerators are discussed in Chapter 15.

Remember that the data acquired during commissioning is not only the basis for the beam modeling in the TPS, it is also the baseline QA dataset for all subsequent periodic QA. That being said, it is critical to document all of the measurements and make sure that this data is easily accessible for future users to know what the baseline value was. Raw data should be saved before any postprocessing such as normalization or smoothing. The data entered into the TPS can be processed (to a certain degree that does not alter the data significantly) but the raw data should be available for future reference. This is another reason not to use "golden beam" data because if the linac does not exactly match the "golden beam" data and during the subsequent annual QA a discrepancy is found, there is no way of knowing if this is a parameter that has changed over time or if this was the baseline value.

16.2.3 Beam Modeling

16.2.3.1 Machine, Beam, and Algorithm Parameters

Each dose calculation algorithm requires unique data to model a treatment beam, but the principle of beam modeling is universal where the goal is to represent the actual treatment beam in the TPS. To be able to accomplish this goal, not only are the dosimetric properties of the beam needed, but the physical dimensions, compositions, and limitations of the linear accelerator treatment head components are also needed. Many TPSs have the standard linear accelerator parameters preprogrammed, but it is important that the user verifies that these parameters are correct. Parameters might include the physical location and composition of the diaphragms, MLC, monitor chamber, wedges, target, as well as the jaw, MLC, and gantry speed and positioning limits.

In modern treatment planning, involving intensity modulation, accurate modeling of the MLC is crucial as the small fields used during IMRT are all shaped by the MLC. This differs from the traditional 3D conformal treatments where the fields were much larger and the MLC model did not affect the dosimetry to such a degree. Depending on the planning system and linac, the MLC modeling can vary. The most important aspect of MLC modeling is to accurately characterize the MLC leaf tip, especially for rounded MLC leaves (LoSasso et al., 1998). The rounded leaf is used so that the beam penumbra is the same across the entire field of view. Therefore, it is important to model the leaf tip on both the central axis and the off-axis. One simple test is to measure the PDD, profiles, and OFs for fields down to 2 cm × 2 cm defined by the MLC alone (Smilowitz et al., 2015). Recommendations are that the agreement between the 50% isodose line from the calculated and measured profiles is within 1 mm (NCS, 2013), or that a distance-to-agreement (DTA) measurement is within 3 mm in the penumbra region (Smilowitz et al., 2015). This measurement is very important in IMRT delivery because multiple abutted fields are used; therefore, inaccurate modeling is compounded. The two common parameters that need to be measured to accurately model the MLC are the leaf-gap width and the MLC transmission (LoSasso et al., 1998; Patel et al., 2005).

The leaf gap represents the amount of transmission through the rounded leaf as the planning system typically calculates the dose assuming a flat MLC tip. The modeled leaf gap increases the penumbra width to the proper amount in the planning system. The leaf transmission accounts for the additional dose that is delivered through the leaves out of the treatment field. The transmission is usually an average of the interleaf and intraleaf transmissions, although some planning systems can model these separately. If modeled separately, it is recommended to use film to differentiate between the interleaf and intraleaf transmissions. The agreement between the measured and calculated transmissions should be within 10% (NCS, 2013) or 3% of the maximum field dose up to 5 cm away from the field edge (Smilowitz et al., 2015).

Another parameter that might be available to model in the planning system is the tongue-and-groove effect (Sykes and Williams, 1998; Huq et al., 2002; Liu et al., 2008). The leaf-gap models the dose profile in a direction parallel to the leaf motion, while the tongue-and-groove models the profile perpendicular to the leaf motion. This parameter accounts for the interlocking design of the MLC where there is partial transmission, as the tongue-and-groove side part of the leaf does not have the same thickness as the central part of an MLC leaf. Inclusion of this parameter in the planning system can reduce the dose calculation error. For all of these measurements, it is recommended to use a high spatial resolution dosimeter, such as film, diode, or microion chamber (Smilowitz et al., 2015).

16.2.3.2 Tuning of Algorithm Parameters

The final step in beam modeling is to tune the various parameters so that the calculated and measured PDDs and profiles match as good as possible. Depending on the algorithm and TPS, the parameters discussed in the previous section might be available to adjust. For example, the amount of transmission allowed through the diaphragms and MLC leaves can be adjusted to affect the penumbra shape and out-of-field dose in the beam model. The energy spectrum of the incident electrons on the target can be tuned to adjust the PDD and the horns of the profiles. The process of matching the measured and calculated profiles and PDDs is usually iterative and is repeated until the desired results are achieved. Remember that when adjusting the beam model to achieve agreement with IMRT fields, the results of the basic beam model for standard fields could change and therefore must be reviewed (Smilowitz et al., 2015).

It is important to realize that no algorithm is perfect and it can be difficult to exactly match the measured data. With this in mind, the user must understand for what purpose the algorithm will be used clinically and tune the parameters to fit that specific need (Fraass et al., 1998). This could be a trade-off in out-of-field dose versus the shape of the penumbra, or it could be that large-field sizes are modeled well but small-field sizes are not. In modern radiation therapy, the trend is moving toward small treatment fields as a result of the use of IMRT and SBRT. It is very important to understand how the linac is going to be used and the model should be tuned to ensure good agreement between the measured and calculated profiles of fields that will be treated, especially in the penumbra region.

Once an acceptable model has been developed, the final step is to compare dosimetric data calculated by the algorithm to measured doses. This data must include point measurements but also clinical treatment plans. This is crucial in modern treatment planning with intensity modulation as the dose delivery and the required calculations are complex. If a linac is to be used to treat IMRT, there are several fields that should be measured to ensure proper modeling in the TPS. The first field to check would be a small field defined by the MLC on the central axis, preferably as small as the smallest allowed field size in the TPS (~2 cm × 2 cm). The OF and profile should be measured and the agreement between measured and calculated outputs should be within 2%, and the penumbra should agree within 3 mm (Smilowitz et al., 2015). A second field would be the same small field but off-axis. The tolerance in this off-axis case is 5% as additional parameters have been added that can confound the results (Smilowitz et al., 2015). There should also be a test of abutting fields to test the leaf-gap model, and a test for the tongue-and-groove effect. Finally, there should be a test to assess the ability of the algorithm to handle tissue inhomogeneity, especially for inverse optimization algorithms, which typically use very simple heterogeneity corrections during the optimization (Fogliata et al., 2007; Gershkevitsh et al., 2008; Aarup et al., 2009).

The last test of a TPS is always an end-to-end test to ensure that all aspects of the TPS are performing appropriately. This test involves scanning a phantom, creating a treatment plan, calculating the dose delivered to the phantom from a treatment plan, and finally delivering the plan to the phantom and measuring the dose. An end-to-end test ensures the functionality and accuracy of all aspects of the treatment planning process. During this testing, it is important to gain an understanding of the accuracy of the algorithm and planning system so that this information can be passed on to other individuals involved in treatment in the clinic. It is recommended to start with simple, flat, homogeneous phantoms before progressing to more anthropomorphic-type phantoms (Ezzell et al., 2009). The most stringent test uses an anthropormorphic phantom with tissue heterogeneities. These types of phantoms are available from external sources and can be irradiated and sent out for independent review to ensure that the entire planning workflow is performing properly (Smilowitz et al., 2015). Additional information on end-to-end tests can be found in Chapter 19.

16.2.4 Calculation of MUs/Time and Plan Normalization

The dose in the TPS is typically calculated relative to a normalization point which can be different for each dose calculation algorithm and TPS. It is very important to understand what point is being used and why. During commissioning, the user needs to verify that the isodose lines calculated by the algorithm are accurate for both simple treatment plans and complex ones. The final step to make a plan deliverable is that the MUs or the treatment time (for Co-60 teletherapy) need to be determined, that is, the relative isodose lines need to be converted into absolute dose. There are many ways of normalizing a treatment plan to determine absolute dose. Some of the common means are to normalize the prescription dose to isocenter, or to normalize the dose so that an amount of the target structure, >95% typically, is covered by the prescription dose. It is important to verify that the isodose lines in both absolute and relative doses agree for any mode of normalization that will be used clinically.

Unlike 3D conformal radiation therapy fields, it can be difficult to verify the MU calculation for IMRT fields due to the use of small, irregular apertures in each field. Other verification methods such as film or chamber array can be used to verify the dose delivered from the IMRT field. Film can be used in a phantom at a known depth to compare the calculated and measured fluences in that plane and the same phantom can be used with an ion chamber, MOSFET, thermoluminescent dosimeter (TLD), optically stimulated luminescent dosimeter (OSLD), or diode to measure absolute point doses. Chamber arrays can be used to measure either planar or volumetric dose depending on the detector arrangement in the phantom. Test cases such as those proposed in the AAPM TG 119 report can be used along with clinical plans that are calculated on actual patient images (Ezzell et al., 2009). Thorough testing that ensures accurate dose delivery must be done before any patient is treated.

16.2.5 Dosimetric Effects of Immobilization Devices and Couch

The dosimetric effects of external patient immobilization and support devices have been discussed in the literature in the last three decades. Recently, the development of image-guided radiation therapy (IGRT) and SBRT led to the introduction of solid, sandwich-like, carbon fiber couch-top design, as well as more complex patient immobilization and motion tracking systems. This in turn sparked a renewed interest and necessity to quantify the dosimetric effects of those devices. Couch tops and patient immobilization devices affect the dose distribution to different degrees depending on the setup geometry and the device design. The introduction of VMAT was another recent technological advance, where the dose perturbations from the immobilization devices may lead to a noticeable increase in surface dose and decrease to target doses. Usually, immobilization devices are included in the planning CT scan. Thereby, the TPS should be able to handle immobilization devices effects within the scope of their accuracy, and therefore they will not be discussed in detail here.

The attenuation for a single open beam varies within several percent depending on the energy and the incidence angle. It can relatively easily be accounted for by adjusting the beam's MUs. However, in case of complex delivery such as IMRT or VMAT, this simplistic approach would not work (Mihaylov et al., 2011). Modern TPSs are capable of accounting for the patient support devices in the dose calculations. The treatment couch can be integrated in the TPS process either through an actual CT scan of the support couch, or through modeling of the couch via contouring. In the first approach, the CT scan of the treatment couch is tailored to the patient/phantom CT scan through TPS fusion modules, third party software, or in-house developed software. The second approach utilizes automated or manual device contouring. In the former case, the couch geometry and density information are automatically generated by virtue of the actual CT scan, while in the latter situation the couch geometry and physical properties can be specified within the model.

Clinical implementation of TPS couch-top modeling requires adequate commissioning and validation measurements of the device in the TPS. As a minimum, the steps should include the following: (1) verification of the couch structure with a CT scan of the device; (2) verification of the daily patient position reproducibility (indexing) so there is consistency between planning and treatment; (3) validation of the couch density through attenuation measurements; and (4) verification of the couch model by performing measurements for TPS calculated doses. For a detailed discussion on the subject, the reader is referred to the guidance document of AAPM TG 176 on dosimetric effects caused by couch tops and immobilization devices (Olch et al., 2014).

16.3 Validation of 3D Dose Calculation Algorithms

16.3.1 Test Cases and Validation for Photon Beam Dose Calculation

In reality, it is inconceivable that all possible clinical scenarios can be covered in the dosimetric tests of a dose calculation algorithm. Thereby, the test cases for the dose calculation algorithm should cover all common basic clinical scenarios, as well as the possible extremes of the use of the dose calculations for all practical purposes. The most obvious and straightforward tests should be performed on the data acquired for the TPS commissioning. The tested dose calculation algorithm should reproduce the input data with an acceptable accuracy for its commissioning (Van Dyk et al., 1993; Fraass et al., 1998; IAEA, 2004; Palta et al., 2008; Smilowitz et al., 2015). The details of those tests are summarized in Table 16.1 and include PDDs, profiles, wedge factors, and OFs. Note that, on the contrary to physical wedges, the nonphysical wedges can be considered as an extension of the open-field geometry and therefore only one field (greater than 15 cm) can be used for each wedge angle (Smilowitz et al., 2015). Model-based algorithms tend to require fewer measurements for their validation since they are based on first principles, where the radiation generation and transport, as well as the energy deposition, are modeled with as few assumptions as possible. In that case, measurements testing more basic situations need to be acquired and the performance of the dose calculation algorithm validated.

Table 16.1 Typical Data Tests Required for TPS Commissioning

Test Number	Parameter	Field Size (cm²)	Depth (cm)
1	Square fields (on- and off-axis)	5 × 5, 10 × 10, 15 × 15, 20 × 20, and 30 × 30	d_{max}, 10, 20. Off-axis: 0.5, 1, 1.5, 5 cm
2	Rectangular fields (on- and off-axis)	5 × 30, 30 × 5	d_{max}, 10, 20. Off-axis: 0.5, 1, 1.5 cm
3	Asymmetric fields (open and wedged)	$x_1 = 2, x_2 = 8, y_1 = 5, y_2 = 5$ $x_1 = 5, x_2 = 5, y_1 = 2, y_2 = 8$ $x_1 = 4, x_2 = 6, y_1 = 5, y_2 = 5$ $x_1 = 5, x_2 = 5, y_1 = 4, y_2 = 6$ Same field size with 45° wedge filter	10
4	Shaped fields	• 20 × 20 cm² field with cord block • 20 × 20 cm² corner and central block • Oval shape • C-shape, central axis under block	10
5	Hard wedges on- and off-axis	5 × 5, 10 × 10, 20 × 20 for 45°	10 Off-axis: 0.5, 1, 1.5, 5 cm
6	Oblique incidence (on- and off-axis)	5 × 5, 10 × 10 for 45°	10
7	Missing scatter	20 × 20	10
8	Buildup region	5 × 5, 10 × 10, 30 × 30	0–3
9	Source-to-surface distance	90–120 cm, 10 × 10, 30 × 30	10

The majority of the reference measurements used for dose calculation commissioning are relative. These include PDDs, beam profiles, OFs, and wedge factors. Patients, however, are treated with absolute rather than relative doses. Therefore, the dose calculation algorithms must be tested against absolute dosimetric measurements. The absolute doses per MU, calculated by the TPS for all available energies, should at the bare minimum acceptably reproduce the measured doses under calibration conditions. Those conditions usually employ square fields, at nominal beam incidence, where dose is measured in water at a certain depth for a given source-to-surface distance (SSD) or source-axis distance (SAD). Comparisons between measured and calculated absolute doses for several different depths in water will additionally validate the accuracy of the dose calculation algorithms. The calculated doses should be compared to measurements performed with a properly calibrated ionization chamber, adequate for absolute reference dosimetry. Additional tests should include absolute dosimetric comparisons for nonsquare fields. Beams with oblique incidence would further validate the dose calculation accuracy. Extending the absolute dosimetric comparisons at different SSDs for some of the above test cases would directly confirm the inverse square law if field sizes are properly scaled. In addition to the absolute dosimetric measurements, the MUs calculated by the TPS can be compared to the MUs calculated by a secondary MU check program, acquired from a vendor different from the TPS vendor. The caveat of such a comparison is that the secondary MU check software requires parameter input of some kind and its validity should be tested by itself. Therefore, the absolute dosimetry mentioned above remains the gold standard in TPS dose calculation validation and shall be performed.

Another class of tests includes a set of simple nonsquare fields. Computed and measured doses need to be compared for elongated and asymmetric fields as well as for wedged asymmetric and elongated fields. In addition, blocked fields should also be checked including central and partial blocking. A special setup, where the edge of a cubic phantom is irradiated at normal and oblique incidences, would validate the dose calculation algorithm in the case of missing tissue. Furthermore, MLC-shaped fields should also be included in the tests, since MLCs are very commonly utilized in modern 3D radiotherapy (Smilowitz et al., 2015).

All modern dose calculation algorithms account for medium heterogeneities in the dose calculations. The first thing that needs to be verified is that the planning system reports the correct mass or electron densities (Smilowitz et al., 2015). After ascertaining that a proper density model is present in the TPS, the actual tissue heterogeneity dose calculation validation tests should be carried over. Performing these dosimetric tests on heterogeneous phantoms, where absolute doses are cross-compared, is a necessity. There are commercially available heterogeneous phantoms, but if they are not available, there are alternative solutions. Instead, a simple heterogeneous phantom, as the one presented in Figure 16.1, can be built from solid water and inexpensive materials, which can be acquired in a local hardware store. The phantom on the figure consists of solid water slabs with density of 1.0 g/cm^3, Styrofoam slabs, which for all practical purposes are treated by a TPS as air, and dry-wall slabs with density of ~0.50 g/cm^3. One of the solid water slabs is drilled for a Farmer-type chamber insert. The ionization chamber can be used for either absolute or relative dosimetry. This particular heterogeneous phantom can be irradiated in different

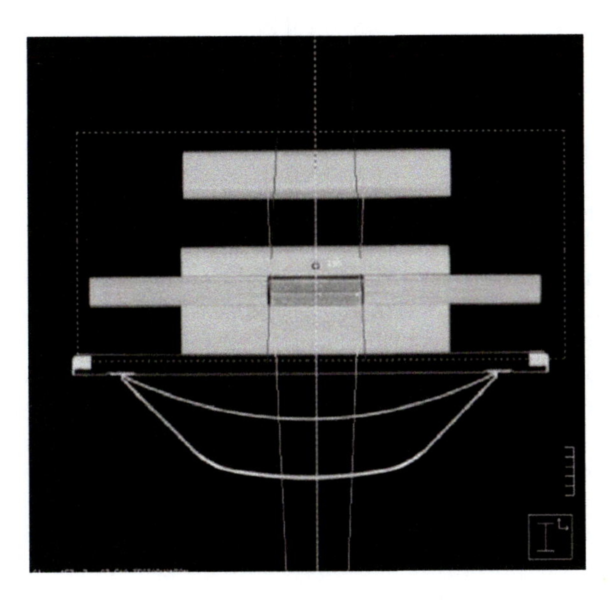

Figure 16.1

User-built heterogeneous phantom. The phantom consists of solid water blocks combined with Styrofoam and dry-wall slabs. The solid water and the Styrofoam are available in every radiation oncology department, while the dry-wall material is inexpensive and can be purchased in every local hardware store.

configurations, thereby testing the radiation transport and the dose calculations with different degrees of heterogeneity in front and behind the point of measurement. All available photon energies should be tested, as well as the accuracy of the dose calculations for different field sizes and shapes. This will give the users deeper insight in the capabilities of the TPS in terms of heterogeneous dose calculations. There is no need to mention that absolute dosimetric measurements should be the choice in these tests for dose deposition in heterogeneous media.

Modern linear accelerators are equipped with MLCs which allow the delivery of IMRT. As beam modifying devices, the MLCs should be tested in a fashion similar to wedges and blocks. The first set of tests should verify the leaf-tip modeling (penumbra) for several leaf positions. Film or electronic portal imaging devices (EPIDs) are the measurement tools of choice because of the required submillimeter leaf-positional accuracy (Pecharroman-Gallego et al., 2011). Calculated dose profiles for several elongated fields should be compared with measured profiles derived from TPS-generated planar dose distributions. The acceptable tolerance is within 5%–10% of the maximum dose. Tongue-and-groove effects can be tested by adding fields which are orthogonal to the direction of leaf motion. In those tests again calculated and measured (by film or EPID) profiles should be compared. Picket-fence fields should be used to verify leaf-positioning accuracy (Sykes and Williams, 1998; Williams and Metcalfe, 2006). Leaf-position errors can easily be identified even with the naked eye. MLC leaf leakage and transmission should be measured with film, and the derived profiles must be compared with TPS-generated profiles. It should be noted that when measuring MLC leakage and transmission, the collimating jaw or backup jaw must be retracted at least 5 cm beyond the MLC edges. Differences of about 10% between measured and calculated profiles for MLC leakage and transmission are acceptable.

16.3.1.1 3D Conformal Radiation Therapy

The test cases of actual patient data need to start from the simplest possible setup. Such a case is outlined in Figure 16.2, where a two opposed fields whole-brain plan is shown. The dose across the entire volume should be fairly uniform which can be inferred from the DVHs. The plan also contains beam blocks which spare the lenses. Measured and calculated absolute doses and profiles should be compared and evaluated. Whole-brain treatments are performed with low-energy photon beams. However, for dose calculation algorithm testing purposes, measurements and calculations should be performed for all available energies.

The next test case is presented in Figure 16.3. This is again a very simple case with only two opposed fields. In this plan, wedges are used to compensate for the missing tissue, and thereby create uniform dose over the target. Physical wedges and enhanced dynamic wedges (EDWs) should be used in separate plans, since they both need to be tested. Measured and calculated absolute and planar, at least, doses should be compared.

A very similar setup is presented in Figure 16.4 with two opposed breast tangential fields. Conceptually, this is not much different from the previous two cases, but given that breast treatments are quite common in radiotherapy it is worthwhile verifying the dose calculation algorithm capabilities for that particular site.

The AP/PA plan in Figure 16.5 is for a lung case. The AP field contains a wedge in order to compensate for missing tissue. In this plan, the dose calculation algorithm is tested in rather heterogeneous media where the tissue densities vary from ~0.25 g/cm³ (lung) to ~1.8 g/cm³ (bone). While the two test cases above do not require posterior beams, the AP/PA setup requires a beam through the treatment couch. If the linac is equipped with an IGRT couch, and the TPS is capable of modeling that couch, this setup will also test the couch model.

A three-field pancreatic cancer case is presented in Figure 16.6. All three fields use wedges in order to generate a uniform dose distribution across the target. In this scenario, approximately two third of the dose is delivered through the posterior

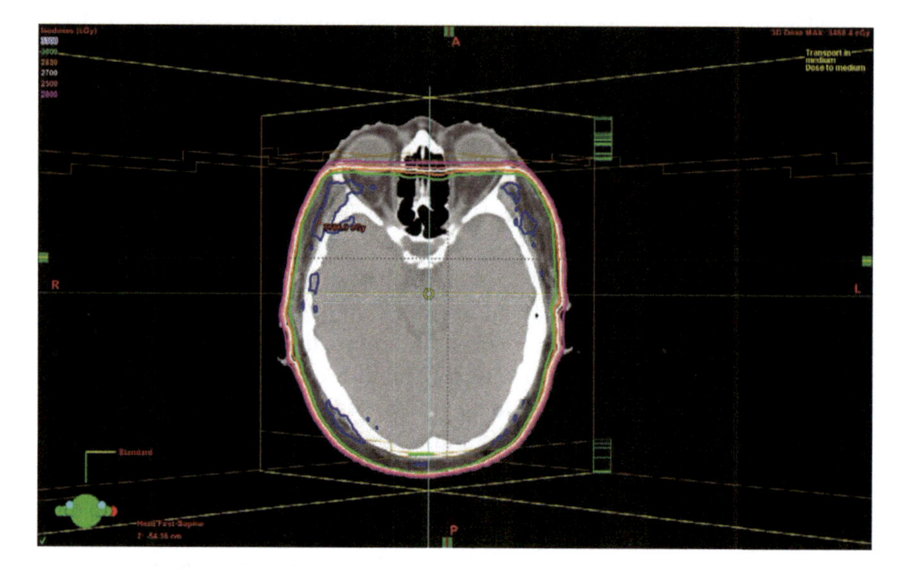

Figure 16.2

Test case of a whole brain treated with two opposed open 6 MV beams.

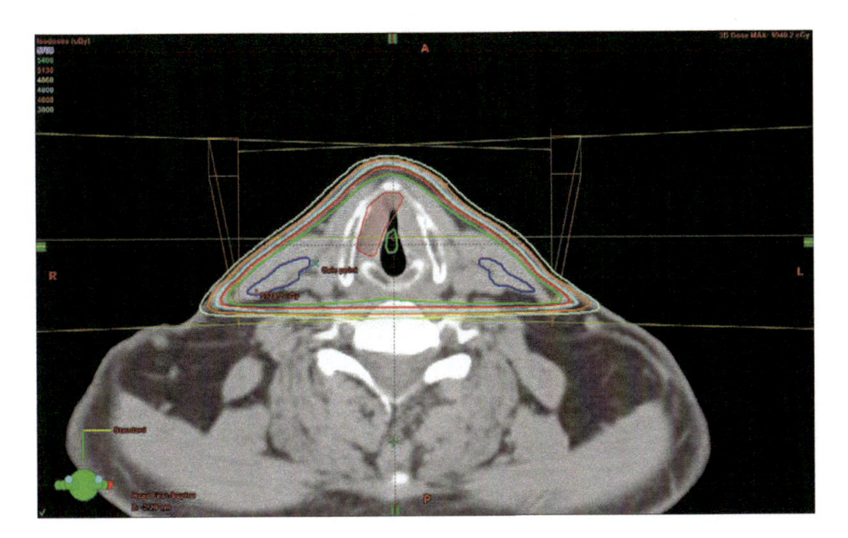

Figure 16.3

Test case of a larynx tumor treated with two opposed wedged 6 MV beams.

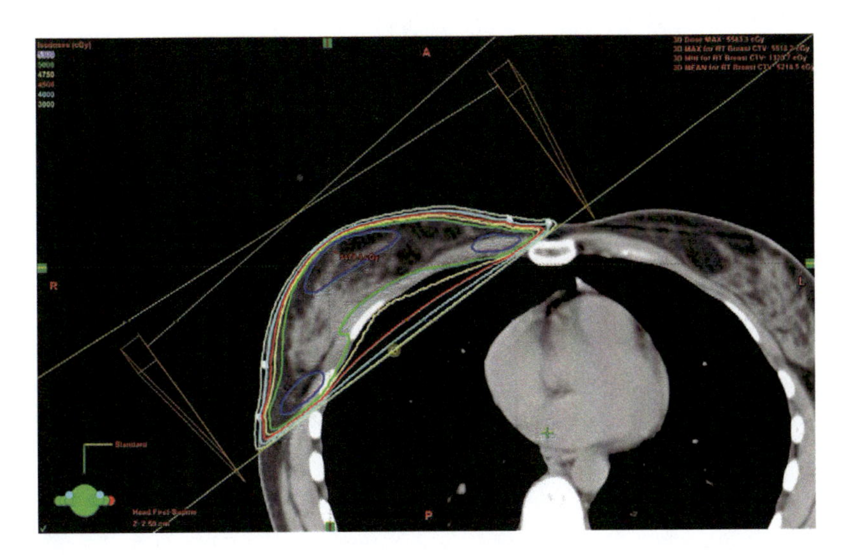

Figure 16.4

A very similar setup with two opposed breast tangential fields.

beams through the treatment couch. This is another good example to test the couch model as well as to demonstrate the necessity of patient indexing (Mihaylov et al., 2011; Olch et al., 2014). Needless to say, plans with both physical and EDWs need to be created and the dose distributions should be verified against measurements.

The next level of complexity would be a four-field box, or a five-field brain plan, similar to the one presented in Figure 16.7. This five-field plan contains a noncoplanar field and the majority of the fields utilize wedges.

The test cases presented above cover different clinical 3D scenarios. They range from simple two-field setups to rather complex five-field setups with wedges and noncoplanar beam arrangements. Those tests cover the majority of the clinical

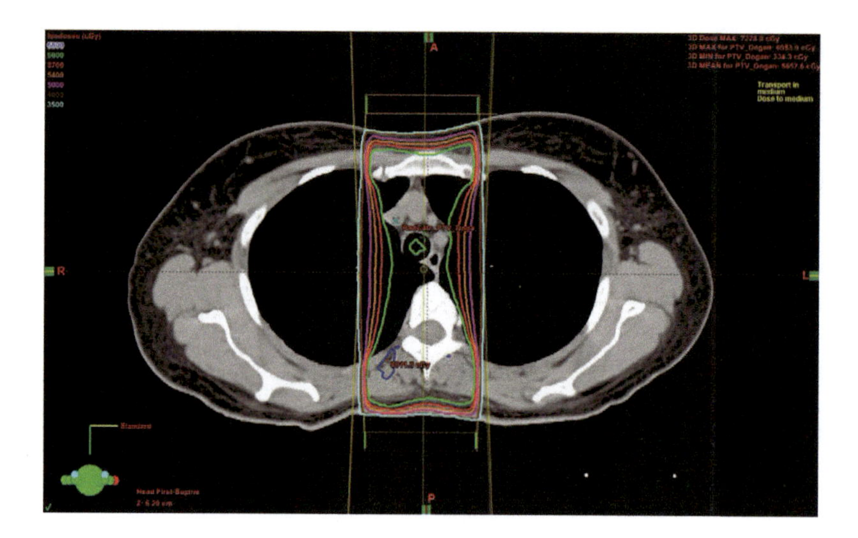

Figure 16.5

Example of an AP/PA lung setup. The AP field contains a wedge in order to compensate for missing tissue.

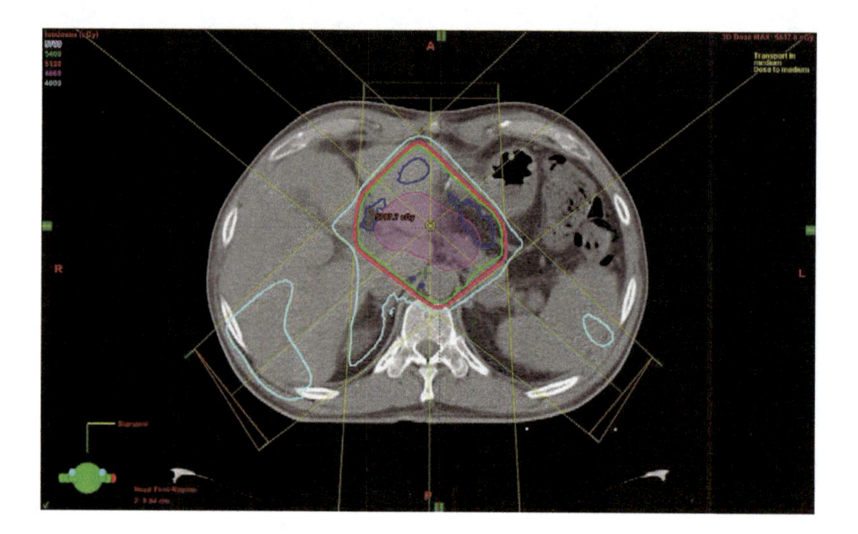

Figure 16.6

Example of a more complex three-field pancreatic cancer setup. All fields are wedged, so that adequate dose conformality to the target can be achieved.

situations where forward planning would be used. The resolution of the dose grids used in the abovementioned test case calculations should be no larger than 3 mm and should at least match the slice thickness of the used CT datasets.

16.3.1.2 Intensity-Modulated Radiation Therapy

Figures 16.8 and 16.9 outline a seven-field mediastinum and a nine-field head-and-neck case, respectively. Verifying the accuracy of IMRT plans is a more involved procedure than 3D plans. While MUs for simple 3D plans can be

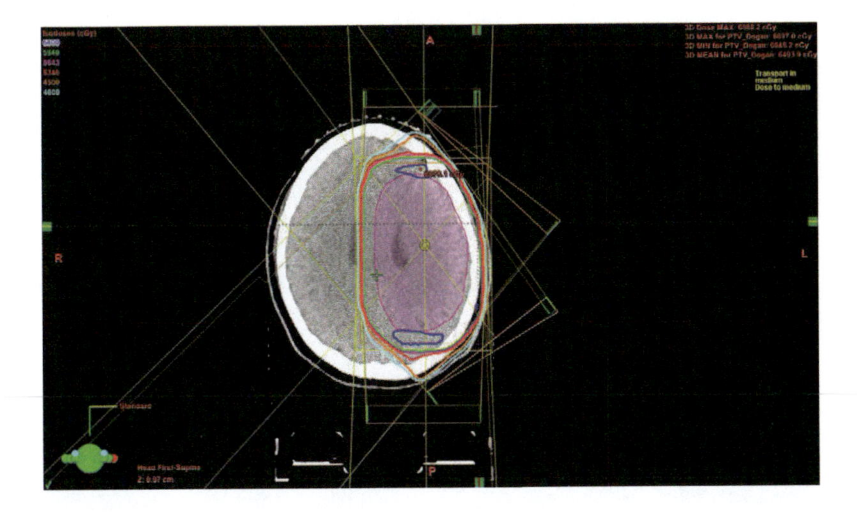

Figure 16.7

Five-field brain case example. All fields contain either wedges or beam blocks. This axial cut does not show the fifth, noncoplanar field.

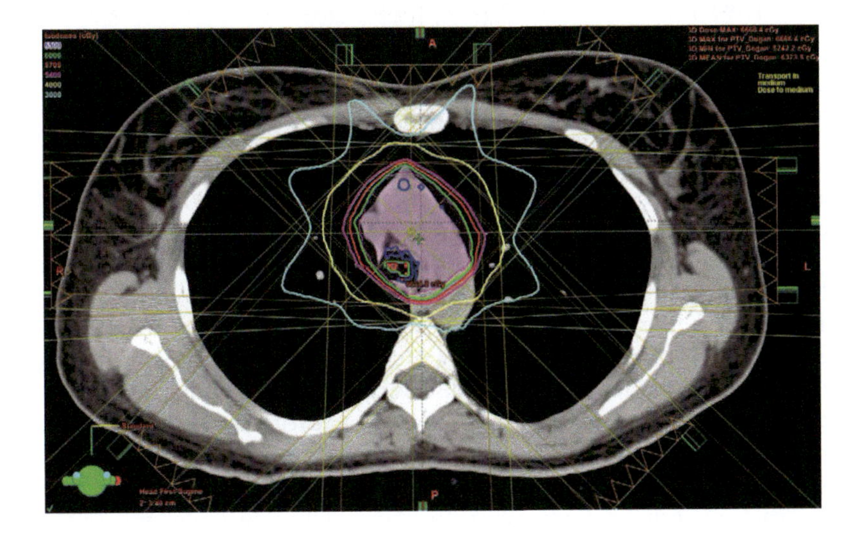

Figure 16.8

Seven-field IMRT setup for a mediastinum case.

checked manually, for IMRT plans the user needs to resort to commercially available secondary MU calculator programs. The gold standard for verifying the MU calculations by the TPS is an ionization chamber measurement. Depending on target size and dose uniformity, chambers of different sizes can be utilized for absolute dose verification with these measurements.

Another feature of IMRT plans is that they very often result in large spatial dose gradients. This in turn brings two issues: spatial resolution of the detector as well as positioning accuracy of the entire dose distribution. Film or EPID measurements should be used for relative dose verification between measured

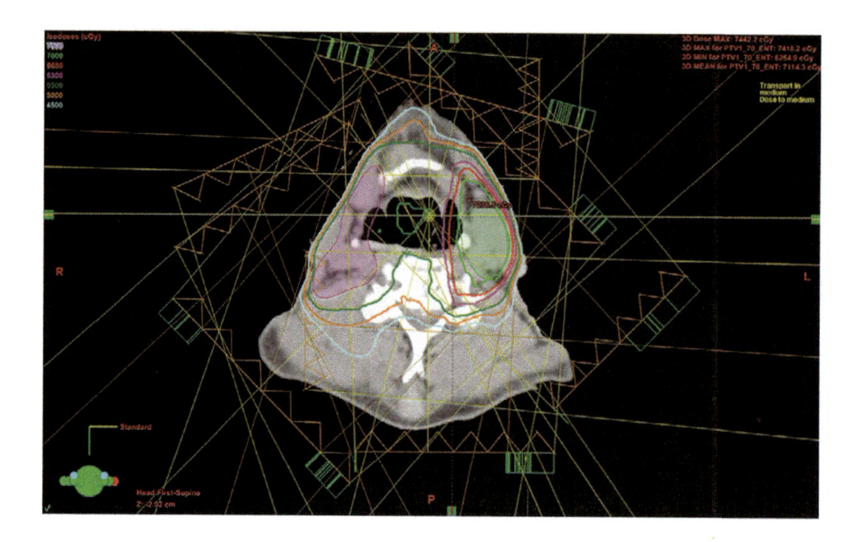

Figure 16.9

Nine-field IMRT setup for a head-and-neck case. The majority of fields are split since this is a simultaneously integrated boost case.

and computed planar dose distributions. Careful alignment of the measured and computed planar doses is required in order to verify that the dose is delivered to the correct spatial location (Low et al., 2011).

More recently, vendors of QA equipment developed ion chamber and diode arrays, which are suitable for verification of IMRT plans. The spatial resolution of those arrays is inferior to EPID or film, but their functionality and the associated dose verification processing tools make them very popular (see Chapter 8 for more details). A common metric used in the evaluation of the planar dose distributions obtained with the arrays is based on gamma analysis (Low and Dempsey, 2003; Low et al., 2013). It is recommended that a gamma evaluation with a criterion of at least 3%/3 mm of the local dose (Low and Dempsey, 2003; Zhen et al., 2011) is used with absolute dosimetry, as discussed in Chapter 14. However, the use of a 2%/2 mm criterion in the gamma analysis may assist in the discovery of deviations during IMRT/VMAT commissioning, which would be obscured with a more relaxed criterion of 3%/3 mm (Smilowitz et al., 2015). In the gamma evaluation, all points with a measured dose below 10% of the reference dose should be discarded to avoid false positives due to low signal to noise in the low dose area. The choice of the actual cutoff value is at the discretion of the user, in part based on the treatment site, the equipment used, and the choice between a 2D and 3D gamma evaluation. The maximum allowed number of points sampled with gamma greater than unity is 10%. If the gamma value averaged over all sample points is >0.5, special attention to the problem is needed. A DTA of 3 mm or better requires that the dose calculations are performed on a grid with a resolution of no more than 3 mm. The slice thickness of the imaging dataset used for dose computation should be considered to meet this criterion. All equipment (arrays) used should be calibrated properly and their limitations in terms of dosimetric and spatial accuracy/precision should be known and taken into consideration.

16.3.1.3 Stereotactic Radiation Therapy

Stereotactic radiotherapy (SRS and SBRT) is usually associated with small fields since the employed high doses are delivered to relatively small targets

(Benedict et al., 2010; Solberg et al., 2012). IMRT delivery techniques are generally also associated with fields of limited sizes. In addition, IMRT fields are often elongated and irregularly shaped. Therefore, all the tests associated with SRS and SBRT are also pertinent to IMRT (Benedict et al., 2001). Small fields are defined as fields that do not exhibit lateral electron equilibrium in the beam center. Usually, fields which are less than 3×3 cm^2 are referred to as small fields, as discussed in Chapter 9. With decreasing field size, the total photon energy spectrum shifts to a spectrum determined by the thickness of the flattening filter. The photon fluence decreases with smaller fields due to the partial covering of the effective spot size.

OFs, depth dose curves, and profiles of small fields should be measured to verify that the TPS is able to predict the data within 5% accuracy (Azimi et al., 2012). An appropriate calculation grid size for dose calculation is advised and special attention should be given to the shape of the depth dose curves for each clinically used grid size. In small-field dosimetry, appropriate detectors, such as diamond detectors, diodes, or PinPoint chambers, need to be used (Pappas et al., 2008; Amin et al., 2011; Underwood et al., 2013; Papaconstadopoulos et al., 2014). The small detectors do not average over the field size for OF measurements, while during profile measurements they sample accordingly the profile edges. In performing OF measurements, the use of small-field detectors should be cross-checked against a reference detector such as a Farmer-type chamber. This can be easily accomplished when OFs for field sizes from 6×6 cm^2 to 3×3 cm^2 are measured with both small field and reference ion chambers. The measured OFs with both detectors should differ by less than a percent. In addition, simple plots of OFs as a function of field size must result in a smooth curve, which is a complimentary consistency test. More information on dose measurements in small fields can be found in Chapter 9.

16.3.2 Test Cases and Validation for Electron Beam Dose Calculation

The basic tests described in Section 16.3.1 are also valid for external beam electron fields. For all electron energies the computed and measured PDDs and profiles should agree. As bare minimum, the TPS should reproduce the commissioning data used to validate the electron dose calculation algorithm.

MUs for electron fields are generally checked by direct comparison with measured data. The points to be checked are the points of maximum dose on the central beam axis. If standard beam inserts are clinically used, tables of OFs should be made available. Another variable that must be checked carefully for its influence on the dose calculation is the SSD. Within the range of SSDs used clinically, for example, from 90 cm up to a maximum of 120 cm, the inverse square law can be applied. If deviations occur between calculated and measured doses, a virtual source position can be inserted in the algorithm to obtain a proper correction, as discussed in detail in Section 15.4.2.2.

Nominal doses per MU can be measured and verified with properly calibrated ionization chambers, adequate for electron beam calibration. Calculations need to be performed for a given setup, nominal electron energy, SSD, depth, and cone size. This setup should be different from the beam calibration setup mentioned in the previous paragraph, thereby truly validating the performance of the dose calculation algorithm. Absolute dosimetric measurements should be performed in a solid water phantom under the same conditions as the calculations. The conversion from raw electrometer reading (charge) into dose (cGy) should be achieved according to a calibration protocol similar to the TG 51 report (Almond et al., 1999; McEwen et al., 2014), where all the correction factors are utilized in the charge-to-dose conversion.

Furthermore, surface slope tests should be performed. In those tests electron beams with oblique incidence should be checked, which will yield the effects of central axis tilt on depth dose and penumbra. Dose calculations in inhomogeneous media need also to be performed for electron beams. A calculation setup similar to the one used for testing tissue heterogeneities in photon beams can be used. The resulting dose distributions should at least qualitatively be evaluated, although measurements may be better suited (Smilowitz et al., 2015).

16.4 Summary

The state-of-the-art technology for radiation oncology continues to change at a rapid rate. Advanced techniques such as inverse planning for IMRT and VMAT, Monte Carlo dose calculations, and 4D techniques are widely available for clinical use and have introduced new challenges for treatment planning and delivery. Increased complexity of modern TPSs results in new demands on the commissioning and QA requirements. Therefore, it is critical to develop a comprehensive treatment planning QA program, which will incorporate all new requirements imposed by the rapidly evolving new technologies. Verification, education, documentation, and communication should be part of the main components of a QA program, which is expected to significantly reduce the magnitude and frequency of errors associated with the treatment planning process, which will be greatly beneficial to the patient.

References

Aarup LR et al. (2009) The effect of different lung densities on the accuracy of various radiotherapy dose calculation methods: Implications for tumour coverage. *Radiother. Oncol.* 91: 405–414.

Able CM and Thomas MD (2005) Quality assurance: Fundamental reproducibility tests for 3D treatment-planning systems. *J. Appl. Clin. Med. Phys.* 6(3): 13–22.

Almond PR et al. (1999) AAPM's TG-51 protocol for clinical reference dosimetry of high-energy photon and electron beams. *Med. Phys.* 26: 1847–1870.

Amin MN et al. (2011) Small field electron beam dosimetry using MOSFET detector. *J. Appl. Clin. Med. Phys.* 12(1): 50–57.

Andersen ES et al. (2012) Bladder dose accumulation based on a biomechanical deformable image registration algorithm in volumetric modulated arc therapy for prostate cancer. *Phys. Med. Biol.* 57: 7089–7100.

Aspradakis MM et al. (2005) Elements of commissioning step-and-shoot IMRT: Delivery equipment and planning system issues posed by small segment dimensions and small monitor units. *Med Dos.* 30: 233–242.

Azimi R et al. (2012) The effect of small field output factor measurements on IMRT dosimetry. *Med. Phys.* 39: 4691–4694.

Bedford JL et al. (2003) Commissioning and quality assurance of the Pinnacle (3) radiotherapy treatment planning system for external beam photons. *Brit. J. Radiol.* 76(903): 163–176.

Benedict SH et al. (2001) Intensity-modulated stereotactic radiosurgery using dynamic micro-multileaf collimation. *Int. J. Radiat. Oncol. Biol. Phys.* 50: 751–758.

Benedict SH et al. (2010) Stereotactic body radiation therapy: The report of AAPM Task Group 101. *Med. Phys.* 37: 4078–4101.

Boda-Heggemann J et al. (2011) kV cone-beam CT-based IGRT: A clinical review. *Strahlenther. Onkol.* 187: 284–291.

Borca VC et al. (2005) Dosimetric evaluation of a commercial 3D treatment planning system using the AAPM Task Group 23 test package. *Med. Phys.* 32: 744–751.

Bortfeld T et al. (1994) Realization and verification of three-dimensional conformal radiotherapy with modulated fields. *Int. J. Radiat. Oncol. Biol. Phys.* 30: 899–908.

Breitman K et al. (2007) Experimental validation of the Eclipse AAA algorithm. *J. Appl. Clin. Med. Phys.* 8(2): 76–92.

Bruinvis IAD et al. (2005) Quality assurance of 3-D treatment planning systems for external photon and electron beams. Netherlands Commission on Radiation Dosimetry, NCS Report 15. (http://radiationdosimetry.org)

Buzdar SA et al. (2013) Accuracy requirements in radiotherapy treatment planning. *J. Coll. Physicians Surg. Pak.* 23: 418–423.

Camargo PRTL et al. (2007) Implementation of a quality assurance program for computerized treatment planning systems. *Med. Phys.* 34: 2827–2836.

Chui CS et al. (1994) Dose calculation for photon beams with intensity modulation generated by dynamic jaw or multileaf collimations. *Med. Phys.* 21: 1237–1244.

Dahele M and McLaren DB (2015) Stereotactic body radiotherapy. *Clin. Oncol.* 27: 249–250.

Dahlin H et al. (1983) User requirements on CT-based computed dose planning systems in radiation therapy. *Acta Oncol.* 22: 397–415.

Das IJ et al. (2008) Accelerator beam data commissioning equipment and procedures: Report of the TG-106 of the Therapy Physics Committee of the AAPM. *Med. Phys.* 35: 4186–4215.

Ezzell GA et al. (2003) Guidance document on delivery, treatment planning, and clinical implementation of IMRT: Report of the IMRT Subcommittee of the AAPM Radiation Therapy Committee. *Med Phys.* 30: 2089–2115.

Ezzell GA et al. (2009) IMRT commissioning: Multiple institution planning and dosimetry comparisons, a report from AAPM Task Group 119. *Med. Phys.* 36: 5359–5373.

Fogliata A et al. (2007) On the dosimetric behaviour of photon dose calculation algorithms in the presence of simple geometric heterogeneities: Comparison with Monte Carlo calculations. *Phys. Med. Biol.* 52: 1363–1385.

Ford EC et al. (2009) Evaluation of safety in a radiation oncology setting using failure mode and effects analysis. *Int. J. Radiat. Oncol. Biol. Phys.* 74: 852–858.

Fraass BA et al. (1998) American Association of Physicists in Medicine Radiation Therapy Committee Task Group 53: Quality assurance for clinical radiotherapy treatment planning. *Med. Phys.* 25: 1773–1829.

Fraass BA et al. (1999) Optimization and clinical use of multisegment intensity-modulated radiation therapy for high-dose conformal therapy. *Semin. Radiat. Oncol.* 9: 60–77.

Franks KN et al. (2015) Stereotactic ablative body radiotherapy for lung cancer. *Clin. Oncol.* 27: 280–289.

Garcia-Molla R et al. (2015) Validation of a deformable image registration produced by a commercial treatment planning system in head and neck. *Phys. Med.* 31: 219–223.

Gershkevitsh E et al. (2008) Dosimetric verification of radiotherapy treatment planning systems: Results of IAEA pilot study. *Radiother. Oncol.* 89: 338–346.

Grau C et al. (2006) The emerging evidence for stereotactic body radiotherapy. *Acta Oncol.* 45: 771–774.

Henderson DR et al. (2015) Stereotactic body radiotherapy for prostate cancer. *Clin. Oncol.* 27: 270–279.

Hu YA et al. (2008) Evaluation of an electron Monte Carlo dose calculation algorithm for electron beams. *J. Appl. Clin. Med. Phys.* 9(3): 1–15.

Hunt AM and Burman CM (2003) Treatment planning considerations in IMRT. In: *A Practical Guide to Intensity-Modulated Radiation Therapy*, pp. 103–121. (Madison, WI: Medical Physics Publishing).

Huq MS et al. (2002) A dosimetric comparison of various multileaf collimators. *Phys. Med. Biol.* 47: N159–N170.

IAEA (1998) *Accidental Overexposure of Radiotherapy Patients in San Jose, Costa Rica.* (Vienna, Austria: International Atomic Energy Agency).

IAEA (2000) Lessons learned from accidental exposures in radiotherapy. Safety Reports Series No. 17. (Vienna, Austria: International Atomic Energy Agency).

IAEA (2004) Commissioning and quality assurance of computerized planning systems for radiation treatment of cancer. Technical Reports Series No. 430. (Vienna, Austria: International Atomic Energy Agency).

IAEA (2007) *Specification and Acceptance Testing of Radiotherapy Treatment Planning Systems, TECDOC-1540.* (Vienna, Austria: International Atomic Energy Agency).

IAEA (2012) *Analisis Probabilista De Seguridad De Tratamientos De Radioterapia Con Acelerador Lineal. TECDOC-1670/S.* (Vienna, Austria: International Atomic Energy Agency).

ICRP (2000) Prevention of accidental exposures to patients undergoing radiation therapy. International Commission on Radiological Protection Publication 8692000303ICRP *Ann. ICRP* 30 (3): 7–70.

ICRU (1976) Determination of absorbed dose in a patient irradiated by beams of X or gamma rays in radiotherapy procedures. Report 24. (Bethesda, MD: International Commission on Radiation Units and Measurements).

ICRU (1987) Use of computers in external beam radiotherapy procedures with high-energy photons and electrons. Report 42. (Bethesda, MD: International Commission on Radiation Units and Measurements).

IPEMB (1996) A guide to commissioning and quality control of treatment planning systems. Report 68. (York, England: Institute of Physics and Engineering in Medicine and Biology).

Jamema SV et al. (2008) Commissioning and comprehensive quality assurance of commercial 3D treatment planning system using IAEA Technical Report Series—430. *Australas. Phys. Eng. Sci. Med.* 31: 207–215.

Keall P (2004) 4-dimensional computed tomography imaging and treatment planning. *Sem. Radiat. Oncol.* 14: 81–90.

Keall P et al. (2006) The clinical implementation of respiratory-gated intensity-modulated radiotherapy. *Med. Dos.* 31: 152–62.
Khan FM (2010) *The Physics of Radiation Therapy.* (Philadelphia, PA: Lippincott Williams & Wilkins).
Kuchnir FT et al. (2000) Commissioning and testing of a commercial intensity modulated treatment planning system. In: *Proceedings 22nd Annual International Conferences IEEE Engineering in Medicine and Biology Society.* Vol 2 (Chicago, IL: IEEE), pp. 1184–1186.
Kumarasiri A et al. (2014) Deformable image registration based automatic CT-to-CT contour propagation for head and neck adaptive radiotherapy in the routine clinical setting. *Med. Phys.* 41: 121712.
Kupelian P and Langen K (2011) Helical tomotherapy: Image-guided and adaptive radiotherapy. In: *IMRT, IGRT, SBRT—Advances in the Treatment Planning and Delivery of Radiotherapy,* pp. 165–80. (Eds. Meyer JL, Dawson LA, Kavanagh BD, Purdy JA and Timmerman R). Basel, Switzerland: Karger Medical and Scientific Publishers.
Langen KM et al. (2010) QA for helical tomotherapy: Report of the AAPM Task Group 148. *Med. Phys.* 37: 4817–4853.
Li J and T Zhang (2008) Implementation of convolution/superposition model of photon dose calculation. In: *Proceedings of the 7th Asian-Pacific Conference on Medical and Biological Engineering,* pp. 442–446. (Eds. Peng Y and Weng X) Berlin, Heidelberg: Springer.
Li XA et al. (2012) The use and QA of biologically related models for treatment planning—Report of Task Group No.166. *Med. Phys.* 39: 1386–1409.
Liu C et al. (2008) Multileaf collimator characteristics and reliability requirements for IMRT Elekta system. *Int. J. Radiat. Oncol. Biol. Phys.* 71: S89–S92.
Lopez O et al. (2009) Preventing accidental exposures from new external beam radiation therapy technologies. International Commission on Radiological Protection Publication 112, *Ann. ICRP:* 39 (4).
Lopez-Tarjuelo J et al. (2014) Acceptance and commissioning of a treatment planning system based on Monte Carlo calculations. *Technol. Cancer Res. Treat.* 13: 129–138.
LoSasso T et al. (1998) Physical and dosimetric aspects of a multileaf collimation system used in the dynamic mode for implementing intensity modulated radiotherapy. *Med. Phys.* 25: 1919–1927.
LoSasso T (2008) IMRT delivery performance with a Varian multileaf collimator. *Int. J. Radiat. Oncol. Biol. Phys.* 71: S85–S88.
Low DA and Dempsey JF (2003) Evaluation of the gamma dose distribution comparison method. *Med. Phys.* 30: 2455–2464.
Low DA et al. (2011) Dosimetry tools and techniques for IMRT. *Med. Phys.* 38: 1313–1338.
Low DA et al. (2013) Does the gamma dose distribution comparison technique default to the distance to agreement test in clinical dose distributions? *Med. Phys.* 40: 071722.
Ma C-M et al. (2002) A Monte Carlo dose calculation tool for radiotherapy treatment planning. *Phys. Med. Biol.* 47: 1671–1689.
Mayles WPM and Evans PM (1999) In vivo dosimetry and portal verification. In: Physics aspects of quality control in radiotherapy. Report 81, pp. 229–232. (York, UK: Institute of Physics and Engineering in Medicine and Biology).

McCullough EC and Holmes TW (1985) Acceptance testing computerized radiation-therapy treatment planning systems – direct utilization of CT scan data. *Med. Phys.* 12: 237–242.

McEwen M et al. (2014) Addendum to the AAPM's TG-51 protocol for clinical reference dosimetry of high-energy photon beams. *Med. Phys.* 41: 041501.

Mihaylov IB et al. (2011) Carbon fiber couch effects on skin dose for volumetric modulated arcs. *Med. Phys.* 38: 2419–2423.

Mijnheer BJ et al. (1987) What degree of accuracy is required and can be achieved in photon and neutron therapy? *Radiother. Oncol.* 8: 237–252.

Mijnheer BJ et al. (2004) Quality assurance of treatment planning systems - Practical examples for non-IMRT photon beams. ESTRO Booklet No. 7. (Brussels, Belgium: European Society for Radiotherapy and Oncology).

Moradi F et al. (2012) Commissioning and initial acceptance tests for a commercial convolution dose calculation algorithm for radiotherapy treatment planning in comparison with Monte Carlo simulation and measurement. *J. Med. Phys.* 37: 145–150.

Morgan AM et al. (2008) Clinical implications of the implementation of advanced treatment planning algorithms for thoracic treatments. *Radiother. Oncol.* 86: 48–54.

Mutic S and Dempsey JF (2014) The ViewRay system: Magnetic resonance-guided and controlled radiotherapy. *Sem. Radiat. Oncol.* 24: 196–199.

NCS (2013) Code of practice for the quality assurance and control for intensity modulated radiotherapy. Netherlands Commission on Radiation Dosimetry Report 22. (http://radiationdosimetry.org).

NCS (2015) Code of practice for the quality assurance and control for volumetric modulated arc therapy. Netherlands Commission on Radiation Dosimetry Report 24. (http://radiationdosimetry.org).

Niranjan A et al. (2003) Radiosurgery: Current techniques. *Tech. Neurosurg.* 9: 119–127.

Oelfke U et al. (2006) Linac-integrated kV-cone beam CT: Technical features and first applications. *Med. Dos.* 31: 62–70.

Olch AJ et al. (2014) Dosimetric effects caused by couch tops and immobilization devices: Report of AAPM Task Group 176. *Med. Phys.* 41: 061501.

Olszewska A et al. (2003) Quality assurance of treatment planning systems: The QUASIMODO project. *Radiother. Oncol.* 68: S73.

Otto K (2008) Volumetric modulated arc therapy: IMRT in a single gantry arc. *Med. Phys.* 35: 310–317.

Paganetti H et al. (2008) Clinical implementation of full Monte Carlo dose calculation in proton beam therapy. *Phys. Med. Biol.* 53: 4825–4853.

Palta JR et al. (2008) Quality assurance of intensity-modulated radiation therapy. *Int. J. Radiat. Oncol. Biol. Phys.* 71: S108–S112.

Pantelis E et al. (2008) Dosimetric characterization of CyberKnife radiosurgical photon beams using polymer gels. *Med. Phys.* 35: 2312–2320.

Papaconstadopoulos P et al. (2014) On the correction, perturbation and modification of small field detectors in relative dosimetry. *Phys. Med. Biol.* 59: 5937–5952.

Papanikolaou et al. (2004) Tissue inhomogeneity corrections for megavoltage photon beams—Report of Task Group No. 65. (Madison, WI: Medical Physics Publishing).

Pappas E et al. (2008) Small SRS photon field profile dosimetry performed using a PinPoint air ion chamber, a diamond detector, a novel silicon-diode array (DOSI), and polymer gel dosimetry. Analysis and intercomparison. *Med. Phys.* 35: 4640–4648.

Patel I et al. (2005) Dosimetric characteristics of the Elekta Beam Modulator™. *Phys. Med. Biol.* 50: 5479–5492.

Pecharroman-Gallego R et al. (2011) Simplifying EPID dosimetry for IMRT treatment verification. *Med. Phys.* 38: 983–992.

Saw CB et al. (2001) Commissioning and quality assurance for MLC-based IMRT. *Med. Dosim.* 26: 125–133.

Scielzo G et al. (2002) Dosimetric evaluation of two radiotherapy treatment planning systems using Report 55 by the AAPM Task Group. *Tumori* 88: 53–58.

Sharma SC et al. (2007) Commissioning and acceptance testing of a CyberKnife linear accelerator. *J. Appl. Clin. Med. Phys.* 8(3): 119–125.

Smilowitz JB et al. (2015) AAPM medical physics practice guideline 5.a.: Commissioning and QA of treatment dose calculations—Megavoltage photon and electron beams. *J. Appl. Clin. Med. Phys.* 16(5): 14–34.

Solberg TD et al. (2012) Quality and safety considerations in stereotactic radiosurgery and stereotactic body radiation therapy: Executive summary. *Pract. Radiat. Oncol.* 2: 2–9.

Song Y et al. (2009) The development of a novel radiation treatment modality—Volumetric modulated arc therapy. *IEEE Eng. Med. Biol. Soc. Ann. Conf.* 2009: 3401–3404.

SSRMP (1997) Quality control of treatment planning systems for teletherapy. Swiss Society for Radiobiology and Medical Physics. Report 7: ISBN 3-908125-23-5.

Suh Y et al. (2008) A deliverable four-dimensional intensity-modulated radiation therapy-planning method for dynamic multileaf collimator tumor tracking delivery. *Int. J. Radiat. Oncol. Biol. Phys.* 71: 1526–1536.

Sykes JR and Williams PC (1998) An experimental investigation of the tongue and groove effect for the Philips multileaf collimator. *Phys. Med. Biol.* 43: 3157–3165.

Thor M et al. (2014) Evaluation of an application for intensity-based deformable image registration and dose accumulation in radiotherapy. *Acta Oncol.* 53: 1329–1336.

Underwood TS et al. (2013) Detector density and small field dosimetry: Integral versus point dose measurement schemes. *Med. Phys.* 40: 082102.

Van Dyk J et al. (1993) Commissioning and quality assurance of treatment planning computers. *Int. J. Radiat. Oncol. Biol. Phys.* 26: 261–273.

Van Dyk J, Barnett RB and Battista JJ (1999) Computerized radiation treatment planning systems. In: *The Modern Technology of Radiation Oncology: A Compendium for Medical Physicists and Radiation oncologists*, Vol. 1, pp. 231–286. (Ed. Van Dyk J, Madison, WI: Medical Physics Publishing).

Van Dyk J (2005) Advances in modern radiation therapy. In: *The Modern Technology of Radiation Oncology: A Compendium for Medical Physicists and Radiation Oncologists*, Vol.2, pp. 16–21. (Ed. Van Dyk J, Madison, WI: Medical Physics Publishing).

Van Dyk J (2008) Quality assurance of radiation therapy planning systems: Current status and remaining challenges. *Int. J. Radiat. Oncol. Biol. Phys.* 71: S23–S27.

van Herk M (2004) Errors and margins in radiotherapy. *Sem. Radiat. Oncol.* 14: 52–64.

Vanderstraeten B et al. (2006) Accuracy of patient dose calculation for lung IMRT: A comparison of Monte Carlo, convolution/superposition, and pencil beam computations. *Med. Phys.* 33: 3149–3158.

Venencia C and Besa P (2004) Commissioning and quality assurance for intensity modulated radiotherapy with dynamic multileaf collimator: Experience of the Pontificia Universidad Católica de Chile. *J. Appl. Clin. Med. Phys.* 5(3): 37–54.

Venselaar J, Welleweerd H and Mijnheer B (2001) Tolerances for the accuracy of photon beam dose calculations of treatment planning systems. *Radiother. Oncol.* 60: 191–201.

Verhaegen F and Seuntjens J (2003) Monte Carlo modelling of external radiotherapy photon beams. *Phys. Med. Biol.* 48: R107–R164.

Wang L et al. (2006) Commissioning and quality assurance of a commercial stereotactic treatment-planning system for extracranial IMRT. *J. Appl. Clin. Med. Phys.* 7(1): 21–34.

Webb S (1994) Optimizing the planning of intensity-modulated radiotherapy. *Phys. Med. Biol.* 39: 2229–2246.

Webb S (1998) Intensity-modulated radiation therapy: Dynamic MLC (DMLC) therapy, multisegment therapy and tomotherapy. An example of QA in DMLC therapy. *Strahlenther. Onkol.* 174 (Suppl. 2): 8–12.

Webb S (2000) Advances in three-dimensional conformal radiation therapy physics with intensity modulation. *Lancet Oncol.* 1: 30–36.

Wen N et al. (2014) IMRT and RapidArc commissioning of a TrueBeam linear accelerator using TG-119 protocol cases. *J. Appl. Clin. Med. Phys.* 15(5): 74–88.

Williams MJ and Metcalfe P (2006) Verification of a rounded leaf-end MLC model used in a radiotherapy treatment planning system. *Phys. Med. Biol.* 51: N65–N78.

Wooten HO et al. (2015) Benchmark IMRT evaluation of a Co-60 MRI-guided radiation therapy system. *Radiother. Oncol.* 114: 402–405.

Zhen H et al. (2011) Moving from gamma passing rates to patient DVH-based QA metrics in pretreatment dose QA. *Med. Phys.* 38: 5477–5489.

Zhou B et al. (2010) GPU-accelerated Monte Carlo convolution/superposition implementation for dose calculation. *Med. Phys.* 37: 5593–5603.

17

Patient-Specific Quality Assurance: Pretreatment 3D Dose Verification

Dietmar Georg, Catharine H. Clark, and Mohammad Hussein

17.1 Introduction

17.1.1 Rationale for Pretreatment Verification

The overall intention in modern radiation oncology is to keep the dose delivered to the patient as close as possible to the prescribed dose, within tight dosimetric tolerance limits, while reducing the dose burden to healthy tissues as much as possible. Verifying a treatment plan and the underlying dose distribution has thus become a common standard and integral part of patient-specific quality assurance (QA). In the pre–intensity-modulated radiation therapy (IMRT) era, such procedures were largely based on independent dose or monitor unit (MU) calculations based on empirical (factor-based) dose calculation methods or *in vivo* dosimetry using point detectors (Dutreix et al., 1997; Georg et al., 2004; Stern et al., 2011; Mijnheer et al., 2013b). With the development and clinical introduction of fluence-modulated treatment techniques in the mid-1990s, point dose calculations and/or point dose measurements became inappropriate for patient-specific pretreatment QA (Ezzell et al., 2003; Mijnheer and Georg, 2008). The reasons are manifold. First, the resulting dose distribution in a patient being treated with IMRT is a superposition of many nonhomogeneous dose distributions of static beams, angles, or arcs; although it is important to underline that IMRT with a static or rotating gantry, as we use it today, often still aims at a homogeneous dose in the planning target volume (PTV). Single-point dose verification per beam is thus no longer representative. Second, as IMRT treatment delivery has become more sophisticated, many more field-defining elements are involved in dose shaping. In other words, multileaf collimators (MLCs) on standard C-arm linear accelerators, or even new treatment delivery units specifically developed for IMRT, for example, the TomoTherapy unit or the robotic linac system CyberKnife, introduced a new level of geometric and dosimetric complexity, which again could no longer be handled with traditional pretreatment QA. Third, treatment plan optimization changed dramatically with IMRT, and particularly in the early days of IMRT, dose calculation accuracy was an issue of concern.

Another issue is the electronic data transfer from imaging to the treatment planning system (TPS), from TPS to record and verify (R&V) system, or oncology information management system, and finally to the delivery unit. Full electronic and network-based data transfer enabled computer-controlled conformal therapy before the IMRT era (Fraass et al., 1995); however, without the development of information technology, IMRT would not be manageable because of the enormous amount of data that needs to be communicated when applying treatment techniques with time variable fluence/dose patterns.

As a result of the rapidly changing technology and techniques in nearly all steps involved in the treatment chain and the associated potential pitfalls, it was obvious to check the end product of the treatment chain in a dry run prior to the first fraction being delivered to the patient. Since traditional empirical point dose calculations were not able to handle IMRT in its entire complexity, patient-specific QA procedures were largely based on experimental dosimetric methods during the first decade of IMRT.

17.1.2 Evolution of Pretreatment Three-Dimensional Verification

Climbing up the learning curve, i.e., using and improving first-generation IMRT systems and utilizing both the upcoming experience with IMRT and the past experience in conformal radiotherapy, paved the way for less workload and machine-intensive patient-specific pretreatment QA procedures. In other words, it became obvious that for the widespread clinical implementation of IMRT experimental treatment plan verification was the "bottleneck." Therefore, the role of calculation techniques, complementing experimental dosimetric methods, was soon explored (Karlsson et al., 2010). Today, it is well understood that an overall QA program for advanced radiotherapy techniques requires several components, i.e., independent dose and MU calculation can be only part of a more comprehensive QA program in a department. For instance, the Netherlands Commission on Radiation Dosimetry (NCS) recommends performing at least 100 experimental verifications of IMRT/volumetric-modulated arc therapy (VMAT) prior to moving to calculation techniques (NCS 2013, 2015).

Another aspect is the exchange or installation of a new treatment unit, where patient-specific QA using experimental dosimetry is also highly recommended for a limited number of patients, for example, 20 or 30. When introducing IMRT/VMAT on a new machine, such a procedure helps to validate the system's performance and to be able to benchmark it with existing or previous ones. The same arguments hold when introducing a new IMRT/VMAT treatment technique or when IMRT/VMAT is clinically implemented for a new indication in a department.

The recent decades in radiation oncology have been largely influenced by technology developments for and around IMRT, and these developments will continue. Some examples are the gradual replacement of static fixed-field IMRT by IMRT in a rotating gantry, mostly for reasons of delivery efficiency, or the use of flattening filter-free photon beams (Georg et al., 2011; Teoh et al., 2011). Today's high-precision radiation oncology techniques go beyond the dogma of a homogeneous dose in the target volume and try to utilize the advances in medical imaging for subtumor volume characterization in order to intentionally deliver an inhomogeneous dose distribution (Thorwarth and Alber, 2010; Bentzen and Gregoire, 2011). This latter concept is called "dose painting."

The implementation of any new treatment technique in radiation oncology increases the overall workload and implies a potential danger for serious errors in the planning and delivery of radiotherapy. The challenge at the department level is to setup an effective net of QA procedures, since the goal of a routine pretreatment verification procedure is to catch errors before the actual treatment begins. So far pretreatment verification procedures have been, and are, performed in a reactive manner, irrespective whether they are based on dosimetric measurements or calculations. Deviations between the expected doses calculated by the TPS or surrogates of the delivered dose are determined, and if large deviations are observed action is taken. This reactive philosophy of doing patient-specific QA is workload intensive. In general, technology in treatment planning and delivery has become more mature, i.e., dose calculation algorithms have become very sophisticated and treatment delivery units with MLCs are even more stable than conventional linear accelerators with block collimators used two decades ago. For that reason, more proactive than reactive QA procedures are currently being explored (Noel et al., 2014).

The aim of this chapter is to provide a thorough overview of measurement- and calculation-based techniques for patient-specific pretreatment QA that provide multidimensional dose information, ideally in "full" three dimension. This implies semi-3D approaches: two-dimensional (2D) methods in multiple planes so that three-dimensional (3D) dosimetric information can be extrapolated. Despite the implementation of a dense QA net, errors may remain that cannot be detected with patient-specific dosimetric procedures; some are related to the limitations of the dose calculation performed with the TPS, some to delivery units, and some are patient related, as will be discussed in Chapter 18. Besides giving this overview on the techniques and their use in modern radiation oncology practice, the final section addresses the impact of recent developments in radiation oncology on current and future concepts for patient-specific pretreatment QA.

Finally, we want to underline that patient-specific treatment QA is not only good clinical practice, but is also highly recommended and necessary from a legal point of view. In several countries, there are legal aspects based on radiation protection legislations when using ionizing radiation in medical procedures for diagnostic or therapeutic procedures, where QA programs, including quality control measures, claim to assess or verify the administered patient dose or activity. Since the legal framework will certainly vary with country, a more detailed discussion of this argument is beyond the scope of this chapter.

17.1.3 Errors That Can Be Detected and Those Which Cannot—Limitations of TPS and Delivery Units

The errors which could occur during the processes of planning and delivering an IMRT plan can have various impacts on the delivery of the plan to the patient (Steers and Fraass, 2016). These can be divided into those originating from the TPS and those from the transfer to and delivery on the linac.

The main limitation of the TPS is that it provides a model of the patient and plan delivery and is only as good as the input data for the dose calculation algorithms, and therefore there are limits to the modeling capability. These include approximations made in the parameters for MLC transmission, dosimetric leaf separation, and others which are dependent on the individual planning system. Those with fewer parameters generally appear easier to implement, but may have limitations in capability. Those with many parameters have many options for optimal modeling, but it may be difficult to know how to best balance them. The capabilities are also very dependent on the manufacturer guidelines and it is not always obvious whether the user has achieved the optimal model for their given delivery system (Clark et al., 2014).

The limitations of the delivery systems include the fact that the feedback systems have time delays and therefore dose may have been delivered fractionally before an interlock is applied. Errors can potentially occur in any of the parameters required by the plan, such as collimator or gantry rotation limitations that are not well modeled in a TPS, for example, when control points (CPs) are too sparse and not within mechanical limits. However, the most common errors are in the MLCs as the motors can be affected due to wear-and-tear, leading to a leaf traveling slower than expected and therefore this leaf lags behind the other leaves. For example, the tolerance on the control software of C-arm linacs with MLC can result in two possible feedback scenarios: if possible, all the other leaves are slowed down and the dose rate is decreased to compensate for the slower leaf, or an interlock may be activated. Figure 17.1 shows an example of single leaf error simulations in a prostate 5-field IMRT case measured with the PTW 2D array detector system. Deliberate changes of 1, 2, and 5 mm were introduced into a single MLC leaf position, and measurements are compared against the original unperturbed dose distribution using the gamma index with 3%/3 mm criteria. In this example, it is seen that a 5 mm error is detected whereas a more likely error of 2 mm is not detectable using this device and the chosen passing criteria. The likelihood of a 5 mm MLC error in practice is very low as the tolerance on most linac MLC control software is commonly 2 mm.

Some of these errors can be detected by pretreatment QA. For example, issues with a plan which is undeliverable or causes interlocks will be seen during the QA delivery. Also some modeling issues may be identified, although it may not be clear where these are coming from. The pretreatment QA can also identify transfer issues, for example, wrong plan or loss of information, and for this reason it is important that the clinical plan is delivered in the same way and through the same software modality as it will be to the patient, for example, in a QA mode. Serious errors can occur if there is a failure in the transfer process between the TPS and the R&V system. The most significant case where this happened was the New York radiotherapy incident in 2005 (Bogdanich, 2010). In this particular case, an error in saving the IMRT plan to the database meant the MLC CP data

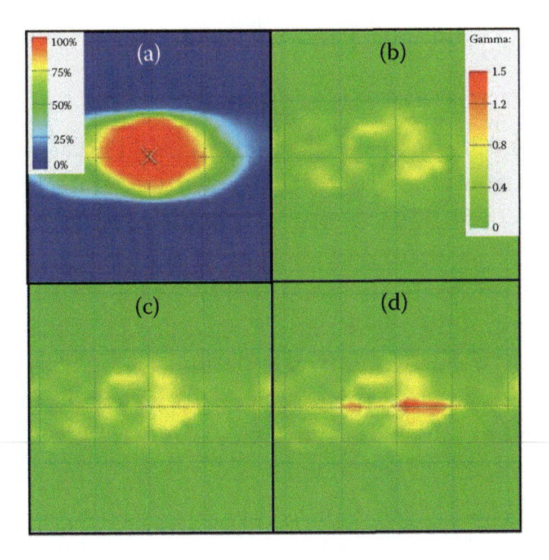

Figure 17.1

Example of the detectability of different deliberately introduced single multileaf collimator leaf position errors showing (a) a prostate IMRT dose distribution and 3%/3 mm gamma index results for deliveries with (b) 1 mm error, (c) 2 mm error, and (d) 5 mm error.

were not passed on to the R&V system. Three fractions were delivered before the error was noticed, resulting in a fatal overdose. Although there was a catalog of contributing factors and missed opportunities, the error itself was ultimately noticed when a QA measurement for the plan was performed after the three fractions were delivered.

Some errors may be seen in the pretreatment QA processes, which do not continue to be present in the patient plan delivery. These could include random variations, for example, in dose rate, which cause changes in MLC speed and manifest themselves as errors seen in the gamma analysis of the plan. Other errors may be made in setup or positioning of the phantom or detector, which would not relate to the patient delivery. These errors can be checked by resetting up the phantom or detector and redelivering the plan.

There are also errors which cannot be detected by pretreatment QA, such as those which may occur randomly and those which are associated with patient setup or anatomy changes as discussed in Chapter 18. Some systematic errors may also be missed due to incorrect QA analysis parameters (Nelms et al., 2011). It is for these reasons that checks should also take place prior to, or during, the delivery to the patient as well as during the pretreatment QA session. Any issues seen in the pretreatment QA should be checked and understood before discussing the clinical implications with the attending physician.

Reports from the Imaging and Radiation Oncology Core (IROC) at the MD Anderson Cancer Center in Houston, TX, revealed that patient-specific IMRT verification by means of dosimetric measurements as currently performed is suboptimal. On the one hand, in-house QA results were not predictive of IROC phantom test results; on the other hand, the various QA devices were found to have differences in their ability to detect dosimetrically acceptable and unacceptable treatment plans (Hussein et al., 2013c; Kry et al., 2014; McKenzie et al., 2014).

17.2 Measurement-Based Techniques

17.2.1 Overview and Examples of What Is Done in Clinical Practice

The complexity of the 3D dose distributions in IMRT treatments requires careful QA. IMRT distributions are characterized by numerous steep dose gradients in order to conform as tightly as possible to the target volume while minimizing the dose to normal tissue. Conventional 3D conformal radiation therapy (3DCRT) treatments are composed of treatment fields with static uniform beam profiles and therefore patient-specific QA consists of simple independent dose and MU verification calculations, which are supplemented by routine machine-specific QA (which includes basic checks such as output constancy, energy, beam flatness, and symmetry). In IMRT, the complex MLC pattern means that an independent dose calculation needs to be multidimensional as well, and a single-point dose calculation is not sufficient. The MLC pattern varies from patient to patient and the number of MUs has a correlation with the complexity of this pattern. Therefore, it is highly recommended, if not necessary, to perform a patient-specific QA measurement to verify the fluence from the IMRT beams at least during the initial phase of implementation of a new treatment technique.

Historically, this has been performed using ion chambers and radiographic/radiochromic films within cubic or semianthropomorphic phantoms. As an example, once an IMRT treatment plan for a patient is complete it is possible in most TPSs to create a verification plan. Essentially, a verification plan is an automatic copy of the same geometry, MLC configuration, and MUs calculated on a computed tomography (CT) scan of the physical phantom to be used for performing the verification measurement. Where this feature is not available, the plan can be manually copied onto the CT data of the phantom, ensuring that the delivery parameters are identical. This verification plan can be used to generate a predicted dose for comparison against a measurement in the same conditions. For ionization chamber measurements, the predicted dose can be typically calculated by creating a contour on the CT dataset that simulates the collecting volume of the chamber; see the example in Figure 17.2. The mean dose to this structure is then recorded. It is often necessary to move the isocenter position relative to the phantom in order to position the ion chamber in regions of the dose distribution that are of interest, for example, in the high-dose PTV region (see the example in Figure 17.3).

It is also necessary to set the position such that the dose across the chamber is homogeneous, and to avoid areas of high-dose gradient as this is sensitive to small setup errors. For comparison against film measurements, a dose plane can be exported from the TPS at the same plane as the film within the phantom. The measured versus predicted plane is commonly compared using metrics such as the gamma index analysis (Low et al., 1998). An example of a gamma index map is shown in Figure 17.4. In this example, the ionization chamber provides an absolute point dose measurement and the film provides a relative 2D measurement of the IMRT fluence. These methods, however, are time consuming, particularly for film measurements, which require extensive resources for calibration, processing, and analysis as well as the costs of single-use films, requiring multiple batches to be purchased, which also involve physical archiving. These resource costs have historically limited the number of patients that could be treated with advanced IMRT.

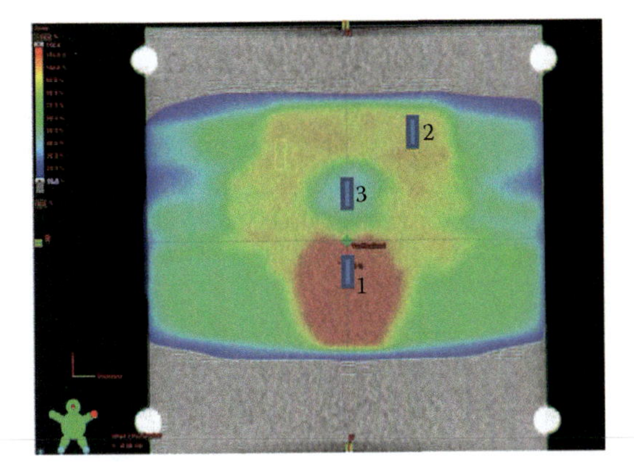

Figure 17.2

Coronal image indicating three regions (shown as blue rectangles) for ionization chamber measuring points in a clinical prostate and nodes IMRT treatment plan. In this example, the rectangles drawn represent the collection volume of a Farmer-type 0.6 cm^3 ionization chamber and are sampling (1) the high-dose PTV, (2) the elective nodal PTV, and (3) a low-dose sparing region.

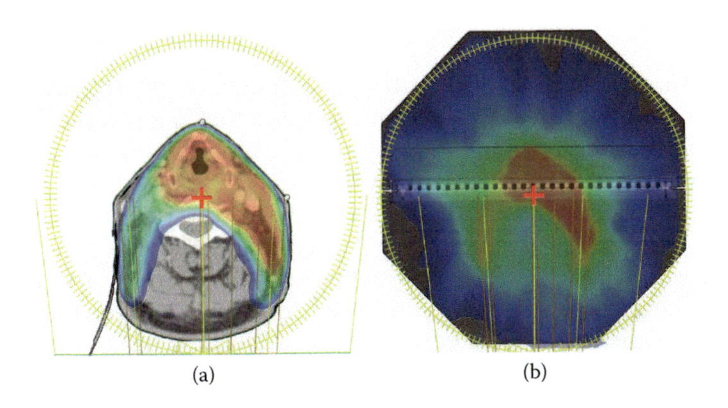

Figure 17.3

Example of (a) in-patient and (b) in-phantom dose distribution. The red cross indicates the location of the isocenter on both images, which has been shifted to move the high-dose region of the dose distribution over the ionization chamber measuring plane.

In recent years, various commercial 2D and 3D ionization chamber or diode detector arrays have become available. These electronic devices have allowed for verification of absolute or relative dose in 2D or 3D with near real-time results. This allows the analysis to be performed online in the IMRT QA measurement session and therefore out of tolerance results can be investigated immediately. Conventional methods such as ionization chamber point dose measurements and film dosimetry are gradually being replaced by detector arrays and are addressed in the following sections and in more detail in Chapter 8.

17.2.2 "Full-3D" versus "Semi-3D" Methods

The only dosimeters available, at the time of writing, that have been able to measure a "full-3D" dose distribution are polymer gel dosimeters and radiochromic

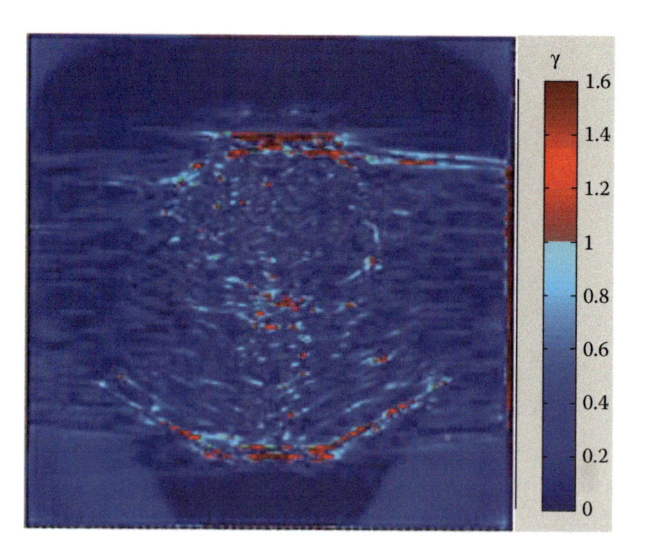

Figure 17.4

Example of a two-dimensional gamma index map.

3D detectors, as discussed in Chapters 5 and 6, respectively. The advantage of these dosimeters is that they are tissue equivalent and can be molded into an anthropomorphic shape. After irradiation, the dosimeter requires scanning using magnetic resonance imaging (MRI), optical CT, or x-ray CT and processing of the measured signal. It has been estimated that the entire process from fabrication to analysis can take up to 45 hours rendering this unsuitable for routine measurement (Baldock et al., 2010). It has the potential to be used as a benchmarking tool for commissioning treatment plans. Currently, this technology has been mainly confined to research institutions and the current processing and analysis timescales have meant there is a limited market. However with research into optimization and more cost-effective scanning techniques, this type of dosimeter may become more available in the future (Baldock et al., 2010). The power of full-3D methods lies certainly in their use for dosimetric end-to-end tests rather than for everyday use. The increasing number of IMRT/VMAT patients and the workload associated with full-3D experimental methods remain conflicting, even if reading of 3D dosimeters is substantially improved.

Several detector systems have been developed which could be referred to as semi-3D. These include arrays which make measurements in two or more plans (either orthogonal or at rotating angles), and then a 3D "measured" dose cube is interpolated from the measure data. These systems allow a 3D gamma calculation to be made between the 3D calculated and 3D "measured" dose. Further details of these systems are given in Section 17.2.4.

17.2.3 Single versus Hybrid Verification Methods

There are two practical approaches to measuring complex dose distributions. The first is to make measurements of individual fields, most commonly at zero gantry angle, i.e., the planar approach. The second approach is to measure the fields in combination as they would be delivered clinically, i.e., the composite approach. Figure 17.5 shows schematically these two approaches. The advantage

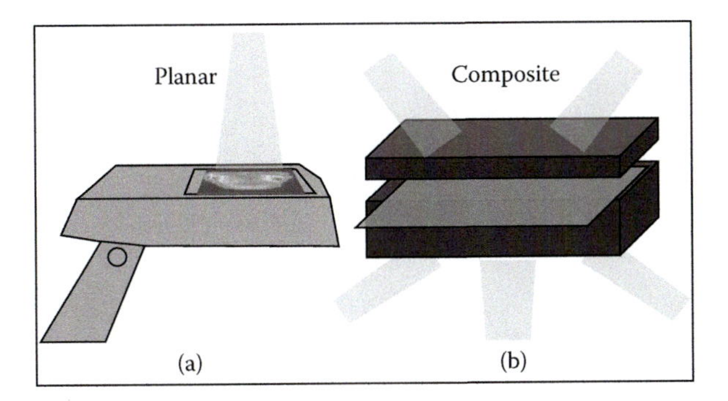

Figure 17.5

Schematic diagram showing two approaches of measurement. (a) The diagram shows the "planar" approach where the gantry angle is orthogonal to the measured plane, i.e., at gantry 0 degrees as typical for electronic portal imaging device measurements. (b) The diagram shows the "composite" approach where the fields are delivered at their planned gantry angle onto a phantom and detector; this example shows film sandwiched in solid water as one such approach.

of the former approach is that if any errors are seen, it is known from which field they came; however, the field is usually delivered differently from the way it is delivered clinically. The latter approach can assess how the fields combine from their clinical gantry angles such that any issues related for instance to the MLC carriage sag can be assessed. Also the total 3D dose distribution is clinically the most relevant quantity. However, if an error is seen, it will not be obvious what the cause was. Either of these approaches can be suitable for static-field IMRT, but for rotational IMRT, i.e., VMAT or tomotherapy, there is generally no simple or meaningful way to collapse the beam angles and therefore the measurements are usually taken using the clinical gantry angles. This also has implications in the way the measurements are taken, i.e., in a single plane or with a planar array which rotates to follow the gantry head and be continuously perpendicular to the beam. Further information on these types of measurements is given in Section 17.2.4.

17.2.4 Brief Overview of Multidimensional Detectors

As an introduction to 2D or 3D arrays, it is a good starting point to consider what would make an ideal detector array. An ideal detector should have a high resolution, ideally comparable to or better than the grid spacing used for the dose calculation in the TPS; be able to measure a true 3D dose distribution; have no angular dependence to beam delivery direction; have linear dose–response, no energy, and dose rate response; be water equivalent; robust in general; perfect short- and long-term reproducibility; be easy to calibrate; and enable real-time measurements. Obviously, all of these criteria cannot be easily achieved in a cost-effective manner. Manufacturers have attempted to fulfill as many of these wish list, but no currently available detector system fulfills all these criteria.

The first generation of commercial detector arrays was designed in a planar configuration using arrays of diodes or vented ionization chambers configured with 5–10 mm detector-to-detector spacing. These devices were originally developed to be able to measure per-beam fluences. In the next development step, adaptations were made in hardware and software to be able to measure composite

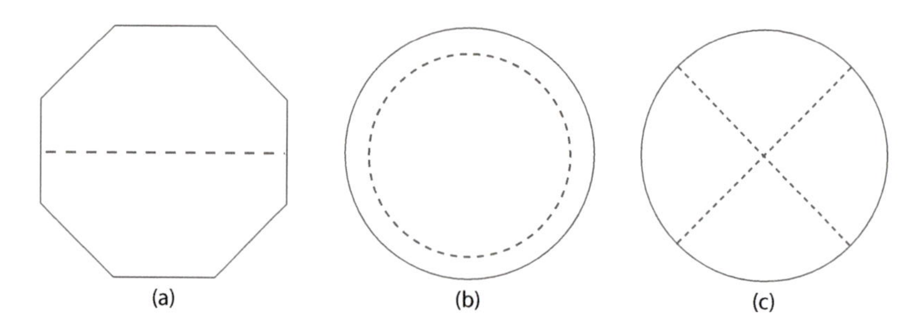

Figure 17.6

Schematic diagram showing different types of detector array configurations. (a) Planar, (b) cylindrical, and (c) cross plane.

dose distributions, with dedicated phantoms that house the planar detector arrays (Van Esch et al., 2007). For VMAT, an inclinometer was added to monitor the gantry angle for each beam delivery (Boggula et al., 2011a). The recent generation of detector arrays and phantoms was specifically designed to perform semi-3D measurements by housing the detectors in a cylindrical phantom in, for example, a cross plane. Figure 17.6 shows schematic diagrams of the main detector array and phantom combination designs. Chapter 8 gives a comprehensive overview of existing 2D and semi-3D dosimetry systems.

On the other hand, electronic portal imaging devices (EPIDs), which had begun being developed in the early 1980s, matured with time and linac manufacturers began offering active matrix flat-panel arrays incorporating amorphous silicon (a-Si) photodiodes (Antonuk, 2002), by 2001. These devices were explored for dosimetry because they offer submillimeter resolution and are currently used routinely for IMRT fluence measurements (Van Esch et al., 2004; van Elmpt et al., 2008). The dosimetric properties of EPIDs are discussed in detail in Chapter 7.

17.2.5 Tolerances and Action Limits

There are various ways of comparing a measured dose distribution from a detector array and the TPS dose distribution. The International Commission on Radiation Units and Measurements (ICRU) recommends that for regions of the dose distribution with low-gradient regions (<20%/cm), the dose difference, normalized to the dose prescription point, should be no more than ±3.5%, and for high-gradient regions (>20%/cm) the dose points should have a distance-to-agreement (DTA) of ≤3.5 mm (ICRU, 2010). However, the most common quantitative technique used is the gamma index analysis which combines dose difference and DTA into a single dimensionless metric (Low et al., 1998). This metric is implemented in all known commercial multidimensional detector systems at the time of writing. A gamma index of less than 1 indicates that the measurement point lies within the dose difference and/or DTA passing criteria. A common acceptance threshold is ≥95% of measured points should pass with a gamma index of <1 for passing criteria of 3% dose difference and 3 mm DTA (IPEM, 2008). However, it is important to stress that this should be taken as a guide only and these acceptance criteria are not universal. For example, it has been shown that for the same passing criteria, different devices and software combinations exhibit varying levels of agreement with each other (Fredh et al 2013; Hussein et al., 2013c, 2017). More information on comparing 2D and 3D dose distributions is provided in Chapter 14.

It is recommended that acceptance limits be chosen that are appropriate for the situation. For example, if performing QA for a stereotactic IMRT delivery, such as in stereotactic ablative radiotherapy (SABR), criteria such as 3%/3 mm may not be appropriate, as in this situation spatial accuracy is critical. In this scenario, it would be more sensible to set a tighter DTA passing criterion such as 1 mm. A good practice is to develop a range of plans with known deliberate changes that can be used to test the appropriateness of a chosen acceptance limit. Various studies have previously been performed to assess the suitability of detector arrays for IMRT and VMAT QA and to evaluate appropriate acceptance criteria (Létourneau et al., 2004, 2009; Spezi and Lewis, 2006; Poppe et al., 2007; Bedford et al., 2009; Li et al., 2009; Yan et al., 2009; Feygelman et al., 2010; Chandaraj et al., 2011; Masi et al., 2011; Nelms et al., 2011; Petoukhova et al., 2011; Van Esch et al., 2011; Zhen et al., 2011; Lang et al., 2012; Myers et al., 2012; Syamkumar et al., 2012; Fredh et al., 2013; Heilemann et al., 2013; Hussein et al., 2013a,b,c; Lin et al., 2013; McGarry et al., 2013; Zhu et al., 2013). Studies have also been made using receiver operator characteristics to establish the sensitivity of acceptance criteria (Fawcett, 2006; Gordon and Siebers, 2013; McKenzie et al., 2014; Bojechko and Ford, 2015).

17.2.6 Fluence Verification with Transmission Detectors and Link to Dose Recalculation Based on Fluence Verification by Measurements

EPIDs have been in use for per-beam IMRT QA measurements since the early 2000s. In recent years, there have been developments in performing 3D *in vivo* dosimetry using these systems (van Elmpt et al., 2008), and it is discussed in detail in Chapters 7 and 18. The general principle is that by measuring the exit fluence through a patient using the EPID, it is possible to back project that fluence and, through convolution with an appropriate energy deposition kernel, perform a 3D dose calculation in the patient CT data. This measurement-based dose distribution can then be compared directly against the TPS dose distribution. Although the pretreatment detector array systems highlight if there has been any error in the transfer of the treatment plan to the delivery system, and for some systems in the dose calculation by the TPS, EPID-based systems additionally allow day-to-day monitoring and can be used to highlight any major changes to the patient anatomy that may have a clinical impact (van Elmpt et al., 2008, 2009; Mijnheer et al., 2013a), as discussed in Chapter 18.

Other systems which measure fluence include those which can be attached directly onto the linac gantry head. These devices can be characterized as transmission devices, in that they are designed so that they can be used during patient treatment with almost negligible attenuation effects. The device measures the actual beam fluence exiting the treatment head and it is possible to compare this measured fluence against the predicted fluence. This process can be automated such that an error could be detected and flagged up almost immediately to the radiation therapist. These systems can therefore be accomplished at detecting transfer errors, such as the New York accident, but not variations in the patient anatomy or setup. In theory, it is also possible to use the measured fluence to calculate a dose distribution in the patient anatomy (Boggula et al., 2011b; Godart et al., 2011). Transmission detectors are just now being clinically introduced for QA purposes and it is an upcoming tool. The systems that are currently commercially available are discussed in Chapter 8. More experience and clinical reports are certainly needed to define and demonstrate their place in the overall QA chain. At present, it is unclear whether it is a tool more related to monitoring the linac and MLC performance, for example, as an alternative to log files, or whether it is patient-specific QA in the sense to include the patient, for example, by dose reconstruction.

17.3 Calculation-Based Techniques

17.3.1 Why and When Doing Dose Verification by Calculations?

Depending on the type of measurement, i.e., just a simple 2D fluence verification or a more burdensome dose verification in a phantom, patient-specific pretreatment QA procedure can be time consuming and workload intensive, both in terms of manpower and machine time. This holds especially for advanced treatment techniques. With the increasing number of patients to be treated with fluence-modulated radiotherapy techniques in a department, measurement-based pre-treatment QA may become a limiting factor, or needs to be incorporated into the routine clinical workflow.

When a treatment technique has been in use for a certain period of time and has become a new "standard treatment technique" for which the personnel feel confident and the utilized technology has proven its reliability during periods with extensive and measurement-based QA, then the practiced QA procedures may be revised; ideally with reduced workload (NCS 2013, 2015). The main advantage of an independent dose calculation method is that it is by far less time consuming than a pretreatment measurement-based procedure, and thus enables streamlining of QA procedures in such a way that the frequency of direct dose measurements is limited or optimized, respectively. Furthermore, such calculation procedures do not require machine time and they can even be automated. These advantages and the previous experience with independent dose calculations cannot be neglected when revisiting the overall QA strategy in a department.

Calculation-based methods and verification of dose and MUs for conventional 3DCRT techniques have been used for a long time as routine QA procedures (Dutreix et al., 1997; van Gasteren et al., 1998; Mijnheer et al., 2001). In 3DCRT, based on uniform intensity beams, independent dose calculation and MU verification are even applied in the clinic during the acceptance testing and commissioning of the TPS besides being part of a patient-specific QA program. Calculations are commonly based on single-point dose verification, mostly by projecting the treatment geometry onto a flat homogeneous semi-infinite water phantom or "slab geometry." Even if the procedure is repeated several times for multiple points, this does not come up to requirements for patient-specific QA for advanced treatment techniques.

In the simplest implementation, independent dose calculation is performed in geometric phantoms, similar to multidosimetric measurements that also are usually not performed in patient geometries but rather in simple geometric phantoms neglecting heterogeneities. However, in principle, independent dose calculations can be performed on CT data of the respective patient. A straightforward option to account for heterogeneity effects, when using geometric phantoms for independent dose calculation, are radiological path length corrections (Georg et al., 2007b).

Besides the workload aspect, calculation-based techniques have an advantage in terms of flexibility. This does not hold only for points-of-interest in the target or organs at risk (OAR), but also in terms of calculation geometry or input data. As far as the latter is concerned, it is not limited just to output data of the TPS; log files from the IMRT delivery unit or measured leaf or fluence patterns determined with transmission ionization chambers or EPIDs could be used as input for independent dose and MU calculation as well. More details on this aspect can be found in Section 17.4.

The clinical use of independent dose or MU calculation procedures will rely on in-house developed or commercial software solutions. Implementing such a system will require commissioning, testing, and definition of acceptance and tolerance levels for decision-making, which in turn requires understanding of the underlying dose calculation algorithm.

17.3.2 Algorithms for Independent Dose Verification

Early dose calculation in radiation oncology was based on single points and purely empirical or factor-based algorithms were employed. Despite the upcoming fluence-based dose calculation models implemented in commercial TPSs in the 3DCRT era, independent dose calculation was still relying on factor-based methods and point dose assessments. These methods reach their limitations for fluence-modulated radiotherapy, where computerized treatment plan optimization is driven by tolerance dose considerations of OARs. In other words, the dose in the OARs is of the same concern as the target dose and often the main limiting factor. Consequently, considering a single-point dose in the target is no longer appropriate and sufficient for patient-specific pretreatment QA, even if the underlying dose calculation is performed in 3D. Requirements and expectations of independent dose calculation procedures for patient-specific pretreatment QA will certainly vary among departments. For the outcome of such an independent calculation, a full 3D dose distribution and comparison of dose–volume histograms (DVHs) is generally desirable, but might be considered "overkill" as a QA procedure. The local demand on the accuracy and the verification process itself, i.e., whether multiple points in the target and OARs are considered as being sufficient or if a DVH-based dose validation is the overall goal, has a big impact on the algorithm and methods used for independent dose calculation in a patient-specific QA program.

Most of the currently existing commercial systems specifically designed for independent dose calculation of advanced treatment techniques, such as IMRT or VMAT, employ an empirical formalism. Input parameters are typically directly measured beam data (e.g., output factors, depth–dose parameters, scatter factors) or quantities derived from measurements (e.g., attenuation parameters). The underlying dose calculation is based on a two-step procedure. First, the fluence or MU map is determined, which is convolved with a scatter kernel in a second step. This corresponds to an advanced Clarkson method for scatter calculations. The dose for each subfield, having a uniform intensity, can then be calculated in the traditional way. Details on the dose calculation and scatter integration can be found in the literature (Kung et al., 2000; Xing et al., 2000; Yang et al., 2003). Linthout et al. (2004) have developed and tested an empirical independent dose calculation procedure for IMRT. Again, the dose contribution of each subfield with a uniform intensity is calculated and added up. Their approach was based on traditional dosimetric parameters such as output factors, depth–dose curves, and off-axis ratios, and utilized for verifying dynamic IMRT delivered with a micro-MLC. An acceptance level of ±5% was proposed when comparing TPS calculations at a single point, with independent calculations in a homogeneous phantom.

To achieve higher accuracy with an independent dose calculation approach, even for the most complex treatment techniques, more sophisticated models than the traditionally used factor-based models need to be used. Furthermore, aspects such as leaf transmission and the effects of rounded leaf ends need to be considered when aiming at dose calculations in 2D or 3D. A general feature and advantage of model-based approaches is that these algorithms are usually more versatile and powerful than empirical models. An effective method for model-based dose calculations is offered by a two-step procedure in which first the energy fluence exiting the treatment head is modeled with a multisource approach, and then in the second step the dose deposition in the patient is handled with energy deposition kernels. This approach utilizes the separation of the radiation sources inside the treatment head of a delivery unit and the patient or the phantom.

Following this rationale, ESTRO Booklet 10 presents a conceptual approach for independent dose verification of IMRT (Karlsson et al., 2010). Moreover, it is even extended toward an effective independent dose calculation with an accuracy similar to that of the primary TPS in order to catch systematic uncertainties of TPSs, and a dose calculation algorithm that requires only a few and easily obtainable input data for model tuning (Olofsson et al., 2003, 2006; Georg, 2007a). A research version of such an independent dose calculation software tool was created and tested that could be configured by a minimum of seven measured output factors in the air and the quality index of the high-energy photon beam (Nyholm et al., 2006). Using these input data, a multisource model (covering effects such as direct radiation from the target, scattered radiation from the flattening filter, the collimator edges, wedges, backscatter to the monitor chamber, and electron contamination) could be tuned, as well as a pencil beam model for which all parameters can be extracted from the quality index. The algorithm, as well as the results of a multi-institutional test, has been described elsewhere (Georg et al., 2007b).

Another published semianalytic approach for independent dose calculation was based on the separation of primary and scatter dose contribution using a two-parameter exponential fit (Baker et al., 2006). The corresponding fit parameters were extracted from measured dosimetric data. This model was extensively tested for prostate IMRT, but results for other IMRT pathologies have not been published so far.

Tissue heterogeneity corrections are often not included in independent dose calculations. In general, the impact of heterogeneity corrections on the accuracy of independent dose calculation accuracy will be biased by beam geometry, treatment techniques (e.g., IMRT with static or rotational gantry), treatment site, and field size and shape. Furthermore, the overall agreement between dose calculation methods in regions with large heterogeneities depends on modeling of photon scatter and lateral electron transport by the different dose calculation algorithms, such as the ones used in the TPS and the independent dose calculation method.

With respect to model-based approaches, the use of a second TPS for independent dose calculation is also an alternative for independent dose verification. The utilization of a second TPS has the advantage that small and systematic uncertainties of the primary TPS can be potentially traced (Sub et al., 2012). Recently, new generation calculation-based QA tools are coming into clinical practice that are based on more advanced dose calculation algorithms, for example, convolution/superposition algorithms, similar to the ones used in TPS. These tools enable efficient and accurate independent dose calculations based on the patient CT data set, allowing DVH verification. Unfortunately, only a few reports have been published on using a second TPS or advanced model-based independent dose calculation software (Anjum et al., 2010; Huang et al., 2013; Fontenot, 2012; Nelson et al., 2014; Clemente-Gutierrez and Perez-Vara, 2015). Nevertheless, the expectations on robustness and time efficiency of such approaches were confirmed.

Monte Carlo (MC)-based dose calculations allow, in principle, the highest accuracy. For the purpose of patient-specific dose verification in 3D, the power and elegance of MC calculations lie in the ability of the source and treatment head model modeling, the accuracy being virtually independent of field shape and delivery technique, and the inclusion of patient-specific imaging information. Although the usefulness of MC calculations as an independent check of dose calculations has been demonstrated in the literature (Leal et al., 2003; Ma et al., 2004; Georg et al., 2007a; Pawlicki et al., 2008), it is still not as widely used as other calculation-based and experimental techniques for patient-specific 3D dose verification. The main reasons are the lack of local MC expertise in a department and the calculation times. Developments in multicore CPU technology will improve MC performance by taking full advantage of parallel computing, which might have an impact on the use of MC tools for patient-specific QA in the future.

17.3.3 Limitations of Calculation Techniques and Implications on Overall QA Strategies

Scientific publications and recommendations that describe the role and accuracy of independent dose or MU calculation software for advanced and fluence-modulated treatment techniques are scarce. Most publications deal with static (step-and-shoot) or dynamic IMRT delivery based on MLCs, and only a few of them with rotational IMRT (VMAT) or other delivery techniques such as robotic linacs. However, the role of calculation-based procedures for patient-specific pretreatment QA is increasing. As long as the individual patient anatomy is not included in verification calculations, the accuracy of independent dose calculation is influenced by treatment site-specific factors, which need to be considered in the analyses or definition of acceptance criteria. Without taking heterogeneities into account acceptance levels of about 2.5%–5% compared to the dose calculations carried out with a TPS and/or measurements have been reported (Stern et al., 2011). If patient-specific anatomic information is not included in verification calculations, even if it is a simple path length correction in the ray line between the source and the point of interest, the accuracy of independent dose checks is influenced by treatment site-specific factors. This is probably one of the most dominant limitations of independent dose calculation procedures that are employed in clinics today. In the thoracic or head-and-neck region, accurate results cannot be achieved in simple dose calculation conditions based on a semi-infinite homogeneous phantom. It has been shown that with radiological depth corrections for head-and-neck treatments, almost as good results as for pelvic treatments could be achieved (Georg et al., 2007b). On the other hand, the impact of radiological depth correction was limited for pelvic treatments. A full 3D verification calculation based on the patient CT data set is largely dependent on the availability of appropriate calculation algorithms in the independent dose calculation software.

When using input data directly from the TPS or the database of the R&V system, independent dose verification does not include a check of the correct file transfer from the TPS to the treatment delivery console, or a verification of the MLC performance for IMRT or VMAT delivery. For this reason, independent dose calculation cannot fully replace experimental methods for fluence-modulated treatment techniques. Furthermore, tools for independent dose and dose calculation cannot replace measurements for commissioning equipment for advanced radiotherapy techniques. Independent dose calculation tools require acceptance testing and commissioning of the software prior to their clinical use. However, when used as a general dose verification tool, without the expectation that small deviations will be detected, independent dose and MU verification can be of great value in a department.

In 3DCRT, the total number of MUs per treatment or beam can be intuitively estimated for a given standard technique, such as a four-field pelvic box. Based on experience, outliers can be detected during plan review. For advanced and fluence-modulated radiotherapy techniques, it is difficult to intuitively estimate whether a given number of MUs makes sense for a given patient and treatment setup.

Patient-specific pretreatment QA relying on independent dose or MU calculations assumes that if the results agree within certain limits, the dose delivery is within those limits. This might not be true due to other sources of errors, such as leaf calibration, patient setup or anatomic variations, or errors related to the underlying CT data used for treatment planning. There are also human errors such as selecting the wrong treatment plan or treatment unit. Even if the independent dose calculation is performed in 3D, the dose might be checked only in one point. In such situations, additional fluence map verification would add complementary information as well as *in vivo* dosimetry, as will be discussed in Chapter 18.

In summary, it is important to underline that an independent dose calculation as the patient-specific QA procedure implies that more stringent machine-specific QA needs to be performed in order to be able to monitor dose delivery influences. As mentioned above, the correct transfer of MLC and gantry-collimator parameters from the TPS to the R&V system or treatment console is not necessarily handled with an independent dose calculation. Furthermore, the patient geometry needs to be verified for high-precision and advanced treatment techniques separately, ideally on a daily basis.

17.4 Impact of Recent Developments in Radiation Oncology on Pretreatment Dose Verification

Until recently flattening filters have been considered as an integral part of the treatment head of a medical accelerator. The increasing use of advanced and precision treatment techniques (e.g., stereotactic radiotherapy, IMRT, and VMAT), where inhomogeneous dose distributions are applied or where varying fluence pattern across the beam are delivered, have resulted in the latest generation linear accelerators that offer a flattening filter-free (FFF) photon mode. The associated higher dose rates and doses per pulse of FFF beams improve treatment delivery efficiency. Delivering fractional dose with dose rates between 10 and 20 Gy per minute obviously raises some questions and concerns in dosimetry related to detector response. As far as patient-specific pretreatment QA is concerned, using arrays of ionization chambers, semiconductor detectors, or film, it has been shown that correction factors do not need to be applied (Lang et al., 2012).

FFF beams are one way of decreasing treatment time due to the higher dose rate, although this becomes a "measurable" advantage only at higher doses per fraction, i.e., above around 5 Gy per fraction. VMAT, on the other hand, exploits the various mechanical (leaf motion, gantry motion, collimator motion) and dosimetric degrees of freedom (e.g., dose rate) of modern linear accelerators and achieves an unprecedented efficiency in beam delivery efficiency. The precise interplay and coordination of each parameter are crucial for correct dose delivery to the patient. Therefore, when determining a deviation between measured and predicted dose in a patient-specific pretreatment QA procedure, the reasons can be manifold. In this context, obtaining the dynamic characteristics of the linac and understanding the linac's behavior is becoming increasingly important. For example, a TPS handles VMAT by approximating the dynamic delivery with multiple CPs, each defined by a specified number of MUs with a stated aperture shape at a fixed gantry angle. The transition from one CP to the next one is assumed to be linear. Inconsistency of the motion of MLC and gantry may cause dosimetric discrepancies that need to be quantified. Therefore, in addition to conventional pretreatment patient-specific QA, there is a need for comprehensive proactive machine-specific QA, for example, by using MLC log files of the current digital linear accelerators (Agnew et al., 2012, 2014; Sun et al., 2012; Calvo-Ortega et al., 2014; Kerns et al., 2014; Childress et al., 2015; Pasler et al., 2015). The utilization of machine log files for patient-specific VMAT QA and reconstructing the dose in the patient anatomy is increasing (e.g., Qian et al., 2010; Clemente-Gutierrez and Perez-Vara, 2015). One such commercial system is available from Mobius Medical Systems (www.mobiusmed.com) called MobiusFx, which utilizes linac log files to reconstruct a 3D dose distribution on the patient's CT scan using the company's own developed collapsed-cone convolution algorithm. Despite the fact that still more research is required to fully explore this technique and to be able to gather wide experience, it is expected that a log file-based approach will become an important patient-specific QA procedure in the near future.

Adaptive radiotherapy (ART) is an upcoming treatment concept that has as a basic assumption that fractionated radiotherapy should no longer solely be based on the anatomic information acquired before treatment, but aims to adapt the treatment and treatment plan, respectively, according to anatomic variations during the course of treatment. ART concepts range from online quasi real-time adaptation, to offline adaptations or predefined patient-specific treatment plan libraries. The currently clinically explored ART techniques are plan library or offline based (Martinez et al., 2001; Castadot et al., 2010). However, in any case, the increased number of treatment plans per patient means an increased QA workload. In this context, more proactive than currently executed reactive QA procedures become of utmost importance.

Biologically motivated ART (BioART) or dose painting aims to account for subtumor volume variations that are linked to tumor biology, for example, hypoxic subvolumes, in addition to geometric/anatomic variations. Advances in hybrid and functional imaging support the definition of such tumor tissue characteristics (e.g., Gregoire et al., 2012). Prescribing and delivering inhomogeneous dose distributions to tumor subvolumes with fluence-modulated delivery techniques implies the intensive use of small fields. Dose calculation and experimental dosimetry in small fields are known to be challenging, and may need additional correction factors

(Azangwe et al., 2014). Consequently, results of patient-specific pretreatment QA when performing BioART may need adapted tolerance levels. For more information on small-field dosimetry, the interested reader is referred to Chapter 9.

The accurate delivery of dynamic and fluence-modulated treatments put high demands on dose calculation, dosimetry, beam delivery, and consequently on patient-specific QA. Measuring the total dose provides little information on how the dose is delivered in terms of temporal aspects. Time-resolved dosimetry that acquires data on the same temporal scale as gun pulse frequency is currently being explored, using detectors such as scintillation detectors (Beierholm et al., 2011), but is still in the research stage. Finally, including time as a parameter in dosimetry methods for dosimetric comparison, such as the gamma index, needs to be refined (Podesta et al., 2014).

MR-guided radiotherapy is an upcoming image-guided radiation therapy (IGRT) technique with its inherent dosimetric challenges for 3D dose verification. These are linked to dosimetry in the presence of magnetic fields and the options for 4D ART enabled by high-contrast soft tissue imaging in almost real time. Dosimetry issues in the presence of magnetic fields are explored and first experience with patient-specific QA is just made. More details on this dynamic field can be found in Chapter 26, which are beyond the scope of this chapter.

17.5 Summary

The complexity of planning and delivery of IMRT and VMAT means that careful verification of plans is needed prior to delivery to the patient. Traditionally, this has been undertaken by making measurements in a phantom, but more recently there has been more capability available for independent 3D calculations. The impact of these recent developments, which may streamline patient-specific QA, means that greater numbers of patients may be treated as QA is less of a block in the process, and adaptive and other advanced approaches to radiotherapy techniques can be more easily carried out.

References

Agnew CE et al. (2012) Implementation of phantom-less IMRT delivery verification using Varian DynaLog files and R/V output. *Phys. Med. Biol.* 57: 6761–6777.

Agnew CE, Irvine DM and McGarry CK (2014) Correlation of phantom-based and log file patient-specific QA with complexity scores for VMAT. *J. Appl. Clin. Med. Phys.* 15(6): 204–216.

Anjum MN et al. (2010). IMRT quality assurance using a second treatment planning system. *Med. Dosim.* 35: 274–279.

Antonuk LE (2002) Electronic portal imaging devices: A review and historical perspective of contemporary technologies and research. *Phys. Med. Biol.* 47: R31–R65.

Azangwe G et al. (2014) Detector to detector corrections: A comprehensive experimental study of detector specific correction factors for beam output measurements for small radiotherapy beams. *Med. Phys.* 41: 072103.

Baker CR et al. (2006) A separated primary and scatter model for independent dose calculation of intensity modulated radiotherapy. *Radiother. Oncol.* 80: 385–390.

Baldock C et al. (2010) Polymer gel dosimetry. *Phys. Med. Biol.* 55: R1–R63.
Bedford JL et al. (2009) Evaluation of the Delta4 phantom for IMRT and VMAT verification. *Phys. Med. Biol.* 54: N167–N176.
Beierholm AR et al. (2011) Characterizing a pulse-resolved dosimetry system for complex radiotherapy using organic scintillators. *Phys. Med. Biol.* 56: 3033–3045.
Bentzen SM and Gregoire V (2011) Molecular imaging-based dose painting: A novel paradigm for radiation therapy prescription. *Semin. Radiat. Oncol.* 21: 101–110.
Bogdanich BW (2010) Radiation offers new cures, and ways to do harm. *New York Times*, January 23: 1–16.
Boggula R et al. (2011a) Evaluation of a 2D detector array for patient-specific VMAT QA with different setups. *Phys. Med. Biol.* 56: 7163–7177.
Boggula R et al. (2011b) Patient-specific 3D pretreatment and potential 3D online dose verification of Monte Carlo-calculated IMRT prostate treatment plans. *Int. J. Radiat. Oncol. Biol. Phys.* 81: 1168–1175.
Bojechko C and Ford EC (2015) Quantifying the performance of in vivo portal dosimetry in detecting four types of treatment parameter variations. *Med. Phys.* 42: 6912–6918.
Calvo-Ortega JF et al. (2014) A Varian DynaLog file-based procedure for patient dose-volume histogram–based IMRT QA. *J. Appl. Clin. Med. Phys.* 15(2): 100–109.
Castadot P et al. (2010) Adaptive radiotherapy of head and neck cancer. *Semin. Radiat. Oncol.* 20: 84–93.
Chandaraj V et al. (2011) Comparison of four commercial devices for RapidArc and sliding window IMRT QA. *J. Appl. Clin. Med. Phys.* 12(2): 338–349.
Childress N, Chen Q and Rong Y (2015) Parallel/opposed: IMRT QA using treatment log files is superior to conventional measurement-based method. *J. Appl. Clin. Med. Phys.* 15(1): 4–7.
Clark CH et al. (2014) A multi-institutional dosimetry audit of rotational intensity-modulated radiotherapy. *Radiother. Oncol.* 113: 272–278.
Clemente-Gutierrez F and Perez-Vara C (2015) Dosimetric validation and clinical implementation of two 3D dose verification systems for quality assurance in volumetric-modulated arc therapy techniques. *J. Appl. Clin. Med. Phys.* 16(2): 198–217.
Dutreix A et al. (1997) *Monitor Unit Calculation for High Energy Photon Beams.* ESTRO Booklet No. 3. (Brussels, Belgium: European Society for Radiotherapy and Oncology).
Ezzell GA et al. (2003) Guidance document on delivery, treatment planning, and clinical implementation of IMRT: Report of the IMRT Subcommittee of the AAPM Radiation Therapy Committee. *Med. Phys.* 30: 2089–2115.
Fawcett T (2006) An introduction to ROC analysis. *Pattern Recognit. Lett.* 27: 861–874.
Feygelman V et al. (2010) Evaluation of a 3D diode array dosimeter for helical tomotherapy delivery QA. *Med. Dosim.* 35: 324–329.
Fontenot JD (2012) Feasibility of a remote, automated daily delivery verification of volumetric-modulated arc therapy treatments using a commercial record and verify system. *J. Appl. Clin. Med. Phys.* 13(2): 113–123.
Fraass BA et al. (1995) A computer-controlled conformal radiotherapy system. I: Overview. *Int. J. Radiat. Oncol. Biol. Phys.* 33: 1139–1157.

Fredh A et al. (2013) Patient QA systems for rotational radiation therapy: A comparative experimental study with intentional errors. *Med. Phys.* 40: 31716.

Georg D et al. (2004) On empirical methods to determine scatter factors for irregular MLC shaped beams. *Med. Phys.* 31: 2222–2229.

Georg D et al. (2007a) Patient-specific IMRT verification using independent fluence-based dose calculation software: Experimental benchmarking and initial clinical experience. *Phys. Med. Biol.* 52: 4981–4992.

Georg D et al. (2007b) Clinical evaluation of monitor unit software and the application of action levels. *Radiother. Oncol.* 85: 306–315.

Georg D, Knöös T and McClean B (2011) Current status and future perspective of flattening filter free photon beams. *Med. Phys.* 38: 1280–1293.

Godart J et al. (2011) Reconstruction of high-resolution 3D dose from matrix measurements: Error detection capability of the COMPASS correction kernel method. *Phys. Med. Biol.* 56: 5029–5043.

Gordon J and Siebers J (2013) Addressing a gap in current IMRT quality assurance. *Int. J. Radiat. Oncol. Biol. Phys.* 87: 20–21.

Gregoire V et al. (2012) Radiotherapy for head and neck tumours in 2012 and beyond: Conformal, tailored, and adaptive? *Lancet Oncol.* 13: e292–e300.

Heilemann G, Poppe B and Laub W (2013) On the sensitivity of common gamma-index evaluation methods to MLC misalignments in Rapidarc quality assurance. *Med. Phys.* 40: 31702.

Huang JY et al. (2013) Investigation of various energy deposition kernel refinements for the convolution/superposition method. *Med. Phys.* 40: 121721.

Hussein M et al. (2013a) A methodology for dosimetry audit of rotational radiotherapy using a commercial detector array. *Radiother. Oncol.* 108: 78–85.

Hussein M et al. (2013b) A critical evaluation of the PTW 2D-ARRAY seven29 and Octavius II phantom for IMRT and VMAT verification. *J. Appl. Clin. Med. Phys.* 14(6): 274–292.

Hussein M et al. (2013c) A comparison of the gamma index analysis in various commercial IMRT/VMAT QA systems. *Radiother. Oncol.* 109: 370–376.

Hussein M, Clark CH and Nisbet A (2017) Challenges in calculation of the gamma index in radiotherapy—Towards good practice. *Phys. Med.* 36: 1–11.

ICRU (2010) Prescribing, recording, and reporting intensity-modulated photon-beam therapy (IMRT). ICRU Report No.83. *J ICRU* 10. (Bethesda, MD: International Commission on Radiation Units and Measurements).

IPEM (2008) Guidance for the clinical implementation of intensity modulated radiation therapy. IMRT Working Party Report 96. (York, England: Institute of Physics and Engineering in Medicine).

Karlsson M et al. (2010) *Dose Verification in Advanced Radiotherapy—Concepts and Models for Independent Dose Calculation.* ESTRO Booklet No. 10. (Brussels, Belgium: European Society for Radiotherapy and Oncology).

Kerns JR, Childress N and Kry SF (2014) A multi-institution evaluation of MLC log files and performance in IMRT delivery. *Radiat. Oncol.* 9: 176–186.

Kry SF et al. (2014) Institutional patient-specific IMRT QA does not predict unacceptable plan delivery. *Int. J. Radiat. Oncol. Biol. Phys.* 90: 1195–1201.

Kung JH, Chen GT and Kuchnir FK (2000) A monitor unit verification calculation in intensity modulated radiotherapy as a dosimetry quality assurance. *Med. Phys.* 27: 2226–2230.

Lang S et al. (2012) Pretreatment quality assurance of flattening filter free beams on 224 patients for intensity modulated plans: A multicentric study. *Med. Phys.* 39: 1351–1356.

Létourneau D et al. (2004) Evaluation of a 2D diode array for IMRT quality assurance. *Radiother. Oncol.* 70: 199–206.

Létourneau D et al. (2009) Novel dosimetric phantom for quality assurance of volumetric modulated arc therapy. *Med. Phys.* 36: 1813–1821.

Li JG, Yan G and Liu C (2009) Comparison of two commercial detector arrays for IMRT quality assurance. *J. Appl. Clin. Med. Phys.* 10(2): 62–74.

Lin MH et al. (2013) Measurement comparison and Monte Carlo analysis for volumetric-modulated arc therapy (VMAT) delivery verification using the ArcCHECK dosimetry system. *J. Appl. Clin. Med. Phys.* 14(2): 220–233.

Linthout N et al. (2004) A simple theoretical verification of monitor unit calculation for intensity modulated beams using dynamic mini-multileaf collimation. *Radiother. Oncol.* 71: 235–241.

Low DA et al. (1998) A technique for the quantitative evaluation of dose distributions. *Med. Phys.* 25: 656–661.

Ma C et al. (2004) Monitor unit calculation for Monte Carlo treatment planning. *Phys. Med. Biol.* 49: 1671–1687.

Martinez AA et al. (2001) Improvement in dose escalation using the process of adaptive radiotherapy combined with three-dimensional conformal or intensity-modulated beams for prostate cancer. *Int. J. Radiat. Oncol. Biol. Phys.* 50: 1226–1234.

Masi L et al. (2011) Quality assurance of volumetric modulated arc therapy: Evaluation and comparison of different dosimetric systems. *Med. Phys.* 38: 612–621.

McGarry CK et al. (2013) Octavius 4D characterization for flattened and flattening filter free rotational deliveries. *Med. Phys.* 40: 091707.

McKenzie EM et al. (2014) Toward optimizing patient-specific IMRT QA techniques in the accurate detection of dosimetrically acceptable and unacceptable patient plans. *Med. Phys.* 41: 121702.

Mijnheer B et al. (2001) *Monitor Unit Calculation for High Energy Photon Beams—Practical Examples.* ESTRO Booklet No. 6. (Brussels, Belgium: European Society for Radiotherapy and Oncology).

Mijnheer B and Georg D (Eds.) (2008) *Guidelines for the Verification of IMRT, European Guidelines for Quality Assurance in Radiotherapy.* ESTRO Booklet No. 9. (Brussels, Belgium: European Society for Radiotherapy and Oncology).

Mijnheer B et al. (2013a) 3D EPID-based in vivo dosimetry for IMRT and VMAT. *J. Phys. Conf. Ser.* 444: 12011.

Mijnheer B Beddar S, Izewska J and Reft C (2013b) In vivo dosimetry in external beam radiotherapy. *Med. Phys.* 40: 070903.

Myers P et al. (2012) Evaluation of PTW Seven29 for tomotherapy patient-specific quality assurance and comparison with ScandiDos Delta 4. *J. Med. Phys.* 37: 72–80.

NCS (2013) Code of practice for the quality assurance and control for intensity modulated radiotherapy. NCS Report 22. (Delft, the Netherlands: Netherlands Commission on Radiation Dosimetry) http://radiationdosimetry.org.

NCS (2015) Code of practice for the quality assurance and control for volumetric modulated arc therapy. NCS Report 24 (Delft, the Netherlands: Netherlands Commission on Radiation Dosimetry) http://radiationdosimetry.org.

Nelms BE, Zhen H and Tomé W (2011) Per-beam, planar IMRT QA passing rates do not predict clinically relevant patient dose errors. *Med. Phys.* 38: 1037–1044.
Nelson CL et al. (2014) Commissioning results of an automated treatment planning verification system. *J. Appl. Clin. Med. Phys.* 15(5): 57–65.
Noel CE et al. (2014) Process-based quality management for clinical implementation of adaptive radiotherapy. *Med. Phys.* 41: 081717.
Nyholm T et al. (2006) Pencil kernel correction and residual error estimation for quality-index-based dose calculations. *Phys. Med. Biol.* 51: 6245–6262.
Olofsson J, Georg D and Karlsson M. (2003) A widely tested model for head scatter influcence on photon beam output. *Radiother. Oncol.* 67: 225–238.
Olofsson J et al. (2006) Evaluation of uncertainty predictions and dose output for model-based dose calculations for megavoltage photon beams. *Med. Phys.* 33: 2548–2556.
Pasler M et al. (2015) Linking log files with dosimetric accuracy—A multi-institutional study on quality assurance of volumetric modulated arc therapy. *Radiother. Oncol.* 117: 407–411.
Pawlicki T et al. (2008) Moving from IMRT QA measurements toward independent computer calculations using control charts. *Radiother. Oncol.* 89: 330–337.
Petoukhova AL et al. (2011) The ArcCHECK diode array for dosimetric verification of HybridArc. *Phys. Med. Biol.* 56: 5411–5428.
Podesta M, Persoon LC and Verhaegen F (2014) A novel time dependent gamma evaluation function for dynamic 2D and 3D dose distributions. *Phys. Med. Biol.* 59: 5973–5985.
Poppe B et al. (2007) Spatial resolution of 2D ionization chamber arrays for IMRT dose verification: Single-detector size and sampling step width. *Phys. Med. Biol.* 52: 2921–2935.
Qian J et al. (2010) Dose reconstruction for volumetric modulated arc therapy (VMAT) using cone-beam CT and dynamic log files. *Phys. Med. Biol.* 55: 3597–3610.
Spezi E and Lewis DG (2006) Gamma histograms for radiotherapy plan evaluation. *Radiother. Oncol.* 79: 224–230.
Steers J and Fraass B (2016) IMRT QA: Selecting gamma criteria based on error detection sensitivity. *Med. Phys.* 43: 1982–1994.
Stern RL et al. (2011) Verification of monitor unit calculations for non-IMRT clinical radiotherapy: Report of AAPM Task Group 114. *Med. Phys.* 38: 504–530.
Sun B et al. (2012) Evaluation of the efficiency and effectiveness of independent dose calculation followed by machine log file analysis against conventional measurement based IMRT QA. *J. Appl. Clin. Med. Phys.* 13(5): 140–154.
Syamkumar SA et al. (2012) Characterization of responses of 2d array seven29 detector and its combined use with octavius phantom for the patient-specific quality assurance in rapidarc treatment delivery. *Med. Dosim.* 37: 53–60.
Teoh M et al. (2011) Volumetric modulated arc therapy: A review of current literature and clinical use in practice. *Brit. J. Radiol.* 84: 967–996.
Thorwarth D and Alber M (2010) Implementation of hypoxia imaging into treatment planning and delivery. *Radiother. Oncol.* 97: 172–175.
van Elmpt W et al. (2008) A literature review of electronic portal imaging for radiotherapy dosimetry. *Radiother. Oncol.* 88: 289–309.
van Elmpt W et al. (2009) 3D in vivo dosimetry using megavoltage cone-beam CT and EPID dosimetry. *Int. J. Radiat. Oncol. Biol. Phys.* 73: 1580–1587.

Van Esch A, Depuydt T and Huyskens DP (2004) The use of an aSi-based EPID for routine absolute dosimetric pre-treatment verification of dynamic IMRT fields. *Radiother. Oncol.* 71: 223–234.

Van Esch A et al. (2007) On-line quality assurance of rotational radiotherapy treatment delivery by means of a 2D ion chamber array and the Octavius phantom. *Med. Phys.* 34: 3825–3837.

Van Esch A et al. (2011) Implementing RapidArc into clinical routine: A comprehensive program from machine QA to TPS validation and patient QA. *Med. Phys.* 38: 146–166.

van Gasteren JJM et al. (1998) Determination and use of scatter correction factors of megavoltage photon beams, NCS Report 12 (Delft, the Netherlands: Netherlands Commission on Radiation Dosimetry) http://radiationdosimetry.org.

Xing L et al. (2000) Monitor unit calculation for an intensity modulated photon field by a simple scatter-summation algorithm. *Phys. Med. Biol.* 45: N1–N7.

Yan G et al. (2009) On the sensitivity of patient-specific IMRT QA to MLC positioning errors. *J. Appl. Clin. Med. Phys.* 10(1): 120–128.

Yang Y et al. (2003) Independent dosimetric calculation with inclusion of head scatter and MLC transmission for IMRT. *Med. Phys.* 30: 2937–2947.

Zhen H, Nelms BE and Tome W (2011) Moving from gamma passing rates to patient DVH-based QA metrics in pretreatment dose QA. *Med. Phys.* 38: 5477–5489.

Zhu LJ et al. (2013) A comparison of VMAT dosimetric verifications between fixed and rotating gantry positions. *Phys. Med. Biol.* 58: 1315–1322.

18

Patient-Specific Quality Assurance

In Vivo *3D Dose Verification*

Ben Mijnheer

18.1 Introduction

18.1.1 *In vivo* Dosimetry: Terminology

In vivo is Latin for "within the living" and denotes the use of a whole, living organism for specific purposes. Consequently, many different types of *in vivo* measurements are possible in radiation oncology. *In vivo* dosimetry in radiotherapy means the measurement of the radiation dose "within the living," i.e., the dose received by a living object during irradiation, as opposed to *ex vivo* or *in vitro* dose measurements in a phantom simulating that object. As discussed in Section 18.2.1, point detectors are often used to determine the entrance and exit *in vivo* dose, whereas electronic portal imaging devices (EPIDs) are frequently used to determine the transit (transmission) *in vivo* dose. These transit *in vivo* measurements using EPIDs can be analyzed by making a comparison at the EPID level or in the patient, as discussed in Chapter 7. Although the first approach is also based on an *in vivo* measurement and may, for instance, provide useful information about anatomical changes in a patient, this approach does not provide quantitative information about the dose "within the living." According to the nomenclature introduced earlier, this approach will therefore not be considered as *in vivo* dosimetry. Only measurements during patient treatment providing quantitative information about the actual dose received by the patient can according to this definition be considered as *in vivo* patient dosimetry and is the topic of this chapter.

Recently, a number of transmission chamber devices that have to be attached to the head of the treatment machine became commercially available, as discussed in Chapter 8. These devices measure the photon fluence exiting the linear accelerator (linac) head. They are designed to detect errors in the delivery of these photon fluences when using them either before treatment or during treatment. Those measurements, however, do not provide information about the patient-specific contribution to the dose distribution, i.e., cannot be used for *in vivo* dosimetry, and are therefore not further discussed in this chapter. It would, however, be interesting to assess their role in relation to EPID-based *in vivo* transit dosimetry systems.

18.1.2 Why Should We Perform *In Vivo* Patient Dosimetry?

Patient-specific quality assurance (QA) is generally considered to be a pretreatment verification method using either a measurement-based or a calculation-based technique as elucidated in Chapter 17. These techniques are able to detect specific types of error, but also have their limitations as discussed in Chapters 8, 14, 17 and 19 of this book. In some recent publications, it is even suggested that the usefulness of pretreatment verification may be questionable in detecting clinically unacceptable plan errors (Ford et al., 2012; Kry et al., 2014; McKenzie et al., 2014; Bojechko et al., 2015). For instance, in the study of Bojechko et al. (2015), it was shown that the effectiveness of EPID dosimetry in detecting incidents during pretreatment verification was only 6% whereas the *in vivo* first-fraction EPID dosimetry detectability of these incidents was 74%, and an additional 20% was discovered if all fractions were measured. There are many reasons for this important remark made by these authors, including the use of inappropriate analysis methodologies such as lax gamma metrics or gamma criteria, dosimeter deficiencies, and shortcomings in the determination of the three-dimensional (3D) dose calculation. The latter type of error can often not be tracked when performing pretreatment verification measurements, because the dimensions and composition of the phantom in which the measuring device is positioned deviate considerably from the geometry and atomic composition of a patient. This is illustrated in Figure 18.1, which shows the dose distribution of a double-arc lung volumetric-modulated arc therapy (VMAT) treatment of a patient either calculated using the patient planning computed-tomography (CT) data or recalculated on a flat polystyrene phantom for the same treatment parameters. Obviously, the shape and position of the isodose lines are different. Consequently, the relationship between pretreatment homogeneous phantom measurements and the dose distribution in a more realistic patient geometry with tissue heterogeneities is not unambiguous. However, probably the most important reason that the actual dose delivered to a patient differs from the planned dose is because the anatomy used for treatment planning deviates from the anatomy during treatment. Anatomy changes may not only occur in the time period between the planning CT scan and the first treatment but also throughout the treatment course. Also setup errors or deviations from the intended clinical procedure can lead to a dose delivery deviating substantially from the planned dose distribution. *In vivo* dosimetry can therefore be considered as a very suitable method to verify on a day-to-day basis the actual dose delivered to a patient during a radiotherapy treatment course. It is able to detect clinically relevant differences between planned and delivered dose, provides a record of the dose received by an

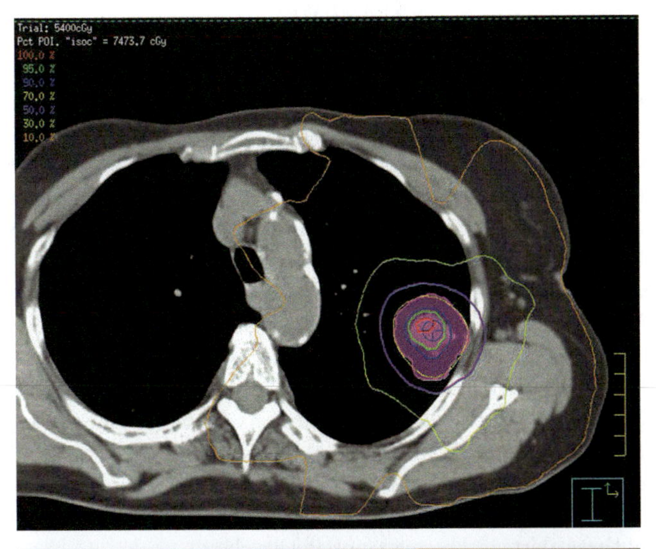

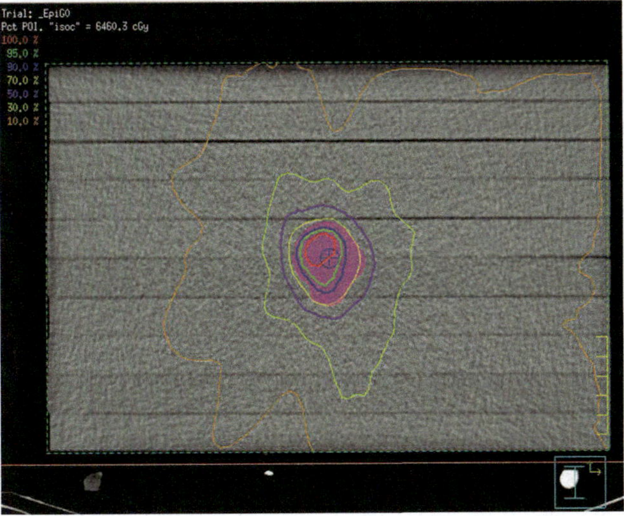

Figure 18.1

Calculated dose distributions for the same double-arc lung volumetric-modulated arc therapy treatment for the patient (top) and phantom (bottom) geometry. Shown are the isodose lines in an axial plane through the isocenter for both situations. The pink area in both figures indicates the planning target volume (PTV).

individual patient, and fulfills legal requirements in some countries. It may replace labor-intensive pretreatment verification measurements, and last but not least, *in vivo* dose verification is able to discover serious errors that otherwise would have remained undetected, as has been reported for *in vivo* dose verification of 3D conformal radiotherapy (Nijsten et al., 2007; Fidanzio et al., 2015) and for intensity-modulated radiotherapy (IMRT) and VMAT delivery (Mans et al., 2010; Mijnheer et al., 2015). A more extensive discussion of the question "Why *in vivo* patient dosimetry?" can be found in several recent publications (IAEA, 2013; Mijnheer, 2013; Mijnheer et al., 2013).

18.1.3 Relationship between *In Vivo* Patient Dosimetry and Other Types of Patient-Specific Quality Assurance

Patient-specific QA in a broader sense includes QA of all steps involved in the treatment process of an individual patient, including physician and physicist plan review, therapist chart review, and patient setup verification. In Chapters 17 and 18, we restrict ourselves to the verification of the correct delivery (i.e., as planned) of the 3D dose distribution to an individual patient with the purpose to distinguish an acceptable plan from an unacceptable plan. In order to perform such a measurement or calculation, the basic question "What is an (un)acceptable plan?" has to be answered. Despite the vast amount of literature concerning patient-specific QA, not many publications or reports are dealing with this question. Differences between actual and intended dose distributions are quantified in many ways using a variety of metrics as discussed in Sections 18.3 and 18.4 and in more detail in Chapter 14. Alerts based on these metrics generally concern dosimetric differences that can be determined with sufficient accuracy, using specific equipment under particular experimental conditions. The clinical relevance of these alerts is, however, not always clear.

Errors that can and cannot be detected by pretreatment verification have been discussed in a more general way in Chapter 17. The examples shown later in this chapter indicate that most deviations between *in vivo* dose measurements and planned dose distributions are due to anatomy changes, variation in patient setup, and the use of a deviating procedure during planning and delivery, for example, the use of wrong CT numbers or lack of bolus material. Some of the clinically relevant errors detected using *in vivo* dosimetry are related to the introduction of a new treatment technique and should have been picked up already at an earlier stage, for example, during the commissioning and testing of that new technique before the first patient was treated. In fact, patient-specific pretreatment verification or *in vivo* dosimetry should never be a substitute for an incomplete commissioning process of new equipment or of a new treatment technique. However, radiotherapy is a dynamic process and sometimes new developments are introduced in the clinic without fully realizing all consequences. In those cases, *in vivo* dosimetry as an end-to-end test provides a safety net to catch errors that may occur in previously untested situations.

When introducing an EPID-based *in vivo* dosimetry program for IMRT and VMAT verification, it is recommended to do this in parallel with the existing pretreatment verification program. With growing confidence in the *in vivo* dosimetry approach, pretreatment verification can then be discontinued, particularly for IMRT and VMAT treatments using relatively simple fields, i.e., not having very high modulation (e.g., see NCS, 2013, 2015). EPID-based *in vivo* transit dosimetry is now the primary dose verification tool at several centers and has replaced nearly all pretreatment dose verification of IMRT and VMAT treatments (Hanson et al., 2014; Fidanzio et al., 2015; Mijnheer et al., 2015).

Finally, it has to be emphasized that patient-specific verification, either pretreatment or *in vivo*, will only enhance patient safety if it is part of a comprehensive quality management program in a radiotherapy department, including a thorough accelerator QA program.

18.1.4 Dimensionality of *In Vivo* Dose Verification

In vivo entrance dose verification measurements, using the point detectors discussed in Chapter 4, have been, and still are, performed in many institutions. In this book, applications of entrance dose measurements to obtain information at specific points of the 3D dose distribution are presented, for example, to assess skin dose (Chapter 4) or the dose outside the treatment volume (Chapter 21). By implementing such an entrance dose verification program, both serious errors and errors of a few percent can be detected as has been discussed, for instance, in the International Atomic Energy Agency (IAEA) Human Health Report No. 8 (IAEA, 2013). Major deviations mentioned in that report concerned, for instance, treatments when a wedge was not inserted into the beam, a source-to-surface distance (SSD) setup was wrongly handled as an isocentric treatment, an incorrect fractionation scheme was used, and wrong treatment data were manually entered into the record-and-verify system rather than being transferred electronically from planning to delivery. Point detectors can be used to discover these types of errors during conventional radiotherapy; a two-dimensional (2D) or 3D dose verification approach is not required, although similar types of errors can also be discovered when using a 2D detector, as shown, for instance, by Fidanzio et al. (2015) and Piermattei et al. (2015). These authors used the transit EPID signal along the central beam axis, obtained during a patient irradiation, to assess the dose at the isocenter in that patient.

In Chapter 2, the dimensionality of dose measurements is discussed with the emphasis on accuracy and precision. This chapter focuses on the 3D aspects of *in vivo* dosimetry because in this way it will be possible to discover errors related to shortcomings in the 3D planning or delivery of patient treatments, particularly when verifying advanced treatment techniques. It should be noted that *in vivo* patient dosimetry is also discussed in other chapters in the book: with point detectors (Chapter 4), during brachytherapy (Chapter 20), and treatments with other modalities orthovoltage x-rays (Chapters 24 and 25), and protons and carbon ions (Chapters 12 and 23).

18.2 Measurement Techniques for *In Vivo* Patient Dosimetry

18.2.1 Instrumentation: Point Detectors, Two-Dimensional Detectors, Electronic Portal Imaging Devices

In several chapters of this book, the characteristics and use of various detectors for 3D dosimetry have been described. In Chapter 4, a large number of point detectors, and their characteristics that make them suitable for certain types of *in vivo* dose measurements, are presented. Detailed information about the use of point detectors for *in vivo* dosimetry of target dose delivery can be found in IAEA Human Health Report No. 8 (IAEA, 2013) and in other papers (e.g., Mijnheer, 2013).

The use of 2D detectors for *in vivo* dosimetry is limited to that of (radiochromic) film and EPIDs. The other planar systems presented in Chapter 8 are mainly used for phantom measurements, for instance, for pretreatment verification, as discussed in Chapter 17. The use of film for *in vivo* dosimetry can be advantageous compared to that of point detectors, for instance, in specific regions of interest such as the eye, or where the patient's contour changes rapidly. Knowledge of the skin dose is of interest when introducing a new treatment technique, and Gafchromic film is often used for that purpose (e.g., Rudat et al., 2014). Measuring the skin dose during total skin electron therapy (TSET) using Gafchromic films has also been reported (Bufacchi et al., 2007). Procedures to perform accurate skin dose measurements using radiochromic film have been described by Devic et al. (2006). Several studies have also shown the usefulness of Gafchromic film during intraoperative radiotherapy (IORT) both in electron beams (Ciocca et al., 2003; Severgnini et al., 2015) and in low-energy x-ray beams (Avanzo et al., 2012). For their use in IORT, radiochromic films are usually wrapped in sterile envelopes, while various methods of analysis can be used. Film is the only 2D detector currently used for out-of-field measurements (see Chapter 21).

As discussed in Chapter 7, EPIDs have a number of favorable characteristics that makes them suitable as detectors for various types of quality control (QC) measurements besides their use for patient setup verification. For the QC of accelerators, they are used for the verification of leaf positions (see Chapter 15). During patient-specific pretreatment verification, they are frequently used to verify the actual photon fluence or dose distribution created by a treatment planning system (TPS; see Chapter 17). An increasing number of centers are using them also for *in vivo* dose verification, as is elucidated in this chapter.

3D dose verification based on the transit dose measured with the onboard detector of a TomoTherapy facility has also been reported (Sheng et al., 2012). Detector sonograms were retrieved and back-projected to calculate entrance fluence, which was then forward-projected on the CT images to calculate the 3D dose distribution. The method was tested with phantom irradiations using ion chamber and film measurements showing root-mean-square errors of about 2.0% for head-and-neck, prostate, and spinal cord stereotactic body radiation therapy (SBRT) patients. The method was also used *in vivo* for the same patient treatments showing agreement compared to planned doses to the majority of planning target volumes (PTVs) and organs at risk (OARs) within 5% for the cumulative treatment course doses. The dosimetric error strongly depends on the error in multileaf collimator (MLC) leaf opening time, with a sensitivity correlating to the gantry rotation period.

18.2.2 Correlation of the Detector Reading with the Dose in the Patient

Performing *in vivo* patient dose measurements implies that the radiation dose measured with a detector has to be correlated with the dose in the patient. This is generally done by first converting the detector signal to dose to water, being the reference material used in dosimetry and in clinical protocols. Recently also other methods of dose reporting, for example, dose to medium instead of dose to water, are proposed, which may result particularly in kilovoltage (kV) photon beams in different dose values as discussed in more detail in Chapters 13 and 24. The next step is to correlate the dose to water at the position of the detector to the dose at specific positions in the patient. This can be done in many ways depending on the type of information one wishes to derive from a particular type of *in vivo* dose measurement. During *in vivo* dosimetry in brachytherapy, a detector is generally positioned as close as possible to the target volume or an OAR, as discussed in Chapter 20 and more extensively by Tanderup et al. (2013). But even in those situations where a dosimeter is placed in a patient body cavity, a correction factor is needed to convert the dose at the position of the detector to the dose in the target volume or OAR if those are the positions where the dose needs to be verified. Such a correction factor depends on the attenuation (dose gradient) of the dose between the two points, as well as the distance between them.

In external beam radiotherapy, entrance and exit dose measurements are generally performed by placing a detector on the skin of the patient at the beam entrance or exit side of the patient. Definitions of entrance and exit dose have been formulated in ESTRO Booklet No. 5 (Huyskens et al., 2001) and in AAPM Report 87 (Yorke et al., 2005), which were also adopted in IAEA Human Health Report No. 8 (IAEA, 2013). For entrance and exit dose determinations, the most common dose reference point chosen in these reports is on the beam central axis at the depth of dose maximum downstream or upstream of the entrance or exit surface, respectively. Detailed information about the calibration of detectors for *in vivo* entrance and exit dosimetry can be found in the above mentioned ESTRO, AAPM, and IAEA documents. It should be noted that the term exit dosimetry is sometimes also used for measurements with a detector placed at a certain distance behind the patient, for instance, when performing EPID dosimetry measurements. In order to avoid confusion, for these type of measurements the term transit or transmission dosimetry has been adopted in this chapter and in Chapter 7 following the proposal by van Elmpt et al. (2008) in their review article on EPID dosimetry.

In vivo transit (transmission) dose measurements using EPIDs are analyzed by making a dose comparison in the patient. In this approach, back-projection techniques have been developed correlating the response of the central pixels of an EPID with the dose at a point in a patient (Nijsten et al., 2007; François et al., 2011; Camilleri et al., 2014; Piermattei et al., 2015). Reconstruction of 2D dose maps at the depth of isocenter in a plane parallel to the EPID has also been reported (Peca and Brown, 2014; Peca et al., 2015). EPID-based 3D *in vivo* dose reconstruction methods have been developed by many groups using different types of dose calculation algorithms. Pencil-beam types of dose calculation models have been described, for instance, by Wendling et al. (2009, 2012) and Gimeno et al. (2014). Van Uytven et al. (2015) and McCowan et al. (2015) used the back-projected fluence to calculate the 3D patient dose distribution via a collapsed-cone convolution method, whereas a Monte Carlo (MC) approach has been described by van Elmpt et al. (2007).

The software for point dose verification developed by François et al. (2011) has been implemented in a commercial system (EPIgray; DOSIsoft, Cachan, France), and the first clinical experience with this system has recently been published (Celi et al., 2016; Ricketts et al., 2016). The approach described by Piermattei et al. (2015), which is clinically used in several Italian hospitals, also became recently available as a commercial product (SOFTDISO; Best Medical, Italy). Software for the verification of EPID-based 3D dose distributions, as well as for point dose verification, has been released by Dosimetry Check (Math Resolutions, Columbia, MD). The commissioning of, and initial experience with, the Dosimetry Check software has been reported (Gimeno et al., 2014; Mezzenga et al., 2014). The EPID-based 3D approach developed in the Netherlands Cancer Institute (NKI) (Wendling et al., 2009) became recently available as Elekta iViewDose. The coming years will show a continuous increase in the use of IMRT, VMAT, and other advanced treatment techniques. It can therefore be expected that in the future more 2D and 3D *in vivo* dose verification tools will become available on the market.

18.3 Clinical Implementation

18.3.1 General Considerations

When starting an *in vivo* dosimetry program in the clinic, a number of issues have to be arranged such as identifying the objectives of the program, formulating a detailed description of the measurement procedures, defining tolerance and action levels, and describing actions to be taken when alerts are raised.

Furthermore, the tasks and responsibilities of the professionals involved in an *in vivo* dosimetry program should be clearly defined. Also the training of personnel in understanding the results of their measurements is important, and it will increase the trust in an *in vivo* dosimetry program in general. These issues have been discussed in detail in the IAEA Human Health Report No. 8 (IAEA, 2013) and in Chapter 9 of another book (Mijnheer, 2013). In this chapter, we elucidate the specific challenges when a 3D *in vivo* dosimetry program will be implemented in the clinic.

The aim of 3D *in vivo* dosimetry is different compared to verifying the dose delivered to a patient at a single point. When using a point detector to verify the entrance dose, a high accuracy can often be obtained if the detector has been calibrated, and all correction factors are well known and properly applied. In other words, the actual entrance dose delivered to the patient can in principle be determined accurately. However, when verifying *in vivo* the 3D dose distribution delivered to a patient, it is complicated to reconstruct the "true" 3D dose distribution inside the patient. Methods to generate an EPID-based dose reconstruction by using the real patient anatomy and patient position are time-consuming and not always very accurate as discussed further on in this chapter. For that reason, the planning CT data are generally used to reconstruct the 3D dose distribution. Consequently, the aim of 3D *in vivo* dosimetry at this moment is often not to measure the "true" dose delivered to a patient with the highest possible accuracy but to detect in a simple and reliable way a deviation between the planned and reconstructed 3D dose distribution, which should be within well-specified criteria. For the large majority of measurements, this will most likely be the case. However, if a flag is out of tolerance then a method should be available to explain the observed difference. This is particularly important when verifying hypofractionated stereotactic radiotherapy treatments, where errors in a single fraction have a larger influence on the total dose than during multifraction treatments having lower dose per fraction. It is obvious that this philosophy of 3D *in vivo* dosimetry is different compared to 3D pretreatment verification, which generally is based on a measurement of the "true" 3D dose distribution in a phantom.

18.3.2 Tolerance and Action Levels

Quantifying differences in 3D dose distributions can be done in many ways, and it is the topic of Chapter 14. The method currently mostly used is gamma evaluation, whereas region-of-interest (ROI) analysis, including dose–volume histogram (DVH) examination, is increasingly used. A variety of metrics are applied when using gamma analysis, such as the gamma pass rate (percentage of γ values <1), mean gamma, and (near) maximum gamma. Also other parameters such as the choice of the ROI, the isodose level, and global or local normalization determine the numerical values of a gamma analysis. Acceptance criteria based on these metrics are discussed in Chapters 14 and 17. These criteria concern, however, pretreatment dose verification analysis but have to be reconsidered when applied to *in vivo* dose verification measurements. They might also be different when comparing dose distributions using 2D gamma analysis in a plane or 3D analysis in a volume.

Currently, not much experience is available on EPID-based *in vivo* dose verification, and alert criteria vary from one institution to another as discussed by van Elmpt et al. (2008) and in Chapter 7. This can be appreciated by considering, for instance, the criteria for isocenter dose verification. These are related to

Table 18.1 Standard Deviations in Isocenter Dose Determinations Measured Using EPID-Based *In Vivo* Dosimetry

Standard Deviation (%)	Type of Treatment	Reference
1.9	VMAT of head-and-neck cancer	Cilla et al. (2016)
~2.5	3DCRT of four different tumor sites	Fidanzio et al. (2015)
3.2	IMRT of most tumor sites	Hanson et al. (2014)
2.9–5.2	IMRT and VMAT of 14 different tumor sites	Mijnheer et al. (2015)
~5.0	3DCRT, IMRT, and VMAT treatments of various tumor sites	Celi et al. (2016)

Note: 3DCRT, three-dimensional conformal radiotherapy; IMRT, intensity-modulated radiotherapy; VMAT, volumetric-modulated arc therapy.

the uncertainty in these *in vivo* point dose determinations, which depend on the measurement method, treatment technique, and treatment site. Standard deviations in isocenter dose determinations as reported by several centers are presented in Table 18.1 and vary between about 2% and 5%. As a consequence, if an action level of ±5% is chosen, which is often done for *in vivo* dosimetry with point detectors during entrance dose determinations, then many alerts can be expected. If a systematic uncertainty is also present, then asymmetric action levels can be used (e.g., Nijsten et al., 2007; Ricketts et al., 2016), whereas a tolerance level of ±7% for the first *in vivo* isocenter dose verification measurement and ±5% for the average of the first five fractions has also been applied (François et al., 2011). Tolerance levels of ±5% for the *in vivo* measured dose at the isocenter for pelvic–abdomen, head-and-neck, and breast irradiations, and ±6% for lung treatments were adopted by Fidanzio et al. (2015). In many other hospitals, isocenter dose deviations >5% are inspected, and an investigation is conducted to trace the origin of the difference (e.g., Camilleri et al., 2014; Hanson et al., 2014; Mijnheer et al., 2015).

When starting to analyze EPID-based *in vivo* patient dose measurements in 2D or 3D, gamma evaluation criteria are often chosen by adapting those applied for pretreatment verification. For instance, at NKI the clinical alert criteria for both pretreatment and *in vivo* dose verification are based on a global 3%/3-mm evaluation for a 3D gamma analysis within the 50% isodose surface. This rather high threshold was chosen to avoid artificial improvement of the gamma evaluation results by including large low-dose volumes, and to avoid buildup areas. Alert criteria at NKI are also more generous in case of palliative treatments (Mijnheer et al., 2015). All groups applying EPID-based *in vivo* dose measurements still have to buildup clinical experience to decide about the optimal action levels to be used in their clinic.

3D EPID-based *in vivo* dosimetry provides a possibility to generate DVHs from the reconstructed 3D dose distribution. Important clinical parameters can be estimated from DVHs, such as the median dose (D_{50}), the near-maximum dose (D_2), and the near-minimum dose (D_{98}) as recommended in ICRU Report 83 (ICRU, 2010). These dose–volume parameters, derived from both the *in vivo* and the planned DVHs, can then be used to understand the importance of observed differences in 3D dose distributions (Mezzenga et al., 2014; Rozendaal et al., 2014). The use of DVHs would give clinicians and medical physicists a tool to interpret dose differences in a more direct way than using gamma evaluation. However, knowledge of the dose level, size, and anatomical location where failure occurs are important properties to evaluate DVH results. In the future radiobiological models, tumor control probability (TCP) and normal tissue complication probability (NTCP) might be used as additional evaluation measures as discussed in more detail in Section 18.5.3.

18.3.3 Follow-Up Actions

When an EPID-based dose verification measurement shows an alert, a number of follow-up actions are needed. One of the first actions is to repeat the measurement, which is obvious if an error in the image acquisition or data analysis is the cause. If the reason for a substantial error is not clear and waiting for the next fraction is not feasible, for instance, in case of a hypofractionated treatment, then it is recommended to perform immediately a phantom measurement. The next step generally is to collect information about a possible variation in patient setup and change in anatomy when the *in vivo* dose measurement was performed. The procedures adapted in two centers are given as examples.

Visual or automatic inspection of the EPID images is applied by the Italian group (Fidanzio et al., 2015; Cilla et al., 2016). If a small patient setup variation or a minor morphological change is the reason for the alert, a warning is sent to the treatment unit staff thus enabling correction of these easier cases. If the cause of the dosimetric discrepancy is more severe or cannot be understood, a new CT scan of the patient is done and used to calculate a hybrid plan (i.e., a plan that uses the new CT data with beam configuration and monitor units of the original plan), and a new *in vivo* dose analysis is performed. A complete replanning procedure is adopted if clinically relevant dosimetric discrepancies are observed between the original and the hybrid treatment plan. In this way, all discrepancies can be eliminated or justified.

At NKI, cone-beam computed tomography (CBCT) verification is performed prior to or during almost all patient treatments with curative intent. Examining the CBCT–planning CT information provided on the day of the EPID-based verification is one of the actions of the medical physicist when an alert is raised (Mijnheer et al., 2015). Radiation oncologists and therapists are consulted in cases where more details are required. If an alert cannot be explained after inspecting the corresponding CBCT scan, then an EPID-based or ion chamber–based 3D dose verification using a phantom irradiation is performed for the same plan but now recalculated using the CT data of the phantom. If the verification result then falls within the alert criteria, and inspection of the log file does not show an unusual behavior of that treatment fraction, no further action is performed.

Once an alert is raised, and follow-up actions can explain the origin of the observed difference, the next question is if the deviation is clinically relevant. This question goes further than the discussion in Chapters 14 and 17 about the interpretation of dose difference results observed during pretreatment verification. The main purpose of pretreatment verification is to identify *dosimetrically unacceptable* IMRT and VMAT patient plans, i.e., considered from the technical point of view. The origin of these unacceptable plans might be related to shortcomings in the dose calculation of the TPS, errors in the transfer of the TPS data to the accelerator, or errors in the delivery of the plan itself. The additional question when analyzing *in vivo* treatment verification data is if the difference in dose distribution is clinically acceptable, i.e., would the treatment outcome in terms of local control or toxicity be affected by this difference. Generally, the alerts are first reviewed by an experienced medical physicist. When the error is understood by the physicist, and it was estimated that there might be negative clinical consequences, the case should be discussed with a radiation oncologist, and a decision for corrective action should then be made by the physicist and radiation oncologist together.

18.3.4 Automation of Electronic Portal Imaging Devices-Based *In Vivo* Dosimetry

An *in vivo* dosimetry program can be implemented in many ways as discussed elsewhere for measurements performed with point detectors (IAEA, 2013; Mijnheer, 2013). EPID-based *in vivo* dosimetry approaches have the advantage over most point detector methods in that the EPID readings can be processed digitally and can be combined with other patient information that is digitally available, thus allowing automation of the whole process. Ideally, an EPID-based *in vivo* dosimetry system should be able to verify each dose delivery automatically, i.e., without human interaction. Such an approach includes automatic tools for input of patient and treatment identification data from the record-and-verify system, image acquisition, data analysis, raising alerts, and scheduling actions when deviations are outside tolerance levels (Olaciregui-Ruiz et al., 2013), and possibly segmentation of imaging data and deformable image registration.

Automation of EPID-based transit dosimetry has been implemented by different groups in various ways, resulting in different workflows. The EPIgray software has an integration with the record-and-verify system but is currently only able to verify the dose at specific points. Follow-up actions such as comparison of dose verification with portal imaging results are a task performed by the physics team and the therapists (François et al., 2011; Celi et al., 2016). The Italian approach also has an integration with the record-and-verify system and the TPS for the verification of the dose at the isocenter. Visual inspection of EPID images, made during a specific fraction, with a reference EPID image obtained during the first treatment fraction, is performed by the therapists in cooperation with a medical physicist. This procedure is recently automated and day-to-day comparisons are now also performed automatically using 2D gamma analysis (Cilla et al., 2016). The Dosimetry Check system has, at the time of writing this chapter, no integration with the record-and-verify network, thus making a laborious process of information manipulation of CT images, treatment plans, and portal imaging necessary (Gimeno et al., 2014).

At the Royal Marsden Hospital NHS Trust (RMH) in London, United Kingdom, 2D gamma maps generated by the EPID *in vivo* patient dosimetry system were only inspected if dose deviations at the isocenter were outside the action level (Hanson et al., 2014). These authors concluded that being able to account for the observed deviations by simply inspecting the results of the automatic *in vivo* 2D analysis illustrates the power of performing a 2D analysis over point dose measurements alone.

Maastro Clinic embarked on 3D EPID-based dosimetry for both pretreatment and *in vivo* dose verification. The entire workflow, from image acquisition to presentation of the 3D reconstructed dose distributions on the kV CBCT images of the day, is a fully automated process (Persoon et al., 2013). The technique was a modification of an earlier developed 3D dose verification procedure method based on in-room megavoltage (MV) CBCT imaging (van Elmpt et al., 2009). A radiation oncologist visually inspects every kV CBCT scan flagged by 3D *in vivo* dosimetry. If the decision for replanning is made, the clinical target volume (CTV) of the primary tumor is redelineated manually on the kV CBCT images, followed by a recalculation of the DVH and gamma metrics for dose intercomparison. The method revealed dose discrepancies and changes over time for lung cancer treatments, and might be used in the future as part of an adaptive strategy for other cancer sites as well.

At NKI, a fully automated EPID-based 3D *in vivo* dosimetry approach has been clinically implemented (Olaciregui-Ruiz et al., 2013). The EPID dosimetry workspace reads the daily delivery schedule from the patient management system (MOSAIQ) and calls the EPID verification software in batch mode as soon as a treatment fraction has been delivered and portal images have been recorded. Results are available a few minutes after delivery of each fraction and alerts are immediately raised when deviations are detected. The only remaining manual action in this workflow is the panel placement for image acquisition. A difference with the Maastro Clinic approach is that currently in clinical routine at NKI the reconstructed 3D dose distribution is shown in the planning CT data. The EPID dosimetry workspace links extra sources of information allowing online visual inspection of CBCT and planning data to identify anatomical changes that influence the dose delivery in such a way that treatment adaptation is necessary.

Automation of EPID-based *in vivo* dosimetry is extremely important otherwise the additional time for analysis will hamper large-scale clinical implementation because of the increase in workload. For instance, at NKI more than 5000 RT treatments per year are verified resulting in alerts in about 30% of the treatments. Furthermore, a solution that automatically provides the correct input data to the dose verification software greatly improves the reliability of the results. Automatic analysis may also allow a more frequent use of *in vivo* dose verification, for example, during those fractions at which patient setup imaging is performed, or even for all fractions. The additionally obtained dose information can then be used to examine interfractional trends. Large-scale implementation of automated *in vivo* patient dosimetry will also allow analysis of the data of cohorts of patients treated for the same disease over a longer period, i.e., to observe any time trend of the results.

Finally, it should be realized that EPID-based *in vivo* dosimetry is an end-to-end test of a patient irradiation that can be obtained with minimum effort if the system is automated. In addition to the regular QC of the system itself, the main workload for the staff in the department will be the follow-up actions after alerts are raised. The result of those actions for a specific case will not only improve the treatment of that particular patient but may also contribute to improvement of the complete irradiation procedure after discussion between the various members of the radiotherapy team. In summary, a structured and automated alert handling approach of EPID-based *in vivo* dosimetry facilitates consistent decision-making and is an excellent tool to check the overall clinical process.

18.4 Clinical Experience

18.4.1 Examples of Errors Discovered by Means of *In Vivo* Dose Verification

In the review paper of van Elmpt et al. (2008), three types of errors were distinguished that can be spotted using EPID dosimetry: *machine-related* errors, *plan-related* errors, and *patient-related* errors. *Machine-related* errors are due to hardware faults and include errors in the presence and direction of a wedge, the presence of a segment, MLC leaf position/speed, leaf sequencing, collimator angle, beam flatness and symmetry, linac output, and gantry angle. Most of these errors can be detected both by pretreatment and by *in vivo* dose verification. *Plan-related* errors include dose calculation errors, wrong TPS commissioning data (such as beam fit errors) and data transfer errors, which include the selection of a wrong patient plan. Most of these errors can be detected by *in vivo* dose verification, but not all of them by pretreatment verification. Errors in the dose calculation can, for instance, only be detected when using a suitable phantom, and cannot be identified if EPID image gray scale or portal dose distributions are used for verification purposes. Because generally the dose is measured in a homogeneous phantom, dose calculation errors related to the tissue heterogeneity of the patient will also not be detected by pretreatment verification (see Figure 18.1). *Patient-related* errors are due to changes in the patient's position or anatomy compared to the situation at planning, but they include also treatment couch interception of a beam, obstructions from immobilization devices, as well as treating a patient with an incorrect plan. Table 18.2 summarizes some of these types of errors traced by various groups through EPID-based *in vivo* dosimetry. These studies concern only papers published after the review article on EPID dosimetry by van Elmpt et al. (2008) and have been discussed further in the updated literature review provided in Chapter 7.

One of the most valuable contributions of EPID-based *in vivo* dosimetry to improving the quality of patient treatments is the verification, and adaptation if necessary, of lung cancer treatments. Many groups have shown that during a course of lung cancer treatment very often anatomical changes occur, leading to considerable dose differences compared to the planned situation (see Table 18.2). In the example shown in Figure 18.2, the data presented in the upper part of the figure indicate an overdosage in the volume surrounded by the 50% isodose surface. Inspection of the CBCT scan made during that day demonstrated a considerable reduction of atelectasis resulting in a higher dose in the PTV and in the lung. As a result of these observations, a new CT scan was made and a new plan was generated.

Anatomical changes also often occur during treatment of head-and-neck cancer and may have dosimetric consequences. Figure 18.3 illustrates such a case where the verification showed a considerable change in the delivered 3D dose distribution resulting in an alert. When the medical physicist inspected the CBCT data of that day, it was noticed that the underdosage was caused by an increase in tissue/fluid that moved the gross tumor volume (GTV) outside the PTV-boost volume and/or by tumor progression. The radiation oncologist decided to immediately acquire a new CT scan, after which a new treatment plan was created.

Figure 18.4 shows dose discrepancies measured with EPID-based dosimetry at the Tom Baker Cancer Centre, Calgary, Alberta, Canada, as a result of a variation in prone position of a rectal cancer patient in a belly board (Peca et al., 2015). The setup error was readily observable in the dose difference maps. It is a typical example of the use of an incorrect patient contour in the dose calculation based on the planning CT when the patient becomes more relaxed during a treatment series.

An error in dose delivery that occurred during a spine SBRT treatment at CancerCare Manitoba, Winnipeg, Manitoba, Canada, was discovered when using their in-house developed 3D EPID-based reconstruction software (Van Uytven et al., 2015) and is shown in Figure 18.5. A 25% decrease in the 3D

Table 18.2 Types of Error Traced by EPID-Based *In Vivo* Dosimetry

Potential Error	Error Type	References
Machine-related	Transfer error	Mans et al. (2010), Mijnheer et al. (2015)
Plan-related	Dose calculation error	Mans et al. (2010), Fidanzio et al. (2015), Mijnheer et al. (2015)
	Immobilization system not included in the treatment plan	Fidanzio et al. (2015)
	Bolus material not taken into account	Mijnheer et al. (2015)
Patient-related: anatomy changes	Changes in atelectasis and pleural effusion	Piermattei et al. (2009), Mans et al. (2010), Persoon et al. (2012), Wendling et al. (2012), Persoon et al. (2013), Fidanzio et al. (2015), Mijnheer et al. (2015)
	Variation in patient contour when the patient becomes more relaxed during treatment	Mans et al. (2010), Fidanzio et al. (2015), Peca et al. (2015)
	Gas pockets in the planning CT scan resulting in an underdose in the PTV during treatment	Camilleri et al. (2014), Cilla et al. (2014), Fidanzio et al. (2015)
	Weight loss resulting in an overdose in the PTV during treatment	Mans et al. (2010), Camilleri et al. (2014), Cilla et al. (2014, 2016)
	Incomplete bladder filling resulting in an overdose in the PTV during treatment	Ricketts et al. (2016)
Patient-related: delivery errors	Bar of the treatment couch in the entrance beam during treatment	Piermattei et al. (2009), Fidanzio et al. (2015)
	Imperfect immobilization allowing the patient to move during treatment	Hanson et al. (2014), Cilla et al. (2016)
	Wrong patient setup during treatment	Fidanzio et al. (2015), Mijnheer et al. (2015)

Note: CT, computed tomography; PTV, planning target volume.

gamma pass rate over the PTV region (using 3%/3 mm) was observed, whereas also the mean dose in the PTV decreased by almost 4%. After detecting these suspiciously high dosimetric delivery differences during EPID dosimetry, analysis of the CBCT scan showed that a pancreatic stent moved into the path of the VMAT delivery, whereas during CT simulation it was out of the irradiated volume. Obviously, such a type of error cannot be detected during pretreatment verification.

Figure 18.6 shows the results of a 2D EPID-based dose verification of an IMRT head-and-neck cancer case as observed at the RMH. The patient had lost a small amount of weight around the lower neck, and the physicist had calculated that it would not affect the 3D dose distribution very much and therefore the treatment was continued. However, the weight loss was having an effect on the patient immobilization in the head-and-neck fixation device, resulting in an error detected with EPID *in vivo* dosimetry. The patient's shoulder was moving into and out of the beam, which was clearly visible in the lateral beams. Although CBCT imaging is performed routinely on head-and-neck cancer patients in that center, the field of view is generally too small to see the shoulder position, and so this variation was not seen. As a result of the EPID dosimetry analysis, the patient had a new immobilization mask made and was rescanned and replanned. A similar major setup error spotted by EPID *in vivo* dosimetry of VMAT treatments of head-and-neck tumors is described by Cilla et al. (2016). By inspecting possible causes, an imperfect immobilization by a thermoplastic mask was noticed allowing the patient to displace the head during treatment delivery.

Figures 18.2 through 18.6 show the effect of nonoptimal patient treatments on the dose delivery during these irradiations. Anatomical changes, daily variation in patient setup, deviations in the position of a metal prosthesis inside the patient, as well as irradiation of unintended body parts such as a shoulder, often occur in clinical practice. These situations may be picked up by the therapists when performing patient setup verification. The impact of EPID-based *in vivo* dosimetry therefore strongly depends on the ability of the therapists to identify and correct setup uncertainties and to notice and handle anatomical changes. However, an *in vivo* dosimetry measurement is able to quantify dose deviations resulting from these deviations from the planned situation. On the basis of that measurement result, it can be decided that a new plan is necessary to solve the

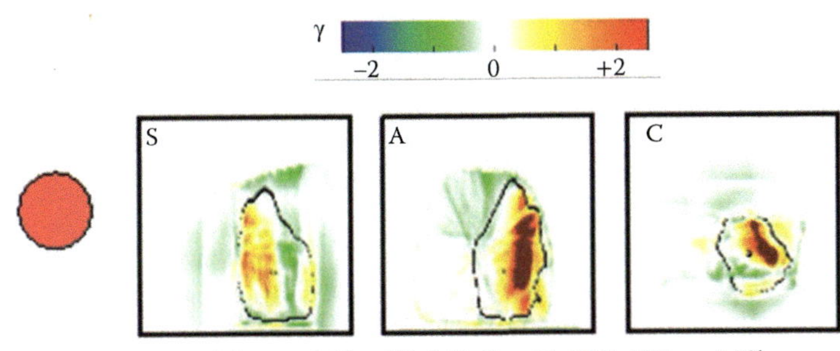

Mean γ : 0.80; γ 1%: 4.14; % $\gamma \leq 1$: 76.2; ΔDisoc: 4.0%

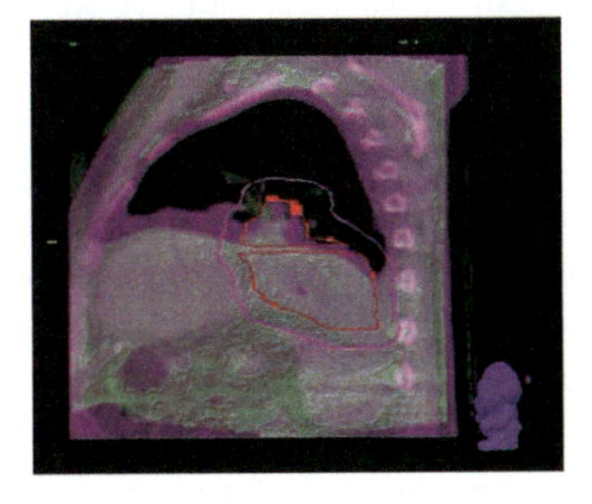
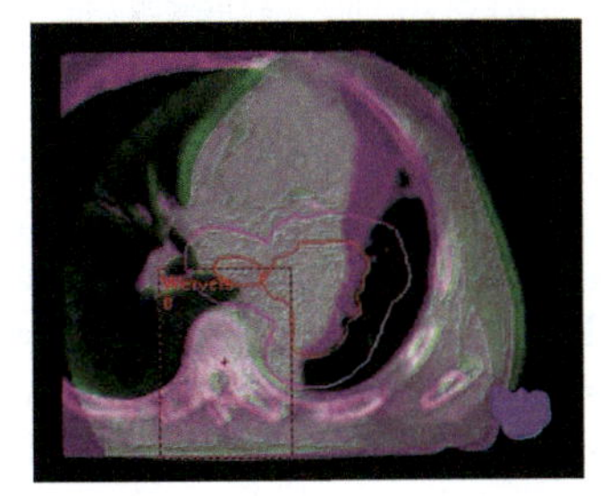
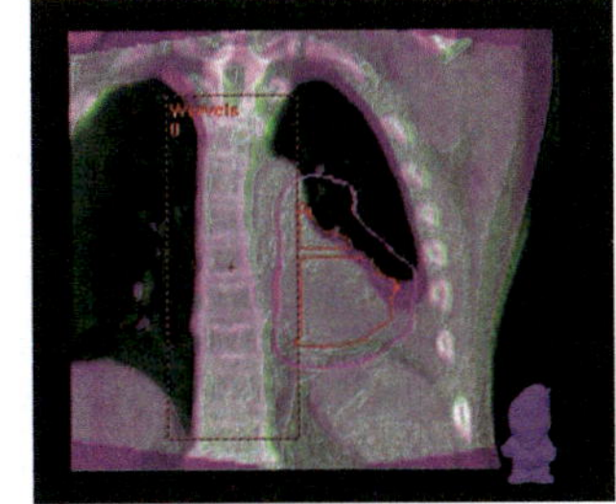

Figure 18.2

Top: the outcome of a three-dimensional (3D) electronic portal imaging device (EPID)-based *in vivo* dose verification of a hypofractionated volumetric-modulated arc therapy treatment of a lung cancer patient at the Netherlands Cancer Institute. Indicated are the results of a 3D gamma evaluation in a sagittal, axial, and coronal plane through the isocenter. The yellow and red color indicate regions where the EPID dose is higher than the planned dose, whereas the green color indicates regions where the EPID dose is equal or lower than the planned dose. The 50% isodose line is shown in black. The red dot indicates that at least one of the four alert criteria is outside the action level. Bottom: A comparison of a CBCT scan (green), made prior to the treatment that day, with the planning computed-tomography scan (purple) in the three orthogonal planes, demonstrating a reduction of atelectasis. Information within the rectangular dashed (clip) box is used for image registration.

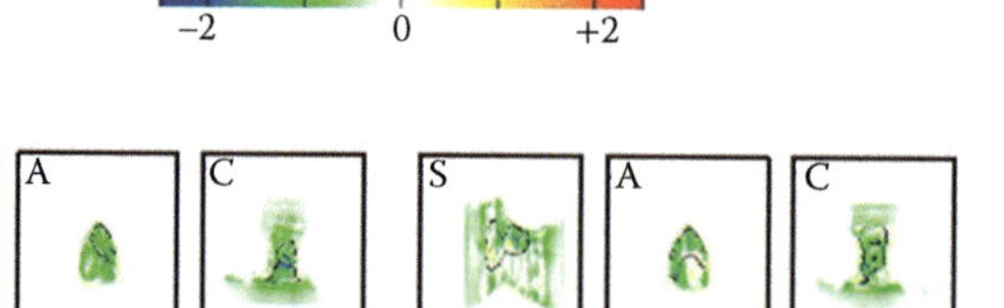

Mean γ : 1.05/0.77; γ 1% : 2.36/2.02: % γ ≤ 1: 47.5/70.2; ΔDisoc : −7.8%

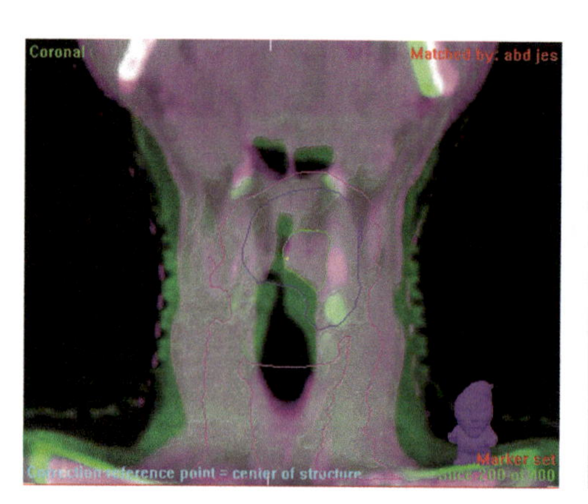

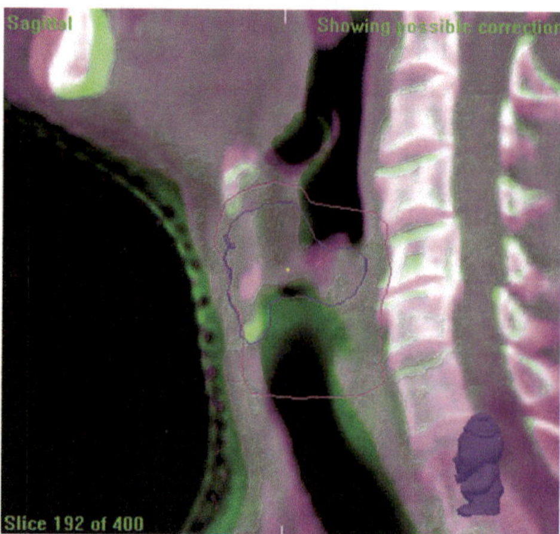

Figure 18.3

Results of a three-dimensional electronic portal imaging device–based *in vivo* dose verification of a two-arc volumetric-modulated arc therapy treatment of a larynx cancer patient at the Netherlands Cancer Institute, presented in the same way as in Figure 18.2. The results of the gamma evaluation are given for the two arcs separately, whereas the dose difference at isocenter is given for the total dose. The contours indicated in the bottom part of the figure are as follows: green = GTV; blue = PTV-boost; pink = PTV-elective.

problem, and/or the medical physicist may give an advice to the therapists to pay more attention to patient-related issues such as immobilization.

The experience in some centers and the examples discussed in this section demonstrate that EPID-based *in vivo* patient dosimetry requires a change in attitude to patient-specific QA. It is no longer purely a physics or technical matter, but it requires in addition clarifying dose differences observed in the clinic, i.e., incorporating many issues that may happen during the daily treatment of patients. As a result, medical physicists will be much more involved in assuring the quality of the actual patient treatment than when only performing a pretreatment dose verification measurement.

18.4.2 *In vivo* Dosimetry in Clinical Practice

The majority of the clinical observations presented in Table 18.2 and in Figures 18.2 through 18.6 concern situations that cause errors of at least 5% at the isocenter or reference point, a large drop in gamma pass rate or a change in another metric when the dose distributions are compared in 2D or 3D. Some of the errors mentioned in the table are related to human errors made in the clinic.

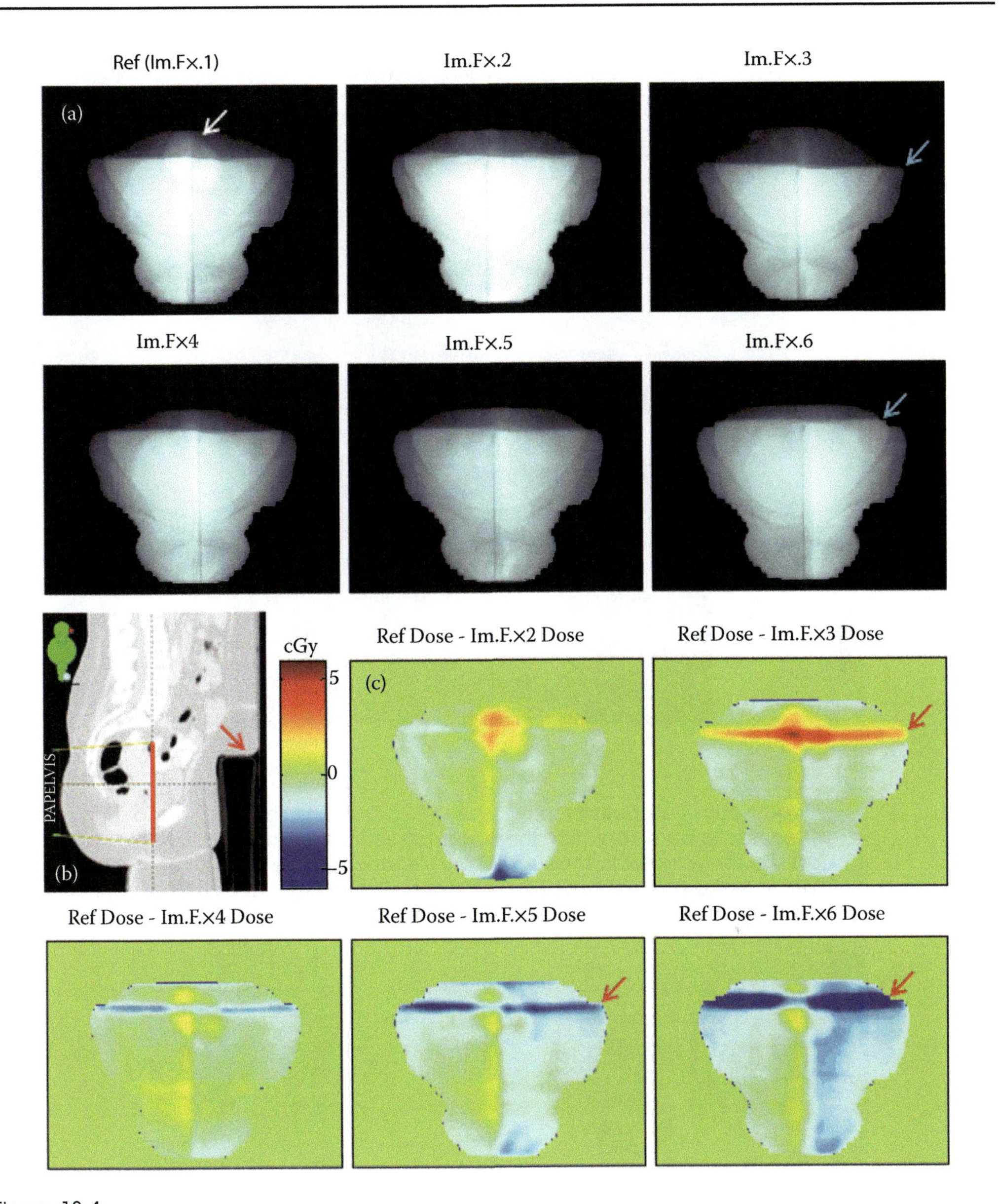

Figure 18.4

Posterior-anterior (PA) field of a rectal cancer treatment. (a) Raw electronic portal imaging device images. (b) Lateral view from the treatment planning system (TPS) showing the imaging plane (red) and the body adapting to the belly board opening. (c) Dose difference maps between the first imaged fraction (Im.Fx.1) and five later treatment days. The large horizontal mismatch is located at the edge of the belly board opening (red arrows), indicating inconsistent setup of the patient with respect to the board in the superior-inferior (SUP-INF) direction. This was verified by visual inspection of the raw images (blue arrows). All dose difference maps are affected by the gas bubble present during the first imaged fraction (white arrow). (From Peca S et al., In DA Jaffray [ed.], *IFMBE Proceedings 51*, Switzerland, Springer International Publishing, 2015. With permission.)

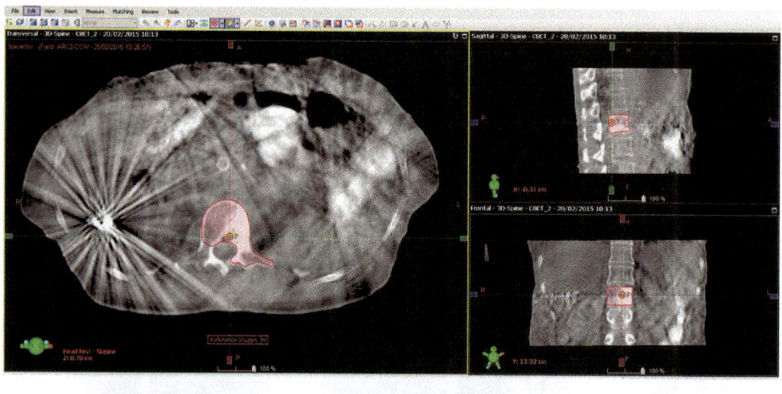

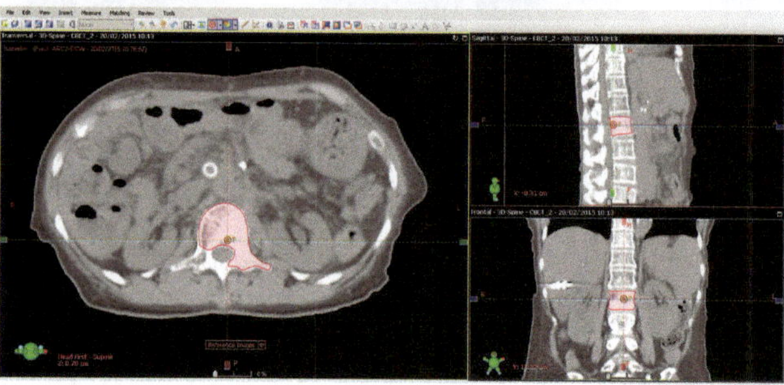

Figure 18.5

Cone-beam computed-tomography (CBCT) scans showing that during a spine stereotactic body radiation therapy treatment, a pancreatic stent moved into the path of the volumetric-modulated arc therapy delivery (top), while during CT simulation (bottom) it was out of the irradiated volume. (Courtesy of Boyd McCurdy.)

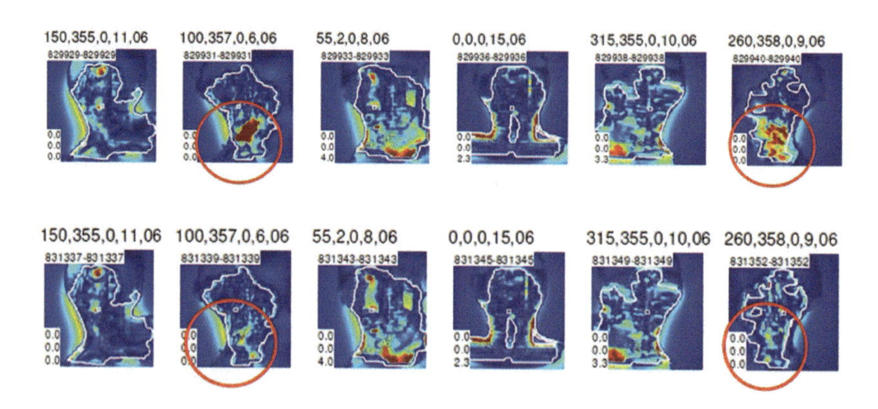

Figure 18.6

Results of two-dimensional electronic portal imaging device–based dose verification in a plane through the isocenter of an intensity-modulated radiotherapy head-and-neck cancer case at two different days. The circles indicate areas in the lateral beams where the patient's shoulder was moving into and out of the beam. (Courtesy of Ian Hanson.)

The discussion in the department of the occurrence of these errors should result in improved procedures, thus avoiding these types of errors in the future.

A general phenomenon when performing *in vivo* transit dosimetry in the pelvic region is the presence of gas pockets during treatment and not in the CT scan. Depending on the position and dimensions of these gas pockets, an error warning may be obtained. Large gas pockets can, however, easily be recognized on an EPID image or CBCT scan, and the measurement should be repeated during the next fraction to ensure the correct dose delivery when the size of the gas pocket is reduced. Another outcome that may occur during transit dosimetry, but does not affect the dose delivered to a patient, is a notification of an underdosage because of traversal of the beam through parts of the couch or immobilization device at the distal side of the patient. These alerts have no clinical consequence and can easily be recognized on EPID images or a CBCT scan.

Finally, it has to be realized that transit dosimetry has another fundamental limitation. Relatively small setup deviations may cause already rather large differences in transit dose values in regions with a strong variation in external contour, for instance, during tangential field irradiations of breast cancer, as discussed by several groups (Nijsten et al., 2009; Mijnheer et al., 2015; Celi et al., 2016). It is interesting to note that the latter group observed a considerable reduction in setup variation by irradiating breast cancer patients in lateral instead of supine position, reducing the fraction of misalignments from 38% of the checked beams to less than 2%. This intrinsic sensitivity of transit dosimetry for setup variation will lead to alerts that are real but maybe not clinically relevant and should therefore be inspected by a medical physicist. A solution for this limitation of transit dosimetry that is simple, fast, and automatic is still lacking.

18.4.3 Correlating *In Vivo* Dosimetry Results with In-Room Imaging Data

An obvious question is what additional information can be obtained with *in vivo* 3D dosimetry if 3D setup verification using in-room imaging, for instance, by means of CBCT is performed. The clinical examples shown in Table 18.2 indicate that not all these errors traced by EPID-based *in vivo* dosimetry can be discovered by CBCT. Anatomy changes due to changes in atelectasis and pleural effusion, as well as contour variations due to weight loss or other reasons, can often also be observed by CBCT. However, *in vivo* 3D dosimetry will quantify the effect of these anatomy changes on the 3D dose distribution. Furthermore, it will depend on the training of the therapists and the availability of a clinical protocol if the resulting deviations in dose delivery are considered to be clinically relevant. Merging the information from *in vivo* 3D dosimetry and CBCT will help in choosing the most suitable follow-up action.

The idea of combining *in vivo* dosimetry with in-room imaging is not new. Figure 18.7 shows an isocentric cobalt-60 unit, constructed in 1960 at NKI, having both options (www.historad.nl/en/welcome). The base part of the construction was a sturdy ring, partly sunk in the floor, which could rotate. On the inner surface of the ring were attached a cobalt source, a small diagnostic x-ray unit, an image intensifier coupled to a Vidicon TV camera, and a counterweight that also acted as a protection shield and contained ionization chambers for transit dosimetry.

As mentioned before, one of the first follow-up actions after observing a deviation between a measured and intended dose distribution is to get information

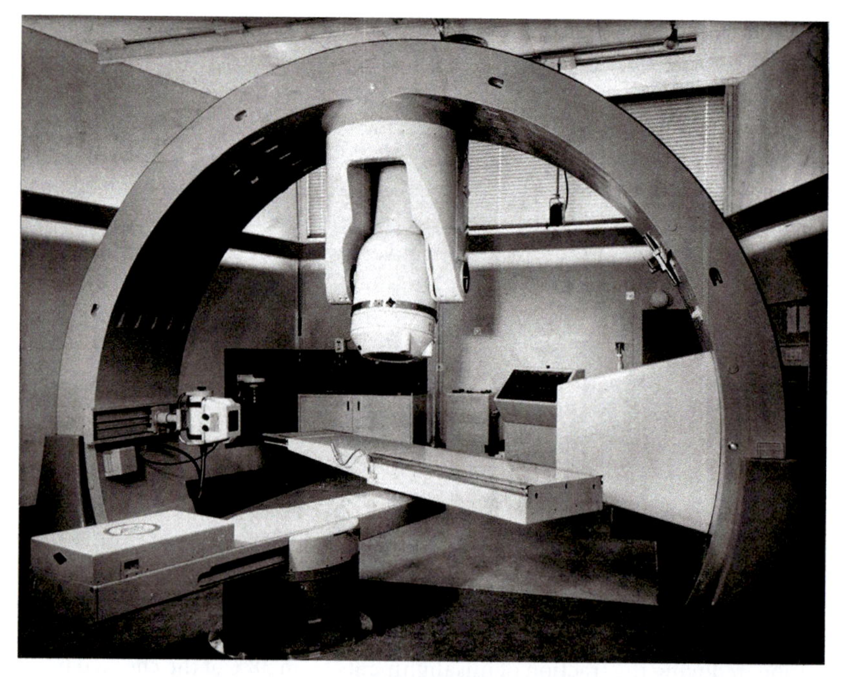

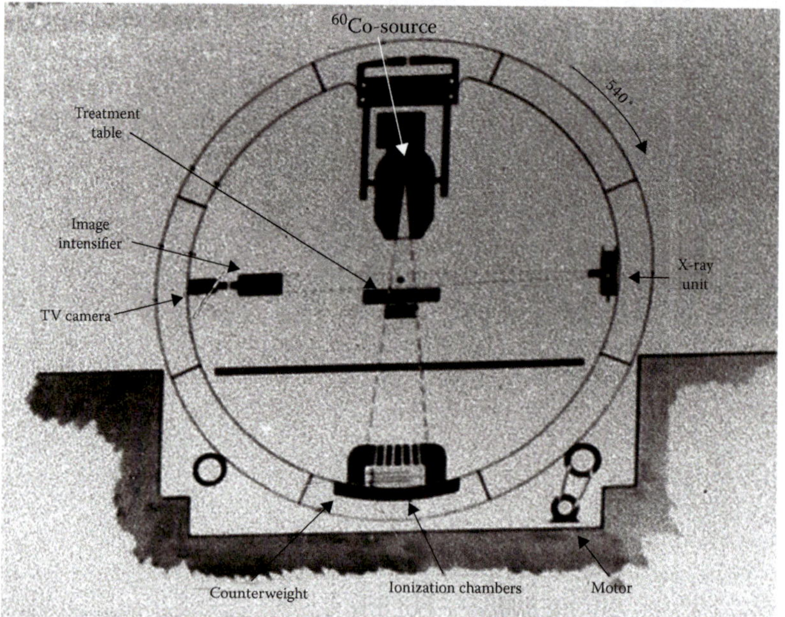

Figure 18.7

Isocentric cobalt-60 unit in 1960 at the Netherlands Cancer Institute, combining in-room imaging with *in vivo* transit dosimetry. (www.historad.nl/en/welcome, the Netherlands Cancer Institute, Amsterdam, the Netherlands.)

about the setup and anatomy of the patient at the time of the *in vivo* dose measurement. This can be a visual inspection of localization portal images made for setup verification just before the start of the treatment, or of the portal images made during the actual treatment. Various groups have also combined 3D in-room image-guided radiotherapy (IGRT) approaches with 3D EPID-based *in vivo* dose verification (e.g., McDermott et al., 2008; Persoon et al., 2013). A common problem with these approaches is how to take changing anatomy into account. In case of treatment adaptation, the radiation oncologist often has to define a new CTV in a CT or CBCT scan, which is time-consuming. Also tumor redelineation is challenging when using CBCT instead of CT data because of the difference in image quality, and sometimes the limited field of view (see Figure 18.6). Redelineation of the CTV on CT images by using deformable image registration in combination with CBCT images introduces also additional dosimetric uncertainties as discussed, for instance, by Veiga et al. (2015). Furthermore, a more fundamental and unsolved problem is how to incorporate disappearing voxels, resulting, for instance, from a shrinking tumor, in deriving the total dose delivered to a changing CTV. Currently, 3D *in vivo* dosimetry is almost entirely restricted to verifying the dose in the PTV. If for the PTV no alert is raised, it can be expected that CTV coverage is in many cases also achieved during a course of radiotherapy as is discussed in more detail in Section 18.5.2.

Finally, it should be noted that patient setup deviations are often already detected and corrected by IGRT procedures prior to the treatment of a patient. Thus, deviations in dose delivery to a patient due to setup uncertainties will generally be small. However, setup corrections resulting from an IGRT procedure might not be correctly performed, for example, when a couch has been shifted in the wrong direction. Also, the patient position may change between image acquisition and the actual treatment. These types of error may then be detected by *in vivo* dosimetry if the dose differences resulting from these setup deviations are large enough. However, small shifts in the patient position that introduce changes in the target dose coverage are not always readily detected by *in vivo* dosimetry, as discussed for instance by Bojechko and Ford (2015). It should therefore be realized that *in vivo* EPID transit dosimetry should not be used in isolation but is most effectively used in combination with image guidance.

18.5 Recent Developments and Future Approaches

18.5.1 Real-Time Dosimetry

Active point detectors such as diodes or metal oxide semiconductor field effect transistors (MOSFETs) can in principle be used for real-time (online) measurements and interrupt a treatment if the observed difference between measured and predicted dose exceeds a predefined level. An example of a possible workflow for the use of real-time *in vivo* dosimetry during image-guided brachytherapy is given in the V20/20 paper on *in vivo* dosimetry in brachytherapy (Tanderup et al., 2013). Also during external beam therapy such a procedure is in principle possible, for instance, when performing *in vivo* dose measurements with diodes or MOSFETs during total body irradiation (TBI) or intraoperative radiation therapy (IORT) (Ciocca et al., 2006). However, despite the fact that *in vivo* dosimetry is often performed during TBI or IORT, information about the use of real-time *in vivo* dosimetry to adapt these treatments is scarce.

An EPID-based real-time delivery verification system, called WatchDog, has been developed by an Australian group (Fuangrod et al., 2013). A reference dataset of predicted EPID images is compared with a cumulative EPID image acquired during treatment. The comparison is performed within the frame time of the imager and allows for both geometric and dosimetric verification as a function of control point and gantry angle. In a recent publication, the authors reported on the first clinical demonstration of their real-time delivery verification system (Woodruff et al., 2015). Such a system can in principle determine in real time whether external beam radiation treatments are being delivered as planned, but do not provide information on the effect of deviating treatments on the dose distribution in the patient.

At NKI, a software package for real-time EPID-based 3D *in vivo* dose verification has been derived from their clinically used offline 3D dose verification system (Spreeuw et al., 2016). Portal images are processed faster than the frame rate of the portal imager by precomputing all input for 3D EPID-based dose reconstruction and speeding up the reconstruction algorithm. After a portal image is acquired, the dose distribution is reconstructed in 3D in the patient planning CT data, and compared with the 3D dose distribution predicted by the TPS using DVH parameters. The complete processing of a single portal frame, including dose verification, takes less than 300 ms on a standard CPU. Whenever dose differences outside tolerance levels are detected an alert is generated, which can be used as a trigger to stop the linac automatically. The system is tested clinically, particularly with respect to establishing criteria to halt the linac.

An interesting new application of EPID-based verification is the possibility to perform time-dependent measurements during VMAT to verify the field shape and intensity of sub-arcs (Liu et al., 2013; Woodruff et al., 2013; Podesta et al., 2014). The models are used to verify pretreatment delivered MLC fluences, and the approach of Woodruff et al. (2015) has been incorporated in their real-time approach. By measuring in a time-dependent manner, dose deviations have been observed by Podesta et al. that might be hidden in integrated dose images. However, their method was until now only tested pretreatment using phantom irradiations and not yet for *in vivo* dose verification.

18.5.2 *In Vivo* Dosimetry during Adaptive Radiotherapy

The dose actually given to a patient may deviate considerably from the planned dose distribution as a result of variation in patient position and change in anatomy during the course of radiotherapy. Adapting the position of a patient before a treatment fraction is delivered, based on a setup verification measurement prior to that treatment, is common practice. However, adaptive radiotherapy (ART), adapting a treatment plan based on in-room imaging to account for anatomical changes during a patient's treatment course, is more challenging. ART can be resource intensive because it needs replanning procedures necessitating both additional use of planning equipment and staff time. Furthermore, any adaptation of a plan requires in principle also an additional patient-specific verification, with a phantom measurement either before the next fraction is delivered or *in vivo* during that fraction, thus enhancing the workload even more.

Portal imaging has been used by several groups to detect anatomy changes. For instance, in a study by McDermott et al. (2006), when analyzing EPID localization images it was found that for 57% of lung and 37% of head-and-neck cancer patients progressive anatomical changes occurred. Persoon et al. (2012) presented a method based on 2D portal transit dosimetry to record dose changes throughout the course of treatment and to allow trend analysis of dose discrepancies. Recently, two fully automated commercial systems for determining daily changes in treatment delivery, using EPID images behind a patient, have been introduced (PerFRACTION, Sun Nuclear Corporation, Melbourne, FL and Adaptivo, Standard Imaging, Middleton, WI). These systems compare the transit image of the first fraction against those of the other fractions in a treatment course, and are therefore able to detect interfraction treatment variations due to anatomical changes. A similar approach is used in the SOFTDISO system developed by the Italian group (Piermattei et al., 2015; Cilla et al., 2016).

An obvious choice for applying ART is the treatment of lung cancer. Many studies have been published in the literature describing anatomical changes during lung cancer treatment such as tumor regression or progression, changes in atelectasis, and pleural effusion. These changes occur frequently and require often an adapted treatment plan. For instance, in a recent study, a new planning CT scan was made in 8% of the patients to mitigate the risk of tumor underdosage during lung cancer treatment (Kwint et al., 2014). In that study, CBCT scans were evaluated to trace these changes using a decision support system (Traffic light protocol). EPID-based *in vivo* dosimetry is used in an increasing number of centers to detect the dosimetric effects of anatomical changes during fractionated radiotherapy of lung cancer (see Table 18.2).

The second site where ART may play a role is in the treatment of head-and-neck cancer. There is, however, variation in the literature regarding the dosimetric impact of anatomical changes on various structures during ART in head-and-neck cancer as summarized recently (Brouwer et al., 2015, Brown et al., 2015). Several studies report no significant difference between planned and delivered doses for the GTV, spinal cord, and brain stem, but did observe a significant increase in parotid gland dose. In contrast, other studies found that changes during treatment significantly decreased the dose to target volumes and significantly increased the dose to surrounding OARs such as the spinal cord and brain stem. Generally speaking, anatomical changes in the head-and-neck region may cause more dose deviations in OARs than in the target volume. The reason is that CTV coverage is usually more robust to anatomical changes because of the incorporation of a CTV–PTV margin.

With the current approaches of *in vivo* dosimetry the dose in the high-dose region, i.e., the PTV, is generally verified, which may show only limited impact of daily anatomical changes. This has recently been confirmed for VMAT treatments of head-and-neck cancer demonstrating that including CBCT information in EPID dose reconstructions only slightly improves the agreement with TPS calculations (Rozendaal et al., 2015). Using EPID-based *in vivo* dosimetry for the verification of OAR doses might therefore be more relevant. However, OARs are often located in regions with a large dose gradient, thus requiring accurate knowledge of the position of an OAR with respect to the beam geometry. Verification of the OAR position is therefore a prerequisite for assessing the actual dose in an OAR. Also even if the position of an OAR is known, accurate knowledge of the complete 3D dose distribution in it is often not needed and only information about the maximum dose is of importance for the prediction of a biological effect in that OAR, particularly for serial organs. Also some OARs are very small having a volume comparable to the voxel size of the TPS, and consequently only point dose information is required. For all those reasons, the use of EPID-based *in vivo* dose verification of OAR doses is still limited and needs more investigation.

In vivo verification of the dose in OARs using other types of detector, for example, diodes or TLDs, is until now also not very often performed. Exceptions are *in vivo* dose verification in OARs during brachytherapy (Chapter 20), out-of-field *in vivo* dose measurements (Chapter 21), and *in vivo* rectal wall dosimetry during external beam radiotherapy of prostate cancer (Hardcastle et al., 2010).

Several centers use as adaptive approach multiple precalculated plans combined with a best fitting plan-of-the-day based on imaging information obtained prior to that treatment (e.g., Martinez et al., 2001; Tuomikoski et al., 2013). The use of EPID-based *in vivo* dosimetry for these types of ART treatments is basically the same as during a non-ART schedule, except that the *in vivo* dosimetry measurement now has to be repeated each time a different plan of the day is used. In order to restrict the workload, tools have to be developed to select patients who are expected to benefit from plan adaptation during a treatment course.

A novel approach for improving dose conformity is real-time adaptation *during* treatment delivery. This can be achieved in several ways, for instance, by real-time electromagnetic transponder-guided MLC tracking as shown by Keall et al. (2015). Another approach will be the real-time image guidance of a patient irradiated with the newly developed MRI cobalt unit (Mutic and Dempsey, 2014) and MRI linac (Lagendijk et al., 2014) as explained in more detail in Chapter 26. Further investigation of EPID-based real-time dose verification methods will be necessary to evaluate their usefulness in real-time QA of these types of ART.

18.5.3 Biologically Based Alert Criteria

As discussed in Chapter 2, there is a vast amount of evidence that for a similar total absorbed dose, lowering the absorbed dose per fraction will reduce the biological effect, whereas increasing the absorbed dose per fraction will increase that effect. The clinical effect of underdosage of the target volume and the related dose accuracy requirements for the PTV, therefore, have to be assessed using a TCP model. In a similar way, an NTCP model has to be used to evaluate the effect of over- or underdosage of an OAR. Late-responding normal tissues often have a low α/β ratio making them especially sensitive to over- or underdosage. Furthermore, the dose distribution in OARs is generally very inhomogeneous, requiring rather complicated calculations to assess accuracy requirements for OARs. All these considerations play an important role in the optimization of a radiotherapeutic treatment and is discussed in detail in Chapter 2.

The same models can be applied for quantifying the biological impact of differences between planned and actual delivered dose distributions. Zhen et al. (2013) investigated the use of TCP and NTCP models to quantify the effects of different types of intentionally induced errors to IMRT patient plans to simulate both TPS errors and machine delivery errors in the IMRT QA process. The changes in TCP and NTCP for various anatomical structures were calculated as the new QA metrics to quantify the clinical impact of these errors on patient treatments. The changes/degradations in TCP and NTCP caused by the errors varied widely depending on dose patterns unique to each plan, and are good indicators of each plan's robustness to that type of error. Sumida et al. (2015) proposed a gamma index–based dose evaluation that integrates the radiobiological parameters of TCP and NTCP calculations. Their results showed that the radiobiological gamma index passing rates for prostate and head-and-neck cases were for some cases different compared to those using the physical gamma indices.

These approaches are promising but still need more tests in daily clinical practice. First, the TCP and NTCP models, as well as the radiobiological parameters to calculate TCP and NTCP values, have an intrinsic uncertainty and therefore caution should be taken when choosing an appropriate TCP/NTCP model. Furthermore, deviations in TCP and NTCP data as a result of differences between planned and measured *in vivo* dose distributions have to be translated into action levels. These are related to the institutional constraints applied during the planning process, but need further investigation of how to implement them in clinical practice. Finally, the measurement uncertainty should also be taken into account when determining the NTCP in an OAR, particularly if geometric uncertainties related to patient setup errors are involved. It should therefore be kept in mind that biologically based evaluation metrics are interesting research quantities, but clinically they still should be used with caution, as discussed in Chapter 2 and elsewhere, for example, in ICRU Report 83 (2010).

18.6 Summary

In vivo patient dosimetry has been in use for a long time in radiotherapy. These measurements generally concern dose verification of conventional three-dimensional conformal radiotherapy (3DCRT) treatments using point (0D) detectors, a technique that is not very suitable for the verification of IMRT and VMAT. More recently, 2D and 3D *in vivo* dose verification using EPIDs has been introduced, which is the main topic of this chapter. After discussing the relationship between *in vivo* dosimetry and other types of patient-specific QA, the clinical implementation of EPID-based dose verification has been elucidated. Issues such as the definition of tolerance/action levels and follow-up actions in case an alert is raised have been discussed in detail. The need for a simple workflow, if possible fully automated, is demonstrated and should be a prerequisite for large-scale implementation. An overview is given of the different types of error that have been detected by various groups using EPID-based *in vivo* dosimetry, and typical examples are shown to illustrate the specific 2D or 3D aspects of these errors. Most of these errors concern errors in the workflow or human mistakes in the clinic, as well as inaccuracies in the dose delivery due to anatomical changes during a treatment course. These types of error cannot be detected by means of pretreatment dose verification, indicating the importance of *in vivo* patient dosimetry for patient-specific QA of RT treatments. The results of some studies concerning the correlation of 3D dose verification with 3D in-room imaging are presented. This topic will become more important in the future with the increasing use of adaptive radiotherapy. Finally, it can be expected that real-time EPID-based *in vivo* dosimetry approaches will be further developed to have a tool to stop a linac in case of serious delivery errors.

Acknowledgments

I thank Agnieszka Olszewska, Anke van Mourik, and Tomas Janssen for providing examples of *in vivo* three-dimensional dose verification at the Netherlands Cancer Institute (NKI), and the EPID dosimetry group at NKI for their useful comments.

References

Avanzo M et al. (2012) In vivo dosimetry with radiochromic films in low-voltage intraoperative radiotherapy of the breast. *Med. Phys. 39: 2359–2368.*

Bojechko C and Ford EC (2015) Quantifying the performance of in vivo portal dosimetry in detecting four types of treatment parameter variations. *Med. Phys. 42: 6912–6918.*

Bojechko C et al. (2015) A quantification of the effectiveness of EPID dosimetry and software-based plan verification systems in detecting incidents in radiotherapy. *Med. Phys.* 42: 5363–5369.

Brouwer CL et al. (2015) Identifying patients who may benefit from adaptive radiotherapy: Does the literature on anatomic and dosimetric changes in head and neck organs at risk during radiotherapy provide information to help? *Radiother. Oncol.* 115: 285–294.

Brown E et al. (2015) Predicting the need for adaptive radiotherapy in head and neck cancer. *Radiother. Oncol.* 116: 57–63.

Bufacchi A et al. (2007) *In vivo* EBT radiochromic film dosimetry of electron beam for Total Skin Electron Therapy (TSET). *Phys. Med.* 23: 67–72.

Camilleri J et al. (2014) Clinical results of an EPID-based *in-vivo* dosimetry method for pelvic cancers treated by intensity-modulated radiation therapy. *Phys. Med.* 30: 690–695.

Celi S et al. (2016) EPID based in vivo dosimetry system: Clinical experience and results. *J. Appl. Clin. Med. Phys.* 17(3): 262–276.

Cilla S et al. (2014) An in-vivo dosimetry procedure for Elekta step and shoot IMRT. Phys. Med. 30: 419–426.

Cilla S et al. (2016) Initial clinical experience with Epid-based in-vivo dosimetry for VMAT treatments of head-and-neck tumors. *Phys. Med.* 32: 52–58.

Ciocca M et al. (2003) In vivo dosimetry using radiochromic films during intraoperative electron beam radiation therapy in early-stage breast cancer. *Radiother. Oncol.* 69: 285–289.

Ciocca M et al. (2006) Real-time in vivo dosimetry using micro-MOSFET detectors during intraoperative electron beam radiation therapy in early-stage breast cancer. *Radiother. Oncol.* 78: 213–216.

Devic J et al. (2006) Accurate skin dose measurements using radiochromic film in clinical applications. *Med. Phys.* 33: 1116–1124.

Fidanzio A et al. (2015) Routine EPID *in-vivo* dosimetry in a reference point for conformal radiotherapy treatments. *Phys. Med. Biol.* 60: N141–N150.

Ford EC et al. (2012) Quality control quantification (QCQ): A tool to measure the value of quality control checks in radiation oncology. *Int. J. Radiat. Oncol. Biol. Phys.* 84: e263–e269.

Francois P et al. (2011) In vivo dose verification from backprojection of a transit dose measurement on the central axis of photon beams. *Phys. Med.* 27: 1–10.

Fuangrod T et al. (2013) A system for EPID-based real-time treatment delivery verification during dynamic IMRT treatment. *Med. Phys.* 40: 091907.

Gimeno J et al. (2014) Commissioning and initial experience with a commercial software for in vivo volumetric dosimetry. *Phys. Med.* 30: 954–959.

Hanson IM et al. (2014) Clinical implementation and rapid commissioning of an EPID based in-vivo dosimetry system. *Phys. Med. Biol.* 59: N171–N179.

Hardcastle N et al. (2010) In vivo real-time rectal wall dosimetry for prostate radiotherapy. *Phys. Med. Biol.* 55: 3859–3871.

Huyskens D et al. (2001) *Practical Guidelines for the Implementation of in vivo Dosimetry with Diodes in External Radiotherapy with Photon Beams (Entrance Dose)*. ESTRO Booklet No. 5. (Brussels, Belgium: European Society for Radiotherapy and Oncology).

IAEA (2013) Development of procedures for in vivo dosimetry in radiotherapy. IAEA Human Health Report No. 8. (Vienna, Austria: International Atomic Energy Agency).

ICRU (2010) Prescribing, recording, and reporting intensity-modulated photonbeam therapy (IMRT). Report 83. (Bethesda, MD: International Commission on Radiation Units and Measurements).

Keall PJ et al. (2015) The first clinical treatment with kilovoltage intrafraction monitoring (KIM): A real-time image guidance method. *Med. Phys.* 42: 354–358.

Kry SF et al. (2014) Institutional patient-specific IMRT QA does not predict unacceptable plan delivery. *Int. J. Radiat. Oncol. Biol. Phys.* 90: 1195–1201.

Kwint M et al. (2014) Intra thoracic anatomical changes in lung cancer patients during the course of radiotherapy. *Radiother. Oncol.* 113: 392–397.

Lagendijk JJW, Raaymakers BW and van Vulpen M (2014) The magnetic resonance imaging–linac system. *Sem. Radiat. Oncol.* 24: 207–209.

Liu B et al. (2013) A novel technique for VMAT QA with EPID in cine mode on a Varian TrueBeam linac. *Phys. Med. Biol.* 58: 6683–6700.

Mans A et al. (2010) Catching errors with in vivo dosimetry. *Med. Phys.* 37: 2638–2644.

Martinez AA et al. (2001) Improvement in dose escalation using the process of adaptive radiotherapy combined with three-dimensional conformal or intensity-modulated beams for prostate cancer. *Int. J. Radiat. Oncol. Biol. Phys.* 50: 1226–1234.

McCowan PM et al. (2015) An in vivo dose verification method for SBRT-VMAT delivery using the EPID. *Med. Phys.* 42: 6955–6963.

McDermott LN et al. (2008) 3D *in vivo* dose verification of entire hypo-fractionated IMRT treatments using an EPID and cone-beam CT. Radiother. Oncol. 86: 35–42.

McDermott LN et al. (2006) Anatomy changes in radiotherapy detected using portal imaging. *Radiother. Oncol.* 79: 211–217.

McKenzie EM et al. (2014) Toward optimizing patient-specific IMRT QA techniques in the accurate detection of dosimetrically acceptable and unacceptable patient plans. *Med. Phys.* 41: 121702.

Mezzenga E et al. (2014) Pre-treatment and in-vivo dosimetry of helical tomotherapy treatment plans using the Dosimetry Check system. *J. Instrum.* 9: C04039.

Mijnheer B (2013) In vivo dosimetry. In: *The Modern Technology of Radiation Oncology*, Vol. 3, pp. 301–336. (Ed. Van Dyk J, Madison, WI: Medical Physics Publishing).

Mijnheer B et al. (2013) *In vivo* dosimetry in external beam radiotherapy. *Med. Phys.* 40: 070903.

Mijnheer BJ et al. (2015) Overview of three year experience with large scale EPID-based 3D transit dosimetry. *Pract. Radiat. Oncol.* 5: e679–e687.

Mutic S and Dempsey JF (2014) The ViewRay system: Magnetic resonance–guided and controlled radiotherapy. *Sem. Radiat. Oncol.* 24: 196–199.

NCS (2013) Code of practice for the quality assurance and control for intensity modulated radiotherapy. Report 22 (The Netherlands: The Netherlands Commission on Radiation Dosimetry) http://www.radiationdosimetry.org.

NCS (2015) Code of practice for the quality assurance and control for volumetric modulated arc therapy. Report 24 (The Netherlands: The Netherlands Commission on Radiation Dosimetry) http://www.radiationdosimetry.org.

Nijsten SMJJG et al. (2007) Routine individualized patient dosimetry using electronic portal imaging devices. *Radiother. Oncol.* 83: 65–75.

Nijsten SMJJG et al. (2009) Prediction of DVH parameter changes due to setup errors for breast cancer treatment based on 2D portal dosimetry. *Med. Phys.* 36: 83–94.

Olaciregui-Ruiz I et al. (2013) Automatic in vivo portal dosimetry of all treatments. *Phys. Med. Biol.* 58: 8253–8264.

Peca S and Brown DW (2014) Two-dimensional in vivo dose verification using portal imaging and correlation ratios. *J. Appl. Clin. Med. Phys.* 15(4): 117–128.

Peca S, Brown D and Smith WL (2015) In vivo EPID dosimetry detects interfraction errors in 3D-CRT of rectal cancer. In: IFMBE Proceedings 51, pp. 531–534. (Ed. Jaffray DA, Switzerland: Springer International Publishing).

Persoon LCGG et al. (2012) Interfractional trend analysis of dose differences based on 2D transit portal dosimetry. *Phys. Med. Biol.* 57: 6445–6458.

Persoon LCGG et al. (2013) First clinical results of adaptive radiotherapy based on 3D portal dosimetry for lung cancer patients with atelectasis treated with volumetric-modulated arc therapy (VMAT). *Acta Oncol.* 52: 1484–1489.

Piermattei A et al. (2009) Integration between in vivo dosimetry and image guided radiotherapy for lung tumors. *Med. Phys.* 36: 2206–2214.

Piermattei A et al. (2015) aSi EPIDs for the in-vivo dosimetry of static and dynamic beams. *Nucl. Instrum. Methods Phys. Res.* A. 796: 93–95.

Podesta M et al. (2014) Time dependent pre-treatment EPID dosimetry for standard and FFF VMAT. *Phys. Med. Biol.* 59: 4749–4768.

Ricketts K et al. (2016) Clinical experience and evaluation of patient treatment verification with a transit dosimeter. *Int. J. Radiat. Oncol. Biol. Phys.* 95: 1513–1519.

Rozendaal RA et al. (2015) Impact of daily anatomical changes on EPID-based in vivo dosimetry of VMAT treatments of head-and-neck cancer. *Radiother. Oncol.* 116: 70–74.

Rozendaal RA et al. (2014) In vivo portal dosimetry for head-and-neck VMAT and lung IMRT: Linking γ-analysis with differences in dose–volume histograms of the PTV. *Radiother. Oncol.* 112: 396–401.

Rudat V et al. (2014) In vivo surface dose measurement using GafChromic film dosimetry in breast cancer radiotherapy: Comparison of 7-field IMRT, tangential IMRT and tangential 3D-CRT. *Radiat. Oncol.* 9: 156.

Severgnini M et al. (2015) *In vivo* dosimetry and shielding disk alignment verification by EBT3 GAFCHROMIC film in breast IOERT treatment. *J. Appl. Clin. Med. Phys.* 16(1): 112–120.

Sheng K et al. (2012) 3D dose verification using tomotherapy CT detector array. *Int. J. Radiat. Oncol. Biol. Phys.* 82: 1013–1020.

Spreeuw H et al. (2016) Online 3D EPID-based dose verification: Proof of concept. *Med. Phys.* 43: 3969–3974.

Sumida I et al. (2015) Novel radiobiological gamma index for evaluation of 3-dimensional predicted dose distribution. Int. J. Radiat. Oncol. Biol. Phys. 92: 779–786.

Tanderup K, Beddar S, Andersen C, Kertzscher G and Cygler J (2013) *In vivo* dosimetry in brachytherapy. *Med. Phys.* 40: 070902.

Tuomikoski L et al. (2013) Implementation of adaptive radiation therapy for urinary bladder carcinoma: Imaging, planning and image guidance. *Acta Oncol.* 52: 1451–1457.
van Elmpt W et al. (2008) A literature review of electronic portal imaging for radiotherapy dosimetry. *Radiother. Oncol.* 88: 289–309.
van Elmpt W et al. (2009) 3D in vivo dosimetry using megavoltage cone-beam CT and EPID dosimetry. *Int. J. Radiat. Oncol. Biol. Phys.* 73: 1580–1587.
van Elmpt WJ et al. (2007) Treatment verification in the presence of inhomogeneities using EPID-based three-dimensional dose reconstruction. *Med. Phys.* 34: 2816–2826.
Van Uytven E et al. (2015) Validation of a method for in vivo 3D dose reconstruction for IMRT and VMAT treatments using on-treatment EPID images and a model-based forward-calculation algorithm. *Med. Phys.* 42: 6945–6954.
Veiga C et al. (2015) Toward adaptive radiotherapy for head and neck patients: Uncertainties in dose warping due to the choice of deformable registration algorithm. *Med. Phys.* 42: 760–769.
Wendling M et al. (2012) In aqua vivo EPID dosimetry. *Med. Phys.* 39: 367–377.
Wendling M et al. (2009) A simple back-projection algorithm for 3D EPID dosimetry of IMRT treatments. Med. Phys. 36: 3310–3321.
Woodruff HC et al. (2015) First experience with real-time EPID-based delivery verification during IMRT and VMAT treatments. *Int. J. Radiat. Oncol. Biol. Phys.* 93: 516–522.
Woodruff HC et al. (2013) Gantry-angle resolved VMAT pretreatment verification using EPID image prediction. *Med. Phys.* 40: 081715.
Yorke E et al. (2005) Diode in vivo dosimetry for patients receiving external beam radiation therapy. AAPM Report 87. (Madison, WI: Medical Physics Publishing).
Zhen H, Nelms BE and Tome WA (2013) On the use of biomathematical models in patient-specific IMRT dose QA. *Med. Phys.* 40: 071702.

19

Audits Using End-to-End Tests

David S. Followill, Catharine H. Clark, and Tomas Kron

19.1 Philosophy

19.1.1 Background

The linear accelerators currently in use by radiation therapy (RT) centers are accompanied by advanced technology and can deliver complex treatment beam arrangements such as, but not limited to, those required for intensity-modulated radiotherapy (IMRT) or stereotactic radiosurgery (SRS). Quality assurance (QA) and an understanding of these complex treatment deliveries are essential to avoid harming patients (Derreumaux et al., 2008; Bogdanich, 2010a,b,c). The European Society for Radiotherapy and Oncology (ESTRO) has issued guidelines on IMRT verification and recommended independent audits by an outside QA agency or comparison with another RT institution (Mijnheer and Georg, 2008) to ensure safe treatment delivery. Similar documents were also issued by the American Association of Physicists in Medicine (AAPM) to provide guidance on commissioning complex treatment delivery systems, clinical implementation of IMRT, and commissioning of treatment planning dose calculations (Ezzell et al., 2003; Smilowitz et al., 2016). In North America, the Imaging and Radiation Oncology Core QA Center in Houston (IROC Houston), formerly the Radiological Physics Center (RPC), located at the University of Texas MD Anderson Cancer Center, has been providing independent remote and on-site QA dosimetry audits for complex IMRT, SRS, proton therapy (PT), and stereotactic body radiation therapy (SBRT) treatments. Some European countries also have national audit networks, for example, in the United Kingdom, the Institute of Physics and Engineering in Medicine's (IPEM) IMRT working party strongly recommended performing audits and external review of IMRT programs when RT departments had no prior experience using these advanced techniques on patients (James et al., 2008). The IPEM's recommendations resulted in national audits being conducted throughout the United Kingdom for IMRT (Budgell et al., 2011), volumetric-modulated arc therapy (VMAT) (Clark et al., 2014), and SBRT for lung (Distefano et al., 2015) to assure safe implementation, modeling, and delivery of these advanced techniques in an independent manner. These national audits benefited from experience of the UK National Cancer Research Institute (NCRI) Radiotherapy Trials Quality Assurance Group's comprehensive dosimetry QA programs for national IMRT clinical trials such as the CHHIP (Khoo et al., 2008) and PARSPORT (Clark et al., 2009a,b). Unfortunately, many of these audits are resource intensive, both for the auditor and the audited, and most other examples are therefore restricted to developed countries such as Japan (Nishio et al., 2006; Mizuno et al., 2008), the United Kingdom (Khoo et al., 2008; Budgell et al., 2011; Clark et al., 2014; Distefano et al., 2015), or Australia (Williams et al., 2012; Kron et al., 2002).

Surveys carried out by the International Atomic Energy Agency (IAEA) and others in the Asia and Pacific region (Tatsuzaki and Levin, 2001; Kron et al., 2015a), Latin America (Zubizarreta et al., 2004), Africa (Abdel-Wahab et al., 2013), and Europe (Rosenblatt et al., 2013) have shown that there is a substantial increase in the number of RT machines capable of delivering complex treatments that are used worldwide. However, this increase in capability has not been paralleled by a corresponding ability to audit complex techniques in these countries. Even though many RT centers throughout the world participate in reference beam output dosimetry audits (Izewska et al., 2003), many do not understand or perform comprehensive QA tests to verify their complex RT modalities and technologies. These advanced technologies typically consist of many interdependent processes that make them difficult to verify properly, therefore, the entire treatment process must be evaluated using what is known as an end-to-end (E2E) QA test.

19.1.2 Definition of E2E QA Audit

Since complex RT patient treatment processes are composed of a series of interdependent events (also known as a chain of events), it is extremely difficult to perform QA for each individual event. A failure mode and effects analysis (FMEA) of IMRT alone has identified up to 216 possible failure modes that could each be verified (Huq et al., 2016). The treatment process for most RT patients, complex or simple, can be simplified into a chain of events as shown in Figure 19.1. This chain of events, although not inclusive of every step a patient goes through, spans from imaging of a patient to treatment planning to final dose delivery. For RT treatments, an E2E QA test is an audit methodology that tests whether all of the components in the treatment process function in a manner such that the desired radiation dose is delivered accurately only to the intended spatial location.

If the E2E test fails to meet the acceptance criterion, then an investigation by the appropriate members of the RT team must be conducted to determine the discrepancy so that it can be corrected. These investigations require the staff to understand the details of the treatment process sufficiently well so as to isolate the reason for the failed test.

19.1.3 Transition to the Use of Advanced Treatment Technologies

19.1.3.1 Imaging Systems

In modern-day RT centers, the sophisticated advanced RT treatments now rely, more than ever, on new complex imaging systems. These systems include

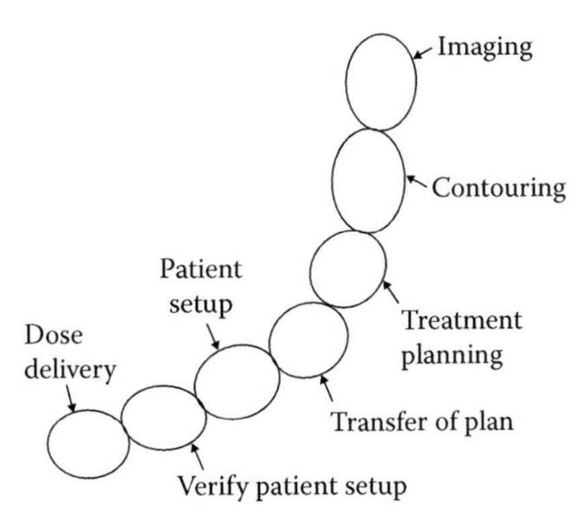

Figure 19.1

Generalized chain of events in the treatment process for complex radiation therapy.

simulators, such as computed tomography (CT) scanners, positron-emission tomography (PET)/CT scanners, magnetic resonance imaging (MRI) scanners, and so on, that acquire the initial patient images used for target and organ at risk (OAR) delineation for treatment planning, as well as onboard imaging systems on the accelerators, such as kilovoltage imagers, cone beam CT, electronic portal imaging devices (EPIDs), that localize the targets and OARs daily during the actual patient dose delivery phase of the RT treatment. Each of the imaging systems has its own unique characteristics which, if they do not function appropriately, may result in the radiation dose being calculated incorrectly by the planning system or delivered to the wrong site within the patient. Such a weak link in the RT treatment chain of events can be discovered using an E2E QA test and subsequent investigation.

19.1.3.2 Treatment Planning Systems

One of the most essential and complex components of the RT treatment chain of events is the treatment planning process, more specifically, the treatment planning computer (TPC). The computer's ability to accurately calculate the dose is critical to delivering safe and efficacious doses to RT patients. There are different dose calculation algorithms used in each make and model of TPC that each require different dosimetry data to model the radiation beams used in the planning process. Some of the more common dose calculation algorithms in use today include pencil beam, convolution superposition, collapsed cone convolution, deterministic radiation transport, analytical anisotropic, and Monte Carlo algorithms. Ensuring that the treatment beams are appropriately modeled in a TPC requires a comprehensive commissioning of the TPC and validation of the beam models to calculate accurate doses. No matter the complexity and rigor of a TPC algorithm, if the dosimetry data used to model the treatment beams are inadequate, then the computer will calculate inaccurate doses. Depending on the delivery modalities to be used for RT patients, a center may have general-use and/or specialized TPCs, each requiring unique QA and dose calculation accuracy (under homogeneous and heterogeneous conditions) validation tests that should be performed before the computer is accepted for clinical use, as discussed in detail in Chapter 16. As part of the commissioning process for RT TPCs, the AAPM's Medical Physics Practice Guidelines (MPPG) (Smilowitz et al., 2016) and IAEA's Technical Report Series 430 (IAEA, 2004) recommend E2E QA tests be performed for each commissioned modality. The E2E QA tests results will help to improve the understanding of the TPC's beam models and dose calculation algorithm strengths and weaknesses.

19.1.3.3 Treatment Delivery Modalities/Systems

Each RT center may employ a variety of complex treatment delivery modalities (listed as follows) depending on the specific make/model of treatment machines and accessories available to the RT team.

1. Three-dimensional conformal radiation therapy (3DCRT)
2. IMRT
3. SRS
4. SBRT
5. Image-guided radiation therapy (IGRT)
6. VMAT
7. PT
8. Target motion management

The listed treatment modalities have unique dose delivery characteristics and details that require unique validation tests before using them clinically. It is important to note that the use of them is also often linked to each other such as IMRT and IGRT. As stated above, the number of possible failure modes associated with each of the treatment modalities can be extremely high numbering in the 100s making it nearly impossible to validate each mode. In addition, understanding the complete treatment process of each modality, the interplay of individual processes, and the complexity of the data flow can be daunting and in some cases remain a "black box" to the RT team due to manufacturer proprietary information. A comprehensive validation test that can provide assurance that the desired RT delivery process can be delivered within acceptable limits of uncertainty is the E2E QA test.

19.1.4 Need for an E2E QA Verification Tool

There are three primary reasons for incorporating an E2E QA verification test into a comprehensive RT QA program: (1) need for treatment delivery accuracy, (2) to ensure RT patient safety, and (3) credentialing to ensure consistency among RT participants in multi-institutional clinical trials. All three of these reasons overlap to some degree. If one can verify accurate delivery, then one has confidence that the patient is getting the prescribed dose and the RT site could be credentialed. Considering it from another perspective, an RT site that becomes credentialed using an E2E QA test has demonstrated that it has the ability to deliver accurate and safe doses to their patients under the specific credentialing conditions. As new technologies and complex treatment modalities are introduced into the RT treatment process, a great deal of more effort is needed to satisfy the requirements of accuracy, safety, and consistency, as mentioned above. Typical risk-based approaches to QA dictate that the frequency and intensity of a QA verification effort should increase when the probability an error occurring increases, the amount of damage that can be caused increases, and the likelihood of the error not being detected is increased. With the clinical implementation of complex RT treatments, the possibility of an error occurring along the chain of events increases if there is not a proper QA program in place. Although we like to believe we can have unlimited resources to verify every possible aspect of the treatment, that just is not possible due to limited personnel effort and budgetary constraints. The E2E QA test serves the role of providing the RT team with a degree of confidence that the RT treatment delivery process in the end can deliver the intended dose to the intended location by verifying the process on a patient substitute prior to a true patient being treated. In this manner if there are errors detected, then these can be corrected, and reverified before clinically treating patients.

Even though E2E QA tools are extremely valuable, RT staff and processes can change requiring reverifications. Typically, the E2E test gives one a snapshot in time of the RT treatment delivery process. Currently, the greatest use of these E2E tests involve credentialing of RT centers to participate in multi-institutional clinical trials. Since the vast majority of the RT centers do participate in trials, credentialing has the dual benefit of ensuring delivery accuracy and patient safety, while ensuring consistency in dose delivery across many participating institutions. All of these benefits depend on a careful evaluation of the QA test results to identify the potential reasons for errors and required solutions. Otherwise, the effort and results of the QA test are worthless. Using the E2E test to simulate a patient just before going clinical with a new or changed treatment process allows the RT center to perform that one last verification that all of the components in a long chain of events are delivering the prescribed treatment.

19.2 Different E2E QA Verification Approaches

19.2.1 North America (IROC Houston QA Center)

19.2.1.1 Clinical Trial Credentialing

A multi-institutional clinical trial may incorporate a credentialing program in an effort to minimize the number of protocol violations and improve the overall quality of the trial (Followill, 2012). As new treatment technologies are introduced into clinical trials, there needs to be confidence that they were correctly implemented clinically and personnel were trained appropriately. These technologies tend to be complex such that there is more reliance on an E2E verification of the complete treatment process. This verification typically includes the process from imaging the patient, planning the treatment, patient setup, and delivery of dose. The National Cancer Institute (NCI), USA, funded clinical trials perform this assessment as part of "credentialing," i.e., the process to vet institutions and individuals prior to entering a patient on study. The purpose of this process is to insure that the RT team and institution have the necessary equipment, resources, radiation dose computational tools, treatment expertise, and understanding of the protocol.

Credentialing studies are often seen as an additional burden in the busy clinic, as it is sometimes perceived that no patient will benefit from the exercise. A good credentialing program however, will identify and correct trial protocol ambiguities, provide education to the staff, and potentially limit the number of treatment plans that must be resubmitted due to protocol violations. The credentialing program will also encourage improvements in treatment delivery that will affect all patients treated at the center, not just those treated on clinical trials.

For trials using more complex technologies, credentialing may require the institution to simulate, plan, and treat a geometric or anthropomorphic E2E QA phantom provided by IROC Houston. When anthropomorphic phantoms are used, the delivered measured dose is compared with the institution's calculated plan to determine the agreement (Molineu et al., 2005, 2013; Ibbott et al., 2006, 2008; Followill et al., 2007; Oldham et al., 2008), and whether an institution will be credentialed or not. These credentialing E2E QA phantom tests not only benefit clinical trials but also serve to verify an institution's clinical implementation of a new RT process and thus ensuring a safe accurate dose delivery to their patients.

19.2.1.2 Phantom-Detector Design (Photons and Protons)

IROC Houston has designed and constructed a family of anthropomorphic phantoms (Figure 19.2) that are used in credential RT centers to participate in advanced technology clinical trials. These QA phantoms are water-filled plastic shells with imageable targets, avoidance structures, and heterogeneities. The phantoms contain thermoluminescent dosimeters (TLDs) for point doses, and radiochromic film in two to three cardinal planes for dose distributions, to measure the dose delivered by the institution. Phantoms have been constructed that represent (1) the head, for SRS brain trials; (2) the head-and-neck, for IMRT trials; (3) the thorax, for IMRT and SBRT treatments of lung tumors; (4) the abdomen, for SBRT treatment of small tumors in organs such as the liver; (5) the spine, for SBRT treatment of metastases; and (6) the pelvis, for IMRT treatments of the prostate and cervix. The abdomen and lung phantoms can be placed on a two-dimensional (2D) reciprocating table to simulate respiratory motion in the anterior–posterior (AP) and superior–inferior (SI) directions. The proton E2E QA phantoms are similar in design and use the same dosimeters. However, they are built using proton equivalent plastics that have the correct relative linear stopping power versus Hounsfield unit (HU) relationship for treatment planning instead of the electron density versus HU relationship used for photon treatment planning. The IROC Houston phantoms provide a consistent test to evaluate each RT center's ability to deliver a specific radiotherapy treatment ensuring that all participating institutions pass the same QA criteria (Ibbott et al., 2006; Kry et al., 2013).

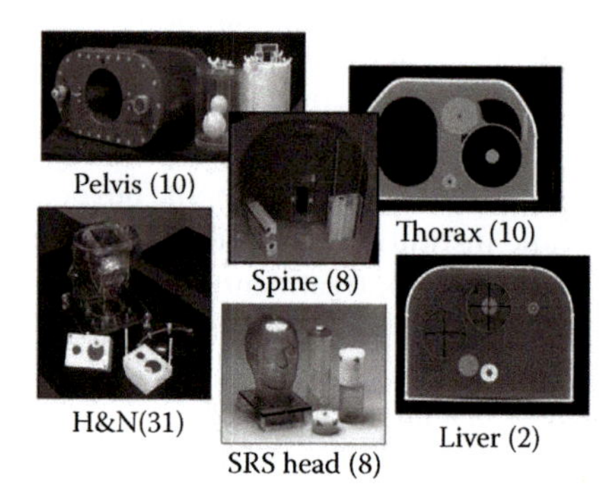

Figure 19.2

Imaging and Radiation Oncology Core Houston family of photon phantoms. SRS, stereotactic radiosurgery.

19.2.1.3 Dosimeter Accuracy and Precision

The IROC Houston E2E QA phantoms all use TLD as absolute dosimeters and radiochromic film as planar dosimeters in the sagittal, coronal, and axial planes. The TLDs are used to verify the delivered doses at points within the targets and OARs, while the films, normalized to a TLD dose point, are used to perform a gamma analysis between the measured dose distribution and the institution's calculated dose distribution. The precision of the IROC Houston TLD program has been previously described by Kirby and colleagues. It has been shown that the accuracy of absolute dose determined in the TLD readout system is equivalent to ion chambers to within ±4% at a 90% confidence interval (Kirby et al., 1992). The TLD system is also precise to within ±3% and is capable of detecting dose errors on the order of ±5%.

The uncertainty of the dose distribution as determined by the radiochromic film and TLD combination at one standard deviation is between 2.6% and 3.5%. This is consistent with a previously published report using the same methodology (Davidson et al., 2008). TLD used as an absolute dosimeter helped to reduce the variation that can occur between the film calibration process and the actual film used in the phantom at the time of irradiation. The uncertainty estimate included the uncertainty of the TLD dose (Kirby et al., 1992), the film uniformity, the film-to-film variation, and the fit of the sensitometric curve (Devic et al., 2004). The TLD uncertainty was included because the film was normalized to the adjacent target TLD housed within the phantom. Film registration to an institution's dose distribution is accomplished by using reference pin prick marks on the film that are correlated to the phantom geometry. The uncertainty in the registration is <1 mm.

19.2.1.4 Establishing Audit Acceptance Criteria

The acceptance criteria used by IROC Houston for its various E2E QA phantoms depend on several considerations. The primary consideration is the precision of the dosimeters used in each phantom. Clearly, acceptance criteria cannot be less than the uncertainty of the dosimeter used. The second consideration taken into account is the RT treatment modality that is being verified. An example of this is the ±5%/3 mm criterion used for the SBRT spine phantom as compared to the ±7%/4 mm criterion used for our IMRT head-and-neck (H&N) phantom. The tighter criteria were established at the behest of the trial principal investigator due to the proximity of the target to a very sensitive OAR. The third consideration involves using the phantom results from the first 10–12 institution irradiations as a pilot study to aid in determining what the 90% confidence limits are of the results. The final consideration involves setting a QA baseline that is stringent enough to insure consistency between RT centers, but does not impact on trial accrual in an overly adverse manner. The last consideration is not scientific but in the clinical trial arena it can have a significant impact.

The E2E QA phantom acceptance criteria for each phantom type (photon and proton), currently in use, are shown in Table 19.1. Molineu et al. investigated whether tightening the acceptance criteria for the IMRT H&N and pelvic phantoms would have a significant impact on the pass rate and found that for these specific phantoms the pass rate would decrease by 7% going from 7%/4 mm to 5%/4 mm (Molineu et al., 2014). When these data were presented to the clinical trial group they were met with resistance to lowering the acceptance criteria.

Table 19.1 IROC Houston Acceptance Criteria for End-to-End QA Phantoms

SRS head	5%/3 mm
IMRT H&N	7%/4 mm
IMRT pelvis	7%/4 mm
SBRT and IMRT lung	5%/5 mm
SBRT spine	5%/3 mm
SBRT liver	7%/4 mm
SBRT lung/spine	7%/5 mm (lung) and 5%/3 mm (spine)
Proton head	5%/3 mm
Proton spine	7%/5 mm
Proton lung	5%/5 mm
Proton pelvis	7%/4 mm
Proton liver	7%/4 mm

Note: IMRT, intensity-modulated radiotherapy; QA, quality assurance; IROC Houston, Imaging and Radiation Oncology Core Houston; SBRT, stereotactic body radiation therapy; SRS, stereotactic radiosurgery.

19.2.1.5 Audit Practical Logistics

IROC Houston issues credentials for NCI-sponsored National Clinical Trial Network (NCTN) groups. The various requirements for credentialing might include any combination of questionnaires, knowledge assessment forms, benchmarks, or phantom E2E QA irradiations. The specific credentialing requirements for each protocol can be found on IROC Houston's website (http://irochouston.mdanderson.org). The website also houses the Credentialing Status Inquiry (CSI) form. Once an institution has reviewed the protocol's credentialing requirements, a CSI form should be completed and submitted to IROC Houston. This form is used both to inquire whether credentialing requirements have been met, and to notify IROC Houston that the institution is requesting credentialing for a specific protocol. IROC Houston will contact the institution to discuss any missing requirements.

If an IROC Houston E2E QA phantom irradiation is required, an online phantom request form, found on IROC Houston's website, must be submitted. Once the form is completed, the institution may be placed on a wait list for the requested phantom depending on availability. Institutions are then prioritized based on a multitude of factors including internal review board (IRB) approval of the protocol, expected protocol accrual by the institution, completion of other credentialing requirements, and readiness by the institution's staff to complete the phantom irradiation. The average wait time is approximately 2–3 weeks. Once the requested phantom becomes available, an IROC Houston staff member will contact the institution's physicist to arrange the logistics of the process. At this time, the physicist at the institution must agree to irradiate and return the phantom within 10 business days. The phantom is shipped along with detailed irradiation instructions. The key instruction is that the E2E phantom is to be treated as an actual patient. Once the phantom has been irradiated, it is returned to IROC Houston along with a completed phantom irradiation form. IROC Houston will allow the TLD to fade for 14 days prior to reading them. During that time interval, the institution must submit their phantom irradiation Digital Imaging and Communication in Medicine-Radiation Therapy (DICOM-RT) data set to be used as part of the film dose distribution gamma analysis. A key component to the IROC film dose distribution analysis is that the film dose is normalized to a TLD dose measurement adjacent to the film. Thus, the dose distribution comparison is an absolute dose comparison, not a relative comparison. Once the dosimeters are read, evaluated, and compared to the institution values, the institution is notified of the results. If discrepancies are found, IROC Houston physicists contact the institution to try and resolve the differences and to possibly schedule a repeat phantom. The institution will receive a formal report of the E2E QA phantom irradiation.

Once the institution has met all requirements, IROC Houston issues a credentialing letter to the institution, the NCI's Cancer Trial Support Unit (CTSU) and other IROC offices of the credentials. In the course of the past 10 years, IROC Houston's E2E phantom program has grown in phantom types and number available for each. This is correlated with the number of advanced technology clinical trials groups requiring an E2E QA phantom as for one of their credentialed requirements.

19.2.2 United Kingdom

19.2.2.1 Clinical Trial Credentialing

The European clinical trial audit was first developed by the European Organization for Research and Treatment of Cancer (EORTC) in the 1980s (Hansson and Johansson, 1991; Dutreix et al., 1993; Kouloulias et al., 2003). These early programs used questionnaires and dosimetry audits. In the United Kingdom, a similar approach was developed for large multi-institutional clinical trials such as CHART (Aird et al., 1995), RTO1 (Mayles et al., 2004; Moore et al., 2006), and START (Venables et al., 2001a,b). In 2003, the Radiotherapy Trials QA Group (RTTQA) was set up, funded through the National Institute for Health Research Clinical Research Network (www.rttrialsqa.org.uk.). The Radiation Oncology Group within the EORTC was established for European trials, with the responsibility for assuring the quality of the data from radiotherapy trials (Kouloulias et al., 2003; Budiharto et al., 2008).

Currently, clinical trial credentialing consists of benchmark cases for contouring and treatment planning, dosimetry audits, facility questionnaires, and RT process documents. Each specific trial is assessed and the appropriate credentialing processes are identified and included within the trial. For complex RT trials, most of the credentialing processes are used; however, for more straightforward delivery techniques only one or two may be used. When trials have similar RT requirements as compared to previous trials for the same anatomic site, then the new trial's credentialing requirements will be assessed in the light of what has been previously accomplished, and prior trial credentialing may be deemed appropriate. This streamlining ("grandfathering") approach significantly reduces the workload for both the RT institutions and the QA group.

19.2.2.2 Clinical QA

The process of implementing new techniques, especially those which may be classified as advanced, is complex. It is not always clear whether the "best RT delivery" has been achieved more for a particular system or combination of systems. Widespread or national audits can help to provide confidence prior to clinical implementation of new technologies/modalities (Clark et al., 2015) and also allow comparison with others using similar systems. In the United Kingdom, there have been several such E2E audits on IMRT (Budgell et al., 2011), VMAT, and tomotherapy (Hussein et al., 2013; Clark et al., 2014), and more recently lung stereotactic radiotherapy (Distefano et al., 2015; Lee et al., 2015) and cranial SRS (Dimitriadis et al., 2015). These specific audits have been varied in terms of how fully E2E they have been, but all have, at a minimum, required planning and delivery, with some requiring imaging and delineation of volumes as well. All of these E2E audits have been a collaborative effort between different auditing groups, including the primary standards laboratories—the National Physical Laboratory (NPL), the IPEM, interdepartmental audit groups (Eaton et al., 2015), and the clinical trials QA group, RTTQA. These harmonized efforts have led to the formation of a national Dosimetry Audit Network (DAN) which aims to inform and streamline dosimetry audits such that they are accessible to all institutions, minimizes redundancy, and appropriate expertise can be drawn upon when needed.

19.2.2.3 Phantom-Detector Design

Traditionally, ion chambers and homogeneous phantoms have been used (Venables et al., 2001b; Moore et al., 2006; Clark et al., 2009b). However, more recently, there has been an increased use of more advanced anthropomorphic phantoms and a variety of detectors. Anthropomorphic phantoms are those which closely mimic the human body in terms of shape and composition. For example, a thorax phantom may have the possibility of including breast shaping, as well as including less dense material for lungs and more dense materials to represent the vertebrae. Recent examples of the use of advanced anthropomorphic phantoms, shown in Figure 19.3, were in a lung stereotactic ablative radiotherapy (SABR) audit using the CIRS thorax phantom (Distefano et al., 2015), and in a cranial SRS audit using the CIRS STEEV phantom (Dimitriadis et al., 2015).

There has also been a greater use of a range of detectors, for example, alanine dosimeters that are small in size and lend themselves well to measurements of higher doses. This has been facilitated by the use of holders for the alanine pellets (5 mm diameter and 2.4 mm thickness), which are externally shaped as a Farmer-type ion chamber, hold nine stacked pellets together (Hussein et al., 2013; Distefano et al., 2015) and can be placed within the phantom at desired locations.

In addition to phantoms holding a single dosimeter, there have also been E2E audit phantoms that have used a range of detectors together in order to perform absolute and relative dosimetry simultaneously. One such example is the recent cranial SRS phantom for which a custom dosimetry insert was designed to house two orthogonal film planes, four stacked alanine pellets, and an Exradin W1 plastic scintillator detector simultaneously.

19.2.2.4 Dosimeter Precision

There are various detectors used in the United Kingdom for E2E audits that commonly include ion chamber, alanine, and film. More recently, the Exradin W1 plastic scintillator detector and glass beads (Jafari et al., 2014) have been investigated as potential dosimeters and used alongside more conventional dosimeters in E2E audits. The RTTQA uses a PTW Semiflex ion chamber, which has

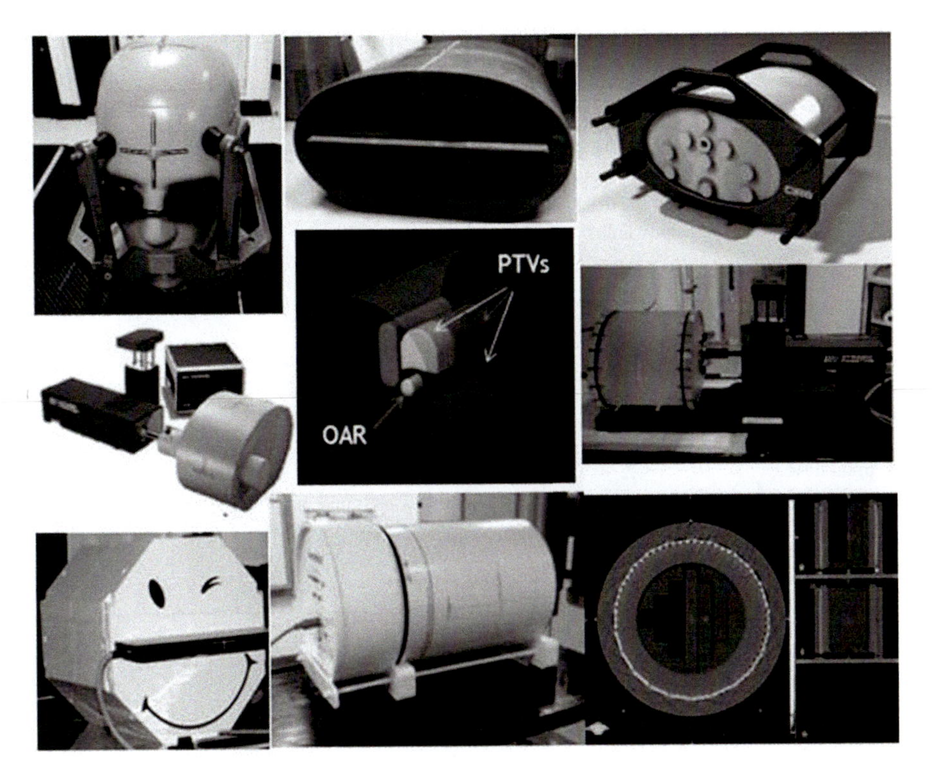

Figure 19.3

Phantoms used in UK dosimetry audits. OAR, organ at risk and PTV, planning target volume.

been calibrated directly against the graphite calorimeter at the NPL, and has an uncertainty of approximately 2% (k = 2) in reference conditions and increases to 2.3% in IMRT fields (Sanchez-Doblado et al., 2007). The alanine dosimeter uncertainty is 1.7% (k = 2) at doses over 10 Gy in reference conditions, and is expected to increase when used in complex IMRT-type field audits (Sharpe et al., 2006). The uncertainty of the dose distributions as determined by radiochromic film is similar to IROC Houston's experience of being approximately 3% as noted in Section 19.2.1.3.

19.2.2.5 Establishing Audit Acceptance Criteria

Establishing audit acceptance criteria can be difficult for a new RT technique or technology, especially where no audit has taken place before. In this case, a more prudent approach has been applied with a national audit being run without prior tolerances (Clark et al., 2014: Distefano et al., 2015; Diez et al., 2016). The full collected dataset is then analyzed at the end of the audit visits and tolerances are applied retrospectively (see Table 19.2). Initial reports can be sent to the institutions with their own results, to be followed up by a full report where the institution can assess their own results in the context of those from the full E2E audit.

Once the E2E audit result data have been collected, then there are several established approaches which can be used to create acceptability criteria.

Table 19.2 Examples of Tolerances Used for the RTTQA ArcCHECK Device in Clinical Trial QA

	Green (Optimal)	Amber (Mandatory)	Red (Suspension)
	Results within good agreement, typically 2 SD of the mean of previous audits	Audit outlier (>2 SD, or achievable by many centers based on national guidelines) Investigation recommended and any mitigating factors noted	Extreme outlier (>3 SD, or gross error based on national guidelines) Immediate investigation required before further participation in the trail
Reference output	<2%	2%–3%	>3%
Clinical plan point dose	<3%	3%–5%	>5%
Clinical plan gamma analysis	>95% 3%/3 mm and >90% 3%/2 mm (IMRT) or >95% 3%/2 mm (VMAT)	90%–95% 3%/3 mm or <90% 3%/2 mm (IMRT) or <95% 3%/2 mm (VMAT)	<90% 3%/3 mm

Note: IMRT, intensity-modulated radiotherapy; RTTQA, Radiotherapy Trials QA Group; SD, standard deviation; VMAT, volumetric-modulated arc therapy.

These include the use of the TG-119 report recommendation of 1.96 σ (Ezzell et al., 2009), and two or three sigma creating "traffic light" levels (see Table 19.2).

19.2.2.6 Audit Practical Logistics

The logistics of an E2E audit depends on the geography of the country concerned as well as the organizational structure of any clinical trial associated with the audit. In the United Kingdom, the distances between hospitals are often not large and hence visits by car are feasible. The first audit for a particular technique or hospital is generally preferred to be a site visit as these allow a good relationship to be developed with an institution, but also issues associated with, but not directly measured by the audit, can be more easily reviewed; for example, barometer calibration. Subsequent audits can be more efficiently carried out remotely by post, where a phantom is sent to the institution to be imaged, planned, and irradiated at the institution's convenience. Other aspects of an E2E audit which must be considered when designing an audit include cost, complexity of phantom/detector combination (in terms of ability for institutions to use themselves), availability of staff, distances to travel (and time), and transportability of phantoms/detectors (by careful audit teams or by courier). There are also logistics in terms of the detector itself, for example, alanine or TLDs are ideal to use as remote audits sent through the post, but need to be returned to the audit center to be read out. Alanine is not cheap to read out and therefore batching of the pellets may be preferable to keep the expenses down, but at the cost of time delay for the report. Again film is easy to send to institutions, but readout is complex and therefore it may be preferable to use a detector array, which also has the possibility of measuring absolute dose. However, the disadvantage of an array is that it will have considerably worse spatial resolution, and is more subject to wear and tear from being sent to institutions (Hussein et al., 2013).

19.2.3 Australia and New Zealand

E2E audits in Australia and New Zealand began in the 1990s in the context of clinical trials under the auspices of the Trans-Tasman Radiation Oncology Group (TROG) (Kron et al., 2002: Ebert et al., 2011). TROG distinguishes between three levels of technology involvement for radiotherapy trials:

1. Level I: No RT question, or RT dose not critical.
2. Level II: RT is standard in both arms.
3. Level III: RT is the research question, specifically the technology/technique being tested.

For trials that fall into the categories of Level II and III, the participation in a suitable dosimetric audit is recommended. If inverse planned treatments (i.e., IMRT, VMAT, or tomotherapy) are used in Level III trials, there is generally a requirement for an independent E2E audit of the treatment in question. In the case of Australia and New Zealand, these E2E audits are traditionally conducted during site visits to the participating institution (Kron et al., 2002; Ebert et al., 2011; Healy et al., 2013). Although this is a resource- and time-consuming process, it is often combined with an educational component about the trial and easily allows for observation of activities that are otherwise difficult to audit when not physically at the institution (Kron et al., 2012). Site visits for TROG participants also made it possible to perform some of the first audits of IGRT in Australia and New Zealand (Middleton et al., 2011; Kron et al., 2012). More recently, TROG has started to investigate remote auditing using electronic portal imaging based on the work of Greer and coworkers (Woodruff et al., 2013; Fuangrod et al., 2014).

In 2012, the Australian government funded the setup of the Australian Clinical Dosimetry Service (ACDS) following demand from the professions and the development of practice standards for radiation oncology (Kron et al., 2015b). The ACDS operates audits at the three levels as shown in Table 19.3 (Williams et al., 2012: Kron et al., 2013).

Table 19.3 Dosimetric Services as Classified by the Australian Clinical Dosimetry Service

Dosimetry Level	ACDS	Detector Type	Mode	System Checked	Comments
Level I	Output under reference conditions	TLD, OSLD	Remote	Every radiation beam	Identical to IROC Houston audit
Level Ib	Output under reference conditions	Ion chamber	On-site	Every radiation beam	Offered for new centers prior to opening
Level II	Dose distribution in physical phantoms	Detector array	Remote	Planning system	Can include heterogeneity and allows clarification of Level III findings
Level III	Anthropomorphic phantom end-to-end	Ion chamber, possibly film array	On-site	Entire treatment chain	Treatment specific—most relevant for clinical trials

Note: IROC Houston, Imaging and Radiation Oncology Core QA Center in Houston; OSLD, optically stimulated luminescent dosimeter; TLD, thermoluminescent dosimeter.

Of particular interest is the introduction of Level II dosimetric audits, which allows direct assessment of the setup of the radiotherapy treatment planning system (TPS) using a 2D detector array and a combination of treatment fields (Lye et al., 2014). For E2E auditing, the ACDS decided to commence its auditing program focusing on dose delivery in the presence of heterogeneities in an anthropomorphic lung phantom, and is currently extending the program to IMRT/VMAT including flattening filter-free (FFF) beams.

19.3 E2E Audit Results

19.3.1 North America (IROC Houston)

19.3.1.1 Results (Historical and Trending)

The IROC Houston E2E QA phantoms, while not anatomically exact, approximate the true anatomy of the various disease sites to be used in the protocols. The phantoms provide a consistent test to evaluate each institution's ability to deliver a specific radiotherapy treatment ensuring that all participating institutions passed the same QA criteria (Ibbott et al., 2008; Followill et al., 2012). Phantoms were used to assess SRS, IMRT, SBRT, VMAT, moving targets, and PT. Since the beginning of the RPC/IROC Houston phantom QA program, over 3100 E2E audit QA phantoms were mailed to participating institutions in North America and elsewhere. The E2E phantom credentialing process continues to be a large component of credentialing and in 2015, 650 phantoms (Figure 19.4) were sent to sites in 15 countries. The historical and 2015 phantom pass rates are seen in Table 19.4. During the time period 2001–2008, the RPC mailed IMRT head-and-neck phantoms to 537 distinct institutions. A total of 763 irradiations were analyzed. Of these, 595 irradiations or 78% successfully met the irradiation criteria. More than 125 institutions failed to meet the irradiation criteria on the first attempt and had to repeat the phantom irradiation. Of those failing to meet the criteria, the majority failed only the dose criterion. The remaining unsuccessful

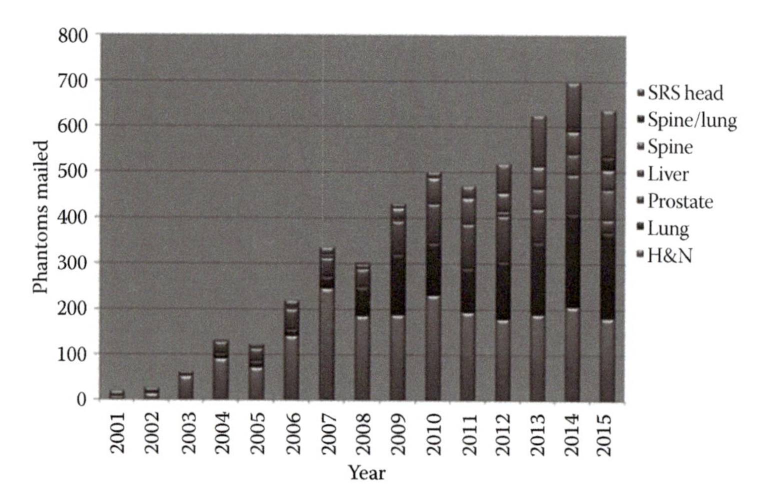

Figure 19.4

End-to-end QA phantoms mailed by Imaging and Radiation Oncology Core QA Center in Houston since 2001.

Table 19.4 IROC Houston End-to-End QA Phantom Pass Rate for its Five Photon Phantoms

Site	Technique	Historical Acceptable (%)	2015 Acceptable (%)
H&N	IMRT	84	90
Pelvis	IMRT	85	69[a]
Lung	3DCRT/IMRT	83	87
Spine	IMRT	75	92
Liver	3DCRT/IMRT	71	79

[a]Prostate phantom only sent to sites just implementing IMRT.

irradiations failed the distance-to-agreement (DTA) criterion or both the dose and DTA criteria.

The most common TPSs used to plan the irradiations of the phantom were the Philips Pinnacle and Varian Eclipse systems. The historical pass rates for these two TPSs were approximately 73% and 85%, respectively. The difference is believed to be because of difficulties at that time in modeling the penumbra at the ends of rounded multileaf collimator (MLC) leaves (Cadman et al., 2002).

19.3.1.2 Accomplishments in Reducing Errors

Some of the key findings from the E2E QA phantom irradiations include the standardization of the use of IMRT, the need for MLC QA, the use of accurate tissue heterogeneity correction dose calculation algorithms, and the need for a single isocenter per target for SBRT, and that image guidance methods are essential to minimize uncertainty. Specifically, the RPC saw the pass rate for the IMRT H&N phantom go from only 66% to 90% at the end of 2015. The RPC/IROC Houston was always proactive in the design and implementation of the phantoms prior to the activation of the protocols as was the case for RTOG 0631, a spine metastases trial. The use of the phantoms influenced the accuracy of trial patient data by requiring institutions to understand the protocols and test their new technologies on plastic patients before a real patient. It was the RPC's lung phantom results that identified that not all tissue heterogeneity correction dose calculation algorithms were the same resulting in target dose differences of up to 15%. As such, the requirement to use only modern accurate heterogeneity correction algorithms was set in place for RTOG lung trials. Another important benefit of the RPC/IROC Houston's phantom irradiation credentialing requirement was that the phantoms were built the same, used the same dosimeters and were evaluated using the same established reading system, and the results were analyzed using the same analysis software and passing criteria. Unlike relying on each institution's own QA tests, which can be highly variable among institutions (see also Chapters 14 and 17), the RPC phantom requirement was a QA audit that was constant between participating institutions helping to ensure radiotherapy treatment delivery consistency for multi-institutional clinical trials not found elsewhere (Followill et al., 2012).

19.3.2 United Kingdom

19.3.2.1 Results (Historical and Trending)

During the past decade, there have been several national UK audits carried out which have helped to support the implementation of, and set the standards for, advanced radiotherapy techniques. These have included a national IMRT audit (Budgell et al., 2011), which was designed to be independent of linear accelerator, TPS and treatment delivery method, and suitable for a plan from any clinical site. The aim was to provide an independent check on the efficient implementation of IMRT in the United Kingdom, identify problems in the modeling and delivery of IMRT, and act as a preclinical independent check for institutions starting to implement IMRT or moving to new treatment sites. For the more complex IMRT plans (mainly head-and-neck), 96.7% of fields achieved a 95.0% pixels passing percentage at a ±4%/4 mm gamma criterion where the dose was normalized to a point in a high-dose low-gradient region. A threshold of 20% was also used where "the dose at the normalization point" was identified for each beam, and the mean of these was used as the nominal "100% dose" for the plan. All the subsequent gamma analyses were confined to areas of dose higher than 20% of this value (Budgell et al., 2011). For the alanine measurements, 94.9% of the beams audited were within the preset 5.0% tolerance from the dose predicted by the TPS. Three of the audit results exceeding the 5.0% tolerance had large deviations of 77.1%, 29.1%, and 14.1%, which were traced to human error associated with carrying out the audit measurements and not affecting patient treatment. Excluding the three measurements outside 10%, the mean difference was 0.05% with a standard deviation (SD) of 1.5%. The results of this audit showed that the overall standard of beam modeling and delivery were within national guidelines.

More recently, the second national IMRT audit took place to focus on the development toward rotational IMRT (VMAT and tomotherapy). The audit also developed a novel approach to audit (Hussein et al., 2013) of using an array of ion chambers such that multiple dose points could be measured simultaneously and initial results could be given during the on-site visit (Clark et al., 2014). Forty two of 43 institutions passed their clinical trial plan with 95.0% of the measured points passing a ±3%/3 mm criterion, suggesting that in the United Kingdom, TPS modeling and delivery can achieve high accuracy for rotational IMRT. However, issues were also identified in particular with the lack of couch modeling in some TPSs.

Other advanced RT techniques which have been recently audited include intraoperative radiotherapy (Eaton et al., 2013), using compact mobile kilovoltage x-ray sources for the treatment of breast and other cancers. A national dosimetry audit for SABR in the lung has also recently been completed, coordinated by the UK SABR Consortium, to provide verification of planning and delivery of the high doses delivered in a few fractions required by this technique (Distefano et al., 2015).

Brachytherapy has been a technique in radiotherapy which in general has had less auditing attention paid to it. However, in 2010, a survey of brachytherapy quality control practices was carried out, linked to the introduction of a new code of practice (Palmer et al., 2012). Within the United Kingdom, a recent coordinated audit approach has brought together three brachytherapy audits. These include a well-type chamber audit of source strength, and an audit measuring absorbed dose in a geometric phantom for the INTERLACE trial. The third study was the first multicenter fully E2E dosimetry audit for high-dose rate (HDR) cervix brachytherapy (Palmer et al., 2015). It used a novel phantom together with radiochromic film dosimetry audit methods, obtaining dose maps using triple-channel film dosimetry, to compare TPS planned and measured (delivered) dose distributions around clinical treatment applicators. The audit visits also took the opportunity to review local procedures. The mean difference between planned and measured dose at Point A was 20.6% for plastic applicators and 23.0% for metal applicators, at standard uncertainty 3.0%. Isodose distributions agreed within 1 mm over a dose range of 2–16 Gy. Mean gamma pass rates exceeded 97.0% for plastic and metal applicators at 3% (local)/2 mm criteria.

19.3.2.2 Accomplishments in Reducing Errors

Historically, the UK dosimetry audits have been effective in identifying a range of errors, most notably at Exeter (Joslin, 1988), as well as showing an overall trend of improvement with time, in terms of fewer out-of-tolerance results, smaller ranges, and SDs of the distribution of results (Thwaites, 1996: Thwaites et al., 2003).

19.3.3 Australia and New Zealand

TROG dosimetric E2E audits have been successful in rolling out QA for several trials that use high tech radiotherapy. Over the years about half of the 80 radiotherapy centers in Australia and New Zealand had site visits to be credentialed for participation in trials. Due to resource constraints trial chairs often have to build in funding for QA in their trial's budget. Given the importance of QA in conducting successful trials as demonstrated in an earlier TROG trial (Peters et al., 2010), and the impact of QA in improving trial statistical power (Pettersen et al., 2008), there is typically no concern with adding the funding. However, it is acknowledged that site visits will become the exception rather than the norm in E2E testing, and TROG has embarked on several methods to try and reduce the need for site visits without compromising quality.

The establishment of the ACDS has helped to support clinical trials QA and participation in Level I dosimetric audits conducted by ACDS (or other suitable groups) is considered a requirement for all institutions participating in radiotherapy clinical trials. The approach of the ACDS has been informed by statistical risk management with the notion that a false positive result is better than a false negative one where an institution that delivers suboptimal radiotherapy passes the audit. This can be seen in particular in the Level II audits where initially approximately 1/3 of institutions did not meet the stringent optimal pass criteria (Lye et al., 2014). The relatively high fail rate was due to systematic issues with a planning system that does not model steep wedges well (Lye et al., 2014). The ACDS also aimed to link their Level II and III audits, which can help to identify why E2E tests in an anthropomorphic phantom fail. It also allows combining data not only from different institutions but also audit types to study performance of TPSs and draw conclusions that would be difficult to draw from a single institution (Dunn et al., 2015).

19.4 Conclusion

E2E tests are an important component of quality radiotherapy. As complexity of radiotherapy techniques and technologies increases they become essential to ensure both safe implementation and safe and efficient practice. Although the approaches taken to E2E testing across the world differ due to historical, resource, and demand reasons, they all share the same goal: to improve radiotherapy practice and ensure consistent and safe application. This is important not only for clinical practice but also for the clinical research that underpins evidence-based medicine.

References

Abdel-Wahab M et al. (2013) Status of radiotherapy resources in Africa: An International Atomic Energy Agency analysis. *Lancet Oncol.* 14: e168–e175.

Aird EGA et al. (1995) Quality assurance in the CHART clinical trial. *Radiother. Oncol.* 36: 235–244.

Bogdanich W (2010a) As technology surges, radiation safeguards lag. *New York Times*, January 26.

Bogdanich W (2010b) Radiation offers new cures, and ways to do harm. *New York Times*, January 23.

Bogdanich W and Rebelo K (2010c) A pinpoint beam strays invisibly, harming instead of healing. *New York Times*, December 28.

Budgell G et al. (2011) A national dosimetric audit of IMRT. *Radiother. Oncol.* 99: 246–252.

Budiharto T et al. (2008) Profile of European radiotherapy departments contributing to the EORTC radiation oncology group (ROG) in the 21st century. *Radiother. Oncol.* 88: 403–410.

Cadman P et al. (2002) Dosimetric considerations for validation of a sequential IMRT process with a commercial treatment planning system. *Phys. Med. Biol.* 47: 3001–3010.

Clark CH et al. (2009a) Pre-trial quality assurance processes for an intensity-modulated radiation therapy (IMRT) trial: PARSPORT, a UK multicentre phase III trial comparing conventional radiotherapy and parotid-sparing IMRT for locally advanced head and neck cancer. *Brit. J. Radiol.* 82(979): 585–594.

Clark CH et al. (2009b) Dosimetry audit for a multi-centre IMRT head and neck trial. *Radiother. Oncol.* 93: 102–108.

Clark CH et al. (2014) A multi-institutional dosimetry audit of rotational intensity-modulated radiotherapy. *Radiother. Oncol.* 113: 272–278.

Clark CH et al. (2015) Radiotherapy dosimetry audit: Three decades of improving standards and accuracy in UK clinical practice and trials. *Br. J. Radiol.* 88(1055): 20150251.

Davidson SE et al. (2008) Technical note: Heterogeneity dose calculation accuracy in IMRT: Study of five commercial treatment planning systems using an anthropomorphic thorax phantom. *Med. Phys.* 35: 5434–5439.

Derreumaux S et al. (2008) Lessons from recent accidents in radiation therapy in France. *Rad. Prot. Dos.* 131: 130–135.

Devic S et al. (2004) Dosimetric properties of improved GafChromic films for seven different digitizers. *Med. Phys.* 31: 2392–2401.

Dimitriadis A et al. (2015) Current practice of cranial stereotactic radiosurgery (CSRS) in the UK. *Radiother. Oncol.* 115: S863.

Distefano G et al. (2015) UK SABR Consortium lung dosimetry audit; absolute dosimetry results. *Radiother. Oncol.* 115: S75–S76.

Dunn L et al. (2015) National dosimetric audit network finds discrepancies in AAA lung inhomogeneity corrections. *Phys. Med. Biol.* 31: 435–441.

Dutreix A et al. (1993) Preliminary results of a quality assurance network for radiotherapy centres in Europe. *Radiother. Oncol.* 29: 97–101.

Eaton DJ et al. (2015) Interdepartmental dosimetry audits—Development of methods and lessons learned. *J. Med. Phys.* 40: 183–189.

Eaton DJ et al. (2013) A national dosimetry audit of intraoperative radiotherapy. *Br. J. Radiol.* 86(1032): 20130447.

Ebert MA et al. (2011) Dosimetric intercomparison for multicenter clinical trials using a patient-based anatomic pelvic phantom. *Med. Phys.* 38: 5167–5175.

Ezzell GA et al. (2003) Guidance document on delivery, treatment planning, and clinical implementation of IMRT: Report of the IMRT subcommittee of the AAPM radiation therapy committee. *Med. Phys.* 30: 2089–2115.

Ezzell GA et al. (2009) IMRT commissioning: Multiple institution planning and dosimetry comparisons, a report from AAPM Task Group 119. *Med. Phys.* 36: 5359–5373.

Followill DS et al. (2007) Design, development, and implementation of the Radiological Physics Center's pelvis and thorax anthropomorphic quality assurance phantoms. *Med. Phys.* 34: 2070–2076.

Followill DS et al. (2012) Credentialing for participation in clinical trials. *Front. Oncol.* 2: 1–8.

Fuangrod T et al. (2014) An independent system for real-time dynamic multileaf collimation trajectory verification using EPID. *Phys. Med. Biol.* 59: 61–81.

Hansson U and Johansson KA (1991) Quality audit of radiotherapy with EORTC mailed in water TL-dosimetry. *Radiother. Oncol.* 20: 191–196.

Healy B et al. (2013) Results from a multicenter prostate IMRT dosimetry intercomparison for an OCOG-TROG clinical trial. *Med. Phys.* 40: 071706.

Huq MS et al. (2016) The report of Task Group 100 of the AAPM: Application of risk analysis methods to radiation therapy quality management. *Med. Phys.* 43: 4209–4262.

Hussein M et al. (2013) A methodology for dosimetry audit of rotational radiotherapy using a commercial detector array. *Radiother. Oncol.* 108: 78–85.

IAEA (2004) Commissioning and quality assurance of computerized planning systems for radiation treatment of cancer. IAEA Technical Report Series No. 430. (Vienna, Austria: International Atomic Energy Agency).

Ibbott GS, Molineu HA, and Followill DS (2006) Independent evaluations of IMRT through the use of an anthropomorphic phantom. *Techn. Cancer Res. Treat.* 5: 481–487.

Ibbott GS et al. (2008) Challenges in credentialing institutions and participants in advanced technology multi-institutional clinical trials. *Int. J. Radiat. Oncol. Biol. Phys.* 71: S71–S75.

Izewska J et al. (2003) The IAEA/WHO TLD postal dose quality audits for radiotherapy: A perspective of dosimetry practices at hospitals in developing countries. *Radiother. Oncol.* 69: 91–97.

Jafari SM et al. (2014) Glass beads and Ge-doped optical fibres as thermoluminescence dosimeters for small field photon dosimetry. *Phys. Med. Biol.* 59: 6875–6889.

James H et al. (2008) Guidance for the clinical implementation of intensity modulated radiation therapy. IPEM Report 96. (York, UK: Institute of Physics and Engineering in Medicine).

Joslin C (1988) *Incident in Radiotherapy Department.* (Exeter, England: Exeter Health Authority).

Khoo VS and Dearnaley DP (2008) Question of dose, fractionation and technique: Ingredients for testing hypofractionation in prostate cancer—The CHHiP trial. *Clin. Oncol.* 20: 12–14.

Kirby TH, Hanson WF and Johnston DA (1992) Uncertainty analysis of absorbed dose calculations from thermoluminescence dosimeters. *Med. Phys.* 19: 1427–1433.

Kouloulias VE et al. (2003) The quality assurance programme of the radiotherapy group of the European Organization for Research and Treatment of Cancer (EORTC): A critical appraisal of 20 years of continuous efforts. *Eur. J. Cancer* 39: 430–437.

Kron T et al. (2002) Dosimetric intercomparison for two Australasian clinical trials using an anthropomorphic phantom. *Int. J. Radiat. Oncol. Biol. Phys.* 52: 566–579.

Kron T et al. (2012) Credentialing of radiotherapy centres for a clinical trial of adaptive radiotherapy for bladder cancer (TROG 10.01). *Radiother. Oncol.* 103: 293–298.

Kron T, Haworth A and Williams I (2013) Dosimetry for audit and clinical trials: Challenges and requirements. *J. Phys. Conf. Ser.* 444: 012014.

Kron T et al. (2015a) Medical physics aspects of cancer care in the Asia Pacific region: 2014 survey results. *Australas. Phys. Eng. Sci. Med.* 38: 493–501.

Kron T et al. (2015b) The development of practice standards for radiation oncology in Australia: A tripartite approach. *Clin. Oncol.* 27: 325–329.

Kry SF et al. (2013) Algorithms used in heterogeneous dose calculations show systematic differences as measured with the Radiological Physics Center's anthropomorphic thorax phantom used for RTOG credentialing. *Int. J. Radiat. Oncol. Biol. Phys.* 85: e95–e100.

Lee J et al. (2015) UK SABR Consortium lung dosimetry audit; relative dosimetry results. *Radiother. Oncol.* 115: S74–S75.

Lye J et al. (2014) A 2D ion chamber array audit of wedged and asymmetric fields in an inhomogeneous lung phantom. *Med. Phys.* 41: 101712.

Mayles WPM et al. (2004) Questionnaire based quality assurance for the RT01 trial of dose escalation in conformal radiotherapy for prostate cancer (ISRCTN 47772397). *Radiother. Oncol.* 73: 199–207.

Middleton M et al. (2011) Successful implementation of image-guided radiation therapy quality assurance in the Trans-Tasman Radiation Oncology Group 08.01 PROFIT Study. *Int. J. Radiat. Oncol. Biol. Phys.* 81: 1576–1581.

Mijnheer B and Georg D (Eds.) (2008) Guidelines for the verification of IMRT. ESTRO Booklet No. 9. (Brussels, Belgium: European Society for Radiotherapy and Oncology).

Mizuno H et al. (2008) Feasibility study of glass dosimeter postal dosimetry audit of high-energy radiotherapy photon beams. *Radiother. Oncol.* 86: 258–263.

Molineu A et al. (2005) Design and implementation of an anthropomorphic quality assurance phantom for intensity-modulated radiation therapy for the Radiation Therapy Oncology Group. *Int. J. Radiat. Oncol. Biol. Phys.* 63: 577–583.

Molineu A et al. (2013) Credentialing results from IMRT irradiations of an anthropomorphic head and neck phantom. *Med. Phys.* 40: 022101.

Molineu A et al. (2014) Is it feasible to tighten the criteria for IROC's anthropomorphic phantoms? *Med. Phys.* 41: 352.

Moore AR et al. (2006) A versatile phantom for quality assurance in the UK Medical Research Council (MRC) RT01 trial (ISRCTN 47772397) in conformal radiotherapy for prostate cancer. *Radiother. Oncol.* 80: 82–85.

Nishio T et al. (2006) Dosimetric verification in participating institutions in a stereotactic body radiotherapy trial for stage I non-small cell lung cancer: Japan clinical oncology group trial (JCOG 0403). *Phys. Med. Biol.* 51: 5409–5417 .

Oldham M et al. (2008) The feasibility of comprehensive IMRT verification using novel 3D dosimetry techniques compatible with the RPC head and neck phantom. *Int. J. Radiat. Oncol. Biol. Phys.* 72: S145.

Palmer AL, Bidmead M and Nisbet A (2012) A survey of quality control practices for high dose rate (HDR) and pulsed dose rate (PDR) brachytherapy in the United Kingdom. *J. Cont. Brachyther.* 4: 232–240.

Palmer AL et al. (2015) A multicentre 'end to end' dosimetry audit for cervix HDR brachytherapy treatment. *Radiother. Oncol.* 114: 264–271.

Peters LJ et al. (2010) Critical impact of radiotherapy protocol compliance and quality in the treatment of advanced head and neck cancer: Results from TROG 02.02. *J. Clin. Oncol.* 28: 2996–3001.

Pettersen MN, Aird E and Olsen DR (2008) Quality assurance of dosimetry and the impact on sample size in randomized clinical trials. *Radiother. Oncol.* 86: 195–199.

Rosenblatt E et al. (2013) Radiotherapy capacity in European countries: An analysis of the directory of radiotherapy centres (DIRAC) database. *Lancet Oncol.* 14: e79–e86.

Sánchez-Doblado F et al. (2007) Uncertainty estimation in intensity-modulated radiotherapy absolute dosimetry verification. *Int. J. Radiat. Oncol. Biol. Phys.* 68: 301–310.

Sharpe PHG and Sephton JP (2006) Therapy level alanine dosimetry at the NPL. In: Proceeding of the 216th PTB Seminar on Alanine dosimetry for clinical applications, PTB-Dos-51 (Braunschweig, Germany: PTB).

Smilowitz JB et al. (2016) AAPM Medical Physics Practice Guidelines 5.a: Commissioning and QA of treatment planning dose calculations—Megavoltage photon and electron beams. *J. Appl. Clin. Med. Phys.* 17(1): 14–34.

Tatsuzaki H and Levin CV (2001) Quantitative status of resources for radiation therapy in Asia and pacific region. *Radiother. Oncol.* 60: 81–89.

Thwaites DI (1996) External audit in radiotherapy dosimetry. In: *Radiation incidents*, pp. 21–28. (Eds. Faulkner K and Harrison RM, London, UK: British Institute of Radiology).

Thwaites DI et al. (2003) The United Kingdom's radiotherapy dosimetry audit network. In: *Standards and Codes of Practice for Medical Radiation Dosimetry.* IAEA STI-pub-1153, pp. 183–90. (Vienna, Austria: International Atomic Energy Agency).

Venables K et al. (2001a) A survey of radiotherapy quality control practice in the United Kingdom for the START trial. *Radiother. Oncol.* 60: 311–318.

Venables K et al. (2001b) The START trial—Measurements in semi-anatomical breast and chest wall phantoms. *Phys. Med. Biol.* 46: 1937–1948.

Williams I et al. (2012) The Australian Clinical Dosimetry Service: A commentary on the first 18 months. *Australas. Phys. Eng. Sci. Med.* 35: 407–411.

Woodruff HC et al. (2013) Gantry-angle resolved VMAT pretreatment verification using EPID image prediction. . 40: 081715.

Zubizarreta EH, Poitevin A and Levin CV (2004) Overview of radiotherapy resources in Latin America: A survey by the International Atomic Energy Agency (IAEA). *Radiother. Oncol.* 73: 97–100.

Brachytherapy Dosimetry in Three Dimensions

J. Adam M. Cunha, Christopher L. Deufel, and Mark J. Rivard

20.1 Introduction

Compared to external beam radiation therapy (EBRT), brachytherapy poses certain challenges yet offers potentially better dose conformality due to using lower photon energies and having geometric advantages. Administration of brachytherapy usually has fewer geometric constraints than EBRT due to lack of potential collisions of equipment with the patient, and since the radioactive sources are placed inside or near the cancerous tumor, the dose to the tumor may be made very high while still protecting nearby healthy tissues. While typically more invasive than EBRT, brachytherapy offers unique geometric possibilities through applicator placement during surgery that exploits anatomic alterations not possible when delivering EBRT. Building upon these concepts, this chapter focuses on aspects of dosimetry as it applies to calculating dose, using dose kernels in three-dimensional (3D) planning, and validating treatment delivery. To this end, Section 20.2 describes the basis for calculating and measuring brachytherapy dose distributions in three dimensions, Section 20.3 covers the practical aspects of performing image-guided 3D treatment planning, and Section 20.4 covers the importance of techniques for commissioning applicators and validation of treatment delivery with *in vivo* dosimetry.

20.2 3D Data Considerations

20.2.1 Scope

The goal of brachytherapy dosimetry is generally to determine the dose distribution in the vicinity of one or more brachytherapy sources. This may be in water, tissue, phantom, or other media. Often there is the assumption that the source position(s) and medium composition remain fixed across the time period between simulation and the dose delivery. Section 20.2.2 examines the necessary parameters to specify the dose distribution from an individual brachytherapy source. This is followed in Section 20.2.3 by a discussion of comparison metrics for 3D dose distributions, which is necessary (for example) when commissioning a clinical treatment-planning system (TPS). The section concludes with a review of the resources data archives available for obtaining brachytherapy source dosimetry information in Section 20.2.4.

20.2.2 Reference Data Acquisition

For clinical brachytherapy dosimetry, the current standard-of-care method for determining the dose distribution in a patient is through the AAPM TG-43 formalism (Nath et al., 1995; Rivard et al., 2004; Pérez-Calatayud et al., 2012). This dose calculation formalism is used to quantify the dose distribution near a brachytherapy source within a water phantom. The brachytherapy source is assumed to be cylindrically symmetric in design and thus have a cylindrically symmetric dose distribution. Consequently, the 3D dose distribution may be specified using only two parameters (e.g., r and θ).

In clinical TPSs, dose was historically obtained using reference data such as along–away tables in which distance from the source origin along the source axis of symmetry and distance from the source origin away from the source axis of symmetry were denoted for a computational lookup table having a Cartesian coordinate system. Based on the frequent use of uniform binning across the ranges for both along and away distances, large interpolation errors in dose calculation close to the source (where the dose was highest) could result. Therefore, a newer technique was developed in which the dose falloff was approximated by accounting for the decreased solid angle from the source with increasing distance. For a small source and at large distances, this dose falloff followed an inverse-square effect. For circumstances when the point of interest was located closer to the source, the source could often be approximated as a line segment due to the spatial distribution of radioactivity within the encapsulation. With this characterization in a polar coordinate system, variations in dose beyond those expected due to solid angle were expressed as anisotropy. Using these geometric approximations of the source design, the formulism of the newer technique was less subject to interpolation errors than the approach using along–away tables (Rivard, 1999; Kouwenhoven et al., 2001; Meli, 2001; Rivard et al., 2002; Song et al., 2003).

The dose distribution in the vicinity of a brachytherapy source may be specified using the TG-43 formalism. This formalism requires specification of brachytherapy dosimetry parameters, which are used in clinical TPSs to permit uniform dose calculation worldwide for a given brachytherapy source model. Acquisition of the dosimetry parameters used in the TG-43 formalism requires calculations or measurements of the dose distribution in the vicinity of a brachytherapy source. The following sections describe the necessary methods for obtaining reference datasets for 3D brachytherapy dose distributions.

20.2.2.1 Monte Carlo Techniques

20.2.2.1.1 Source Codes

Numerous computer source codes have been written to accomplish radiation transport simulations, and several of these have been applied to the field of radiotherapy (Rogers, 2006). Specific to brachytherapy dosimetry, several codes have been used (Williamson, 2006) since the first 3D dose characterization of a brachytherapy source by Krishnaswamy (1971). Popular codes for acquiring brachytherapy source reference data include EGS4, GEANT4, MCNP, PENELOPE, and PTRAN (Rivard et al., 2009a; Beaulieu et al., 2012). These codes simulate emission of radiation from the brachytherapy source and propagation in the surrounding medium. Monte Carlo methods are used to randomly sample the possible photon energies, emission angles, and radiological interactions following nuclear disintegration, and to make estimates of macroscopic quantities such as absorbed dose through the summation of microscopic phenomena. This topic is explored further by Thomadsen et al. (2008).

20.2.2.1.2 Simulation Reference Data

Monte Carlo codes require fundamental reference data to perform dosimetric estimates. These data include knowledge of the emission spectrum of the radionuclide contained within the brachytherapy source, cross-section libraries for radiological interactions (generally photoelectric effect and incoherent scattering for the brachytherapy energy regime), physical properties (elemental composition and mass density) of all materials of interest, and geometric specification of the 3D environment. As a practical example for a high-dose-rate (HDR) ^{192}Ir brachytherapy source, the photon and beta spectrum need to be specified; ^{192}Ir undergoes nuclear disintegration via beta decay to ^{192}Pt (95.13%) or electron capture to ^{192}Os (4.87%) (Ballester et al., 2009; Rivard et al., 2010a; Granero et al., 2011). The Evaluated Nuclear Data File/B version VI Release 8 (IAEA, 2016) is the current accepted standard for the radiological interaction cross-section libraries of Monte Carlo codes. From here, photoatomic cross-section libraries are prepared and formatted for accessibility by a particular Monte Carlo code. The final components (physical properties) are generally taken from reference reports such as by the International Commission on Radiation Units and Measurements (ICRU, 1989, 1992), especially for phantom compositions. The mass density of liquid water at 22°C is chosen for comparison to measurements performed at a similar reference temperature. However, the mass density does not change by more than 0.5% from 20°C to 37°C (USGS, 2016). While there are extant standards for the compositions of human tissues (ICRP, 1975, 2002; ICRU, 1992), there is large variability among patients even for the same body location (Landry et al., 2010). The importance of tissue specification becomes increasingly important with diminishing photon energy, as for low-dose-rate (LDR) low-energy sources. For estimating air-kerma strength or reference air-kerma rate, the simulation occurs in vacuum and the photon energy fluence in each energy bin (0.1 keV is generally adequate) is multiplied by the photon energy and mass energy absorption coefficients in each energy bin. Good practice guidelines for performing reference Monte Carlo dose calculations are summarized in the AAPM Task Group 43 update (Rivard et al., 2004).

20.2.2.1.3 Source Design

Brachytherapy may be distinguished from other types of radiation therapy based on geometric conditions. If the ionizing radiation source is farther than approximately 0.1 m from the clinical target volume, the treatment modality may be considered as EBRT instead of brachytherapy (DIN, 2000). Unlike nuclear medicine, in which a liquid solution containing radioactivity is injected with biological agents devised to localize the radiation emissions, brachytherapy sources are sealed sources that generally are macroscopic and contained within a metallic or plastic capsule. An exception is microsphere brachytherapy for hepatic cancers where ^{90}Y sources contained in resin or glass spheres of approximately 30 μm in diameter are injected into the liver and become lodged in the microvasculature (Dezarn et al., 2011).

The cylindrical symmetry of modern brachytherapy sources is based on the historical design of ^{226}Ra needles and ^{222}Ra tubes (Aronowitz, 2002). Practically, there are deviations from this goal, due to either minor issues such as the eyelet of a needle or major issues such as nonuniformity of the radioactivity distribution (Mitch et al., 2006). The latter causes azimuthal asymmetry in the dose distribution, which can neither be accounted for with conventional brachytherapy TPSs nor easily measured by the clinical medical physicist.

The encapsulation material for brachytherapy sources is typically metallic for leakage protection and sterilization purposes: commonly titanium is used for low-energy LDR sources (often referred to as seeds), and stainless steel for high-energy HDR sources. Of course, there are infinite design possibilities, and over two dozen LDR seed models have been devised in the past century. There can be substantial volume within the capsule not occupied by either the radioactive pellet or a radiopaque marker. This void permits mobility of the internal components and instability in the external dose distribution (see, e.g., Rivard, 2001). Given that typical clinical use simultaneously employs several sources, the dosimetric impact of dynamic internal components is less than for a single source. High-energy HDR sources with radionuclides such as ^{192}Ir or ^{60}Co are designed with proportionately smaller voids within their capsule, and the internal components are not as mobile. In fact, some manufacturers crimp the HDR capsule to affix the location of the radioactive source within the encapsulation.

Brachytherapy sources are positioned within the body through several methods. For LDR seeds that are permanently implanted interstitially, there is a risk that the seeds may migrate away due to dynamic tissue properties and body-fluid circulation. To prevent this, seeds may be stranded within a suture-like material (Lee et al., 2003; Reed et al., 2007) or linked together (Al-Qaisieh et al., 2004). Less commonly, LDR seeds are rigidly positioned within an applicator such as for ocular brachytherapy (Chiu-Tsao et al., 2012). Using an entirely different approach, HDR sources are attached to a delivery cable and administered within needles or applicators via a computer-controlled stepping motor (Kubo et al., 1998). The source-positioning accuracy using this method is approximately 1 mm, but it can vary if the geometry of the tubes connecting the remote afterloading system to the patient is altered (Jursinic, 2014).

Brachytherapy source strength is specified in terms of reference air-kerma rate (ICRU, 1985) or air-kerma strength (Nath et al., 1987). Use of other antiquated units such as apparent activity is discouraged (Williamson et al., 1999) and has resulted in numerous medical errors and treatments that overdosed the patient (NRC, 2009). Source strength is determined by the vendor who provides a calibration certificate for each shipment. The clinical medical physicist is required to perform an assay of the source strength preceding patient treatment (Butler et al., 2008). The necessary instrumentation to perform this task includes a reentrant well-type air ionization chamber, a chamber insert to reproducibly position the source within the chamber, an electrometer for measuring the source strength with either integrated charge (applicable for LDR sources) or electrical current (applicable for HDR sources), and a barometer and thermometer (sometimes combined as one device). All these instruments should have calibration traceability to a national metrology institute specializing in radioactivity such as the Laboratoire National Henri Becquerel (LNHB) in Europe or the National Institute of Standards and Technology (NIST) in the United States.

20.2.2.1.4 Phantom Design

Derivation of 3D dose distributions is reliant upon establishing the medium of interest. For photon-emitting sources, the medium for specifying absorbed dose is water for spheres of 15- and 40-cm radii for low-energy (<50 keV) and high-energy (≥50 keV) sources, respectively. These dimensions provide adequate radiation backscatter to approximate photon scatter conditions for an infinite phantom for distances up to 10 cm from the centrally positioned source (Pérez-Calatayud et al., 2004; Melhus and Rivard, 2006).

The computational phantom is segmented into regions to provide spatial resolution for the evaluated dose distribution. This segmenting generally follows the polar coordinate system of the TG-43 formalism, with partitions made for each radius and angle to be sampled. Depending on the Monte Carlo code and method used to estimate the dose distribution, there can be volume averaging associated with each segmented partition. For locations close to the source, where the dose rate is highest and perhaps most clinically important, this volume averaging can substantially reduce the estimated value of the true quantity. For angular resolution, 1° sampling is generally adequate where higher resolution can reveal interesting properties of the source angular radiation emissions.

20.2.2.2 Measurement Techniques, Detector Types, and Features

In addition to estimating the 3D radiation dose distribution in the vicinity of a brachytherapy source using computational tools such as Monte Carlo methods, it is crucial to include a practical evaluation of reality using experimental methods. It cannot be overstated how important it is to measure the radiation dose distribution for a source instead of relying only upon design drawings and assumptions of fabrication processes. There have been instances where the manufactured source geometry has differed substantially from the intended source design (Wierzbicki et al., 1998; Rivard, 2001).

Several types of radiation detectors have been used for this purpose and include thermoluminescent dosimeters (TLDs), TLD powder, radiochromic film, ionization chambers, diodes and solid-state devices, diamond detectors, and scintillators. These radiation detectors have different attributes and weaknesses (Williamson and Rivard, 2005, 2009). The useful properties of radiation dosimeters relevant to brachytherapy are included in Chapters 4 through 6 and 8.

The most established method is TLD as evaluated by the Interstitial Collaborative Working Group (Anderson et al., 1990). TLDs for radiation dosimetry were developed in the 1980s (Cameron, 1991) and now come in several sizes such as (1 mm^3) microcubes, 1 mm × 3 mm × 3 mm chips, and powder (Tailor et al., 2008). Their main attributes are high signal-to-noise ratio and the ability for precision positioning in a solid phantom material. However, their regular use in the clinic is diminishing in favor of film and solid-state devices. Also, there are recent concerns on the necessary correction factors to account for detector energy response (Carlsson Tedgren et al., 2012; Horowitz and Moscovitch, 2013; Massillon-JL et al., 2014).

Radiochromic film is a newer detector type that has improved spatial resolution and detector volume averaging over TLDs. It is versatile and easily accessible for practicing medical physicists and dosimetrists, and film technology has advanced significantly over the last decade. Radiochromic-type EBT3 film can be used in water, is nearly water equivalent, and has a flat dose–response curve. The film is also flexible and may be curved around phantoms or source applicators. Further, modern compositions demonstrate minimal variations in detector response over the photon energy range available for brachytherapy dosimetry (Sutherland and Rogers, 2010; Lewis et al., 2012). A complete discussion of the dosimetry and calibration of EBT3 film in the context of external beam dosimetry was recently completed by Crijns et al. (2013). The topic of application of modern EBT3 film in brachytherapy has been covered recently (Palmer et al., 2013a), whereas its use in low-energy synchrotron beams has been discussed in Chapter 25. It is important when using radiochromic film to account for any dose–response nonlinearity of the film and also any discrepancies between the red, green, and blue signals of the film scanner. In that study, the variance across the film was under 1.4% for 1 Gy with the largest deviations happening at the edges of the film. In general, it is recommended to use films that are a few millimeters larger than the area needed for measurements to avoid these edge effects.

Air ionization chambers are the gold standard for measurement of EBRT radiation fields due to the well-known corrections necessary to convert detector response to absorbed dose. However, for brachytherapy dosimetry, there can be significant volume averaging of dose gradients and large corrections for phantom displacement by the detector. Further, detectors can exhibit large response variations with radiation angular incidence. Therefore, air ionization chambers are infrequently used for accurate brachytherapy dosimetry, especially in proximity to the source.

Diodes have the highest signal-to-noise ratio of any radiation detector used for measuring brachytherapy dose distributions (Williamson and Rivard, 2005). This permits dose measurements of LDR sources over a large range of distances where the dose rate can vary by a few orders of magnitude. However, diode response is sensitive to temperature and tiny changes in voltage potential. Further, they generally do not have uniform angular response, which complicates correlating measured output with absorbed dose. Diodes serve especially well for relative measurements of similar dose rates from low-energy sources. Other solid-state devices such as metal oxide silicon field effect transistors (MOSFETs) and diamond detectors have similar properties as diodes (Rustgi, 1998; Williamson and Rivard, 2005; Lambert et al., 2007). However, these detectors are now used more frequently for *in vivo* dose measurements than for reference measurements of 3D dose distributions (Tanderup et al., 2013).

Plastic scintillators are advantageous for brachytherapy dosimetry in that they are often water equivalent or tissue equivalent depending on composition customization, have small active volumes and correspondingly small volume averaging, and have large signal-to-noise ratio and dynamic range (resultant with good response linearity) for measurements close to the source. They can also be moved via a computer-controlled stage within a water phantom for measurements without phantom corrections (Lambert et al., 2006; Therriault-Proulx et al., 2011). However, their potential for spatial resolution for discerning a 3D brachytherapy dose distribution is confounded by the stem effect, which is the contribution of signal, i.e., light, from within the optic fiber connecting the scintillator to the detector (Kertzscher et al., 2011).

Other methods for measuring brachytherapy dose distributions include polymer gels (Massillon-JL et al., 2009) and Fricke gel chemical dosimeters (De Deene et al., 2001; MacDougall et al., 2002), which have a unique attribute of rendering the 3D dose distribution around a brachytherapy source as discussed in more detail in Chapters 5 and 6. However, the signal-to-noise ratio and reproducibility of these detectors are inferior to other detector types. An additional method for discerning a brachytherapy dosimetry parameter, specifically the dose-rate constant, is the photon spectrometry technique (Chen and Nath, 2007). This method combines measurements of the emitted photon spectrum with Monte Carlo simulations of the measurement system but is unfortunately without traceability to a national metrology laboratory for evaluation of source strength (Williamson et al., 1998). Consequently, it is not used for consideration of candidate values for measurements of the dose-rate constant.

20.2.3 3D Data Comparison Metrics

With availability of a brachytherapy source 3D dose distribution, there are several methods for comparing it to other 3D dose distributions. As examples, this task would be necessary for commissioning the source within a clinical TPS, comparing two different source models of the same radionuclide, or even comparing sources containing different radionuclides.

There are several established methods for comparing dose distributions, which were largely developed in EBRT for treatment-plan comparisons as discussed in detail in Chapter 14. Dahlin, Hogstrom, and others proposed using distance-to-agreement (DTA) to indicate regions where a specified tolerance was exceeded (Dahlin et al., 1983; Hogstrom et al., 1984). A complementary and contemporaneous approach using dose differences was suggested by Mah et al. (1989). These metrics of DTA and percentage dose differences were combined into a single method by Van Dyk, Cheng, Low, Venselaar, and others (Van Dyk et al., 1993; Cheng et al., 1996; Low et al., 1998; Venselaar et al., 2001). The modern terminology for this method is the gamma index method, which allows users to quantify the level of agreement between two 3D dose distributions and knowingly respond.

For EBRT treatments such as intensity-modulated radiotherapy (IMRT), passing criteria for the gamma index are typically 3% and 3 mm. In regions of steep dose gradients for EBRT, the gamma index method should default to the DTA comparison (Low et al., 2013). Beyond the gamma index method, another approach for comparing dose distributions with high gradients was examined by Moran et al. (2005). For EBRT, use of this method for treatment-plan evaluation has become mature and considerations have been made for the presence of noise in the datasets (Low and Dempsey et al., 2003), processor time savings for comparing the datasets over three dimensions (Ju et al., 2008), and pooling of institution plan data to guide the comparison criteria (Ruan et al., 2012).

In brachytherapy, the dose gradients are even steeper than for EBRT, especially for single source dose distributions. Poon and Verhaegen (2009) appear to have been the first group to evaluate 3D brachytherapy dose distributions using the gamma index method, where they selected 3%/2 mm criteria for comparing treatment plans derived using analytical calculations with plans from Monte Carlo simulations. Soon afterward, application of this method to general comparisons of 3D brachytherapy dose distributions was suggested with consideration of how the criteria should vary depending on media and brachytherapy source photon energy (Rivard et al., 2010). For a shielded applicator, dose difference criteria of 1% or 2% and DTA criteria of 1 or 2 mm were evaluated with $(0.5 \text{ mm})^3$ grid size for comparing the reference Monte Carlo results to the treatment-plan dose distribution (Yang et al., 2011). Nearly 100% passing was achieved over the sampling region (8-cm deep and 16-cm diameter) with largest discrepancies occurring at the applicator edge in contact with the patient with dose gradients >60% per millimeter. For another shielded applicator, Petrokokkinos et al. (2011) observed that 5%/2 mm criteria were not sufficient for explaining differences between reference Monte Carlo results and the treatment-plan dose distribution, and that the percentage difference, sometimes exceeding 30%, was the crucial metric. In the same year, Bannon and colleagues evaluated the potential for high-resolution characterization of a curvilinear brachytherapy source using line segments in a brachytherapy TPS using TG-43 dosimetry parameters (Bannon et al., 2011). No other method is available for describing nonstraight sources in conventional TPSs. Using dose superposition of discrete line segments that exceeded the 2%/2 mm criterion, passing rates of 99% were observed.

The principle of the gamma index analysis has also been considered for evaluating agreement between measured and planned doses for *in vivo* dosimetry (Tanderup et al., 2013), and is extended to account for uncertainties in position and time (Espinoza et al., 2015). Additional discussion of quantitative comparisons of 3D dose distributions is given in Chapter 14.

20.2.4 Brachytherapy Dosimetry Data Availability

Brachytherapy dosimetry data are available to clinical medical physicists from several sources. The first resource to be considered is the scientific literature. Based on dosimetric prerequisites and publication requirements set forth by the AAPM, dosimetry investigators who characterize the 3D brachytherapy dose distribution will try to have their results published (Williamson et al., 1998; Rivard et al., 2004; Li et al., 2007; Pérez-Calatayud et al., 2012). These data serve as the basis for subsequent evaluations by professional societies such as the AAPM and GEC-ESTRO (Groupe Européen de Curiethérapie–European Society for Radiotherapy and Oncology) for formulating consensus datasets for promoting uniform worldwide use.

Clinical medical physicists interested in a new brachytherapy source model will often have marketing materials provided to them from the source vendor. These materials may refer to unpublished research on brachytherapy dose distributions and dosimetry parameters, and may also include peer-reviewed scientific publication(s) for the same source. When more than one dataset is available, the clinical user should always make preference toward data that have undergone scientific peer review. Methods on how the clinical medical physicist should evaluate brachytherapy dosimetry data in the absence of societal consensus data are detailed elsewhere (Rivard et al., 2004; Thomadsen et al., 2008).

Brachytherapy sources are listed on the Brachytherapy Source Registry if they have met certain dosimetric prerequisites and robust practices are in place for vendor calibration standards (Williamson et al., 1998; Rivard et al., 2004; Li et al., 2007; Rivard et al., 2010a). This Registry is managed jointly by the AAPM and the MD Anderson Imaging and Radiation Oncology Core (IROC) Quality

Assurance Center, Houston, Texas. The initial motivation for developing this Registry was to identify brachytherapy sources that met the dosimetric prerequisites and delineate the sources that may be included in cooperative clinical trials. For source models not included on the Registry, clinical users are advised to either file an application with the AAPM for having that source be included on the Registry or limit use of the source to situations where a clinical trial is overseen by an appropriate regulatory body such as the local institutional review board or other committee formally designated to approve, monitor, and review biomedical and behavioral research.

Another societal resource to clinical medical physicists seeking reference data for brachytherapy dose distributions is the ESTRO website managed by the BRAchytherapy PHYsics Quality assurance System (BRAPHYQS) working group of GEC-ESTRO (ESTRO, 2016). This website is most convenient in that data are included in spreadsheet format to facilitate the task of source commissioning through quantitative comparisons with TPS results.

Yet another online resource for brachytherapy dose distributions is the Carleton University website (Carleton Laboratory for Radiotherapy Physics, 2013). This website includes brachytherapy dosimetry results for ^{125}I, ^{103}Pd, ^{192}Ir, and ^{169}Yb sources, all from Monte Carlo simulations using the same code (BrachyDose) and methods. These data have high spatial resolution and therefore permit evaluation of dosimetry parameters close to the source. Further, the dose distributions for some high-energy sources are parsed into dose from primary photons originating from the source, single-scattered photons, and multiple-scattered photons. These data support the calculation of dose distributions in heterogeneous media and differing radiation scatter conditions using collapsed-cone methods (Williamson et al., 1991; Russell et al., 2005).

20.3 3D Treatment Planning

Modern brachytherapy treatment planning is considerably more complex than it was just a few decades ago, thanks in large part to advances in imaging technology. Brachytherapy TPSs that use 3D images have become an important part of how the brachytherapy team targets cancer and minimizes normal tissue complications. Such systems feature sophisticated algorithms for 3D dose calculation and volume optimization, permit 3D organ and applicator reconstruction, and have volume-based dosimetric reporting tools for plan selection and patient follow-up. Whereas Section 20.2 introduced 3D source reference data in brachytherapy, Section 20.3 focuses on applying that data in the TPS. This topic is divided into five main areas: algorithms for calculating dose (Section 20.3.1), mapping dose to image datasets (Section 20.3.2), dose optimization methods for 3D anatomy-based brachytherapy (Section 20.3.3), 3D volume-based reporting of plan dosimetry (Section 20.3.4), and incorporating radiobiological models into the dose calculations (Section 20.3.5).

20.3.1 Treatment Planning System Dose Calculation Algorithms

The dose calculation algorithms used by most, if not all, commercial TPSs may be categorized as either correction-based or transport-based algorithms. Correction-based algorithms (as the moniker suggests) utilize correction factors in order to calculate the dose to a point inside the patient from a source with a given source strength. The most common correction-based algorithm used in brachytherapy is defined by the AAPM TG-43 report (Rivard et al., 2009b, 2010; Beaulieu et al., 2012). The TG-43 formalism may be used to perform a 3D dose calculation to all regions inside a water medium according to the source type, source strength, and position of the source relative to the reference point. The correction factors adopted by the TG-43 formalism have physical interpretations, for example, the radial dose function represents attenuation and scatter in water along a distance orthogonal to the source axis (removing the effects for solid angle) and are determined by a combination of Monte Carlo and empirical methods. The accuracy of the TG-43 dose calculation is fundamentally limited by the accuracy of the underlying correction factors. Because the TG-43 formalism is an all-water calculation, the patient and applicator heterogeneities are ignored, which may or may not be clinically significant depending upon the application. Correction-based TG-43-style calculations that account for applicator heterogeneities have been proposed by several authors, including Astrahan et al. who developed the Plaque Simulator™ TPS (Astrahan, 2005; Rivard et al., 2009b; Deufel and Furutani, 2015). TG-43-style approaches are relevant to any situation where the source position is fixed relative to the applicator, for example, eye plaques for treatment of ocular melanoma or vaginal cylinders with internal shields.

Brachytherapy has steadily moved toward the adoption of transport-based algorithms in order to accurately handle patient and applicator heterogeneities. Transport-based algorithms calculate the dose in 3D geometry using the fundamental equations for radiation transport. Transport algorithms have the difficult task of accurately accounting for photon and electron scatter, which can contribute up to 50% of the total dose for ^{192}Ir sources at large distances, and even larger percentages of the total dose for low-energy sources (Carlsson and Ahnesjo, 2000). There are different approaches to solving the transport equation in a time-efficient manner. Monte Carlo methods are perhaps the most well-known class of transport algorithm and take a stochastic approach. Graphics processor unit (GPU) methods have helped to hasten the Monte Carlo approach for brachytherapy applications using parallel computing (Hissoiny et al., 2012). Discrete Boltzmann equation, for example, ACUROS, and collapsed-cone algorithms simplify the transport calculation in order to reduce dose calculation times. Discrete Boltzmann equation solutions discretize the energy transport in space, whereas collapsed-cone methods use point kernel dose superposition that separate primary and scatter terms (Carlsson and Ahnesjo, 2000; Russell et al., 2005; Beaulieu et al., 2012). Discrete Boltzmann and collapsed-cone methods are currently available only for HDR ^{192}Ir source models. Gynecological applicators with internal shields are a type of treatment where fast 3D transport algorithms are of great clinical benefit (Mikell et al., 2013). The TPS is able to display dosimetric penumbra and shield transmission information relative to organs inside the body, permitting identification of the proper shield size and orientation to shape the dose distribution to treat the diseased site while minimizing dose to nearby healthy tissues. Future developments in commercial TPSs will hopefully include transport calculations for common low-energy brachytherapy sources, including ^{125}I, ^{103}Pd, and ^{131}Cs. Additionally, the future of brachytherapy treatment planning includes more regular use of Monte Carlo, or comparable methods, in HDR brachytherapy calculations.

20.3.2 Organ and Applicator Delineation

3D image sets, whether they originate from computed tomography (CT), magnetic resonance imaging (MRI), positron emission tomography (PET), or ultrasound, allow the user to visualize the internal anatomy of the patient in three dimensions when contouring structures and delineating applicators. 3D images permit soft-tissue windowing, which helps define the boundaries of an organ or target in the axial, and sometimes on nonaxial, image planes. Moreover, 3D images facilitate volume-based dose reporting, i.e., DVH, methods for evaluating dosimetry, and have led to the adoption of complex dose optimization methods for plan creation

3D organ and applicator delineation is similar in brachytherapy and EBRT, and the reader may find the discussion of contouring basics in other sections of this book. However, the consequences of interobserver variability in contour delineation are more dramatic for brachytherapy than for EBRT. Brachytherapy dose distributions have steeper dose gradients due to the proximity of sources inside and adjacent to the target or organs at risk (OARs). These higher dose gradients increase the sensitivity of the volume dose calculations and statistics for how contours are drawn. Interobserver tests have been performed to quantify how the dose statistics in brachytherapy treatment planning varies among observers. For example, Hellebust et al. (2013) compared the contours on MR images from ten different observers using six different patient scans, and found that gross tumor volume (GTV) and high-risk clinical target volume (HR-CTV) coverage (D_{90}) varied with a standard deviation of 8%–10%, whereas doses for the rectum, bladder, and sigmoid (D_{2cm3}) varied with standard deviations between 5% and 11%. Although 3D organ delineation permits better assessment of the dose to tissues inside the patient, it also raises the stakes in the consequences of contour errors or variability. 2D planning methods, in comparison, had much more formulistic methods of calculating target, for example, ICRU-defined reference points for a tandem-and-ovoids applicator, and likely less interobserver variability.

Another challenge with 3D organ delineation is how to handle variation in organ location and size within the radiation course for fractionated brachytherapy. The relative locations of applicator, target, and OARs may change from fraction to fraction, and if the motion is significant, unique plans may need to be generated for each fraction (Morgia et al., 2013; Nesvacil et al., 2013). The location of hot spots in target and OARs may change between plans for fractionated treatments, and it has been debated as to how clinical users should sum dose between fractions. A *worst-case scenario* approach is the easiest method to use, since the cumulative DVH parameter, for example, D2cc, becomes simply the sum of the DVH parameters over all plans, for example, D2cc_Fx1 + D2cc_Fx2 + and so on (Anderson et al., 2013). An argument for using the worst-case scenario is that hot spots are often in the same location, as well as this is how dose has been tracked historically and OAR dose limits were established using this approach. A disadvantage of the "worst-case" approach is that it does not distinguish between a treatment course where the same spot received high dose in all fractions and a treatment course where the hot spot was relocated in each fraction. A more sophisticated method uses deformable image registration (DIR) to morph the 3D dose distributions from all treatment fractions onto a single-patient image set. It can even be used to add the dose between EBRT and brachytherapy sessions (Vasquez Osorio et al., 2015). This approach is still considered investigational but has the potential to improve our knowledge of how hot spots and other doses produce toxicity (Andersen et al., 2013; Petric et al., 2013; Kim et al., 2014; Sabater et al., 2014).

In HDR brachytherapy, the applicator must be delineated in order for the TPS to determine where sources will be positioned inside the patient. Just as with organ delineation, the high-dose gradients make the treatment-plan statistics extremely sensitive to the accuracy of the applicator delineation. 3D treatment plans are generally optimized using computerized methods and may have a higher level of modulation, or variation in treatment times, between dwell positions than 2D plans (Schindel et al., 2013). 3D treatment-planning methods therefore enhance the effect that applicator movement has on the shape of the dose distribution. For a sequence of 20 HDR cervix cancer treatments, Schindel and colleagues found that accuracy in applicator reconstruction of the tandem-and-ovoids applicator should be better than 1.5 mm to prevent greater than 10% variation in the reported rectal D2cc dose. Consequently, it is important that the treatment team use appropriately high-resolution imaging and aim to identify the applicator as accurately as possible. The team should also immobilize the applicator with respect to the patient to prevent relative movement between the times of simulation and treatment.

20.3.3 Optimization

Brachytherapy optimization is defined as the process of determining how much and where to position source strength in the vicinity of a patient. Treatment planning optimization has changed substantially in recent years due to the ability to visualize soft tissue inside the patient and identify the relative positions of target, organs, and sources. Additionally, there have been several technological improvements in how accurately sources can be positioned with LDR and HDR treatment devices, and HDR brachytherapy affords a level of source strength modulation that was previously unavailable. Thus, optimization is an integral part of the brachytherapy treatment process.

Most popular algorithms employed by radiotherapy were developed to solve problems in other fields, including mathematics, economics, and logistics. Several classes of optimization algorithms are available and readers interested in an overview may refer to several recent publications (Ezzell, 1994; Thomadsen et al., 2008; Adamczyk et al., 2013; De Boeck et al., 2014).

Phenomenological approaches to optimization are based on human experience, such as standard loading, and are popular with many clinical users. The primary advantages of phenomenological algorithms are efficient plan generation, along with a connection to past practice and past outcomes. In 3D planning, the volumetric doses to structures can be evaluated and reported, and the treatment planner can adjust the source strengths as needed. Since the phenomenological approach is not strictly quantitative in its approach, the primary disadvantage is suboptimally quantitative, for example, DVH, statistics for complicated plans with many source positions and volumetric constraints. The main advantage of phenomenological approaches is that treatment plans can be created very quickly, though modern computing power now allows for extremely fast deterministic optimization as well.

Deterministic optimization algorithms offer a much more quantitative approach to brachytherapy treatment planning. Deterministic algorithms search for a solution according to a set of rules that can be mathematically programmed and, given the same initial conditions, will repeatedly generate the same result. The treatment planner specifies constraints for the healthy tissue and target, as well as any limits on maximum dwell times or any desire for smoothing among neighboring dwell positions. Several classes of deterministic algorithm exist. For situations where the optimization can be written in terms of linear and quadratic objective functions, constrained linear programming routines have been used to identify globally optimum solutions in radiotherapy applications (Jozsef et al., 2003; Alterovitz et al., 2006).

Heuristic optimization algorithms use a different quantitative approach to brachytherapy treatment planning than deterministic algorithms. Heuristic optimization incorporates stochastic elements and has methods that steer algorithms toward solutions that are most likely to be optimal. However, heuristic algorithms do not guarantee that the solution is the best of all possible options. The Nelder–Mead Simplex routine is a heuristic direct search algorithm widely used in the radiotherapy industry for optimization, mostly because of its convergence speed and because it does not require the computation of derivatives (Yao et al., 2014). The Nelder–Mead Simplex routine finds a solution that is nearly best, but is unlikely to find the global optimum. Furthermore, the solution may depend on initial conditions. Simulated annealing and generic algorithms are other common heuristic approaches. Simulated annealing searches travel in a downhill gradient direction except that inferior solutions are retained with some probability in order to reduce local minimum trapping. Genetic algorithms start with many initial solutions (parents), perturb solutions slightly with random changes (offspring), and select the best perturbed solutions among both parents and offspring for a new round of breeding.

Deterministic and heuristic optimization algorithms may be used to generate a set of solutions that explore the trade-off between target and OARs for a given patient treatment. A set of quantitatively optimal solutions is termed a Pareto surface (Milickovic et al., 2002; Alterovitz et al., 2006). Every plan along the Pareto surface is nondominated, i.e., there is no other plan that produces better quantitative statistics for all of the target and OAR structures. Once the Pareto surface is mapped, the physician can choose which plan is best for the patient among all possible optimal plans. Pareto optimization is a promising approach, and research is ongoing with regard to efficient plan generation and methods for communicating the Pareto surface trade-offs.

Other algorithms used in brachytherapy optimization include nonnegative least squares, which was originally developed by Lawson and Hanson and proposed by Deufel and Furutani for quality assurance (QA) of optimization systems (Deufel and Furutani, 2014). The least squares algorithm uses a linear system of equations to determine source strengths and source positions to deliver prescription dose to a target surface. Since the problem only used knowledge of the target surface, the solution is globally optimal only for target surface coverage. Geometrical optimization is another form of a deterministic algorithm that determines relative dwell times or source strength according to the distances of the source position relative to all other sources (Edmundson, 1990). Geometrical optimization generates a modified peripheral loading, and therefore the solution may not be clinically useful for complex treatments since it does not use any information about the relative positions of targets and critical structures with respect to the applicator.

Popular commercial TPSs make use of one or more of the previously described optimization algorithms. For example, Varian VariSeed™ optimization for LDR prostate uses simulated annealing. Varian BrachyVision™ for HDR treatment planning permits geometric optimization as well as a modified Nelder–Mead Simplex routine. Nucletron OncentraSeed™ uses an inverse-planning simulated annealing (IPSA) algorithm (Pouliot et al., 2005). OncentraBrachy™ offers both IPSA (Lessard and Pouliot, 2001) and a linear programming hybrid inverse-planning and optimization (HIPO) algorithm (Lahanas et al., 2003).

20.3.4 Volume-Based Dose Reporting

3D TPSs are able to calculate and report volume doses for target and normal tissue structures. The TPS uses closed coplanar loops on axial slices of the volumetric image to generate the boundaries of the volumetric structure, and an array of points inside a structure is used for reporting dose to that structure. The voxel resolution for structures is generally user configurable. Once the dose to each volume element of a structure is known, dose–volume histograms (DVHs) or other volumetric statistics can be computed.

Plans that incorporate volume optimization of source strength and source positions should employ volume-based dose reporting. 3D dose calculation and dose–volume analysis of target and normal tissues is recommended by the American Brachytherapy Society (ABS) for customizing patient plans for prostate, cervical, vaginal, and breast cancers (Beriwal et al., 2012; Davis et al., 2012; Small Jr. et al., 2012; Viswanathan et al., 2012; Yamada et al., 2012; Shah et al., 2013). Plans that have standard loadings may also benefit from volume-based dose reporting since the location of critical structures relative to the treatment delivery device may be different among patients and also among different treatment fractions for the same patient.

Brachytherapy has a unique set of challenges when reporting volume dose to structures, owing to the high-dose gradients in brachytherapy plans. An overview of the topic is presented by several authors (Karouzakis et al., 2002; Kirisits et al., 2007; Purdy, 2008; Ebert et al., 2010). Brachytherapy dose distributions typically have very steep dose gradients due to the proximity of sources inside and adjacent to the target. These dose gradients make dose calculations and statistics sensitive to the alignment of the dose grid, dose grid size, image slice size, and slice interpolation/extrapolation. Straube et al. (2005) noted differences among commercial TPSs and traced them back to how the systems placed structure reference points. The reported volume dose differences were worse for small structures in high-dose gradients (Straube et al., 2005). A general recommendation is to obtain high-resolution images and use the highest possible resolution for the dose calculation grid setting in the TPS.

20.3.5 Radiobiological Modeling

3D treatment planning can also incorporate estimates for the radiobiological effects of dose. This topic is not new to brachytherapy and the reader may examine several recent publications on the topic (Baltas et al., 2010; Bentzen and Yarnold, 2010; de Leeuw et al., 2011; Gagne et al., 2012a,b; Li et al., 2012; Giantsoudi et al., 2013). The reader should be aware of additional considerations when applying radiobiological models to brachytherapy. A first consideration is that high fraction doses are typically administered in brachytherapy and that cell kill will depend on the shape of the cell-survival curve at the higher doses. This consideration is also present in stereotactic radiosurgery (SRS) and stereotactic body radiation therapy (SBRT) treatments, but generally more so in brachytherapy where sources are in very close to tumor and normal tissue. A brachytherapy plan usually contains regions of dose >200% prescription. Another consideration is that brachytherapy plans have steep dose gradients inside the target. Modern TPSs assign a single dose to a voxel, which can lead to overestimates or underestimates for the cell in that voxel due to volume averaging. This is another consequence of the limitations of 3D volume dose reporting in high-dose-gradient regions. This topic is well described by Dale et al. (1997).

20.4 3D Measurement Techniques

20.4.1 Motivation for 3D Measurements in Brachytherapy

Brachytherapy by its nature generates conformal dose distributions; however, this advantage over other forms of radiation therapy can create situations where the planned dose distribution differs significantly from the delivered dose distribution. For example, misalignment between the images/coordinate system used for treatment planning and those used for treatment delivery will cause significantly more loss of target coverage or increased OAR dose than similar misalignment for EBRT. This is a result of the high conformality of brachytherapy plans in conjunction with the rapid dose falloff due to the inverse distance-squared component of the brachytherapy sources.

Brachytherapy can be administered through a number of different routes, including interstitial, intracavitary, intraluminal, and superficial, but for the purposes of dosimetry the most discerning parameter is whether the sources are confined to a fixed geometry, for example, cylindrical applicators for vaginal cancers, or whether they are placed in a geometry that can vary with every application and is influenced by variables determined at the time of application, for example, permanent seed brain implants. Figure 20.1 shows a qualitative evaluation of the geometric potential dependence of several brachytherapy procedures on the axis of implant geometry reproducibility. Where a procedure falls on this axis influences the types of dosimetry verification that should be present for the treatment of the patient. For example, the geometry of a cylindrical applicator for gynecologic brachytherapy will be static for each treatment in which it is used, thus barring applicator failure. In this case, the dose geometry relative to the applicator is very likely static and only global applicator movement is a concern. In the case of a freehand, i.e., no template, prostate HDR brachytherapy implant, the geometry will be different for every implant, but stability is expected, and depends upon, from the time of planning image acquisition to treatment. On the far end of the spectrum is a procedure such as postsurgery placement of seeds on the resection bed of an excised brain tumor. In this case, a treatment plan may not be

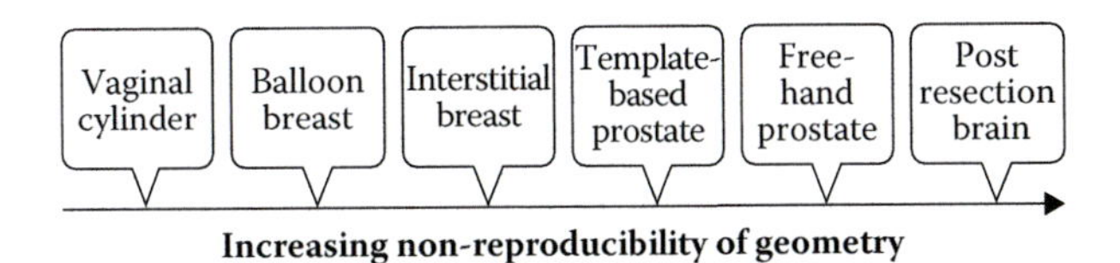

Figure 20.1

Reproducibility of the source geometry during patient treatment influences what type of dosimetry verification should be used.

generated beforehand and dosimetry, in the context of quantitatively evaluating the dose distribution, only occurs after the procedure.

For afterloader-based brachytherapy, analysis of the dose distribution from applicators and interstitial catheters used in the clinic is a critical component of safe brachytherapy practice. When possible, each applicator should be commissioned prior to first clinical use even under circumstances where the same applicator model has been used previously. Performing a dosimetric analysis of each applicator can reveal flaws in device fabrication that could adversely affect treatment dose distributions. In fact, it can also reveal consequences of design decisions made by the manufacturer that are features of the applicator. A prime example of this is the effect of source channel size in curved gynecological applicators. Hellebust et al. (2007) demonstrated that the marker string used to identify the dwell positions in the ring applicator incorrectly models the source position by following the outside edge of the channel rather than staying in the center of the channel. This resulted in erroneous simulated dwell positions in the ring by distances as great as 2.5 mm.

Figure 20.2 depicts a typical brachytherapy workflow for both permanent and temporary, afterloaded, implants. Temporary implants with reproducible geometry, such as eye plaques, may be planned before implantation. Applicators with variable geometry are typically planned after implant placement, but before radioactive source loading. Historically, clinical practice of permanent seed implant brachytherapy had three dosimetric components: plan generation using forward- or inverse-planning techniques; seed calibration, which mainly consisted of spot checks of source strength for a subset of the seeds obtained from a vendor; and postimplant dosimetry. Dosimetry of brachytherapy plans occurs only in the penultimate step (see Figure 20.2) in permanent implant brachytherapy (Davis et al., 2012). At the time of the postimplant dosimetry, a CT scan is used to evaluate the plan quality. However, at this point only underdosed volumes can be remedied by scheduling another implant to fix the deficiency. Unfortunately, overdosed regions cannot be fixed since the seeds are already imbedded in patient tissue. For afterloader-based brachytherapy, dosimetry traditionally occurs only at the time of source calibration and applicator commissioning with no dosimetric verification of individual dose plans.

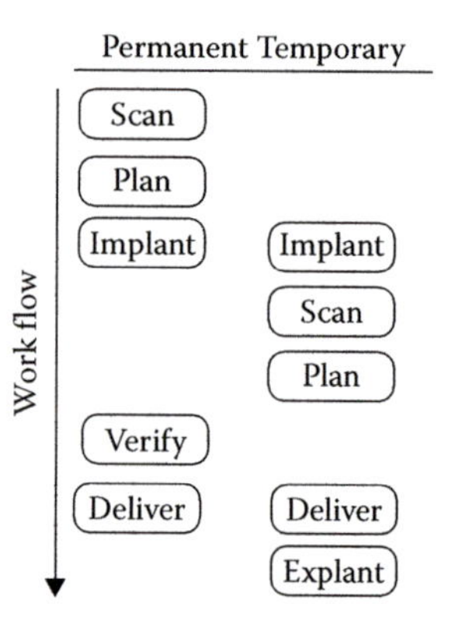

Figure 20.2

General workflow for current state-of-the-art brachytherapy for permanent implants and temporarily implanted brachytherapy.

3D techniques used for dosimetry in brachytherapy include *in vivo* dosimetry and source characterization, which is directly related to the earlier discussion of 3D data considerations and will not be addressed further, and phantom dosimetry/applicator commissioning. The remainder of this chapter introduces the concept of dimensionality for the classification of dosimetric systems (Section 20.4.2), followed by a review of state-of-the-art applicator commissioning (Section 20.4.3), the introduction of new technologies for *in vivo* dosimetry of both permanent implants and afterloader-based brachytherapy (Section 20.4.4), and finally, an analysis of uncertainties in brachytherapy source calibration is discussed (Section 20.4.5).

20.4.2 On the Dimensionality of 3D Dosimetry

A 3D measurement consists of measuring a volume of interest to a desired spatial resolution. The methodologies for obtaining this measurement have a range of geometric configurations, and it is useful to define a classification scheme for the measurement geometry. There are two fundamental aspects of the 3D measurement: the first is the inherent dimensionality of the measurement device, and the second is the dimensionality of the placement of the measurement device. For example, a diode measurement is 0D; the measurement is valid for one point in space. But if the diode is translated throughout a volume taking a matrix of measurements, the result is a set of values that span the 3D space. Thus we define the dimensionality as $D_{0,3}$; the measurement point is inherently 0D but has three dimensions available for placement of that measurement. In this case, the spatial resolution of this measurement is a function of the number of points that are measured and the placement precision of the detector relative to the source.

Film on the other hand offers a 2D measurement that can be configured in multiple parallel or intersecting planes. This configuration gives a hybrid resolution that is discrete in the space between the films but continuous in the plane of each sheet of film. Thus we have a measurement dimensionality of $D_{2,3}$, an inherently 2D measurement with three spatial dimensions and six degrees of freedom: three translations and three rotations for configuration of the 2D planes. The resolution of this measurement is a function of the inherent film resolution and the discreteness of the film placement. A subset of the film geometry is a set of films placed parallel to each other at a given separation spacing. The dimensionality of this arrangement is $D_{2,1}$. Another subset is the film-based ruler used for afterloader source position QA. In this case the film is one piece that is used for measurement along an axis and would be classified as $D_{1,1}$.

Gel-based dosimetry is inherently 3D since the entire volume is filled by the radiation-sensitive material. The entire phantom is placed at one point in space and generally not translated to generate multiple measurements giving a placement dimensionality of 0; thus the dimensionality of gel dosimetry is $D_{3,0}$.

20.4.3 Applicator Commissioning

Since the geometry of an applicator may change with each patient implantation, there is no one best way to perform the dosimetric analysis of an applicator at the time of commissioning and routine QA. However, at a minimum, dosimetric commissioning should determine the actual dwell positions inside the applicator using a dosimetry technique that measures the true position of the radiation source as delivered by the afterloader. As discussed earlier in this chapter, this information needs then to be translated into the treatment plan.

Rigid applicators, for example, cylindrical applicators for vaginal brachytherapy, provide reproducible dose distributions because the geometry of the applicator's internal components does not vary between patient implants. Because of this fixed geometry, a dosimetric analysis can be performed once during applicator commissioning and can be applied on each applicator use thereafter. Applicator modeling and commissioning was discussed for gynecological applicators in a 2010 GEC-ESTRO recommendation (Hellebust et al., 2010). It is recommended for applicator reconstruction that clinics use the digital image library (LIB), a computer-aided design (CAD) file that is provided for the user by the applicator vendor. The user maps the CAD representation of the applicator to the CT or MR image of the applicator using reference points, and this registration places the dwell positions in the proper place according to the manufacturer's specification. However, for the two largest afterloader brachytherapy TPS vendors (Nucletron/Elekta and Varian) the LIB method is only available for applicators sold by each vendor and for use on their own systems. In addition, applicators do not have an infinite lifetime and deterioration of the materials, especially in plastic applicators, can cause changes in the dose geometry of the applicator. Thus it is still recommended that a complete dosimetric measurement for each applicator be part of the commissioning and annual QA (Nath et al., 1997).

Commissioning and annual QA can be performed with most of the dosimeters listed in Section 20.2.2.2, but dose gradients exceeding 12%/mm (Hellebust et al., 2010) or even +60%/mm (Yang et al., 2011) make quantitative QA technically challenging. For example, proper use of film for dosimetry requires setup precision and reproducibility of the applicator/film geometry. In fact, a complete system for performing an audit of the planned and delivered dose distributions based on applicators to be used in treatment has been proposed (Palmer, 2013b) in which they noted that applicator-based plan QA for brachytherapy is akin to the quality control used for advanced EBRT techniques such as IMRT and volumetric-modulated arc therapy (VMAT). Applicator commissioning using 3D dosimeters is able to detect missing or misplaced sources. Palmer and colleagues performed a gamma analysis between the planning system expectation and radiochromic-type EBT3 film in a tandem-and-ring applicator setup and reported a favorable passing rate of 99.2% using 3%/3 mm criterion. Of note is the inclusion of a simulated error in the planning system where the authors intentionally removed one dwell position in the ring applicator. The result was observed on the film where the passing rate dropped to 49.9% (Palmer 2013b). This is a key observation since it is important for the commissioning and annual QA of brachytherapy applicators to identify any difference in the expected source position from the actual source positions. Hellebust et al. (2007) and Awunor et al. (2013) found offsets from the vendor-supplied source position of up to 6.1 mm. However, even more critical is that they found variance between two different units of the same applicator part. Four CT/MR compatible Interstitial Ring Ø26 mm were analyzed. The locations of dwell positions in two applicators of the same model were found to differ by up to 2.5 mm. Therefore, different applicators of the same model may require separate reconstruction criteria. Similar methods can be applied to any individual interstitial or intravascular applicators (Song et al., 2006), breast implants (Bernard et al., 2005), or eye plaques (Poder et al., 2013).

20.4.4 *In Vivo* Dosimetry

As presented in the Foreword of the IAEA Human Health Report No. 8, Development of procedures for *in vivo* dosimetry in radiotherapy

(IAEA, 2013): "There is no general consensus among radiotherapy centres on the cost effectiveness of *in vivo* dosimetry, and until recently its routine implementation was not widespread. ... However, a recent series of major accidents in radiotherapy, which would have been prevented if *in vivo* dosimetry systems had been in place, has strengthened the reasoning in favour of *in vivo* dosimetry. It is now more broadly considered that preventing the severe consequences of serious errors justifies the effort and costs of *in vivo* dosimetry programmes."

The main goal of *in vivo* dosimetry is to have a record of the dose distribution *as delivered* to the patient. Evaluation of the as-delivered dose ideally happens in real time so that errors in execution can be caught and treatment interrupted if the delivered dose is not within predefined limits of acceptance. However, in the case of afterloaded brachytherapy, even a record of the as-delivered dose is more than current clinical practice requires. Part of the reason for minimal regulatory requirements for requiring *in vivo* dosimetry in brachytherapy is that it is simply harder than for EBRT. However, the trend is toward increasing use of *in vivo* dosimetry and regulations are likely coming; the difficulties are numerous. Lack of exit dose because of the rapid inverse-squared dose falloff means detectors outside the body have to be extremely sensitive. In addition, there are a number of problem scenarios that are unique to brachytherapy as follows:

- *Displacement of the dosimeter relative to plan position:*
    - Organ-induced displacement
    - Manipulation error
- *Displacement of source position(s) relative to plan position(s):*
    - Displacement of one or more catheters or applicators, including rotation for some applicators
    - Organ-induced displacement
    - Setup error when connecting the applicator/catheters to the afterloader
- *Combination of source and sensor displacements: if errors align, no effect on dose measured.*
- *If the implant is used as to also provide the dosimeter access, any organ-related change with respect to the implant as a whole that does not impact the relative distances between catheters will yield a false-negative result, for example, catheter retraction in prostate HDR brachytherapy.*

There have been several recent review articles on the topic of *in vivo* dosimetry and brachytherapy. Readers interested in going into depth on this topic may refer to Lambert et al. (2007) and Tanderup et al. (2013), as well as the 2016 World Congress of Brachytherapy where *in vivo* dosimetry was a major part of the conference program.

20.4.5 Measurement Uncertainty in Brachytherapy Dosimetry

Uncertainties in dosimetry are a critical component of any measurement and their evaluation in the context of brachytherapy is not new. Analyses of the uncertainties involved in dosimetry for brachytherapy applications are available piecemeal in the literature, but the joint AAPM/GEC-ESTRO Task Group 138 Report (DeWerd et al., 2011) provides a complete analysis of all the uncertainties propagated from the source manufacturer to the clinic calibration (Figure 20.3), with those associated with clinical procedures covered in a joint GEC-ESTRO/AAPM

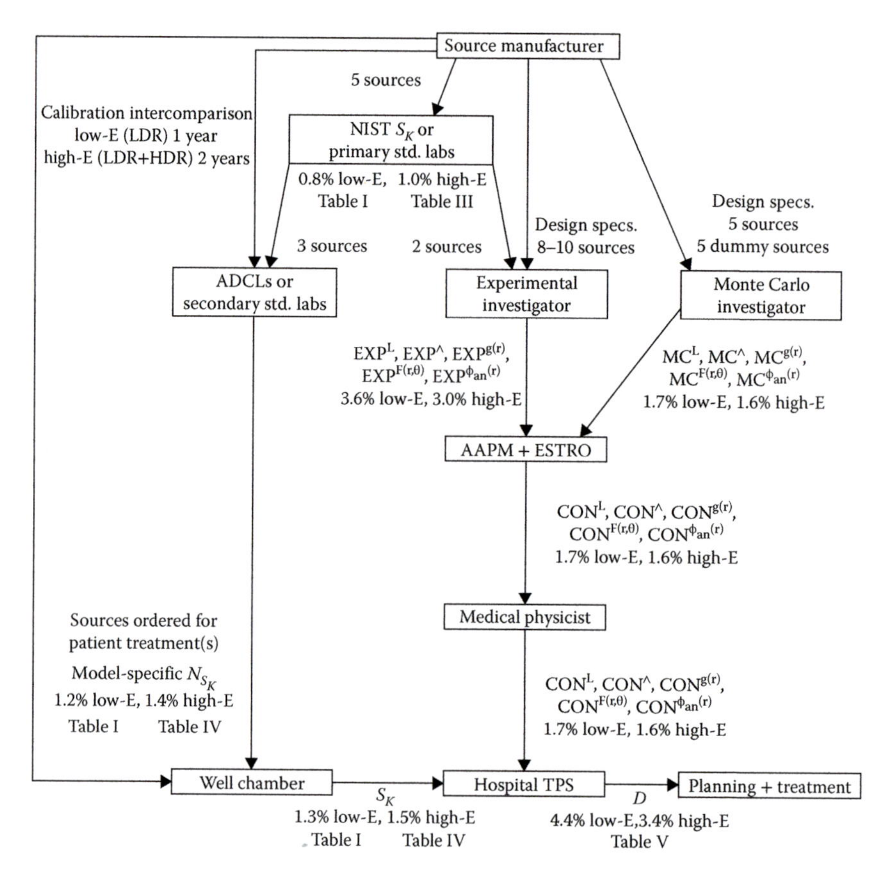

Figure 20.3

End-to-end uncertainties in the brachytherapy source dosimetry calculation. (From DeWerd LA et al., *Med. Phys.*, 38, 782–801, 2011.)

report (Kirisits et al., 2014). We refer the reader to those references for detailed discussion.

20.5 Summary

Brachytherapy is an established treatment modality in radiation therapy, with rigorous methods in place for the calculation and measurement of 3D dose distributions. In many aspects, brachytherapy is superior to EBRT. The placement of brachytherapy sources in close contact or within the target region is physically advantageous over EBRT, and brachytherapy reference dose calculations are often more accurate due to the well-known radionuclide disintegration scheme. Nevertheless, the measurement of brachytherapy dose distributions is more challenging than for EBRT, mainly due to the higher dose gradients and larger detector response corrections at lower photon energy. As a result the methods for validating brachytherapy treatment delivery are currently inferior to and less commonly employed than those existing for EBRT. Given its great potential and geometric possibilities for clinical administration, 3D dosimetry for brachytherapy will continue to be an area of innovative research and development.

References

Adamczyk M et al. (2013) Evaluation of clinical benefits achievable by using different optimization algorithms during real-time prostate brachytherapy. *Phys. Med.* 29: 111–116.

Alterovitz R et al. (2006) Optimization of HDR brachytherapy dose distributions using linear programming with penalty costs. *Med. Phys.* 33: 4012–4019.

Al-Qaisieh B et al. (2004) The use of linked seeds eliminates lung embolization following permanent seed implantation for prostate cancer. *Int. J. Radiat. Oncol. Biol. Phys.* 59: 397–399.

Andersen ES et al. (2013) Simple DVH parameter addition as compared to deformable registration for bladder dose accumulation in cervix cancer brachytherapy. *Radiother. Oncol.* 107: 52–57.

Anderson LL, Nath R and Weaver KA (1990) *Interstitial Brachytherapy: Physical, Biological, and Clinical Considerations. Interstitial Collaborative Working Group (ICWG).* (New York, NY: Raven Press).

Anderson C et al. (2013) Critical structure movement in cervix brachytherapy. *Radiother. Oncol.* 107: 39–45.

Aronowitz JN (2002) Buried emanation: The development of seeds for permanent implantation. *Brachytherapy.* 1: 167–178.

Astrahan MA (2005) Improved treatment planning for COMS eye plaques. *Int. J. Radiat. Oncol. Biol. Phys.* 61: 1227–1242.

Awunor OA, Dixon B and Walker C (2013) Direct reconstruction and associated uncertainties of ^{192}Ir source dwell positions in ring applicators using Gafchromic film in the treatment planning of HDR brachytherapy cervix patients. *Phys. Med. Biol.* 58: 3207–3225.

Ballester F et al. (2009) Evaluation of high-energy brachytherapy source electronic disequilibrium and dose from emitted electrons. *Med. Phys.* 36: 4250–4256.

Baltas D et al. (2010) A radiobiological investigation on dose and dose rate for permanent implant brachytherapy of breast using ^{125}I or ^{103}Pd sources. *Med. Phys.* 37: 2572–2586.

Bannon EA, Yang Y, Rivard MJ (2011) Accuracy assessment of the superposition principle for evaluating dose distributions of elongated and curved ^{103}Pd and ^{192}Ir brachytherapy sources. *Med. Phys.* 38: 2957–2963.

Beaulieu L et al. (2012) Report of the Task Group 186 on model-based dose calculation methods in brachytherapy beyond the TG-43 formalism: Current status and recommendations for clinical implementation. *Med. Phys.* 39: 6208–6236.

Bentzen SM and Yarnold JR (2010) Reports of unexpected late side effects of accelerated partial breast irradiation–radiobiological considerations. *Int. J. Radiat. Oncol. Biol. Phys.* 77: 969–973.

Beriwal S et al. (2012) American Brachytherapy Society consensus guidelines for interstitial brachytherapy for vaginal cancer. *Brachytherapy.* 11: 68–75.

Bernard S et al. (2005) Optimization of a breast implant in brachytherapy PDR: Validation with Monte Carlo simulation and measurements with TLDs and Gafchromic films. *Radiother. Oncol.* 76: 326–333.

Cameron J (1991) Radiation dosimetry. *Environ. Health Perspect.* 91:45–48.

Carleton Laboratory for Radiotherapy Physics (2013) Database of TG-43 brachytherapy dosimetry parameters. Taylor REP and Rogers DWO, http://www.physics.carleton.ca/clrp/seed_database (accessed on July 16, 2017).

Carlsson ÅK and Ahnesjo A (2000) The collapsed cone superposition algorithm applied to scatter dose calculations in brachytherapy. *Med. Phys.* 27: 2320–2332.

Carlsson Tedgren Å et al. (2012) Determination of absorbed dose to water around a clinical HDR ^{192}Ir source using LiF:Mg,Ti TLDs demonstrates an LET dependence of detector response. *Med. Phys.* 39: 1133–1140.

Chen Z and Nath R (2007) Photon spectrometry for the determination of the dose-rate constant of low-energy photon-emitting brachytherapy sources. *Med. Phys.* 34: 1412–1430.

Cheng A et al. (1996) Systematic verification of a three-dimensional electron beam dose calculation algorithm. *Med. Phys.* 23: 685–693.

Crijns W et al.(2013) Calibrating page sized Gafchromic EBT3 films. *Med. Phys.* 40: 012102.

Dahlin H et al. (1983) User requirements on CT based computerized dose planning systems in radiotherapy. *Acta Radiol. Oncol.* 22: 398–141.

Dale RG et al.(1997) Calculation of integrated biological response in brachytherapy. *Int. J. Radiat. Oncol. Biol. Phys.* 38: 633–642.

Davis BJ et al. (2012) American Brachytherapy Society consensus guidelines for transrectal ultrasound-guided permanent prostate brachytherapy. *Brachytherapy* 11: 6–19.

De Boeck L, Belien J and Egyed W (2014) Dose optimization in high-dose-rate brachytherapy: A literature review of quantitative models from 1990 to 2010. *Oper. Res. Health Care.* 3: 80–90.

De Deene Y, Reynaert N and DeWagter C (2001) On the accuracy of monomer/polymer gel dosimetry in the proximity of a high-dose-rate ^{192}Ir source. *Phys. Med. Biol.* 46: 2801–2825.

de Leeuw AAC et al.(2011). The effect of alternative biological modelling parameters (α/β and half time of repair $T_{½}$) on reported EQD2 values in the treatment of advanced cervical cancer. *Radiother. Oncol.* 101: 337–342.

Deufel CL and Furutani KM (2014) Quality assurance for high dose rate brachytherapy treatment planning optimization: Using a simple optimization to verify a complex optimization. *Phys. Med. Biol.* 59: 525–540.

Deufel CL and Furutani KM (2015) Heterogeneous dose calculations for Collaborative Ocular Melanoma Study eye plaques using actual seed configurations and Task Group Report 43 formalism. *Brachytherapy* 14: 209–230.

DeWerd LA et al. (2011) A dosimetric uncertainty analysis for photon-emitting brachytherapy sources: Report of AAPM Task Group No. 138 and GEC-ESTRO. *Med. Phys.* 38: 782–801.

Dezarn WA et al. (2011) Recommendations of the American Association of Physicists in Medicine on dosimetry, imaging, and quality assurance procedures for ^{90}Y microsphere brachytherapy in the treatment of hepatic malignancies. *Med. Phys.* 38: 4824–4845.

DIN (2000) *Begriffe in der radiologischen Technik—Teil 8: Strahlentherapie, DIN 6814-8:2000-12.* [Terms and Definitions in the Field of Radiology—Part 8: Radiotherapy]. (Berlin, Germany: Deutsches Institut für Normung).

Ebert MA et al. (2010) Comparison of DVH data from multiple radiotherapy treatment planning systems. *Phys. Med. Biol.* 55: N337–N346.

Edmundson GK (1990) Geometry based optimization for stepping source implants. In: *Brachytherapy HDR and LDR*, pp. 184–192. (Columbia, MD: Nucletron).

Espinoza A et al. (2015) The evaluation of a 2D diode array in "magic phantom" for use in high dose rate brachytherapy pretreatment quality assurance. *Med. Phys.* 42: 663–673.

ESTRO: BRAPHYQS Working Group (2016) A database of TG-43 brachytherapy dosimetry parameters: http://www.estro.org/about/governance-organisation/committees-activities/tg43 (accessed on July 16, 2017).

Ezzell GA (1994) Quality assurance of treatment plans for optimized high dose rate brachytherapy-planar implants. *Med. Phys.* 21: 659–661.

Gagne NL et al. (2012a) BEDVH—A method for evaluating biologically effective dose volume histograms: Application to eye plaque brachytherapy implants. *Med. Phys.* 39: 976–983.

Gagne NL, Leonard KL and Rivard MJ (2012b) Radiobiology for eye plaque brachytherapy and evaluation of implant duration and radionuclide choice using an objective function. *Med. Phys.* 39: 3332–3342.

Giantsoudi D et al. (2013) A gEUD-based inverse planning technique for HDR prostate brachytherapy: feasibility study. *Med. Phys.* 40: 041704.

Granero D et al. (2011) Dosimetry revisited for the HDR ^{192}Ir brachytherapy source model mHDR-v2. *Med. Phys.* 38: 487–494.

Hellebust TP et al. (2007) Reconstruction of a ring applicator using CT imaging: Impact of the reconstruction method and applicator orientation. *Phys. Med. Biol.* 52: 4893–4904.

Hellebust TP et al. (2010) Recommendations from Gynaecological (GYN) GECESTRO Working Group: Considerations and pitfalls in commissioning and applicator reconstruction in 3D image-based treatment planning of cervix cancer brachytherapy. *Radiother. Oncol.* 96: 153–160.

Hellebust TP et al. (2013) Dosimetric impact of interobserver variability in MRI-based delineation for cervical cancer brachytherapy. *Radiother. Oncol.* 107: 13–19.

Hissoiny S et al. (2012) Sub-second high dose rate brachytherapy Monte Carlo dose calculations with bGPUMCD. *Med. Phys.* 39: 4559–4567.

Hogstrom KR et al. (1984) Dosimetric evaluation of a pencil-beam algorithm for electrons employing a two-dimensional heterogeneity correction. *Int. J. Radiat. Oncol. Biol. Phys.* 10: 561–569.

Horowitz YS and M Moscovitch (2013) Highlights and pitfalls of 20 years of application of computerised glow curve analysis to thermoluminescence research and dosimetry. *Rad. Prot. Dosim.* 153: 1–22.

IAEA (2013) Development of procedures for in-vivo dosimetry in radiotherapy. Human Health Reports No. 8, STI/PUB/1606. (Vienna, Austria: International Atomic Energy Agency).

IAEA (2017) Nuclear Data Services (Vienna, Austria: International Atomic Energy Agency. Available at: https://www-nds.iaea.org/public/download-endf/ENDF-B-VI-8/ (accessed on July 16, 2017).

ICRP (1975) Report on the Task Group on Reference Man. Publication No. 23. (Pergamon, Oxford, UK: International Commission on Radiological Protection).

ICRP (2002) Basic anatomical and physiological data for use in radiological protection: Reference values. Publication No. 89. (Pergamon, Oxford, UK: International Commission on Radiological Protection).

ICRU (1985) Dose and volume specification for reporting intracavitary therapy in gynecology. Report No. 38. (Bethesda, MD: International Commission on Radiation Units and Measurements).

ICRU (1989) Tissue substitutes in radiation dosimetry and measurement. Report No. 44. (Bethesda, MD: International Commission on Radiation Units and Measurements).

ICRU (1992) Photon, electron, proton and neutron interaction data for body tissues. Report No. 46. (Bethesda, MD: International Commission on Radiation Units and Measurements).

Jozsef G, Streeter OE and Astrahan MA (2003) The use of linear programming in optimization of HDR implant dose distributions. *Med. Phys.* 30: 751–760.

Ju T, Simpson T, Deasy JO and Low DA (2008) Geometric interpretation of the γ dose distribution comparison technique: Interpolation-free calculation. *Med. Phys.* 35: 879–887.

Jursinic PA (2014) Quality assurance measurements for high-dose-rate brachytherapy without film. *J. Appl. Clin. Med. Phys.* 15(1): 246–261.

Kertzscher GC, Andersen E, Edmund J and Tanderup K (2011) Stem signal suppression in fiber-coupled Al2O3:C dosimetry for ^{192}Ir brachytherapy. *Radiat. Meas.* 46: 2020–2024.

Karouzakis K et al. (2002) Brachytherapy dose–volume histogram computations using optimized stratified sampling methods. *Med. Phys.* 29: 424–432.

Kim H, Huq MS et al. (2014) Mapping of dose distribution from IMRT onto MRI-guided high dose rate brachytherapy using deformable image registration for cervical cancer treatments: preliminary study with commercially available software. *J. Contemporary Brachytherapy.* 6: 178–184.

Kirisits C (2007) Accuracy of volume and DVH parameters determined with different brachytherapy treatment planning systems. *Radiother. Oncol.* 84: 290–297.

Kirisits C (2014) Review of clinical brachytherapy uncertainties: Analysis guidelines of GEC-ESTRO and the AAPM. *Radiother. Oncol.* 110: 199–212.

Kouwenhoven E, van der Laarse R and Schaart DR (2001) Variation in the interpretation of the AAPM TG-43 geometry factor leads to unclearness in brachytherapy dosimetry. *Med. Phys.* 28: 1965–1966.

Krishnaswamy V (1971) Calculation of the dose distribution about californium-252 needles in tissue. *Radiology* 98: 155–160.

Kubo HD, et al. (1998) High dose-rate brachytherapy treatment delivery: Report of the AAPM Radiation Therapy Committee Task Group No. 59. *Med. Phys.* 25: 375–403.

Lahanas M, Baltas D and Zamboglou N (2003) A hybrid evolutionary algorithm for multi-objective anatomy-based optimization in high-dose-rate brachytherapy. *Phys. Med. Biol.* 48: 399–415.

Lambert J et al. (2006) A plastic scintillation dosimeter for high dose rate brachytherapy. *Phys. Med. Biol.* 51: 5505–5516.

Lambert J et al. (2007) In vivo dosimeters for HDR brachytherapy: A comparison of a diamond detector, MOSFET, TLD, and scintillation detector. *Med. Phys.* 34: 1759–1765.

Landry G et al. (2010) Sensitivity of low energy brachytherapy Monte Carlo dose calculations to uncertainties in human tissue composition. *Med. Phys.* 37: 5188–5198.

Lee W et al. (2003) Limited resection for non-small cell lung cancer: Observed local control with implantation of I-125 brachytherapy seeds. *Ann. Thorac. Surg.* 75: 237–243.

Lessard E and J Pouliot (2001) Inverse planning anatomy based dose optimization for HDR-brachytherapy of the prostate using fast simulated annealing algorithm and dedicated objective function. *Med. Phys.* 25: 773–779.

Lewis D et al. (2012) An efficient protocol for radiochromic film dosimetry combining calibration and measurement in a single scan. *Med. Phys.* 39: 6339–6350.

Li Z et al. (2007) Dosimetric prerequisites for routine clinical use of photon emitting brachytherapy sources with average energy higher than 50 keV. *Med. Phys.* 34: 37–40.

Low D et al. (1998) A technique for the quantitative evaluation of dose distributions. *Med. Phys.* 25: 656–661.

Low D and Dempsey JF (2008) Evaluation of the gamma dose distribution comparison method. *Med. Phys.* 30: 2455–2464.

Low D et al. (2013) Does the γ dose distribution comparison technique default to the distance to agreement test in clinical dose distributions? *Med. Phys.* 40: 071722.

MacDougall ND, Pitchford WG and Smith MA (2002) A systematic review of the precision and accuracy of dose measurements in photon radiotherapy using polymer and Fricke MRI gel dosimetry. *Phys. Med. Biol.* 47: R107–R121.

Mah E et al. (1989) Experimental evaluation of a 2D and 3D electron pencil beam algorithm. *Phys. Med. Biol.* 34: 1179–1194.

Massillon-JL G et al. (2009) The use of gel dosimetry to measure the 3D dose distribution of a ^{90}Sr/^{90}Y intravascular brachytherapy seed. *Phys. Med. Biol.* 54: 1661–1672.

Massillon-JL G et al. (2014) Influence of phantom materials on the energy dependence of LiF:Mg,Ti thermoluminescent dosimeters exposed to 20–300 kV narrow x-ray spectra, ^{137}Cs and ^{60}Co photons. *Phys. Med. Biol.* 59: 4149–4166.

Melhus CS and Rivard MJ (2006) Approaches to calculating AAPM TG-43 brachytherapy dosimetry parameters for ^{137}Cs, ^{125}I, ^{192}Ir, ^{103}Pd, and ^{169}Yb sources. *Med. Phys.* 33: 1729–1737.

Meli JA (2001) Let's abandon geometry factors other than that of a point source in brachytherapy dosimetry. *Med. Phys.* 28: 1965–1966.

Mikell JK et al. (2013) Commissioning of a grid-based Boltzmann solver for cervical cancer brachytherapy treatment planning with shielded colpostats. *Brachytherapy* 12: 645–653.

Milickovic N et al. (2002) Multiobjective anatomy-based dose optimization for HDR-brachytherapy with constraint free deterministic algorithms. *Phys. Med. Biol.* 47: 2263–2280.

Mitch M, Mitchell M and Seltzer S (2006) Illustrating manufacturing variability in prostate brachytherapy seeds through x-ray spectrometry and radiochromic film measurements. *Med. Phys.* 33: 2218.

Moran JM, Radawski J and Fraass BA (2005) A dose gradient analysis tool for IMRT QA. *J. Appl. Clin. Med. Phys.* 6(2): 62–73.

Morgia M et al. (2013) Tumor and normal tissue dosimetry changes during MR-guided pulsed-dose-rate (PDR) brachytherapy for cervical cancer. *Radiother. Oncol.* 107: 46–51.

Nath R et al. (1987) Specification of brachytherapy source strength. AAPM Report No. 21 (Task Group 32 report). (New York, NY: American Institute of Physics).

Nath R et al. (1995) Dosimetry of interstitial brachytherapy sources: Recommendations of the AAPM Radiation Therapy Committee Task Group No. 43. *Med. Phys.* 22: 209–234.

Nath R et al. (1997) Code of practice for brachytherapy physics: Report of the AAPM Radiation Therapy Committee Task Group No. 56. *Med. Phys.* 24: 1557–1598.

Nesvacil N et al. (2013). A multicentre comparison of the dosimetric impact of inter- and intra-fractional anatomical variations in fractionated cervix cancer brachytherapy. *Radiother. Oncol.* 107: 20–25.

NRC (2009) Reportable medical events involving treatment delivery errors caused by confusion of units for the specification of brachytherapy sources. NRC Information Notice 2009-17. (Washington, DC: Nuclear Regulatory Commission, http://pbadupws.nrc.gov/docs/ML0807/ML080710054.pdf).

Palmer AL, Nisbet A and Bradley D (2013a) Verification of high dose rate brachytherapy dose distributions with EBT3 Gafchromic film quality control techniques. *Phys. Med. Biol.* 58: 497–511.

Palmer AL et al. (2013b) Design and implementation of a film dosimetry audit tool for comparison of planned and delivered dose distributions in high dose rate (HDR) brachytherapy. *Phys. Med. Biol.* 58: 6623–6640.

Pérez-Calatayud J, Granero D and Ballester F (2004) Phantom size in brachytherapy source dosimetric studies. *Med. Phys.* 31: 2075–2081.

Pérez-Calatayud J et al. (2012) Dose calculation for photon-emitting brachytherapy sources with average energy higher than 50 keV: Report of the AAPM and ESTRO. *Med. Phys.* 39: 2904–2929.

Petric P et al. (2013) Uncertainties of target volume delineation in MRI guided adaptive brachytherapy of cervix cancer: A multi-institutional study. *Radiother. Oncol.* 107: 6–12.

Petrokokkinos L et al. (2011) Dosimetric accuracy of a deterministic radiation transport based ^{192}Ir brachytherapy treatment planning system. Part II: Monte Carlo and experimental verification of a multiple source dwell position plan employing a shielded applicator. *Med. Phys.* 38: 1981–1992.

Poder J and Corde S (2013) I-125 ROPES eye plaque dosimetry: Validation of a commercial 3D ophthalmic brachytherapy treatment planning system and independent dose calculation software with GafChromic® EBT3 films. *Med. Phys.* 40: 121709.

Poon E and Verhaegen F (2009) A CT-based analytical dose calculation method for HDR ^{192}Ir brachytherapy. *Med. Phys.* 36: 3982–3994.

Pouliot J, Lessard E and Hsu I-C (2005) Advanced 3D Planning. In: *Brachytherapy Physics*, 2nd ed., pp. 393–414. (Eds. Thomadsen BR, Rivard MJ and Butler WM, Madison, WI: Medical Physics Publishing).

Purdy JA (2008) Quality assurance issues in conducting multi-institutional advanced technology clinical trials. *Int. J. Radiat. Oncol. Biol. Phys.* 71: S66–S70.

Reed DE et al. (2007) A prospective randomized comparison of stranded vs. loose ^{125}I seeds for prostate brachytherapy. *Brachytherapy.* 6: 129–134.

Rivard MJ (1999) Refinements to the geometry factor used in the AAPM Task Group No. 43 necessary for brachytherapy dosimetry calculations. *Med. Phys.* 26: 2445–2450.

Rivard MJ (2001) Monte Carlo calculations of AAPM Task Group Report No. 43 dosimetry parameters for the MED3631-A/M ^{125}I source. *Med. Phys.* 28: 629–637.

Rivard MJ et al. (2002) Comment on: Let's abandon geometry factors other than that of a point source in brachytherapy dosimetry. *Med. Phys.* 29: 1919–1920.

Rivard MJ et al. (2004) Update of AAPM Task Group No. 43 Report: A revised AAPM protocol for brachytherapy dose calculations. *Med. Phys.* 31: 633–674.

Rivard MJ, Venselaar JLM and Beaulieu L (2009a) The evolution of brachytherapy treatment planning. *Med. Phys.* 36: 2136–2153.

Rivard MJ et al. (2009b) An approach to using conventional brachytherapy software for clinical treatment planning of complex, Monte Carlo-based brachytherapy dose distributions. *Med. Phys.* 36: 1968–1975.

Rivard MJ, et al. (2010) Influence of photon energy spectra from brachytherapy sources on Monte Carlo simulations of kerma and dose rates in water and air. *Med. Phys.* 37: 869–876.

Rogers DWO (2006) Fifty years of Monte Carlo simulations for medical physics. *Phys. Med. Biol.* 51: R287–R301.

Ruan D et al. (2012) Evolving treatment plan quality criteria from institution-specific experience. *Med. Phys.* 39: 2708–2712.

Russell KR, Tedgren ÅC and Ahnesjö A (2005) Brachytherapy source characterization for improved dose calculations using primary and scatter dose separation. *Med. Phys.* 32: 2739–2752.

Rustgi SN (1998) Application of a diamond detector to brachytherapy dosimetry. *Phys. Med. Biol.* 43: 2085–2094.

Sabater S et al. (2014) Dose accumulation during vaginal cuff brachytherapy based on rigid/deformable registration vs. single plan addition. *Brachytherapy* 13: 343–351.

Schindel J et al. (2013) Dosimetric impacts of applicator displacements and applicator reconstruction-uncertainties on 3D image-guided brachytherapy for cervical cancer. *J. Contemporary Brachytherapy* 5: 250–257.

Shah C et al. (2013) The American Brachytherapy Society consensus statement for accelerated partial breast irradiation. *Brachytherapy.* 12: 267–277.

Small Jr W et al. (2012) American Brachytherapy Society consensus guidelines for adjuvant vaginal cuff brachytherapy after hysterectomy. *Brachytherapy.* 11: 58–67.

Song H, Luxton G and Hendee WR (2003) Calculation of brachytherapy doses does not need TG-43 factorization. *Med. Phys.* 30: 997–999.

Song H, et al. (2006) Application of Gafchromic® film in the dosimetry of an intravascular brachytherapy source. *Med. Phys.* 33: 2519–2524.

Straube W et al. (2005) DVH analysis: Consequences for quality assurance of multi-institutional clinical trials. *Med. Phys.* 32: 2021–2022.

Sutherland JG and Rogers DWO (2010) Monte Carlo calculated absorbed-dose energy dependence of EBT and EBT2 film. *Med. Phys.* 37: 1110–1116.

Tailor R, Tolani N and Ibbott GS (2008) Thermoluminescence dosimetry measurements of brachytherapy sources in liquid water. *Med. Phys.* 35: 4063–4069.

Tanderup K et al. (2013) *In vivo* dosimetry in brachytherapy. *Med. Phys.* 40: 070902.

Therriault-Proulx F et al. (2011) A phantom study of an *in vivo* dosimetry system using plastic scintillation detectors for real-time verification of ^{192}Ir HDR brachytherapy. *Med. Phys.* 38: 2542–2551.

Thomadsen BR et al. (2008) Anniversary paper: Past and current issues, and trends in brachytherapy physics. *Med. Phys.* 35: 4708–4723.

U.S. Geological Survey (USGS) (2017) The USGS Water Science School, Washington DC, USA. (http://water.usgs.gov/edu/density.html see also: http://www.csgnetwork.com/waterinformation.html) (accessed on July 16, 2017).

Van Dyk J et al. (1993) Commissioning and quality assurance of treatment planning computers. *Int. J. Radiat. Oncol. Biol. Phys.* 26: 261–273.

Vasquez Osorio EM et al. (2015) Improving anatomical mapping of complexly deformed anatomy for external beam radiotherapy and brachytherapy dose accumulation in cervical cancer. *Med. Phys.* 42: 206–220.

Venselaar J, Welleweerd H and Mijnheer B (2001) Tolerances for the accuracy of photon beam dose calculations of treatment planning systems. *Radiother. Oncol.* 60: 191–201.

Viswanathan AN et al. (2012) American Brachytherapy Society consensus guidelines for locally advanced carcinoma of the cervix. Part II: high-dose-rate brachytherapy. *Brachytherapy.* 11: 47–52.

Wierzbicki JG et al. (1998) Calculated dosimetric parameters of the IoGold ^{125}I source model 3631-A. *Med. Phys.* 25: 2197–2199.

Williamson JF, Baker RS and Li Z (1991) A convolution algorithm for brachytherapy dose computations in heterogeneous geometries. *Med. Phys.* 18: 1256–1265.

Williamson JF et al. (1998) Dosimetric prerequisites for routine clinical use of new low energy photon interstitial brachytherapy sources. *Med. Phys.* 25: 2269–2270.

Williamson JF et al. (1999) On the use of apparent activity (A_{app}) for treatment planning of ^{125}I and ^{103}Pd interstitial brachytherapy sources: Recommendations of the American Association of Physicists in Medicine radiation therapy subcommittee on low-energy brachytherapy source dosimetry. *Med. Phys.* 26: 2529–2530.

Williamson JF and Rivard MJ (2005) Quantitative dosimetry methods for brachytherapy. In: *Brachytherapy Physics*, 2nd ed., pp. 233–294. (Eds. Thomadsen BR, Rivard MJ and Butler WM, Madison, WI: Medical Physics Publishing).

Williamson JF (2006) Brachytherapy technology and physics practice since 1950: A half-century of progress. *Phys. Med. Biol.* 51: R303–R325.

Williamson JF and Rivard MJ (2009) Thermoluminescent detector and Monte Carlo techniques for reference-quality brachytherapy dosimetry. In: *Clinical dosimetry for radiotherapy: AAPM Summer School*, pp. 437–499. (Eds. Rogers DWO and Cygler JE, Madison, WI: Medical Physics Publishing).

Yamada Y et al. (2012) American Brachytherapy Society consensus guidelines for high-dose-rate prostate brachytherapy. *Brachytherapy.* 11: 20–32.

Yang Y et al. (2011) Treatment planning of a skin-sparing conical breast brachytherapy applicator using conventional brachytherapy software. *Med. Phys.* 38: 1519–1525.

Yao R et al. (2014) Optimization for high-dose-rate brachytherapy of cervical cancer with adaptive simulated annealing and gradient descent. *Brachytherapy.* 13: 352–3560.

21

Dose Outside the Treatment Volume in External Beam Therapy

Stephen F. Kry, Rebecca M. Howell, and Bryan P. Bednarz

21.1 Introduction

Dose outside the planning target volume (PTV) is defined as "nontarget dose." In radiation therapy, any "nontarget" dose should be minimized, as it offers no therapeutic benefit. Nontarget dose is delivered by nontarget radiation and can be in-field or out-of-field. "In-field nontarget dose" is that within a primary field border, such as entrance and exit dose along the beam path. "Out-of-field nontarget dose" is not only outside of the PTV but also outside of any primary field edge; dose deposited by leakage, scatter, or secondary radiation, i.e., stray radiation.

The algorithms in commercial treatment planning systems accurately calculate the high dose regions and areas within the primary beam path. However, beyond a few centimeters outside the treatment field edge, the accuracy of dose calculation algorithms is substantially reduced (Howell et al., 2010; Taylor et al., 2010; Joosten et al., 2011; Huang et al., 2013). In these cases, alternate methods are required to assess the dose. Such assessments require additional care to avoid potentially large errors associated with a variety of complications unique to performing dose measurements outside the treatment field. Furthermore, because such dose assessments are uncommon, the physicist will often not have an *a priori* estimate of the expected dose, and, as a result, large dosimetric errors are more easily missed. This is particularly true for modern radiotherapy procedures, which are highly complex, thereby making the estimation of low out-of-field doses even more challenging.

Knowing the dose outside the treatment volume can be important for several reasons. Such low doses are of substantial concern in the treatment of pregnant patients, as the fetus is particularly radiosensitive (Stovall et al., 1995). Similarly, cardiovascular implantable electronic devices (pacemakers and defibrillators) may suffer damage or detect erroneous signals in the presence of radiation. High linear energy transfer (LET) particles have been shown to be particularly damaging to these implantable devices, although relatively high cumulative doses from leakage and scatter can also induce malfunctions, and noise or oversensing events can be induced by specific dose rates (Marbach et al., 1994; Hurkmans et al., 2012). In addition to these effects, many tissues in the patient are also sensitive to radiation. The lens of the eye is known to be sensitive to cataract induction at low doses, and there is recent strong evidence that cataract formation may be best described by a linear, no-threshold model (Ainsbury et al., 2009), meaning even very low doses of radiation are of some concern. Similarly, dose to the heart may also be of concern (Carr et al., 2005; Darby et al., 2005), as even relatively low doses outside of the treatment field yield increased cardiac toxicity (Huddart et al., 2003).

More broadly and most thoroughly studied, radiation is a well-documented risk factor for cancer induction in virtually any tissue (NCRP, 2011). About 10% of long-term survivors develop a second cancer (Meadows et al., 2009; de Gonzalez et al., 2011; Takam et al., 2011), although only a fraction of these second cancers are attributable to radiation treatment, as age, genetics, and environmental factors also contribute to the risk (de Gonzalez et al., 2011; Travis et al., 2012). While most induced cancers occur in intermediate or high dose regions (Diallo et al., 2009), even low doses increase the risk of cancer (Brenner et al., 2000; Chaturvedi et al., 2007). Low radiation doses outside the treated volume are actually an increasing concern because the risk of late effects from secondary radiation is more evident today than in the past. This is because the success of cancer screening and modern therapies have increased the number of cancer patients who survive and live long enough for the adverse radiation effects on healthy tissues to manifest. Consequently, properly assessing nontarget doses, as a means to document and minimize them, is an increasingly important issue, as discussed for instance in a review article by Xu et al. (2008).

21.2 Characteristics of Out-of-Field Radiation and Dose

There are three main sources of out-of-field radiation dose to a patient (Figure 21.1): (1) patient scatter, primary radiation that enters the target patient but is then scattered to locations outside the treatment field; (2) head scatter, primary radiation scattered off structures in the accelerator head, so that the radiation still leaves the open field but is no longer directed at the target; and (3) head leakage, radiation that penetrates through the shielding components of the head and deposits dose in the patient away from the treatment field. In close proximity to the field edge, the dominant source of radiation dose is patient scatter, while at greater distances from the field edge head leakage becomes the dominant source of radiation. Head leakage becomes dominant at ~20 cm from the field edge although the exact distance depends on a host of treatment and patient parameters, such as treatment energy, depth in the patient, and size of the field, because all these

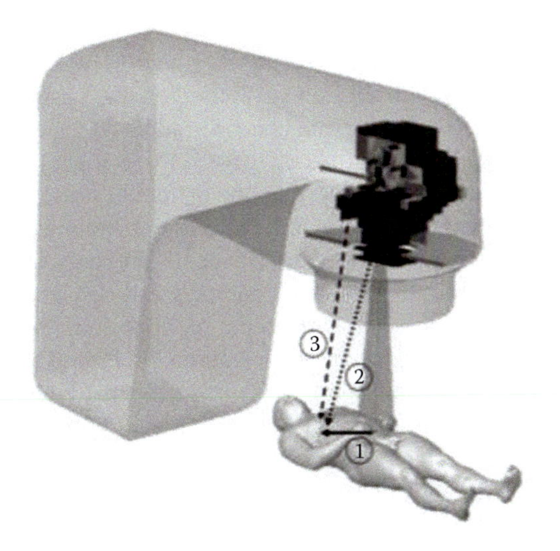

Figure 21.1

Sources of secondary radiation from a medical linear accelerator. (1) Patient scatter. (2) Collimator scatter. (3) Head leakage.

parameters affect the amount of patient scatter (Kase et al., 1983). The relative contributions of head leakage and patient scatter will also affect the spectrum of the out-of-field radiation. While the mean energy at d_{max} for a 6 MV beam is around 1.6 MeV (for a Varian accelerator), the average energy outside of the treatment field is typically much lower because scattered radiation has a low energy. Typical mean energies of the photon spectrum outside the treatment field of a 6 MV beam are between 0.2 and 0.6 MeV (Scarboro et al., 2011).

The out-of-field dose depends strongly on the distance from the field edge because patient scatter and collimator scatter decrease with increasing distance from the edge. In general, the out-of-field dose decreases approximately exponentially with distance from the field edge. However, the behavior is less predictable, at least close to the field, for highly modulated fields produced in intensity-modulated radiation therapy (IMRT) where the dose falloff can depend substantially on the objectives and complexity of the treatment plan. The out-of-field dose also depends on the field size. The larger the field size, the larger the volume that is irradiated resulting in more patient scatter. The field-size effect becomes less relevant at larger distances from the field edge where patient scatter contributions are reduced. In addition, out-of-field dose for small field sizes varies considerably between different accelerator models due to variations in shielding (Joosten et al., 2011). In contrast, the out-of-field dose varies only slightly with beam energy, being roughly the same magnitude for beam energies ranging from 4 to 25 MV (Stovall et al., 1995). The out-of-field dose also varies minimally with depth in the patient (Stovall et al., 1995; Kry et al., 2006), except for near the surface of the patient due to contamination of electrons (Starkschall et al., 1983; Kry et al., 2006). While these parameters only minimally impact the total out-of-field dose, the relative contributions of head leakage and patient scatter may vary. For example, as treatment energy increases, the amount of patient scatter decreases, but the amount of head leakage increases, largely offsetting each other.

An additional source of out-of-field dose results from neutron production in the head of the accelerator when the primary photon beam exceeds ~10 MV. Neutrons are primarily produced in high-Z components of the linac head, primarily the primary collimator, but can also be produced in the target, flattening filter (FF), jaws, and multileaf collimators (MLCs) (Mao et al., 1997; Howell et al., 2009a). Neutrons are not produced within the patient because the photon-beam energies used for clinical therapy are well below the thresholds for neutron production in the low-Z tissues in the body. The neutron energy spectra are largely consistent among different photon-beam energies and accelerators from different manufacturers (Howell, 2009b). The maximum neutron energy is ~10 MeV, below which there is a fast neutron peak centered between 0.1 and 1 MeV, and below that is a low-energy tail that arises from neutrons being scattered elastically throughout the treatment vault (Howell et al., 2009a; Sanchez-Doblado et al., 2012).

Unlike photons, the out-of-field neutron dose does not depend on distance from the treatment field or size of the primary radiation field; the patient is exposed to a large, relatively uniform bath of neutrons. However, the neutron fluence, and therefore the dose, does depend on treatment energy. The number of neutrons typically increases by a factor of 10 from 10 MV to 15 MV, and by an additional factor of 2 from 15 MV to 18 MV (Howell et al., 2009b). Additionally, the neutron dose is highly dependent on depth in the patient. The strong depth dependency results from neutrons being rapidly thermalized as they enter hydrogenous tissue of the patient. The neutron dose equivalent, for relevant clinical energies, decreases by approximately 15% per centimeter of depth in tissue (Kry et al., 2009), meaning the majority of neutron contribution occurs predominantly in superficial tissue.

Proton or carbon ion beam therapies also produce secondary radiation outside of the treatment field. The dose equivalent outside of the treatment field from these therapies is primarily due to neutrons and to a much lesser extent protons and light-charged ions (Newhauser et al., 2005; Fontenot et al., 2008). These particles result from nuclear interactions within beam line components inside the treatment head, and, unlike photon therapy, also arise in an important quantity from interactions within the patient tissues. Neutrons generated via interactions in the treatment head are commonly referred to as external neutrons, whereas those created in the patient are referred to as internal neutrons. Because of the high particle energies involved, neutrons produced during particle therapy have a higher energy distribution than seen during photon therapy; neutrons are produced with energy up to the incident particle energy, i.e., hundreds of MeV. This is important because most neutron detectors cannot detect neutrons with energies above ~20 MeV. Neutron production varies with treatment parameters. Higher treatment energies lead to more neutron production. Additionally, and in general terms, treatments that irradiate more beam line components have higher external neutron production (Mesoloras et al., 2006; Jarlskog and Paganetti 2008; Zheng et al., 2008; Yonai et al., 2010). Therefore, external neutron production is much more pronounced during passive scattering delivery than pencil beam scanning delivery. Internal neutron production is largely driven by treatment field size. Larger fields that irradiate more tissue result in higher internal neutron production, which is similar for both passive scattering and scanning beam delivery.

21.3 Typical Out-of-Field Dose Levels

Typical out-of-field photon doses for 6 MV and 18 MV conventional beams are described in the American Association of Physicists in Medicine (AAPM) TG-36 report (Stovall et al., 1995), and detailed comparisons for advanced therapies are described in the AAPM TG-158 report (Kry et al., 2017).

Compared to conventional treatments, IMRT treatments have an advantage near the edge of the treatment field due to improved conformality, but a disadvantage away from the treatment field due to increased head leakage resulting from increased modulation and longer beam-on times. That is, more conformal treatments decrease high and intermediate absorbed doses, but increase low doses to tissue further away from the treatment field. Figure 21.2 presents data that combines IMRT treatments for both segmental and dynamic IMRT delivery for various treatment sites. The data presented in the figure represent 6 MV treatments with the exception of a single 18 MV data set, excluding neutrons (Kry et al., 2007a). The large variation in data presented reflects differences in target volumes and modulation, as well as differences between different types of accelerators (Joosten et al., 2011). Variations within individual data series typically reflect different measurement locations in a phantom, i.e., different depths and different locations relative to heterogeneities. Volumetric modulated arc therapy

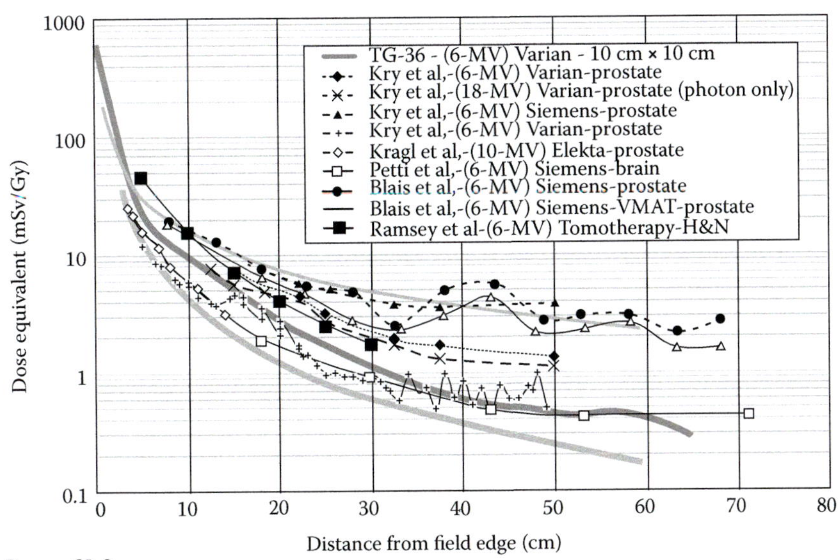

Figure 21.2

Out-of-field dose data from various intensity-modulated radiation therapy (IMRT) treatments representing different treatment sites, treatment energies, and delivery systems (Kry SF et al., *Int. J. Radiat. Oncol. Biol. Phys.*, 62, 2005; Petti PL et al., *Med. Phys.*, 33, 2006; Ramsey C et al., *J. Appl. Clin. Med. Phys.*, 7, 3, 2006; Kragl G et al., *Z. Med. Phys.*, 21, 2011; Blais AR et al., *Med. Phys.*, 39, 2012). The shown upper and lower bounds describe the empirical range of doses associated with different IMRT techniques (including helical tomotherapy (Ramsey C et al., *J. Appl. Clin. Med. Phys.*, 7, 3, 2006). For comparison, the "TG-36" line represents out-of-field dose data from the AAPM Task Group 36 report (Stovall M et al., *Med. Phys.*, 22, 1995) for a 6 MV treatment beam energy and conventional (no MLC) 10 cm × 10 cm field.

(VMAT) is a form of IMRT and therefore, the out-of-field dose characteristics of VMAT are similar to those of other IMRT treatment techniques depending primarily on field size and number of monitor units (MUs). Single arc therapy has fewer MUs than traditional IMRT, but multiple arc therapy can use comparable or more MUs during treatment delivery. When fewer MUs are required, lower out-of-field doses have been reported for VMAT compared to IMRT as a result of less leakage (Lafond et al., 2011; Blais et al., 2012), but near the treatment field where patient scatter dominates there is no dose difference between VMAT and other forms of IMRT.

There are more specialized cases that require additional consideration. Due to their unique delivery approaches, stereotactic radiosurgery (SRS) and stereotactic body radiotherapy (SBRT) can have different out-of-field dose distributions, often being higher than traditional therapies. Many SRS/SBRT treatments use noncoplanar fields, which can substantially increase the dose throughout the patient especially when beams are angled along the patient axis (Shepherd et al., 1997). Special delivery units also have distinct characteristics. For example, CyberKnife units tend to produce more out-of-field dose than other modalities. Gamma Knife treatments typically produce out-of-field doses that are larger than linac-based systems, but smaller than those from CyberKnife units (Petti et al., 2006; Zytkovicz et al., 2007).

The flattening filter free (FFF) modality has been used for decades with CyberKnife and TomoTherapy units, but is now also a standard feature on modern C-arm accelerators (Xiao et al., 2015). Removal of the FF results in decreased head leakage because the photon beam is delivered more efficiently. Collimator scatter is also reduced because the FF is no longer introducing scatter to the primary beam. However, patient scatter may be increased if the same electron energy is incident on the Bremsstrahlung target for FF and FFF modes, as in the Varian implementation of FFF, because the softer photon spectrum will experience more scatter. Overall, FFF beams produce comparable or less dose near the target volume and substantially less dose further from the target volume (Cashmore, 2008; Kry et al., 2010; Kragl et al., 2011). Furthermore, for higher energy treatments, neutron production in the FFF mode is reduced by as much as 70% over flattened beams due to more efficient photon delivery and the elimination of the FF as a source of neutrons (Kry et al., 2007b, 2008; Mesbahi, 2009).

The nontarget dose equivalents from proton and carbon ion therapy are generally less than that from photon therapy. Near the treatment field where there is an abundance of scattered photon dose, the dose equivalent from particle therapy is much less. However, farther from the treatment field (~10+ cm), particle therapy has much less of an advantage, and may produce higher dose equivalents due to secondary neutrons (Polf and Newhauser, 2005; Mesoloras et al., 2006; Wroe et al., 2007; Zheng et al., 2007). Low proton energies typically have lower dose equivalents at all distances from the field edge because of the lower neutron production associated with lower treatment energies. Similarly, scanning beams have lower external neutron production and are therefore broadly advantageous.

21.4 Photon Dosimetry

21.4.1 General Considerations

Measurement of nontarget photon or electron doses outside of the treatment field poses many unique challenges because the radiation field is different outside of the treatment field than within it. These differences require special consideration of four general measurement conditions that are substantially different for in-field and out-of-field measurements. Specific implications of these out-of-field specific conditions are presented in this section for several types of dosimeter.

1. *Dose at the surface:* Outside the treatment field, the dose at the surface is elevated by contaminant electrons, so there is a build-down effect rather than the buildup effect observed at the surface inside the treatment field. The dose is elevated by a factor of 2 or more at the patient surface, and decreases to a depth of approximately d_{max}, below which the dose becomes approximately constant with depth (Starkschall et al., 1983; Kry et al., 2006). Therefore, if a dosimeter is placed on the patient surface, it will substantially overestimate the dose, unless a very superficial dose estimate is desired. If a superficial dose estimate is not desired, the dosimeter should be covered by bolus of a thickness of approximately d_{max}. If measurements are made inside a phantom, depth is only a minor consideration because it has little impact on the dose.
2. *Energy spectrum:* The average photon energy is much lower outside the treatment field than within it. Consequently, compared to its response inside the treatment field where the calibration usually occurs, a dosimeter that is not tissue equivalent will over-respond due to increased photoelectric interactions with the softer radiation. While this means that measurements will tend to be conservative overestimates, this effect can be sizeable to the point of providing unacceptable accuracy unless it is accounted for.
3. *Dosimeter dynamic range:* Because dose levels outside the treatment field are low, the MUs for phantom measurements often must be scaled up to achieve an appropriate reading by the dosimeter. This is clearly not possible for *in vivo* measurements, and an appropriate dosimeter must be selected for such an application.
4. *Presence of other particles:* It is important to know and consider if measurements are being made in a mixed field, for example photon/neutron, as dosimeters can respond very differently to different types of radiation.

21.4.2 Point Dosimeters

Extensive study of many point detectors has been done in an out-of-field setting. While basic characteristics of many common detectors can be found in Chapter 4, details of their use in out-of-field applications are detailed here.

21.4.2.1 Luminescent Dosimeters (LDs)

These detectors are generally well suited for out-of-field dosimetry (Knezevic et al., 2013). These detectors have several advantages; they are unobtrusive for *in vivo* measurements, and for thermoluminescent dosimeters (TLDs) in particular, many commercial phantoms are preconfigured to accommodate them. Compared to calibration within a 6 MV treatment field, outside the field TLD will over-respond by 5%–12% for LiF-based TLD, for example, TLD-100 (Scarboro et al., 2011; Edwards and Mountford, 2004), while Al_2O_3-based optically stimulated luminescence dosimeters (OSLDs) will over-respond by 5%–31% (Scarboro et al., 2012). The low dose outside the treatment field is generally well handled by luminescent dosimeters which are sensitive to low doses of radiation. However, there are nonlinearities in the dose response of LDs. While the dose–response is quite linear at low doses, this also requires that the calibration be done at a low dose; if the detector is calibrated at 2 Gy, there will typically be a few percent nonlinear response, depending on the particular type of LD, which will be included in the calibration that may warrant attention for the measured dose. As a final consideration, TLD-100 has been shown to dramatically over-respond, by a factor of 10 or more, outside of the treatment field in an 18 MV beam because of the presence of neutrons (Kry et al., 2007c). A neutron insensitive detector, for example, TLD-700 or nanoDots, should be used to measure the photon component of the out-of-field dose from treatments using photon beams with energy higher than 10 MV.

21.4.2.2 Diodes

Diodes can be suitable for out-of-field dosimetry, but some caution is warranted. The most concerning aspect of diode detectors is their overresponse to the low-energy photons outside the treatment field compared to within it. Even just 1 cm outside of the treatment field, the energy response was found to be up to 70% different than in-field for an EDD-5 diode (Edwards and Mountford, 2004). Diodes are typically recommended for certain energy ranges and it is therefore important to choose appropriately low-energy diodes for out-of-field dosimetry. Even so, large corrections or uncertainties are likely. Unlike most other detectors, diodes have a dose rate dependence, often called as source-to-surface (SSD) dependence. While this effect is a relatively small 2%–7% for different types of diodes (Saini and Zhu, 2004), it is particularly hard to manage for out-of-field dosimetric applications because the dose rate varies by several orders of magnitude across the wide range of doses encountered out-of-field. Similarly, diodes show an angular dependence of 2%–5% (Yorke et al., 2005). This is again a small effect, but one particularly hard to manage in an out-of-field context because head leakage radiation and patient scatter will be incident from different directions, preventing consistent orientation of the detector.

21.4.2.3 MOSFETs

These detectors can be reliable for out-of-field dosimetry, but attention is required. Depending on the calibration and irradiation conditions, as well as the particular MOSFET detector being considered, MOSFET dosimeters can over-respond to low-energy x-rays by 50%–600%, drastically impacting the quality of the measurement unless care is taken to ensure accurate calibration or correction of the signal (Wang et al., 2004; Panettieri et al., 2007; Cheung et al., 2009). In addition, MOSFETs do exhibit small angular dependence of 2%–6% (Panettieri et al., 2007), which, like diodes, is more difficult to manage because of the various angles of incidence of out-of-field radiation. Final considerations for this detector are that MOSFETs have a limited accumulated dose lifespan, although this is typically not a concern for low doses associated with out-of-field measurements. Also they are generally unable to measure doses below ~0.1 cGy, which is a relatively low dose even for out-of-field applications.

21.4.2.4 Ion Chambers

Ion chambers have been used to measure out-of-field doses, and have advantages in terms of having relatively little energy dependence. One drawback for this type of detector in an *in vivo* setting is that it requires a high voltage, necessitating the use of biased electronics on or near the patient.

21.4.3 Multidimensional Detectors

Film is the only multidimensional detector currently evaluated in an out-of-field setting. EBT Gafchromic film has been used to measure doses outside of the treatment field (Chiu-Tsao and Chan, 2009; Van den Heuvel et al., 2012) although this process is not without challenges. EBT2 film, in particular, is only indicated for doses >1cGy (ISP users guide: http://www.ashland.com/Ashland/Static/Documents/ASI/Advanced%20Materials/ebt2.pdf). Doses outside the treated volume may easily be less than this, requiring either upscaling of the delivered MUs or limiting the measurement location to relatively small distances from the field edge. Additionally, depending on the range of measurement locations, doses can span many orders of magnitude. Creation of a calibration curve that can span this range can be challenging, although some successes were reported (Van den Heuvel et al., 2011).

There is always a conceptual advantage to perform 3D dosimetry as this provides spatially complete data. While such information is informative, it is often not essential because the treatment planning system can calculate dose out to 3 cm beyond the field edge with reasonable accuracy. Beyond this distance, the dose gradients become relatively predictable; the dose decreases primarily in only one direction—away from the treatment field. It does not change substantially with changes in depth or lateral changes parallel to the field edge (Scarboro et al., 2010). Moreover, many of the sensitive structures of interest are either well described by a point, for example, lens of the eye or embryo, or the concern is the maximum dose to the volume of interest or device, for example, fetus or cardiovascular implantable electronic device. In either case, a point dosimeter is a suitably accurate and practical choice.

21.5 Neutron Dosimetry

Most neutron detectors are sensitive to neutrons only over a particular energy range. Thus, neutron dosimetry requires an understanding of the detector response function as well as some general knowledge of the energy range of the neutron spectra being measured. Neutron detectors can be divided into two categories: fast neutron detectors (detectors that respond directly to fast neutrons) and thermal neutron detectors (detectors that respond only to thermal neutrons).

21.5.1 Fast Neutron Dosimeters

Fast neutron dosimeters include bubble detectors, track etch detectors, and tissue equivalent proportional counters (TEPCs), although TEPCs are generally reserved for study of microdosimetric properties (see Chapter 12).

Bubble detectors are sealed tubes filled with a polymer gel that, upon neutron interaction, produce a bubble that remains fixed within the polymer (Lewis et al., 2012). Typical bubble detectors have a somewhat uniform response (per unit dose equivalent) to neutrons over the energy range of ~100 keV–10 MeV (Ing et al., 1997). At energies less than and greater than this range, the response decreases rapidly. While this energy range encompasses the majority of neutrons produced during photon therapy, room-scattered neutrons are underestimated by bubble detectors because these neutrons have an energy that is too low to be detected optimally by bubble detectors. Consequently, these detectors are of little use in a vault maze or outside a treatment room where the average neutron energy is low. Caution should also be used before performing measurements with bubble detectors in a phantom. Inside a phantom, neutrons lose energy rapidly and therefore the bubble detectors will under-respond to them. Similarly, bubble detectors are not appropriate for measuring neutrons from proton therapy because they are insensitive to the high-energy neutrons with energies greater than 10 MeV.

Track etch detectors are typically made of polymers. When a neutron interacts in these materials, the recoil nuclei from the constituent materials leave behind microscopic damage trails or tracks. Chemical processing (etching) is used to enlarge the tracks. The track density can then be scored and neutron dose determined by using a calibration factor. Different polymers sensitive to different energy ranges of neutrons are available and can be used. Consequently, the response of the detector is dependent on the energy spectrum of the neutrons. Sensitivity decreases rapidly to neutrons with energy greater than 10 MeV. However, these detectors remain sensitive to low energy and thermal neutrons. While the detector will respond to a wide range of neutron energies, the response is not uniform across these energies, requiring attention to differences between the incident spectrum and the calibration spectrum. Because these detectors are intended for personnel monitoring, they are calibrated on the surface of a phantom. Therefore, using them in air or at depth in phantom will increase the measurement uncertainty, as the fluence and spectrum are both substantially altered by the absence of scattering material or additional buildup (Kry et al., 2009). Calibration of these detectors for *in vivo* measurements is possible through application of multiple detector calibration factors that vary as a function of position in the phantom/patient, thereby accounting for the changes in the neutron spectrum. This approach was implemented recently with good success (Sanchez-Doblado et al., 2012; Hälg et al., 2014).

21.5.2 Thermal Neutron Detectors

Many neutron detectors respond overwhelmingly to only thermal neutrons and have a very low cross-section to fast neutrons. This includes LiI, BF_3, ^3He, TLD-600 and activation foils containing ^{197}Au or ^{115}In. These detectors can provide some of the most precise measurements of neutron fluence and energy spectrum. However, this is a complicated process because the neutrons that are detected (thermal) contribute a negligible amount of dose. Therefore, the fast neutrons, which are of interest, must be thermalized so they can be measured by the detector. This is done through the use of a neutron moderator. Fast neutrons are thermalized by the moderator that surrounds the thermal neutron detector. The detector then measures these thermalized neutrons, and the signal is related back to the fluence of fast neutrons by a calibration process.

Incomplete or incorrect consideration of the detector response or neutron spectrum being measured can easily result in errors of several orders of magnitude. Moreover, because the presence of a patient or phantom strongly influences the energy of the neutrons, the calibration of the detector, which is typically done in air, is usually incorrect if it is used *in vivo* or even placed on the surface of a phantom or patient. Such applications are difficult and require careful consideration of the energy of the neutrons at the measurement location compared to the energy of the neutrons during calibration. Neutron dosimeters with location-specific calibration factors have been proposed as a solution to this very difficult problem and can be effective, but such corrections are not trivial (Sanchez-Doblado et al., 2012).

Because primary proton and photon beams are pulsed, passive detectors are generally preferred for measurements; active detectors such as ^6LiI detectors and BF_3 proportional counters are subject to pulse-pileup effects.

Most of the neutron dosimetry systems substantially perturb the radiation field. Neutron dosimetry is thus often conducted in air, which provides the most accurate measurements of neutron spectrum or dose. However, the dose equivalent in air, or at the surface of a patient, substantially overestimates the dose to organs within the patient because of the very sharp dose gradient for neutrons. Such a value can be used as a conservative estimate of the neutron dose equivalent at depth in the patient. Alternately, an estimate of the dose in the patient can be made by propagating the dose in air/at the surface to dose in a phantom or patient based on neutron percent depth dose-equivalent curves (d'Errico et al., 2001; Kry et al., 2009).

21.5.3 Multidimensional Neutron Detectors

Because of the challenges inherent in neutron dosimetry, there are no current multidimensional neutron dosimeters. High-precision neutron dosimetry typically requires large volume detectors that fundamentally prevent meaningful spatial resolution. Even smaller, low-precision dosimeters, such as bubble detectors, are relatively large volume and can only be considered as point dosimeters. However, point dosimetry often suffices for neutron dosimetry. For many applications, such as treatment vault shielding or personnel dose, it suffices to know the relatively uniform neutron dose equivalent throughout the room, which can be determined from a small collection of point measurements. If good spatial information on neutron dose equivalent is desired, for example, the dose equivalent throughout a patient, a typical approach is to use a Monte-Carlo neutron transport model and to calculate neutron dose equivalent. Of course, this Monte-Carlo model must be well validated against good quality neutron measurements, typically in air.

21.6 Phantoms

Phantoms are essential for characterizing the out-of-field dose to patients during radiation therapy treatments. Several physical phantoms have been developed for taking out-of-field dose measurements ranging in size and complexity. These phantoms include large scanning tanks that permit measurements far from the treatment isocenter (Stovall et al., 1995), plastic slab phantoms put together in an anthropomorphic shape (Mutic and Low, 1998; Klein et al., 2006), adult anthropomorphic phantoms (Stovall et al., 2006; Scarboro et al., 2010; Taddei et al., 2013), and pediatric anthropomorphic phantoms (Stovall et al., 2006). Anatomically, realistic anthropomorphic phantoms are often utilized over slab phantoms or water tanks because they are conceptually more elegant and allow for measurements at locations that correspond to organ positions. However, the dosimetric differences between the relatively simple versus the highly complex phantom possibilities have never been quantified, so it is unclear if there is a notable dosimetric difference, or simply a conceptual advantage.

Most investigations that use physical phantoms to measure out-of-field dose only consider photon contributions. Neutron measurements in phantoms, when appropriate, are very challenging. The presence of the dosimeter inside a phantom alters the neutron energy spectrum and ultimately the response of the detector. For example, a neutron detector calibrated in air will be based on a large proportion of fast neutrons. However, inside a patient, there is an abundance of thermal neutrons because of the rapid thermalization in hydrogenous tissue. Particularly for thermal neutron detectors, this change in spectrum will result in a drastically different signal, often higher because of the additional thermal neutrons. Consequently, the signal does not correctly describe the neutron dose equivalent, which is typically lower because thermal neutrons do not deposit a great deal of dose. Unless the detector is appropriately calibrated to account for these spectral changes (Hälg et al., 2012; Sanchez-Doblado et al., 2012), the dosimeter will provide false readings and can substantially overestimate the neutron dose.

21.7 Summary

Dose outside the treatment volume is detrimental to the patient, and may be particularly concerning under special circumstances, such as the treatment of a pregnant patient or one with an implantable electronic device. Measuring this dose poses many unique challenges because the radiation field outside the treatment volume is much different than inside it. The average energy is lower, dose rate is lower, and the dose distribution (e.g., dose variation with depth) is much different. These factors must be considered when conducting measurements.

Measurements with the greatest potential for error are those of neutrons, particularly dose or dose-equivalent measurements in or on a phantom or patient. It is critical to consider the energy spectrum of neutrons being measured in comparison to the energy response of the dosimeter. The neutron spectrum changes dramatically throughout a patient or phantom, which can easily mean that the calibration spectrum cannot be accurately applied to the measurement condition. Improved neutron dosimetry in a patient or phantom is a clear area where additional data and/or methodologies would be beneficial.

References

Ainsbury EA et al. (2009) Radiation cataractogenesis: A review of recent studies. *Rad. Res.* 172: 1–9.

Blais AR et al. (2012) Static and rotational step-and-shoot IMRT treatment plans for the prostate: A risk comparison study. *Med. Phys.* 39: 1069–1078.

Brenner DJ et al. (2000) Second malignancies in prostate carcinoma patients after radiotherapy compared with surgery. *Cancer.* 88: 398–406.

Carr ZA et al. (2005) Coronary heart disease after radiotherapy for peptic ulcer disease. *Int. J. Radiat. Oncol. Biol. Phys.* 61: 842–850.

Cashmore J (2008) The characterization of unflattened photon beams from a 6 MV linear accelerator. *Phys. Med. Biol.* 53: 1933–1946.

Chaturvedi AK et al. (2007) Second cancers among 104,760 survivors of cervical cancer: Evaluation of long-term risk. *J. Natl. Cancer Inst.* 99: 1634–1643.

Cheung T, Butson MJ and Yu PKN (2009) Energy dependence corrections to MOSFET dosimetric sensitivity. *Australas. Phys. Eng. Sci. Med.* 32: 16–20.

Chiu-Tsao ST and Chan MF (2009) Use of new radiochromic devices for peripheral dose measurement: potential in-vivo dosimetry application. *Biomed. Imag. Interv. J.* 5: 1–12.

Darby SC et al. (2005) Long-term mortality from heart disease and lung cancer after radiotherapy for early breast cancer: Prospective cohort study of about 300,000 women in US SEER cancer registries. *Lancet Oncol.* 6: 557–565.

de Gonzalez AB (2011) Proportion of second cancers attributable to radiotherapy treatment in adults: A cohort study in the US SEER cancer registries. *Lancet Oncol.* 12: 353–360.

d'Errico F et al. (2001) Depth dose-equivalent and effective energies of photoneutrons generated by 6-18 MV x-ray beams for radiotherapy. *Health Phys.* 80: 4–11.

Diallo et al. (2009) Frequency distribution of second solid cancer locations in relation to the irradiated volume among 115 patients treated for childhood cancer. *Int. J. Radiat. Oncol. Biol. Phys.* 74: 876–883.

Edwards CR and Mountford PJ (2004) Near surface photon energy spectra outside a 6 MV field edge. *Phys. Med. Biol.* 49: N293–301.

Fontenot J et al. (2008) Equivalent dose and effective dose from stray radiation during passively scattered proton radiotherapy for prostate cancer. *Phys. Med. Biol.* 53: 1677–1688.

Hälg RA et al. (2012) Field calibration of PADC track etch detectors for local neutron dosimetry in man using different radiation qualities. *Nucl. Instr. Meth. Phys. Res.* Section A 694: 205–210.

Hälg RA et al. (2014) Measurements of the neutron dose equivalent for various radiation qualities, treatment machines and delivery techniques in radiation therapy. *Phys. Med. Biol.* 59: 2457–2468.

Howell RM et al. (2009a) Effects of tertiary MLC configuration on secondary neutron spectra from 18 MV x-ray beams for the Varian 21EX linear accelerator. *Med. Phys.* 36: 4039–4046.

Howell RM et al. (2009b) Secondary neutron spectra from modern Varian, Siemens, and Elekta linacs with multileaf collimators. *Med. Phys.* 36: 4027–4038.

Howell RM et al. (2010) Accuracy of out-of-field dose calculations by a commercial treatment planning system. *Phys. Med. Biol.* 55: 6999–7008.

Huang JY et al. (2013) Accuracy and sources of error of out-of field dose calculations by a commercial treatment planning system for intensity-modulated radiation therapy treatments. *J. Appl. Clin. Med. Phys.* 14(2): 186–197.

Huddart RA et al. (2003) Cardiovascular disease as a long-term complication of treatment for testicular cancer. *J. Clin. Oncol.* 21: 1513–1523.

Hurkmans CW et al. (2012) Management of radiation oncology patients with a pacemaker or ICD: A new comprehensive practical guideline in The Netherlands. *Radiat. Oncol.* 7: 198.

Ing H, Noulty RA and McLean TD (1997) Bubble detectors—A maturing technology. *Radiation Measurements* 27: 1–11.

Jarlskog CZ and Paganetti H (2008) Risk of developing second cancer from neutron dose in proton therapy as function of field characteristics, organ, and patient age. *Int. J. Radiat. Oncol. Biol. Phys.* 72: 228–235.

Joosten A et al. (2011) Variability of a peripheral dose among various linac geometries for second cancer risk assessment. *Phys. Med. Biol.* 56: 5131–5151.

Kase KR et al. (1983) Measurements of dose from secondary radiation outside a treatment field. *Int. J. Radiat. Oncol. Biol. Phys.* 9: 1177–1183.

Klein EE et al. (2006) Peripheral doses from pediatric IMRT. *Med. Phys.* 33: 2525–2531.

Knezevic Z et al. (2013) Photon dosimetry methods outside the target volume in radiation therapy: Optically stimulated luminescence (OSL), thermoluminescence (TL) and radiophotoluminescence (RPL) dosimetry. *Radiation Measurements* 57: 9–18.

Kragl G et al. (2011) Flattening filter free beams in SBRT and IMRT: Dosimetric assessment of peripheral doses. *Z. Med. Phys.* 21: 91–101.
Kry SF et al. (2005) Out-of-field photon and neutron dose equivalents from step-and-shoot intensity-modulated radiation therapy. *Int. J. Radiat. Oncol. Biol. Phys.* 62: 1204–1216.
Kry SF et al. (2006) A Monte Carlo model for calculating out-of-field dose from a Varian 6 MV beam. *Med. Phys.* 33: 4405–4413.
Kry SF et al. (2007a) A Monte Carlo model for out-of-field dose calculation from high-energy photon therapy. *Med. Phys.* 34: 3489–3499.
Kry SF et al. (2007b) Reduced neutron production through use of a flattening-filter-free accelerator. *Int. J. Radiat. Oncol. Biol. Phys.* 68: 1260–1264.
Kry SF et al. (2007c) The use of LiF (TLD-100) as an out-of-field dosimeter. *J. Appl. Clin. Med. Phys.* 8(4): 169–175.
Kry SF et al. (2008) Energy spectra, sources, and shielding considerations for neutrons generated by a flattening filter-free Clinac. *Med. Phys.* 35: 1906–1911.
Kry SF et al. (2009) Neutron spectra and dose equivalents calculated in tissue for high-energy radiation therapy. *Med. Phys.* 36: 1244–1250.
Kry SF, Vassiliev ON and Mohan R (2010) Out-of-field photon dose following removal of the flattening filter from a medical accelerator. *Phys. Med. Biol.* 55: 2155–2166.
Kry SF et al. (2017) AAPM TG-158: Measurement and calculation of doses outside the treated volume from external-beam radiation therapy. *Med Phys.* (In press).
Lafond C et al. (2011) Evaluation and analyze of out-of-field doses in head and neck radiation therapy for different delivery techniques: from 3DCRT to VMAT. *Int. J. Radiat. Oncol. Biol. Phys.* 81: S907.
Lewis BJJ et al. (2012) Review of bubble detector response characteristics and results from space. *Radiat. Prot. Dos.* 150: 1–21.
Mao XS et al. (1997) Neutron sources in the Varian Clinac 2100C/2300C medical accelerator calculated by the EGS4 code. *Health Physics.* 72: 524–529.
Marbach JR et al. (1994) Management of radiation oncology patients with implanted cardiac pacemakers: Report of AAPM Task Group No. 34. *Med. Phys.* 21: 85–90.
Meadows AT et al. (2009) Second neoplasms in survivors of childhood cancer: Findings from the childhood cancer survivor study cohort. *J. Clin. Oncol.* 27: 2356–2362.
Mesbahi A (2009) A Monte Carlo study on neutron and electron contamination of an unflattened 18-MV photon beam. *Appl. Radiat. Isot.* 67: 55–60.
Mesoloras G et al. (2006) Neutron scattered dose equivalent to a fetus from proton radiotherapy of the mother. *Med. Phys.* 33: 2479–2490.
Mutic S and Low DA (1998) Whole-body dose from tomotherapy delivery. *Int. J. Radiat. Oncol. Biol. Phys.* 42: 229–232.
NCRP (2011). Second primary cancers and cardiovascular disease after radiation therapy. Report 170. (Bethesda, MD: National Council on Radiation Protection and Measurements).
Newhauser W et al. (2005) Monte Carlo simulations of a nozzle for the treatment of ocular tumours with high-energy proton beams. *Phys. Med. Biol.* 50: 5229–5249.
Panettieri et al. (2007) Monte Carlo simulation of MOSFET detectors for high-energy photon beams using the PENELOPE code. *Phys. Med. Biol.* 52: 303–316.

Petti PL et al. (2006) Peripheral doses in CyberKnife radiosurgery. *Med. Phys.* 33: 1770–1779.
Polf JC and Newhauser WD (2005) Calculations of neutron dose equivalent exposures from range-modulated proton therapy beams. *Phys. Med. Biol.* 50: 3859–3873.
Ramsey C et al. (2006) Out-of-field dosimetry measurements for a helical tomotherapy system. *J. Appl. Clin. Med. Phys.* 7(3): 1–11.
Saini AS and Zhu TC (2004) Dose rate and SDD dependence of commercially available diode detectors. *Med. Phys.* 31: 914–924.
Sanchez-Doblado FC et al. (2012) Estimation of neutron-equivalent dose in organs of patients undergoing radiotherapy by the use of a novel online digital detector. *Phys. Med. Biol.* 57: 6167–6191.
Scarboro SB et al. (2010) Effect of organ size and position on out-of-field dose distributions during radiation therapy. *Phys. Med. Biol.* 55: 7025–7036.
Scarboro SB et al. (2011) Variations in photon energy spectra of a 6 MV beam and their impact on TLD response. *Med. Phys.* 38: 2619–2628.
Scarboro SB et al. (2012) Energy response of optically stimulated luminescent dosimeters for non-reference measurement locations in a 6 MV photon beam. *Phys. Med. Biol.* 57: 2505–2515.
Shepherd SF et al. (1997) Whole body doses from linear accelerator-based stereotactic radiotherapy. *Int. J. Radiat. Oncol. Biol. Phys.* 38: 657–665.
Starkschall G, George FJS and Zellmer DL (1983) Surface dose for megavoltage photon beams outside the treatment field. *Med. Phys.* 10: 906–910.
Stovall M et al. (1995) Fetal dose from radiotherapy with photon beams: Report of AAPM Radiation Therapy Committee Task Group No. 36. *Med. Phys.* 22: 63–82.
Stovall M et al. (2006) Dose reconstruction for therapeutic and diagnostic radiation exposures: Use in epidemiological studies. *Radiat. Res.* 166 (1 Pt 2): 141–157.
Taddei PJ et al. (2013) Analytical model for out-of-field dose in photon craniospinal irradiation. *Phys. Med. Biol.* 58: 7463–7479.
Takam R et al. (2011) Out-of-field neutron and leakage photon exposures and the associated risk of second cancers in high-energy photon radiotherapy: Current status. *Radiat. Res.* 176: 508–520.
Taylor ML et al. (2010) Stereotactic fields shaped with a micro-multileaf collimator: Systematic characterization of peripheral dose. *Phys. Med. Biol.* 55: 873–881.
Travis LB et al. (2012) Second malignant neoplasms and cardiovascular disease following radiotherapy. *J. Natl. Cancer Inst.* 104: 357–370.
Van den Heuvel F, Crijns W and Defraene G (2011) Companding technique for high dynamic range measurements using Gafchromic films. *Med. Phys.* 38: 6443–6448.
Van den Heuvel F et al. (2012) Out-of-field contributions for IMRT and volumetric modulated arc therapy measured using gafchromic films and compared to calculations using a superposition/convolution based treatment planning system. *Radiother. Oncol.* 105: 127–132.
Wang B, Kim CH and Xu XG (2004) Monte Carlo modeling of a high-sensitivity MOSFET dosimeter for low- and medium-energy photon sources. *Med. Phys.* 31: 1003–1008.

Wroe A, Rosenfeld A and Schulte R (2007) Out-of-field dose equivalents delivered by proton therapy of prostate cancer. *Med. Phys.* 34: 3449–3456.

Xiao Y et al. (2015) Flattening filter-free accelerators: A report from the AAPM Therapy Emerging Technology Assessment Work Group. *J. Appl. Clin. Med. Phys.* 16(3): 12–29.

Xu XG, Bednarz B and Paganetti H (2008): A review of dosimetry studies on external-beam radiation treatment with respect to second cancer induction. *Phys. Med. Biol.* 53: R193–R241.

Yonai S et al. (2010) Measurement of absorbed dose, quality factor, and dose equivalent in water phantom outside of the irradiation field in passive carbon-ion and proton radiotherapies. *Med. Phys.* 37: 4046–4055.

Yorke ED et al. (2005) Diode in-vivo dosimetry for patients receiving external beam radiation therapy. AAPM Report No. 87. (Madison, WI: Medical Physics Publishing).

Zheng Y et al. (2007) Monte Carlo study of neutron dose equivalent during passive scattering proton therapy. *Phys. Med. Biol.* 52: 4481–4496.

Zheng Y et al. (2008) Monte Carlo simulations of neutron spectral fluence, radiation weighting factor and ambient dose equivalent for a passively scattered proton therapy unit. *Phys. Med. Biol.* 53: 187–201.

Zytkovicz A et al. (2007) Peripheral dose in ocular treatments with CyberKnife and Gamma Knife radiosurgery compared to proton radiotherapy. *Phys. Med. Biol.* 52: 5957–5971.

22

Imaging Dose in Radiation Therapy

Jonathan Sykes, Parham Alaei, and Emiliano Spezi

22.1 Introduction

Previous chapters in this book have concentrated on the instrumentation and measurement techniques for dosimetry of the therapeutic beams. In this chapter, a variety of measurement and calculation techniques will be reviewed for characterizing the radiation dose from x-ray imaging systems used in radiation therapy (RT). X-ray imaging systems are now used extensively throughout a patient's treatment for all complex RT, and in many cases for simple palliative RT as well. Nearly all patients will undergo a multislice computed tomography (CT) examination for localizing the target volume and nearby organs at risk. In addition, a variety of x-ray imaging systems are available to image the patient at the point of treatment, either immediately prior to beam delivery to ensure accurate patient alignment (e.g., AAPM, 2009; Korreman et al., 2010; Moore et al., 2014), or during beam delivery to monitor intrafraction motion (e.g., Ng et al., 2012). The electronic portal imaging device is probably the most basic of these systems and can be used to locate and track anatomy with little or no additional radiation dose to the patient. However, poor image quality limits the device to applications where bony anatomy or radiographic markers are sufficient surrogates for the target anatomy. Megavoltage cone beam CT* (MV-CBCT) can be used to provide 3D visualization of soft tissues with little additional hardware but at the cost of increased imaging dose. The addition of a kilovoltage (kV) x-ray system to the gantry has become the mainstay of x-ray image guidance, providing CBCT with superior image quality compared to MV-CBCT and the option of planar radiographic or fluoroscopic imaging (Jaffray et al., 1999; Jaffray and Siewerdsen, 2000). Other image guidance systems that utilize x-ray imaging include: Accuray CyberKnife® and the Brainlab ExacTrac system, both of which utilize dual/stereoscopic kV x-ray imaging systems; Accuray TomoTherapy®, which uses helical megavoltage CT (MVCT); and the Mitsubishi VERO radiotherapy unit, which has a dual x-ray imaging system mounted on a rotating gantry capable of stereoscopic kV imaging, as well as CBCT. These systems have been discussed in detail in Chapter 10.

While all these systems can be used to perform image guidance and enable greater accuracy in the delivery of RT, they all lead to additional x-ray dose to the patient. In imaging intensive CBCT-based image-guided RT (IGRT) regimes this dose can be of the order of 1–2 Gy for a course of RT (Spezi et al., 2012). It is therefore important to quantify this dose in order to justify the risks of using x-ray imaging against the benefits for a particular IGRT protocol.

In this chapter, the measurement of imaging dose is reviewed for the various imaging modalities with particular emphasis on kV-CBCT imaging.

22.2 Dose Measurement for CT

The established and standardized method for measuring fan beam CT dose, since 1981, has been to measure the computed tomography dose index (CTDI) (Shope et al., 1981). CTDI is measured using a 100-mm-long pencil ionization chamber in a cylindrical phantom. The phantom is made of polymethylmethacrylate (PMMA) and is either 32 cm diameter and 15 cm length for measurement of dose in the body or 16 cm diameter and 15 cm length for measuring dose in the head (Figure 22.1). The $CTDI_{100}$, defined in Equation 22.1, was designed to measure the dose to air over a volume which encapsulates the slice width plus the tails of the fan beam profile on either side. The measurement is performed for a single axial (i.e., not helical) rotation of the x-ray tube without table shift. For a narrow slice (≤ 1 cm), a 100 mm long chamber is sufficient to capture enough of the profile without significant loss of accuracy.

$$CTDI_{100} = \frac{1}{nT} \int_{-50\text{mm}}^{50\text{mm}} D(z)\mathrm{d}z \qquad (22.1)$$

where n is the number of detector rows and T is the thickness of each row (mm).

* MV-CBCT was developed and commercialized by Siemens Medical Systems, but is no longer commercially available since Siemens stopped producing radiation therapy treatment machines.

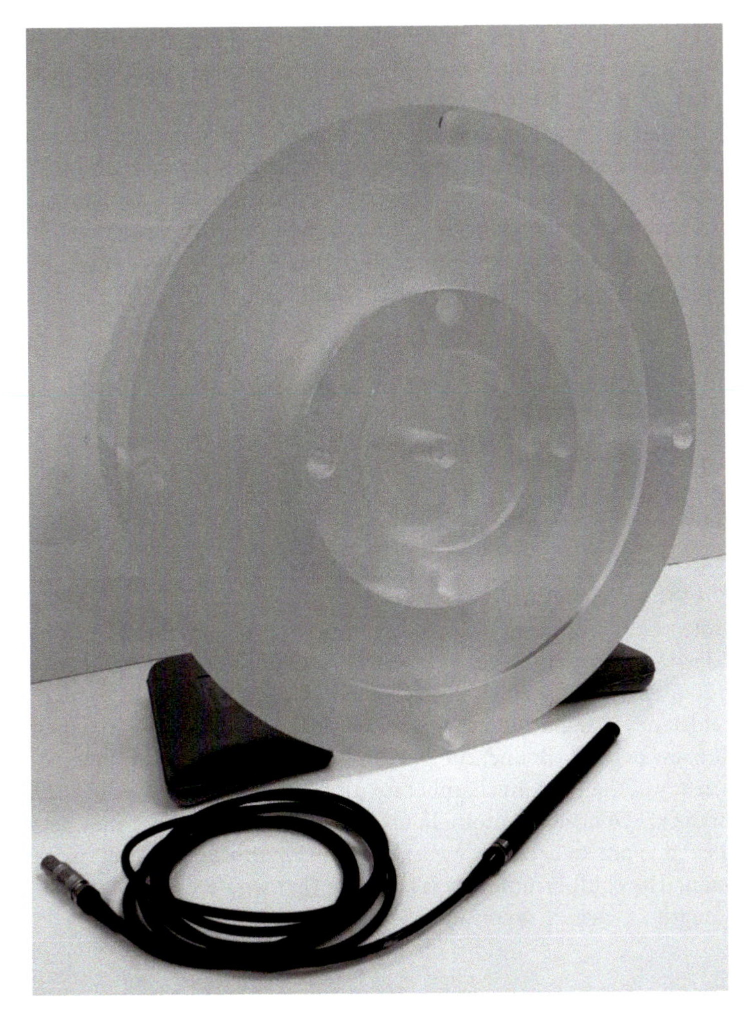

Figure 22.1

Computed tomography dose index (CTDI) phantom with 100 mm long pencil chamber.

Capturing the entire profile, including the tails due to scatter, is equivalent to measuring the dose for a spiral acquisition. Although there are some variants of the CTDI measurement, the $CTDI_{100}$ is generally accepted as a standard (Kim et al., 2011) and is the standard adopted by the International Electrotechnical Commission (IEC) (2009).

To estimate the average dose in the axial plane, $CTDI_w$, defined in Equation 22.2, is calculated from measurements at the center and the periphery of the phantom.

$$CTDI_w = \frac{1}{3}CTDI_{center} + \frac{2}{3}CTDI_{periphery} \tag{22.2}$$

where $CTDI_{center}$ is the CTDI measured at the center of the phantom and $CTDI_{periphery}$ is the average of the dose measured in at least four peripheral points on the CTDI phantom. $CTDI_w$ alone tells us what average dose in a plane would be given by a series of contiguous axial scans (stepped table feed) or helical scans with a pitch of unity. For a helical scan with pitch not equal to unity, $CTDI_{vol}$ can be calculated using Equation 22.3. The $CTDI_{vol}$, when multiplied by the scan length gives the dose length product (DLP), defined in Equation 22.4. If $CTDI_{vol}$ is expressed in mGy, DLP is generally given in mGy·mm. It is an estimate of the total absorbed dose in the scanned volume, which can then be related to the stochastic effects of radiation exposure and used as a first-order estimate in public health monitoring (Smith-Bindman et al., 2009).

$$CTDI_{vol} = \frac{CTDI_{center}}{Pitch} \quad (22.3)$$

where Pitch is the table increment per revolution as a fraction of the detector width (nT), and

$$DLP = CTDI_{vol} \times L \quad (22.4)$$

where L is the scan length.

The introduction of multislice detectors with typical detector widths up to 4 cm and with some up to 16 cm has raised discussion about the validity of CTDI (Dixon, 2003; Mori et al., 2005; Boone, 2007). Boone showed that the loss in collection efficiency using a 100 mm chamber with wider beam widths up to 4 cm was only a few percent (Boone, 2007). However, it was also shown that the collection efficiency of the 100 mm chamber for a 1 cm beam width was only 82% at the center of the CTDI head phantom. In other words, while $CTDI_{100}$ may be a useful dose index, it is not an accurate measure of the equilibrium dose at the center of a long scan. The equilibrium dose is the dose that would be delivered for a long scan of length L, where L is considerably longer than 100 mm.

22.3 Dose Measurement for CBCT for Radiotherapy Applications

22.3.1 CTDI for CBCT

The CTDI concept starts to break down with increasing cone angle for a number of reasons:

1. The 100 mm chamber length is not long enough to cover the entire beam profile which is typically comparable to or longer than the beam width in radiotherapy applications.
2. The phantom size is not sufficient to capture the entire beam width and its scatter.
3. The weightings of central and peripheral dose in the calculation of $CTDI_w$ are not necessarily a good estimate of the average dose across the axial plane given the radiological shape of the bow-tie filters used by CBCT systems, and the partial arc scanning utilized by some protocols.

These problems were largely ignored in the studies published soon after the introduction of CBCT in radiotherapy practice (Sykes et al., 2005; Amer et al., 2007; Walter et al., 2007; Osei et al., 2009; Sawyer et al., 2009; Hyer and

Hintenlang, 2010; Falco et al., 2011), and researchers employed "CTDI-like" techniques to measure the CBCT dose. They continued to use a 100 mm chamber in a cylindrical PMMA phantom with the same diameters as the body and head CTDI phantoms, 32 and 16 cm, respectively, but lengthened. This method is still used by Varian Medical Systems for measuring CBCT dose for the OBI system (Varian Medical Systems, 2012). To account for the beam length being similar in length or longer than the standard 15 cm, CTDI phantom dose measurements have also been reported with additional scatter material (Amer et al., 2007). In some cases, this has been achieved using two or three CTDI phantoms placed end to end (Palm et al., 2010). In recognition that the $CTDI_{100}$ only measures an average of the central 100 mm portion of the CT, some authors introduced the term cone beam dose index (CBDI) (Amer et al., 2007; Osei et al., 2009; Hyer and Hintenlang, 2010).

One method to ensure that the entire dose profile is acquired, at least for cone angles typical of diagnostic CT scanners, is to measure the dose at the center of a lengthened CTDI-type phantom using a 300 mm long cylindrical ion chamber (Geleijns et al., 2009; Hu and Mclean, 2014) or radiochromic film (Hu and Mclean, 2014).

An alternative to measuring a CTDI-like quantity is to measure the point dose using a 0.6 cm^3 Farmer-type chamber, as suggested by Fahrig et al. (2006), in cylindrical PMMA phantoms (Islam et al., 2006; Song et al., 2008; Sykes et al., 2010). Song and collaborators measured the dose at the center of two CTDI phantoms placed end to end using a Farmer-type chamber, and termed this measurement CBCTDI (Song et al., 2008). Sykes and colleagues used the same technique to measure the dose for both the Elekta XVI and Varian OBI systems (Sykes et al., 2010). Typical doses for the two systems, taken from Sykes et al. (2010), are given in Table 22.1.

The equivalence of measuring DL $(z = 0)$, the dose at the center of a helical scan of length L and CTDIL, the integral dose for a single static slice between $-L/2$ and $L/2$, was demonstrated by Dixon (2003). Furthermore, the equivalence of DL $(z = 0)$ was measured at the center of a helical scan (scan length = L and pitch = 1) and a CBCT scan (aperture a), where $a = L$. This was confirmed experimentally by Mori and colleagues who used a photodiode stepped through the beam to measure profiles of various beam widths (Mori et al., 2005). They showed that dose profiles acquired along the central axis of both helical CT and CBCT scans were equivalent. An alternative to the stepping diode, which would require many

Table 22.1 Cone Beam Dose Measurements (Similar to $CTDI_w$) for Standard Imaging Protocols on the Varian OBI and Elekta Synergy CBCT Systems

Varian OBI Imaging Protocols	Exposure Parameters (kV/mAs/bt)	Doses (mGy)	Elekta Synergy Imaging Protocols	Exposure Parameters (kV/mAs/bt)	Dose (mGy)
Low dose head	100/72/bt	2.8	Low dose head	100/36	1.4
Standard dose head	100/145/bt	5.6	Medium dose head	100/144	5.4
High-quality head	100/720/bt	27.8	High dose head	100/288	10.7
Pelvis	125/655/bt	24.9	Pelvis M10	100/819/bt	12.7
Pelvis spotlight	125/360/bt	20.2	Pelvis M15	100/819/bt	14.0
			Pelvis M20	100/819/bt	15.3

Note: Manufacturer's recommended protocol settings may change over time based on the introduction of new technology or feedback from customers. (bt signifies a bow-tie filter was used.)

repeat CBCT scans, is the use of a CT dose profiler (RTI Electronics AB, Mölndal, Sweden) as demonstrated by Palm et al. (2010). Based on their respective previous findings, Boone and Dixon proposed a unified and self-consistent approach to both multidetector helical (moving) and stationary CBCT systems by measuring a point dose for a single beam width (Dixon and Boone, 2010). Their theory suggests that this single measurement is sufficient to calculate the dose for a stationary or moving scan of any other beam width. This theory eventually formed the basis of the American Association of Physicists in Medicine (AAPM) Task Group (TG) 111 report on the evaluation of radiation dose in x-ray CT (AAPM, 2010).

22.3.2 AAPM Recommendations on CBCT Dosimetry

AAPM TG-111 report (AAPM, 2010) presents the theoretical underpinnings of measuring dose in axial, helical fan beam or CBCT with table translation, as well as in stationary phantom CBCT. For axial or helical scanning, the report notes that there is an equilibrium dose constant which is independent of the collimation or the pitch. The dose for any particular scan can then be determined as the product of the equilibrium dose constant and a factor (pnT/a) where p is the pitch, nT the total width of the detector, i.e., n rows of width T, and a is the width of collimation. This considerably reduces the number of measurements that need to be made as long as a is known. The equilibrium dose constant is the dose measured at the central scan plane ($z = 0$) for a scan of length L_{eq} and with pitch $p = a/nT$. L_{eq} has to be sufficiently large to ensure that further increment of L does not significantly increase the measured dose.

For scan lengths shorter than L_{eq} the dose DL, at the center of the scan on the rotation axis, asymptotically approaches D_{eq} as described by Equations 22.5 and 22.6:

$$D_L(z=0)=h(L)D_{eq} \tag{22.5}$$

and

$$h(L)=1-\alpha\exp\left(\frac{-4L}{L_{eq}}\right) \tag{22.6}$$

where α, L_{eq}, and D_{eq} can be determined experimentally by fitting the curve of dose versus the number of scan lengths.

For CBCT, where there is no table movement and typically only one rotation, the dose for a given scan with collimation width a is simply the dose measured at the center of the scan at the midpoint on the central axis. AAPM states that the CBCT dose would also reach an equilibrium for large collimation widths, but a_{eq} would be greater than 400 mm. Because this is not possible, even on RT image guidance systems, measurement of the dose equilibrium is therefore not clinically relevant. That said, measurement of the dose for a few collimation widths would allow a fit to Equation 22.6, which could then be used for the calculation of the dose for any collimation width used in clinical practice. Since the 100 mm long chamber does not cover the entire beam profile, it can underestimate the dose by 2%–5% for wide angle CBCT (Osei et al., 2009). This is one reason why the AAPM TG-111 methods are based on measuring the point dose (e.g., using a Farmer-type chamber) in a geometrical phantom (typically cylindrical) that is sufficiently long to provide full scatter conditions for the irradiated scan length.

AAPM TG-111 also introduces the concepts of integral dose (E_{tot}) and planar average equilibrium dose $\left(\overline{D_{eq}}\right)$, and shows that $E_{tot} = \rho\pi R^2 L\overline{D_{eq}}$ where R is the radius of the volume, L is the scan length, and ρ is the mass density of the phantom. According to this report, the integral dose serves as a simplified indicator of patient risk: the presumption is that cancer risk increases the larger the dose and irradiation volume containing radiosensitive tissue.

The practicalities of measuring $\overline{D_{eq}}$, the average dose over the scan plane are discussed briefly. Recognizing that when $D_{eq}(r) = A + Br^2$, i.e., $D_{eq}(r)$ has a parabolic form, the measurement of $D_{eq}(r)$ at two points, such as in the center and in the periphery of the phantom as conventionally measured for $CTDI_{vol}$, would be reasonable. Note, however, that this leads to $E_{tot} = \frac{1}{2} \cdot D(r=0) + \frac{1}{2} \cdot D(r=R-1)$ instead of the more commonly used formula $CTDI_{vol} = \frac{1}{3} \cdot CTDI(r=0) + \frac{2}{3} \cdot CTDI(r=R-1)$. Recognizing that $D_{eq}(r)$ does not always follow a parabolic form, AAPM TG-111 notes that more detailed measurements or the use of Monte Carlo (MC) modeling might be required. This would be the case for the large fields of view of the Elekta XVI and Varian OBI CBCT systems, and for the medium field of view of the Elekta system where the detector panel is shifted laterally to extend the field of view. This creates a central cylinder which is exposed from all 360° while material in the remaining volume is only exposed from 180°. In addition, the very different bow-tie filter designs between the Elekta and Varian systems will lead to different radial dose profiles.

AAPM TG-111 notes that measuring the free-in-air dose equilibrium pitch product* is an important measurement to make at commissioning as it can be used for quality assurance purposes to assess constancy of exposure and to infer the equilibrium dose measured in a phantom given a scanner with the same phantom factor (ratio of dose equilibrium in phantom to free-in-air dose equilibrium).

While AAPM TG-111 presents the theoretical underpinnings of the measurement of CBCT dose, it does not offer much in the way of standardization of CBCT dose measurement. The report suggests, but does not dictate, the use of a Farmer-type chamber. It also discusses various phantom designs with different dimensions and cross-sectional shapes (circle or ellipse) and different materials, but does not make any recommendations for a standard phantom and chamber as with the CTDI concept. The methods used by Islam et al. (2006), Song et al. (2008), and Sykes et al. (2010) are in many ways closely aligned with those of AAPM TG-111.

22.3.3 IAEA Recommendations for CBCT Dosimetry

The IAEA has published an update on the status of CT dosimetry for wide-cone beam scanners (IAEA, 2011) which includes recommendations on CT and CBCT dosimetry based on the IEC 60601-2-44 report (IEC, 2009). This is a more pragmatic approach to dosimetry of CBCT scanners than the AAPM recommendations and can be performed with current dosimetry equipment. They noted that even for a 10 mm wide beam $CTDI_{100}$, measured at the center of the phantom, only 82% and 63% of the dose is collected for the head and body phantoms, respectively (Boone, 2007). This illustrates that the $CTDI_{100}$ was never as accurate

* The product of the pitch with the free-in-air dose equilibrium corrects for variation in dose due to choice of pitch.

as one might desire with doses underestimated for long scan lengths and overestimated for short scan lengths. However, the $CTDI_{100}$ accuracy stayed constant for beam widths between 10 and 40 mm, and only decreased significantly for beam widths greater than 40 mm. The IAEA recommends a two-tier approach to the $CTDI_{100}$ with measurements for beam widths of less than 40 mm following the existing method, but for those greater than 40 mm they exploit Equation 22.7 which states that the CTDI for a beam width greater than 40 mm is related to the CTDI for a beam width less than 40 mm by the ratio of the $CTDI_{free\text{-}in\text{-}air}$ at the two beam widths.

$$CTDI_{100,(N\times T)>40} = CTDI_{100,ref} \times \left(\frac{DTDI_{free\text{-}in\text{-}air,N\times T}}{CTDI_{free\text{-}in\text{-}air,ref}} \right) \tag{22.7}$$

where $CTDI_{100,ref}$ is the $CTDI_{100}$ measured in a phantom for the reference beam of $(N\times T)_{ref}$ using an integration of 100 mm, N is the number of detector rows, T is the thickness of a single detector row and where $(N\times T)_{ref}$ is typically 20 mm, $DTDI_{free\text{-}in\text{-}air,N\times T}$ is the $CTDI_{free\text{-}in\text{-}air,ref}$ for a beam width of $N\times T$, and $CTDI_{free\text{-}in\text{-}air,ref}$ is the $CTDI_{free\text{-}in\text{-}air}$ for the reference beam width.

The measurement of $CTDI_{free\text{-}in\text{-}air,ref}$ is itself done in two tiers with a single chamber (100 mm length) position used for beam widths less than 60 mm; and two or three positions, each stepped by 100 mm to cover beam widths larger than 60 mm. Note that the IAEA formalism results in a dose index equivalent to the $CTDI_{100}$ but for a wider beam width, and therefore, retains the fundamental problem that the collection efficiency of the 100 mm pencil chamber is considerably less than 100% as described previously. The IAEA formulation is not equivalent to measuring the dose that would be given for a particular scan using, for example, a Farmer-type chamber at the center of a large phantom or using a 300 mm pencil chamber ($CTDI_{300}$) (Hu and Mclean, 2014).

22.3.4 Comparison of AAPM and IAEA Results

Hu and McLean measured CBCT dose using both the AAPM and IAEA methods of CT dosimetry and compared the results with those obtained with the previous $CTDI_{w,100}$ method as used by Varian to measure CBCT dose for the OBI system (Hu and Mclean, 2014). They used three CTDI phantoms stacked end to end, and a variety of dosimeters including a PTW 30009 100 mm pencil chamber, a PTW 30017 300 mm pencil chamber, an IBA Farmer-type chamber, and Gafchromic XR QA2 film. Taking a standard head protocol with a 184 mm beam width as an example they found that $CTDI_{w,100}$ was 5.53 mGy in a 15 cm phantom with a 100 mm pencil chamber and 4.28 mGy using the IAEA correction factor. In a 45 cm phantom, $CTDI_{w,300}$ was 5.48 mGy with a 300 mm pencil chamber. Using the AAPM formalism with a 0.6 cm^3 Farmer-type chamber, the dose was 5.49 mGy. For a pelvis scan with a 206 mm beam width, the $CTDI_{w,100}$ in a 15 cm phantom was 18.06 and 15.88 mGy with and without the IAEA correction, respectively, and in the 45 cm phantom $CTDI_{w,300}$ was 22.37 mGy. The AAPM formalism gave 22.7 mGy.

As Hu and McLean comment, there is good agreement between the $CTDI_{w,300}$ and the Farmer-type 0.6 cm^3 chamber measurements, despite one measuring the combined central and peripheral integrated dose profiles, and the other the peak central dose. As expected, they showed a clear difference between $CTDI_{w,300}$ and $CTDI_{w,100}$ measured in 45 and 15 cm phantoms due to the loss of collection efficiency of the 100 mm chamber and the lack of scatter material. No comment was made in the article about the IAEA corrected $CTDI_{w,100}$ measurement, which seems to reduce the dose further from what could be considered the gold standard using the $CTDI_{w,300}$ measurement.

22.4 Measurement (and Calculation) of Dose for Planar kV Imaging

In radiation oncology, kV x-ray imaging systems are used for IGRT in both radiographic and fluoroscopic modes (Yin et al., 2009). Radiographic modes are typically used to ensure correct placement of the isocenter and patient alignment using two orthogonal views, usually anterior–posterior and lateral. The use of implanted radiographic markers can make this a highly accurate method of aligning the target volume (Schiffner et al., 2007). Fluoroscopic modes can be used to monitor respiration prior to treatment, for example, to monitor diaphragm position, to ensure correct positioning for deep inspiration breath hold (DIBH) techniques in treatment of breast cancer (Borst et al., 2010), or to monitor target motion during treatment at sites such as prostate, lung, liver, and pancreas (Shirato et al., 2004; Ng et al., 2012). In this section, both radiographic and fluoroscopic modes will be considered equivalent for the purpose of measuring the radiation dose. The difference between the two is simply that in radiographic mode the x-ray tube is switched on once, while in fluoroscopic mode the tube is pulsed once per image frame acquired and can therefore be considered as a sequence of radiographic images.

In diagnostic radiology, there are two primary concerns: (1) the deterministic effects of radiation exposure, for example, skin erythema from long exposures where the skin dose exceeds 2000 mGy, and (2) the stochastic risk of secondary induced cancer from radiation exposure. The first effect can be determined from a measurement using an ion chamber, thermoluminescent dosimeters (TLDs), or optically stimulated luminescent dosimeters (OSLDs). The second effect can be estimated from the effective dose. AAPM TG-75 report (Murphy et al., 2007) provides an example of the estimation of effective dose for a particular radiographic exposure. This example uses the entrance dose given as dose area product (DAP) multiplied by a factor F, which is specific to a technique and is derived using MC calculations of dose in a mathematical phantom. The MC technique is described in NRPB Report 186 (Jones and Wall, 1985) and the F-factors for a variety of radiographic procedures have been calculated by Le Heron (1992).

It is not standard practice to characterize the 3D dose distribution in a homogeneous phantom in diagnostic radiology. Measuring the DAP is sufficient in the diagnostic community to estimate the effective dose for a procedure as described above.

In RT, there are two purposes of measuring the radiation dose from radiographic/fluoroscopic procedures: (1) to calculate effective dose* from the imaging to be combined with the effective dose from the treatment beam/source as recommended by AAPM TG-75 (Murphy et al., 2007), and (2) to calculate the

* Effective dose, as defined by Jacobi (1975) is "the mean absorbed dose from a uniform whole-body irradiation that results in the same total radiation detriment as from the non-uniform, partial-body irradiation in question" (Equation 22.10).

additional dose to critical organs which may already be receiving a threshold dose from the treatment beam/source.

For the calculation of effective dose, it is not necessary to calculate dose for an individual patient and therefore use of effective dose calculations based on the entrance dose are deemed to be sufficient. Ideally, the calculation of dose to individual organs should be done based on the individual patient anatomy (Ding et al., 2008b; Spezi et al., 2009; Alaei et al., 2010; Pawlowski and Ding, 2014; Poirier et al., 2014) discussed in Section 22.8. Unfortunately, these techniques are not widely available. In the absence of such methods, it may be sufficient to estimate dose to organs based on knowledge of the basic depth dose distribution of kV x-rays in water. In order to measure the dose distribution of kV x-rays, the RT medical physicists can turn to techniques for characterizing the dose in superficial x-ray units. For the absolute dose calibration, there are a number of protocols to choose from, for example, the Institution of Physics and Engineering in Medicine and Biology (IPEMB) code of practice (IPEMB, 1996), the IAEA TRS-398 code of practice (IAEA, 2000), and AAPM Report 61 (Ma et al., 2001). Methods of relative dosimetry for kV x-ray beams have been reviewed comprehensively by Hill et al. (2014) in the context of superficial x-ray treatment systems, and include details of measurement using ion chambers, diamond detectors, diodes, metal–oxide–semiconductor field-effect transistors (MOSFETs), optical fibers, OSLDs, plastic scintillator detectors, TLDs, radiochromic film, and gel dosimeters.

Typical doses for radiographic and fluoroscopic systems vary widely depending on the technique and the anatomical site being imaged. In particular, fluoroscopic imaging doses will depend on the length of the procedure and the pulse-repetition frequency. Tien and colleagues compared the entrance skin exposure (ESE) for two centers treating brain, thorax, abdomen, and pelvic regions with the CyberKnife system. They found the average ESE to be 17, 53, 41, and 68 cGy, respectively, for these regions (Tien et al., 2014). In a white paper from Accuray (Accuray, n.d.), effective doses of 0.24, 3.56, and 16 mSv were reported for image guidance with intrafraction tracking of motion for the head, chest, and pelvis with a total number of projection images of 54, 138, and 196, respectively.

Ding and Munro (2013) compared radiation dose values from an orthogonal pair of MV portal images, kV radiographs, and CBCT in the head, thorax, and pelvic regions. They performed MC simulations to determine the portal imaging doses and concluded that kV radiographs deliver the least dose among imaging modalities with typical doses in the order of a fraction of a cGy.

22.5 Dose Measurement for MV Portal Imaging

Portal imaging using the MV photon beam and an image receptor, for example, radiographic film, which was subsequently replaced by electronic portal imaging devices, was the principle imaging tool prior to development of CBCT, and is still being utilized heavily for patient position verification. This is often accomplished by taking AP and lateral radiographs to verify isocenter location, but may involve imaging individual beam portals, in which case the images are often double exposed by imaging both with the beam-limiting devices (blocks, multileaf collimators [MLCs]) in place and without them.

The portal imaging dose has traditionally been estimated by equating each monitor unit (MU) delivered for an image to one cGy, which is obviously not accurate as the patient dose varies depending on the imaging field size and patient size. Jones and Shrimpton (1991) measured the portal film dose for 100 patients and reported average doses of up to 150 cGy per course of linac-based treatment. More recently, Kudchadker et al. (2004) evaluated radiation exposure from portal films in pediatric patients and reported mean total doses per course of RT to be between 17 and 46 cGy, with most of the dose due to open-field dose from the double-exposure technique. Additional *in vivo* studies provide further data for dose using electronic portal imaging systems both at the surface and internally. Walter et al. (2007) measured the electronic portal imaging dose to patients and reported a skin dose of 5.8–6.9 cGy and a rectal dose of approximately 3 cGy for a pair of portal images. Stock et al. (2012) measured the electronic portal imaging dose in an anthropomorphic phantom and reported doses ranging between 3 and 5 cGy. Ding and Munro (2013) reported MC calculated doses for MV electronic portal imaging to be in the order of 2–4 cGy per orthogonal pair.

The reduction of MV portal imaging dose is achievable by: (1) limiting the collimator size in double-exposure imaging and (2) using fewer MUs in imaging smaller volumes, such as in the head and neck region and for pediatric cases.

22.6 Dose Measurement for MVCT and MV-CBCT

MVCT scanning in RT is exclusive to TomoTherapy systems (Accuray, Sunnyvale, CA). The TomoTherapy Hi Art system utilizes a 3.5 MV x-ray beam and a row of xenon detectors to acquire fan beam CT images for patient position adjustments. During imaging, a 4 mm jaw width (as projected to isocenter) is used for all image acquisitions. The MVCT images are acquired at couch speeds of 4, 8, and 12 mm/rotation, corresponding to pitch values of 1, 2, and 3, referred to as fine, normal, and coarse image acquisition, respectively. The images are then reconstructed as fine, normal, and coarse, corresponding to slice thicknesses of 1 or 2, 2 or 4, and 3 or 6 mm, respectively, the thinner of each set is obtained from interpolation. Thus, the acquisition pitch determines the slice thickness, imaging dose, and duration of image acquisition (Shah et al., 2008). The imaging dose is also dependent on the length of imaged volume and patient size. The reported measured doses range from less than 1 cGy to over 2 cGy in cylindrical and anthropomorphic phantoms (Shah et al., 2008) utilizing multiple scan average dose (MSAD) measurements. Similar types of measurements in a cylindrical acrylic phantom reported doses between 0.2 and 1 cGy for pitches between 4 and 1 (Meeks et al., 2005). Another set of measurements using TLDs in an anthropomorphic phantom indicated imaging doses of <1 cGy for coarse setting (Shah et al., 2012).

Imaging dose from MV-CBCT employed in Siemens linacs has been measured by several groups (Gayou et al., 2007; Morin et al., 2007a,b; Isambert et al., 2009; Quinn et al., 2011; Halg et al., 2012). The dose from this imaging modality generally increases with higher MU protocol, which produces better quality images. Due to fixed gantry start/stop angles, there is also a steep dose gradient within the patient with higher dose on the anterior portions of the body, assuming supine position. This has been illustrated by Miften and colleagues using a treatment-planning system (Miften et al., 2007). The reported doses from MV-CBCT imaging range from a fraction of cGy up to 12 cGy, depending on the protocol used.

22.7 Dosimeters for All Modalities

Virtually any dosimeter used in RT can be used for imaging dose measurements provided it has been calibrated for the quality of the imaging beam and characterized for its behavior in such a beam quality. Ion chambers can be used to measure dose from any imaging beam, regardless of beam quality, with the caveat that they need to be calibrated for the imaging beam quality if different than the therapeutic one. For example, to use an ion chamber in the kV energy range, the chamber must be calibrated for that beam quality with traceability to a dosimetry standard laboratory.

TLDs have been extensively used for dose measurements in anthropomorphic phantoms and kV-CBCT beams (Sykes et al., 2005; Amer et al., 2007; Saw et al., 2007; Wen et al., 2007; Kan et al., 2008; Marinello et al., 2009; Osei et al., 2009; Palm et al., 2010; Cheng et al., 2011; Dufek et al., 2011; Halg et al., 2012). They have also been used for skin dose measurements from imaging beams. One of the limitations of TLDs is the energy dependence of their response; hence, to use them for measurements in kV beams they either have to be calibrated for the same beam quality or their response corrected using a correction factor (Kron et al., 1998; Nunn et al., 2008).

Among other dosimeters commonly available, OSLDs can also be used for imaging dose measurements. The use of OSLDs in kV x-ray beams has been studied by several groups (Winey et al., 2009; Ding and Malcolm, 2013; Giaddui et al., 2013).

Other dosimeters such as MOSFETs (Cheung et al., 2003; Ehringfeld et al., 2005), radiographic, and radiochromic films have also been used for MV imaging dose measurements and could be used for kV measurements as well (Marinello et al., 2009; Isambert et al., 2009; Alvarado et al., 2013; Giaddui et al., 2013; Nobah et al., 2014).

22.8 Dose Calculation Methods

22.8.1 Dose Calculation Algorithms for MV-CBCT

In case of MV-CBCT, all the current algorithms available in treatment-planning systems can be utilized to compute the imaging dose as the same 6 MV therapeutic beam is used for imaging. This has been done by several groups, indicating its feasibility and the ability of including MV-CBCT dose in treatment planning (Miften et al., 2007; Morin et al., 2007a,b). The Siemens units can also perform MV-CBCT using an "Imaging beam line" (IBL), which employs a degraded 4.2 MeV beam and a carbon target to produce the imaging beam (Faddegon et al., 2008). The IBL beam data has been collected and modeled in one commercial treatment-planning system utilizing a convolution/superposition algorithm (Flynn et al., 2009). Figure 22.2 shows a dose distribution from MV-CBCT using a treatment-planning system.

22.8.2 Dose Calculation Algorithms for kV-CBCT

Calculating kV-imaging dose using available algorithms poses greater challenges as these algorithms have been developed to calculate the dose from MV beams which predominantly interact with tissue through Compton interactions. Interactions in the kV range are predominantly through photoelectric effect, which is not modeled accurately with these algorithms, with the exception of

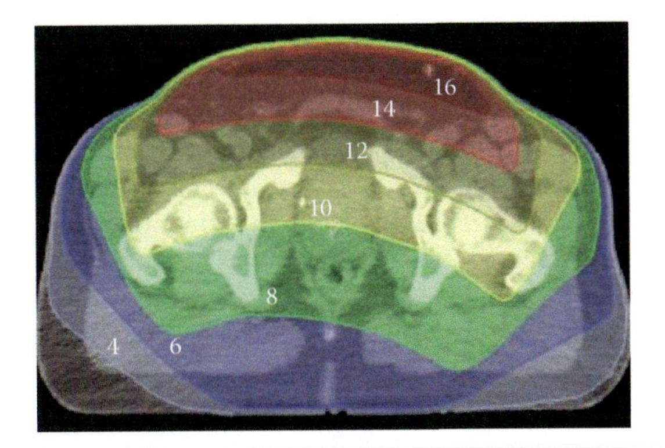

Figure 22.2

Distribution of dose deposited in the pelvis by a single fraction of MV-CBCT imaging for a prostate patient, with 10 cGy at isocenter. The isodose lines are labeled in cGy. (Reproduced from Miften M et al., *Med. Phys.*, 34, 3760–3767, 2007. With permission.)

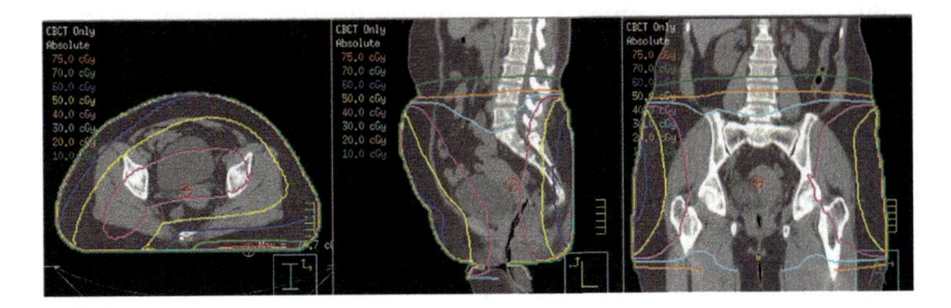

Figure 22.3

Isodose distribution showing the imaging dose from 25 fractions of pelvic imaging for one patient using the Elekta XVI pelvis imaging protocol (120 kVp, 1 mAs, 650 projections) calculated using the Pinnacle treatment-planning system. (Reproduced from Alaei P and Spezi E, *Phys. Med.*, 31, 647–658, 2015. With permission.)

MC methods. A common algorithm applied in treatment-planning systems, convolution/superposition, has been used for dose calculation from kV-CBCT by addition of kV energy deposition kernels (Alaei et al., 2010) producing reasonable results in soft tissue and lung but underestimating the dose in and around bone (Alaei et al., 2001, 2010). Convolution-based algorithms do not account for atomic number changes in the medium, which are needed for accurate dose calculations in the kV range. A proposed algorithm (Ding et al., 2008b) overcomes this problem by introducing a correction factor to account for atomic number changes. This algorithm is currently not available commercially. Figure 22.3 demonstrates such a dose calculation using the convolution-based algorithm implemented in the Pinnacle treatment-planning system (Philips, Milpitas, CA).

22.8.3 Dose Calculation Using MC Methods

Frequently used methods for estimating dose from CBCT are based on calculating the dose to either the CTDI phantoms or to simplified humanoid computational phantoms. In the diagnostic world of radiation protection, it may be sufficient to relate the dose for CT scan protocols used by a particular hospital to the radiation risk for the purpose of justification and for reporting dose.

However, in radiotherapy where many CBCT scans might be performed during the course of treatment, it may be necessary to calculate the dose to specific critical organs to ensure that the combined treatment and concomitant imaging dose does not exceed the dose criteria specified in the treatment protocol. In such cases, individualized patient dose calculations may be required. The MC method is regarded as the most accurate approach to model ionizing radiation transport for radiotherapy and imaging applications (Verhaegen and Seuntjens, 2003; Spezi and Lewis, 2008), and it is an ideal tool for CBCT patient dosimetry. The calculation of concomitant dose from both kV- and MV-CBCT units has been carried out extensively with the EGSnrc code system (which includes the BEAMnrc and DOSXYZnrc codes) and, to a lesser extent, with other MC codes such as MCNP and Geant4 (Chow et al., 2008; Ding et al., 2008a; Gu et al., 2008; Ding and Coffey, 2009, 2010; Downes et al., 2009; Spezi et al., 2009, 2011, 2012; Walters et al., 2009; Qiu et al., 2011, 2012; Deng et al., 2012a,b; Zhang et al., 2012; Ding and Munro, 2013; Son et al., 2014). As reported by Alaei and Spezi (2015), several groups developed MC models for CBCT imaging systems and calculated 3D dose distributions using patient-specific CT scans or virtual phantoms. This is the result of several works aimed at improving particle transport models for x-ray photon beams in the diagnostic energy range (Kawrakow, 2013). The commissioning of an MC model for a CBCT unit is in principle similar to the commissioning of a treatment-planning system for external beam radiotherapy. First, the model of the unit's head, including source, filters, and beam collimators, has to be built. Second, the model has to be calibrated for absolute dose calculation, and dose profiles obtained in reference conditions must be validated against experimental measurements. The process for the absolute dose calibration of a CBCT MC model was described by Ding et al. (2008a) and Downes et al. (2009). The MC calibration factor, defined in Equation 22.8, is specific to each CBCT beam and is derived by measuring, in reference conditions, the absolute dose to a point in a phantom with known geometry, and by calculating the MC dose to the same point.

$$F_{\mathrm{MCcal}} = \frac{D_{\mathrm{exp}}}{D_{\mathrm{MCcal}}} \tag{22.8}$$

where D_{exp} is the measured dose in units of Gy and D_{MCcal} is the MC dose, calculated in the same reference conditions, in units of Gy per incident particle.

Once the computational model is commissioned, 3D dose calculation can be carried out by sampling the photons incident on the patient with one of the following methods using: (1) full MC simulation of the beam line (Qiu et al., 2011, 2012); (2) a phase space file representing the invariant parts of the unit or fixed field sizes (Chow et al., 2008; Ding et al., 2008a; Ding and Coffey, 2009, 2010; Downes et al., 2009; Walters et al., 2009; Spezi et al., 2012); (3) a source model representing the main sources of radiation (Spezi et al., 2011; Deng et al., 2012a,b; Zhang et al., 2012; Ding and Munro, 2013; Montanari et al., 2014); and (4) an x-ray spectrum (Gu et al., 2008; Ding et al., 2010). Several groups (Chow et al., 2008; Downes et al., 2009; Spezi et al., 2009, 2011, 2012) have developed a computational model for the Elekta XVI CBCT unit using the EGSnrc/BEAMnrc code system and Beampp (a C++ implementation of the BEAMnrc MC code). A number of other groups have developed MC models of the Varian OBI CBCT scanner (Gu et al., 2008; Qiu et al., 2011, 2012; Zhang et al., 2012; Ding and Munro, 2013;

Montanari et al., 2014). All publications reported 3D dose data calculated on voxelized geometries representing human anatomy, based on patient CT scans or virtual phantoms, and present doses to various organs of interest. While it is not feasible to list all the organ dose data in this chapter, we summarize in Table 22.2 the typical kV-CBCT doses for three anatomical sites. Note that the performance of the CBCT systems in Table 22.2 should not be judged on the basis of the data reported, since there is no reason to assume that the acquisition settings have been optimized to give the same trade-off between imaging dose and image quality.

The following observations are generally applicable to understanding patient dose from CBCT. As shown in Figure 22.4, MC simulations have demonstrated that bony structures can receive two to four times the dose delivered to the soft tissue (Ding et al., 2008a; Downes et al., 2009; Spezi et al., 2012). Furthermore, the use of computational phantoms based on micro-CT images pointed out that the average dose to bone surface cells can be up to 80% higher than the average dose to organs at risk in a typical head and neck CBCT scan (Walters et al., 2009). This is caused by the increased mass-energy absorption coefficient due to the photoelectric interaction within the materials of higher atomic number. It has also

Table 22.2 Typical MC Calculated Patient Doses, in cGy, for Three Anatomical Sites for the Elekta Synergy CBCT System and the Varian OBI CBCT System

	Pelvis/Abdomen	Head and Neck	Chest
Elekta XVI (Spezi et al., 2012)	1.5–3.3	0.1–0.2	1.2–3.4
Varian OBI (Ding and Coffey, 2009; Montanari et al., 2014)	1–5	0.2–0.5	2–9

Note: Doses reported are for the body, i.e., not for a specific organ.

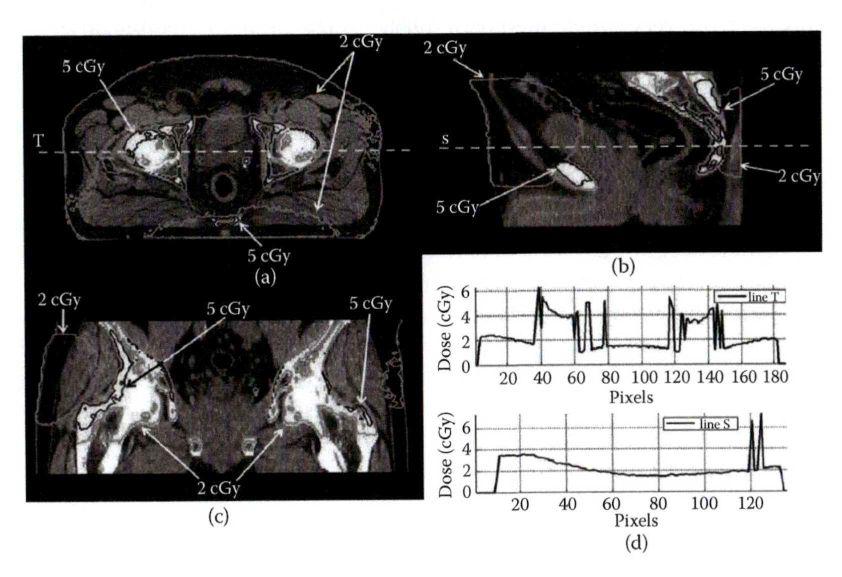

Figure 22.4

Patient dose from XVI CBCT pelvis scan simulated using the M10 collimator and F1 bow-tie filter. (a) Transverse, (b) sagittal, and (c) coronal dose contours are shown. Absolute dose profiles in the transverse and sagittal plane are shown in (d). (Reproduced from Downes P et al., *Med. Phys.* 36, 4156–4167, 2009. With permission.)

been shown that the addition of the bow-tie filter significantly reduces the dose by 22% in the pelvis and 45% in the chest (Spezi et al., 2012). This is primarily due to the attenuation of the dose to the peripheral tissues but also due to the beam hardening effect. Moreover, the bow-tie filter reduces the scattered dose from the periphery of the patient to the imager which has the additional advantage of increasing image quality.

Downes et al. (2009) also showed that the CBCT imaging dose has a left–right asymmetry due to the increased number of exposures at the start and stop gantry angles as the gantry rotation accelerates and decelerates at the beginning and end of each scan. Unlike CT imaging where the patient is normally central in the CT scanner, in radiotherapy the isocenter is typically set to the center of the target volume which may itself be offset from the center of the patient's cross section. Chow (2009) studied this effect and found for the pelvis phantom variation in the mean dose of up to 20% for up to 10 cm anterior–posterior shifts. Dose variations for the chest and head and neck were typically between 7% and 17%. It has been found that kV-CBCT doses are highly (inversely) correlated with patient size, expressed in weight or body mass index (BMI) (Zhang et al., 2012; Alaei et al., 2014). In particular, doses to pediatric patients were found to be of the order of two times that of an adult (Deng et al., 2012a; Zhang et al, 2012).

22.9 Estimating Effective Dose and Risk

The $CTDI_{vol}$ measurement is an estimate of the average dose in the central axial plane of the scan and is typically calculated as one-third of the central dose and two-thirds of the peripheral dose, as discussed in Sections 22.2 and 22.3. This is independent of the scan length and therefore does not relate to the total dose to the patient and the subsequent risk of radiation-induced malignancy. A commonly used and very simple method to relate $CTDI_{vol}$ to total imaging dose is to multiply it by the length of the scan. This is known as the dose length product and has been discussed in Section 22.2. Unlike the above, the quantity referred to as integral dose (total energy absorbed in a volume) can be used as a surrogate to estimate patient risk, assuming that the risk increases with the dose and volume irradiated. AAPM TG-111 (AAPM, 2010) presents the methodology to relate integral dose to scan length and other quantities as elucidated in Section 22.3.2.

For a more accurate assessment of radiation risk, the dose to individual organs and their respective organ sensitivities are needed. The effective dose, defined in Equation 22.10, measured in units of sievert (Sv), is a summation of tissue equivalent doses, shown in Equation 22.9, and tissue-specific weighting factors defined in ICRP Report 103 (ICRP, 2007). The effective dose can be related to radiation risk using, for example, data presented in the Biological Effects of Ionizing Radiation (BEIR) Report published by the National Academies Concerning Radiation Health Risks (BEIR, 2006). The equivalent dose H_T for tissue/organ T is given by

$$H_T = \sum_R W_R \cdot D_{T,R} \tag{22.9}$$

where W_R is the weighting factor for radiation type R and $D_{T,R}$ is the absorbed dose for tissue T by radiation type R. The effective dose E is then given by

$$E = \sum_T W_T \cdot H_T \tag{22.10}$$

where W_R *and* W_T are the weighting factors as given in ICRP Report 103 (ICRP, 2007) and H_T is the equivalent dose for tissue or organ type T.

One method of measuring organ dose to calculate effective dose is to use an anthropomorphic phantom for CBCT. This has been performed by a number of groups using small radiation dosimeters such as TLDs (Sykes et al., 2005; Amer et al., 2007; Wen et al., 2007; Osei et al., 2009; Sawyer et al., 2009; Palm et al., 2010; Stock et al., 2012), fiber optic-coupled water-equivalent plastic scintillators (Hyer et al., 2010), silicon-photodiode dosimeters (Koyama et al., 2010), and MOSFETs (Perks et al., 2008; Kim et al., 2010).

An alternative method of estimating the effective dose is to use the ImPACT CT patient dose calculator (http://www.impactscan.org/ctdosimetry.htm). The ImPACT dose calculator, designed originally for fan beam CT, uses a library of MC calculated dose calculations (Jones and Shrimpton, 1991) for organ doses in a humanoid mathematical phantom. The library covers numerous commercial CT scanners each characterized by the ratio of peripheral to central $CTDI_w$ and central to in-air $CTDI_w$ for both the head and body phantoms. To calculate the effective dose, the operator selects the scanner type and the start and stop positions of the scan. The software will provide individual organ doses with their weighting factors and equivalent doses, as well as the total effective dose. Ideally when using the ImPACT calculator, the CT scanner for which the dose is to be calculated will be one of the scanners in the ImPACT library. If not, the CT scanner can be matched to the closest one in the library using ImPACT factors derived from a linear combination of the ratios of the central and peripheral normalized $CTDI_{100}$ to $CTDI_{air}$. This method has been employed by several authors to match a CBCT scanner with fan beam CT scanners in the ImPACT library (Amer et al., 2007; Sawyer et al., 2009). Hyer and Hintenlang (2010) compared organ doses from the ImPACT dose calculator with previously published MC calculated organ doses (Hyer et al., 2010). They found that many organs agreed within 40%, with generally better agreement for the pelvis scan. However, some discrepancies of more than 100% were also found. They concluded that the ImPACT dose calculator is not suitable for calculating CBCT dose.

Gu et al. have modeled both kV- and MV-CBCT systems using MCPNX (Gu et al., 2008) and applied these models to calculate organ doses to the VIP-Man phantom that was developed from the National Library of Medicine's Visible Human Project (Xu et al., 2000). They concluded that the effective dose for the head and neck and prostate was 8.53 and 6.25 mSv, respectively, for a 125 kVp kV-CBCT exposure of 1350 mAs.

22.10 Combining Dose from RT and Imaging

For radiotherapy, the risk of concomitant imaging needs to be considered in the context of the existing risk of secondary cancer induction from radiotherapy treatment. In addition, the dose to critical organs already receiving high doses from the treatment needs to be assessed to ensure the additional imaging dose does not exceed organ dose limits. The imaging dose needs to be considered both within the treated volume and also peripheral to the volume.

Qiu et al. (2012) performed MC dose calculations for relatively large volume gynecological intensity-modulated RT (IMRT) treatments with field length of ~15 cm, and for CBCT scans of length ~24 cm. They concentrated on modeling the in-field dose, discussing out-of-field dose only briefly. In-field doses for organs at risk were calculated using organ equivalent doses calculated using linear, linear-exponential, and plateau radiobiological models. The greatest increment in dose, from imaging one CBCT per fraction, was 2.5% for the bowel with the linear model but this reduced to 1.3% for the plateau model. For dose in the peripheral region, the CBCT dose was compared with the linac scatter and leakage doses. In the peripheral low dose regions, where there is low risk of secondary malignancies, the incremental dose from CBCT was found to be an order of magnitude less than the IMRT scatter dose and less than or equal to the linac leakage dose.

Chow et al. (2008) concentrated on in-field dose and compared CBCT dose with the treatment dose for a prostate IMRT case. The planning target volume (PTV) dose rose by 0.6 Gy (0.8%) for a 78 Gy/39 fractions treatment, which suggests the CBCT dose was ~1.5 cGy per scan. The femoral heads saw the largest increase in dose of 2.5 cGy (5%).

Perks et al. (2008) measured the peripheral dose at the center and on the surface of an anthropomorphic phantom. They measured the dose for a prostate IMRT treatment using MOSFETs and kV-CBCT using TLDs. The dose from the IMRT dropped from the prescription dose of 2 Gy (per fraction) down to 1 cGy at 16 cm and 0.4 cGy at 21 cm distance from the field edge. In comparison, the CBCT dose was 0.5 and 0.2 cGy at the same positions, respectively (7 and 12 cm, respectively, from the imaged volume edge). They used an S20 collimator which arguably provides a longer field of view than necessary for prostate IGRT. The nominal dose per scan was 6 cGy which they acknowledge was twice that normally used in their clinic. To put this into context, 6 cGy is four times the United Kingdom's diagnostic reference level for imaging the abdomen/pelvis, and is arguably three to five times higher than necessary for adequate image quality for CBCT image guidance (Sykes, 2010).

Harrison and colleagues published two articles on the subject of combined treatment and imaging doses covering anatomical sites of larynx, breast, and prostate (Harrison et al., 2006, 2007). They compared imaging dose from 2D portal imaging and 3D CT imaging with the treatment dose. While this work was not based on CBCT, the differences between CT and CBCT doses are likely to be minimal so the work provides a good perspective on the relative impact of kV imaging on the combined treatment and imaging dose. For the prostate, they measured dose using TLDs in the Alderson-Rando phantom. Neutron doses were also calculated for the 15 MV beams. They calculated the dose to multiple organs both in-field and out-of-field for combinations for a 37 fraction two phase prostate treatment with 26 CT images and 4 portal images. The excess relative risk (ERR) was found to be <0.1 for most organs with bone surfaces, small intestine, and muscle having ERR < 0.3. Increases in total dose due to portal imaging of up to 20% were found for bone marrow and bone surfaces. They employed similar methods for the larynx and breast and concluded that the dose to critical organs increased by 5%–20% with increases of up to 30% for bone surfaces and bone marrow. They noted that by far the largest component of dose to these organs was from scatter and leakage from the MV beam.

Alaei et al. (2014) used a treatment-planning system to compute the imaging dose for head and neck and pelvic treatments and added the imaging dose to the therapeutic one. They showed that high-dose imaging procedures add an appreciable dose to the therapeutic one received by patients. This could become an issue of concern if an organ at risk is proximal, but outside, the treated volume but within the volume irradiated by the imaging beam. They also demonstrated the inverse relationship between imaging dose and BMI.

The studies above computed the imaging dose retrospectively. It is, however, beneficial to do so prospectively and account for imaging dose at the time of treatment planning. To this extent, both Alaei et al. (2014) and Grelewicz and Wiersma (2014) combined the kV imaging and MV therapy beams to perform inverse planning, hence accounting for imaging dose during optimization.

Previous work focused on the dose from kV-CBCT. Combining MV-CBCT imaging dose with the therapeutic one is more straightforward and can easily be accomplished using treatment-planning systems as shown by Miften et al. (2007), Morin et al. (2007a,b), and Akino et al. (2012). Whereas the combination of kV-CBCT beams with therapeutic MV ones requires MC codes or other software not commonly available, combining MV-CBCT dose with the therapeutic one can be accomplished routinely in a clinical setting.

One issue with calculating the imaging dose prospectively is that it is not always known what imaging will be required for a particular patient. For instance, an IGRT protocol may have as its basis the use of imaging for a few initial fractions and then weekly thereafter, but if the weekly images show a change in patient setup or anatomy then this may trigger further imaging.

22.11 Clinical Consequences and Benefits

22.11.1 Detrimental Effects of Radiation Exposure

To date there have been no large-scale epidemiologic studies of the cancer risks associated with x-ray imaging. The evidence we have is derived from measurement and calculation of organ doses and applying organ-specific cancer incidence or mortality data derived from studies of atomic-bomb survivors (Brenner and Hall, 2007). The estimated attributable lifetime risk of death from cancer due to a single, typical, CT scan is ~0.01% increasing to 0.1% for exposures in early childhood (Brenner and Hall, 2007). However, these risks are calculated for the general population and not specifically for patients undergoing treatment with radiotherapy. Therefore, the risk of imaging alone is small compared to the risk of treatment failure and other morbidities associated with radiotherapy treatment. It appears sensible that, if possible, the risk from the imaging dose should be incorporated into the overall risk calculation including the treatment dose.

The primary risk of radiation exposure from radiotherapy, including any concomitant imaging but excluding the risks of treatment failure and comorbidities, is the induction of a secondary primary malignancy (SPM). As Tubiana noted in his review (Tubiana, 2009), these rarely occur before 10 years after treatment. However, with increased long-term survival rates, the incidence of these malignancies is likely to increase. Tubiana found from cancer registries that the incidence could be as high as 20%. He also noted that SPMs tended to occur in tissues receiving more than 2 Gy. Data derived from the US Surveillance, Epidemiology, and End Results (SEER) cancer registry by Berrington de Gonzalez et al. (2011) found that 9% of 5-year survivors developed a solid tumor and that the relative risk was highest for tissues that typically received more than 5 Gy. From the previous section, we know that the imaging dose is typically small in comparison to the treatment dose. Nevertheless, it adds to the radiation dose burden and contributes to the increased risk of SPM induction.

Very low doses are also associated with complications. Perks et al. (2008) collected a number of such effects in their paper including prolonged azoospermia at doses >2.5 Gy (Howell and Shalet, 1998), loss of ovarian function at doses <2 Gy (Wallace et al., 2003), and hypothyroidism or thyroid nodules with median dose equivalents as low as 0.09 Sv (Imaizumi et al., 2006). Cataract formation can also occur with ERR of 1.98 ERR/Gy with no lower threshold and with measurable hazard ratios for doses as low as 60 mGy (Chodick et al., 2008).

22.11.2 Clinical Benefits

Although there is little doubt of the clinical benefit of CBCT imaging for quality assurance (QA) of patient setup, there is, as yet, little published evidence on improved outcomes attributable to the use of CBCT imaging or other image-guided modalities. Chow et al. (2008) calculated the normal tissue complication probability (NTCP) increase from CBCT imaging during IMRT treatment of the prostate to be 0.5%, although they recognized that the NTCP model used was relatively crude and did not take into account the relative biological effectiveness (RBE) of kV imaging. Nevertheless, they found that NTCP decreased by 3% when the clinical target volume (CTV) to PTV margin was reduced from 10 to 5 mm showing a net benefit of using CBCT imaging for every fraction of treatment. Kron et al. (2010) showed that even when daily online IGRT (CBCT) was used with an adaptive strategy for bladder cancer, the integral dose to both the whole irradiated volume and the irradiated volume minus the CTV was less than the conventional treatment. This is because on average the irradiated volume is smaller in the adaptive strategy than that required otherwise to ensure the bladder is covered the majority of the time. The exception to this was for patients with smaller treatment volumes. Zelefsky et al. (2012) compared cohorts of patients in which one group received prostate IMRT with IGRT, and the other group received the same treatment but without IGRT. They found that biochemical tumor control was significantly better for patients with high-risk prostate carcinoma when IGRT was employed. In addition, late urinary toxicity was almost halved in the group with IGRT. While this study was performed using MV portal imaging and gold seed markers, Moseley et al. (2007) have demonstrated the equivalence of kV-CBCT and gold seed marker versus MV portal image-based IGRT. More recently, Bujold et al. (2012) reviewed the literature and concluded that IGRT has enabled treatments such as hypofractionated stereotactic ablative radiotherapy of the lung, spine, and liver. They also concluded that "an improvement in relapse rate in prostate cancer, Hodgkin disease, and head-and-neck cancers using IGRT has been consistently reported," and that "there is a suggestion that prostate and head-and-neck cancer patients might have lower toxicity with IGRT, especially when combined with other technical advances like IMRT."

22.12 Closing Remarks

There are several x-ray imaging options available for IGRT in modern clinical practices. There is a wide variation in the protocols used for these imaging options due to a number of factors. In some cases, requirements for high throughput of patients on machines and availability of suitably trained staff to interpret and act on these images limits the amount of imaging that can be performed. Variable perceptions of risk of imaging dose also contribute to the utilization of x-ray-based IGRT. Many centers will use preset image acquisition protocols defined by the manufacturers. These protocols may well be based on the experiences of early adopters of the equipment; however, there is a lack of evidence and consensus on the minimum image quality required to perform IGRT specific to the particular anatomical site or size of patient. For this reason, the imaging dose may well be higher than necessary. Further work is required in order to optimize image acquisition protocols and the frequency of imaging to achieve the aims of IGRT. To achieve this, standardization of the way radiation dose for imaging is reported needs to be improved.

Further work is also required to understand the risks and benefits attributable to using x-ray-based imaging in RT. The use of dose calculation models to calculate the imaging dose for individual patients may have immediate benefit in estimating the total dose to critical organs/structures of concern, but if implemented for all patients would provide useful data for future analysis. To achieve this, dose calculation algorithms need to be made accessible and integrated efficiently into the clinical workflow so that there are minimal overheads. Ideally, imaging dose calculations would be automated, running in the background, and storing the required dosimetric data in the patient record.

References

AAPM (2009) The role of in-room kV x-ray imaging for patient setup and target localization. Report of AAPM Task Group 104 of the Therapy Imaging Committee. (One Physics Ellipse, College Park, MD: American Association of Physicists in Medicine).

AAPM (2010) Comprehensive methodology for the evaluation of radiation dose in x-ray computed tomography. Report of AAPM Task Group 111: The future of CT dosimetry. (One Physics Ellipse, College Park, MD: American Association of Physicists in Medicine).

Accuray (n.d.) Estimation of the imaging dose for the CyberKnife® robotic radiosurgery system. (no longer available online).

Akino Y et al. (2012) Megavoltage cone beam computed tomography dose and the necessity of reoptimization for imaging dose-integrated intensity-modulated radiotherapy for prostate cancer. *Int. J. Radiat. Oncol. Biol. Phys.* 82: 1715–1722.

Alaei P, Gerbi BJ and Geise RA (2001) Lung dose calculations at kilovoltage x-ray energies using a model-based treatment planning system. *Med. Phys.* 28: 194–198.

Alaei P, Ding G and Guan H (2010) Inclusion of the dose from kilovoltage cone beam CT in the radiation therapy treatment plans. *Med. Phys.* 37: 244–248.

Alaei P, Spezi E and Reynolds M (2014) Dose calculation and treatment plan optimization including imaging dose from kilovoltage cone beam computed tomography. *Acta Oncol.* 53: 839–844.

Alaei P and Spezi E (2015) Imaging dose from cone beam computed tomography in radiation therapy *Phys. Med.* 31: 647–658.

Alvarado R et al. (2013) An investigation of image guidance dose for breast radiotherapy. *J. Appl. Clin. Med. Phys.* 14(3): 25–38.

Amer A et al. (2007) Imaging doses from the Elekta Synergy x-ray cone beam CT system. *Br. J. Radiol.* 80(954): 476–482.

BEIR (2006) Health risks from exposure to low levels of ionizing radiation: *BEIR VII Phase 2.* (Washington, DC: The National Academies Press).

Berrington de Gonzalez A et al. (2011) Proportion of second cancers attributable to radiotherapy treatment in adults: A cohort study in the US SEER cancer registries. *Lancet Oncol.* 12(4): 353–360.

Boone JM (2007) The trouble with CTD 100. *Med. Phys.* 34: 1364–1371.

Borst GR et al. (2010) Clinical results of image-guided deep inspiration breath hold breast irradiation. *Int. J. Radiat. Oncol. Biol. Phys.* 78: 1345–1351.

Brenner DJ and Hall EJ (2007) Computed tomography—An increasing source of radiation exposure. *New England J. Med.* 357: 2277–2284.

Bujold A et al. (2012) Image-guided radiotherapy: Has it influenced patient outcomes? *Sem. Radiat. Oncol.* 22: 50–61.

Cheng HCY et al. (2011) Evaluation of radiation dose and image quality for the Varian cone beam computed tomography system. *Int. J. Radiat. Oncol. Biol. Phys.* 80: 291–300.

Cheung T, Butson MJ and Yu PKN (2003) MOSFET dosimetry in-vivo at superficial and orthovoltage x-ray energies. *Australas. Phys. Eng. Sci. Med.* 26: 82–84.

Chodick G et al. (2008) Original contribution risk of cataract after exposure to low doses of ionizing radiation: A 20-year prospective cohort study among US radiologic technologists. *Amer J. Epidemiol.* 168: 620–631.

Chow JCL (2009) Cone-beam CT dosimetry for the positional variation in isocenter: A Monte Carlo study. *Med. Phys.* 36: 3512–3520.

Chow JCL et al. (2008) Evaluation of the effect of patient dose from cone beam computed tomography on prostate IMRT using Monte Carlo simulation. *Med. Phys.* 35: 52–60.

Deng J et al. (2012a) Kilovoltage imaging doses in the radiotherapy of pediatric cancer patients. *Int. J. Radiat. Oncol. Biol. Phys.* 82: 1680–1688.

Deng J et al. (2012b) Testicular doses in image-guided radiotherapy of prostate cancer. *Int. J. Radiat. Oncol. Biol. Phys.* 82: 39–47.

Ding GX, Duggan DM and Coffey CW (2008a) Accurate patient dosimetry of kilovoltage cone-beam CT in radiation therapy. *Med. Phys.* 35: 1135–1144.

Ding GX, Pawlowski JM and Coffey CW (2008b) A correction-based dose calculation algorithm for kilovoltage x rays. *Med. Phys.* 35: 5312–5316.

Ding GX and Coffey CW (2009) Radiation dose from kilovoltage cone beam computed tomography in an image-guided radiotherapy procedure. *Int. J. Radiat. Oncol. Biol. Phys.* 73: 610–617.

Ding GX and Coffey CW (2010) Beam characteristics and radiation output of a kilovoltage cone-beam CT. *Phys. Med. Biol.* 55: 5231–5248.

Ding A, Gu J, Trofimov AV and Xu XG (2010) Monte Carlo calculation of imaging doses from diagnostic multidetector CT and kilovoltage cone-beam CT as part of prostate cancer treatment plans. *Med. Phys.* 37: 6199–6204.

Ding A et al. (2010) Monte Carlo calculation of imaging doses from diagnostic multidetector CT and kilovoltage cone-beam CT as part of prostate cancer treatment plans. *Med. Phys.* 37: 6199–6204.

Ding GX and Malcolm AW (2013) An optically stimulated luminescence dosimeter for measuring patient exposure from imaging guidance procedures. *Phys. Med. Biol.* 58: 5885–5897.

Ding GX and Munro P (2013) Radiation exposure to patients from image guidance procedures and techniques to reduce the imaging dose. *Radiother. Oncol.* 108: 91–98.

Dixon RL (2003) A new look at CT dose measurement: Beyond CTDI. *Med. Phys.* 30: 1272–1280.

Dixon RL and Boone JM (2010) Cone beam CT dosimetry: A unified and self-consistent approach including all scan modalities—with or without phantom motion. *Med. Phys.* 37: 2703–2718.

Downes P et al. (2009) Monte Carlo simulation and patient dosimetry for a kilovoltage cone-beam CT unit. *Med. Phys.* 36: 4156–4167.

Dufek V, Horakova I and Novak L (2011) Organ and effective doses from verification techniques in image-guided radiotherapy. *Radiat. Prot. Dosim.* 147: 277–280.

Ehringfeld C et al. (2005) Application of commercial MOSFET detectors for in vivo dosimetry in the therapeutic x-ray range from 80 kV to 250 kV. *Phys. Med. Biol.* 50: 289–303.

Faddegon BA et al. (2008) Low dose megavoltage cone beam computed tomography with an unflattened 4 MV beam from a carbon target. *Med. Phys.* 35: 5777–5786.

Fahrig R et al. (2006) Dose and image quality for a cone-beam C-arm CT system. *Med. Phys.* 33: 4541–4550.

Falco MD et al. (2011) Preliminary studies for a CBCT imaging protocol for offline organ motion analysis: Registration software validation and CTDI measurements. *Med. Dos.* 36: 91–101.

Flynn RT et al. (2009) Dosimetric characterization and application of an imaging beam line with a carbon electron target for megavoltage cone beam computed tomography. *Med. Phys.* 36: 2181–2192.

Gayou O et al. (2007) Patient dose and image quality from mega-voltage cone beam computed tomography imaging. *Med. Phys.* 34: 499–506.

Geleijns J et al. (2009) Computed tomography dose assessment for a 160 mm wide, 320 detector row, cone beam CT scanner. *Phys. Med. Biol.* 54: 3141–3159.

Giaddui T et al. (2013) Comparative dose evaluations between XVI and OBI cone beam CT systems using Gafchromic XRQA2 film and nanoDot optical stimulated luminescence dosimeters. *Med. Phys.* 40: 062102.

Grelewicz Z and Wiersma RD (2014) Combined MV + kV inverse treatment planning for optimal kV dose incorporation in IGRT. *Phys. Med. Biol.* 59: 1607–1621.

Gu J et al. (2008) Assessment of patient organ doses and effective doses using the VIP-man adult male phantom for selected cone-beam CT imaging procedures during image guided radiation therapy. *Radiat. Prot. Dos.* 131: 431–443.

Halg RA, Besserer J and Schneider U (2012) Systematic measurements of whole-body imaging dose distributions in image-guided radiation therapy. *Med. Phys.* 39: 7650–7661.

Harrison RM et al. (2006) Organ doses from prostate radiotherapy and associated concomitant exposures. *Brit. J. Radiol.* 79 (942): 487–496.

Harrison RM et al. (2007) Doses to critical organs following radiotherapy and concomitant imaging of the larynx and breast. *Brit. J. Radiol.* 80(960): 989–995.

Hill R et al. (2014) Advances in kilovoltage x-ray beam dosimetry. *Phys. Med. Biol.* 59: R183–R231.

Howell S and Shalet S (1998) Gonadal damage from chemotherapy and radiotherapy. *Endocrinol Metab Clin North Am.* 27(4): 927–943.

Hu N and Mclean D (2014) Measurement of radiotherapy CBCT dose in a phantom using different methods. Australas. Phys. Eng. Sci. Med. 37: 779–789.

Hyer DE and Hintenlang DE (2010) Estimation of organ doses from kilovoltage cone-beam CT imaging used during radiotherapy patient position verification. *Med. Phys.* 37: 4620–4626.

Hyer DE et al. (2010) An organ and effective dose study of XVI and OBI cone-beam CT systems. *J. Appl. Clin. Med. Phys.* 11(2): 181–197.

IAEA (2000) Absorbed dose determination in external beam radiotherapy: An international code of practice for dosimetry based on standards of absorbed dose to water. Technical Reports Series No. 398. (Vienna, Austria: International Atomic Energy Agency).

IAEA (2011) Status of computed tomography dosimetry for wide cone beam scanners. Human Health Reports No. 5. (Vienna, Austria: International Atomic Energy Agency).

ICRP (2007) The 2007 recommendations of the International Commission on Radiation Protection. Publication 103. *Annals of ICRP* 37(2–4).

IEC (2009) IEC 60601-2-44 Medical Electrical Equipment—Part 2-44: Particular requirements for the basic safety and essential performance of x-ray equipment for computed tomography (Third Edition–2009). (Geneva, Switzerland: International Electrotechnical Commission). (https://webstore.iec.ch/publication/2661&preview=1).

Imaizumi M et al. (2006) Radiation dose-response relationships for thyroid nodules and autoimmune thyroid diseases in Hiroshima and Nagasaki atomic bomb survivors 55–58 years after radiation exposure. *JAMA*. 295(9): 1011–1022.

IPEMB (1996) The IPEMB code of practice for the determination of absorbed dose for x-rays below 300 kV generating potential (0.035 Mm Al-4 Mm Cu HVL; 10–300 kV generating potential). Institution of Physics and Engineering in Medicine and Biology. *Phys. Med. Biol.* 41: 2605–2625.

Isambert A et al. (2009) Dose délivrée au patient lors de l'acquisition d'images par tomographie conique de haute énergie. *Cancer/Radiother.* 13: 358–364.

Islam MK et al. (2006) Patient dose from kilovoltage cone beam computed tomography imaging in radiation therapy. *Med. Phys.* 33: 1573–1582.

Jacobi W (1975) The concept of the effective dose–a proposal for the combination of organ doses. *Radiat. Env. Biophys.* 12: 101–109.

Jaffray DA et al. (1999) A radiographic and tomographic imaging system integrated into a medical linear accelerator for localization of bone and soft-tissue targets. *Int. J. Radiat. Oncol. Biol. Phys.* 45: 773–789.

Jaffray DA and Siewerdsen JH (2000) Cone-beam computed tomography with a flat-panel imager: Initial performance characterization. *Med. Phys.* 27: 1311–1323.

Jones DG and Wall BF (1985) Organ doses from medical x-ray examinations calculated using Monte Carlo techniques. Report NRPB-R186. (Chilton, Didcot, UK: National Radiological Protection Board).

Jones DG and Shrimpton PC (1991) Survey of CT Practice in the UK Part 3: Normalised organ doses calculated using Monte Carlo techniques. Document NRPB-R250. (Chilton, UK: National Radiological Protection Board).

Kan MW, Leung LH, Wong W and Lam N (2008) Radiation dose from cone beam computed tomography for image-guided radiation therapy. *Int. J. Radiat. Oncol. Biol. Phys.* 70: 272–279.

Kawrakow I (2013) The EGSnrc Code System: Monte Carlo simulation of electron and photon transport. NRCC Report PIRS-701. (http://irs.inms.nrc.ca/software/egsnrc/documentation.html).

Kim S et al. (2010) Radiation dose from cone beam CT in a pediatric phantom: Risk estimation of cancer incidence. *Am. J. Roentgenol.* 194: 186–190.

Kim S et al. (2011) Computed tomography dose index and dose length product for cone-beam CT: Monte Carlo simulations of a commercial system. *J. Appl. Clin. Med. Phys.* 12(2): 84–95.

Korreman S et al. (2010) Report on 3D CT-based in-room image guidance systems: A practical and technical review and guide. *Radiother. Oncol.* 94: 129–144.

Koyama S et al. (2010) Radiation dose evaluation in tomosynthesis and C-Arm cone-ceam CT examinations with an anthropomorphic phantom. *Med. Phys.* 37: 4298–4306.

Kron T et al. (1998) Dose response of various radiation detectors to synchrotron radiation. *Phys. Med. Biol.* 43: 3235–3259.

Kron T et al. (2010) Adaptive radiotherapy for bladder cancer reduces integral dose despite daily volumetric imaging. *Radiother. Oncol.* 97: 485–487.

Kudchadker RJ et al. (2004) An evaluation of radiation exposure from portal films taken during definitive course of pediatric radiotherapy. *Int. J. Radiat. Oncol. Biol. Phys.* 59: 1229–1235.

Le Heron JC (1992) Estimation of effective dose to the patient during medical x-ray examinations from measurements of the dose-area product. *Phys. Med. Biol.* 37: 2117–2126.

Ma CM et al. (2001) AAPM protocol for 40–300 kV x-ray beam dosimetry in radiotherapy and radiobiology. *Med. Phys.* 28: 868–893.

Marinello G et al. (2009) Radiotherapie des cancers de la prostate: Evaluation in vivo de la dose delivre par tomographie conique de basse energie (kV). *Cancer/Radiother.* 13: 353–357.

Meeks SL et al. (2005) Performance characterization of megavoltage computed tomography imaging on a helical tomotherapy unit. *Med. Phys.* 32: 2673–2681.

Miften M et al. (2007) IMRT planning and delivery incorporating daily dose from mega-voltage cone-beam computed tomography imaging. *Med. Phys.* 34: 3760–3767.

Montanari D et al. (2014) Comprehensive evaluations of cone-beam CT dose in image-guided radiation therapy via GPU-based Monte Carlo simulations. *Phys. Med. Biol.* 59: 1239–1253.

Moore CJ et al. (2014) Developments in and experience of kilovoltage x-ray cone beam image-guided radiotherapy. *Brit. J. Radiol.* 79 Spec No 1 S66–S78.

Mori S et al. (2005) Enlarged longitudinal dose profiles in cone-beam CT and the need for modified dosimetry. *Med. Phys.* 32: 1061–1069.

Morin O et al. (2007a) Dose calculation using megavoltage cone-beam CT. *Int. J. Radiat. Oncol. Biol. Phys.* 67: 1201–1210.

Morin O et al. (2007b) Patient dose considerations for routine megavoltage cone-beam CT imaging. *Med. Phys.* 34: 1819–1827.

Moseley DJ et al. (2007) Comparison of localization performance with implanted fiducial markers and cone-beam computed tomography for on-line image-guided radiotherapy of the prostate. *Int. J. Radiat. Oncol. Biol. Phys.* 67: 942–953.

Murphy MJ et al. (2007) The management of imaging dose during image-guided radiotherapy: Report of the AAPM Task Group 75. *Med. Phys.* 34: 4041–4063.

Ng JA et al. (2012) Kilovoltage intrafraction monitoring for prostate intensity modulated arc therapy: First clinical results. *Int. J. Radiat. Oncol. Biol. Phys.* 84: e655–e661.

Nobah et al. (2014) Radiochromic film based dosimetry of image-guidance procedures on different radiotherapy modalities. *J. Appl. Clin. Med. Phys.* 15(6): 229–239.

Nunn AA et al. (2008) LiF:Mg, Ti TLD response as a function of photon energy for moderately filtered x-ray spectra in the range of 20–250 kVp relative to 60Co. *Med. Phys.* 35: 1859–1869.

Osei EK et al. (2009) Dose assessment from an online kilovoltage imaging. *J. Radiol. Prot.* 29: 37–50.

Palm A, Nilsson E and Herrnsdorf L (2010) Absorbed dose and dose rate using the Varian OBI 1.3 and 1.4 CBCT system. *J. Appl. Clin. Med. Phys.* 11(1): 229–240.

Pawlowski JM and Ding GX (2014) An algorithm for kilovoltage x-ray dose calculations with applications in kV-CBCT scans and 2D planar projected radiographs. *Phys. Med. Biol.* 59: 2041–2058.

Perks JR et al. (2008) Comparison of peripheral dose from image-guided radiation therapy (IGRT) using kV cone beam CT to intensity-modulated radiation therapy (IMRT). *Radiother. Oncol.* 89: 304–310.

Poirier Y et al. (2014) Experimental validation of a kilovoltage x-ray source model for computing imaging dose. *Med. Phys.* 41: 041915.

Qiu Y et al. (2011) Mega-voltage versus kilo-voltage cone beam CT used in image guided radiation therapy: Comparative study of microdosimetric properties. *Rad. Prot. Dos.* 143: 477–480.

Qiu Y et al. (2012) Equivalent doses for gynecological patients undergoing IMRT or RapidArc with kilovoltage cone beam CT. *Radiother. Oncol.* 104: 257–262.

Quinn A et al. (2011) Megavoltage cone beam CT near surface dose measurements: Potential implications for breast radiotherapy. *Med. Phys.* 38: 6222–6227.

Saw CB et al. (2007) Performance characteristics and quality assurance aspects of kilovoltage cone-beam CT on medical linear accelerator. *Med. Dos.* 32: 80–85.

Sawyer LJ, Whittle SA, Matthews ES, Starritt HC and Jupp TP (2009) Estimation of organ and effective doses resulting from cone beam CT imaging for radiotherapy treatment planning. *Brit. J. Radiol.* 82(979): 577–584.

Schiffner DC et al. (2007) Daily electronic portal imaging of implanted gold seed fiducials in patients undergoing radiotherapy after radical prostatectomy. *Int. J. Radiat. Oncol. Biol. Phys.* 67: 610–619.

Shah AP et al. (2008) Patient dose from megavoltage computed tomography imaging. *Int. J. Radiat. Oncol. Biol. Phys.* 70: 1579–1587.

Shah A, Aird E and Shekhdar J (2012) Contribution to normal tissue dose from concomitant radiation for two common kV-CBCT systems and one MVCT system used in radiotherapy. *Radiother. Oncol.* 105: 139–144.

Shirato H et al. (2004) Intrafractional tumor motion: Lung and liver. *Sem. Rad. Oncol.* 14: 10–18.

Shope TB, Gagne RM and Johnson GC (1981) A method for describing the doses delivered by transmission x-ray computed tomography. *Med. Phys.* 8: 488–495.

Smith-Bindman R et al. (2009) Radiation dose associated with common computed tomography examinations and the associated lifetime attributable risk of cancer. Arch. Intern. Med. 169: 2078–2086.

Son K, Cho S et al. (2014) Evaluation of radiation dose to organs during kilo-voltage cone-beam computed tomography using Monte Carlo simulation. *J. Appl. Clin. Med. Phys.* 15(2): 295–302.

Song WY et al. (2008) A dose comparison study between XVI® and OBI® CBCT systems. *Med. Phys.* 35: 480–486.

Spezi E and Lewis G (2008) An overview of Monte Carlo treatment planning. *Rad. Prot. Dos.* 131: 123–129.

Spezi E et al. (2009) Monte Carlo simulation of an x-ray volume imaging cone beam CT unit. *Med. Phys.* 36: 127–136.

Spezi E et al. (2011) A virtual source model for kilo-voltage cone beam CT: Source characteristics and model validation. *Med. Phys.* 38: 5254–5263.

Spezi E et al. (2012) Patient-specific three-dimensional concomitant dose from cone beam computed tomography exposure in image-guided radiotherapy. *Int. J. Radiat. Oncol. Biol. Phys.* 83: 419–426.

Stock M et al. (2012) IGRT induced dose burden for a variety of imaging protocols at two different anatomical sites. *Radiother. Oncol.* 102: 355–363.

Sykes JR et al. (2005) A feasibility study for image guided radiotherapy using low dose, high speed, cone beam x-ray volumetric imaging. *Radiother. Oncol.* 77: 45–52.

Sykes JR (2010) PhD thesis: Quantification of geometric uncertainties in image guided radiotherapy. University of Leeds, UK.

Sykes JR et al. (2010) Evaluation report x-ray tomographic image guided radiotherapy systems. (http://nhscep.useconnect.co.uk/CEPProducts/Catalogue. aspx).

Tien CJ, Lee SW and Dieterich S (2014) Estimated clinical impact of fractionation scheme and tracking method upon imaging dose in CyberKnife robotic radiosurgery. *Austin J. Nucl. Med. Radiother.* 1: 1–5.

Tubiana M (2009) Can we reduce the incidence of second primary malignancies occurring after radiotherapy? A critical review. *Radiother. Oncol.* 91: 4–15.

Varian Medical Systems (2012) Dose in CBCT—OBI advanced imaging On-Board Imager kV imaging systems v1.4 and v1.5.

Verhaegen F and Seuntjens J (2003) Monte Carlo modelling of external radiotherapy photon beams. *Phys. Med. Biol.* 48: R107–R164.

Wallace WHB, Thomson AB and Kelsey TW (2003) The radiosensitivity of the human oocyte. *Hum. Reprod.* 18: 117–121.

Walter C et al. (2007) Phantom and in-vivo measurements of dose exposure by image-guided radiotherapy (IGRT): MV portal images vs. kV portal images vs. cone-beam CT. *Radiother. Oncol.* 85: 418–423.

Walters BRB et al. (2009) Skeletal dosimetry in cone beam computed tomography. *Med. Phys.* 36: 2915–2922.

Wen N et al. (2007) Dose delivered from Varian's CBCT to patients receiving IMRT for prostate cancer. *Phys. Med. Biol.* 52: 2267–2276.

Winey B, Zygmanski P and Lyatskaya Y (2009) Evaluation of radiation dose delivered by cone beam CT tomosynthesis employed for setup of external breast irradiation. *Med. Phys.* 36: 164–173.

Xu XG, Chao TC and Bozkurt A (2000) VIP-Man: An image-based whole-body adult male model constructed from color photographs of the visible human project for multi-particle Monte Carlo calculations. *Health Phys.* 78: 476–486.

Yin F-F et al. (2009). *The Role of In-Room kV X-Ray Imaging for Patient Setup and Target Localization Report of AAPM Task Group 104. Data Management.*

Zelefsky MJ et al. (2012) Improved clinical outcomes with high-dose image guided radiotherapy compared with non-IGRT for the treatment of clinically localized prostate cancer. *Int. J. Radiat. Oncol. Biol. Phys.* 84: 125–129.
Zhang Y et al. (2012) Personalized assessment of kV cone beam computed tomography doses in image-guided radiotherapy of pediatric cancer patients. *Int. J. Radiat. Oncol. Biol. Phys.* 83: 1649–1654.

23

Dose Verification of Proton and Carbon Ion Beam Treatments

Katia Parodi

23.1 Introduction to Ion Therapy with Pencil-Beam Scanning

Ion beam therapy is a rapidly emerging treatment modality that makes use of the favorable properties of ion interaction with tissue. When ions are slowed down in matter, they transfer most of their energy in Coulomb interactions with the atomic electrons, and the energy transfer rate is relatively low at the initial therapeutic beam energies of 50–230 MeV for protons and 90–430 MeV/u for ^{12}C ions, but rapidly increases toward the end of the beam penetration depth (Schardt et al., 2010; Newhauser and Zhang, 2015). This energy loss process is responsible for the main feature of the ion depth–dose deposition, exhibiting a pronounced sharp maximum, the Bragg peak, at an energy-dependent depth (Wilson, 1946).

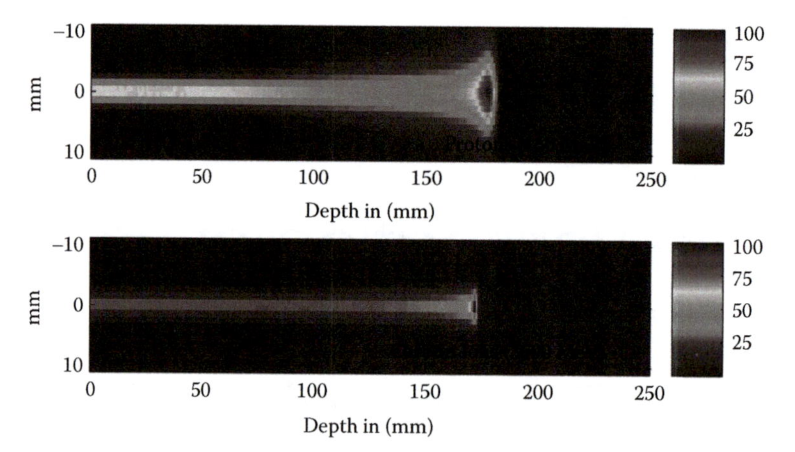

Figure 23.1

Two-dimensional distributions of dose deposition calculated in water with the FLUKA Monte Carlo code (Böhlen et al. 2014; Ferrari et al. 2005) for a proton (top) and carbon ion (bottom) pencil-like beam of initial lateral Gaussian distribution (3-mm full width at half maximum [FWHM]), entering a water target from the left. The gray scale levels indicate percentages of the maximum dose, as specified by the gray scale bar. (Adapted from Parodi K, Habilitation thesis, Ruprecht-Karls-Universität Heidelberg, Heidelberg, Germany, 2008.)

Additional processes that influence the exact shape of the three-dimensional (3D) dose distribution of an ion beam are (1) energy loss straggling, affecting the longitudinal Bragg peak width; (2) multiple Coulomb scattering, affecting the lateral beam spread; and (3) nuclear interactions, which modify the particle energy fluence spectrum. Energy straggling and multiple Coulomb scattering are especially responsible for the broader Bragg peak and marked lateral spread of proton beams compared to carbon ions at the same penetration depth (Figure 23.1). Nuclear reactions prevent primary particles from reaching the Bragg peak and are especially responsible for the complex mixed radiation field in the case of carbon ion beams, where secondary projectile fragments of reduced charge give rise to a characteristic dose tail distal to the Bragg peak (Figure 23.2). More information about the interaction of light ions with matter can be found in Chapter 12.

In addition to the various physical processes influencing ion interaction with tissue and detection systems, the pattern of dose deposition, and hence its verification, is also highly dependent on the beam delivery technique. Most of the clinical experience of particle therapy relies on the more traditional (semi-) passively scattered delivery, where extended treatment fields are formed almost instantaneously. Here, scattering foils are used, often in combination with active beam deflection or wobbling, for lateral beam broadening. Longitudinal range modulation is obtained with fast spinning wheels or binary range shifters of different thickness, followed by patient-specific mechanical shaping via collimators and compensators (Chu et al., 1993). However, higher degrees of tumor dose conformity and flexibility in treatment planning can be achieved with the introduction of active beam scanning, where narrow, monoenergetic, pencil-like beams are laterally scanned over arbitrarily shaped tumor volumes by exploiting fast (10–100 m/s) horizontal and vertical magnetic deflection (Pedroni et al., 2004; Furukawa et al., 2010). Adjustment of each pencil beam to the intended penetration depth for optimal 3D placement of the Bragg peak is achieved by

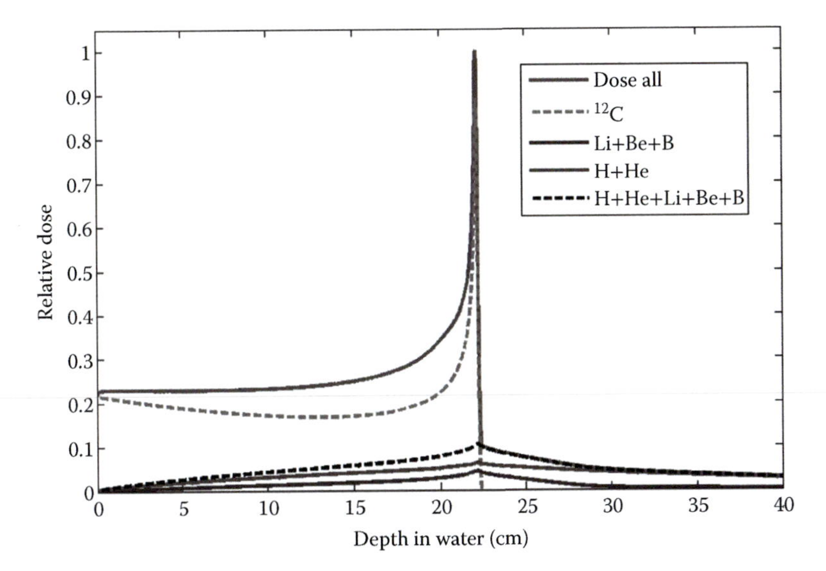

Figure 23.2

Depth-dose distribution in water calculated by the FLUKA Monte Carlo code for a perfectly monoenergetic carbon ion beam at 350 MeV/u. The total dose (gray solid line) is compared with the separate contributions of the primary ^{12}C beam (gray dashed line) and of the lighter projectile fragments for charge $3 \leq Z \leq 5$ (black solid line) and $Z < 3$ (dark gray solid line). While the surviving primary beam stops in the Bragg peak, the lighter ions are solely responsible for the dose tail beyond the Bragg peak, as shown by the integral dashed black line. (Adapted from Parodi K, Habilitation thesis, Ruprecht-Karls-Universität Heidelberg, Heidelberg, Germany, 2008.)

energy variation prior to scanning, which can be performed either directly at the accelerator level, for example, at synchrotron accelerators with typical energy switching times of few seconds, or just after beam extraction, for example, from cyclotrons, with very fast beam wedges able to degrade the beam energy by 5 mm water-equivalent depth in about 50 ms (Pedroni et al., 2004).

Starting from the pioneering developments and first clinical experience with several flavors of beam scanning at the end of the last century (Kanai et al., 1983; Haberer et al., 1993; Pedroni et al., 1995), pencil beam scanning is currently entering routine clinical usage and represents the state-of-the-art of modern solutions of particle therapy worldwide. Therefore, only this beam delivery modality is considered in this chapter. Similar to the challenges inherent to the sophistication of modern beam delivery techniques in photon therapy, the wide range of pencil beam parameters (energy, spot size, fluence) and the dynamic forming of dose delivery as superposition of several thousands (depending on target volume) of these tiny beams pose strict requirements to the clinical 3D dosimetry of ion therapy, as addressed in the following sections.

23.2 Dosimetric Measurements for Pencil Beam Characterization and Treatment Planning System Basic Data Generation

Once stable ion beams are produced by the accelerator at a new particle therapy facility, several dosimetric characterizations are required for acceptance testing and commissioning. In particular, individual monoenergetic, pencil-like beams need to be extensively characterized in water and air for different physical quantities.

The most critical parameter affecting the accuracy of the dose deposition is the knowledge of the beam range in water, which is related to the ion beam energy. Although the physical definition of beam range refers to the position at which the particle fluence drops to 50% of its initial value, in clinical practice of ion beam therapy the term "range" refers to a certain falloff percentage (typically 80% for monoenergetic pristine Bragg peaks, 90% for extended dose distributions) of the depth–dose curve or, in some cases, to the Bragg peak position itself. In order to properly assess this quantity, acquisitions of laterally integrated pencil beam depth–dose curves can be obtained at a fine (down to 10 μm) spatial resolution with water columns. This is done by measuring with large area (typically ≈4 cm radius) plane parallel ionization chambers (ICs) the ratio of ionization produced before and after a variable water column. In this approach, the first fixed position chamber serves to provide normalization to the incoming beam current, thus eliminating the influence of unavoidable beam current fluctuations. Due to the high accuracy (~100 μm) and precision (~10 μm) achievable with this measuring approach (Schardt et al., 2008; Kurz et al., 2012), special attention has to be devoted to the absolute calibration of the detector (i.e., knowledge of the water-equivalent thickness of all components traversed by the beam prior to water) and the controlled water condition (i.e., water type and temperature). As the purpose of such measurements is the validation of the beam range for the entire energy interval of therapeutic interest, implying hundreds of energy settings, for example, 255 for the Heidelberg Ion-Beam Therapy Center (HIT) synchrotron accelerator (Parodi et al., 2012) to be verified, the acquisition can be optimized in speed by properly measuring only few points around the expected Bragg peak position, sufficient for determination of the practical beam range. Once an accurate relationship between initial beam energy and range has been experimentally assessed, water-column measurements spanning the whole range of therapeutic beam energies can be repeated at coarser time intervals to verify the beam energy stability. For more frequent quality assurance (QA), faster tests can be designed and performed at few representative beam energies, for example, by using a fixed stack of several, up to 180 (Grant et al., 2014), independent vented plane parallel ICs interleaved with 2 to 3 mm thick absorbers, which provide a coarser Bragg curve in one single acquisition (Brusasco et al., 2000). Range QA can also be performed by measuring the transmission of monoenergetic treatment fields (e.g., lines or planes) after properly designed absorbers (e.g., beam wedges) with simple dosimetric films or detector arrays (Mirandola et al., 2015).

Whereas knowledge of the beam range in water is essential for selection of the appropriate beam energies at the beginning of the planning process, 3D dosimetric characterization of the pencil beams in water and air is the next fundamental information to be fed to the treatment planning system (TPS) for proper planning of ion therapy treatments. According to the typical beam model of pencil beam algorithms, which separate the lateral and depth dependence of the 3D dose distribution, typical dosimetric measurements for TPS basic data generation include laterally integrated depth–dose curves and lateral beam profiles in water, complemented by 2D beam-spot characterization in air at different treatment room positions.

For collection of laterally integrated depth–dose curves, water columns are again the instrument of choice, typically providing a finer (submillimeter) sampling in the more rapidly changing parts of the Bragg curve, for example, around the peak, while coarser (up to centimeters) sampling in the plateau region and in the long ranging tail for carbon ions. Different from the measurements required for range assessment, more stringent conditions apply for acquisition of laterally integrated Bragg curves for TPS basic data generation, as several parameters can influence the shape of the acquired data. In particular, the beam-spot size should be minimized to prevent lateral loss of particles. This is, however, somewhat unavoidable at high beam energies due to considerable beam broadening (protons) and large-angle secondary emission (carbon ions) in water, rapidly exceeding the lateral dimension of currently available chambers regardless of the initial beam size (Gillin et al., 2010; Grevillot et al., 2011; Parodi et al., 2012). Moreover, low beam currents and sufficient chamber voltage should be used to minimize distortions of the peak-to-plateau ratio due to depth-dependent changes of saturation effects and charge collection efficiency (Kurz et al., 2012). Finally, the users should be aware of additional effects that can have a minor influence on the shape of the Bragg curves. In particular, the typically neglected energy dependence of stopping power ratios in the conversion of the measured ionization of air into dose to water should be considered depth dependent, thus impacting the peak-to-plateau ratio, as discussed in detail in Chapter 12. Moreover, the assumption of energy-independent calibration of the detector water-equivalent thickness, typically measured at high energies, can introduce Bragg peak position shifts up to 0.2 mm if not properly accounted for (Kurz et al., 2012).

Characterization of the lateral beam broadening in water, which provides the basic data for lateral parametrization of pencil beams under the typical assumption of single or double Gaussian distributions, can be assessed from 1D or 2D dose distribution measurements (Pedroni et al., 2005; Schwaab et al., 2011). Due to the wide dynamic range of dose from the central core to the low-dose envelope far away from the pencil beam center, as well as the varying contributions of different energies and even particle species, especially for primary carbon ions, ICs are still considered the detectors of choice. However, in this case small-volume, cylindrically shaped "pinpoint" ICs with radii typically less than 1.5 mm (Karger et al., 2010) are used to provide the best possible approximation of desired point measurements.

Due to considerable time consumption associated with these kinds of measurements, the chambers are typically arranged in special patterns with multiple locations not shadowing each other for parallel acquisitions in one beam delivery (Karger et al., 2010). The sampling frequency is then increased by laterally moving the system through the irradiation field, typically consisting of monoenergetic single spot data or line scans (Sawakuchi et al., 2010a; Schwaab et al., 2011). In this way, 1D lateral profiles can be collected at different depths in water and can be used for TPS basic data generation, either directly or indirectly via validation of dedicated Monte Carlo models of the beamline used to generate the data (Gillin et al., 2010; Parodi et al., 2013). More time-efficient measurements can be obtained using 2D detection systems, such as detector arrays, electronic portal imaging devices (EPIDs), or films (Sawakuchi et al., 2010a; Martisíková et al., 2012; Mirandola et al., 2015). However, issues related to measurement in water surrogates (for detector arrays and EPIDs), dose linearity (films), and possible linear energy transfer (LET) dependences (diode-based arrays, EPIDs, and films) need to be properly understood and taken into account for proper interpretation of the measurements, in order to provide the correct data for lateral beam modeling in the TPS. In any case, such faster measurements can complement the more time-demanding IC measurements done for basic data generation, for example, to assess the constancy of the lateral beam parameters for exemplary energies.

Characterization of beam spots in air considerably simplifies the measuring conditions due to the reduced influence of air on the particle fluence spectrum with respect to water. The purpose of these measurements is to characterize the beamline-specific divergence and dimension of the beam prior to entrance in the patient, which depends on the geometrical position in the treatment room. To this aim, 2D detectors are typically the method of choice, and spatial resolution is a critical parameter in order to enable single acquisitions at each geometrical position. Moreover, the detector itself should be of low material budget, to avoid additional beam broadening due to the measuring process itself. Hence, films or scintillation screens read by charge-coupled devices are typical solutions adopted at different facilities, and measurements are pursued in few positions chosen at the isocenter as well as several tens of centimeters in front and beyond it. Such data can be acquired in parallel to the beam-spot characterization (position and shape) in the upstream position-sensitive beam monitors, to collect additional data for monitoring the beam stability over time (Mirandola et al., 2015). Given the relative simplicity of the setup, these approaches can also be used for more frequent constancy checks of the lateral beam shape in air at isocenter, unless other setups are preferred to combine several checks in one single measurement.

The information resulting from all earlier-mentioned measurements enables a complete 3D dosimetric description of pencil-like beams when entering a water medium placed at an arbitrary position in the treatment room (Figure 23.3). However, the laterally integrated dose measurement is only a relative acquisition (ratio of ionizations), and therefore does not convey absolute information on the amount of dose deposition. Moreover, even if using absolute measurements deduced from the lateral dose profiles acquired for a certain number of delivered ions for individual pencil beams, the relationship between these numbers of ions and the monitor units (MUs) read out by the upstream beam-monitor system (typically consisting of transmission ICs) has to be established for linking the TPS predictions to the beam-delivery system. Hence, dosimetric measurements are needed for calibration of individual pencil beams to connect the upstream beam-monitor readout to the dose to water deposited at isocenter. The typical MU calibration approach consists in delivering square monoenergetic treatment fields of sufficient size (up to 10×10 cm^2) to avoid field-size effects, i.e., reduced dose at the center of the field due to lateral loss of particles no longer compensated by neighboring beam spots (Sawakuchi et al., 2010a). At isocenter a calibrated cylindrically shaped IC (typically a Farmer-type chamber) is placed, embedded in water or a water-equivalent medium (Karger et al., 2010). To rule out (or at least reduce) dependencies on the exact depth positioning of the chamber, these measurements are typically performed in the entrance region of the Bragg curve. The energy-dependent calibration factor CF, i.e., the link between the number of irradiated particles N per measured number of MUs, and the measured dose D when delivering a square field scanned at Δx, Δy spot separation, can be obtained from the following formula (Karger et al., 2010):

$$CF(E) = \frac{\mathrm{N}}{MU} = \frac{D\Delta x\,\Delta \mathrm{y}}{MU}\left(\frac{1}{\rho}\frac{\mathrm{d}E}{dx}\right)^{-1} \tag{23.1}$$

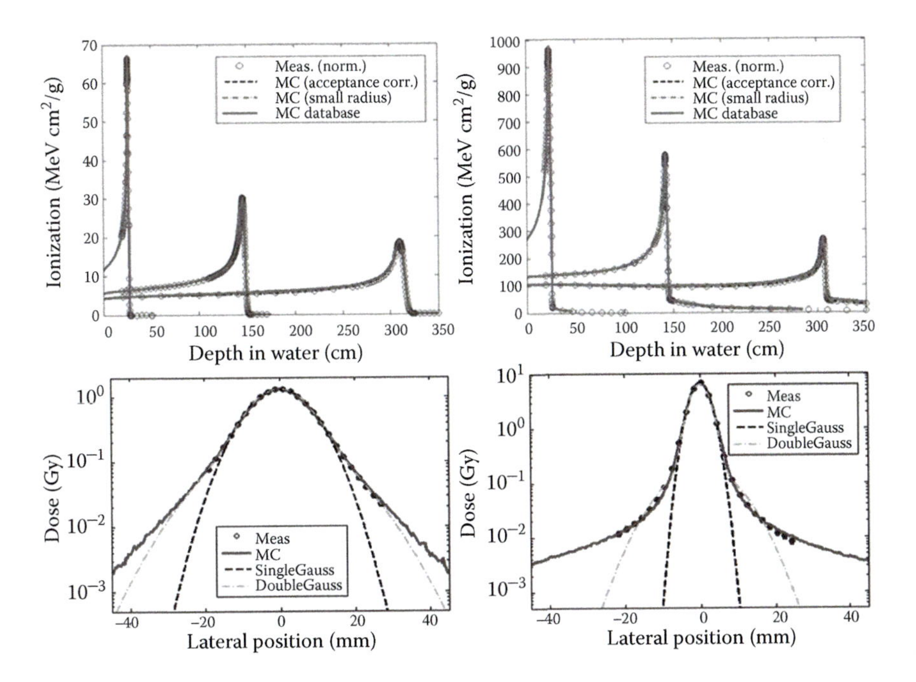

Figure 23.3

Top row: Examples of water-column depth–dose distributions in water (symbols) and corresponding FLUKA Monte Carlo simulations, including (dash and dash-dotted line) or not (solid line) the acceptance of the large area plane parallel ionization chamber (IC) for three energies spanning the entire therapeutic range for protons (left) and carbon ions (right). (Adapted from Parodi K et al., *Phys. Med. Biol.*, 57, 3759–3784, 2012.) *Bottom row*: Pinpoint ionization chamber (IC) measurements (symbols) of lateral profiles sampled in water at ≈16.5 cm shortly before the Bragg peak in comparison to FLUKA Monte Carlo simulations (solid lines) and related single Gaussian (dashed lines) and double Gaussian (dash dotted) parametrization for two intermediate energies of protons (157.53 MeV/u) and carbon ion beams (299.94 MeV/u). (Adapted from Parodi K et al., *J. Radiat. Res.*, 54(Suppl. 1), i91–i96, 2013.)

where the last term in parentheses indicates the inverse of the mass-stopping power. The only unknown quantity is the linear energy loss for a single particle, dE/dx. Since the beam is no longer purely monoenergetic when entering the chamber, this quantity is typically obtained from calculations, for example, based on Monte Carlo simulations of laterally integrated depth–dose distributions for many beam particles (to reduce statistical fluctuations), evaluated at the water-equivalent penetration depth at which the chamber is positioned. The latter information has to be deduced on the basis of the geometrical chamber location, including the effective point of measurement, as well as the water-equivalent thickness of the material in which it is embedded, if resorting to solid water-like surrogates (Mirandola et al., 2015). In this way, the whole procedure is self-consistent, as uncertainties in the determination of the dose or average stopping power will compensate when comparing dose measurements with calculations based on the same computational platform (Karger et al., 2010). Given the energy dependence of CF, its measurement has to be repeated at least for few representative energies spanning the whole range of therapeutic interest, to provide an appropriate trend, which can enable interpolation of the calibration factor to all available beam-energy settings. Being the delivery of a sufficiently large scanned field needed for each acquisition, the measurement is not optimized for time efficiency. Hence, for constancy checks, faster measurements relying on individual pencil beams depositing their entire energy within a large area, plane parallel Bragg peak IC of lower dosimetric absolute accuracy (similar to the one used in the water column) can be considered (Gillin et al., 2010; Karger et al., 2010).

As a final step for determination of absorbed dose delivered to patients, the relationship between the water system, where all TPS computations are performed, and the patient model, based on the planning x-ray computed tomography (CT) image, needs to be established. To this aim, the patient model can be stretched into a water-equivalent system along the beam-penetration depth using the concept of water-equivalent thickness (Hong et al., 1996; Krämer et al., 2000). The underlying idea is that the quantity of clinical interest is the dose to water, and this is mainly determined by the correct longitudinal location of the Bragg peak position, neglecting particle fluence changes between water and real tissue due to different scattering and nuclear interaction processes. In this approximation, a certain thickness of arbitrary material can be converted into an equivalent thickness of water if the (approximately) energy-independent stopping-power ratio of the material relative to water is known. Therefore, for each used CT imaging protocol, a relationship between the CT gray scale values (Hounsfield units, HUs) and the ion stopping–power ratio (often referred to as relative water-equivalent path length, rWEPL) needs to be experimentally established. This can again be achieved with water-column measurements after traversing a sample of a certain thickness d, by assessing the Bragg peak shift Δ when using a material x of known HUs in comparison to the same thickness of water (Rietzel et al., 2007):

$$\mathrm{rWEPL} = 1 + \frac{\Delta}{d} \tag{23.2}$$

The formula could be simplified to the measurement of the Bragg peak shift δ after traversing the sample of known HUs with respect to the reference Bragg curve in water without traversing any sample, i.e., rWEPL = δ/d, provided that the water-equivalent thickness of an air layer of the same sample thickness d can be considered negligible with respect to the desired measurement accuracy (or otherwise corrected for). These acquisitions need to be repeated for different tissue-equivalent substitutes or real animal tissue samples in regions of sufficient HU homogeneity, and for an initial beam energy where the stopping-power ratio can be reasonably considered energy independent. The collected measuring points, often complemented by stoichiometric calculations, serve to determine the piecewise linear relationship between the CT scanner- and protocol-dependent HU values and the ion relative stopping power or rWEPL (Schneider et al., 1996; Rietzel et al., 2007). In this way, any arbitrary ray traversing the patient CT can be converted into the water-equivalent system by multiplying each traversed depth l in a certain image voxel with the corresponding rWEPL value and summing up. The dose to water in the water-equivalent system can then be estimated on the basis of the pencil beam algorithms, using all earlier-mentioned data for characterization of the pencil beams in water and air, together with the MU calibration (Hong et al., 1996; Krämer et al., 2000). The dose calculation in water can be finally stretched back to the patient system, using the inverse operation based on the empirically determined HU-rWEPL calibration. The validity of the HU-rWEPL calibration over time can be guaranteed by checking the constancy of the HU values of known materials at the CT scanner for the given imaging protocols, together with the already mentioned verification of the stability of the ion-beam range in water.

Although all the described dosimetric measurements enable 3D calculations of absorbed dose, they do not offer sufficient beam-quality characterization for biological calculations. While clinical practice of proton therapy still relies on a constant relative biological effectiveness (RBE) of 1.1 (Paganetti, 2014), the complex RBE dependence with radiation quality (LET), dose, tissue type, and so on cannot be neglected in carbon ion therapy. To this aim, most TPSs used in scanned carbon ion therapy rely on biophysical models (Scholz et al., 1997), which help properly weight the energy deposit of each heavy charged particle of the mixed radiation field with its corresponding biological response, and finally take into account the nonlinear superposition of all these components for single as well as multiple treatment fields (Krämer and Scholz, 2006). Typical information needed for such biological calculations is the depth-dependent characterization of the fluence energy spectra of all ions with charge lower than or equal to the primary projectile in water (Krämer et al., 2000; Parodi et al., 2012). Although work is ongoing to enable such measurements with clinically affordable setups, for example, based on small silicon detectors operated in water or water-equivalent materials (Hartmann, 2013), the current experience still relies on nuclear physics measurements of beam attenuation and fragmentation, typically acquired with complex and bulky detector setups, for example, with large space requirements for time-of-flight fragment identification, at basic research institutes (Hättner et al., 2013), complemented by Monte Carlo calculations (Mairani et al., 2010). In addition to the ongoing experimental campaigns aiming at extending our knowledge of nuclear fragmentation in different materials of biological interest, there are also several attempts to provide a deeper understanding of the link between measurable physical quantities at the micro- (nano) dosimetric level and biological response (Hawkins, 1996; Inaniwa et al., 2010; Casiraghi and Schulte, 2015). If successful and widely accepted, these emerging approaches could promote a valuable paradigm shift, enabling to obtain the needed TPS information for biological calculations on the basis of micro- (or nano)dosimetric measurements to be performed in water tanks with certified medical devices (Inaniwa et al., 2015).

23.3 Dosimetric Measurements for TPS Commissioning

Once the basic data described in Section 23.2 are properly fed to the TPS, several tests need to be performed to enable commissioning of the system. Although a large variety of tests can be designed at the different facilities, the common trend is to consider treatment plans of different complexity to probe the abilities of the planning system in calculating dose distributions that correspond well to the measured ones.

Depending on the purpose of the dosimetric test, different detectors can be used. For example, field-size dependencies, which reflect the TPS ability to accurately model the lateral beam shape approximately at the target entrance, are typically measured in a similar setup as the MU calibration described in Section 23.2. A Farmer-type chamber is inserted in water or a water-like slab at a shallow depth to enable measurements in the Bragg peak entrance region for scanned square fields of different lateral size and initial beam energy. Conversely, different scanning patterns of still monoenergetic beams can be best tested with 2D detectors of high spatial resolution perpendicular to the beam direction, similar to the setups indicated in Section 23.2 for the assessment of the lateral beam extension in water and air.

Dosimetric measurements of 3D-extended treatment fields in water are typically performed with small-volume pinpoint chambers, again assembled in a special arrangement for multiple acquisition during irradiation, similar to the setup described for the point measurements of lateral beam distributions. In fact, due to the complexity of scanning beam delivery, the dose needs to be probed at several positions of the treatment field, as inaccurate delivery of few pencil beams could result in the total dose being correct at several positions, but incorrect at a few others (Karger et al., 2010). Moreover, in such a situation of volumetric dose delivery, the beam application can be quite time-consuming (from several tens of seconds up to minutes) due to the time needed for energy switching, which is especially pronounced at synchrotron-based facilities, in addition to lateral beam scanning. Hence, repeated single-point measurements would be too time-demanding, while the possibility of moving a single IC across the treatment field during irradiation, as done in passively scattered delivery, is not an option due to the slow dynamic formation of the total dose distribution. To this aim, the usage of 24 chambers coupled to 2 electrometers has been reported by Karger et al. (1999) for the optimization of data acquisition. These authors used a geometrical IC arrangement shifted in three heights each of two rows and four different side positions, to avoid neighboring chambers to interfere with each individual measurement. Other proposed configurations include two arrays of 13 ICs, which are arranged orthogonally to each other (Lomax et al., 2004). Depending on the application at fixed beam ports or gantries, the chambers can be operated in water or in solid materials, respectively. In the latter case, usage of polymethyl methacrylate (PMMA) is quite common due to its convenient mechanical properties and costs (Henkner et al., 2015). In addition to the limited freedom in the chamber positioning with respect to water, special care has to be given to the nonwater equivalence of PMMA, thus calling for proper fluence corrections for accurate dosimetry (Palmans et al., 2002; Karger et al., 2010).

The TPS capability to design treatment plans built by the proper lateral and longitudinal superposition of 3D pencil beams can be verified by optimizing the dose delivery to differently shaped target volumes, such as cubes or spheres of different size placed at different penetration depths in water. Whereas large volumes can hint on the capability to handle a large number of beams and properly account for divergence effects in case of large lateral beam deflections, small volumes are optimal to probe the accuracy of the TPS in modeling lateral beam broadening in depth, which is especially critical at large penetration and thus complementary to the field-size dependency probed in the entrance region of monoenergetic square fields. Moreover, whereas spherical targets pose a challenging optimization problem requesting a relatively high-intensity modulation, cubical targets are of special interest for testing geometrical properties of the optimized plan. For example, fine sampling of distal or lateral dose falloffs via repeated shifts of the ICs, or a combination of longitudinal water-column measurements with lateral 2D detection, for example, using films or EPIDs, enable probing the TPS capability to provide the geometrically correct target dose extension, and to reproduce the beam penumbra. In addition to regular body shapes,

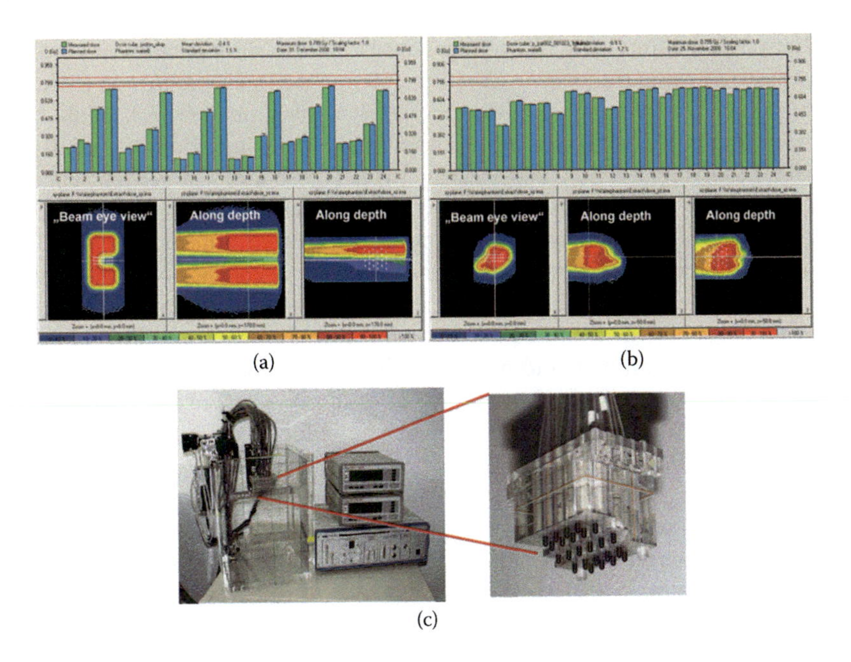

Figure 23.4

Comparison between experimental data (bars on the left of the ionization chamber [IC] numbering) and FLUKA Monte Carlo (MC) calculation (bars on the right of the IC numbering) for a U-shaped field (a) and a test patient case (b) delivered with protons and measured in water with multiple ICs, with the device shown in panel (c). The MC dose distributions in three orthogonal views with the location of the ICs marked by small crosses are illustrated below the bar diagrams in both panels (a) and (b). (Adapted from Karger CP et al., *Med. Phys.*, 26, 2125–2132, 1999; Parodi K et al., *Phys. Med. Biol.*, 57, 3759–3784, 2012.)

test cases may include patterns introducing strong dose gradients, for example, U-shaped targets as indicated in Figure 23.4a, or nonflat entrance surfaces, such as wedges or spherical shapes. Since the pencil beam algorithms underlying all existent TPSs are particularly challenged in the presence of tissue heterogeneities, additional measurements with materials different from water are strongly encouraged. These tests span from regular slab phantom geometries introducing strong interfaces, such as bone-air, perpendicular or parallel to the beam direction, up to more realistic phantoms with small tissue-equivalent inserts as well as anthropomorphic phantoms. Although in these cases ICs could be embedded in appropriate tissue-like media (Jäkel et al., 2000), such measurements are challenged by the extreme nonreference condition, violating electron equilibrium and making accurate dosimetry difficult to be achieved. Hence, a better choice is to perform the dosimetric measurements in water after traversal of the tissue heterogeneities (Bauer et al., 2014). Since all these test cases can generate treatment fields with remarkable dose gradients, the spatial resolution achievable with the typically employed small-volume ICs can be insufficient. Hence, additional measurements with 2D detectors of high spatial resolution placed perpendicular to the treatment field can be used to complement the pointwise acquisitions, however, only in terms of relative dosimetry due to remaining issues of these detectors from the already mentioned LET dependence (Spielberger et al., 2002).

The verified ability of a TPS to reproduce absorbed dose measurements can be considered sufficient to ensure accurate patient treatment for proton therapy, within the ca. 2% uncertainty (1SD) of IC reference dosimetry (IAEA, 2000) and the even larger 10–15% uncertainties underlying the assumption of a constant RBE of 1.1 (Paganetti et al., 2002; Paganetti, 2014). Besides a larger uncertainty of ca. 3% (1SD) for IC reference dosimetry of carbon ion beams (IAEA, 2000), assessment of absorbed dose alone cannot be considered sufficient for a TPS to be used in carbon ion therapy, where the complex dependency of RBE on several parameters cannot be described as a single multiplication factor. Although the knowledge of RBE-weighted dose, i.e., absorbed dose multiplied by the local RBE value, is also estimated to be affected by a considerable uncertainty in the range of 20% (Karger and Jäkel, 2007), which can be ascribed to experimental uncertainties and corresponding accuracy of the biological models as well as their input parameters, a TPS for carbon ion therapy must also be commissioned with respect to its ability to perform reasonable biological dose predictions. To this aim, phantoms for "biological dosimetry" can be designed, which include special suspensions of cells to be irradiated and followed up, in order to compare measurable endpoints, for example, cell survival, with the TPS predictions for the biological system under consideration, and related model parameters input to the TPS (Krämer et al., 2003). Encouraging results could be achieved so far, showing, for example, a TPS agreement with biological survival data within 3 percentage points in the target region, i.e., within 11% in terms of biological dose, for the more complex simultaneous optimization of a multifield carbon ion irradiation (Gemmel et al., 2008). Obviously, these are quite cumbersome and time-consuming experiments, requiring special preparation of the phantom and prolonged observation and analysis of the cells with respect to reference nonirradiated colonies. Nevertheless, many facilities had to undergo similar validation measurements at least once for approval prior to the start of clinical carbon ion treatment (Facoetti et al., 2015).

23.4 Dosimetric Measurements for Pretreatment Dose Verification

Similar to the current practice of intensity-modulated photon radiotherapy, ion-beam therapy plans of patient-specific treatment need to be verified dosimetrically prior to the first treatment session, unless an independent and certified computational engine is available to cross check the TPS calculation. In particular, the requirement of certification rules out usability of the various Monte Carlo research platforms that are available at many centers for independent verification of the TPS predictions (Sawakuchi et al., 2010b; Grevillot et al., 2011; Bauer et al., 2014). Interest in the development of independent and potentially certifiable dose calculation engines for scanned ion-beam therapy is rapidly growing because of the main motivation of enabling a considerable reduction of the amount of patient-specific QA measurements (Meier et al., 2015). However, the current situation in most clinical centers is still to rely on dosimetric measurements for pretreatment dose verification of patient plans.

Different from the current trends in intensity-modulated photon radiotherapy, where verification of the integral dose delivered by beams from several directions is performed by using semi-3D detector systems, for example, rotating 2D detectors arrays, plan verification in ion therapy is still focused on the absorbed dose delivered by each individual treatment field. To this aim, each planned field, specifying energy, lateral position, and number of ions (or MUs) for every pencil-like beam impinging from a given main direction, i.e., the treatment field angle, is recalculated in the geometry and medium of the dosimetric measurement. The latter medium is recommended to be water, and the measurement is typically performed with multiple small-volume ICs, similar to the setup described in Section 23.2 for dosimetric measurements of 3D extended treatment fields, to enable the highest dosimetric accuracy. In principle, the separate verification of the few (≈1-3) beam directions typically delivered in ion-beam therapy could simplify the dosimetric setup, having the water tank fixed at isocenter and using the fixed beam port, or only one beam gantry angle, depending on the facility. However, since the quality of the pencil beams, especially in terms of beam shape and size, can be dependent on the angle of the beam gantry in case of rotating beam lines, special phantoms have also been designed, which embed the ICs in water-equivalent solid media or PMMA (Henkner et al., 2015). As already mentioned in Section 23.3, in such cases special care is needed to take fluence corrections into account, to prevent introducing a bias, for example, up to 5% for protons beams (Palmans et al., 2002), in the comparison between TPS and measurement due to an inaccurate dosimetric assessment. Additional factors to be considered in the comparison are potential dose averaging effects in the measurement, especially for positions in high-dose gradients, due to the finite size of the ICs, regardless of the surrounding medium.

In general, scanned treatment fields exhibit complex intensity modulation, even in the case of homogeneous dose delivery, because of preirradiation of more proximal locations of the target volume from beams of higher penetration depth. Moreover, the dose distribution tends to be highly distorted when recalculated in the water system (or another homogeneous medium), due to the variable water-equivalent thickness of the original patient model for which the plan is optimized (Figure 23.4b). Hence, dosimetric measurements for pretreatment plan verification have to probe the dose distribution at many positions, by repeating the delivery of the same field and performing a parallel acquisition of multiple ICs placed at different locations in the treatment field, for example, within the high-dose target area and in the distal falloff region sensitive to the beam range (Karger et al., 2010). Automated data analysis systems enable to directly determine the measured dose by applying all required conversion factors of measured ionization into dose to water (IAEA, 2000) and compare it to the treatment plan calculation at the same measuring position. Although corrections for the earlier-mentioned dose averaging effects can be taken into account in this step, dosimetric measurements in high-dose gradients are also challenged by possible setup misalignments, where minor offsets could easily introduce large discrepancies between measurement and TPS. Therefore, for determining the acceptability of the verified plan, many analysis tools rule out certain chamber positions that are more prone to uncertainties according to the dose gradients expected on the basis of the TPS calculation. In particular, taking into account all possible sources of uncertainties from the dose delivery, the calculation engine, and the dose measuring process itself (Karger et al., 1999, 2010), approaches already implemented into clinical practice evaluate the mean and standard deviation between measured and calculated absorbed dose values normalized to the maximum beam dose. Plans are accepted where such a deviation remains below a threshold of typically 3–5%, when including only the accepted measuring positions.

Optionally, additional more relaxed tolerances, for example, up to ±7%, can be applied to the maximum deviation acceptable for single IC positions outside the high-dose gradient region (Molinelli et al., 2013).

Given the unavoidable tradeoff in the number of accessible measuring positions within a reasonable time slot for verification of each field of an entire treatment plan for each individual patient, which still represents a major bottleneck in clinical routine operation, efforts are ongoing to augment the measurement sampling frequency without prolonging the needed acquisition time. Although 2D detector arrays based on ICs would be unaffected by LET dependencies, they are not directly usable in water and still suffer from coarse spatial resolution, which makes an interpolation of the dose distribution more challenging than in photon therapy, due to the additional "depth" dimension of ion beams with their characteristic Bragg peak. Nevertheless, a dedicated solution using a commercial 2D IC array placed after a specially designed accordion-type water phantom, which allows an easy change of the measurement depth, has recently been implemented successfully for patient-specific QA of scanned carbon ion therapy (Furukawa et al., 2013). Here, a 3D gamma index analysis with 3%/3 mm acceptance criteria for at least 90% of the evaluated points is used to compensate for uncertainties due to chamber volume (75 mm^3) effects, which are especially critical at the longitudinal steep dose gradients of carbon ion beams. Alternatives such as EPIDs have been demonstrated to offer excellent spatial resolution but are prone to dosimetric uncertainties due to the already mentioned dependencies, besides again not being suitable for direct usage in water. Nevertheless, also in this case promising results have been recently reported for verification of ion fluences in the entrance region perpendicular to the beam direction, separately measured for each energy component of the treatment field (Martišíková et al., 2013). Such solutions of relatively easy setup could complement the coarser dosimetry with multiple pinpoint ICs, and may contribute to the simplification of patient-specific QA in the near future. Ultimately, true 3D dosimetric solutions, for example, using polymeric detectors being under investigation at several centers, could contribute to a future paradigm shift moving the current verification of single fields to the verification of the entire dose delivery for a given patient, similar to current trends of modern photon therapy and also performed in all earlier-mentioned biological dosimetry experiments (cf. Section 23.3). However, besides the cumbersome readout, these tools are also challenged by LET dependencies (Karger et al., 2010), which could suggest again only a measurement complementary to IC dosimetry, possibly including a refined model to correct for the quenched measured signal, or to predict the detector response when projecting the plan on the considered dosimetric system.

23.5 3D *In Vivo* Treatment QA and Range Verification

Accurate verification of the actual dose delivered to the patient or, at least, *in vivo* confirmation of the beam range, is still an unmet challenge of ion beam therapy. The stopping of the primary ions in the Bragg peak placed within the tumor prevents applicability of transmission detection similar to EPID dose reconstruction in photon therapy (cf. Chapters 7 and 18). Nevertheless, several methods have been proposed and tested during phantom experiments and even in the first clinical trials, which utilize different physical principles.

The most extensively investigated technique is positron-emission tomography (PET), which exploits the minor amount of β^+-activity formed in nuclear interactions between the primary ions and the nuclei of the irradiated tissue (Kraan, 2015; Parodi, 2015). Although nuclear β^+-activation is not directly proportional to the dose delivery, which is dominated by electromagnetic Coulomb interaction, it still exhibits a spatial correlation that can be used for range monitoring. The degree of correlation is highly dependent on the used ion species. For proton beams, only target nuclei of the tissue can be activated, as long as the beam energy is above the threshold for nuclear interaction. This activation results in a spatial pattern of positron emitters, which is sensitive to the tissue composition and ceases being created few millimeters before the Bragg peak, depending on the energy threshold of the considered β^+-activation channel and the medium stopping properties. For carbon ion beams, a stronger spatial correlation is provided by the formation of positron-emitting isotopic projectile fragments (^{10}C, ^{11}C), which accumulate toward the end of their range shortly before the Bragg peak. In terms of signal strength, the activity production at the same physical dose is typically higher by a factor of about 3 in carbon-rich PMMA (Parodi et al., 2002) for protons than for carbon ions, due to the higher fluence required to compensate for the lower stopping power. On the other hand, the amount of detectable signal strongly depends on the implementation of PET imaging to measure the annihilation photon pairs resulting from the β^+-decay. Already investigated clinical workflows include imaging during beam delivery (in-beam) (Enghardt et al., 2004), just afterward, with a variable delay of 0–5 min after end of irradiation for PET installations directly integrated in the treatment site (onboard) (Nishio et al., 2010), or in the treatment room (in-room) (Zhu et al., 2011), and up to 15 min for scanners located outside the treatment room (off-line) (Parodi et al., 2007; Bauer et al., 2013). Commonly recognized merits of in-beam, onboard, and in-room installations are the possibility to detect the important activity contribution from the short-lived ^{15}O ($T_{1/2} = 2$ min) and retain the best correlation of the measured signal with the irradiation, due to a reduced influence of physiological processes (washout effects), which may dislocate the activity from its place of production. On the other hand, geometrical constraints for integration in the beam delivery have so far restricted clinical in-beam and onboard implementations to dual-head scanners of limited detection efficiency and imaging performances, in comparison to the more traditional full ring arrangements utilized in in-room and off-line PET scanners. However, the latter off-line approach is only capable to image long-lived positron emitters (especially ^{11}C, with $T_{1/2} = 20$ min). Its main limitations include the already discussed loss of correlation between measured and actually produced signal, as well as the considerably reduced activity strength due to physical and biological decay in the long time elapsed between end of irradiation and imaging, making acquisition times up to 30 min necessary for collecting sufficient counting statistics.

To infer information on the actual treatment delivery, the measured PET signal can be compared to a prediction based on the planned treatment and the registered time course of irradiation and imaging, typically relying on detailed Monte Carlo calculations, or to a measurement of a previous fraction to verify reproducibility. The main clinical investigations reported so far could indicate some potential of PET imaging in verifying *in vivo* the beam range with few millimeter accuracy in favorable anatomical locations featuring low perfused bony structures and reliable positioning, for example, in the skull base region (Parodi et al., 2007; Fiedler et al., 2010; Knopf et al., 2011; Zhu et al., 2011; Nischwitz et al., 2015). Also its capability to validate the HU-range calibration curve, when using the same rWEPL approach as the planning system for the prediction of the PET activity, has been reported (Fiedler et al., 2012). Moreover, with all investigated workflows it was proven possible to identify cases of patient mispositioning or interfractional anatomical changes (Parodi, 2004; Nishio 2010; Bauer et al., 2013) as shown in Figure 23.5, even with attempts to quantify the dosimetric consequences of the observed discrepancies between measured and predicted PET activity patterns (Parodi, 2004). However, regardless of the used workflow, a major common challenge was identified in the extremely low-counting statistics for the typically applied fraction doses of a few Gy (RBE), resulting in activity concentrations up to two orders of magnitude lower than usual values imaged in conventional nuclear medicine. Since all the clinically investigated detector solutions relied either on commercial full ring tomographs or on dedicated dual-head cameras based on commercially available detector components of whole-body or small-animal PET scanners optimized for much higher statistics conditions, ongoing research is aiming at the development of next-generation instrumentation specifically tailored to this unconventional application of PET imaging. Proposed detector designs feature dual-head cameras capable of ultrafast time-of-flight photon detection for suppression of limited angle artifacts (Crespo et al., 2007), or dedicated ring scanners with special detector

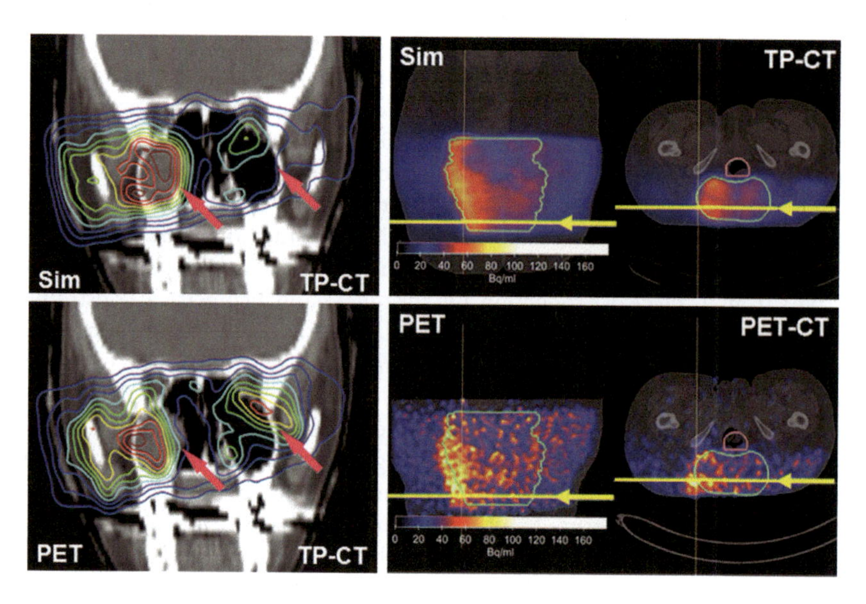

Figure 23.5

Comparison between Monte Carlo (MC) simulated positron-emission tomography (PET) distributions ("Sim," top row) and the corresponding PET measurements ("PET," bottom row), superimposed onto the planning computed tomography (CT) ("TP-CT") or a coregistered CT ("PET-CT") for carbon ion treatment of a skull-base (left) and sacral (right) tumor. Regardless the different instrumentation (left: dual-head, right: full ring) and clinical workflow (left: in-beam, right: off-line), in both cases PET imaging was able to indicate an inaccurate delivery due to anatomical change (left, red arrows denote regions of marked disagreement with evident measured over-range) or mispositioning (right, yellow arrows indicate beam direction, again with measured over-range). (Adapted from Parodi K, PhD thesis, Dresden University of Technology, Germany, 2004; Bauer J et al., *Radiother. Oncol.*, 107, 218–226, 2013.)

arrangements, for example, slanted or axially shifted as investigated by Tashima et al. (2012), to enable an opening for the beam passage.

An alternative nuclear-based technique under extensive investigation worldwide is based on the detection of the energetic, up to several MeV, photons promptly emitted in the deexcitation stage of nuclei having undergone nuclear interaction. This prompt gamma imaging technique holds the promise of overcoming major limitations of PET imaging by providing a signal that is produced in real time, i.e., sub-nanosecond time scale, and is not affected by physiological washout processes. Owing to the typically lower energy threshold of the reaction cross sections in comparison to β^+-activation, a promising spatial correlation could be observed between the location of the Bragg peak and the distal falloff of the prompt gamma signal, measured perpendicular to the beam direction (Min et al., 2006; Testa et al., 2008). Depending on the available counting statistics, the real-time nature of the signal could enable monitoring of the beam range for each individual pencil beam delivered to the target, especially for the most distal beams of typically higher fluence. However, contrary to PET, this new technique cannot borrow or simply adapt existing instrumentation from nuclear medicine, given the much higher photon energies in comparison to those used in medical single-photon imaging. Hence, no off-the-shelf solution exists and major efforts are being invested in the development of different detector concepts. These include providing mechanical (Smeets et al., 2012) or electronic collimation (Polf et al., 2015) of the prompt gammas, resolving the additional characteristic emission energy for nuclear spectroscopy (Verburg and Seco, 2014), converting the energetic photons in Compton electrons for easier charged particle detection (Kim et al., 2012) or exploiting the different arrival times of the prompt gammas to infer the range-dependent stopping time of ions penetrating in tissue (Hueso-González et al., 2015). For all these technologies, initial promising measurements could be reported in several phantom experiments, and clinical testing of two prototypes of a mechanically collimated single slit camera is currently ongoing with passively scattered (Richter et al., 2016) and actively scanned (Xie et al., 2017) delivery.

In addition to nuclear-based techniques, additional approaches are being investigated, which do not rely on nuclear interactions but rather make use of ionization, which is the main mechanism responsible for dose deposition. The possibility to directly visualize the Bragg peak from the detection of pressure waves resulting from the localized ion energy deposition, first pioneered for a passively scattered proton treatment field delivered to a liver patient in the late 1990s, is now receiving considerable attention worldwide, although being likely restricted to suitable anatomical locations of low signal attenuation and good accessibility for a transducer (Parodi and Assmann, 2015). However, despite its intriguing promise of real-time *in vivo* verification of pencil beam range colocalized to the patient anatomy, this so-called "ionoacoustics" approach is still at the infancy of its development. Other methods under consideration, such as implanted diodes and radiography/tomography, involve the measurement of exploratory beams of higher energies than the therapeutic ones, in order to reach larger penetration depths than for treatment, or completely traverse the patient. A first approach is based on the usage of multiple diodes, which need to be placed at the distal edge of the tumor region in accessible anatomical locations like the prostate, for example, attached to a rectal balloon (Bentefour et al., 2015). From the time-dependent signal recorded by an exploratory low-dose scout beam formed with passive range modulation, it is possible to infer information on the beam range and calculate a corresponding correction for the lower energy treatment plan, in case of observed deviations. Initial promising results have been recently reported for a prototype system tested in an anthropomorphic phantom (Bentefour et al., 2015), and first clinical evaluation should be started soon. Although initially conceived for the monitoring of passively scattered proton treatment of prostate tumors, the method could also be envisioned for application to beam scanning (Hoesl, 2016).

Another proposed method exploits the approximate energy independence of the stopping-power ratio relative to water or rWEPL, which represents the key information for calculation of the beam range in the patient at the planning stage (cf. Section 23.2). Hence, low-dose ion-based radiographic or tomographic transmission images of the patient at the treatment site, calibrated to water-equivalent thickness or rWEPL, respectively, could provide information on the patient stopping properties at the treatment site and could be used to refine the treatment plan prior to delivery (Schneider et al., 2005). To this aim, different detector setups have been proposed for measurement of residual range or energy of passively scattered or actively scanned proton and carbon ion beams (Schulte and Penfold, 2012; Parodi, 2014). However, also in this case so far only phantom experiments have been reported and further improvements of the prototype hardware and software solutions are warranted prior to clinical application.

In summary, despite the various techniques being proposed and in some cases already clinically investigated, *in vivo* verification of the delivered dose or, at least, the beam range in ion therapy is still a demanding challenge, particularly at the level of individual pencil beam delivery in modern beam scanning. Hence, the years to come will allow drawing conclusions on the success of ongoing efforts that encompass a large spectrum of detector developments, exploiting different signals produced in the interaction of ion beams in tissue.

23.6 Neutron Dosimetry and Spectrometry

Since the start of ion beam therapy, it has been well known that neutrons can be produced in many nuclear interactions of the ion beam with the traversed material, including the beamline and the patient itself. Modern beam scanning delivery can considerably reduce neutron production, by minimizing the amount of material encountered by the beam on its path to the patient (Schneider et al., 2002; Yonai et al., 2014). However, the patient itself is a considerable source of neutrons, which cannot be suppressed. Neutrons, being a highly penetrating type of radiation, typically escape from the place of production without interaction. Hence, although the contribution of neutrons is typically negligible in the treatment area, i.e. "in-field," it can still pose a hazard to the patient by performing biologically effective dose depositions in healthy tissue and critical organs far away from the tumor area, i.e., "out-of-field." Moreover, radiation background of neutrons bouncing back and forth in the treatment room, or penetrating through the walls to other rooms, can represent a considerable hazard for the personnel. Hence, neutron dosimetry has always been an important task to confirm the adequateness of implemented shielding measures. To this aim, a wide variety of proportional counter detectors are typically used to measure neutron ambient dose-equivalent $H^*(10)$, i.e., the dose equivalent that would be generated in the associated aligned and expanded radiation field at a depth of 10 mm on the radius of the International Commission on Radiation Units and Measurements

(ICRU) sphere oriented opposite to the direction of incident radiation (Vana et al., 2003). Other types of detectors used for neutron dosimetry are discussed in Chapter 21.

Recently, the focus on the long-term effects of radiation, especially in terms of enhanced carcinogenic risk, has gained considerable attention, particularly due to the increasing use of protons for pediatric patients in view of their typically lower integral dose than conventional photon radiation within the treated area. Promising results of several computational studies foresee a reduction up to one order of magnitude, depending on treatment facility, beam particle and energies, as well as anatomical location, for the lifetime risk of radiation-induced secondary malignant neoplasms in comparison to modern photon therapy (see, e.g., Newhauser et al., 2009). However, challenges remain for the measurement and prediction of neutron stray radiation and its related exposure of patient organs, given the complicated radiation field. Hence, recent international efforts have promoted intercomparison of different commercially available dosimetry systems in measurement campaigns of out-of-field doses at different facilities, with the specific goal to provide a detailed characterization of the neutron energy spectra beyond the more unspecific equivalent dose, with a special focus on scanning beams (Kaderka et al., 2012; Farah et al., 2015; Mares et al., 2016). Typical detectors for neutron spectroscopy feature Bonner spheres consisting of a central proportional counter core, for example, ^{3}He, surrounded by different thicknesses of moderating material, for example, polyethylene, also complemented by lead shells to extend the detection range to high-energy neutrons (Farah et al., 2015). Spectral information can be deduced from simultaneous acquisition of different diameter spheres, taking into account their respective fluence response function typically calculated by means of Monte Carlo simulations, validated in well-known neutron sources, to unfold the underlying spectra (Barros et al., 2014). These neutron detectors can provide all necessary characterization of the neutron ambient dose and underlying spectra at different distances and angles from the beam, which is essential to complement the detailed in-field dosimetric characterization reviewed in the previous sections in this chapter. Hence, these data can provide crucial information to validate and complement the prediction of Monte Carlo calculations, which are extremely sensitive to the accurate modeling of the beam delivery, target, and treatment room with related shielding, thus enabling a better accounting of stray radiation and estimation of its impact on the patient organ exposure for a more informed decision on the optimal treatment strategy that also accounts for out-of-field secondary effects.

23.7 Conclusion and Outlook

Radiation therapy with proton and carbon ion beams is a rapidly emerging modality that promises superior selectivity of the dose deposited to the tumor, with excellent sparing of healthy tissue. Especially its most modern form of implementation using scanned beam delivery, which is nowadays entering clinical routine, poses challenges for accurate characterization of the individual pencil beams and in-field dosimetric verification of the dynamically built, intensity-modulated extended fields delivered to representative phantoms. Despite the established dosimetric instruments and protocols, which already enable safe clinical application at operating facilities, research is still ongoing to reduce the remaining 2–3% uncertainties of reference IC dosimetry, for instance via comparison to calorimetry (see Chapter 12). Additional investigations aim to overcome LET-dependence issues of high-resolution 2D and 3D detectors and to improve the understanding of radiation-induced biological effects. Moreover, considerable efforts in instrumentation development are being devoted to the challenging problem of *in vivo* verification of the delivered dose or beam range, trying to exploit multiple secondary emission products formed during the therapeutic irradiation, or low-dose pretreatment exposures at higher energies than the therapeutic ones. Finally, coordinated campaigns are ongoing at different facilities to improve our understanding of out-of-field stray radiation, including detailed characterization of neutron energy spectra. All these aspects will contribute substantially to the improvement of the delivered treatment and the establishment of adaptive treatment workflows, thus impacting the clinical quality of this rapidly emerging form of modern radiation therapy.

Acknowledgment

I thank former colleagues from the Heidelberg Ion Beam Therapy Center and Helmholtzzentrum für Schwerionenforschung Darmstadt, particularly Stephan Brons, Andrea Mairani, Florian Sommerer, Julia Bauer, and Dieter Schardt, as well as collaborators from Helmholtzzentrum München, particularly Werner Rühm, Vladimir Mares, and Sebastian Trinkl, for several fruitful discussions on topics reviewed in this contribution.

References

Barros S et al. (2014) Comparison of unfolding codes for neutron spectrometry with Bonner spheres. *Radiat. Prot. Dosimet* 161: 46–52.

Bauer J et al. (2013) Implementation and initial clinical experience of offline PET/CT-based verification of scanned carbon ion treatment. *Radiother. Oncol.* 107: 218–226.

Bauer J et al. (2014) Integration and evaluation of automated Monte Carlo simulations in the clinical practice of scanned proton and carbon ion beam therapy. *Phys. Med. Biol.* 59: 4635–4659.

Bentefour el H et al. (2015) Validation of an in-vivo proton beam range check method in an anthropomorphic pelvic phantom using dose measurements. *Med. Phys.* 42: 1936–1947.

Bohlen TT et al. (2014) The FLUKA code: Developments and challenges for high energy and medical applications. *Nuclear Data Sheets.* 120: 211–214.

Brusasco C et al. (2000) A dosimetry system for fast measurement of 3D depth-dose profiles in charged-particle tumor therapy with scanning techniques. *Nucl. Instrum. Meth. Phys. Res. B.* 168: 578–592.

Casiraghi M and Schulte RW (2015) Nanodosimetry-based plan optimization for particle therapy. *Comput. Math. Meth. Med.* 2015: 908971.

Chu WT, Ludewigt BA and Renner TR (1993) Instrumentation for treatment of cancer using proton and light-ion beams. *Rev. Sci. Instrum.* 64: 2055–2122.

Crespo P et al. (2007) Direct time-of- flight for quantitative, real-time in-beam PET: A concept and feasibility study. *Phys. Med. Biol.* 52: 6795–6811.

Enghardt W et al. (2004) Charged hadron tumour therapy monitoring by means of PET. *Nucl. Instrum. Meth. Phys. Res.* A. 525: 284–288.

Facoetti A et al. (2015) In vivo radiobiological assessment of the new clinical carbon ion beams at CNAO. *Radiat. Prot. Dosimet* 166: 379–382.
Farah J et al. (2015) Measurement of stray radiation within a scanning proton therapy facility: EURADOS WG9 intercomparison exercise of active dosimetry systems. *Med. Phys.* 42: 2572–2584.
Ferrari A et al. (2005) FLUKA: A multi-particle transport code. CERN-2005-10, INFN/TC_05/11, SLAC-R-773 (Geneva, Switzerland: CERN European Organization for Nuclear Research).
Fiedler F et al. (2010) On the effectiveness of ion range determination from in-beam PET data. *Phys. Med. Biol.* 55: 1989–1998.
Fiedler F et al. (2012) Online irradiation control by means of PET. In: *Ion Beam Therapy Fundamentals, Technology, Clinical Applications*, pp. 527–543. (Ed. Linz U, Berlin, Germany: Springer-Verlag).
Furukawa T et al. (2010) Performance of the NIRS fast scanning system for heavy-ion radiotherapy *Med. Phys.* 37: 5672–5682.
Furukawa T et al. (2013) Patient-specific QA and delivery verification of scanned ion beam at NIRSHIMAC. *Med. Phys.* 40: 121707.
Gemmel A et al. (2008) Biological dose optimization with multiple ion fields. *Phys. Med. Biol.* 53: 6991–7012.
Gillin M et al. (2010) Commissioning of the discrete spot scanning proton beam delivery system at the University of Texas M D Anderson Cancer Center, Proton Therapy Center, Houston. *Med. Phys.* 37: 154–163.
Grant RL et al. (2014) Relative stopping power measurements to aid in the design of anthropomorphic phantoms for proton radiotherapy. *J. Appl. Clin. Med. Phys.* 15(2): 121–126.
Grevillot L et al. (2011) A Monte Carlo pencil beam scanning model for proton treatment plan simulation using GATE/GEANT4. *Phys. Med. Biol.* 56: 5203–5219.
Haberer T et al. (1993) Magnetic scanning system for heavy ion therapy. *Nucl. Instrum. Meth. Phys. Res. A.* 330: 296–305.
Hartmann B (2013) A novel approach to ion spectroscopy of therapeutic ion beams using a pixelated semiconductor detector. PhD thesis, Ruprecht-Karls-University, Heidelberg, Germany.
Hattner E et al. (2013) Experimental study of nuclear fragmentation of 200 and 400 MeV/u (12)C ions in water for applications in particle therapy. *Phys. Med. Biol.* 58: 8265–8279.
Hawkins RB (1996) A microdosimetric-kinetic model of cell death from exposure to ionizing radiation of any LET, with experimental and clinical applications. *Int. J. Radiat. Biol.* 69: 739–755.
Henkner K et al. (2015) A motorized solid-state phantom for patient-specific dose verification in ion beam radiotherapy. *Phys. Med. Biol.* 60: 7151–7163.
Hoesl M et al. (2016) Clinical commissioning of an in vivo range verification system for prostate cancer treatment with anterior and anterior oblique proton beams. *Phys. Med. Biol.* 61: 3049–3062.
Hong L et al. (1996) A pencil beam algorithm for proton dose calculations. *Phys. Med. Biol.* 41: 1305–1330.
Hueso-Gonzalez F et al. (2015) First test of the prompt gamma ray timing method with heterogeneous targets at a clinical proton therapy facility. *Phys. Med. Biol.* 60: 6247–6272.

IAEA (2000) Absorbed dose determination in external beam radiotherapy—an international code of practice for dosimetry based on standards of absorbed dose to water. Technical Report Series No 398. (Vienna, Austria: International Atomic Energy Agency).

Inaniwa T et al. (2010) Treatment planning for a scanned carbon ion beam with a modified microdosimetric kinetic model. *Phys. Med. Biol.* 55: 6721–6737.

Inaniwa T et al. (2015) Reformulation of a clinical-dose system for carbon-ion radiotherapy treatment planning at the National Institute of Radiological Sciences, Japan. *Phys. Med. Biol.* 60: 3271–3286.

Jakel O et al. (2000). Quality assurance for a treatment planning system in scanned ion beam therapy. *Med. Phys.* 27: 1588–1600.

Kaderka R et al. (2012), Out-of-field dose measurements in a water phantom using different radiotherapy modalities. *Phys. Med. Biol.* 57: 5059–5074.

Kanai T et al. (1983) Three-dimensional beam scanning for proton therapy. *Nucl. Instrum. Meth. Phys. Res. A.* 214: 491–496.

Karger CP et al. (1999) A system for three-dimensional dosimetric verification of treatment plans in intensity-modulated radiotherapy with heavy ions. *Med. Phys.* 26: 2125–2132.

Karger CP and Jakel O (2007) Current status and new developments in ion therapy. *Strahlenther. Onkol.* 183: 295–300.

Karger CP et al. (2010) Dosimetry for ion beam radiotherapy. Phys. Med. Biol. 55: R193–R234.

Kim CH, Park JH, Seo H and Lee HR (2012) Gamma electron vertex imaging and application to beam range verification in proton therapy. *Med. Phys.* 39: 1001–1005.

Knopf AC et al. (2011) Accuracy of proton beam range verification using post-treatment positron emission tomography/computed tomography as function of treatment site. *Int. J. Radiat. Oncol. Biol. Phys.* 79: 297–304.

Kraan AC (2015) Range verification methods in particle therapy: Underlying physics and Monte Carlo modeling. *Front. Oncol.* 5: 150.

Kramer M et al. (2000) Treatment planning for heavy-ion radiotherapy: Physical beam model and dose optimization. *Phys. Med. Biol.* 45: 3299–3317.

Kramer M, Wang JF and Weyrather W (2003) Biological dosimetry of complex ion radiation fields. *Phys. Med. Biol.* 48: 2063–2070.

Kramer M and Scholz M (2006) Rapid calculation of biological effects in ion radiotherapy. *Phys. Med. Biol.* 51: 1959–1970.

Kurz C, Mairani A and Parodi K (2012) First experimental-based characterization of oxygen ion beam depth dose distributions at the Heidelberg Ion Beam Therapy Center. *Phys. Med. Biol.* 57: 5017–5034.

Mares V et al. (2016) A comprehensive spectrometry study of a stray neutron radiation field in scanning proton therapy. *Phys. Med. Biol.* 61: 4127–4140.

Mairani A et al. (2010) The FLUKA Monte Carlo code coupled with the local effect model for biological calculations in carbon ion therapy. *Phys. Med. Biol.* 55: 4273–4289.

Martisikova M et al. (2012) Characterization of a flat-panel detector for ion beam spot measurements. *Phys. Med. Biol.* 57: 485–497.

Martišikova M et al. (2013) High-resolution fluence verification for treatment plan specific QA in ion beam radiotherapy. *Phys. Med. Biol.* 58: 1725–1738.

Meier G et al. (2015) Independent dose calculations for commissioning, quality assurance and dose reconstruction of PBS proton therapy. *Phys. Med. Biol.* 60: 2819–2836.
Min CH et al. (2006) Prompt gamma measurements for locating the dose falloff region in the proton therapy. *Appl. Phys. Lett.* 89: 183517.
Mirandola A et al. (2015) Dosimetric commissioning and quality assurance of scanned ion beams at the Italian National Center for Oncological Hadrontherapy. *Med. Phys.* 42: 5287–5300.
Molinelli S et al. (2013), Dosimetric accuracy assessment of a treatment plan verification system for scanned proton beam radiotherapy: One-year experimental results and Monte Carlo analysis of the involved uncertainties. *Phys. Med. Biol.* 58: 3837–3847.
Newhauser WD et al. (2009) The risk of developing a second cancer after receiving craniospinal proton irradiation. *Phys. Med. Biol.* 54: 2277–2291.
Newhauser WD and Zhang R (2015) The physics of proton therapy. *Phys. Med. Biol.* 60: R155–R209.
Nischwitz SP et al. (2015) Clinical implementation and range evaluation of in vivo PET dosimetry for particle irradiation in patients with primary glioma. *Radiother. Oncol.* 115: 179–185.
Nishio T et al. (2010) The development and clinical use of a beam on-line PET system mounted on a rotating gantry port in proton therapy. *Int. J. Radiat. Oncol. Biol. Phys.* 76: 227–286.
Paganetti H et al. (2002) Relative biological effectiveness (RBE) values for proton beam therapy. *Int. J. Radiat. Oncol. Biol. Phys.* 53: 407–421.
Paganetti H (2014) Relative biological effectiveness (RBE) values for proton beam therapy. Variations as a function of biological endpoint, dose, and linear energy transfer. *Phys. Med. Biol.* 59: R419–R472.
Palmans H et al. (2002) Fluence correction factors in plastic phantoms for clinical proton beams. *Phys. Med. Biol.* 47: 3055–3071.
Parodi K, Enghardt W and Haberer T (2002) In-beam PET measurements of $\beta+$ radioactivity induced by proton beams. *Phys. Med. Biol.* 47: 21–36.
Parodi K (2004) On the feasibility of dose quantification with in-beam PET data in radiotherapy with 12C and proton beams. Forschungszentrum Rossendorf Wiss-Techn-Ber FZR-415, PhD thesis, Dresden University of Technology, Germany.
Parodi K et al. (2007) Patient study on in-vivo verification of beam delivery and range using PET/CT imaging after proton therapy. *Int. J. Radiat. Oncol. Biol. Phys.* 68: 920–934.
Parodi K (2008) Positron-emission-tomography for in-vivo verification of ion beam therapy—Modelling, experimental studies and clinical implementation. Habilitation thesis, Ruprecht-Karls-Universitat Heidelberg, Heidelberg, Germany.
Parodi K et al. (2012) Monte Carlo simulations to support start-up and treatment planning of scanned proton and carbon ion therapy at a synchrotron-based facility. *Phys. Med. Biol.* 57: 3759–3784.
Parodi K, Mairani A and Sommerer F (2013) Monte Carlo-based parametrization of the lateral dose spread for clinical treatment planning of scanned proton and carbon ion beams. *J. Radiat. Res.* 54(Suppl. 1): i91–i96.
Parodi K (2014) Heavy ion radiography and tomography. *Phys. Med.* 30: 539–543.

Parodi K and Assmann W (2015) Ionoacoustics: A new direct method for range verification. *Mod. Phys. Lett. A.* 30: 1540025.

Parodi K (2015) On- and off-line monitoring of ion beam treatment. Nucl. Instrum. *Meth. Phys. Res. A.* 809: 113–119.

Pedroni E et al. (1995) The 200 MeV proton therapy project at PSI: Conceptual design and practical realization. *Med. Phys.* 22: 37–53.

Pedroni E et al. (2004) The PSI gantry 2: A second generation proton scanning gantry. *Z. Med. Phys.* 14: 25–34.

Pedroni E et al. (2005) Experimental characterization and physical modelling of the dose distribution of scanned proton pencil beams. *Phys. Med. Biol.* 50: 541–561.

Polf JC et al.(2015) Imaging of prompt gamma rays emitted during delivery of clinical proton beams with a Compton camera: Feasibility studies for range verification. *Phys. Med. Biol.* 60: 7085–7099.

Richter C et al. (2016) First clinical application of a prompt gamma based in vivo proton range verification system. *Radiother. Oncol.* 118: 232–237.

Rietzel E, Schardt D and Haberer T (2007) Range accuracy in carbon ion treatment planning based on CT-calibration with real tissue samples. *Radiother. Oncol.* 23: 2–14.

Sawakuchi GO et al. (2010a) Experimental characterization of the low-dose envelope of spot scanning proton beams. *Phys. Med. Biol.* 55: 3467–3478.

Sawakuchi GO et al. (2010b) An MCNPX Monte Carlo model of a discrete spot scanning proton beam therapy nozzle. *Med. Phys.* 37: 4960–4970.

Schardt D et al. (2008) Precision Bragg-curve measurements for light-ion beams in water. Scientific Report 2007, GSI Helmholtzzentrum fur Schwerionenforschung, GSI Report 2008, 1:373.

Schardt D, Elsasser T and Schultz-Ertner D (2010) Heavy-ion tumor therapy: Physical and radiobiological benefits. Rev. *Mod. Phys.* 82: 383–425.

Schneider U et al. (2005) Patient specific optimization of the relation between CT-Hounsfield units and proton stopping power with proton radiography. *Med. Phys.* 32: 195–199.

Schneider U, Pedroni E and Lomax A (1996) The calibration of CT Hounsfield units for radiotherapy treatment planning. *Phys. Med. Biol.* 41: 111–124.

Schneider U et al. (2002) Secondary neutron dose during proton therapy using spot scanning. *Int. J. Radiat. Oncol. Biol. Phys.* 53: 244–251.

Scholz M et al. (1997) Computation of cell survival in heavy ion beams for therapy. *Radiat. Environ. Biophys.* 36: 59–66.

Schulte RW and Penfold SN (2012) Proton CT for improved stopping power determination in proton therapy. *Trans. Am. Nucl. Soc.* 106: 55–58.

Schwaab J et al. (2011) Experimental characterization of lateral profiles of scanned proton and carbon ion pencil beams for improved beam models in ion therapy treatment planning. *Phys. Med. Biol.* 56: 7813–7827.

Smeets J et al. (2012) Prompt gamma imaging with a slit camera for real-time range control in 525 proton therapy. *Phys. Med. Biol.* 57: 3371–3405.

Spielberger B et al. (2002) Calculation of the x-ray film response to heavy charged particle irradiation. *Phys. Med. Biol.* 47: 4107–4120.

Tashima H et al. (2012) A single-ring OpenPET enabling PET imaging during radiotherapy. *Phys. Med. Biol.* 57: 4705–4718.

Testa E et al. (2008) Monitoring the Bragg peak location of 73 MeV/u carbon ion beams by means of prompt gamma-ray measurements. *Appl. Phys. Lett.* 93: 093506.
Vana N, Hajek M and Berger T (2003) Ambient dose equivalent H*(d)—An appropriate philosophy for radiation monitoring onboard aircraft and in space? In: Proceedings of the IRPA regional congress on radiation protection in Central Europe. (Bratislava, Slowak Republic: International Radiation Protection Association [IRPA]).
Verburg JM and Seco J (2014) Proton range verification through prompt gammaray spectroscopy. *Phys. Med. Biol.* 59: 7089–7106.
Wilson R (1946) Radiological use of fast protons. *Radiology.* 47: 487–491.
Xie Y et al. (2017) Prompt gamma imaging for in vivo range verification of pencil beam scanning proton therapy. *Int. J. Radiat. Oncol. Biol. Phys.* 99: 210–218.
Yonai S, Furukawa T and Inaniwa T (2014) Measurement of neutron ambient dose equivalent in carbon-ion radiotherapy with an active scanned delivery system. *Radiat. Prot. Dosimet* 161: 433–436.
Zhu X et al. (2011) Monitoring proton radiation therapy with in room PET imaging. *Phys. Med. Biol.* 56: 4041–4057.

Section V
Emerging Technological Developments

24

Dosimetry of Small Animal Precision Irradiators

Frank Verhaegen and Dietmar Georg

24.1 Introduction

A modern trend in radiation biology preclinical research is to mimic, from a technological point of view, the radiation conditions of human radiotherapy as closely as possible. This has led, in the last decade, to a range of commercial and in-house-made irradiators, which often come equipped with various imaging devices to allow image-guided irradiation (Verhaegen et al., 2011; Tillner et al., 2014, 2016). Several of these devices include a gantry that allows a kilovolt (kV) x-ray radiation source to irradiate target structures from different angles, with very small beams (down to 0.5 mm diameter). This novel equipment has enabled high-precision radiation experiments which may lead to new fundamental insights or allow improved translational studies. The advanced radiation research platforms offer opportunities to precisely target tumors, or substructures therein, and can also be used for innovative normal tissue-response studies. The availability of these new platforms is expected to lead to major advances in radiation oncology, in combination with other anticancer agents (Butterworth et al., 2015).

The novel radiation platforms, however, pose new challenges for the mechanical targeting accuracy, dosimetric accuracy, and imaging characteristics. Essential for the radiation treatment planning is the availability of low noise, high spatial resolution images, most frequently cone beam CT (CBCT) images in the current implementations, but magnetic resonance imaging (MRI)- and positron emission tomography (PET)-based plannings are also contemplated (Bolcaen et al., 2014; Trani et al., 2015). Older animal irradiation systems often used larger, uniform radiation fields, from static radiation sources, without employing image guidance. The radiation sources were either x-ray tubes up to a few hundred kV, or ^{137}Cs or ^{60}Co gamma sources. In these systems, the radiation commissioning and dosimetry were uncomplicated, often requiring only a limited set of dose measurements following, for instance, the recommendations of one of the kV dosimetry protocols (IAEA, 2000; Ma et al., 2001) or the megavolt dosimetry protocol of the American Association of Physicists in Medicine (AAPM) Task Group 51 (Almond et al., 1999). It should be noted, though, that the radiobiology field is notorious for not reporting the irradiation conditions used, which leads to the conjecture that large uncertainties may exist in the dosimetry of many radiobiology studies. A recent literature survey (Desrosiers et al., 2013) discovered that only 48% of studies reported their irradiation geometries and only 7% referred to any published dosimetry standard or guidelines, whereas only 4% reported any dose uncertainty in their works. The published standards are not directly applicable to the potentially very small fields from the novel irradiation platforms.

In this chapter, we explore the possible approaches to measuring and calculating dose distributions with a special focus on the small beams employed to irradiate small structures in animal models.

24.2 Small Animal Precision Irradiators

Several prototype small-field irradiators were developed in the past, such as an ^{192}Ir brachytherapy source (Stojadinovic et al., 2007) without image guidance, a micro-CT scanner fitted with an iris collimator (Graves et al., 2007), and a 320 kV industrial x-ray tube fitted with a fixed imaging panel (Song et al., 2010). Currently used systems comprise two commercial radiation research platforms: XRAD-SMART from PXi (North Branford, CT) (Lindsay et al., 2014) and SARRP from Xstrahl, Inc. (Camberley, UK) (Wong et al., 2008), and an in-house built system (Tillner et al., 2014). All these systems use gantries to rotate a high dose rate x-ray tube with maximum voltage of 225 kV around the specimen and image guidance. They can deliver static or arced beams. They are all capable of producing very small beams, down to 0.5 mm diameter, from manually placed fixed-field collimators. Their x-ray tubes have a dual focus, allowing both x-ray CT imaging and irradiation with the same tube, with the small and large focus, respectively. These two modes employ different filtrations. The position and spatial distribution of the large focal spot for irradiation has implications for the dosimetry in beams approaching the dimensions of the focal spot (see Section 24.6). An overview of these systems can be found in a recent review (Tillner et al., 2014). The x-ray imaging systems are usually CBCT imaging panels. Figure 24.1 shows an example of a commercial irradiator.

To irradiate small targets stably with small x-ray beams moving on gantries is no trivial task. Several of the systems mentioned employ corrections of the animal stage during gantry motion to compensate for gantry flex and other causes of systematic error. With this compensation mechanism, a targeting accuracy of better than 0.5 mm was reported (Clarkson et al., 2011; Tillner et al., 2014). A study by Lindsay et al. (2014) comparing similar irradiators at three different institutes showed that the gantry flex can be different among irradiators, but that mechanical calibration, which is needed very infrequently, removes most of the systematic targeting error. After mechanical calibration not only the mechanical

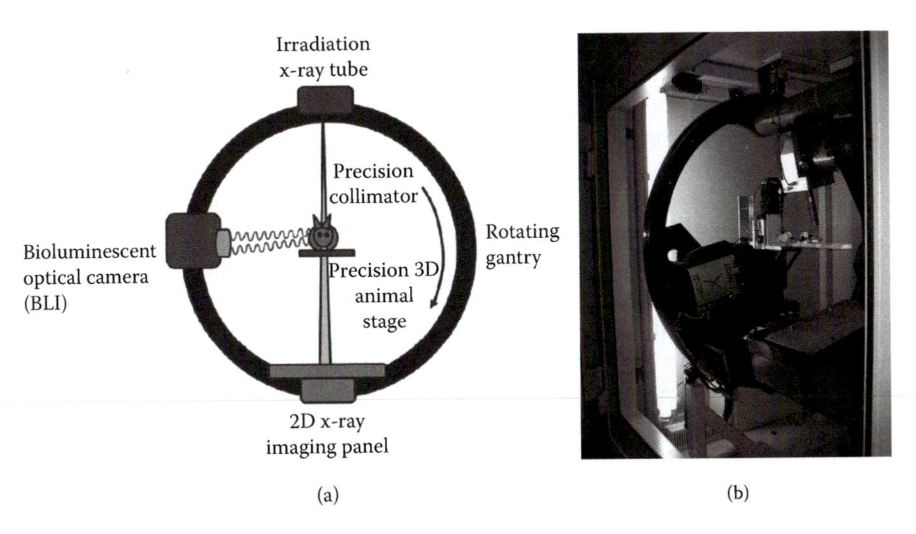

Figure 24.1

(a) Schematic overview of a possible arrangement of a small animal precision irradiation platform, including on the same gantry a dual-focus x-ray tube, a 2D x-ray imaging panel for CBCT imaging, and a bioluminescent camera for imaging optical markers. (b) Commercial implementation of XRAD 225Cx small animal irradiator (PXi) as installed at the University of Maastricht.

isocenter corresponds to the radiation isocenter, but also the CBCT panel is aligned with the irradiation device.

24.3 Beam Calibration

Most institutes employ the highest photon energy their irradiator can provide, for example, 225 kV for the animal irradiations on the commercial platforms. In this case, absolute calibration only needs to be performed for this photon energy. However, some groups (Bazalova et al., 2014) have used lower photon energies, which require absolute calibration at the specific energies. For absolute dosimetry calibration in kV x-ray beams, several protocols exist, for example, the AAPM's Task Group 61 (TG-61) report (Ma et al., 2001), or the International Atomic Energy Agency's Technical Report Series, TRS-398 (IAEA, 2000). In this chapter, we mostly focus on the former since this is the most widely used kV x-ray dosimetry protocol in the small animal irradiation literature.

Absolute calibration of a small animal irradiation platform starts with the measurement of the first and second half-value layers (HVLs) (HVL1 and HVL2) of the x-ray qualities to be used for irradiation. HVL1 and HVL2 are used to look up correction and conversion factors in the protocol needed to derive absolute dose in various conditions. These are defined as the thickness of a high-purity filter (Al or Cu) which is needed to reduce the measured air kerma rate by a factor of two (HVL1) or the additional thickness (HVL2) to reduce the air kerma by another factor of two. The beam should be scatter-free, and collimated with a diameter not exceeding 4 cm. The beam diameter (or other characteristic dimension) should enable covering the sensitive volume of the radiation detector. TG-61 recommends a distance of at least 50 cm between detector and filter, which itself should be about 50 cm away from the radiation source. Due to the internal dimensions of research platforms this is not possible, therefore, the filter is placed approximately midway between the source and detector. Scatter from the filter in the detector should be avoided, so a small detector needs to be used. A radiation detector with a relatively flat energy response is recommended, i.e., with a response variation of no more than 5% over the range of the encountered photon energies. Furthermore, the use of a monitor ion chamber is recommended to take into account the fluctuation of the tube output. However, since the current implementations of the research platforms are not equipped with a monitor chamber, one has to rely on the mAs reading of the devices, and perform repeated measurements. If HVL cannot be determined experimentally, they can be derived from calculation models, for example, SpekCalc (Poludniowski et al., 2009), from which also photon spectra can be extracted, which are difficult to measure.

Absolute calibration of a kV radiation beam should be performed with an ionization chamber (IC) which has been calibrated at a primary or, more commonly, a secondary radiation standards laboratory. TG-61 recommends the use of cylindrical ICs, such as the Farmer-type chamber, but for absolute dosimetry below 70 kV thin window parallel plate ion chambers are recommended. Corrections to the reading for reference temperature/pressure, ion recombination, and polarity effects need to be performed. Ideally, the IC should have a few calibration points in the relevant kV energy range, so that the HVL or photon spectrum can be used to derive the calibration factor in the specific beam(s). TG-61 recommends a calibration field size of 10×10 cm^2, which is impractical for the precision irradiators. Instead, often a field of 4×4 cm^2 is used.

TG-61 provides guidelines for two different dosimetry methods, the "in-air" and "in-phantom" methods. For the former method, which is recommended for $40 \leq kV \leq 300$, air kerma is derived, free in air, with an IC. This quantity is then converted to dose to water at a phantom surface, positioned at the same location as the chamber's effective point of measurement. The latter method, which is recommended for $100 \leq kV \leq 300$, entails a measurement by an IC at a reference depth (2 cm) in a water phantom. Above 100 kV both formalisms are consistent. For both methods, various correction and conversion factors are required. Some of these need to be inter/extrapolated from the TG-61 tables for the smaller field size employed in the calibration than 10×10 cm^2. Electron contamination can be ignored for small animal precision irradiators with open-ended applicators (Verhaegen et al., 1999). TG-61 also recommends dose conversion to other depths than the reference depth, and conversion to dose to other materials than water. More details can be obtained from the TG-61 protocol (Ma et al., 2001).

24.4 Dose Measurements

In this section, we discuss dosimetry methods which can be used mostly for relative dosimetry, but to some extent also for absolute dosimetry. A few general characteristics that all small-field precision irradiators exhibit are sharp beam penumbras and a steep drop in beam output for small fields (Newton et al., 2011; Granton et al., 2012; Lindsay et al., 2014). A number of issues related to dose measurements in small-field animal irradiators will also be valid for measurements in synchrotron microbeams, as will be discussed in detail in Chapter 25.

24.4.1 Small Volume Detectors

ICs are certainly the gold standard for absolute dosimetry, but also for relative dosimetry due to their small energy dependence. Standard small volume ICs around 0.1 cm^3 were used for basic beam data acquisition, for example, HVLs, output factors, depth dose, and cross-beam profiles, in dedicated or in-house-developed small water phantoms (Frenzel et al., 2014; Lindsay et al., 2014). For relative dosimetry in small fields, volume-averaging effects are important and, therefore, smaller volume detectors are required. Liquid-filled ICs, scintillation detectors, or diamond detectors offer advantages due to their extremely small sensitive volume. Very small volume air-filled ICs (volume < 0.01 cm^3) were reported to show a pronounced time-depending behavior, for example, long irradiation times are required until the detector stabilizes, which in turn hinders dosimetry (McEwen, 2010; Kuess et al., 2014). Diamond detectors have been explored to a much lesser degree for small-field dosimetry in the kV range but cannot be used generally unless the well-known dose-rate dependence is accounted for (Lansley et al., 2010; Kuess et al., 2014). MOSFET detectors were used in phantoms somewhat resembling a mouse shape* (De Lin et al., 2008). Another interesting and upcoming detector for small-field dosimetry in general is the scintillation detector. For the lowest energy x-rays, pronounced energy dependence was reported (Peralta and Rego, 2014). However, kV beam dosimetry with modern plastic scintillators has only received interest recently (Boivin et al., 2016) and further research is required to draw final conclusion on their practicability for dosimetry related to small animal irradiators. More information on small-field dosimetry can be found in Chapter 9.

24.4.2 Radiochromic Film

Radiochromic external beam therapy (EBT)-type films are extensively used for dosimetric purposes in radiation physics, mostly for quality assurance (QA) with respect to fluence-modulated radiotherapy, and recently also for small-field dosimetry. In this context and in light of the upcoming animal research with dedicated irradiation units, radiochromic films thus have found another key application. They are versatile and powerful, and the dosimetric applications reach from basic beam data acquisition, for example, profiles, output factors, depth dose curves, to the verification of treatment plans in dedicated mouse phantoms (van Hoof et al., 2012; Kuess et al., 2014; Lindsay et al., 2014), such as the PlastiMouse and PlastiRat phantoms† (van Hoof et al., 2013).

There are some pitfalls when using radiochromic films, which can turn into drawbacks that lead to uncertainties when not accounted for. The most important ones are the signal dependency on film orientation, postirradiation darkening, film nonuniformity, and the sensitive surface that requires careful film handling. On the other hand, radiochromic films have distinct advantages such as near tissue or water equivalence, great flexibility in applications, very little energy dependency, and the readout options with affordable document scanners. Small animal irradiation units are operated around 200 kV. Even for lower energy kV beams radiochromic films are successfully used for dosimetry purposes, although the energy response of EBT films can vary significantly between the MV and kV range (Brown et al., 2012; Villarreal-Barajas and Khan, 2014;

* CIRS Inc., Norfolk, VA (www.cirsinc.com).
† SmART Scientific Solutions BV, Maastricht, the Netherlands (www.smartscientific.com).

Steenbeke et al., 2016). As far as these document scanners are concerned, several scanner-dependent corrections might need to be applied, especially due to warm-up effects and nonuniformity of the transmission light source and related scatter effects (Fuss et al., 2007). The latter effect is certainly negligible for the small fields for small animal precision irradiators, given that film dosimetry is performed with a template in the center of the document scanner. As far as film evaluation is concerned, either the red channel of the red–green–blue (RGB) information after film scanning is used for dosimetry, or a novel triple channel method that enables to directly correct for uniformity (van Hoof et al., 2012; Dreindl et al, 2014).

At present, the most commonly used radiochromic film is EBT3-type film. Compared to earlier types of film (first generation EBT film and EBT2-type films) it was improved due to a symmetric structure of film layers that in turn reduced face-up/face-down signal dependency and a matte film surface that reduced Newton ring artifacts. More details on the use of radiochromic films and the importance of the dosimetry process consistency can be found in film dosimetry-specific literature (Fuss et al., 2007; van Hoof et al., 2012; Dreindl et al., 2014), and in Chapter 8. Very recently a new type of EBT film, i.e., EBT XD, was launched with an extended dose range for verifying hypofractionated stereotactic radiotherapy (Palmer et al., 2015).

24.4.3 Three-Dimensional Dosimetry Systems

Polymer gels, one of the few "true" three-dimensional dosimeters, have been explored for dosimetry for more than a decade, mostly with respect to verifying fluence-modulated radiotherapy with high-energy photon beams. Polymer gels consist of hydrogel and vinyl monomers in which polymerization reactions take place after irradiation that in turn are proportional to the absorbed dose. These polymerization reactions can be read out via MRI, x-ray CT, optical CT, or ultrasound methods (De Deene et al., 2015), but the most commonly used method is MRI. The accessibility to an MRI scanner, as well as the gel fabrication and stability, has hampered its widespread clinical use. Standard polymer gel dosimeters are susceptible to atmospheric oxygen that inhibits the polymerization processes. This drawback can be overcome by adding a metallo-organic complex to the gel recipe, thus removing the problem of oxygen inhibition and enabling polymer gels to be more easily manufactured. Detailed information on polymer gel dosimetry can be found in Chapter 5. As far as the dosimetric application of polymer gels in kV x-ray beams is concerned, reports in the literature are scarce and mostly limited to synchrotron radiation (Rahman et al., 2012).

PRESAGE dosimeters are rather new systems that offer distinct advantages over polymer gels, such as linear dose response and the readout options with an optical CT scanner (Sakhalkar et al., 2009). PRESAGE dosimeters are solid plastic dosimeters that contain a leuco dye and are thus in essence a radiochromic type dosimeter with an effective atomic number close to water. They have been used for dosimetry in high-energy photon beams for verifying fluence-modulated radiotherapy as discussed in Chapter 6. Moreover, their potential for dosimetry in kV beams provided by precision small animal irradiators is demonstrated by various groups. For commissioning a small animal precision irradiator, PRESAGE dosimeters were used to extract depth dose data, output factors, and profiles (Newton et al., 2011), achieving results comparable to those acquired with radiochromic films. Figure 24.2 shows an example of an application of

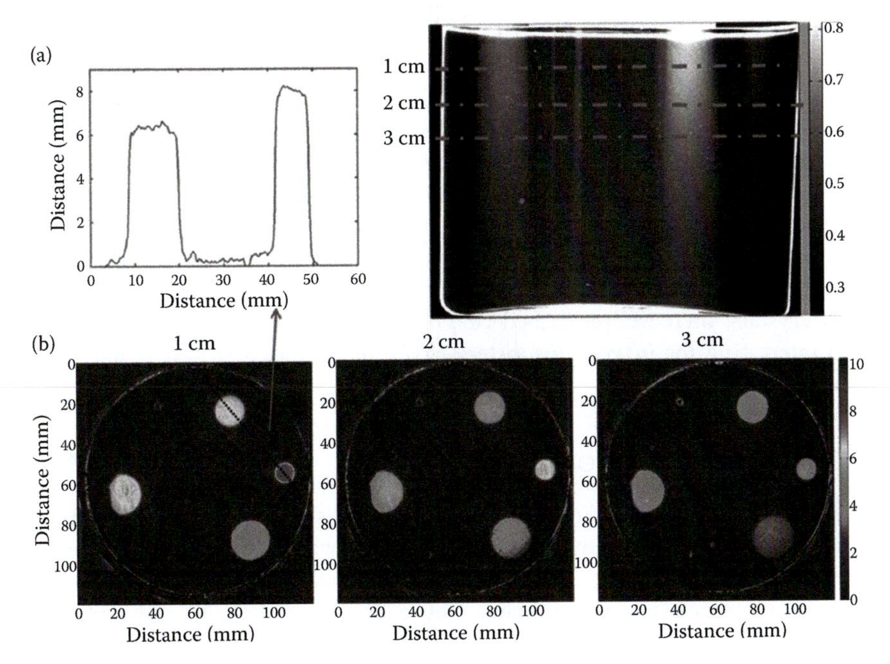

Figure 24.2

Illustration of the use of a PRESAGE 3D cylindrical dosimeter (a) irradiated with five circular 225 kVp x-ray fields (diameter 20, 15, 10, 2.5, and 1 mm) incident on the top surface. (b) Dose distributions are shown at three depths along with a line profile through the center of the 15 and 10 mm fields. (Reproduced from Newton J et al., *Med. Phys.*, 38, 6754–6762, 2011. With permission.)

PRESAGE dosimetry for kV x-ray fields. Dosimetry close to the phantom surface is, however, challenging due to optical refraction artifacts during readout.

When using small irradiation fields and irradiations from different directions, geometric consistency checks can be nicely performed with such 3D dosimeters. The same group (Bache et al., 2015) recently presented a method to use 3D printing technology to develop anatomically accurate rat phantoms, including a high atomic number spine, printed from a mold based on the CT image. Part of the phantom was filled with uniform PRESAGE plastic dosimeter medium.

Very recently, a novel silicon dosimeter was proposed for true 3D dosimetry, i.e., the FlexyDos3D dosimeter, which can be evaluated as well with an optical CT scanner (De Deene et al., 2015). It was described to be simple in fabrication, nontoxic, and can be molded in an arbitrary shape with high geometrical precision, which is an asset to mimic animals, if needed. The dosimeter formulation can be variably optimized in terms of dose sensitivity. Although it should also be applicable for kV x-rays, according to the authors' knowledge it is not yet characterized in this energy range so far.

24.4.4 Other Passive Detectors

Electron paramagnetic resonance (EPR) dosimetry is an upcoming method for reference beam dosimetry and audits, respectively, in radiation therapy. EPR dosimetry is mostly used so far for high-energy photon beams where alanine is the most commonly used material (Zeng et al., 2004), but recently its low-energy response was also investigated (Anton and Buermann, 2015). A disadvantage for small-field dosimetry is the relatively large size of standard EPR pellets

(5 mm diameter and 3.5 mm thickness for alanine). Recently, the energy dependence of lithium formate and alanine dosimeters was explored for kV photons (Waldeland et al., 2010; Adolfsson et al., 2015; Anton and Buermann, 2015). One of these studies (Waldeland et al., 2010) reported less energy dependence for lithium formate in the kV range compared to the MV range. In general, EPR dosimetry is so far not systematically introduced on a large scale to kV dosimetry.

Another passive but more widely used type of dosimetry system are thermoluminescence detectors (TLDs), for which the energy response in the kV and MV range is studied in detail, for example, Nunn et al. (2008). The small size of TLD crystals, their well-described characteristics, and availability make them also interesting for *in vivo* dosimeters in animal research (Kuess et al., 2014; Karagounis et al., 2016).

24.5 Dose Calculation

The sections on dose measurement in this chapter discussed dosimetry in simple uniform phantoms, or phantoms with a shape somewhat representative of a small animal, with the possible exception of the PlastiMouse phantom which is a fairly accurate representation of a complete mouse (van Hoof et al., 2013). kV photon dosimetry is very sensitive to the (biological) material being irradiated; large differences exist in the mass energy–absorption coefficients of water and human tissues such as adipose, muscle, and bone (Berger et al., 2010). This is mostly due to the dominance of the photoelectric effect over the Compton effect for kV photon energies. To calculate the probability for the latter, only the electron density is required, while the probability of photoelectric effect depends very strongly on the effective atomic number Z^{3-4} of the tissues. Cortical bone and skeletal muscle, which often occur in each other's proximity anatomically, differ by more than a factor of six in their energy absorption around 30 keV. For higher photon energies, the differences are reduced but can still be substantial. An additional challenge is that the tissue composition of rodents is virtually unknown, therefore in dose calculations human tissues are assigned to animals. It is currently unknown to what degree this causes systematic dose calculation errors.

Nevertheless, due to strong compositional heterogeneity of animal tissues, the atomic number is assumed to vary between 6 and 14 (as in human tissues), 3D dose calculation in a voxel geometry derived from high-resolution CT imaging is perhaps the most accurate method to derive 3D dose distributions in animals. Several methods to perform dose calculations in small animal specimens are discussed in the literature, ranging from simple analytical models (Stojadinovic et al., 2007), superposition convolution (Jacques et al., 2011; Cho and Kazanzides, 2012), to Monte Carlo simulation (Tryggestad et al., 2009; Granton et al., 2012; van Hoof et al., 2013; Noblet et al., 2016). A recent review on dose calculations for small animal precision irradiation studies discussed many of the issues in detail (Verhaegen et al., 2014). Figure 24.3 illustrates a 3D dose calculation in a mouse lung target using Monte Carlo photon simulation techniques in a dedicated small animal treatment-planning system, SmART-Plan (van Hoof et al., 2013), clearly demonstrating the influence of tissue heterogeneities.

The development of realistic mathematical rodent phantoms can be combined with detailed dose calculations to derive 3D dose estimates, but caution is needed to interpret the models (Segars et al., 2004; Mauxion et al., 2013).

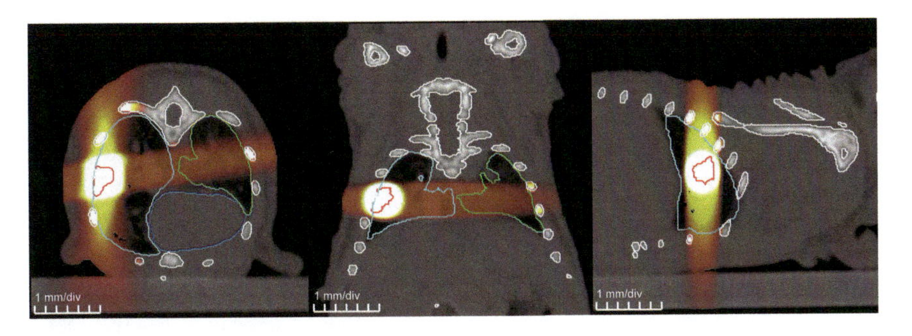

Figure 24.3

A 3D Monte Carlo dose calculation of small kV x-ray beams in a mouse lung target with SmART-Plan. The presence of various tissues and the high dose to bone can be noted. (Courtesy of Stefan van Hoof, MAASTRO Clinic, Maastricht, the Netherlands.)

Photon scatter for narrow kV photon beams irradiating small animal geometries is not studied in detail. This phenomenon is important because it may degrade the CBCT imaging quality, and therefore compromise the conversion of the geometry into a voxelized phantom. Scatter is also not, or only approximately, modeled in most dose calculation algorithms for small animals, with the exception of Monte Carlo codes. Therefore, photon scatter may also degrade the dose calculation accuracy.

A final issue of relevance in this section is the dose reporting method for kV photon beams. Calculated dose can be reported as dose to medium in medium, $D_{m,m}$, or dose to water in medium, $D_{w,m}$. For both, particle transport is performed in the proper animal media, but for the former dose scoring is also done in medium, whereas for the latter, scoring is done in water. Both are absorbed doses, but fundamentally different quantities. This is a somewhat complex and potentially confusing matter which is discussed more in detail in the literature (Enger et al., 2012; Thomson et al., 2013; Verhaegen et al., 2014), as well as in Chapter 13. As recommendation we may mention that the dose reporting method in studies should always be mentioned clearly.

24.6 Focal Spot Issues

The position and intensity distribution of the focal spot in an x-ray tube become important for radiation fields having sizes comparable to the focal spot itself. Since most precision irradiators have focal spot sizes of the order of a few millimeters, and the beams used for precision irradiation are about the same dimension, this issue is important. Figure 24.4 shows the geometry of the x-ray target producing the photon focal beam distribution. Each x-ray tube will have its own particular distribution which, moreover, may drift and change relative amplitudes slowly over time. The focal spot distribution will influence the absolute output of the x-ray tube for the smallest fields. It is therefore imperative that the photon beam output of a precision irradiator for the smallest field is measured regularly and accurately, for example, with calibrated radiochromic film. The use of radiation dosimeters with dimensions of the active volume exceeding

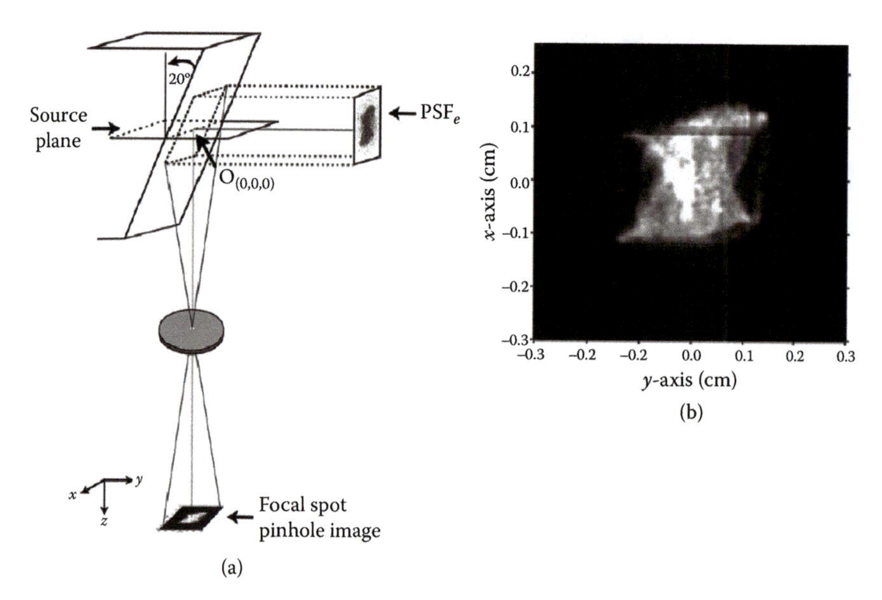

Figure 24.4

(a) Schematic representation of an electron beam impinging from the right (PFS$_e$: phase space file for electrons) on the angled x-ray target, producing a photon beam. The figure also indicates how the photon focal beam distribution can be measured, with a pinhole camera. (b) Typical focal spot distribution of the photon source emanating from an x-ray tube used for precision irradiation of small animals. For CBCT imaging, the same x-ray tube may be used with a smaller spot size, due to the lower tube currents needed. (Reproduced from Granton and Verhaegen. *Phys. Med. Biol.* 58: 3377–3395, 2013. With permission.)

the beam dimensions should be avoided. A comparison (Lindsay et al., 2014) of the absolute measured output at 4 cm deep in a solid water phantom of three irradiators of the same type, using radiochromic film, revealed differences of up to 180% for a 2.5 mm diameter field, and even for a 5 mm diameter field differences of about 25% were noted. Results for the smallest collimator available on their device (1 mm diameter) were not reported, but could differ substantially. In another irradiator (SARRP, from Xstrahl Inc.), the smallest field size available is 0.5 mm diameter, therefore even more caution is warranted in this case. While some of these differences may be due to small differences in collimator geometry and positioning, a source of systematic and random errors for the smallest fields, the main cause is most likely the focal spot distribution variation between x-ray tubes. The influence of small drifts of the focal spot on beam output was also demonstrated (Granton and Verhaegen, 2013). Adding a monitor IC to these devices would seem warranted.

A fast analytical method was developed (Granton and Verhaegen, 2013) to avoid time-consuming Monte Carlo simulations of the primary electron beam producing the photon beam. This method uses measured focal spot distributions with a pinhole camera (Figure 24.4b) to project photons down to the level of the exit plane of the photon collimator. At that level, a new photon phase space file is created, from which photons are sampled for further Monte Carlo simulation in the animal specimen. In the same study (Granton and Verhaegen, 2013), the authors showed that scatter from the photon collimator only had a minor influence on beam output.

24.7 Portal Dosimetry

The two commercially available image-guided precision irradiators are equipped with imaging panels that can capture the photon beam after passing through the specimen. In the XRAD-SMART PXi system (Lindsay et al., 2014), this is done with the CBCT imaging panel, which is always opposite to the x-ray tube at any gantry angle. In the SARRP Xstrahl system (Wong et al., 2008), a fixed imaging panel at the bottom of the setup serves the same purpose, which means the x-ray tube must be facing downward. Analogously to the use of electronic portal imaging devices mounted on linear accelerators for human radiotherapy (van Elmpt et al., 2008; Mans et al., 2010; Persoon et al., 2012), the imaging panel onboard of small animal irradiators could be used as a portal dosimeter to capture the photon beam, with or without an animal specimen present. In the latter case, the panel could be used to verify the photon fluence distribution of the treatment beam. In case the specimen is present, the captured photon fluence could be used to ascertain the planned radiation dose was correctly delivered.

Factors which could cause discrepancies between planned and delivered dose could be related to plan transfer from planning system to irradiation device, errors in beam delivery, or anatomical changes in the specimen (motion, shifts, geometry changes in tumor, etc.). Provided the photon fluence can be predicted, for example, by the dose planning system, it can be compared to the measured photon fluence, from which errors could be deduced. The portal images or the photon fluence can also be converted to absolute dose distributions. This dose verification method is currently not implemented in the commercially available irradiators. Recently, the first efforts were published to develop a portal dosimetry method for small animal irradiators (Granton et al., 2012). This work was based on detailed Monte Carlo simulations of the irradiator, including the focal photon spot, the animal specimen, and the imaging panel. For the latter, a complex response model was included. Figure 24.5 shows a comparison between a predicted and measured portal image of the head region of a mouse. The gamma function shows that in some pixels, mostly at the bottom right (shoulder region), the combined difference criteria of intensity difference and distance-to-agreement

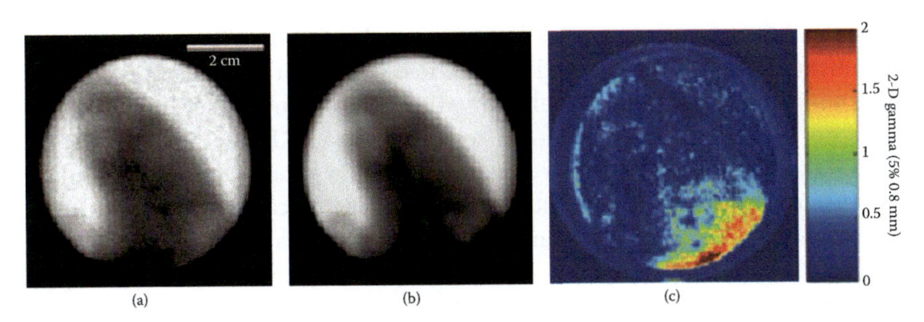

Figure 24.5

Comparison of a simulated portal image of a mouse (a) head at top left, shoulder at bottom right, to a measured image (b) for a 2 cm diameter photon beam. Panel (c) shows a gamma function comparison (intensity difference criterion 5%, distance-to-agreement criterion 0.8 mm). Pixels in the gamma function exceeding a value of unity violate the combined difference criteria of intensity difference and distance-to-agreement (Persoon et al., 2011). (Reproduced from Granton PV et al., *Med. Phys.* 39, 4155–4166, 2012. With permission.)

were violated. This probably means that changes occurred in the mouse position between the acquisition of the CBCT image and the delivery of the dose.

For the smallest fields, the usefulness of comparing predicted and acquired 2D image/dose distributions at the imaging panel may be reduced. However, in principle, even full 3D dose reconstruction in the specimen is possible, employing photon back-projection techniques, again in analogy to portal dosimetry in human radiotherapy (van Elmpt et al., 2008). Studies are needed to establish this technique solidly.

24.8 QA Aspects

After acceptance testing, commissioning, and subsequent clinical implementation, a QA program is set up for radiation delivery devices, such as medical linear accelerators or orthovoltage treatment units. Moreover, QA and quality control (QC) procedures that cover the whole dosimetry chain from beam calibration, imaging and treatment planning until dose delivery need to be put in place. These end-to-end tests are mandatory when using ionizing radiation in human medicine (see Chapter 19). Pure research areas are commonly not covered by medical radiation protection legislation but mostly by more general radiation protection legislations with less specific recommendations. Moreover, there are no national or international QA/QC recommendations from professional bodies for research beam delivery devices.

Nevertheless, when performing nonclinical research with small animal precision irradiation systems a similar QA/QC program as for radiation therapy needs to be implemented to guarantee safe and reproducible beam delivery also including geometric and imaging aspects. The conceptual ideas on dosimetric and geometric QA/QC for medical systems used in radiation therapy can be certainly applied as a starting point. However, the individual design of the small animal irradiator and its imaging system, which can vary as described above, needs to be considered.

Based on the experience of multicenter trials in radiation oncology, for preclinical animal research in a multiinstitutional setting, dosimetry and geometric intercomparisons/audits are also recommended. Audits not only reveal uncertainties that might deteriorate final results of a study, but are also helpful to centers with less experience. As a first step, Lindsay et al. (2014) reported on a multiinstitutional dosimetric and geometric comparison in the frame of commissioning an image-guided small animal irradiator. Another group (Rankine et al., 2013) reported on end-to-end testing with a PRESAGE 3D dosimetry method to assess the isocenter precision under gantry rotation, and the accuracy of coincidence of the imaging and therapeutic mechanical coordinate systems. With the increasing number of small animal irradiation systems being installed and the related research activities carried out around the world, the importance for QA/QC recommendations of such devices has just been realized. Published documents from several working groups are expected within the coming years.

24.9 Summary and Outlook

The field of image-guided precision small animal irradiation is young, and offers many new exciting possibilities to discover important new ways of treating cancer patients (and possibly others) with radiation combined with other agents.

To exploit fully the potential benefits of the new research platforms, it is important that highly accurate radiation dosimetry methods, and the imaging needed to enable the dosimetry, are implemented promptly. In this chapter the challenges and opportunities were discussed, which are the development of precise and accurate dosimeters in points, planes and volumes, in realistic irradiation geometries, exposed to small beams of kV x-rays. Highly realistic animal phantoms may need to be developed. Onboard imaging panels could be employed to perform dose verification, but much work is needed to establish this technique for small animals.

The efforts to enable 3D dose calculation in heterogeneous animal geometries with dedicated planning systems were also discussed. More developments are needed as the radiation platforms will increase their degrees of freedom, for example, by adding variable collimators, beam shutters, synchronous couch/gantry motion, and motion-gated irradiation. There will also be a need to develop 4D dosimetry techniques (van der Heyden et al., 2017), to register the real-time dose distributions administered to living animals. And finally, in the near future, small animal image-guided irradiation platforms may be developed that allow irradiation with small proton or other particle beams. This will present new challenges and opportunities to develop novel dosimetry techniques.

References

Adolfsson ES et al. (2015) Measurement of absorbed dose to water around an electronic brachytherapy source. Comparison of two dosimetry systems: Lithium formate EPR dosimeters and radiochromic EBT2 film. *Phys. Med. Biol.* 60: 3869–3882.

Almond PR et al. (1999) AAPM's TG-51 protocol for clinical reference dosimetry of high-energy photon and electron beams. *Med. Phys.* 26: 1847–1870.

Anton M and Buermann L (2015) Relative response of the alanine dosimeter to medium energy x-rays. *Phys. Med. Biol.* 60: 6113–6129.

Bache ST et al. (2015) Investigating the accuracy of microstereotactic-body-radiotherapy utilizing anatomically accurate 3D printed rodent-morphic dosimeters. *Med. Phys.* 42: 846–855.

Bazalova M et al. (2014) Modality comparison for small animal radiotherapy: A simulation study. *Med. Phys.* 41: 011710.

Berger MJ et al. (2010) XCOM: Photon Cross Section Database (version 1.5). (http://physics.nist.gov/xcom, Gaithersburg, MD: National Institute of Standards and Technology (NIST)).

Boivin J et al. (2016) A systematic characterization of the low-energy photon response of plastic scintillation detectors. *Phys. Med. Biol.* 61: 5569–5586.

Bolcaen JB et al. (2014) MRI-guided 3D conformal arc micro-irradiation of a F98 glioblastoma rat model using the small animal radiation research platform (SARRP). *J. Neurooncol.* 120: 257–266.

Brown et al. (2012) TAD Dose-response curve of EBT, EBT2, and EBT3 radiochromic films to synchrotron-produced monochromatic x-ray beams. *Med. Phys.* 39: 7412–7417.

Butterworth KT, Prise KM and Verhaegen F (2015) Small animal image-guided radiotherapy: Status, considerations and potential for translational impact. *Br. J. Radiol.* 88 (1045): 20140634.

Cho NP and Kazanzides P (2012). A treatment planning system for the small animal radiation research platform (SARRP) based on 3D slicer. (http://hdl.handle.net/10380/3364) accessed date August 2016.
Clarkson R et al. (2011) Characterization of image quality and image-guidance performance of a preclinical microirradiator. *Med. Phys.* 38: 845–856.
De Deene Y et al. (2015) FlexyDos3D: A deformable anthropomorphic 3D radiation dosimeter: Radiation properties. *Phys. Med. Biol.* 60: 1543–1563.
De Lin M et al. (2008) Application of MOSFET detectors for dosimetry in small animal radiography using short exposure times. *Radiat. Res.* 170: 260–263.
Desrosiers M et al. (2013) The importance of dosimetry standardization in radiobiology. *J. Res. Natl. Inst. Stand. Technol.* 118: 403–418.
Dreindl R, Georg D and Stock M (2014) Radiochromic film dosimetry: Considerations on precision and accuracy for EBT2 and EBT3 type films. *Z. Med. Phys.* 24: 153–163.
Enger SA et al. (2012) Dose to tissue medium or water cavities as surrogate for the dose to cell nuclei at brachytherapy photon energies. *Phys. Med. Biol.* 57: 4489–4500.
Frenzel T et al. (2014). Partial body irradiation of small laboratory animals with an industrial x-ray tube. *Z. Med. Phys.* 24: 352–362.
Fuss M et al. (2007) Dosimetric characterization of GafChromic EBT film and its implication on film dosimetry quality assurance. *Phys. Med. Biol.* 52: 4211–4225.
Granton PV et al. (2012) A combined dose calculation and verification method for a small animal precision irradiator based on onboard imaging. *Med. Phys.* 39: 4155–4166.
Granton PV and Verhaegen F (2013) On the use of an analytic source model for dose calculations in precision image-guided small animal radiotherapy. *Phys. Med. Biol.* 58: 3377–3395.
Graves EE et al. (2007) Design and evaluation of a variable aperture collimator for conformal radiotherapy of small animals using a microCT scanner. *Med. Phys.* 34: 4359–4367.
IAEA (2000) Absorbed dose determination in external beam radiotherapy. Technical Report Series Report TRS 398. (Vienna, Austria: International Atomic Energy Agency).
Jacques R et al. (2011) Real-time dose computation: GPU-accelerated source modeling and superposition/convolution. *Med. Phys.* 38: 294–305.
Karagounis IV, Abatzoglou IM and Koukourakis MI (2016) Technical note: Partial body irradiation of mice using a customized PMMA apparatus and a clinical 3D planning/LINAC radiotherapy system. *Med. Phys.* 43: 2200.
Kuess P et al. (2014) Dosimetric challenges of small animal irradiation with a commercial x-ray unit. *Z. Med. Phys.* 24: 363–372.
Lansley SP et al. (2010) Comparison of natural and synthetic diamond x-ray detectors. *Australas. Phys. Eng. Sci. Med.* 33: 301–306.
Lindsay PE et al. (2014) Multi-institutional dosimetric and geometric commissioning of image-guided small animal irradiators. *Med. Phys.* 41: 031714.
Ma C-M et al. (2001) AAPM protocol for 40-300 kV x-ray beam dosimetry in radiotherapy and radiobiology. *Med. Phys.* 28: 868–893.
Mans A (2010) 3D Dosimetric verification of volumetric-modulated arc therapy by portal dosimetry. *Radiother. Oncol.* 94: 181–187.

Mauxion T et al. (2013) Improved realism of hybrid mouse models may not be sufficient to generate reference dosimetric data. *Med. Phys.* 40: 052501.
McEwen MR (2010) Measurement of ionization chamber absorbed dose k(Q) factors in megavoltage photon beams. *Med. Phys.* 37: 2179–2193.
Newton J et al. (2011) Commissioning a small-field biological irradiator using point, 2D, and 3D dosimetry techniques. *Med. Phys.* 38: 6754–6762.
Noblet C et al. (2016) Validation of fast Monte Carlo dose calculation in small animal radiotherapy with EBT3 radiochromic films. *Phys. Med. Biol.* 61: 3521–3535.
Nunn AA et al. (2008) LiF: Mg,Ti TLD response as a function of photon energy for moderately filtered x-ray spectra in the range of 20-250 kVp relative to 60Co. *Med. Phys.* 35: 1859–1869.
Palmer AL et al. (2015) Evaluation of Gafchromic EBT-XD film, with comparison to EBT3 film, and application in high dose radiotherapy verification. *Phys. Med. Biol.* 60: 8741-8752.
Peralta L and Rego F (2014) Response of plastic scintillators to low-energy photons. *Phys. Med. Biol.* 59: 4621–4633.
Persoon LC et al. (2012) Interfractional trend analysis of dose differences based on 2D transit portal dosimetry. *Phys. Med. Biol.* 57: 6445–6458.
Persoon LC et al. (2011) A fast three-dimensional gamma evaluation using a GPU utilizing texture memory for on-the-fly interpolations. *Med. Phys.* 38: 4032–4035.
Poludniowski G et al. (2009) SpekCalc: A program to calculate photon spectra from tungsten anode x-ray tubes. *Phys. Med. Biol.* 54: N433–N438.
Rahman WN et al. (2012) Polymer gels impregnated with gold nanoparticles implemented for measurements of radiation dose enhancement in synchrotron and conventional radiotherapy type beams. *Australas. Phys. Eng. Sci. Med.* 35: 301–309.
Rankine LJ et al. (2013) Investigating end-to-end accuracy of image guided radiation treatment delivery using a micro-irradiator. *Phys. Med. Biol.* 58: 7791–7801.
Sakhalkar HS et al. (2009) A comprehensive evaluation of the PRESAGE/optical-CT 3D dosimetry system. *Med. Phys.* 36: 71–82.
Segars WP et al. (2004) Development of a 4-D digital mouse phantom for molecular imaging research. *Mol. Imaging Biol.* 6: 149–159.
Song KH et al. (2010) An x-ray image guidance system for small animal stereotactic irradiation. *Phys. Med. Biol.* 55: 7345–7362.
Steenbeke F et al. (2016) Quality assurance of a 50-kV radiotherapy unit using EBT3 GafChromic film: A feasibility study. *Technol. Cancer Res. Treat.* 15: 163–170.
Stojadinovic S et al. (2007) MicroRT-small animal conformal irradiator. *Med. Phys.* 34: 4706–4716.
Thomson RM, Tedgren AC and Williamson JF (2013) On the biological basis for competing macroscopic dose descriptors for kilovoltage dosimetry: Cellular dosimetry for brachytherapy and diagnostic radiology. *Phys. Med. Biol.* 58: 1123–1150.
Tillner F et al. (2014) Pre-clinical research in small animals using radiotherapy technology—A bidirectional translational approach. *Z. Med. Phys.* 24: 335–351.
Tillner F et al. (2016) Precise image-guided irradiation of small animals: A flexible non-profit platform. *Phys. Med. Biol.* 61: 3084–3108.

Trani D et al. (2015) What level of accuracy is achievable for preclinical dose painting studies on a clinical irradiation platform? *Radiat. Res.* 183: 501–510.
Tryggestad E et al. (2009) A comprehensive system for dosimetric commissioning and Monte Carlo validation for the small animal radiation research platform. *Phys. Med. Biol.* 54: 5341–5357.
van der Heyden B et al. (2017) The influence of respiratory motion on dose delivery in a mouse lung tumor irradiation using the 4D MOBY phantom. *Br. J. Radiol.* 90 (1069): 20160419.
van Elmpt W et al. (2008) A literature review of electronic portal imaging for radiotherapy dosimetry. *Radiother. Oncol.* 88: 289–309.
van Hoof SJ, Granton PV and Verhaegen F (2013) Development and validation of a treatment planning system for small animal radiotherapy: SmART-Plan. *Radiother. Oncol.* 109: 361–366.
van Hoof SJ et al. (2012) Evaluation of a novel triple-channel radiochromic film analysis procedure using EBT2. *Phys. Med. Biol.* 57: 4353–4368.
Verhaegen F, Granton P and Tryggestad E (2011) Small animal radiotherapy research platforms. *Phys. Med. Biol.* 56: R55–R83.
Verhaegen F et al. (1999) Monte Carlo modelling of radiotherapy kV x-ray units. *Phys. Med. Biol.* 44: 1767–1789.
Verhaegen F et al. (2014) A review of treatment planning for precision image-guided photon beam pre-clinical animal radiation studies. *Z. Med. Phys.* 24: 323–334.
Villarreal-Barajas JE and Khan RF (2014) Energy response of EBT3 radiochromic films: Implications for dosimetry in kilovoltage range. *J. Appl. Clin. Med. Phys.* 15(1): 331–338.
Waldeland E et al. (2010). The energy dependence of lithium formate and alanine EPR dosimeters for medium energy x rays. *Med. Phys.* 37: 3569–3575.
Wong J et al. (2008) High-resolution, small animal radiation research platform with x-ray tomographic guidance capabilities. *Int. J. Radiat. Oncol. Biol. Phys.* 71: 1591–1599.
Zeng GG et al. (2004) An experimental and Monte Carlo investigation of the energy dependence of alanine/EPR dosimetry: I. Clinical x-ray beams. *Phys. Med. Biol.* 49: 257–270.

25

3D Dosimetry in Synchrotron Radiation Therapy Techniques

Elke Bräuer-Krisch

25.1 Introduction

Stereotactic synchrotron radiotherapy (SSRT) and microbeam radiation therapy (MRT) are novel approaches in radiation therapy to treat brain tumors and potentially other tumors using synchrotron radiation (Bräuer-Krisch et al., 2010a; Balosso et al., 2014). SSRT is based on a local drug uptake of high-Z elements in tumors followed by stereotactic irradiation with 80-keV photons to enhance the dose deposition only within the tumor (Biston et al., 2004; Adam et al., 2006). Phase I SSRT clinical trials started in 2014 combining conventional radiotherapy (RT) with one fraction of the treatment delivered at the European Synchrotron Radiation Facility (ESRF) (Balosso et al., 2014). Medical physics aspects such as the integration of the iodine concentration in the treatment planning system (TPS) to accurately determine the dose in the tumor were successfully benchmarked. Also, a dosimetry protocol for absolute dose measurements was established (Prezado et al., 2011; Vautrin, 2011; Obeid et al., 2014). For MRT, the medical physics aspects are particularly challenging due to the very small-field sizes. For instance, in case of plane parallel beams, field sizes having a full width at half maximum (FWHM) of 25 μm by several centimeters in height, down to 50 μm × 50 μm spot sizes for pencil beams are used. Potential dosimeters for MRT were investigated over the last 15 years (Bräuer-Krisch et al., 2010b), with a more recent review article (Bräuer-Krisch et al., 2015) summarizing the most important developments in high-resolution dosimetry for MRT.

In MRT, quasi-parallel, highly collimated arrays of x-ray microbeams of medium energy between 50 and 350 keV are applied. Important features of such highly brilliant synchrotron sources are the very small beam divergence and the extremely high dose rate. The minimal beam divergence allows the insertion of a multislit collimator (MSC) to produce spatially fractionated beams of typically 25–75 μm wide microplanar beams separated by wider, 100–400 μm center-to-center (ctc) spaces with a very sharp penumbra (see Figure 25.1). Peak entrance doses of several hundreds of gray (Gy) are extremely well tolerated by normal tissues and at the same time provide a higher therapeutic index for various tumor models in rodents (Laissue et al., 1998; Dilmanian et al., 2002; Miura et al., 2006; Smilowitz et al., 2006). The hypothesis of a selective radio-vulnerability of the tumor vasculature versus normal blood vessels by MRT (Blattmann et al., 2002) was recently more solidified (Bouchet et al., 2013a).

25.2 Three-Dimensional Small-Field Dosimetry and Micrometer-Sized Field Dosimetry

A third generation synchrotron source at ESRF provides a quasi-nondiverging beam with a very high intensity photon flux, where usually a monochromatic beam with a small bandwidth is extracted, by the insertion of monochromators. Filtered "white" beams, with a broad spectrum ranging from 50 to 350 keV in the case of MRT, represent an excellent tool to determine the resolution of any suitable detector system. Monochromatic beams can also provide a tool to characterize the energy dependence of the detectors in ranges typically from 20 to 100 keV. Microbeams at the ESRF can be used for both detector development (Rosenfeld et al., 2001; Bräuer-Krisch et al., 2003; Lerch et al., 2011; Petasecca et al., 2012) and experimental studies in biology to understand the underlying processes involved in MRT (Serduc et al., 2006; Bouchet et al., 2013b; Fernandez-Palomo et al., 2013;

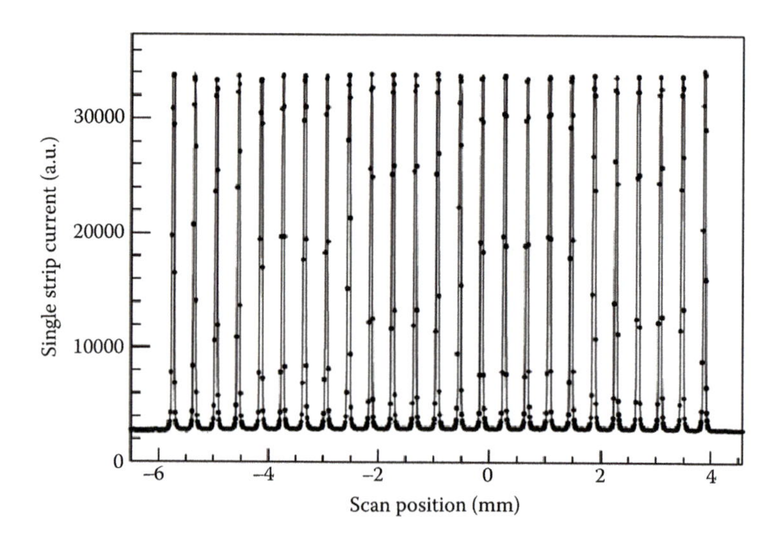

Figure 25.1

Acquisition of the signal of a 10-μm single silicon strip detector scanning horizontally across a microbeam field with 50-μm full width at half maximum (FWHM) and 400-μm center-to-center distance.

Smith et al., 2013), with both research directions supporting the technique to move forward to phase I clinical trials.

The following examples for micrometer-sized field dosimetry for the specific application in MRT mainly emerged through short- and long-term scientific collaborations. The specific combination of dose range, dose rate, medium- to low-energy photons, and extremely high resolution has further pushed technological developments to solve a wide range of the specific medical physics problems in MRT.

25.2.1 Radiochromic Films

25.2.1.1 *Spatial Resolution and Film Readers*

Dose assessment using radiochromic films is traditionally based on linear absorbance measurements often using "white" light sources such as those used in flatbed color scanners as discussed in Chapter 8. The transmission (or reflection) image is analyzed for dose assessment in three wide color channels (red–green–blue [RGB]), and the data obtained either in one channel, usually the red one, or in all of them (Kalef-Ezra and Karava, 2008; Micke et al., 2011; Lewis et al., 2012). Alternatively, spot spectrophotometers, densitometers, and microdensitometers with light sources of appreciable spectral content in the region of intense light absorption are also often used. Low power lasers, such as He/Ne (632.8 nm) and diode (e.g., 650–670 nm) lasers, and broadband red light-emitting diode (LED) sources coupled with band-pass filters are frequently used along with either a photodiode or a photomultiplier.

Flatbed scanners equipped with "white" light sources and arrays of charge-coupled devices (CCDs) are useful in the study of synchrotron radiation fields for various applications such as alignment procedures, uniform film irradiations, and dose mapping. Measurements of the modulation transfer function (MTF) of both instruments, the microdensitometer and a commercial high-resolution flatbed scanner, were used to determine the real spatial resolution of the two systems. The real resolution of the flatbed scanner is only ~100 μm, whereas the microdensitometer has shown a resolution of 20 μm, adequate for applications in MRT. More recently, Bartzsch et al. (2015) have used a modified Zeiss microscope with a resolution better than 5 μm.

25.2.1.2 *Environmental Effects*

Gafchromic films were originally manufactured by International Specialty Products (ISP), Wayne, NJ (www.ispcorp.com), which is now a part of ASHLAND, Bridgewater, NJ (www.ashland.com). When using Gafchromic films, such as HD-810, HD-V2, EBT2, or EBT-XD, the molecular motions influence the structure of the polymer. Therefore, the shape of the absorption spectrum is influenced by the temperature during film irradiation, storage, and reading, usually shifting toward lower wavelength with increasing reading temperature. Humidity and ultraviolet (UV) exposure (even by sunlight or light from fluorescent lamps) may also influence the film response by a degree which depends on the coating used, among other things. Radiochromic films undergo postexposure signal intensification, with the polymerization rate decreasing with time. Adequate time has to elapse between irradiation and measurement to achieve accurate measurements. Thus sticking to a fixed carefully designed protocol is crucial to obtain reproducible and accurate dosimetric results with radiochromic films.

25.2.1.3 Energy Response

Taking into account that (a) most of the imparted energy during the therapeutic use of synchrotron radiation is related to photons of energy less than about 150 keV, and (b) the photon spectrum varies with depth as well as between the peak and the valley region (Siegbahn et al., 2009), some corrections may be necessary to cope with the energy dependence. Measurements carried out by Bartzsch (2014) indicated a 50% decrease in the response of HD-810 films in water with decreasing photon energy from 100 to 40 keV, and about 15% increase in the response of HD-V2 films in the same energy region. Similarly, simulations by Hermida-Lopez et al. (2014) indicated that the EBT2 and EBT3 films exhibit an energy-dependent response in water in the energy region from 10 to 100 keV, with a 10% and 40% maximum reduction at 40 keV, respectively. These authors predicted that the EBT3 films would have a constant response within 2.3% over the entire energy region. However, in practice, one cannot exclude the potential existence of intrinsic energy dependence, a factor usually not taken into account when using radiation transport codes. Thus, radiochromic film energy response curves have to be assessed experimentally.

Muench et al. (1991) showed that the response of HD-810 films to 60-kVp x-rays, (28 keV effective photon energy, kV_{eff}), is lower by about 30% than that to 4-MV x-rays. Kron et al. (1998) reported that MD-55 films underestimated the dose by a factor of two when irradiated with a monoenergetic 26-keV synchrotron-generated x-ray beam. Nariyama (2005) studying the energy response of MD-55 and HD-810 films reported a measurable dose up to 50 and 400 kGy, respectively, and an under-response relative to ^{60}Co gamma rays to low-energy photons. In HD-810 films, an almost constant under-response by 20% was observed in the energy region 30–100 keV, relative to ^{60}Co gamma rays, and a gradual increase in MD-55-2 film response from about 5% to almost 40% as the energy decreases in this energy region. Similarly, Cheung et al. (2004) studying MD-55-2 and HD-810 film observed a gradual decrease in response (up to about 40%) with decreasing energy from 100 to 30 keV, and a large over-response (up to a factor 5 at 50 kV_{eff}) in XR-type T films.

Oves et al. (2008) observed that in LiPCD-loaded EBT films a 0.76 and 0.81 response to 75- and 125-kVp x-rays relative to 6-MV x-rays. Brown et al. (2012) reported the responses of EBT, EBT2, and EBT3 films to 35-keV synchrotron-produced monochromatic beams of 0.76, 1.24, and 0.98 relative to 4-MV x-rays, respectively. Similarly, comparing the output factors of x-ray machines in the energy range from 50 to 125 kVp measured with EBT3 films and a parallel plate ionization chamber, Gill and Hill (2013) reported differences up to only 3.3% in 2.0 cm fields. The differences were consistent with the estimated total uncertainty. On the other hand, Villarreal-Barajas and Khan (2014) irradiating EBT3 films with 70–300-kVp x-ray beams, reported a gradual reduction of the response with decreasing energy from 0.94 at 168 keV_{eff} down to 0.79 at 32 keV_{eff} using the red channel of RGB images, and even lower using the blue one (0.83 and 0.74, respectively). Moreover, Massillon-JL et al. (2012) found a dose-dependent reduction in the response of EBT3 films to 50-kVp x-rays (20 keV_{eff}) up to 11% relative to 6-MV x-rays. In conclusion, even for films such as EBT3 that are often referred as dosimeters with no energy dependence, extra care has to be taken for dosimetry in synchrotron beams used for therapeutic purposes.

25.2.1.4 Microbeam Fields

Several investigators have exposed films in MRT using HD-810, EBT, and HD-V2 films, the most appropriate type for typical dose values in the range of 100–500 Gy peak entrance dose and valley dose values between 1 and 20 Gy. The films themselves come with submicrometer resolution, permitting the determination of relative output factors, transversal dose profiles, and depth-dose distributions with the adequate instrumentation to scale the measured signal to optical density, which is linear with dose over a wide range. The measured valley doses in such profiles have, in general, been 10%–15% higher than those predicted by Monte Carlo (MC) simulations (Nariyama, 2005; Siegbahn et al., 2006; Crosbie et al., 2008; Martinez-Rovira et al., 2012; Bartzsch et al., 2015).

For beam sizes between 25 and 100 μm FWHM, the use of a 3CS Microdensitometer (Joyce-Loebl [JL] Automation) or a modified Zeiss Axio Vert. A1 microscope (www.zeiss.fr) with EC Plan-Neofluar objective lenses, are currently good options, both meeting the requirements in terms of spatial resolution (Figure 25.2). The old technology of a microdensitometer unfortunately is lacking sufficient stability of the instrument in time and requires several attempts to identify stable conditions for reliable data for dose measurements.

Bartzsch et al. (2015) have irradiated HD-810 and HD-V2 films with MRT beams, and films were evaluated using an inverted optical microscope coupled with a CCD-camera with a nominal spatial resolution below 5 μm. For calibration purposes, films can be irradiated homogeneously at 2.0 cm depth in a solid water phantom at a well-known dose, previously measured with an ionization chamber and employing as close as technically possible the IAEA TRS 398 protocol (IAEA, 2000). Percentage depth dose (PDD) curves for the peak and valley dose, as well as peak-to-valley dose ratios (PVDRs), can be measured and compared to MC-calculated dose distributions in MRT. It should, however, be mentioned that the associated uncertainties are high and the differences between the MC-calculated dose and the measured dose can be as high as 15%, with measurement uncertainties depending on the film type. HD-810 has shown a high fluctuation in sensitivity per sheet of up to 8%, which is considerably reduced to about 3% for the HD-V2 and EBT-XD films. Comparing the signal of the two film types, it was found that the peak dose values of the HD-V2 films were slightly higher than those of the HD-810 film, with an opposite situation in the case of the valley dose, resulting in higher PVDR values when the HD-V2 films were used. These differences could partially be attributed to differences in the energy response due to a slight energy shift of the spectrum in the peak and valley regions.

25.2.2 Si-Based Single and Multiple Strip Detector Systems

The fast development of silicon nanotechnology has led to well-established manufacturing of silicon radiation detectors for medical applications. Current available technology allows the feasibility of direct coupling of silicon sensors to their associated readout electronics with extremely high spatial resolution, thus providing an attractive solution as an active beam monitor for MRT. High spatial and temporary resolution detectors are used for instance for QA in MRT, volumetric-modulated arc therapy (VMAT) and stereotactic radiosurgery (SRS). Commercialized Si-strip detector technology has as well successfully been used in brachytherapy over the last 10 years (Wong et al., 2010, 2011).

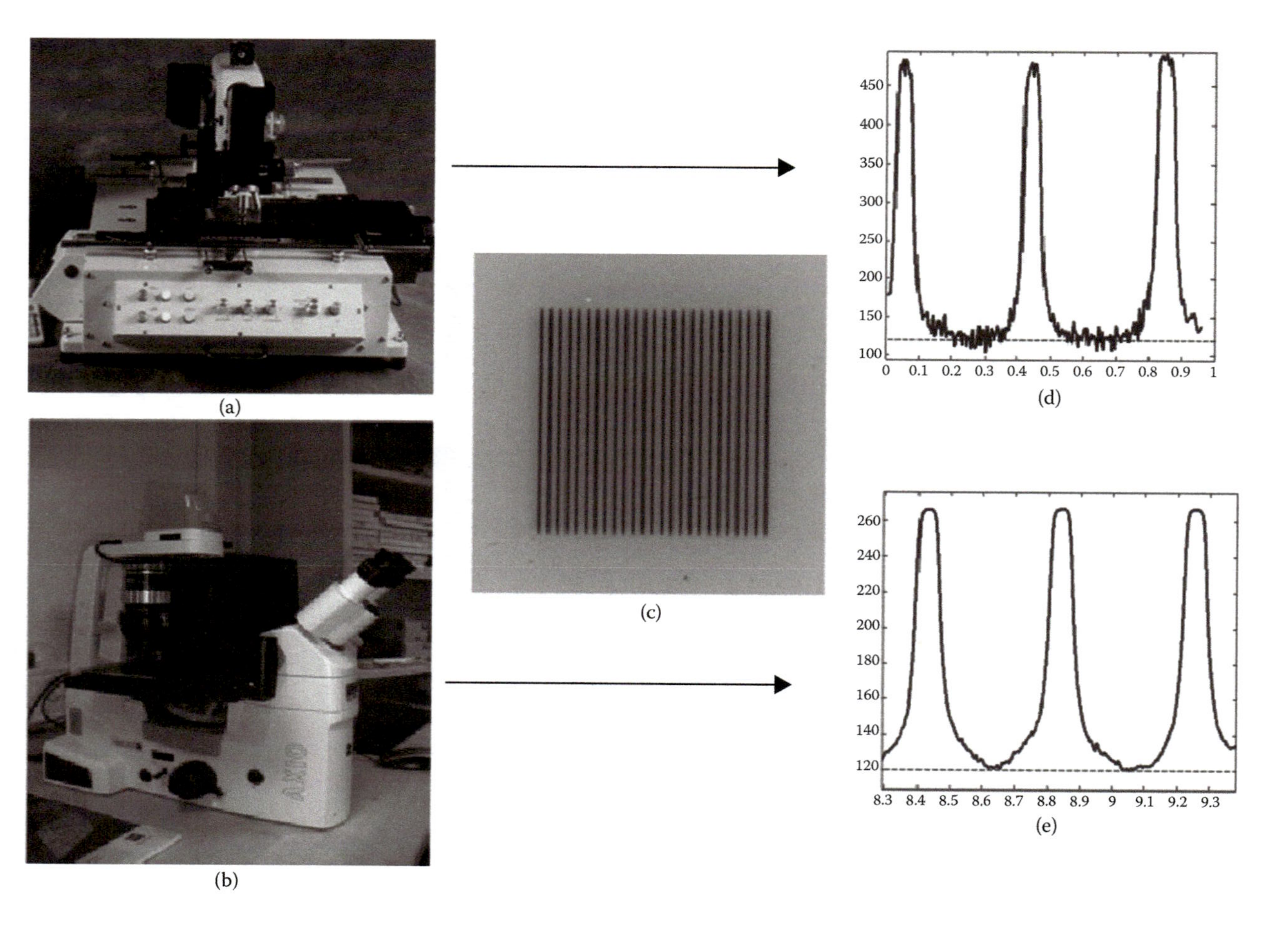

Figure 25.2

(a) JL microdensitometer. (b) Modified Zeiss microscope. (c) Gafchromic film exposed with microbeams (50-μm FWHM, 400-μm center-to-center). (d) Lateral microbeam profiles read out with the microdensitometer. The pen deflection can directly be correlated to the optical densities (1 pixel step size = 1.25 μm). (e) Lateral microbeam profiles read out with the modified Zeiss microscope.

For MRT, and within the framework of a European project (3DMiMic collaboration, www.sintef.no/3dmimic), a novel silicon sensor with multiple strips (or channels) has been proposed to monitor the x-ray beams of an entire microbeam array. Due to the extremely high dose rate in MRT, one key issue was the large amount of charge generated by the x-ray beams. A 10-μm-thick sensor with short strip length reduces the available ionization volume and leads to a minimal beam perturbation and reduced heat load on the silicon sensor (Petasecca et al., 2012).

Isolation structures, such as "guard-ring" and "p-stop," provide a high recombination probability of the generated charge and surround each individual strip to limit the amount of charge reaching the readout electronics. In addition, the guard isolation reduces the charge sharing between adjacent strips for better identification of an x-ray peak intensity position measurement. The high concentration of dopants increases the recombination rate of the generated charge upon ionization thus resulting in a lower total collected charge and generated signal.

The expected radiation damage to a silicon detector for the x-ray photon energies involved in MRT is mostly related to the charging up of the surface oxide layers. In order to assure a proper operation periodical recalibration of the sensor will take into account efficiency losses due to bulk damage without compromising the operation of the system.

In the single channel measurements, the generated current in one single strip and the isolated guard-ring structure can be measured by two separate electrometers. The device can be scanned across a microbeam array to determine absolute peak and valley dose for different field sizes at the desired depth (see Figure 25.1). Such dose measurements require careful alignment of the detector in the peak and in the valley. Recent results using a single Si-strip detector demonstrated an agreement of the measured PVDR within 10% with the expected MC-calculated dose for a 2 cm × 2 cm field size (Bräuer-Krisch et al., 2015).

A 256-channel readout system was developed to monitor the full array of the microbeam field prior to and during treatment, comparable to the function of a large area detector in conventional radiation therapy to monitor the integrated dose per port. The readout system integrates the input, which is then sampled at an adjustable rate to provide various readout capabilities. The maximum charge currently integrated per sample is 9.6 pC with a material resistivity of 5 Ω·cm and the mean current per channel recorded at a sampling rate of 1 kHz. With a thickness of only 250 μm, the detector is transparent enough not to perturb the microbeam array. It is currently the only solution to perform such high-resolution online dosimetry for potential clinical trials in MRT, and it can easily be integrated into a patient safety system to interrupt the beam, in case, the expected dose delivery would be exceeded in any of the ports.

25.2.3 MicroDiamond Detectors

MicroDiamond detectors developed by PTW (www.ptw.de) are 1 μm thick and approximately 1.1 mm wide and are therefore very suitable for relative dose measurements in conventional radiation therapy with IMRT, Gamma Knife, and CyberKnife beams, as discussed in Chapter 9. An example of an MRT-relative microbeam dose profile measured with a PTW microDiamond chamber in edge-on mode at the ESRF is shown in Figure 25.3, demonstrating the excellent resolution of this type of detector. Recent PVDR measurements with a microDiamond detector for MRT application at the Australian Synchrotron

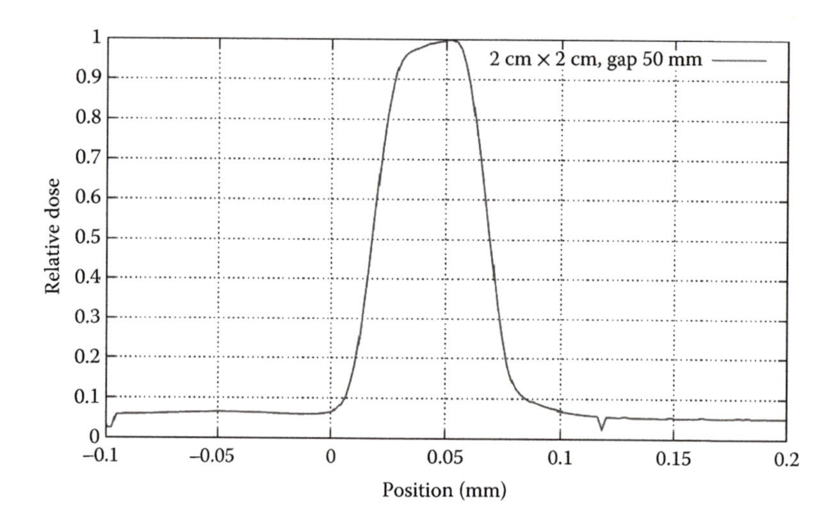

Figure 25.3

Lateral beam profile of a nominal 50-μm FWHM microbeam measured at the European Synchrotron Radiation Facility (ESRF) using a PTW microDiamond detector (measured FWHM: 50.4 μm).

were in good agreement with Gafchromic film measurements (Livingstone et al., 2016).

25.2.4 PRESAGE® Dosimeters

MRT combines several challenging dosimetry problems requiring measurements with both high spatial resolution and high dynamic range at a very high dose rate of up to 20,000 Gy/s. In order to validate the entire chain from the initial x-ray computed tomography (CT) scan through planning to the final treatment, measurements over a large field-of-view (FOV) in three dimensions (3D) can potentially be performed by means of 3D optical CT microscopy using the radiochromic plastic polymer known as PRESAGE.

High-spatial resolution often comes at the price of limiting both the region of space sampled and the dimensionality of the information obtained. For example, single detectors have limited sensitive areas and must be translated through the region-of-interest, involving a series of separate irradiations, rather than necessarily mimicking a single patient treatment.

As discussed in more detail in Chapters 5 and 6, in the field historically known as gel dosimetry, two readout modalities have emerged as leading candidates for quantitative dose imaging: (a) magnetic resonance imaging (MRI) of both radiochromic Fricke gels (Appleby et al., 1987; Schreiner, 2004) and polymer gels (Maryanski et al., 1993), and (b) optical CT (Gore et al, 1996; Doran et al., 2001). Although the MRI-based techniques have been used successfully for dosimetry of SSRT protocols at the ESRF (Boudou et al., 2007), they have proved unsuccessful for MRT, both because the gels themselves are not sufficiently robust to very high dose rates, and because the available spatial resolution is not high enough to characterize microbeams of the order 50 μm (Berg et al., 2004; Bayreder et al., 2008).

PRESAGE is a solid plastic chemical dosimeter based on clear polyurethane mixed with a leucomalachite green reporter dye and a number of organic and/

or metallic initiators (Adamovics and Maryanski, 2006). A radiochromic reaction is induced after exposure to ionizing radiation, resulting in a local change in optical density of the plastic, which is discussed in a comprehensive way in Chapter 6. Effectively, the PRESAGE acts as a "3D radiochromic film" and the response to radiation is linear with dose at the often employed wavelength of 633 nm, compatible with both helium–neon (He–Ne) laser and LED light sources (see Figure 25.4).

Favorable characteristics of PRESAGE include excellent spatial resolution, high dynamic range (Al Nowais et al., 2010), dose rate independence, and the ability to record the dose distribution in 3D, allowing flexible and realistic dosimetry applications. To date, the highest resolution measurements have been made via fluorescent microscopy with pixel sizes down to 78 nm (Annabell et al., 2012). The corresponding disadvantages relate primarily to the fact that PRESAGE is a chemical dosimeter with a relatively complex composition. A number of the constituents, particularly the polyurethane base, are supplier dependent, with batches whose properties do not remain constant over time. The manufacturer has also investigated a number of different formulations over the course of the research program at ESRF. The samples investigated have displayed differing sensitivities to radiation and ambient temperature, with variable degrees of time-evolution of their optical density postirradiation. The inter- and intrabatch variabilities still need to be investigated until the optimum formulation is found and characterized. Thus, although the relative dosimetry is reliable (Doran et al., 2013), moving from the current results to absolute dosimetry will be challenging.

After an initial feasibility study (Doran et al., 2010, 2013), the development of a microimaging scanner has involved several upgrades to reach acquisition times of less than 3 minutes with the addition of a new camera (Zyla sCMOS, Andor Technology PLC, Belfast, UK, www.andor.com) using a large pixel array and fast frame rate. Other additions to the system are motorized positioning stages and a sample mounting system, which allows reproducible positioning of individual samples. This potentially means that absolute changes in optical density can be measured by registration of pre- and postirradiation optical CT scans.

For such important benchmarking experiments, the PRESAGE dosimetry system is currently an excellent choice for ultimate confirmation of a 3D valley dose distribution prior to approval of a treatment plan.

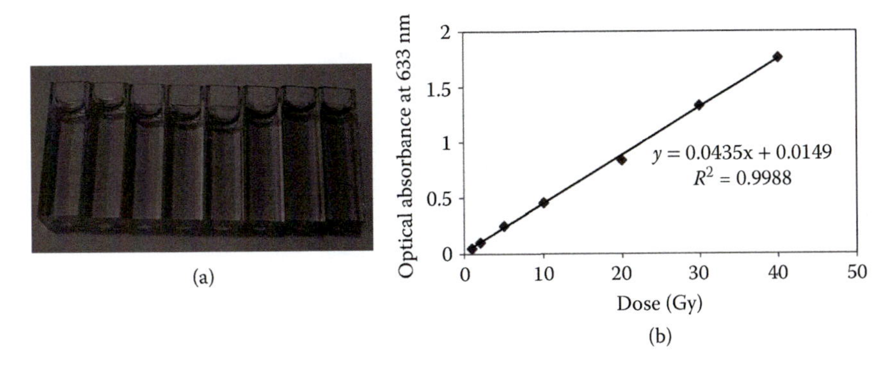

Figure 25.4

(a) Cuvettes of PRESAGE irradiated with a range of doses. (b) Optical absorbance of the cuvettes as measured at 633 nm by a spectrophotometer.

25.2.5 High-Resolution Thermoluminescent Dosimeters

Other potential high-resolution types of dosimeters are thermoluminescent dosimeters (TLDs). At the Institute of Nuclear Physics IFJ PAN in Krakow, Poland, a two-dimensional thermoluminescence (TL) dosimetry system, consisting of LiF:Mg,Cu,P (MCP-N)-based TL foils, a TLD reader with a CCD camera, and a large size planchet heater, was developed to perform high-resolution dosimetry (Ptaszkiewicz et al., 2008). Absolute dose measurements in the low dose range from mGy, up to 20 Gy can be performed after careful calibration of the system. This is particularly important for MRT dosimetry due to the energy dependence of the TL detectors for mean energies around 100 keV (Crosbie et al., 2015).

25.2.6 Fluorescent Nuclear Track Detectors

Fluorescent nuclear track detectors (FNTDs) are a new type of luminescent detectors for dosimetric applications in radiation therapy (Akselrod et al., 2006). These single crystals are made of Al_2O_3:C,Mg and were originally used for neutron and heavy charged particle dosimetry. With the help of an optical readout system the luminescence intensity is linear with dose over a wide range up to 30 Gy and can be quantified with submicrometer resolution (see Figure 25.5). This qualifies these dosimeters for absolute valley dose measurements in MRT, whereas peak dose measurements at 30 Gy can usually only be delivered during storage ring filling modes at lower current (Sykora and Akselrod, 2010). Because of their physical properties and excellent spatial resolution, FNTDs also found successful application in ion beam dosimetry (Osinga et al., 2013). A potential advantage of the FNTD technology system over some other systems is the fact that they are commercially available (Akselrod et al., 2014; www.landauer.com).

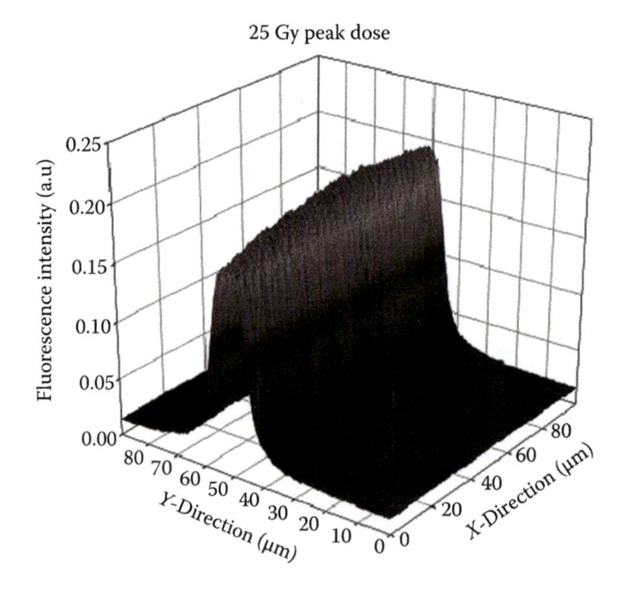

Figure 25.5

High resolution 3D dose distribution of a central microbeam from an exposed fluorescent nuclear track detector (Al_2O_3) using a confocal laser scanning microscope.

25.2.7 Other Techniques

In the context of very high-resolution dosimetry, other techniques developed at synchrotron sources, like samarium-doped glasses (Okada et al., 2011, 2013; Vahedi et al., 2012), and optical fiber dosimetry (Sporea et al., 2014) showed very good potential, but are currently still lacking either sufficient linearity or resolution for accurate dose determination below 50-μm beam sizes. With the interesting field of radiation therapies at synchrotron radiation sources pushing the detector development to very high-resolution dosimetry, the conventional radiation therapy medical physics issues in small-field dosimetry can as well profit from these recent developments.

Most of the high-resolution dosimeters presented in this chapter are used for relative dose measurements, which are typically linked to a calibration protocol for absolute dosimetry. For SSRT as well as for MRT, the sample or detector has to be scanned through the beam due to the limited beam size from the source. This approach was validated in SSRT by Prezado et al. (2011) and in case of MRT, ion recombination corrections were taken into account for ion chamber measurements at dose rates up to 20,000 Gy/s (Bräuer-Krisch et al., 2015; Fournier et al., 2016). A good alternative is the use of graphite calorimeters for absolute dosimety at synchrotron sources; these are currently validated for dose rates up to 50 Gy/s (Harty et al., 2014).

25.3 Look at the Future and Closing Remarks

Preclinical research in MRT has clearly demonstrated the great potential of spatially fractionated radiation therapy using microscopically small beam sizes (Bräuer-Krisch et al. 2010a; Grotzer et al., 2015). Differential effects occur at high peak doses, and for beam sizes below 100 μm, where the normal tissue tolerance is very high (Laissue et al., 2001) compared with pathological tissues. For larger beams, for example, 0.5–5 mm wide, the dose–volume effect (Curtis et al., 1963, 1967) has not yet been exploited a lot in radiation therapy in the clinical environment, with a few exceptions using GRID therapy with megavoltage x-rays and multileaf collimators to spatially fractionate the beam (Penagaricano et al., 2010). Submillimeter beams, called minibeams to differentiate them from microbeams, were first proposed for photons (Dilmanian et al., 2006). They are currently also proposed for proton and heavy ion therapy (Dilmanian et al., 2012, 2015; Prezado et al., 2013; Kłodowska et al., 2015; Martinez-Rovira et al., 2015; Peucelle et al., 2015a,b; Girst et al., 2016), although the experimental evidence for their biological efficacy is fragmentary. Several of these approaches are aiming for homogeneous dose coverage of the tumor, whereas the advantages of the spatial fractionation are exploited in normal tissue. A tighter periodicity between peak and valley doses leads to a better tumor control. In other words, narrow microbeams at tighter spacing are more effective for tumor growth suppression than wide microbeams (Uyama et al., 2011; Griffin et al., 2012).

Research in MRT, which uses beams <100 μm, has led to the conclusion that superior tumor control can be achieved at equal normal tissue sparing when compared to conventional radiation therapy even if the spatially fractionated feature of the beams is maintained in the tumor tissue (Laissue et al., 1998). The different mechanisms explaining a superior tumor control are of mixed origin. First, the high radiation dose delivered in the microbeams induces direct and immediate tumor cell death along the beam path, whereas sublethal doses are given in conventional RT. As a result, debulking of 1/4–1/2 of the tumor may occur in

MRT depending on the irradiation configuration used. Furthermore, differential effects between tumor and normal tissue vasculature have been demonstrated during the last decades. For instance, MRT induces tumor endothelial cell death and decreases blood volume and vessel density. Furthermore, MRT deprives the lesion of oxygen and nutrients, which leads to tumor necrosis (Bouchet et al., 2013a,b, 2015). Other mechanisms such as bystander effects/cell–cell communication (Fernandez-Palomo et al., 2013, 2015) and differences in gene expression (Bouchet et al., 2015) have also been reported. The immune system also certainly plays a very important role, as well as the specific surface of the interface between heavily (peaks) and lightly (valley) irradiated tissue slices. The latter effect is much larger for microbeams than for minibeams (Laissue et al., 2013). A limitation of MRT is that it can currently only be performed at third generation x-ray sources such as the ESRF. Clinical trials with microbeams are proposed in the future to validate the scientific findings, which are hoped to spur the scientific community to develop smaller but adequate sources for improved cancer treatment, and to transfer these approaches safely into the clinic.

References

Adam JF et al. (2006) Prolonged survival of Fischer rats bearing F98 glioma after iodine-enhanced synchrotron stereotactic radiotherapy. *Int. J. Radiat. Oncol. Biol. Phys.* 64: 603–611.

Adamovics J and Maryanski MJ (2006) Characterisation of PRESAGE: A new 3-D radiochromic solid polymer dosemeter for ionising radiation. *Radiat. Prot. Dos.* 120: 107–112.

Akselrod MS et al. (2014) FNTD radiation dosimetry system enhanced with dual-color wide-field imaging. *Radiat. Meas.* 71: 166–173.

Akselrod GM et al. (2006) A novel Al2O3 fluorescent nuclear track detector for heavy charged particles and neutrons. *Nucl. Instr. Meth. Phys. Res.* B 247: 295–306.

Al Nowais S et al. (2010) An investigation of the response of the radiochromic dosimeter PRESAGE TM to irradiation by 62 MeV protons. *J. Phys. Conf. Ser.* 250: 012034.

Annabell N et al. (2012) Evaluating the peak-to-valley dose ratio of synchrotron microbeams using PRESAGE fluorescence. *J. Synchrotron Radiat.* 19: 332–339.

Appleby A, Christman EA and Leghrouz A (1987) Imaging of spatial radiation dose distribution in agarose gels using magnetic resonance. *Med. Phys.* 14: 382–384.

Balosso J et al. (2014) Monoenergetic synchrotron beams: First human experience for therapeutic purpose. *Radiother. Oncol.* 111(1): SP-0205.

Bartzsch S (2014) PhD thesis: Microbeam radiation therapy—Physical and biological aspects of a new cancer therapy and development of a treatment planning system. University of Heidelberg, Germany.

Bartzsch S et al. (2015) Micrometer-resolved film dosimetry using a microscope in microbeam radiation therapy. *Med. Phys.* 42: 4069–4079.

Bayreder C et al. (2008) The spatial resolution in dosimetry with normoxic polymer-gels investigated with the dose modulation transfer approach. *Med. Phys.* 35: 1756–1769.

Berg A et al. (2004) High resolution MR based polymer dosimetry versus film densitometry: A systematic study based on the modulation transfer function approach. *Phys. Med.* Biol. 49: 4087–4108.
Biston MC et al. (2004) Cure of Fisher rats bearing radioresistant F98 glioma treated with cis-platinum and irradiated with monochromatic synchrotron x-rays. *Cancer Res.* 64: 2317–2323.
Blattmann H et al. (2002) Microbeam irradiation in the chorio-allantoic membrane (CAM) of chicken embryo. *Strahlenther. Onkol.* 178: 118.
Bouchet A et al. (2013a) Synchrotron microbeam radiation therapy induces hypoxia in intracerebral gliosarcoma but not in the normal brain. *Radiother. Oncol.* 108: 143–148.
Bouchet A et al. (2013b) Early gene expression analysis in 9L orthotopic tumor-bearing rats identifies immune modulation as a hallmark of response to synchrotron microbeam radiation therapy. PLOS One. 31: e81874.
Bouchet A et al. (2015) Identification of AREG and PLK1 pathway modulation as a potential key of the response of intracranial 9L tumor to microbeam radiation therapy. *Int. J. Cancer* 136: 2705–2716.
Boudou C et al. (2007) Polymer gel dosimetry for synchrotron stereotactic radiotherapy and iodine dose-enhancement measurements. *Phys. Med.* Biol. 52: 4881–4892.
Brauer-Krisch E et al. (2003) MOSFET dosimetry for microbeam radiation therapy at the European Synchrotron Radiation Facility. *Med. Phys.* 30: 583–589.
Brauer-Krisch E et al. (2010a) Effects of pulsed, spatially fractionated, microscopic synchrotron x-ray beams on normal and tumoral brain tissue. *Mutat. Res.* 704: 160–166.
Brauer-Krisch E et al. (2010b) Potential high resolution dosimeters for MRT. *AIP Conf. Proc.* 1266: 89–97.
Brauer-Krisch E et al. (2015) Medical physics aspects of the synchrotron radiation therapies: Microbeam radiation therapy (MRT) and synchrotron stereotactic radiotherapy (SSRT). *Phys. Med.* 31: 568–583.
Brown TAD et al. (2012) Dose-response curve of EBT, EBT2, and EBT3 radiochromic films to synchrotron-produced monochromatic x-ray beams. *Med. Phys.* 39: 7412–7417.
Cheung T, Butson MJ and Yu PKN (2004) Experimental energy response verification of XR type T radiochromic film. *Phys. Med.* Biol. 49: N371–376.
Crosbie JC et al. (2008) A method of dosimetry for synchrotron microbeam radiation therapy using radiochromic films of different sensitivity. *Phys. Med.* Biol. 53: 6861–6877.
Crosbie JC et al. (2015) Energy spectra considerations for synchrotron radiotherapy trials on the ID17-biomedical beamline at the European Synchrotron Radiation Facility. *J. Synchrotron Radiat.* 22: 1035–1041.
Curtis H (1963) The microbeam as a tool in radiobiology. Adv. Biol. *Med. Phys.* 175: 207–224.
Curtis H (1967) The use of a deuteron microbeam for simulating the biological effect of heavy cosmic-ray particles. *Radiat. Res. Suppl.* 7: 250–257.
Dilmanian FA et al. (2002) Response of rat intracranial 9L gliosarcoma to microbeam radiation therapy. *Neuro-Oncol.* 4: 26–38.
Dilmanian FA et al. (2006) Interlaced x-ray microplanar beams: A radiosurgery approach with clinical potential. *PNAS* 103(25): 9709–9714.

Dilmanian FA et al. (2012) Interleaved carbon minibeams: An experimental radiosurgery method with clinical potential. *Int. J. Radiat. Oncol. Biol. Phys.* 84: 514–519.

Dilmanian FA et al. (2015) Minibeam therapy with protons and light ions: Physical feasibility and potential to reduce radiation side effects and to facilitate hypofractionation. *Int. J. Radiat. Oncol. Biol. Phys.* 92: 469–474.

Doran SJ et al. (2013) Establishing the suitability of quantitative optical CT microscopy of PRESAGER radiochromic dosimeters for the verification of synchrotron microbeam therapy. *Phys. Med.* Biol. 58: 6279–6297.

Doran SJ et al. (2001) A CCD-based optical CT scanner for high-resolution 3D imaging of radiation dose distributions: Equipment specifications, optical simulations and preliminary results. *Phys. Med.* Biol. 46: 3191–3213.

Doran SJ et al. (2010) An investigation of the potential of optical computed tomography for imaging of synchrotron-generated x-rays at high spatial resolution. *Phys. Med.* Biol. 55: 1531–1547.

Fernandez-Palomo C et al. (2013) DNA double strand breaks in the acute phase after synchrotron pencil beam irradiation. *JINST* 8: C07005.

Fernandez-Palomo C et al. (2015) Use of synchrotron medical microbeam irradiation to investigate radiation-induced bystander and abscopal effects in vivo. *Phys. Med.* 31: 584–595.

Fournier P et al. (2016) Absorbed dose-to-water protocol applied to synchrotron-generated x-rays at very high dose rates. *Phys. Med.* Biol. 61: N349–N361.

Gill S and Hill R (2013) A study on the use of GafchromicTM EBT3 film for output factor measurements in kilovoltage x-ray beams. *Australas. Phys. Eng. Sci. Med.* 36: 465–471.

Girst S et al. (2016) Proton minibeam radiotherapy reduces side effects in an in vivo mouse ear model. *Int. J. Radiat. Oncol. Biol. Phys.* 95: 234–241.

Gore JC et al. (1996) Radiation dose distributions in three dimensions from tomographic optical density scanning of polymer gels: I. Development of an optical scanner. *Phys. Med.* Biol. 41: 2695–2704.

Griffin RJ et al. (2012) Microbeam radiation therapy alters vascular architecture and tumor oxygenation and is enhanced by a galectin-1 targeted antiangiogenic peptide. Radiat. Res. 177: 804–812.

Grotzer MA et al. (2015) Microbeam radiation therapy: Clinical perspectives. *Phys. Med.* 31: 564–567.

Harty PD et al. (2014) Absolute x-ray dosimetry on a synchrotron medical beam line with a graphite calorimeter. *Med. Phys.* 41: 052101.

Hermida-Lopez M et al. (2014) Influence of the phantom material on the absorbed-dose energy dependence of the EBT3 radiochromic film for photons in the energy range 3 keV-18 MeV. *Med. Phys.* 41: 112103.

IAEA (2000) Absorbed dose determination in external beam radiotherapy: An international code of practice for dosimetry based on standards of absorbed dose to water. Technical Reports Series No. 398. (Vienna, Austria: International Atomic Energy Agency).

Kalef-Ezra J and Karava K (2008) Radiochromic film dosimetry: Reflection vs transmission scanning. *Med. Phys.* 35: 2308–2311.

Kłodowska M et al. (2015) Proton microbeam radiotherapy with scanned pencil-beams—Monte Carlo simulations. *Phys. Med.* 31: 621–626.

Kron T et al. (1998) Dose response of various radiation detectors to synchrotron radiation. *Phys. Med.* Biol. 43: 3235–3259.

Laissue JA et al. (1998) Neuropathology of ablation of rat gliosarcomas and contiguous brain tissues using a microplanar beam of synchrotron-wiggler-generated X rays. *Int. J. Cancer.* 78: 654–660.

Laissue JA et al. (2001) The weaning piglet cerebellum: A surrogate for tolerance to MRT (microbeam radiation therapy) in paediatric neuro-oncology. *Proc. SPIE.* 4508: 65–73.

Laissue JA et al. (2013) Response of the rat spinal cord to x-ray microbeams. *Radiother. Oncol.* 106: 106–111.

Lerch MLF et al. (2011) Dosimetry of intensive synchrotron microbeams. *Radiat. Meas.* 46: 1560–1565.

Lewis D et al. (2012) An efficient protocol for radiochromic film dosimetry combining calibration and measurement in a single scan. *Med. Phys.* 39: 6339–6350.

Livingstone J et al. (2016) Characterization of a synthetic single crystal diamond detector for dosimetry in spatially fractionated synchrotron x-ray fields. *Med. Phys.* 43: 4283–4293.

Martinez-Rovira I, Sempau J and Prezado Y (2012) Development and commissioning of a Monte Carlo photon beam model for the forthcoming clinical trials in microbeam radiation therapy. *Med. Phys.* 39: 119–131.

Martinez-Rovira I, Fois G and Prezado Y (2015) New approaches in GRID therapy by using non-conventional sources. *Med. Phys.* 42: 685–693.

Maryanski MJ et al. (1993) NMR relaxation enhancement in gels polymerized and cross-linked by ionizing radiation: A new approach to 3D dosimetry by MRI. Magn. Reson. Imaging 11: 253–258.

Massillon-JL G et al. (2012) Energy dependence of the new Gafchromic EBT3 film: Dose response curves for 50 kV, 6 and 15 MV x-ray beams. *Int. J. Med. Phys. Clin. Eng. Radiat. Oncol.* 1: 60–65.

Micke A, Lewis DF and Yu X (2011) Multichannel film dosimetry with nonuniformity correction *Med. Phys.* 38: 2523–2534.

Miura M et al. (2006) Radiosurgical palliation of aggressive murine SCCVII squamous cell carcinomas using synchrotron-generated x-ray microbeams. Br. J. Radiol. 79: 71–75.

Muench PJ et al. (1991) Photon energy dependence of the sensitivity of radiochromic film and comparison with silver halide film and LiF TLDs used for brachytherapy dosimetry. *Med. Phys.* 18: 769–775.

Nariyama N (2005) Responses of GafChromic films for distribution of extremely high doses from synchrotron radiation. J. Appl. Radiat. Isot. 62: 693–697.

Obeid L et al. (2014) Absolute perfusion measurements and associated iodinated contrast agent time course in brain metastasis: A study for contrast-enhanced radiotherapy. *J. Cereb. Blood Flow Metab.* 34: 638–645.

Okada G et al. (2011) Spatially resolved measurement of high doses in microbeam radiation therapy using samarium doped fuorophosphate glasses. *Appl. Phys. Lett.* 99: 121105.

Okada G et al. (2013) Examination of the dynamic range of Sm-doped glasses for high-dose and high-resolution dosimetric applications in microbeam radiation therapy at the Canadian synchrotron. *Opt. Mat.* 35: 1976–1980.

Osinga J-M et al. (2013) High-accuracy fluence determination in ion beams using fluorescent nuclear track detectors. *Radiat. Meas.* 56: 294–298.

Oves SD et al. (2008) Dosimetry intercomparison using a 35-keV x-ray synchrotron beam. *Eur. J. Radiol.* 68: S121–S125.

Penagaricano JA et al. (2010) Evaluation of spatially fractionated radiotherapy (GRID) and definitive chemoradiotherapy with curative intent for locally advanced squamous carcinoma of the head and neck: Initial response rate and toxicity. *Int. J. Radiat. Oncol. Biol. Phys.* 76: 1369–1375.

Petasecca M et al. (2012) X-Tream: A novel dosimetry system for synchrotron microbeam radiation therapy. *J. Instrum.* 7: 1–15.

Peucelle C, Martinez-Zovira I and Prezado Y (2015a) Spatial fractionation of the dose using neon and heavier ions: A Monte Carlo study. *Med. Phys.* 42: 5928–5936.

Peucelle C et al. (2015b) Proton minibeam radiation therapy: Experimental dosimetry evaluation. *Med. Phys.* 42: 7108–7113.

Prezado Y et al. (2011) Dosimetry protocol for the forthcoming clinical trials in synchrotron stereotactic radiation therapy (SSRT). *Med. Phys.* 38: 1709–1717.

Prezado Y and Fois GR (2013) Proton minibeam radiation therapy: A proof of concept. *Med. Phys.* 40: 031712.

Ptaszkiewicz M et al. (2008) TLD dosimetry for microbeam radiation therapy at the European Synchrotron Radiation Facility. *Radiat. Meas.* 43: 990–993.

Rosenfeld AB et al. (2001) Feasibility study of online high-spatial-resolution MOSFET dosimetry in static and pulsed x-ray radiation fields. *IEEE Trans. Nucl. Sci.* 48: 2061–2068.

Schreiner LJ (2004) Review of Fricke gel dosimeters. *J. Phys. Conf. Ser.* 3: 9–21.

Serduc R et al. (2006) In vivo two-photon microscopy study of short-term effects of microbeam irradiation on normal mouse brain microvasculature. *Int. J. Radiat. Oncol. Biol. Phys.* 64: 1519–1527.

Siegbahn EA et al. (2006) Determination of dosimetrical quantities used in microbeam radiation therapy (MRT) with Monte Carlo simulations. *Med. Phys.* 33: 3248–3259.

Siegbahn EA et al. (2009) MOSFET dosimetry with high spatial resolution in intense synchrotron-generated x-ray microbeams. *Med. Phys.* 36: 1128–1137.

Smilowitz HM et al. (2006) Synergy of gene-mediated immunoprophylaxis and microbeam radiation therapy for advanced intracerebral rat 9L gliosarcomas. *Neuro-Oncol.* 78: 135–143.

Smith RW et al. (2013) Proteomic changes in the rat brain induced by homogenous irradiation and by the bystander effect resulting from high energy synchrotron x-ray microbeams. *Int. J. Radiat. Biol.* 89: 118–127.

Sporea D et al. (2014) Investigation of UV optical fibers under synchrotron irradiation. *Opt. Express* 22: 31473–31485.

Sykora GJ and Akselrod MS (2010) Novel fluorescent nuclear track detector technology for mixed neutron-gamma fields. *Radiat. Meas.* 45: 594–598.

Uyama A et al. (2011) A narrow microbeam is more effective for tumor growth suppression than a wide microbeam: An in vivo study using implanted human glioma cells. *J. Synchrotron Radiat.* 18: 671–678.

Vahedi S et al. (2012) X-ray induced $Sm3+$ to $Sm2+$ conversion in fluorophosphate and fluoroaluminate glasses for the monitoring of high-doses in microbeam radiation therapy. *J. Appl. Phys.* 112: 073108.

Vautrin M (2011) PhD thesis: Planification de traitement en radiotherapie stereotaxique par rayonnement synchrotron. Developpement et validation d'un module de calcul de dose par simulations Monte Carlo. Universite de Grenoble, Switzerland.

Villarreal-Barajas JE and Khan RFH (2014) Energy response of EBT3 radiochromic films: Implications for dosimetry in kilovoltage range. J. Appl. Clin. *Med. Phys.* 15(1): 356–366.

Wong J et al. (2010) A silicon strip detector dose magnifying glass for IMRT dosimetry. *Med. Phys.* 37: 427–439.

Wong J et al. (2011) The use of silicon strip detector dose magnifying glass on stereotactic radiotherapy QA and dosimetry. *Med. Phys.* 38: 1226–1238.

26

3D Dosimetry in Magnetic Fields

Geoffrey S. Ibbott, Gye Won (Diane) Choi, Hannah Jungeun Lee, Yvonne Roed, and Zhifei Wen

26.1 Introduction

The benefits of visualizing a patient's anatomy during the treatment position led to the development of image-guided radiation therapy (IGRT). Modern linear accelerators are generally equipped with onboard imaging systems that allow the patient to be imaged, potentially at any time, and often just before delivering the radiation. This allows the patient's position to be verified and small adjustments to be made to the treatment parameters to compensate for changes in position or anatomy. Most IGRT modalities use kilovoltage x-rays to generate images: some examples include projection images and cone-beam computed tomography (CBCT) images. Megavoltage images (portal images) made using the treatment beam and an electronic imaging system are frequently used for treatment verification. These images allow positioning of the patient based on anatomical structures and can reduce the uncertainty in aligning the treatment beam with the location of the target.

The feasibility of using magnetic resonance imaging (MRI) for radiation therapy treatment simulation, position verification, and response assessment is currently being investigated, leading to the development of MR-image-guided radiation therapy (MR-IGRT). Compared to x-ray imaging, MRI offers a number of benefits. First, MRI does not use radiation to generate images and thus removes the concern for harmful effects from the imaging dose. This is beneficial especially for lengthy courses of fractionated treatments, during which the imaging dose can accumulate to a significant level over the course of the treatment. Even more important is that imaging radiation very often encompasses normal tissue structures that can be avoided by the treatment beam. MRI also offers superior soft-tissue contrast, allowing the physician to delineate the target volume and surrounding structures with less uncertainty and potentially smaller margins. MR imaging protocols exist that acquire images rapidly, requiring less time than a CBCT image, and offering the potential to acquire dynamic images in multiple planes. Since there appears to be no biological interaction between MR imaging and the treatment radiation, it is possible to image the patient during treatment and to monitor the location of the target volume in real time. Functional imaging capabilities of MRI also allow the assessment of treatment response, during and posttreatment. Such assessment can allow treatment to be adapted according to the radiation response, providing a more personalized treatment for each patient.

Several MR-IGRT treatment machines have been designed and one system is currently commercially available. ViewRay (Oakwood village, OH) produces the MRIdian system, which combines three ^{60}Co teletherapy heads to deliver radiation in a 0.35 T magnetic field. A version in which the ^{60}Co sources are replaced with a linear accelerator has recently become commercially available. An MR-Linac, which combines a 7 MV linear accelerator with a diagnostic-quality MRI scanner, was developed as a collaboration between the University Medical Center Utrecht (the Netherlands) and Elekta (Crawley, UK). In Figure 26.1, a sketch of the MR-Linac indicates that the linear accelerator (manufactured by Elekta) is mounted on a ring around a 1.5 T Philips (Best, the Netherlands) MRI system (Raaymakers et al., 2009). This system is presently installed at several sites and is undergoing preclinical evaluation. Other systems in development are by the Cross Cancer Institute in Canada (Fallone et al., 2009) and the Ingham Institute in Australia (Keall et al., 2014). Despite the benefits of MR imaging, MR-IGRT faces some challenges to be resolved before

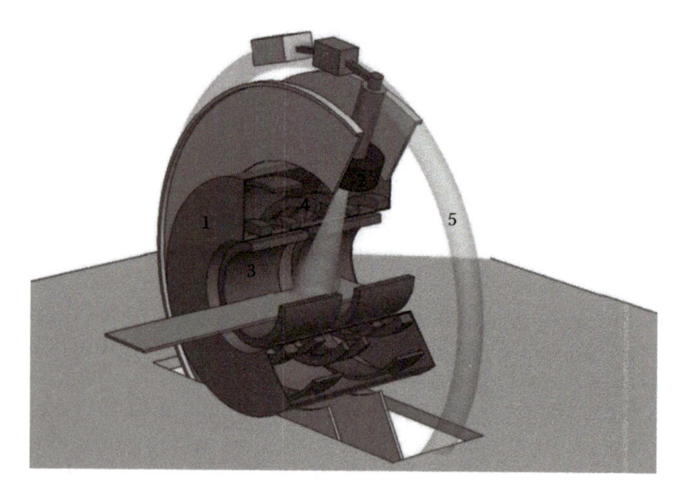

Figure 26.1

Sketch of the MR-Linac system concept: (1) 1.5 T MRI, (2) 6 MV linear accelerator, (3) split gradient coil, (4) superconducting coils, and (5) low magnetic field toroid in the fringe field. (Reproduced from Raaymakers et al., *Phys. Med. Biol.*, 54, N229–N237, 2009. With permission.)

reaching the full potential. A major concern is the change in the delivered dose distribution when strong (>1 T) magnetic fields are used for MR imaging. In photon-beam radiation therapy, dose is delivered by secondary electrons that carry the energy transferred from the incoming photons. Secondary electrons are mostly forward peaked and deposit energy downstream, as they travel deeper into the patient, in segments of straight lines. In a magnetic field, however, the paths of secondary electrons are altered due to the Lorentz force. When a secondary electron is generated in a dense medium and travels into a less dense medium, the electron may be able to follow a circular path, back into the original medium. Due to the lack of electron equilibrium, dose is increased in the upstream medium near the interface. This phenomenon is referred to as the electron return effect (ERE) and is shown to pose clinical concerns (Raaijmakers et al., 2005). Figure 26.2 shows the trajectory of an electron traveling from one medium to the next; the path changes in the presence of the magnetic field. The figure clearly demonstrates the ERE (Raaijmakers, 2008). In a Monte Carlo simulation and a film measurement (Figure 26.3), the ERE is observed to enhance dose at the interface by a factor of 1.3–1.4, over about 1 cm range (Raaijmakers et al., 2005).

The ERE is a clinical concern not only because it significantly enhances dose at interfaces but also because the intensity and pattern of the ERE are dependent on several factors. These include the media comprising the interface, the shape and orientation of the interface, the strength and direction of the magnetic field, and the energy of the beam. Investigations by Raaijmakers and colleagues show that the dose can be enhanced substantially at the interface, depending on the materials, beam energy, and magnetic field strength (Raaijmakers et al., 2005). Some measures such as using opposing beams or multiple beam angles have been shown to reduce the dose enhancement due to the ERE in relatively simple geometries (Raaijmakers et al., 2005; van Heijst et al., 2013). However, it remains a concern for the treatment of heterogeneous treatment sites, and in regions containing airways and air-filled cavities.

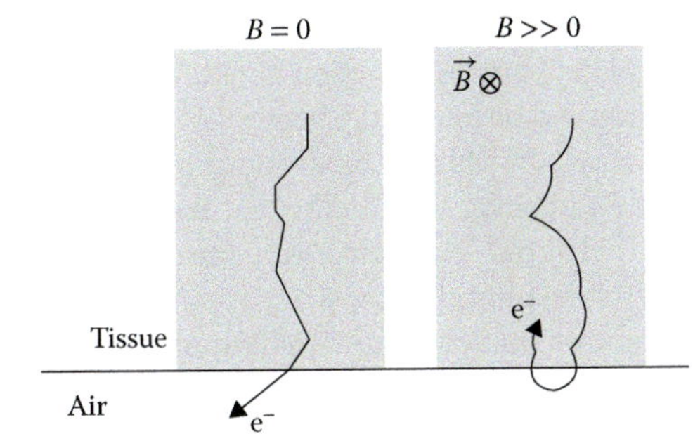

E (MeV)	r (mm)
0.5	2.2
1.0	3.4
2.0	5.6
4.0	10.0
6.0	14.5

Figure 26.2

Left: Examples of the trajectory of a secondary electron traveling in zero ($B = 0$) and strong ($B >> 0$) magnetic field. In a magnetic field, the electron may travel back to the original medium (redrawn after Raaijmakers et al., 2005). *Right*: Radius of helical electron trajectory (r) as a function of electron energy (E) with $B = 1.5$ T. (Reproduced from Raaijmakers AJ et al., *Phys. Med. Biol.*, 50, 1363–1376, 2005. With permission.)

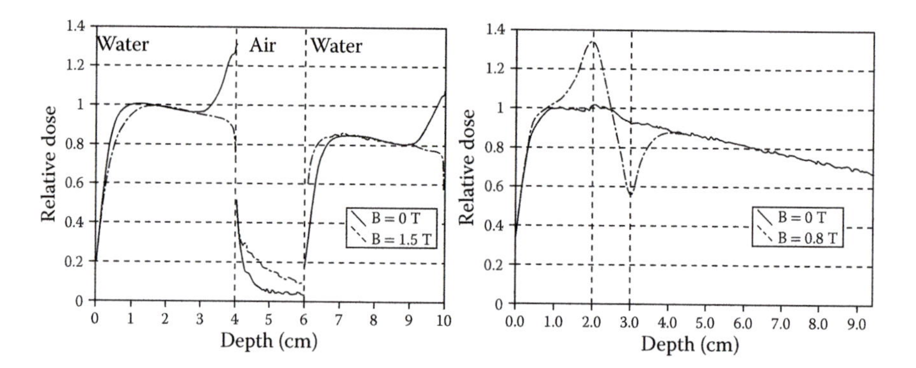

Figure 26.3

Left: Monte Carlo simulation result showing the dose profile along a beam central axis through water-air-water interfaces in the presence and absence of a magnetic field. *Right*: Film measurement of the dose profile along the central beam axis through plastic-air-plastic interfaces. (Reproduced from Raaijmakers AJ et al., *Phys. Med. Biol.*, 50, 1363–1376, 2005. With permission.)

26.2 Issues with 0D and 2D Dosimetry

Conventional quality assurance (QA) procedures in radiation therapy frequently rely on zero and two dimensional (0D, 2D) dosimetry, using point and planar dosimeters to compare the measured dose to the planned dose at a few points of interest and in a few planes. For example, the Imaging and Radiation Oncology Core-Houston (IROC-Houston, formerly the Radiological Physics Center) QA Center uses 0D and 2D dosimetry with thermoluminescence dosimeters (TLDs) and radiochromic film to credential institutions to participate in national cooperative group clinical trials (Ibbott et al., 2008). IROC-Houston ships an anthropomorphic QA phantom to each institution requesting credentialing. The phantom contains structures representing a target volume and organs at risk, as well as dosimeters. The clinic conducts its conventional workflow to obtain CT-simulator images, develop a treatment plan, and deliver the plan to the phantom, after which the phantom is sent back to IROC-Houston for analysis. IROC-Houston uses the planar and point dosimeters that were located in the phantom to compare the delivered dose distribution with the institution's calculated treatment plan. More information about credentialing procedures performed by IROC-Houston can be found in Chapter 19.

In 2010, IROC-Houston reported that 20%–30% of institutions participating in credentialing procedures had failed to deliver treatments that matched their own treatment plans within the IROC criteria. These data were measured using phantoms representing sites such as the head-and-neck, pelvis, spine, and lung (Ibbott, 2010). A more recent report shows that even for irradiations performed since 2012, about 10% of irradiations did not meet the passing criteria (Molineu et al., 2014). Considering the fact that the delivered dose was evaluated in only two or three planes, it is conceivable that a more comprehensive investigation would reveal either better or worse agreement between the plan and the delivered dose. Work in this area has demonstrated that, in some cases, a 3D evaluation identifies regions of disagreement that, because they appeared outside the plane of the 2D dosimeter, were not detected with film analysis (Oldham et al., 2012; Lafratta et al., 2015). For MR-IGRT treatments, where the magnetic field is expected to change the delivered dose distribution in a complicated fashion over the irradiated volume, the need for a more thorough means of measuring the dose to perform a detailed comparison is apparent.

26.3 The Benefits of 3D Dosimetry

In an effort to improve upon the small number of sampling points available with 2D dosimetry, three dimensional (3D) dosimeters were developed to provide volumetric dose information. Materials that have been investigated recently include polymerizing gels, radiochromic gels, and radiochromic plastic materials such as PRESAGE® (Heuris Pharma, Skillman, NJ) developed by Adamovics and Maryanski (2003). Gel dosimeters can be analyzed with one of several readout methods including MRI, x-ray CT, and optical-CT (Baldock et al., 2010). Radiochromic plastics do not generate a signal that is visible with MR or x-ray CT, and therefore are customarily analyzed with optical-CT methods (Jackson et al., 2015). With the capacity to measure and compare volumetric dose distribution, 3D dosimetry has the potential to perform as a complement to, or possibly a substitute for, 0D and 2D dosimetry. The preparation, characteristics, and readout techniques of these dosimeters are described elsewhere in this chapter, and in more detail in Chapters 5 and 6, and will not be discussed further here.

26.4 Dosimeters in Magnetic Fields

26.4.1 General Issues

Before 3D dosimeter systems could be employed as viable QA tools in magnetic field environments, it was essential to first determine if and how the magnetic field affected the response of the dosimeters. There are relatively few investigations on how a strong magnetic field affects the conventional dosimeters. In 2009, Meijsing et al. reported the magnetic field effects on the response of a Farmer NE2571 ion chamber. The study consisted of two parts—GEANT4 Monte Carlo simulation and experiments using the Farmer NE2571 chamber in the magnetic field produced by an electromagnet. Depending on the orientation of the chamber and the strength of the magnetic field, varying from 0 to 1.2 T, the response of the ion chamber varied 10%–15% (Meijsing et al., 2009). More recent data was reported by O'Brien et al. that characterize a number of cylindrical ion chamber models in a 1.5 T magnetic field (O'Brien et al., 2016). These authors showed that chamber design and construction played important roles in the magnitude of the effect. Even more important was the orientation of the ion chamber axis to the magnetic field; the influence of the magnetic field was considerably decreased when the chamber axis was parallel to the magnetic field. O'Brien and colleagues also showed that even small air gaps around an ionization chamber altered the reading of the instrument, suggesting that the use of solid water-equivalent phantoms could lead to measurement errors (O'Brien et al., 2015; O'Brien and Sawakuchi, 2017).

The complications experienced with ion chambers can be largely addressed through selection of an appropriate chamber design, by avoiding air gaps, and by taking care to align the chamber parallel to the magnetic field. This generally means making measurements with a suitable ion chamber in a water phantom which, because of the design of most MR-guided treatment units, is awkward at best, and potentially dangerous due to the high voltages in use. Most motor-driven water phantoms are unsuitable for use in such environments. Consequently, a 3D solid volumetric dosimeter is attractive to avoid some of these complications. In addition, some 3D dosimeters such as Fricke gels exhibit a response that is measurable with MR imaging. This leads to the potential of analyzing the dosimeter with the MR component of the MR-IGRT system, without needing to move the dosimeter to another device. In fact, as will be shown later, imaging is possible during irradiation, allowing for a real-time display of accumulated dose.

The effects of a magnetic field on TLDs, optically stimulated luminescence dosimeters (OSLDs), and a plastic scintillator detector were also investigated (Therriault-Proulx et al., 2016; Wang et al., 2016; Wen et al., 2016a,b). These simple 0D dosimetry tools were shown to exhibit no significant magnetic field effects. Key in these experiments was the elimination of air gaps around the dosimeters.

A recent report indicated discrepancies of up to 15% in measurements with radiochromic film when exposed in a 0.35 T magnetic field (Reynoso et al., 2016). However, measurements at 1.5 T have shown minimal influences, provided great care is taken to avoid even very small air gaps around the film (Reyhan et al., 2015; Alqathami et al., 2016; Lee et al., 2016a; Wen et al., 2016a).

26.4.2 Evaluation of PRESAGE in Magnetic Fields

26.4.2.1 Change in Sensitivity

The effect of a magnetic field on PRESAGE was investigated using cuvettes that fit into a polymethylmethacrylate (PMMA) phantom that was designed to fit between the pole pieces of an electromagnet. Doses of approximately 100, 400, 700, and 1000 cGy were delivered, with either $B = 0$ or $B = 1.5$ T. The net optical density (OD) change due to irradiation was calculated by taking the difference between the average OD of the irradiated cuvettes and that of unirradiated cuvettes handled identically. Since PRESAGE is a relative dosimeter, dose-response curves were plotted as net OD versus monitor unit (MU) settings. Analysis of variance (ANOVA) was performed in R-statistical software to test whether the magnetic field had a significant influence on the dose–response curve. Figure 26.4 illustrates the relative dose-response curves of PRESAGE cuvettes, showing a decrease in the slope and implying an underresponse in the presence of the magnetic field.

Despite the underresponse in comparison to measurements with $B = 0$ T, the response of the PRESAGE dosimeters in $B = 1.5$ T was strictly linear, with $R^2 > 0.99$. The result from an analysis with statistical software showed that the effect of the magnetic field on the sensitivity of PRESAGE was statistically significant with a p-value of 9.34×10^{-4}. Because PRESAGE dosimeters are relative dosimeters and the response remained strictly linear with dose, the change in the PRESAGE response in the magnetic field can be accounted for by applying a small correction factor for the sensitivity. For a comparison of PRESAGE responses in the presence or absence of a magnetic field, this small change in sensitivity was accounted for by scaling by the calibration and did not affect the overall results.

26.4.2.2 Complications in the Measurement of ERE Caused by the Edge Effect

If a 3D dosimeter is to measure the ERE accurately, it must provide reliable measurements at the edge of the dosimeter, where the ERE is expected to either enhance or reduce the dose at the dosimeter–air interface. Dosimetry using

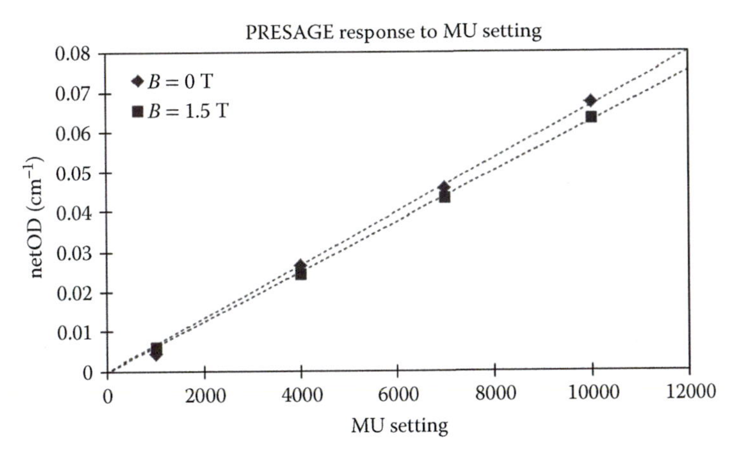

Figure 26.4

Relative dose response of PRESAGE® cuvettes irradiated in $B = 0$ and $B = 1.5$ T magnetic fields.

PRESAGE dosimeters suffers from an edge artifact, which severely degrades dose dependence in the signal at the edge of the dosimeter. The artifact arises from differences in the index of refraction between the dosimeter and the solution used in an optical-CT scanner, and was described in detail (Chisholm et al., 2015). The artifact can be minimized by carefully matching the index of refraction of the solution to that of the dosimeter, but is complicated by variations in refractive index sometimes seen from one part of a dosimeter to another.

The 3D dosimetry community does not have a consistent protocol to identify the region of edge artifact. Conventionally, the region of edge artifact was determined by visually inspecting dose distributions or profiles. This method is subjective and can be inconsistent even for a single observer. Another method was proposed by Choi in which a statistical approach is employed (Choi, 2016).

26.4.2.3 Measurement of ERE with Film and PRESAGE Dosimeter

A PRESAGE dosimeter was employed to measure the ERE in the radiation beam produced by an MR-Linac (Choi, 2016). The PRESAGE dosimeter was manufactured to fit into an ABS (acrylonitrile–butadiene–styrene) plastic phantom designed for the measurement (Choi et al., 2016). The phantom was 3D-printed to provide adequate buildup and scatter around the dosimeter, and to ensure a tight fit. The PRESAGE dosimeter was designed with a coaxial cylindrical cavity, around which the ERE could be expected to be observed in the presence of a magnetic field. Two additional short cylindrical PRESAGE dosimeters were produced as "dummies," such that the sum of their heights was equal to the height of the experimental PRESAGE dosimeter. A piece of EBT3 radiochromic film, whose response in the presence of a magnetic field was assessed as described above, was cut and sandwiched in between the dummies (Figure 26.5). This ensured that once spatially registered, the dose distribution measured by the film would correspond to a cross-sectional plane in the dose distribution measured by the PRESAGE dosimeter.

The dosimeters were irradiated with a single field oriented to be perpendicular to the magnetic field and to the cavity in the center of the PRESAGE dosimeter (Figure 26.6). A dose of 10 Gy was delivered to the film and PRESAGE

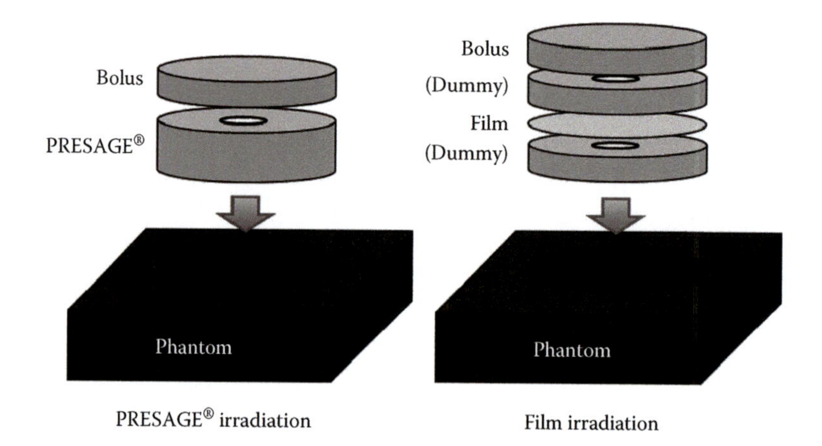

Figure 26.5

Left: ABS plastic phantom and setup for PRESAGE® dosimeter irradiation. *Right:* Setup for film irradiation.

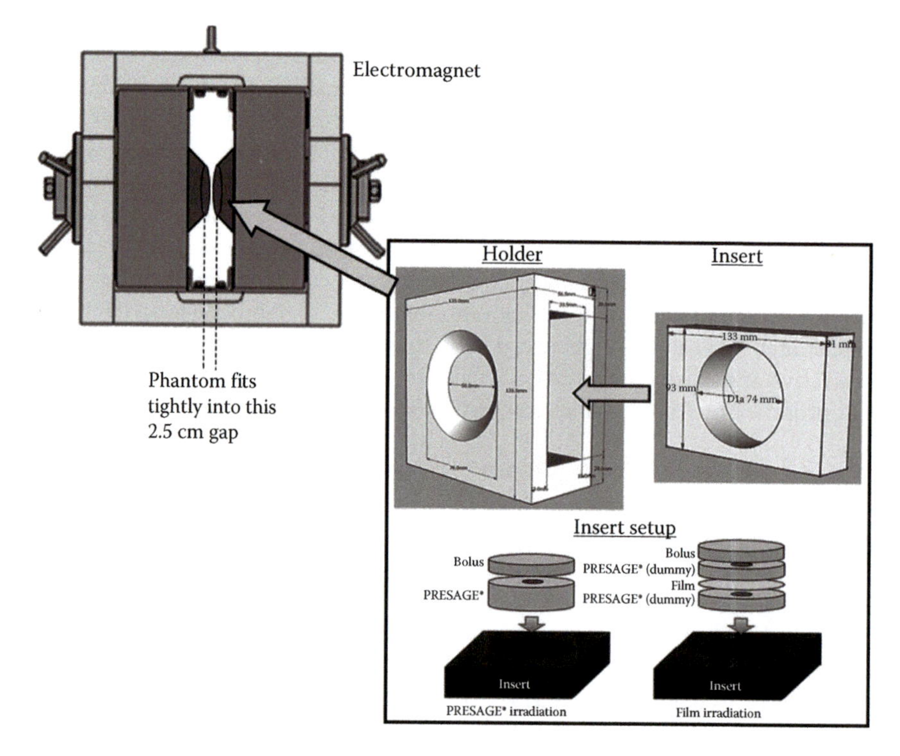

Figure 26.6

The construction and alignment of an ABS plastic phantom (holder and insert) for irradiating PRESAGE® and film dosimeters.

dosimeters. For both dosimeters, a prescan was performed immediately before irradiation. The postscans for the film and the PRESAGE dosimeter were obtained 24 hours and 2 hours postirradiation, respectively.

After film and PRESAGE dosimeters were read out using a flatbed scanner and an optical-CT scanner, respectively, the images were spatially registered for gamma analysis. The rotational orientation was marked using three fiducial markers on the PRESAGE dosimeter and corresponding pen markings on the film. The ABS plastic phantom was also marked to match the irradiation orientation of the film and PRESAGE dosimeters. The fiducial markers on the PRESAGE dosimeter were attached only during the prescan imaging and were detached for the irradiation. For dose comparison, a local gamma analysis was performed to compare the average of three film irradiations to the dose distribution measured by the PRESAGE dosimeter. The gamma analysis was performed with and without a minimum dose constraint.

Figure 26.7 shows a representative measurement with the PRESAGE dosimeter and the local gamma index map calculated from the average of the film measurements. No dose constraint was applied. The figures are oriented to indicate the beam entering the dosimeter from the top of the image. The edge artifact remaining in the PRESAGE dosimeter measurements (a region of less than 2 mm in width) was excluded from the analysis. However, the finite edge artifact resulted in streaking around the cavity, which affected the image quality inside the dosimeter body. Although not shown in the figure, local gamma maps were also calculated with a dose constraint of $D > 1$ Gy, which only excluded the points outside the beam from the analysis. The passing rates calculated for the local gamma analyses were 95.56% with no dose constraint and 94.17% with constraint.

Figure 26.8 shows line profiles taken through the region of edge artifact, across the cavity, obtained along the same axis for EBT3 and two PRESAGE dosimeters. The line profiles agree well with one another, and it is apparent that the PRESAGE dosimeters have captured the dose enhancement and build-down due to the ERE. A significant level of noise can be seen in both PRESAGE dosimeters.

26.4.3 BANG™ Gel Dosimeters for Measurement of Penumbra in a Magnetic Field

Gel-based dosimeters offer great promise for measurement of dose distributions in the presence of magnetic fields. Following irradiation, gel dosimeters

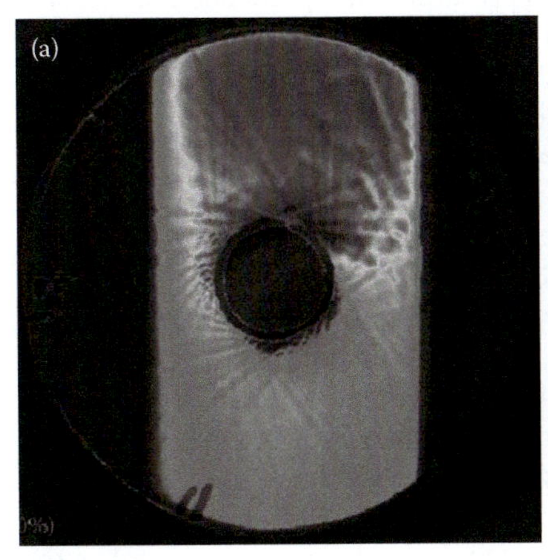

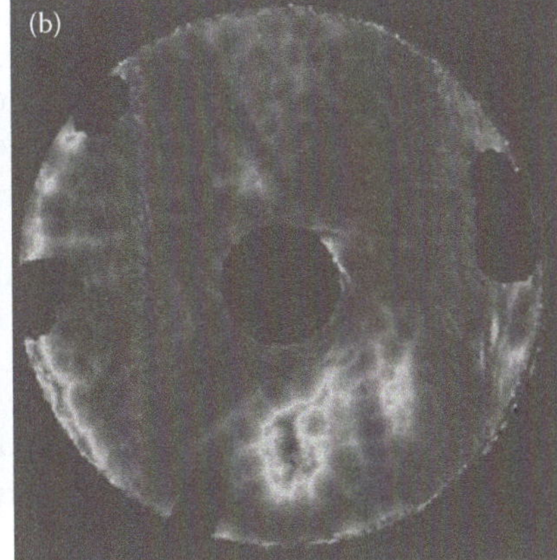

Figure 26.7

(a) Dose distribution measured with the PRESAGE® dosimeter and (b) gamma index map demonstrating the comparison between the external beam therapy (EBT) film and the PRESAGE dosimeter.

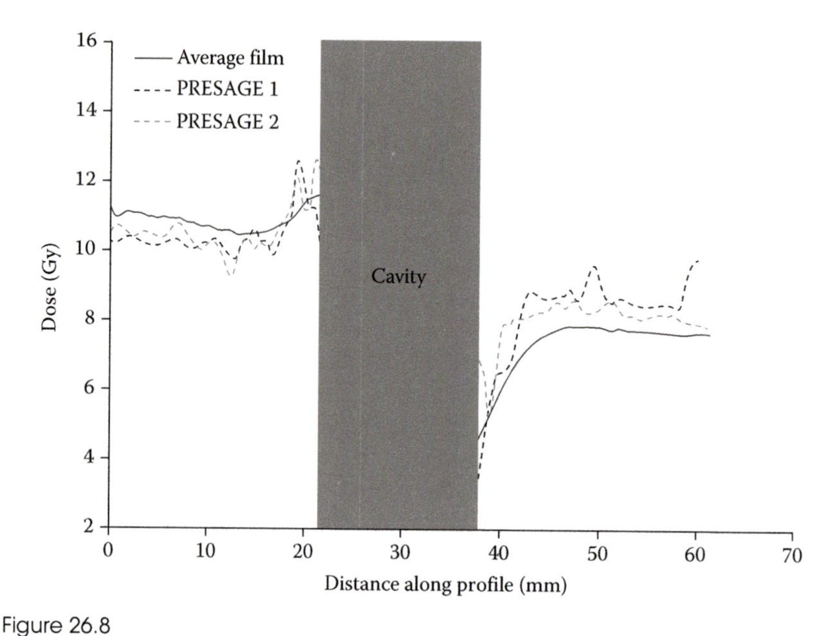

Figure 26.8

Line profiles of averaged film, PRESAGE® dosimeter 1, and PRESAGE dosimeter 2, taken along the same axis.

respond to the deposition of energy in ways that can be detected using MRI. The specifics of dosimeter response and appropriate MRI pulse sequences were discussed previously in Chapter 5. A potential benefit of gel dosimetry is that once the dosimeter is positioned and treatment is delivered, the dosimeter can be analyzed without moving it from the treatment position. This feature was explored by Roed et al. who have irradiated BANG gels (MGS Research Inc., Madison, CT) in the radiation field produced by an MR-Linac (Roed et al., 2016). These authors designed a phantom of unit density silicone that accommodated a BANG gel dosimeter in a specially designed glass flask, which protected the dosimeter from oxygen, and was coated with a UV-absorbing paint. Dosimeters were irradiated at the edges of the radiation field so that the influence of the magnetic field on the beam penumbra could be examined. Dosimeters were irradiated in the MR-Linac beam with the magnet operating at 1.5 T field strength. Analysis of the BANG gel dosimeters was conducted both with the MRI component of the MR-Linac and with a 3 T diagnostic MRI system. Imaging with the MR-Linac was conducted both during irradiation and immediately following irradiation. During irradiation, a 2D-balanced fast-field echo (bFFE) pulse sequence was used, with TR = 4.5 ms, TE = 2.2 ms, and a SENSE factor of 3. The temporal resolution was 517 ms/frame. Following irradiation, a T2 imaging sequence was used, with TR = 1000 ms and TE = 20, 40, 60, 80, and 100 ms.

Imaging during irradiation demonstrated development of the signal as the dosimeter polymerized in response to the dose deposition. Figure 26.9 indicates a logarithmic growth in signal with time following initiation of irradiation, with continuation of the signal growth immediately after irradiation was terminated.

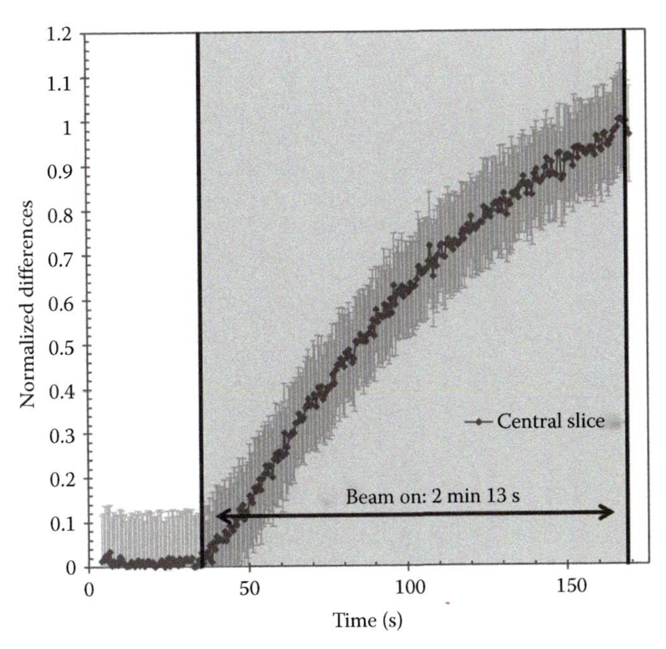

Figure 26.9

Real-time measurements of bFFE signal from a BANG™ gel dosimeter during irradiation.

T2 imaging was performed approximately 22 hours after irradiation. Images were converted to maps of spin–spin relaxation rate (R2) which was previously related to dose through a calibration procedure conducted in the absence of a magnetic field, and which demonstrated a linear response with dose. The BANG gel dosimeters clearly demonstrated the influence of the magnetic field on the penumbra of the radiation field. In the transverse plane of the linac, perpendicular to the magnetic field, the penumbra was noticeably distorted, reflecting the redirection of electrons toward the $+X$ direction as shown in Figure 26.10. Excellent agreement is seen between the BANG gel dosimeter measurements and a measurement of profile taken under identical conditions with radiochromic film.

26.4.4 Fricke Dosimeters in a Magnetic Field

Preliminary work was conducted by Lee et al. to investigate the feasibility of using 3D Fricke-type gel dosimeters both for analysis of dose distributions and for "real-time" dose observations (Lee et al., 2016b). An MR-Linac with a 1.5 T magnetic field was used for these studies. Fricke-type dosimeters were prepared in 97% w/w Milli-Q water with 3% w/w gelatin (300 Bloom), 1 mM ferrous ion, 0.05 mM xylenol orange, 50 mM sulfuric acid, and 1 mM sodium chloride. The dosimeters were prepared in a plastic flask approximately 8.5 cm in diameter and 6 cm in height. The dosimeters were stored at 4°C prior to irradiation and imaging.

To demonstrate postirradiation and real-time dosimetry, the dosimeters were irradiated in air, with a part of each dosimeter outside the treatment field to act as a reference. A pair of perpendicular fields was used—an "Anterior" field

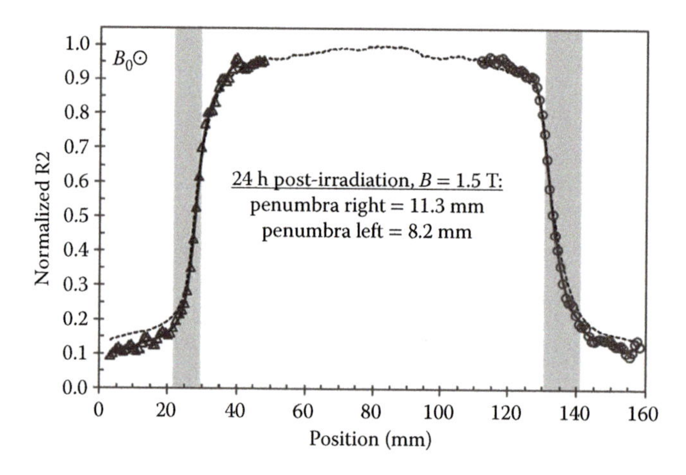

Figure 26.10

Measurements with BANG™ gel of the penumbra of the radiation beam from an MR-Linac. In this figure, the +X direction is to the viewer's right. The symbols indicate BANG gel measurements, while the dotted line is a measurement of beam profile made with radiochromic film. The shaded region indicates the 80%–20% penumbra region.

delivered 10 Gy to the center of the dosimeter through the top of the dosimeter and a "Lateral" field delivered 20 Gy through the side of the dosimeter. This arrangement constructed an overlapping region of high dose with regions of lower dose on either side. MR imaging was performed with the MR-Linac to observe the change in paramagnetic properties pre- and postirradiation using a T1-weighted sequence of TR = 500 ms and TE = 20 ms. MRI during irradiation was done in the MR-Linac using a balanced fast-field echo sequence with TR = 5 ms and TE = 1.7 ms.

Lee et al. (2016b) observed that a significant increase in pixel value from unirradiated to irradiated regions of about 30 Gy. The increase in pixel value and corresponding dose was also visible during irradiation. Figure 26.11 demonstrates this increase in pixel value with dose, indicating that the signal increases in a linear fashion. Visibly, the dosimeter underwent a color change from yellow to purple with the formation of the xylenol orange—ferric complex. Following irradiation, the dosimeter demonstrated the 3D dose distribution as indicated in Figure 26.12.

Perhaps the most comprehensive investigation of 3D dosimetry for measurement of dose distributions from an MR-guided treatment unit was published by Rankine et al. (2017). This study involved the use of a PRESAGE dosimeter to evaluate the dose distributions from a ViewRay MR-cobalt system. The authors evaluated several simple dose distributions, and also several intensity-modulated radiation therapy (IMRT) distributions selected from those recommended by the American Association of Physicists in Medicine (AAPM) (Ezzell et al., 2009). As shown above, the 3D dosimetry system demonstrated excellent agreement with other methods including treatment-planning system calculations and ionization chamber measurements. This test was admittedly less challenging, given the low field strength of the magnetic field, but is still a good demonstration of the value of 3D dosimetry.

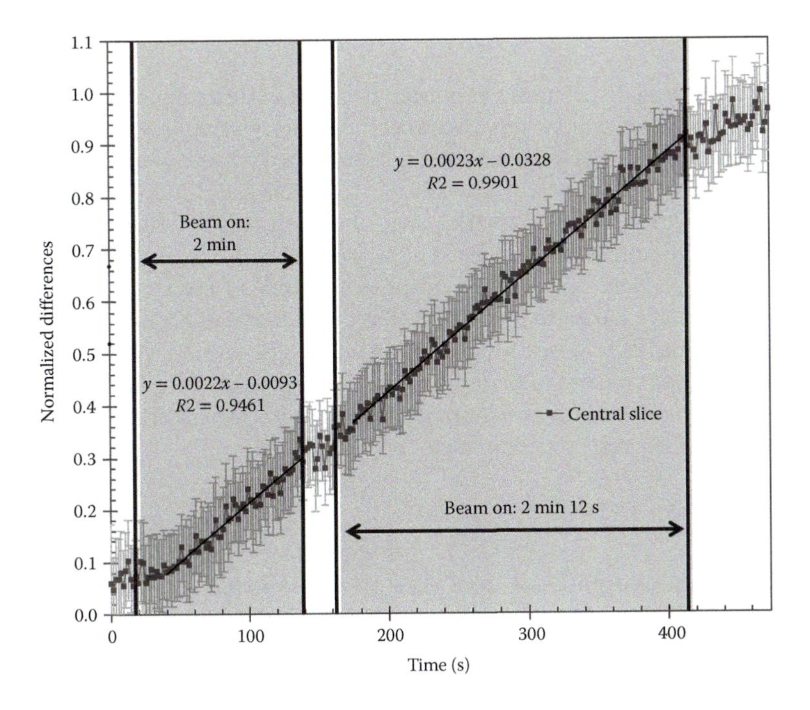

Figure 26.11

Real-time measurements of balanced fast-field echo signal from a Fricke-type gel dosimeter during irradiation. The increasing curve indicates the irradiated portion of the dosimeter.

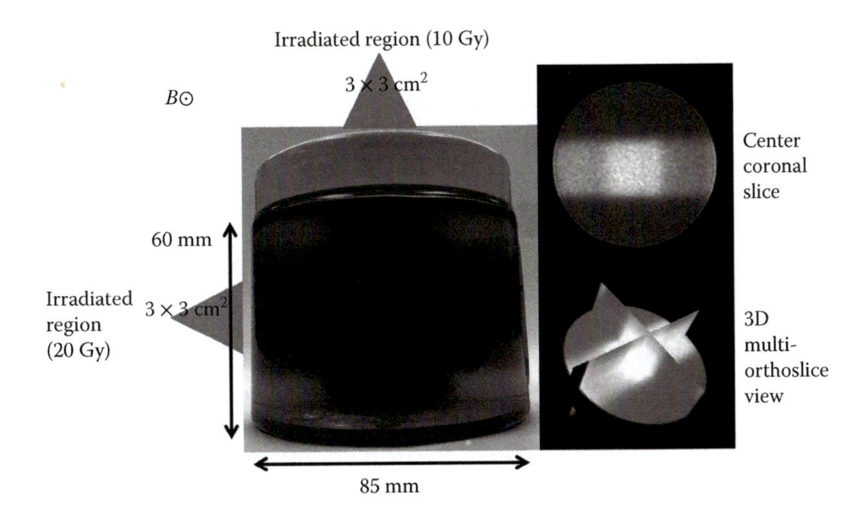

Figure 26.12

An irradiated Fricke gel dosimeter. The dosimeter is shown on the left. Images of the measured dose distribution are shown on the right.

26.5 Summary

The complex dose distributions produced by modern treatment equipment and delivery techniques require more advanced dosimetry systems to provide confidence that the delivered distribution is consistent with the planned distribution. 3D dosimetry techniques are valuable to enable acquisition of volumetric information with a single irradiation. The emerging field of MR-image-guided radiotherapy requires the presence of strong magnetic fields that can affect the performance of most conventional dosimetry systems. However, several novel 3D dosimeters are shown to perform well in the presence of magnetic fields and to provide quantitative dose distributions in volumetric fashion. While the available data are preliminary, these results indicate the potential for 3D dosimeters, including both gels and radiochromic polyurethane, to provide reliable measurements in clinically relevant circumstances.

References

Adamovics J and Maryanski MJ (2003) New 3D radiochromic solid polymer dosimeter from leuco dyes and a transparent polymeric matrix. *Med. Phys. 30:* 1349.

Alqathami M et al. (2016) Development of novel radiochromic films for radiotherapy dosimetry. *Med. Phys.* 43: 3664.

Baldock C et al. (2010) Polymer gel dosimetry. *Phys. Med. Biol.* 55: R1–R63.

Chisholm K et al. (2015) Investigations into the feasibility of optical-CT 3D dosimetry with minimal use of refractively matched fluids. *Med. Phys.* 42: 2607–2614.

Choi G et al. (2016) Using 3D dosimeters for the investigation of the electron return effect (ERE) in MR-guided radiation therapy: A feasibility study. *Med. Phys.* 43: 3856.

Choi GW (2016) MSc thesis: Measurement of the electron return effect using PRESAGE dosimeter. The University of Texas, Houston, TX.

Ezzell GA et al. (2009) IMRT commissioning: Multiple institution planning and dosimetry comparisons, a report from AAPM Task Group 119. *Med. Phys.* 36: 5359–5373.

Fallone BG et al. (2009) First MR images obtained during megavoltage photon irradiation from a prototype integrated linac-MR system. *Med. Phys.* 36: 2084–2088.

Ibbott GS (2010) QA in radiation therapy: The RPC perspective. *J. Phys. Conf. Ser.* 250: 1–7.

Ibbott GS et al. (2008) Challenges in credentialing institutions and participants in advanced technology multi-institutional clinical trials. *Int. J. Radiat. Oncol. Biol. Phys.* 71: S71–S75.

Jackson J et al. (2015) An investigation of PRESAGE(R) 3D dosimetry for IMRT and VMAT radiation therapy treatment verification. *Phys. Med. Biol.* 60: 2217–2230.

Keall PJ et al. (2014) The Australian magnetic resonance imaging-linac program. *Semin. Radiat. Oncol.* 24: 203–206.

Lafratta R et al. (2015) Comparison of 2D and 3D gamma calculations for an IMRT QA phantom. *J. Phys. Conf. Ser.* 573: 012055.

Lee H et al. (2016a) Orientation-dependent response of Gafchromic® EBT-2 and EBT-3 irradiated in the presence of a magnetic field. *Med. Phys. Int.* 4: 200.

Lee H et al. (2016b) Comparison between Fricke-type 3D radiochromic dosimeters for real-time dose distribution measurements in MR-guided radiation therapy. *Med. Phys.* 43: 3660.

Meijsing I et al. (2009) Dosimetry for the MRI accelerator: the impact of a magnetic field on the response of a Farmer NE2571 ionization chamber. *Phys. Med. Biol.* 54: 2993–3002.

Molineu PA, Kry S and Followill D (2014) Is it feasible to tighten the criteria for IROC's anthropomorphic phantoms? *Med. Phys.* 41: 352.

O'Brien DJ et al. (2015) Small air-gaps affect the response of ionization chambers in the presence of a 1.5 T magnetic field. *Med. Phys.* 42: 3724.

O'Brien DJ et al. (2016) Reference dosimetry in magnetic fields: Formalism and ionization chamber correction factors. *Med. Phys.* 43: 4915–4927.

O'Brien DJ and Sawakuchi G (2017) Monte Carlo study of the chamber-phantom air gap effect in a magnetic field. *Med. Phys.* 44: 3830–3838.

Oldham M et al. (2012) A quality assurance method that utilizes 3D dosimetry and facilitates clinical interpretation. *Int. J. Radiat. Oncol. Biol. Phys.* 84: 540–546.

Raaijmakers AJ, Raaymakers BW and Lagendijk JJ (2005) Integrating a MRI scanner with a 6 MV radiotherapy accelerator: Dose increase at tissue-air interfaces in a lateral magnetic field due to returning electrons. *Phys. Med. Biol.* 50: 1363–1376.

Raaijmakers AJE (2008) PhD thesis: MR-guided radiotherapy: Magnetic field dose effects. The University of Utrecht, the Netherlands.

Raaymakers BW et al. (2009) Integrating a 1.5 T MRI scanner with a 6 MV accelerator: Proof of concept. *Phys. Med. Biol.* 54: N229–N237.

Rankine et al. (2017) Three-dimensional dosimetric validation of a magnetic resonance guided intensity modulated radiation therapy system. *Int. J. Radiat. Oncol. Biol. Phys.* 97: 1095–1104.

Reyhan ML, Chen T and Zhang M (2015) Characterization of the effect of MRI on Gafchromic film dosimetry. *J. Appl. Clin. Med. Phys.* 16(6): 325–332.

Reynoso FJ et al. (2016) Magnetic field effects on Gafchromic-film response in MR-IGRT. *Med. Phys.* 43: 6552–6556.

Roed Y et al. (2016) Real-time imaging of 3-dimensional dose distributions with polymer gels using a magnetic resonance-guided linear accelerator. *Int. J. Radiat. Oncol. Biol. Phys.* 96(2S): E633.

Therriault-Proulx F et al. (2016) Impact of a magnetic field on the response from a plastic scintillation detector. *Med. Phys.* 43: 3876.

van Heijst TC et al. (2013) MR-guided breast radiotherapy: Feasibility and magnetic-field impact on skin dose. *Phys. Med. Biol.* 58: 5917–5930.

Wang J et al. (2016) Effect of a strong magnetic field on TLDs, OSLDs, and Gafchromic films using an electromagnet. *Med. Phys.* 43: 3873.

Wen Z et al. (2016b) Study on the magnetic field effect on the Exradin W1 plastic scintillation detector. *Med. Phys.* 43: 3417.

Wen Z et al. (2016a) Investigation on the magnetic field effect on TLDs, OSLDs, and Gafchromic films using an MR-Linac. *Med. Phys.* 43: 3632.

Index

A

B

C

E

K

L

M

N

Q

T

U